Motives *of* Honor, Pleasure, *and* Profit

Motives *of* Honor, Pleasure, *and* Profit

PLANTATION MANAGEMENT IN THE COLONIAL CHESAPEAKE, 1607–1763

LORENA S. WALSH

Published for the
OMOHUNDRO INSTITUTE OF EARLY AMERICAN HISTORY AND CULTURE,
Williamsburg, Virginia,
by the UNIVERSITY OF NORTH CAROLINA PRESS,
Chapel Hill

The Omohundro Institute of Early American History and Culture is sponsored jointly by the College of William and Mary and the Colonial Williamsburg Foundation. On November 15, 1996, the Institute adopted the present name in honor of a bequest from Malvern H. Omohundro, Jr.

Colonial Williamsburg Studies in Chesapeake History and Culture

Manufactured in the United States of America

Library of Congress Cataloging-in-Publication Data
Walsh, Lorena Seebach, 1944–
Motives of honor, pleasure, and profit : plantation management in the Colonial Chesapeake, 1607–1763 / Lorena S. Walsh.
p. cm.
Includes bibliographical references and index.
ISBN 978-0-8078-3234-9 (cloth : alk. paper)
1. Plantations—Chesapeake Bay Region (Md. and Va.)—Management—History. 2. Tobacco industry—Chesapeake Bay Region (Md. and Va.)—Management—History. I. Title.
HD1471.U52.C488 2010
338.109752'09032—dc22 2009033897

The paper in this book meets the guidelines for permanence and durability of the Committee on Production Guidelines for Book Longevity of the Council on Library Resources.

The University of North Carolina Press has been a member of the Green Press Initiative since 2003.

This volume received indirect support from an unrestricted book publication grant awarded to the Institute by the L. J. Skaggs and Mary C. Skaggs Foundation of Oakland, California.

14 13 12 11 10 5 4 3 2 1

TO PETER,

who kept steadfast

through it all

FOREWORD

The passage of time can turn the most ardent love match into a marriage of convenience. In 1943 a union eagerly entered into by the College of William and Mary and a newcomer to town, the Colonial Williamsburg Foundation, brought forth the Institute of Early American History and Culture. So ably has that progeny served the historical community since then that the scholars and teachers who read Institute publications, subscribe to the *William and Mary Quarterly,* and attend its conferences seldom think to ask what benefit the sponsors—the College and the Foundation—might still derive from their creation sixty-some years later. For William and Mary the answer is plain enough: the Institute's worldwide renown enhances the College's own academic reputation by association. In addition, as a practical matter, Institute fellows and members of the staff fill out the ranks of classroom teachers at the College.

The dividend paid to Colonial Williamsburg in return for its annual subsidy is not so easily reckoned. The Institute's star power, while real enough, shines less brightly in the universe of American museums. That hasn't mattered when times were flush; nor has it ever weighed so heavily in hard times that the Foundation withheld its generous support. But Institute director Ron Hoffman for his part and I for mine could never quite shake off the impression that budget-balancers at Colonial Williamsburg regarded their employer's alliance with the College and the Institute as a marriage of convenience, sometimes bordering on inconvenience, albeit a partnership too venerable and too much honored to call into question.

Benefit of doubt does not do justice to the country's oldest and most enduring collaboration between an organization famous for its scholarship and a history museum equally famous for teaching American history to audiences that number in the millions every year. Both employ professional historians. Understanding how their work is mutually supportive is the secret to answering the seldom asked question about the Institute's value to Colonial Williamsburg, or, to a larger point, the value of academic scholarship generally to history museums, historical societies, historic sites, and other such informal educational institutions as cater to a general public eager to learn about America's past.

This book by Lorena Walsh and Lorena's long career as a museum historian, when set side by side, illustrate the advantage that Colonial Williamsburg enjoys from its association with the Institute—that, coupled with the

Foundation's practice of staffing its own research and collections divisions with archaeologists, architectural historians, curators, and a few straight-up historians like Lorena who actively participate in the larger intellectual life of their respective disciplines. The honor roll of notable historians who have worked full time for the Research Division at Colonial Williamsburg includes the likes of Pierce Middleton, Thad Tate, and John Selby in years gone by and more recently Mick Nicholls, John Hemphill, Kevin Kelly, David Konig, Ann Smart Martin, Rhys Isaac, my successor as vice president of the division, Jim Horn, and, of course, Lorena herself, almost the longest serving and certainly the most widely published of them all. For these colleagues and for others as well, the Foundation's involvement with the Institute has been indispensable. Their accomplishments fulfill for Colonial Williamsburg the vows exchanged at the founding of the Institute more than half a century ago.

How so? The book you hold in your hands gives no outward clue. From its cover and its contents it looks and reads like most monographs on the Institute's book list by authors who almost always are university professors. As it should. Lorena has written this and other books, articles, and conference papers during her tenure at the Foundation as her contributions to the scholar's task of bringing new knowledge to bear on questions about the past that are defined and debated mainly by academic historians. Their intellectual world is partly her world too. The question is: How does her being a respected player in that league repay her colleagues at Colonial Williamsburg? What special value do her credentials in the academic community and those of the Foundation's other scholar-authors add to Williamsburg's educational programs for the general public? Pretend for a minute that these men and women were indeed professors, that, instead of working for a museum, they were faculty members in history departments someplace else. Would the influence of their ideas be any different if their publications were simply available in the Foundation's excellent library? After all, most historical organizations borrow the scholarship they need for exhibitions, tours, and other public programs from the usual retail sources. Only a handful find reason to employ research historians on staff. The few exceptions include the National Park Service, the Smithsonian Institution, Historic St. Mary's City in Maryland (where Lorena began her career alongside Lois Carr and Russell Menard), Colonial Williamsburg, and some of the country's leading state historical societies.

Their rationale for investing in this costly in-house talent runs parallel to the thinking at some academic institutions where advanced research is believed to contribute to superior classroom teaching. But the analogy isn't perfect. Because most museum historians are not teachers in the usual sense, only by understanding what they do do can the Institute's influence be fully appreciated. Lorena's "invisible" career — the one unseen by her academic colleagues — provides as good an example as anyone could ask for.

All the time she was collecting research materials and writing *Motives of Honor, Pleasure, and Profit* and, before that, *From Calabar to Carter's Grove,* she also performed duties expected of every member of the Department of Historical Research. But in her case, her own extensive research projects and her far-flung network of professional contacts meant that she brought something extra to those staff assignments. Sometimes it was an Atlantic-spanning perspective that placed the local Williamsburg story in a sweep of events large enough to excite visitors' imaginations. When the Foundation rewrote its educational master plan in 1998, for instance, Lorena was just then working through the argument she presents here in Chapters 5 and 6 about the capitalistic behavior of Chesapeake planters and their shrewd participation in an international trading system. Profit taking, she reminded us even then, bore directly on the treatment of enslaved laborers and that in turn on measures that they themselves took to cope or resist. Those were aspects of slave life that interpreters at Colonial Williamsburg were eager to portray. For years Lorena corresponded with colleagues who were still sorting out these issues; she regularly attended Institute colloquia; she frequently read papers, and commented on others' papers, at scholarly conferences, professional meetings that the Institute often organized or sponsored. To work in the same town with the Omohundro Institute of Early American History and Culture is to be connected to an information exchange unrivaled in any other branch of American historical scholarship. Lorena tapped into the flow of ideas that pass back and forth along Institute channels. Some of those ideas she drew off for her own work; others she funneled into the curricular planning process at Colonial Williamsburg, often years ahead of publication.

Her professional contacts and her own research came into play again when the Foundation's strategic planning efforts led forward to bricks-and-mortar projects that required her expertise. One was a reconstructed slave quarter at Carter's Grove Plantation; another a typical middling planter's farmstead built anew on the outskirts of Williamsburg. Both were designed under hers and Kevin Kelly's watchful eyes to illustrate different agricultural economies, larger and smaller workforces, and management practices appropriate to each. Readers can now find her thinking on those matters in the pages of this book, but years after museum visitors first encountered it in the form of living historical landscapes. The newly discovered knowledge on which these and other exhibits were based was never simply ready and waiting on a library shelf somewhere. Nor did Lorena provide it all from her own research. Other Williamsburg colleagues were contributors too. And the list of credits didn't stop there. Museums that employ their own book-writing scholars enjoy the added benefit of plugging into that whole informal network of researchers whose give-and-take exchanges with one another redefine what historians think is most important about the past and whose discoveries shared with one another reshape the contours of American history long before publications trickle their new ideas into the profession's intellectual mainstream.

Finally, my previous statement that museum historians are seldom teachers needs clarification. Every member of the Department of Historical Research at Colonial Williamsburg is senior faculty when it comes to training the museum's guides and interpreters. The research experience that these five or six professional historians bring into the classroom invigorates their instruction just as working scholars are said to make better teachers at colleges and universities. Here too the Institute has been a steadfast partner. Interpreter training sometimes covers topics on which Institute editors and fellows are recognized authorities and therefore prime guest lecturers. Their payback to Colonial Williamsburg on those occasions is personal, unstinting, and much appreciated.

Long-range planning, project development, interpreter training and mentoring—all have been enriched by the many knowledgeable friends of friends that Colonial Williamsburg meets through the Institute's good offices. And now here comes *Motives of Honor, Pleasure, and Profit,* a keystone study of Chesapeake society and economy that was written and illustrated at one end of Williamsburg and edited and published at the other. Lorena Walsh's latest contribution to both American historical scholarship and American history museumship makes as clear as clear can be that the passing years have done little to diminish the ardor that wed the College of William and Mary, the Institute of Early American History and Culture, and the Colonial Williamsburg Foundation so long ago.

Cary Carson
Vice President (retired), Research Division
Colonial Williamsburg Foundation

PREFACE

The conventional introductory opening, "This study has been a long time in the making," is in this instance a decided understatement, since I began research for a book on the management of large Chesapeake plantations when I joined the Department of Historical Research at the Colonial Williamsburg Foundation in 1981. Over the years, a fellowship at the Mills-Hagley Foundation Regional Economic History Research Center and responsibilities for directing three long-term, museum-focused research projects funded by the National Endowment for the Humanities undertaken by Colonial Williamsburg, Historic Saint Mary's City, and Historic Annapolis, Inc., interrupted writing. So, too, did completion of studies already in progress on women's history and on consumer behavior as well as an unanticipated additional research project on the transatlantic Chesapeake slave trade that has contributed to David Eltis et al., *Voyages: The Trans-Atlantic Slave Trade Database,* http://www.slavevoyages.org, launched at Emory University in 2008.

My original goal was to reconstruct the management practices of elite Virginia and Maryland planters from their correspondence and account books, to generate wherever possible concrete measures (standardized over time and place) of the success or failure of these practices, and to identify from the individual case studies typical practices and common patterns. I then envisioned the end product as a relatively narrow examination of the activities of large planters that would be placed in the wider context of then ongoing individual and collaborative research by a number of scholars on the social and economic history of the Chesapeake region. As time went on, the scope of the manuscript has both contracted and expanded. The initial plan called for inclusion of the Albemarle Sound region, since commitment to tobacco culture made adjacent North Carolina counties arguably part of the "Tobacco Coast." A lack of source materials for large North Carolina plantations in the colonial era comparable to those available for Virginia and Maryland led to dropping that region from the study. A decision to deal comprehensively with most of the better-documented agricultural operations, rather than to select only a few for exposition, required contracting the span of time covered in the present volume to 1607–1763. (A second volume dealing with 1764–1820 is planned.)

On the other hand, the failure of several projected jointly authored books on the Chesapeake economy to materialize necessitated doing additional research and writing on topics I had not planned to tackle on my own. Invitations to contribute to yet other collaborative studies of Chesapeake environmental and economic history, and to comparative investigations of diet, health, and migration also slowed progress on the manuscript but have led to new insights into diverging subregional economies and the interaction of agricultural technologies and the environment and to a reassessment of the development of slavery in the region that have been incorporated in this volume. Finally, when it came time to revise the manuscript for publication, readers and editors insisted on further extending the scope to include the Atlantic context of regional events.

Meanwhile, the issues provoking debate among scholars about the economy of colonial North America changed as my study progressed. When I began research in the early 1980s, debates over whether southern planters made economically rational decisions or drew on a precapitalist, nonprofit maximizing mentality loomed large. Similar debates over the rationales behind economic decision making prevailed among scholars of New England. Some argued for an early emergence of market-oriented, entrepreneurial behavior that fostered intensive economic development in the North, while others argued for, and indeed celebrated, limited market involvement, lauding a striving for "competency" instead of profit as a morally superior alternative to rapacious capitalism. Scholars on both sides of the rationality debates universally maintained that the labor system from which southern slaveholders reaped (or failed to reap) profit was profoundly immoral. Initially, the most heated debates were over whether slavery was profitable. Many regarded attempts to overturn the conventional wisdom that the institution was unprofitable as tantamount to endorsing it as a positive good. A related concern was whether slaveholders contributed to economic development in the South by embracing risk-taking entrepreneurial behavior or retarded it by clinging to pre-industrial ideologies and investment patterns. Contrary to the thrust of many of the studies focused on New England farmers, however, in the context of a contrast between a nonprogressive, overwhelmingly agricultural South and a dynamic, industrializing North, capitalist economic development was usually viewed a positive good. Rationality or irrationality, efficiency or inefficiency, morality or immorality, and public benefits or harms were employed inconsistently and often contradictorily in emotionally charged arguments and counterarguments. Given my own bias toward rational behavior, my working hypothesis was that profit-maximizing strategies did prevail among large Chesapeake planters. I recognized, however, that confirmation required the construction and testing of standardized quantitative measures of agricultural productivity and of response to changing market opportunities. Although my interest in generating measures of productivity and hence of planter performance continued unabated, the more I learned about elite slaveowners' management methods and about the human cost at which successful outcomes were bought, the less sympathy I could muster for the subjects of the study. My enthusiasm for completing the last chapters and for revising the study for publication for a time flagged.

Nonetheless, as my research progressed, and as programs for interpreting African American history at Colonial Williamsburg expanded, the experiences of individual enslaved workers and the histories of plantation communities that could sometimes be reconstructed from slaveholders' records took on new significance. The amount of information I uncovered about the lives of enslaved Africans and African Americans was unexpected, and making these unanticipated results available to the public took precedence over polishing a manuscript summarizing more-expected findings for the already extensively studied (if not well-measured) planter elite. The results of more extensive and more intensive research and analysis than originally planned on the lives and work of enslaved Africans and African Americans has been published separately in Lois Green Carr and Lorena S. Walsh, "Economic Diversification and Labor Organization in the Chesapeake, 1650–1820," in Stephen Innes, ed., *Work and Labor in Early America* (Chapel Hill, N.C., 1988), 144–188; and in Walsh, "Rural African Americans in the Constitutional Era in Maryland, 1776–1810," *Maryland Historical Magazine,* LXXIV (1989), 327–341; "Slave Life, Slave Society, and Tobacco Production in the Tidewater Chesapeake, 1620–1820," in Ira Berlin and Philip D. Morgan, eds., *Cultivation and Culture: Labor and the Shaping of Slave Life in the Americas* (Charlottesville, Va., 1993), 170–199; "'A Place in Time' Regained: A Fuller History of Colonial Chesapeake Slavery through Group Biography," in Larry E. Judson, Jr., ed., *Working toward Freedom: Slave Society and Domestic Economy in the American South* (Rochester, N.Y., 1994), 1–32; "Work and Resistance in the New Republic: The Case of the Chesapeake, 1770–1820," in Mary Turner, ed., *From Chattel Slaves to Wage Slaves: The Dynamics of Labour Bargaining in the Americas* (Bloomington, Ind., 1995), 97–122; *From Calabar to Carter's Grove: The History of a Virginia Slave Community* (Charlottesville, Va., 1997); "Slavery and Agriculture at Mount Vernon," in Philip J. Schwarz, ed., *Slavery at the Home of George Washington* (Mt. Vernon, Va., 2001), 47–77; and "The Differential Cultural Impact of Free and Coerced Migration to Colonial America," in David Eltis, ed., *Coerced and Free Migration: Global Perspectives* (Stanford, Calif., 2002), 117–151.

By the time I finally returned in 1998 to revising earlier drafts of this study for publication, the cold war had ended, capitalism had triumphed as the dominant form of global economic organization, and most historians had accepted that large southern planters were rational, profit-maximizing businessmen, and that even most small to middling colonial tobacco growers were thoroughly engaged in production for international markets. The grounds of debate shifted from issues of rational or irrational (or precapitalist) decision making to more arcane disputes among economic historians over levels of productivity (or lack thereof) in colonial and early national period agriculture. Consequently, some of the profit-seeking behaviors I had anticipated having to document, demonstrate, and argue for could be presented in a more concise, straightforward fashion. On the other hand, new information and new interpretations about the development of slavery in seventeenth-century North America have appeared in recent years in such profusion and at such a quick pace that I have been forced to repeatedly rethink and (much to the consternation of Institute editors) to repeatedly rewrite portions of the study dealing with the evolution of slavery as a labor system.

My interpretations of seventeenth- and eighteenth-century agriculture and of colonial material culture have been profoundly influenced by living for a year and a half in Tanzania, East Africa, in 1967–1969 while my husband was conducting doctoral dissertation research in agricultural economics and I taught at a rural secondary school, and, between 1996 and 2006, by our teaching six separate three-month stints as volunteers at Tumaini University in Iringa, Tanzania. Seeing in the late 1960s how rural families who lived in small, impermanent, earthfast houses and who owned a minimum of material goods organized their dwellings and adjoining yards and carried out everyday household functions made the material world of the seventeenth-century European settlers I subsequently studied much less of a foreign terrain. Residence in Tanzania also afforded invaluable insights into how hard farmers using only "traditional" agricultural methods must work just to provide their families with basic necessities as well as into the stark lack of options that a less developed economy and rural poverty can impose.

Unlearning as well as learning was an important part of the experience. For example, we arrived in Rungwe, Tukuyu District, in early 1968, naively convinced of the invariable superiority of modern Western agriculture. We were absolutely confident of our ability to grow more and better vegetables than did local residents. Just like the Price family, whose experiences in the Belgian Congo in the 1960s Barbara Kingsolver so aptly related in *The Poisonwood Bible* (New York, 1998), we were soon disabused of our presumptions when our best efforts ended in total failure with the onset of the first long rains. Discarding what we thought we knew, we began to learn about alternative methods from our Tanzanian neighbors. We gained a profound respect for the wisdom of "traditional" farmers, who from a lack of formal schooling may not be able to readily explain why they use the methods they do but who from many generations of hard-won experience have developed techniques of ground preparation, seed selection, cultivation, and harvesting that indisputably work. Watching the steps involved in slash-and-burn clearing at first hand in a field just beyond our garden fence and observing crops raised in hills in unfenced fields with hand tools, without any aid from draft animals, afforded a better-informed understanding of the tasks routinely recorded in colonial and early national period farm diaries and plantation work logs or alluded to in plantation overseers' reports.

Between 1996 and 2006, we were privileged to learn not only more about traditional technologies but also more about the rationales underlying them while visiting the home villages and family farms of students we taught at Tumaini University. Interviews with tobacco growers in Kitapilimwa, Iringa District, for example, led us to appreciate more fully how smallholder farmers make local adaptations in the seeding, cultivation, harvesting, housing, and curing methods of this labor-intensive, but not capital-intensive, crop, just as they did in the Chesapeake and in the Caribbean in the seventeenth century. Everywhere, we gained a greater appreciation for the reasons behind smallholders' aversion to risk, given that failed experiments can result in hunger and acute suffering, especially for dependents. On the other hand, we have also seen how quickly small farmers do respond to opportunities afforded by new markets or improved means of transport by working harder, adopting more efficient methods, and expanding culture of cash crops when the costs of doing so involve primarily a sacrifice of leisure and there is a reasonable chance that the increased effort will be rewarded.

We also observed the frequent failure of agricultural development schemes devised by western "experts" that did not take local climate and topography into account or that depended on technologies too capital-intensive for subsistence farmers to adopt. These experiences have generated a healthy skepticism about the practicality and merits of agendas pushed by elite agricultural "improvers," present and past. Hence, when approaching historical documents, my working assumption is that practitioners of traditional farming methods almost always had a compelling reason for doing, or not doing, something and that most self-proclaimed improvers probably knew a great deal less than they supposed.

Abraham and Stella Mayenje, Ebenezer Laisser, Kastory and Vummy Madege, Maureen Fondo, Twisa and Hellen Mwambona, Wallace and Wema Lupenza, Yekonia and Upendo Koko, and Emmanuel and Mama Mwachang'a, along with their extended families, graciously opened their homes and hearts, for which we are profoundly grateful, as we are to Issa Maalim, Muhingo Rweyemanu, Paulina Teveli, Rosemary Kachecheba, and to Sister Jane Eliza and all the Teresina sisters.

From Deo Sabokwigina, Isidore Ntirampeba, and other Burundian refugees at Tumaini University we gained an appreciation of the profound resiliency of the human spirit. Coming to know individuals who witnessed, survived, and have overcome but not forgotten events that no human being should ever have to endure has contributed to some understanding of the physic and spiritual resources those who survived the trauma of enslavement and the Middle Passage in the seventeenth and eighteenth centuries could draw upon.

I have accumulated many debts in the long course of research and writing this study. First I must thank the many archivists and librarians who have facilitated the research. I am especially indebted to staffs at the Rockefeller Library, Colonial Williamsburg Foundation, and Swem Library, College of William and Mary, for extending long-term borrowing privileges, for expert assistance with manuscript collections, and for expediting interlibrary loans of obscure materials. Also helpful were the staffs at the Virginia Historical Society, the Virginia State Library, the Special Collections libraries of the University of Virginia, Georgetown University, and Duke University, the Maryland State Archives, the Maryland Historical Society, and the Manuscript Division of the Library of Congress.

Many friends and colleagues have read parts of this work and have offered helpful advice and criticism. I mention with special thanks Lois Carr, Emory Evans, David Galenson, John McCusker, and Jean B. Russo. The late Robert Gallman and the late John Hemphill gave careful and critical readings of early chapters, and I regret profoundly no longer having the benefit of their expertise and encouragement. Mendy Gladden, Paul Clemens, and an anonymous reader for the Omohundro Institute of Early American History and Culture offered extensive and multifaceted suggestions for revising the manuscript. Theirs was advice I did not always take, but the study is surely more comprehensive, and I hope more comprehensible, as a result. Virginia Montijo copyedited the manuscript cordially and with care. Richard Anderson created the illustrations of gentry houses, and he and Cary Carson jointly produced those of more ordinary structures. Marianne Martin in Colonial Williamsburg's Visual Resources Library assisted with locating and procuring copies of illustrations. Arlene Kasarjian and Tim Allison provided research assistance so long ago they may not remember it.

Colleagues at the Colonial Williamsburg Foundation, the College of

William and Mary, and the Omohundro Institute of Early American History and Culture have offered encouragement, advice, and support over the years, as have former colleagues at Historic Saint Mary's City and the Maryland State Archives. They include Joanne Bowen, Marley Brown, Rex Ellis, P. M. G. Harris, Cathy Hellier, Ronald Hoffman, Kevin Kelly, Russell Menard, Henry Miller, Philip Morgan, Michael Nicholls, Edward Papenfuse, Lou Powers, Wayne Randolph, Linda Rowe, Robert Watson, Jr., Robert C. Watson, and Jim Whittenberg. I have also profited from discussions about research, writing, and interpretation of Chesapeake history with younger colleagues that began while they were graduate students at the College of William and Mary. I learned a great deal from the dissertations of John Coombs, Ann Martin, Turk McCleskey, Lynn Nelson, and Julie Richter; some of their subsequent publications have prompted significant rethinking of portions of this study.

The contributions of two other colleagues require more than passing mention. I owe a special debt to Stanley Engerman, who over the course of many years graciously and efficiently read first, intermediate, and later versions of the entire manuscript, always offering apt advice and, above all, continued encouragement when I despaired of bringing even the first half of the project to a satisfactory conclusion. He saved me many a misstep, and, along with other readers, bears no responsibility for remaining errors.

The most profound debt is to Cary Carson, formerly Director of Research at Historic Saint Mary's City and subsequently at the Colonial Williamsburg Foundation. His unwavering commitment to practicing, inspiring, and facilitating excellence in museum programming and to forging a seamless partnership between public and academic history is virtually unique among museum administrators. At both institutions Cary recruited staffs of historians, archaeologists, and architectural historians whom he challenged to apply multidisciplinary approaches to major museum research topics and to find multidimensional answers to historical and artifactual problems. Staff meetings routinely became ongoing colloquia (often followed up at informal lunchtime seminars and end-of-workday gatherings) that were usually more stimulating than comparable meetings of ivory tower academicians, since abstract documentary research was linked to concrete artifacts and structures and findings and interpretations to widely disseminated public programs as well as to more esoteric academic publications.

Carson encouraged staff historians to learn and to apply the approaches and methodologies of social history, demography, material culture studies, and quantitative history as well as the more standard techniques of documentary analysis. The goal was to further the mission, that many late-twentieth-century history museums embraced, of portraying an inclusive account of all the people who once lived in a region, and not just reflexively commemorating standard textbook political and military events and the accomplishments of a few elite men. He also challenged staff to write programs and publications for the museum-going public of a quality, depth, and breadth that would merit their concurrent adoption in undergraduate and graduate college history courses throughout the country. Equally critically, Cary was willing to bend typically inflexible institutional rules in ways that enabled staff committed to a combination of public and academic history to function productively in both spheres.

Over the years, portions of this study benefited from discussion at the Washington Area Early American Seminar Series, University of Pennsylvania Economic History Seminar, Omohundro Institute of Early American History and Culture Colloquium, University of Delaware History Colloquium, University of Minnesota History Workshop, and Champlain–St. Lawrence Seminar in Early American Studies; at annual meetings of the Economic History Association, Social Science History Association, Southern Historical Association, Association of Caribbean Historians, and Organization of American Historians; and at conferences at the University of Maryland, St. Mary's College of Maryland, the Huntington Library, Katholieke Universiteit, Brussels, the Center for the History of Freedom at Washington University in St. Louis, and the University of Rochester.

Part of Chapter 2 and preliminary statements of major arguments were published in Russell R. Menard, Lois Green Carr, and Lorena S. Walsh, "A Small Planter's Profits: The Cole Estate and the Growth of the Early Chesapeake Economy," *William and Mary Quarterly*, 3d. Ser., XL (1983), 171–196; Carr, Menard, and Walsh, *Robert Cole's World: Agriculture and Society in Early Maryland* (Chapel Hill, N.C., 1991); and Walsh, "Plantation Management in the Chesapeake, 1620-1820," *Journal of Economic History*, XLIX (1989), 393–406; "Summing the Parts: Implications for Estimating Chesapeake Output and Income Subregionally," *William and Mary Quarterly*, 3d Ser., LVI (1999), 53–94; "Mercantile Strategies, Credit Networks, and Labor Supply in the Colonial Chesapeake in Trans-Atlantic Perspective," in David Eltis, Frank D. Lewis, and Kenneth L. Sokoloff, eds., *Slavery in the Development of the Americas* (Cambridge, 2004), 89–119; and "The Role of Aristocratic Management Strategies in the Economic Development of the British North American Chesapeake," in Paul Janssens and Bartolomé Yun-Casalilla, eds., *European Aristocracies and Colonial Elites: Patrimonial Management Strategies and Economic Development, 15th–18th Centuries* (Aldershot, Hampshire, 2005), 199–211.

CONTENTS

ILLUSTRATIONS

Plates

Maps

Figures

TABLES

ABBREVIATIONS AND SHORT TITLES

Ag. Hist.
Agricultural History
AHR
American Historical Review
Arch. Soc. Va. Qtly. Bull.
Archaeological Society of Virginia, *Quarterly Bulletin*
Byrd Correspondence
Marion Tinling, ed., *The Correspondence of the Three William Byrds of Westover, Virginia, 1684–1776,* 2 vols. (Charlottesville, Va., 1977)
CSP
Great Britain, Public Record Office, *Calendar of State Papers,* Colonial Series, *America and West Indies,* ed. W. Noel Sainsbury et al. (London, 1864–)
CWL
John D. Rockefeller, Jr., Library, Colonial Williamsburg Foundation, Williamsburg, Va.
CWR
Department of Training and Historical Research, Colonial Williamsburg Foundation, Williamsburg, Va.
EJCCV
H. R. McIlwaine, ed., *Executive Journals of the Council of Colonial Virginia,* 6 vols. (Richmond, Va., 1925–1966)
Exp. Econ. Hist.
Explorations in Economic History
Force Tracts
Peter Force, comp., *Tracts and Other Papers, Relating Principally to the Origin, Settlement, and Progress of the Colonies in North America, from the Discovery of the Country to the Year 1776,* 4 vols. (1836; rpt. Gloucester, Mass., 1963)
HSP
Historical Society of Pennsylvania, Philadelphia
JAH
Journal of American History
JEH
Journal of Economic History

Jour. So. Hist.
Journal of Southern History
LC
Library of Congress, Washington, D.C.
MCGC
H. R. McIlwaine, ed., *Minutes of the Council and General Court of Colonial Virginia, 1622–1632, 1670–1676* (Richmond, Va., 1924)
MHM
Maryland Historical Magazine
MHS
Maryland Historical Society, Baltimore
MSA
Maryland State Archives, Annapolis
NCHR
North Carolina Historical Review
Northern Neck Hist. Mag.
Northern Neck of Virginia Historical Magazine
NYPL
New York Public Library
Res. Econ. Hist.
Research in Economic History
RVC
Susan Myra Kingsburg, ed., *The Records of the Virginia Company of London,* 4 vols. (Washington, D.C., 1906–1935)
So. Stud.
Southern Studies
UVa
Special Collections Library, University of Virginia, Charlottesville
Va. Hist. Reg.
Virginia Historical Register and Literary Advertiser
VHS
Virginia Historical Society, Richmond
VMHB
Virginia Magazine of History and Biography
VSL
The Library of Virginia, Richmond
WMQ
William and Mary Quarterly
WPREHRC
Working Papers from the Regional Economic History Research Center

Motives *of* Honor, Pleasure, *and* Profit

INTRODUCTION

The founders of the first permanent English settlement on the North American mainland hoped to establish a commercial outpost—a *plantation*—that would produce handsome profits from the mining of precious metals and iron, from trade with the local population, from the gathering or raising of exotic tropical crops, and from the production of naval stores, wine, silk, dyestuffs, and other staple commodities that Englishmen heretofore had to buy from other nations. Spurred by the example of the early European colonizers of the New World, the Portuguese and the Spanish, investors in the London Company of Virginia hoped to reap immense fortunes similar to those being made by their rivals from a modest initial outlay of capital. The colonists at Jamestown could find no gold or silver, however, nor were there any rich, populous native American empires with which they could either trade, or, following the example of the Spanish, plunder. Ambitious projects for raising tropical crops and even for producing more mundane staples also failed, and the fate of most of the first contingents of settlers was, not to amass wealth, but instead to suffer hunger, starvation, disease, and premature, sometimes violent, death. Stay-at-home investors in England earned nothing more than repeated demands from surviving settlers for additional, costly supplies.

After a decade of bitter disappointment, the London Company of Virginia awarded its shareholders the only asset the Company possessed, dividends of large tracts of undeveloped land. It was hoped that by this means individual investors would become more motivated to finance the settlement and development of *particular plantations* over which they were granted a measure of direct control. Some of the investors indeed responded to the Company's initiative, sending fresh shipments of supplies and several thousand more settlers to establish smaller, but yet still corporately organized, agricultural and extractive enterprises in the Chesapeake. These ventures, although somewhat more successful than those of the parent company, also failed to produce the desired windfall profits, and all of the particular plantations were abandoned soon after the devastating native American attack of 1622.[1]

1. An excellent summary of the successes and failures of the Jamestown settlement is David W. Galenson, "The Settlement and Growth of Colonies: Population, Labor, and Economic Development," in Stanley L. Engerman and Robert E. Gallman, eds., *The Cambridge Economic History of the United States,* I, *The Colonial Era* (Cambridge, 1996), 135–141.

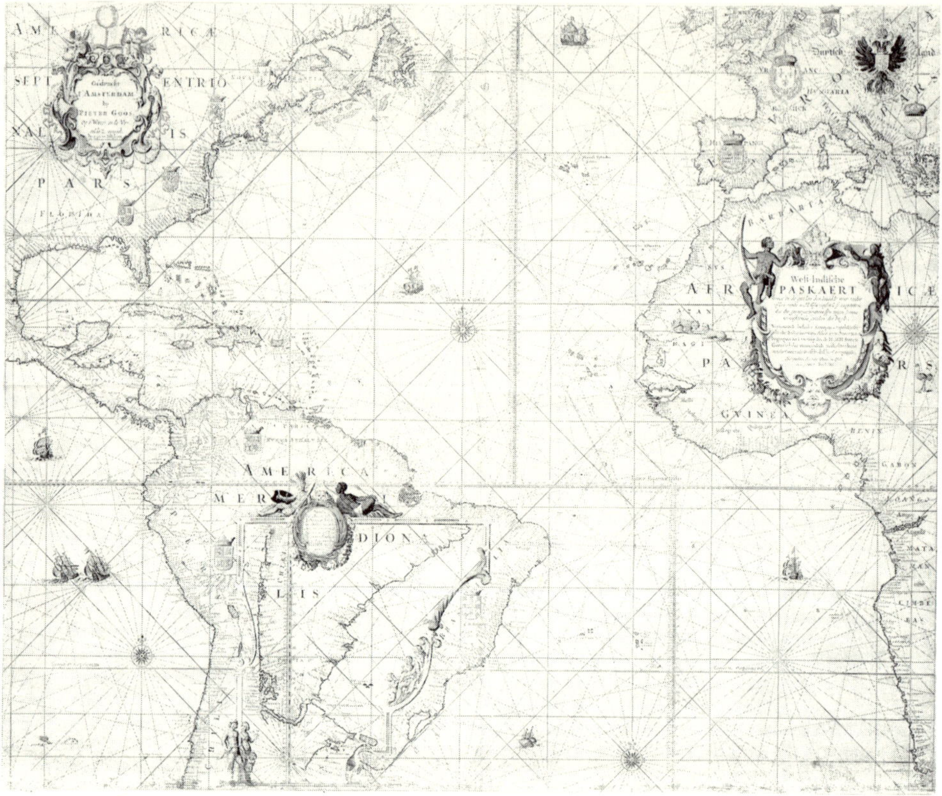

MAP 1. *West-Indische Paskaert.* By Johannes van Keulen. Netherlands, 1710. Black-and-white line engraving with period color. Courtesy, The Colonial Williamsburg Foundation.
The first pan-Atlantic paskaert (or "passage map") was published in the Netherlands in 1621, as rival European powers established colonies and contested trade in the four continents bordering the Atlantic Ocean.

The colony did eventually prosper, however, once settlers found a viable export product and once private property replaced the corporate forms of organization that had offered too few incentives for striving either for the public good or for private gain. A minority of hardened colonists, who survived the disastrous early years and who learned how to grow enough food to sustain themselves and to raise an export crop the proceeds of which enabled them to buy essential imported manufactures, began to acquire farmsteads of their own. These individual holdings also soon came to be termed *plantations,* and their owners, whatever their scale of operation, came to be known as *planters.* The emergence of such privately held plantations is the subject of Chapter 1.

Still, had these early Virginia planters achieved nothing more than feeding and housing themselves, the colony would likely have gone the way of other foundering and ultimately unsuccessful New World colonizing ventures. All that changed when tobacco emerged as a viable export commodity. Although this New World product was frowned upon in official circles in England, it quickly found a ready market among European consumers. In the absence of other commercially sustainable products, for the next two centuries the fortunes of Chesapeake planters became inextricably tied to markets for the addictive "sot weed."

This study uses the terms *planter* and *plantation* according to their seventeenth-century meanings. Anyone who made his (or occasionally her) living primarily from agriculture is considered a planter, and any viable farm, according to contemporary practice a holding of at least fifty to one hundred acres, is considered a plantation. No modern categorizations are employed that use some arbitrarily selected number of acres or of bound laborers to separate smaller, presumably family operated "farms" that might or might not have been profit-making enterprises from larger units termed "plantations" worked by bigger groups of bound laborers, typically supervised by overseers or farm managers, and unambiguously managed for profit.

After the English settlers overcame the initial difficulties of establishing viable settlements in the Chesapeake, the prospect of making a fortune from tobacco lured many more English settlers to the region. From the 1630s into the 1670s, the Chesapeake remained a good "poor man's country" where men and women possessed of modest capital or who were willing to finance migration through a labor contract might, if they survived the hostile Chesapeake disease environment, achieve a greater competency than they would have had they remained at home.

The motives that prompted Englishmen to emigrate, as Robert Wintour, a Catholic gentleman from Gloucestershire who intended to move to the new Maryland settlement, explained in a treatise of 1635, were honor, pleasure, and, above all, profit. Settlers could gain honor from strengthening the English state by becoming "Founders of a nation" and from converting the Indians to Christianity and civilized European ways. Pleasure would come from owning one's own land, producing one's own food in plenty, harvesting fish and game from one's own estate, basking on cold days before a good fire made of one's own wood in one's own chamber or parlor, having servants to serve their master meat on Sundays, and enjoying the "friendly neighbourhood" of other like-minded freeholders and gentlemen settled in the area. Such pleasures could be realized in the Chesapeake colonies because of the large profits that could be made by transporting indentured servants to raise a variety of export and food crops that would provide a handsome subsistence and could be traded for necessities and amenities that could not be produced locally.[2]

Unfortunately, Wintour himself had little chance to garner profits or to enjoy the pleasures of landownership and a comfortable subsistence. He arrived in Maryland in January 1637, bringing with him seven indentured servants and equipment for planting, fishing, trading, building, shoemaking, and defense—as well as books, pictures, and a cittern to provide pleasure in leisure hours. By the next year, Wintour had established his own plantation, but he never had a chance to marry, raise a family, or even, it seems, to grow tobacco, for, sometime during his second summer in the Chesapeake, Wintour died. In his brief career in Maryland, Wintour achieved the honor he aspired to, being appointed a justice of the peace, a captain in the militia, and named a member of the council. And he might have realized a small profit trading beads and knives for beaver skins with Indian neighbors. His start-up costs were high, however, and the debts he incurred in the first year and a half slightly exceeded the value of his movable property. Many other hopeful emigrants experienced a similar fate, but most of those who survived the first year or two of "seasoning" eventually realized some of the benefits that had induced Robert Wintour to leave England. As he had perceptively recognized, for men of limited means "the best of the business . . . [was] at the first," before rich men took over the colony and eliminated opportunities for settlers with little wealth. The era between 1640 and 1679 when small and middling planters were the primary producers in the region is surveyed in Chapter 2.[3]

But, as plantations proliferated and tobacco production expanded, high profits diminished to increasingly modest, sometimes paltry, returns between

2. Robert Wintour's account is reproduced and analyzed in John D. Krugler, ed., *To Live Like Princes: A Short Treatise Sett Downe in a Letter Written by R. W. to His Worthy Friend C. J. R. concerning the Plantation Now Erecting under the Right Honorable Lord Baltemore in Maryland . . .* (Baltimore, 1976), 27–37, esp. 28, 32, and 36. The account is further discussed in Krugler, *English and Catholic: The Lords Baltimore in the Seventeenth Century* (Baltimore, 2004), 135–151.

3. Krugler, ed., *To Live Like Princes,* 40. Wintour's subsequent career is documented in Historic Saint Mary's City biographical files, typescripts, MSA, and in Edward C. Papenfuse et al., *A Biographical Dictionary of the Maryland Legislature, 1635–1789,* 2 vols. (Baltimore, 1979), II, 905. Studies of the period that emphasize the opportunities available to early Chesapeake settlers include Russell R. Menard, "From Servant to Freeholder: Status Mobility and Property Accumulation in Seventeenth-Century Maryland," *WMQ,* 3d Ser., XXX (1973), 37–64; Lois Green Carr and Russell R. Menard, "Immigration and Opportunity: The Freedman in Early Colonial Maryland," in Thad W. Tate and David L. Ammerman, eds., *The Chesapeake in the Seventeenth Century: Essays on Anglo-American Society* (Chapel Hill, N.C., 1979), 206–242; and Lois Green Carr, Russell R. Menard, and Lorena S. Walsh, *Robert Cole's World: Agriculture and Society in Early Maryland* (Chapel Hill, N.C., 1991).

1680 and 1729. Planters exhausted any technological improvements they could make in raising tobacco, and most that could be achieved in packing it, and, unless they were able to expand their labor forces, faced increasingly hard times. Opportunities for newcomers declined as well because long-established planters—who fared better than freed servants or recently arrived free immigrants—were usually able to set up their children with portions of land, bound laborers, and already developed farms. Few propertyless immigrants were able to overcome the advantages that such inheritances conferred on the more privileged native born, and, with diminishing opportunities to acquire freehold land, fewer poor Europeans chose to migrate to the Chesapeake. In the region overall, African slaves quickly replaced indentured servants as the primary form of bound labor. Since initially only rich planters could afford to buy more expensive Africans, the wealthy gained from the shift in labor systems, while ordinary planters suffered hard times. Growing inequality among whites went hand in hand with increasing degradation and exploitation of enslaved blacks.

Distinct subregions emerged as planters in more marginally productive areas abandoned tobacco and turned to an array of diversified agricultural activities and to the extraction of natural resources that they began to sell to other colonies instead of trying to produce staples for European markets. Meanwhile, Virginia planters in the lower tidewater developed sweet-scented tobacco, a new variety that for a time commanded high returns in the English market. The strategies and fortunes of planters able to raise the more favored strain began to diverge from the strategies and fortunes of those living in areas where only ordinary grades of tobacco could be raised. Chapters 3 and 4 analyze planters' struggles to maintain or to expand their agricultural revenues in the face of stagnant or falling prices for their staple, decreasing availability of good land in long-settled areas, and severe shortages of labor among all but the planter elite.[4]

Then, beginning in the 1730s, tobacco prices again improved, and so did plantation revenues, now often augmented with returns from grains and other farm and forest products and from import replacement activities.

4. An earlier treatment of these topics appeared in Lorena S. Walsh, "Summing the Parts: Implications for Estimating Chesapeake Output and Income Subregionally," *WMQ*, 3d Ser., LVI (1999), 53–94. See also Carville V. Earle, *The Evolution of a Tidewater Settlement System: All Hallow's Parish, Maryland, 1650–1783* (Chicago, 1975); Paul G. E. Clemens, *The Atlantic Economy and Colonial Maryland's Eastern Shore: From Tobacco to Grain* (Ithaca, N.Y., 1980); Gloria L. Main, *Tobacco Colony: Life in Early Maryland, 1650–1720* (Princeton, N.J., 1982); and David Scott Hardin, "'Alterations They Have Made at This Day': Environment, Agriculture, and Landscape Change in Essex County, Virginia, 1600–1782" (Ph.D. diss., University of Maryland, 1995).

More parents were living to see at least one of their children come of age, so consequently elite planters were increasingly preoccupied with expanding operations in order to provide handsome portions for their offspring. Not just distinct subregional strategies but also individual family strategies emerged for maintaining and augmenting family fortunes. Either through adopting a more complex agricultural calendar that allowed them to expand and diversify sources of revenues or through astute manipulation of credit, most elite planting families did succeed across the mid-eighteenth century in replicating the levels of wealth and status they had achieved over several generations. Middling and small planters as well as large producers benefited from the opening of new markets and from the quickening pace of trade. Westward migration afforded an outlet for an ever-growing white population, and large-scale importation of enslaved Africans, followed by the onset of natural increase among the enslaved, solved problems of labor shortage that had bedeviled ordinary farmers in earlier decades. Chapters 5, 6, and 7 discuss the so-called Golden Age of Chesapeake agriculture, examining individual successes and failures in a maturing colonial society and economy. The study then concludes with a reassessment of the Golden Age, followed by a brief look at what lay ahead for Virginia and Maryland planters as a rapidly developing imperial crisis dashed their hopes for a renewal of normalcy and prosperity at the end of the Seven Years' War.[5]

The primary focus of this book is on the operation of individual plantations and on the decisions that individual planters made about how they would manage their farms and agricultural workers and market their produce. This approach reveals the wide range of coping strategies large planters adopted and demonstrates how these strategies changed over time. However, the study also attempts to place the successes or failures of individual plantation managers in the broader context of the history of agriculture and society in the Chesapeake, incorporating the many analyses of the region's economy, society, and material culture that have appeared over the past half a century.

The extent to which Chesapeake planters succeeded in building, maintaining, or improving their fortunes and, furthermore, whether they were generally efficient, rational farm managers or inefficient, wasteful husbandmen has been a topic for debate among several generations of historians. Most European travelers who visited Virginia and Maryland in the seventeenth,

5. After 1763, the economic and political conditions in which Chesapeake farmers had to operate changed profoundly. Those developments, as well as the expansion of the plantation regime into the piedmont, are intended to be the subjects of another study on the management of large Chesapeake plantations up to 1820.

eighteenth, and early nineteenth centuries castigated Chesapeake planters for their slovenly husbandry that relied on extensive farming methods that failed to measure up to more advanced European practices, and many early-twentieth-century scholars accepted these negative contemporary assessments. Moreover, the international tobacco market was subject to recurring booms and busts, and many historians have also accepted at face value planters', merchants', and government officials' numerous and bitter complaints about the deleterious effects of overdependence on tobacco that proliferated during economic downturns, but they have given less credence to more optimistic assessments made during economic booms. Finally, a long-standing historiography also accepts planters' laments about debts owed to British merchants on the eve of and after the American Revolution as evidence for increasing planter indebtedness well before the war arising from planters' naivete in dealing with British merchants and from wildly extravagant living. Chesapeake planters have thus often stood indicted for irrational overproduction of their staple as well as for employing poor farming methods, and for being notoriously bad financial managers to boot. In contrast, other economic historians have developed a greater appreciation for the obstacles Chesapeake farmers faced in adopting European farming methods and for the entrepreneurial strategies they devised to remedy the problems of too great dependence on a single staple; they have concluded that, given the mix of resources that the settlers commanded, the majority made rational use of abundant land and of scarce labor and capital. The accounts of individual management practices included in this study provide powerful support for arguments for rational decision making, but they also reveal the tremendous social costs that slave-based plantation agriculture entailed.[6]

In the second half of the twentieth century, historians of the New England colonies also intensely debated whether farmers in that region aimed only to achieve a modest competency for themselves and their families and

6. The classic study of Chesapeake farmers as wasteful husbandmen is Avery Odelle Craven, *Soil Exhaustion as a Factor in the Agricultural History of Virginia and Maryland, 1606–1860* (1926; rpt. Gloucester, Mass., 1965). Contrary arguments can be found in Aubrey C. Land, "Economic Base and Social Structure: The Northern Chesapeake in the Eighteenth Century," *JEH,* XXV (1965), 639–654; Paul G. E. Clemens, "The Operation of an Eighteenth-Century Chesapeake Tobacco Plantation," *Ag. Hist.,* XLIX (1975), 517–531; Carville V. Earle, "The Myth of the Southern Soil Miner: Macrohistory, Agricultural Innovation, and Environmental Change," in Earle, *Geographical Inquiry and American Historical Problems* (Stanford, Calif., 1992), 258–299; Edward C. Papenfuse, Jr., "Planter Behavior and Economic Opportunity in a Staple Economy," *Ag. Hist.,* XLVI (1972), 297–312; and Carr, Menard, and Walsh, *Robert Cole's World.* On the mentalité of the Chesapeake planter, see T. H. Breen, *Tobacco Culture: The Mentality of the Great Tidewater Planters on the Eve of Revolution* (Princeton, N.J., 1985).

to govern their own working lives and so had little interest in production for market, or whether they eagerly embraced the market opportunities (and their attendant risks) that became available. Given Chesapeake farmers' heavy dependence on tobacco, their involvement from the outset in production for distant markets has seldom been questioned. However, scholars of Virginia and Maryland have also been interested in how farm families allocated resources, mobilized family labor, and developed strategies, more or less successful, for endowing succeeding generations. Most studies exploring these questions have concentrated on the small to middling planters who predominated in the region in the seventeenth century, families who operated on a scale similar to northern farmers and who made little or no use of slave labor. Early Chesapeake elites have not been entirely neglected, but most emphasis has been placed on how they consolidated political and cultural rather than economic power.[7]

7. Among the studies of New England farmers that have emphasized an absence of market orientation are Percy Wells Bidwell and John I. Falconer, *History of Agriculture in the Northern United States, 1620–1860* (1925; rpt. New York, 1941); James A. Henretta, "Families and Farms: Mentalité in Pre-Industrial America," *WMQ*, 3d Ser., XXXV (1978), 3–32; Michael Merrill, "Cash Is Good to Eat: Self-Sufficiency and Exchange in the Rural Economy of the United States," *Radical History Review*, IV (1977), 42–71; and Christopher Clark, *The Roots of Rural Capitalism: Western Massachusetts, 1780–1860* (Ithaca, N.Y., 1990). Scholars arguing for market orientation include Charles S. Grant, *Democracy in the Connecticut Frontier Town of Kent* (New York, 1961); Richard L. Bushman, *From Puritan to Yankee: Character and the Social Order in Connecticut, 1690–1765* (Cambridge, Mass., 1967); Stephen Innes, *Creating the Commonwealth: The Economic Culture of Puritan New England* (New York, 1995); and, especially, Winifred Barr Rothenberg, *From Market-Places to a Market Economy* (Chicago, 1992), which relies heavily on analysis of farm account books and diaries. A good summary of the literature is found in Daniel Vickers, "The Northern Colonies: Economy and Society, 1600–1775," in Engerman and Gallman, eds., *Cambridge Economic History of the United States*, I, 209–248. See also Allan Kulikoff, "The Transition to Capitalism in Rural America," *WMQ*, 3d Ser., XLVI (1989), 120–144; and Daniel Vickers, "Competency and Competition: Economic Culture in Early America," ibid., XLVII (1990), 3–29.

For analyses of Chesapeake elites, see Bernard Bailyn, "Politics and Social Structure in Virginia," in James Morton Smith, ed., *Seventeenth-Century America: Essays in Colonial History* (Chapel Hill, N.C., 1959), 90–115; David W. Jordan, "Political Stability and the Emergence of a Native Elite in Maryland," and Carole Shammas, "English-Born and Creole Elites in Turn-of-the-Century Virginia," both in Tate and Ammerman, eds., *The Chesapeake in the Seventeenth Century*, 243–273, 274–296; Martin H. Quitt, "Immigrant Origins of the Virginia Gentry: A Study of Cultural Transmission and Innovation," *WMQ*, 3d Ser., XLV (1988), 629–655; and Allan Kulikoff, "The Colonial Chesapeake: Seedbed of Antebellum Southern Culture?" *Jour. So. Hist.*, XLV (1979), 513–540.

Economic and social comparisons of the Chesapeake with New England are summarized in Jack P. Greene, *Pursuits of Happiness: The Social Development of Early Modern British Colonies and the Formation of American Culture* (Chapel Hill, N.C., 1988).

From the late 1950s on, there have also been heated debates among economic and social historians about the efficiency and profitability of slave-based agriculture in the antebellum cotton South. In contrast to the research on small New England farmers, most attention in this controversy has centered on large plantations. In *Time on the Cross: The Economics of American Negro Slavery* (1974), Robert William Fogel and Stanley L. Engerman contested and have subsequently largely succeeded in overturning long-held assumptions that by the 1850s and 1860s slave labor and agricultural production based on slave labor were economically inefficient and that slavery was economically moribund. Some opponents of cliometric approaches judged the application of sterile calculations of economic efficiency and of profit or loss to a system of exploitation now considered immoral and indefensible to be callous and the finding that slavery was economically profitable morally irresponsible. The controversies generated by *Time on the Cross* subsequently led to a proliferation of quantitative studies of plantation profitability and of the economics of slavery generated by both supporters and opponents of Fogel and Engerman's approach. That antebellum slave-based plantation agriculture was economically viable and profitable has now been widely accepted, aided by a growing recognition that technological efficiency is, not an inherent good, but morally neutral. This outcome was not accidental, as planters responded appropriately to incentives and paid close attention to shifting markets. As Fogel has more recently put it, "Although American slavery was deeply immoral and politically backward, it might, nevertheless, have been a highly efficient form of economic organization that was able to sustain high rates of economic growth and yield substantial profits to its ruling class." On the other hand, Fogel and Engerman's finding that southern slave agriculture was more productive than northern free agriculture continues to generate debate and further research into both forms of economic organization.[8]

8. The publication of Alfred H. Conrad and John R. Meyer, "The Economics of Slavery in the Ante Bellum South," *Journal of Political Economy*, LXVI (1958), 95–130, initiated the debate about the profitability of antebellum plantation agriculture. The most influential and controversial study was Robert William Fogel and Stanley L. Engerman, *Time on the Cross: The Economics of American Negro Slavery*, 2 vols. (Boston, 1974). Further research was reported in Fogel, *Without Consent or Contract: The Rise and Fall of American Slavery* (New York, 1989); Fogel, Ralph A. Galantine, and Richard L. Maning, eds., *Without Consent or Contract: The Rise and Fall of American Slavery: Evidence and Methods* (New York, 1992); Fogel and Engerman, eds., *Without Consent or Contract: The Rise and Fall of American Slavery: Technical Papers*, I, *Markets and Production* (New York, 1992); Fogel and Engerman, eds., *Without Consent or Contract: The Rise and Fall of American Slavery: Technical Papers*, II, *Conditions of Slave Life and the Transition to Freedom* (New York, 1992).

Studies by critics include Paul A. David et al., *Reckoning with Slavery: A Critical Study in*

Chesapeake scholars have addressed the issues raised by Fogel and Engerman primarily for the years following the American Revolution when the future viability of slavery became a subject for contemporary debate. And many have argued that, between 1775 and the invention of the cotton gin, slavery, at least in the Chesapeake, had become unprofitable, although few have attempted to prove their assertions with direct measurements. The profitability of slave-based agriculture in the colonial era has received much less scrutiny, as has the issue of the comparative efficiency of farms worked by slaves, those using indentured servants, and those employing only free family labor. This neglect is in part a result of the dearth of documentary evidence for large plantations in the critical formative years of the late seventeenth and early eighteenth centuries. Edmund Morgan, for example, in *American Slavery, American Freedom* did argue that slave labor was more profitable than servant labor in the seventeenth century, but he lacked economic data to conclusively prove the point. Moreover, he also argued that slavery was adopted in part for reasons that transcended economics; racial bondage was a means for eliminating the political and social unrest caused by unruly white servants and freedmen.[9]

Other colonial economic historians are not convinced that slaves were more profitable than servants or that slavery was a more efficient form of labor extraction than was indentured servitude. They argue that rich plant-

the Quantitative History of American Negro Slavery (New York, 1976); and Herbert G. Gutman, *Slavery and the Numbers Game: A Critique of "Time on the Cross"* (Urbana, Ill., 1975). Robert William Fogel, *The Slavery Debates, 1952–1990: A Retrospective* (Baton Rouge, La., 2003), provides a summary of the issues (quotation on 30); as does Gavin Wright, "Slavery and American Agricultural History," *Ag. Hist.,* LXXVII (2003), 527–552; and Wright, *Slavery and American Economic Development* (Baton Rouge, La., 2006).

Other debates about slave productivity in the colonial era include Peter Mancall, Joshua L. Rosenbloom, and Thomas Weiss, "Slave Prices and the South Carolina Economy, 1722–1809," *JEH,* LXI (2001), 616–639; Mancall, Rosenbloom, and Weiss, "Agricultural Labor Productivity in the Lower South, 1720–1800," *Exp. Econ. Hist.,* XXXIX (2002), 390–424; David Eltis, Frank D. Lewis, and David Richardson, "Slave Prices, the African Slave Trade, and Productivity in the Caribbean, 1674–1807," *Economic History Review,* LVIII (2005), 673–700; Eltis, Lewis, and Richardson, "Slave Prices, the African Slave Trade, and Productivity in Eighteenth-Century South Carolina: A Reassessment," *JEH,* LXVI (2006), 1054–1065; and Mancall, Rosenbloom, and Weiss, "Slave Prices, the African Slave Trade, and Productivity in Eighteenth-Century South Carolina: A Reply," ibid., 1066–1071. For a reassessment of the profitability of free and slave labor, see Elizabeth B. Field-Hendrey and Lee A. Craig, "The Relative Efficiency of Free and Slave Agriculture in the Antebellum United States: A Stochastic Production Frontier Approach," in David Eltis, Frank D. Lewis, and Kenneth L. Sokoloff, eds., *Slavery in the Development of the Americas* (Cambridge, 2004), 236–257.

9. Edmund S. Morgan, *American Slavery, American Freedom: The Ordeal of Colonial Virginia* (New York, 1975).

ers instead turned to slaves reluctantly as a last resort when the supply of indentured servants became inadequate to meet the region's labor needs. Increasing use of slave labor then further discouraged servant emigration, leaving other planters wishing to expand their workforces with no option but to become slaveowners. Arguments that emphasize planters' lack of choice among different forms of bound labor or that posit noneconomic as well as economic motives for embracing slavery have also deflected attention from investigations of actual plantation performance. If, however, as argued here, the planter elite deliberately chose slaves over servants well before diminished supplies of the latter forced their hand, then questions of the comparative efficiency of slave, servant, and family labor become more central to explaining how and why slavery transformed the economy and society of the colonial Chesapeake.[10]

Measurements of planters' economic performance and analysis of whether or not they made optimal use of available resources is one of the main concerns of this study. It addresses these issues by analyzing all the planter correspondence, account books, and farm diaries that could be located in regional archives. The management practices of the thirty-two best-documented Virginia and Maryland planters provide the core of the analysis, supplemented with insights from less complete collections of plantation correspondence and financial records and by travelers' accounts and merchants' letters. The main advantages of plantation account books and farm diaries is that the individual notations of purchases and sales recorded in them can be re-

10. The most widely accepted interpretation of the transition from servants to slaves has been Russell Menard, "From Servants to Slaves: The Transformation of the Chesapeake Labor System," *So. Stud.*, XVI (1977), 355–390, which argues that Chesapeake planters "did not abandon indentured servitude because they preferred slaves; rather, a decline in the traditional labor supply forced planters to recruit workers from new sources, principally but not exclusively from Africa" (355). Allan Kulikoff, *Tobacco and Slaves: The Development of Southern Cultures in the Chesapeake, 1680–1800* (Chapel Hill, N.C., 1986), presented a similar interpretation. Menard's contention that the transition from servants to slaves occurred during the 1680s continues to hold true for the region as an aggregate.

However, John C. Coombs, "Building 'The Machine': The Development of Slavery and Slave Society in Early Colonial Virginia" (Ph.D. diss., College of William and Mary, 2004), challenges Menard's argument that most planters preferred servants over slaves. Small to middling planters might have favored servants, but Coombs persuasively demonstrates that was not the case among the Chesapeake elite. See also Anthony S. Parent, Jr., *Foul Means: The Formation of a Slave Society in Virginia, 1660–1740* (Chapel Hill, N.C., 2003), who accepts Menard's account of the timing of the transition from servants to slaves but who also presents a harsh critique of the actions and motivations of the Virginia elite. David W. Galenson, "Labor Market Behavior in Colonial America: Servitude, Slavery, and Free Labor," in Galenson, ed., *Markets in History: Economic Studies of the Past* (Cambridge, 1989), 52–96, provides some statistics for elite Marylanders.

assembled into annual summaries of plantation revenues and expenditures. These summaries, in turn, permit an assessment of the profitability of individual farms and of the effectiveness of individual plantation management strategies over a period of years. Correspondence between planters and their farm managers and overseers provide insights into day-to-day farm and labor management practices as well as data on the size of the crops produced. Letters exchanged between planters and merchants supply valuable information on markets and on buyers' and sellers' perceptions of the overall economic climate. These sources also allow comparisons of what planters stated to be their goals (or the goals that can be inferred from an analysis of how they allocated their resources) with measures of their actual performance derived from financial records and from crop outputs recorded in farm diaries and accounts.

Such labor-intensive reconstructions of plantation balance sheets are necessary because Chesapeake planters' accounting practices—like those of contemporary West Indian sugar planters—were not designed for making assessments of profit or loss on an annual or any other regular basis. Nor were the planters themselves interested in such measurements. Many elite planters did have a good sense of the market value of their estates, and some did make estimates of their annual incomes. Farm expenditures, on the other hand, were seldom analyzed systematically. In only two account books covering years before 1764 did planters distinguish between current expenses and capital investments, and only in the 1750s did a few others begin to keep systematic records of the annual proceeds of commercial crops other than tobacco. And, although many kept systematic records of tobacco outputs per plantation and per laborer, the main purpose was to calculate the proportion of the crop due to overseers, not to monitor the performance of laborers. Most planters tended instead to evaluate a plantation's performance only on the basis of how much they got for the year's tobacco crop and whether or not the net proceeds enabled them to buy whatever capital and consumer goods they and their families wanted.[11]

Present-day scholars are more interested in whether slave-based agriculture produced average, lesser, or greater returns than other potential investments and in whether or not plantation managers were optimizing available resources. Contemporary planters did not think in these terms, and, indeed,

11. For sugar planters' mentalité, see Stuart B. Schwartz, "Introduction," in Schwartz, ed., *Tropical Babylons: Sugar and the Making of the Atlantic World, 1450–1680* (Chapel Hill, N.C., 2004), 1–26; and B. W. Higman, *Plantation Jamaica, 1750–1850: Capital and Control in a Colonial Economy* (Kingston, Jamaica, 2005).

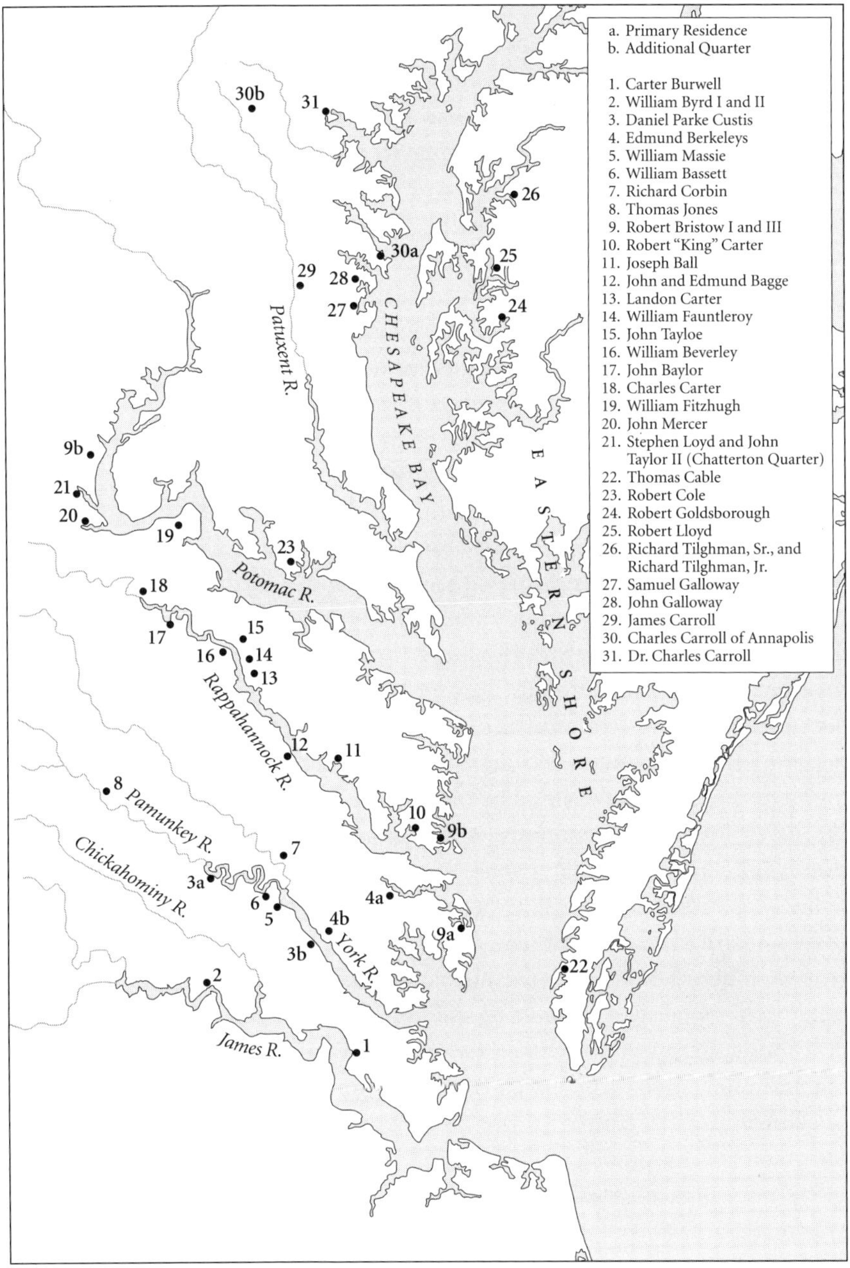

MAP 2. Home Plantations of Elite Planters

the concept of "homo economicus," or pure economic man, is only an analytical abstraction. The concept is useful for analyzing the decisions economic actors made but ought not to be mistaken for an approximation of a reality that was assuredly messier and far more complex. Laziness, inattention, ineptitude, addiction to drink or gaming, or unrestrained passions that could be freely exercised upon largely defenseless subordinates could all undercut rational behavior and decision making. And wars, sickness, droughts, floods, and many other chance misfortunes entirely outside the planter's control could negate the most well-calculated business decisions or the most assiduous application of the best available management practices.

Planters' diaries, letters, and sometimes even their account books provide as much or more information about personal foibles and chance misfortunes, about cooperation or conflict within immediate or extended families, about changing expectations for what constituted acceptable standards of living and levels of provision for subsequent generations, and about relationships with and resistance offered by the enslaved as they do about plantation outputs, revenues, and expenditures. The stories of the rise (and sometimes fall) of personal and family fortunes among the privileged minority who are the focus of this study as well as the more sketchy accounts that can be pieced together from these materials documenting the sufferings, resistance, and occasional minor victories of the bound workers on whose labors the prosperity of these fortunate few depended are often equally or more interesting than the insights provided by bloodless balance sheets revealing owners' profits or losses. Consequently, in the chapters that follow, the organization is both narrative and analytical and both qualitative and quantitative. Whenever surviving records are complete enough to permit the calculation of quantitative measures of planter performance, these calculations are presented. When the qualitative information in the plantation records is especially rich, when it is relevant to estate-building strategies or management decisions, and especially when it has not been published elsewhere, it is presented in narrative form. Given the dearth of systematic empirical work on pre-Revolutionary plantation agriculture, narratives for individual planters are presented in some detail to provide as complete a picture as possible of actual management practices and their outcomes in order to build a basis for assessing plantation efficiency. Since the primary focus of analysis is on issues of profitability, however, the accompanying motives of honor and pleasure are discussed only in the context of estate management and of the creation of a physical environment designed for pleasurable activities. To engage the now extensive literature on honor as a cultural construct or to reconstruct the social activities of large planters would be outside the scope of this study.

Since as a rule only men (and a very few women) who were both wealthy and well educated kept written records of their farming activities, this study examines primarily the performance of the Chesapeake planter elite. I used all the substantive collections of surviving seventeenth- and eighteenth-century plantation records that I could locate, but the result is hardly a representative sample even of large Chesapeake planters. The survival of personal records has surely been haphazard, affected, among other things, by the later political prominence of particular planting families. Political leaders of the Revolutionary and early national eras were not infrequently indifferent or inattentive plantation managers. Their initial elevated status usually depended, however, on their inheriting assets from one or more prior generations of elite planters who did achieve above average success in agriculture. The biases resulting from selective record keeping and subsequent record survival cannot be easily measured since there are virtually no period population censuses and only a smattering of local tax lists.[12]

The numbers of large planters in Virginia and Maryland and their proportion among all free heads of household cannot, then, be determined with any precision before the compilation of comprehensive tax lists enumerating holdings in land, slaves, and other capital assets that were first made in the new American Republic in the 1780s. For the colonial era, few tithable lists (that give the number of heads of households and of taxable adult free and bound laborers for individual counties) or rent rolls (that list all landholdings in a colony at a given point in time) have survived. Often the only way to quantify the proportions of various economic groups is through intensive local studies wherein biographies of resident free adults are painstakingly reconstituted from a wide variety of county and provincial records. The amount of land an individual held, for example, must usually be pieced together by tabulating grants, surveys, patents, purchases, and sales as well as tracing the descent of tracts conveyed by gift or inheritance. The size of his or her workforce is usually known only at the death of a servant or slaveholder, and then only if the number, age, and sex of bound workers was specified in a will or probate inventory. Since many colonists did not make wills, the law did not require that estates of solvent decedents be appraised, and the records of a number of colonial Chesapeake counties have been destroyed, much of the essential information was never recorded or is now lost. Moreover, the most

12. By contrast, see William Kauffman Scarborough, *Masters of the Big House: Elite Slaveholders of the Mid-Nineteenth-Century South* (Baton Rouge, La., 2003), for example, who was able from the manuscript 1850 and 1860 federal censuses to compile a list of the 338 planters who owned at least 250 slaves in the 1850s in the fifteen slave states.

comprehensive data on movable property comes from probate inventories, which value the wealth of decedents rather than of the living population, a group that was older and wealthier than average and hence not representative of the whole population. Most useful are censuses of the living population, again painstakingly reconstituted from local records. This approach requires several years of work in the records of one well-documented county to produce results for just a few points in time, however, and is too labor intensive to be feasible for an entire colony.[13]

A further problem in estimating the size of the planter elite over the course of a century and a half is that the number of acres and of bound laborers that would place a man in the top 1 or 2 percent of resident householders changed dramatically between the early seventeenth and the mid-eighteenth century. Once slaves replaced indentured servants as the primary form of bound labor, the value of large planters' estates rose appreciably. The capital assets owned by men of the local economic elite in the 1640s, 1650s, and 1660s were often no greater than those typically accumulated by middling planters a century later. Robert Wintour, for example, who transported seven indentured servants in 1637 had assets sufficient to merit appointment to the Maryland council, whereas by the 1730s a number of Chesapeake councillors owned one hundred bondpeople or more. Thus, one cannot apply a single set of economic benchmarks across the whole period.

Given all these difficulties, most colonial Chesapeake historians who study economic and social elites have opted to use members of the legislature or major county officeholders as a proxy for the planter elite, if only because lists of provincial and county officers can be easily obtained from surviving records. Moreover, numerous county studies confirm the validity of equating the local political elite with the local economic elite, since resident large planters (conventionally defined as those owning one thousand or more acres of land) almost invariably served in one or more of the major county offices of justice of the peace, sheriff, and delegate to the lower house of the legislature. In all counties across most of the seventeenth century, and in newly

13. Examples of economic ranking based on the 1780s tax lists are Jackson Turner Main, "The One Hundred," *WMQ*, 3d Ser., XI (1954), 354–384; and Main, "The Distribution of Property in Post-Revolutionary Virginia," *Mississippi Valley Historical Review*, XLI (1954–1955), 241–258. Examples of reconstituted censuses include Darrett B. Rutman and Anita H. Rutman, "'More True and Perfect Lists': The Reconstruction of Censuses for Middlesex County, Virginia, 1668–1704," *VMHB*, LXXXVIII (1980), 37–74; Lorena S. Walsh, "The Historian as Census Taker: Individual Reconstitution and the Reconstruction of Censuses for a Colonial Chesapeake County," *WMQ*, 3d Ser., XXXVIII (1981), 242–260; Walsh, "Staying Put or Getting Out: Findings for Charles County, Maryland, 1650–1720," ibid., XLIV (1987), 89–103; and James R. Perry, *The Formation of a Society on Virginia's Eastern Shore, 1615–1655* (Chapel Hill, N.C., 1990).

settled areas in the eighteenth century, however, the number of high offices to be filled regularly exceeded the supply of wealthy candidates, resulting in the appointment or, in the case of the lower house of the legislature, election of a number of men of more modest means. Clearly there were fewer large planters than there were major officers.[14]

Exceedingly crude estimates of the total number of elite Chesapeake planters are offered here for selected years based on the somewhat arbitrary assumption that large planters with land and labor holdings well above the prevailing mean constituted 1 percent of the total white male population and that truly great planters with land and labor holdings well above those of ordinary large planters (thus meeting the qualifications for provincial officers) were one-tenth of the top 1 percent (Table 1). The total number of men operating at the level of the individual planters discussed in subsequent chapters rose from roughly fifty in 1640 to around eight hundred by 1760. Great planters holding thousands of acres and exceptionally large labor forces likely increased from fewer than a dozen in 1640 to around eighty at the close of the Seven Years' War. Complaints of colonial governors across the seventeenth and first quarter of the eighteenth century of a dearth of men of requisite wealth and character available to fill a dozen council seats and of the low wealth and educational levels of many county officers further reinforces

14. For major officeholders considered as representative of the economic gentry, see Jack P. Greene, "Foundations of Political Power in the Virginia House of Burgesses, 1720–1776," *WMQ*, 3d Ser., XVI (1959), 485–506; Quitt, "Immigrant Origins of the Virginia Gentry," ibid., XLIV (1988), 629–655; Martin H. Quitt, *Virginia House of Burgesses, 1660–1706: The Social, Educational, and Economic Bases of Political Power* (New York, 1989); and Jordan, "Political Stability," in Tate and Ammerman, eds., *The Chesapeake in the Seventeenth Century*, 243–273. For the coincidence of economic and political elites on the county level in Virginia, see David Alan Williams, "Political Alignments in Colonial Virginia Politics, 1698–1750" (Ph.D. diss., Northwestern University, 1959), chap. 3; Robert Anthony Wheeler, "Lancaster County, Virginia, 1650–1750: The Evolution of a Southern Tidewater Community" (Ph.D. diss., Brown University, 1972); Cathleene B. Hellier, "'The Bigwigs': The County Court of York County, Virginia, 1700–1705" (seminar paper, 1984, CWR); Linda Hunter Rowe, "Peopling the Power Structure: Urban Oriented Officeholders in York County, Virginia, 1699–1780" (master's thesis, College of William and Mary, 1989); Gwenda Morgan, *The Hegemony of the Law: Richmond County, Virginia, 1692–1776* (New York, 1989), chap. 2; and Darrett B. Rutman and Anita H. Rutman, *A Place in Time: Middlesex County, Virginia, 1650–1750* (New York, 1984), 143–163. For Maryland, see Lois Green Carr, "County Government in Maryland, 1689–1709" (Ph.D. diss., Harvard University, 1968), 467–483, 610–655, 676–691; Kulikoff, *Tobacco and Slaves*, 267–276; and Lorena S. Walsh, "The Development of Local Power Structures: Maryland's Lower Western Shore in the Early Colonial Period," in Bruce C. Daniels, ed., *Power and Status: Officeholding in Colonial America* (Middletown, Conn., 1986), 53–71. For the paucity of wealthy officers in frontier counties, see Richard R. Beeman, *The Evolution of the Southern Backcountry: A Case Study of Lunenburg County, Virginia, 1746–1832* (Philadelphia, 1984), chap. 2.

TABLE 1. Elite Chesapeake Planters, 1640–1760

Year	Top 1% white males: no.	Top 1/10 of 1% white males: no.	Total white population
1640	54	5	8,000
1680	250	25	55,600
1700	315	32	85,200
1730	566	57	171,400
1750	613	61	227,200
1760	843	84	312,400

Note: Adult males were assigned the following proportions of the total white population: 1640, 67% (estimated from percentage of adult men in 1625 Virginia census); 1680, 45% (estimated from 1704 Maryland census); 1700, 37% (from 1704 Maryland census); 1730, 33% (estimated from 1704 and 1755 Maryland censuses); 1750, 27% (from 1755 Maryland census); 1760, 27% (assuming no change from 1755).
Sources: Population estimates are based on John J. McCusker and Russell R. Menard, *The Economy of British America, 1607–1789* (Chapel Hill, N.C., 1985), 135–136.

a contention that the number of great planters was indeed limited. More concretely, research in county tax lists for the 1760s reveals that large planters then made up only between 2 and 7 percent of all adult heads of household. Consequently, estimating the number of elite planters as the top 1 percent of all males appears not unreasonable.[15]

In contrast to the planter elite, the management decisions of middling and small planters who left no written accounts of their farming operations can be studied primarily through probate records. From the contents of their estate inventories, from the levels of movable wealth they left at death, and from the wills in which they laid out the disposition of their worldly assets, scholars have inferred a great deal about their farming strategies and about

15. For shortages of qualified men to serve as councillors, see David W. Jordan, "Maryland's Privy Council, 1637–1715," in Aubrey C. Land, Lois Green Carr, and Edward C. Papenfuse, eds., *Law, Society, and Politics in Early Maryland* (Baltimore, 1977), 65–87; and Williams, "Political Alignments," 248–252. The percentage of large planters among householders is based on unpublished research kindly provided by Kevin Kelly, historian, CWR. John C. Coombs has identified all elite Virginia slaveowners who appear in seventeenth-century county and provincial records. Taking into account loss of information from counties whose records have not survived, Coombs estimates that, during the 1680s and 1690s, no more than fifty men had achieved the level of wealth and influence associated with the term "great planter" ("Beyond the 'Origins Debate': Rethinking the Rise of Virginia Slavery," in Douglas Bradburn and John C. Coombs, eds., *Early Modern Virginia: New Essays on the Old Dominion* [Charlottesville, Va., forthcoming]).

their successes and failures as husbandmen. The main disadvantage of probate inventories is that, although they provide a detailed snapshot of plantation stocks at one point in time — shortly after the death of the farm owner — they reveal little or nothing about the income flows during previous years that produced that level of wealth. Throughout this work, I have incorporated the results of other studies of Chesapeake agriculture derived from probate records in order to provide a broader context for assessing the operations of the elite planters on whom I have primarily focused as well as for forming a more firmly based assessment of the performance of small to middling planters in the region. However, I undertook no new aggregate analyses of probate materials.

The survey of plantation management practices for the years before 1680, a period for which no individual plantation account books or substantive collections of planter correspondence survive, necessarily concentrates on the few corporately sponsored James River settlements for which there are documentary or archaeological records and, for the mid-seventeenth century, on the one middling Maryland plantation for which a detailed contemporary guardian account has been preserved in provincial probate records. Thereafter, Chapters 3 through 7 employ a geographical framework of analysis. Discussions of the normative characteristics of large plantation agriculture for 1680–1729 and 1730–1763 begin with a survey of the lower James / York River peninsula, encompassing the area surrounding the initial Jamestown settlement. Next planter practices in Virginia's later-settled Rappahannock and Potomac basins are discussed before proceeding to an analysis of developments on Maryland's tobacco-centered lower western and less tobacco-centered upper eastern shores. Finally, the contrasting strategies of colonists living in peripheral tobacco areas — the lower Delmarva Peninsula and counties on the south bank of the James River — are considered. Chapters 4 and 7 conclude with overall assessments of agricultural changes and adaptations on large plantations in the Chesapeake circa 1730 and 1760.

In the years covered by this study, planting was not an occupation that more privileged men chose from a long list of potential vocations because they had an affinity or aptitude for farming. The great majority of Chesapeake residents had few ways for making a living outside agriculture. Given the virtual absence of towns, almost everyone lived in the countryside and earned their bread from the land. Even professionals such as ministers, lawyers, physicians, and merchants as well as established artisans at a minimum produced most of the food that their families ate. Many professionals and craftsmen were themselves also actively engaged in commercial farming or in directing bound or hired laborers who worked full-time in agriculture.

Although the initial research strategy relied heavily on calculating measures of economic performance, it soon became clear that a narrow econometric study of plantation profitability that abstracted from moral issues would be incomplete and unsatisfactory. All planters operating on a scale similar to those whose farm records have survived were primarily managing, not their own labor and that of their families, but, from the outset, that of voluntarily or involuntarily bound workers. Struggles with uncooperative weather and soils, volatile and unpredictable markets, and crop pests and diseases were only part of the challenges plantation managers faced. They also needed to enlarge their labor forces in order to produce ever greater quantities of tobacco and eventually other cash crops and also to find ways to force or induce the laborers they already owned to work harder and more carefully. At first, the English settlers planned to employ native Americans as willing or unwilling workers as well as European tenant farmers and indentured European men, women, and children. Chesapeake Indians, however, proved unwilling to work voluntarily for the settlers, nor were the English able to enslave them effectively. Schemes for establishing a reliable body of tenant farmers based on Old World models quickly foundered, given the ready availability of land in the New World. A system of indentured servitude, harsher than the servitude in husbandry then prevailing in England but one that still offered some legal protections for servants and economic opportunity for those who survived their terms of service, helped initially to alleviate the labor supply problem and to finance the migration of thousands of poor to middling English settlers to the Chesapeake.

By the 1630s, however, the wealthiest and most politically powerful planters also began looking to the examples of the older Spanish and Portuguese colonies and to the recently founded more southerly English colonies of Bermuda, Providence Island, Saint Christopher, and Barbados. On these islands, English settlers were, like the Iberians, turning to Africans as an alternative source of bound labor, an alien people whom they could conveniently enslave and hold in hereditary bondage on the basis of their non-Christian status and perceived barbarity. In the 1630s and 1640s, members of Virginia's provincial elite began buying African slaves from Dutch traders, and, after Parliament prohibited commerce with the Dutch in 1660, from West Indian merchants and planters with whom they maintained close trading connections. Ordinary planters had to accept whatever bound laborers were offered locally and that they could afford—almost exclusively European indentured servants. The wealthy planter elite who had the luxury of choice among bound laborers increasingly chose to purchase slaves who could be forced to serve for life over servants whose labor they could command for only a term of years. They decided more expensive but more exploitable enslaved workers would produce higher returns in the long run than would cheaper but less exploitable white servants indentured for shorter terms.[16]

By the 1660s, all provincial officials in Virginia had acquired one or more slaves, and, for some, enslaved Africans had become their main source of plantation labor. In the next two decades, the majority of county officeholders in Virginia would follow suit, and elite Marylanders soon emulated their example. These councillors and burgesses then set about passing laws to protect their rights to hold human property. Before the 1690s, few nonofficeholders had either the wealth or the mercantile connections required to purchase slaves. Then, as the supply of indentured servants further diminished and the supply of Africans to the Chesapeake increased dramatically with the abolition of the Royal African Company's monopoly of the African slave trade in 1698, more ordinary planters began buying slaves as well. Soon after the turn of the eighteenth century, slavery was well entrenched in the region, and, as the black population began growing through natural increase, ownership of slaves trickled further down the economic scale.[17]

From the middle of the seventeenth century, among elite planters, plantation management thus included the necessity of compelling unwilling captives to produce the agricultural surpluses on which the planters' prosperity depended. They did so through a combination of a few positive incentives and a larger measure of violence, or they had to find subordinates—general managers and overseers—willing and able to extract such unwilling labor. More or less successful methods of co-option or coercion are consequently a critical aspect of any assessment of planter performance, as are the ways in which the enslaved resisted and frustrated their captors. Few slaveowners had moral qualms about the institution of human bondage until after the end of the Seven Years' War. Until then, most planters accepted slavery as at worst a distasteful necessity, and some rationalized, even romanticized, their role as slaveowners as a positive good. Although it is possible to view some of the

16. Galenson, "Labor Market Behavior in Colonial America," in Galenson, ed., *Markets in History,* provides a good overview of the shift from servants to slaves. The most recent interpretation of Europeans' justifications for African slavery is David Eltis, *The Rise of African Slavery in the Americas* (Cambridge, 2000). For the early adoption of slavery in Barbados, even before the adoption of sugar cultivation, see John J. McCusker and Russell R. Menard, "The Sugar Industry in the Seventeenth Century: A New Perspective on the Barbadian 'Sugar Revolution,'" in Schwartz, ed., *Tropical Babylons,* 289–330.

17. Coombs, "Building 'The Machine,'" chaps. 3, 7.

elite planters' struggles as heroic, and some of their managerial and entrepreneurial successes as admirable, from a present-day perspective their activities as slaveowners cannot be judged other than morally reprehensible.[18]

Outputs of major export crops per laborer (called in the period "shares per hand"), one of the main measures employed in this study to assess planter performance, encapsulate these moral ambiguities. Because labor, not land, was the scarcest form of agricultural capital in the seventeenth and eighteenth centuries, farmers did not calculate annual outputs on the basis of the number of bushels of corn or wheat or pounds of tobacco produced per acre. Rather they calculated returns by the number of bushels or pounds each laborer working full time in the crop raised in a year. (Sugar planters in the West Indies similarly calculated outputs in terms of the weight of sugar produced by the enslaved.)

Some of the factors that affected the size of the tobacco, grain, and fodder crops an agricultural laborer produced involved struggles with nature — the fertility of the soil, the vagaries of weather and growing seasons, the presence of pests and diseases that could decimate crops and kill livestock. Other factors were economic, some of which planters could control and others that they could not. The ups and downs of commodity markets, the costs of agricultural inputs, the availability and price of shipping, and the ability or inability to borrow money are examples of factors over which individuals had little or no control. On the other hand, planters could choose an optimal crop mix, manage the agricultural calendar to raise maximum outputs of complementary crops and to effectively use labor in off seasons, and adopt more productive farming practices. If one were to judge plantation efficiency on tobacco outputs alone, it would be impossible to argue that planters became more efficient in the eighteenth century, since tobacco outputs per worker peaked in the 1690s and then subsequently dropped to mid-seventeenth-century levels. Instead, increased efficiency came in the mid-eighteenth century from large planters' maintaining tobacco production at reasonable if not extraordinary levels while doubling, trebling, and sometimes even quadrupling outputs per worker of either corn or wheat or both. Adoption of labor-saving technologies, advantages of scale in the production of grains, and better use of the workforce year round all contributed to these achievements.

18. For the ideologies of seventeenth- and eighteenth-century elite Virginia planters, see Parent, *Foul Means*. Rhys Isaac, *Landon Carter's Uneasy Kingdom: Revolution and Rebellion on a Virginia Plantation* (Oxford, 2004), describes the heroic side of planters' struggles with nature and the psychological cost of managing slaves but at the same time also emphasizes the levels of brutality and expropriation inherent in the system.

The amount of output per hand also depended, however, on a planter's ability to extract steady and reliable labor from sometimes unwilling or sickly indentured servants and increasingly from even more alienated and unwilling captive Africans and native-born slaves. Economists today conventionally measure work performance in terms of adult male equivalents. That is, they assume that adult female and child workers do not and are not expected to produce as much as an able, healthy, adult man. Colonial and early republican tobacco planters did not make such assumptions. They did allow that very old, chronically sick, or partially disabled men and women as well as children in their early teens just starting work in the fields could produce only half the output of an able-bodied adult. But aside from the very old and the very young, they acknowledged no differences in the product expected from enslaved males and females. And apparently there was in fact little difference, since no one argued that able-bodied women ought not to be considered full producers when it came to apportioning the proceeds of crops between overseers and plantation owners. Outputs per hand thus include a measure of exploitation of the productive powers of enslaved women that by the early eighteenth century would come to be considered inappropriate for white female laborers as well as hard driving of adolescent workers.[19]

This measure also incorporates the whippings, appropriation of enslaved women's reproductive powers, threats of family sale or separation, denial of workers' time for leisure or for proper child care, and the paring down of expenditures for food, clothing, and housing that were the primary means by which planters sought to maximize outputs for given inputs. Crop shares thus also capture a planter's success or failure to effectively employ violence and to expropriate an optimal proportion of workers' product. These are aspects of plantation management all too easy to overlook when evaluating

19. William Tatham, *An Historical and Practical Essay on the Culture and Commerce of Tobacco* (1800), in G. Melvin Herndon, ed., *William Tatham and the Culture of Tobacco* (Coral Gables, Fla., 1969), 101–102, stated that enslaved women working in tobacco were counted only as half sharers and children as quarter sharers, a breakdown that many Chesapeake historians have accepted uncritically. Here Tatham, a tobacco buyer rather than a planter, was simply wrong. In the course of this research, hundreds of contemporary observations of divisions of crops between labor owner and overseer were examined. Occasionally, old or sick men and women were rated as half sharers, as were children aged between nine and twelve who were tending their first (and less often second) crop. In not one of these contemporary observations was a healthy enslaved woman or older girl ever rated as anything other than a full sharer, nor any child as a quarter sharer. For the exploitation of enslaved women, see Jennifer L. Morgan, *Laboring Women: Reproduction and Gender in New World Slavery* (Philadelphia, 2004), 144–146; and Susan Dwyer Amussen, *Caribbean Exchanges: Slavery and the Transformation of English Society, 1640–1700* (Chapel Hill, N.C., 2007), 134–135, 231–236.

sanitized plantation balance sheets. Planters who inherited rather than purchased slaves were in one sense victims of a system of racial slavery that they had not themselves directly created. However, they took few steps to ameliorate the system and none to end it until after the American Revolution. Rather they ratcheted up work requirements and implemented new measures of psychological as well as physical coercion over their bondpeople.

The overall conclusion of this study is that most large Chesapeake planters did follow economically rational plans. They maximized one of the two scarcest resources in the colonies, labor, and accumulated the other, capital, by engaging in farm building. Land, the most abundant resource, they exploited extensively, although with less destructive long-term impact than later would-be "improving" farmers whose attempts to copy Old World models of agriculture often produced more harm than benefit. The rudimentary technologies planters employed were also suitable to the mix of resources they commanded. When it became cost effective to use more advanced technologies such as plowing or more complex methods such as manuring or crop rotations, planters adopted them. And when new markets developed for alternative staples such as grain, livestock, and naval stores, planters diversified their crop mix to take advantage of these opportunities.

The aim of most great planters was surely to labor for profit, or, more accurately, to force others to labor so that they might profit. The history of Chesapeake plantation management is thus not just a story of struggles with nature or of economic success or failure. It is also a study of individual actions in pursuit of profit for oneself and for one's family that in aggregate produced an immense social failure. As the actions of nineteenth-century robber barons and corrupt twenty-first century corporate executives have forcibly reminded us, individual pursuit of profit, both in the past and in the present, does not always promote economic and social welfare for the majority.

CHAPTER 1

The Plantation Economy Begins 1607–1639

In the 1590s and early 1600s, northern Europeans—English, French, and Dutch—sought to establish colonies in the Americas, as had the Spanish and Portuguese a century earlier. Their motivations were many, including religion and national pride, but especially the lure of supposedly easily obtained riches in the form of plunder, precious metals, furs, and a host of semitropical and tropical commodities that could not be produced in northern climates. Tropical products had become familiar to the northern Europeans through voyages of discovery along the coasts of Africa and the Americas, from booty that privateers captured from Iberian vessels, and through legal trade with Spain and illegal trade with Spanish colonists in the Caribbean. If the more northerly countries could successfully establish export-producing colonies of their own, their nations would be enriched, and wealth-draining exchanges with rival Spain diminished. Such efforts were undertaken by private groups of merchants and gentlemen, usually with approval of, but seldom with direct material support from, their national governments. Lacking experience in planting colonies, promoters and projectors could form no reliable estimates of the costs of establishing viable settlements or of the length of time required for returns to begin coming in. Initially, investors in Europe were deceived by overly sanguine reports of early explorers and promoters and, when they discovered that the expenses would be far greater than they first anticipated, were often unwilling to provide the additional capital necessary for creating permanent, export-producing settlements that were a prerequisite for commerce with the Americas. Discouraged by the unexpectedly long period that had to pass before settlements could become self-supporting, adventurers also underestimated the long-range potential of the colonies. They tended to give up as hopeless projects that were just reaching a stage at which they might achieve success.[1]

1. Richard Pares, *Merchants and Planters, Economic History Review,* Supplement IV (1960), chap. 1; K. G. Davies, *The North Atlantic World in the Seventeenth Century,* Europe and the

Those who actually ventured their lives across the seas often quickly lost them. Survivors faced unanticipated hardships from disease, hunger, hostile native inhabitants or European rivals, poor leadership, low morale, and lack of essential supplies from the home country. Both European investors and settlers at first did not realize the difficulties that would be encountered in procuring or producing basic food and shelter and failed to understand how fundamentally familiar European ways of doing things would have to be changed in order to survive in unfamiliar environments. Woefully ignorant of American conditions, investors in England directed colonists, struggling just to stay alive, to produce wine; citrus and other tropical fruits; olives; sugarcane; ginger; dyewoods and dye-yielding plants such as madder (a red dye plant), woad (a blue dye plant), and indigo (all of which England had to import); rape and other oil-producing plants used in the textile industry; medicines; silk, silkgrass (likely several tropical plants having a fiber with the qualities of silk), flax, hemp, and cotton; lumber, pitch, tar, soap ash, and potash along with moderate quantities of tobacco. Even after most of the early trials failed, projectors continued to demand that colonists grow impractical crops because they remained convinced that climate was constant in any latitude across the globe. Despite repeated experience of colder winters than their theories about climate predicted, for example, they clung to the belief that Mediterranean and tropical crops would grow in the Chesapeake region since it shared the same latitude as southern Spain and Italy.[2]

World in the Age of Expansion, IV (Minneapolis, Minn., 1974); Kenneth R. Andrews, *The Spanish Caribbean: Trade and Plunder, 1530–1630* (New Haven, Conn., 1978), chaps. 6–10; Andrews, *Trade, Plunder, and Settlement: Maritime Enterprise and the Genesis of the British Empire, 1480–1630* (Cambridge, 1984), introduction; Andrews, N. P. Canny, and P. E. H. Hair, eds., *The Westward Enterprise: English Activities in Ireland, the Atlantic, and America, 1480–1650* (Liverpool, 1978); Irene Winchester Duckworth Hecht, "The Virginia Colony, 1607–1640: A Study in Frontier Growth" (Ph.D. diss., University of Washington, 1969), chaps. 1, 2; Robert Brenner, *Merchants and Revolution: Commercial Change, Political Conflict, and London's Overseas Traders, 1550–1653* (Princeton, N.J., 1993), 92–102; Karen Ordahl Kupperman, *Providence Island, 1630–1641: The Other Puritan Colony* (Cambridge, 1993), chap. 4; Kupperman, "Errand to the Indies: Puritan Colonization from Providence Island through the Western Design," *WMQ,* 3d Ser., XLV (1988), 70–99; Kupperman, *The Jamestown Project* (Cambridge, Mass., 2007), 208–209, 211–214; John Wood Sweet, "Introduction: Sea Changes," and Eric Griffen, "The Specter of Spain in John Smith's Colonial Writing," both in Robert Appelbaum and Sweet, eds., *Envisioning an English Empire: Jamestown and the Making of the North Atlantic World* (Philadelphia, 2005), 1–24, 111–134. A useful summary is David W. Galenson, "The Settlement and Growth of the Colonies: Population, Labor, and Economic Development," in Stanley L. Engerman and Robert E. Gallman, eds., *The Cambridge Economic History of the United States,* I, *The Colonial Era* (Cambridge, 1996), 135–207.

2. Kupperman, *Providence Island,* chap. 4; Kupperman, "The Puzzle of the American Climate in the Early Colonial Period," *AHR,* LXXXVII (1982), 1262–1289; William Cronon, *Changes in*

In the absence of adequate supplies from abroad, colonists were overwhelmed with the simultaneous need to learn how to raise enough food to sustain themselves and at the same time to find viable export products, learn how to grow and process them, and also to build houses, roads, forts, and other public works. Settlers were expected as well to lay out towns and to build churches; to set up iron- and saltworks, saw-, oil, and gristmills; and to initiate other industries within a year or two of arrival. Not surprisingly, many settlements failed. In those which succeeded, individual settlers absorbed the costs of repeated failures, finding and experimenting with crops that would thrive in local environments, improving processing technologies, and developing means for obtaining a regular supply of additional laborers. Equally critical to successful colonization was a handing over of power from overseas investors to resident planters and the development of local institutions that incorporated planters into the colony's political structure. Only those settlers who gained a measure of direct control over their own destinies would come to think of New World settlements as their home instead of a place of temporary abode where one either made a quick fortune, died, or moved on.[3]

In the more southerly settlements that eventually fulfilled promoters' visions for staple-producing colonies, plantations soon became the predominant form of agricultural organization. Settlers relied almost entirely on Europe for supplies of manufactures and, until international grain markets expanded in the eighteenth century, usually produced only enough food for subsistence. They learned to cultivate high-yielding local crops such as maize and cassava, limiting or abandoning altogether traditional northern European small grains. Those island colonies that could not produce enough food traded staples for supplies of grain and meat from less specialized settlements. All colonies imported European livestock, gradually adopting simpler, less labor-intensive modes of livestock husbandry than were common in Europe. And they learned new, labor-saving land-clearing techniques from native Americans as well as hoe culture (well known to both Indians and Africans),

---

the Land: Indians, Colonists, and the Ecology of New England (New York, 1983), chap. 3. For an example of the promotional literature, see William Strachey, *The Historie of Travell into Virginia Britania (1612),* ed. Louis B. Wright and Virginia Freund, Hakluyt Society, 2d Ser., CIII (London, 1953). For madder, woad, rapeseed, and silkworms, see Joan Thirsk, *Alternative Agriculture: A History from the Black Death to the Present Day* (Oxford, 1997), 72–96, 104–130. For dye plants and silkgrass, see Kupperman, *Providence Island,* 86–87, 101–102.

3. Kupperman, *Providence Island,* 114–116. See Cronon, *Changes in the Land,* chap. 3, for similar misconceptions about the ease of establishing European settlements in New England and the difficulties settlers faced in the first few years before European-style subsistence patterns were reproduced.

which was employed in the production of both food and staple export crops. They used land, which was abundantly available, extensively and concentrated on maximizing the use of scarce labor. Once subsistence was achieved (and sometimes even before), all available effort went into the production of staples. Free migrants who could afford an ocean crossing and bring enough supplies to sustain themselves until they had established viable farms were few. Corporate investors, merchants, and established planters quickly developed methods for financing the transportation of poor European workers, primarily young, single men, through a term of obligatory service upon arrival. Occasionally, labor was also obtained by enslaving native Americans and, hesitantly at first, following the example of the Spanish and Portuguese, by employing enslaved Africans captured from Iberian ships or by raiding or trading for them in Iberian colonies.

Particular Plantations Established

That England's early attempts at establishing colonies in North America were undertaken by English merchants and gentlemen who financed and organized the projects as joint-stock corporate ventures is hardly surprising. English merchants had for some time pursued overseas trades with the Low Countries, the Baltic, Russia, the eastern Mediterranean, and the East Indies using companies chartered by the crown that provided an element of official support and that spread costs and risks among numerous investors. Dutch and French merchants and courtiers were employing similar forms of organization for overseas trade and colonizing ventures. England's James I, like Elizabeth I before him, supported corporate ventures. But he wanted to end the continuing sea war against Spain that had been precipitated by Elizabeth I's encouragement of English freebooters' depredations against Iberian shipping and settlements. He also wanted to maintain England's claims to parts of the Atlantic seaboard sandwiched between Spanish settlements in Florida to the south and in Canada to the north, where the French had planted a colony to sustain their claims to parts of the Americas in 1604.

James lacked both the funds and the desire to undertake colonial ventures directly. He lent tacit support to merchants and other leading men from London, Bristol, and Plymouth, however, who were determined to establish footholds on the North American mainland. The conclusion of peace with Spain in 1604 diminished the possibility that any colonies invested in by merchants, courtiers, and gentlemen would be threatened by the most powerful nation in Europe. In 1606, a royal charter divided temperate North America into two spheres of English interest. West Country merchants dominated the Plymouth Company, which was granted the right to settle a colony in the area between Chesapeake Bay and present-day Maine, and London merchants, organized as the Virginia Company of London, could establish a settlement further to the south. A royal council composed of investors in the two companies was to oversee their activities. In 1607, both the Plymouth and London Companies sent out contingents of settlers. The Plymouth Company's northern settlement at the mouth of the Kennebec River was abandoned after little more than one year.[4]

The settlement the London Company planted at Jamestown experienced more than its share of the problems posed by investors' and settlers' unrealistic expectations, poor leadership, hostile relations with the indigenous inhabitants, periodic starvation, and appallingly high mortality. Only the chance arrival in June 1610 of a fleet bringing fresh supplies and additional settlers prevented the straggling survivors from abandoning a project that had consumed much capital and many lives but that had produced almost no substantive returns. Attempts over the next few years to move most colonists off unhealthy Jamestown Island and to impose order and to promote industry by martial law failed owing to continued high death rates and periodic fighting with the Indians, well-founded fears of a Spanish attack (despite

4. Charles M. Andrews, *The Colonial Period of American History,* 4 vols. (New Haven, Conn., 1934–1938, 1964), I, chap. 2; Theodore K. Rabb, *Enterprise and Empire: Merchant and Gentry Investment in the Expansion of England, 1575–1630* (Cambridge, 1967), chap. 1; Kupperman, *The Jamestown Project,* 191–192; J. H. Elliott, "The Iberian Atlantic and Virginia," in Peter C. Mancall, ed., *The Atlantic World and Virginia, 1550–1624* (Chapel Hill, N.C., 2007), 552–554. The most recent accounts of the early Jamestown story are James Horn, *A Land as God Made It: Jamestown and the Birth of America* (New York, 2005); J. H. Elliott, *Empires of the Atlantic World: Britain and Spain in America, 1492–1830* (New Haven, Conn., 2006), chaps. 1–4; and Kupperman, *The Jamestown Project.*

Initially, the term "Virginia Company" referred collectively to the two joint-stock companies James I chartered in 1606 for the purpose of establishing settlements on the coast of North America. At that time, Englishmen referred to the entire coastal area between Spanish Florida and the areas to the north claimed by France as "Virginia." The two companies, called the Virginia Company of London (or the London Company) and the Virginia Company of Plymouth (or the Plymouth Company), operated with identical charters but with differing territorial bounds that overlapped in the center. The London Company had a primary interest in the more southerly area between the thirty-fourth and forty-first parallels (or from approximately Cape Fear to Long Island Sound). The Plymouth grantees had a primary interest in the more northerly area, although they could establish colonies anywhere between the thirty-eighth and forty-fifth parallels (or roughly between the upper Chesapeake Bay and a point midway through the modern state of Maine). After their settlement on the Kennebec failed in 1608, the Plymouth investors became inactive, and in 1609 the London group came commonly to be called the Virginia Company and the colony they established to be called Virginia. In 1620, some of the Plymouth adventurers reorganized as the Plymouth Council for New England and obtained a new charter giving them rights in the area of New England north of the fortieth parallel (Wesley Frank Craven, *The Virginia Company of London, 1606–1624* [Williamsburg, Va., 1957], 1–14).

the peace then prevailing), and the continued inability of the settlers to feed themselves or to produce profitable commodities. Leaders of the Virginia Company (as the London adventurers were then commonly called) began casting about for a modified form of organization, one that would provide more initiative for individual investors. Their hope was to raise additional capital by exempting new investors in subsidiary joint-stock ventures from paying the crushing debts the Company had already incurred and to create a means for recruiting additional, better-qualified settlers. The story of Jamestown's earliest years has been told many times and will not be recounted here. Rather, discussions of problems that plagued the Virginia colony's first nine years will be introduced where this evidence helps to explain the successes and failures of the particular plantations that replaced Jamestown and three other "boroughs" the Company had established as the main focus for investment, expanded settlement, and agricultural experimentation.

The transition from a single corporate settlement, directed and financed by overseas investors and managed for their benefit, to numerous privately owned plantations operated by individual settlers for individual profit began in 1616. In that year, the Virginia Company decided to award a dividend of undeveloped land to those Englishmen who had invested in its joint stock. Early hopes for quick profits from the Virginia settlement had been blasted by failure to discover lucrative trading commodities, by poor management in England and in the colony, and by the inability of a majority of the settlers to survive in the deadly environment around Jamestown. Returns from the Company's seven-year monopoly of the colony's trade (1609–1616) were so small that recompense to investors could be made only in shares of uncleared land. Finding some means to reward shareholders was essential if the Company was to persuade disillusioned backers to continue the venture. It was hoped that individual investors might become more motivated to finance the peopling and development of "particular" plantations that they would be allowed to manage independently within the nominal proprietary sphere of the faltering parent company.

The Virginia Company explained its dividend scheme in a printed public announcement. Investors would be granted a first allotment of fifty acres for each share of Company stock they had already purchased in land along the James River over which the Company had effective control. To secure this land, however, the investors would have to make a further payment of £12 10s to meet the costs of surveying their dividends and of sending over a new governor, the Company being in no position to cover these expenses. Further dividends, amounting to a total of at least two hundred acres per share, were promised once the area of settlement should be enlarged. To encourage additional investment in the colony, new adventurers were also offered fifty acres of land for each £12 10s paid. Initially, takers were few. As the Company freely admitted, earlier endeavors had "brought nothing home but adverse successe and bad reports."[5]

Between 1612 and 1618, most English investors interested in colonial ventures devoted their attention and monies to the development of Bermuda, an outpost fortuitously reconnoitered in 1609 by Company leaders shipwrecked there on their way to Virginia. A subsidiary joint-stock organization, the Somers Island Company (composed almost entirely of members of the parent body) was formed in 1612 to finance the settlement of this second colony. Although a singular run of ill luck had left the Virginia project "in a manner cleane defeated," fortune smiled more brightly on the second attempt. Initial difficulties, inevitable in starting up New World outposts, were more quickly overcome in Bermuda. Costs of development were less for a small group of islands. There was no resident native population to threaten the English settlers, the environment proved exceptionally healthy for Europeans, and the indigenous food supply was plentiful. Profits from a variety of semitropical products were quickly realized, occasionally augmented by windfalls from ambergris (a substance produced in the intestines of sperm whales used in making perfume) discovered on the beaches and goods salvaged from wrecked foreign vessels.[6]

Although the first settlers were transported to the island at the expense of the Company's common stock, most of Bermuda's development was underwritten by individual adventurers rather than by the Somers Island Company as a whole. Bermuda investors got quick title to land in proportion to the size of their contribution to the Company's joint stock. Although the common

5. Council for Virginia, *A Brief Declaration* (1616), in Alexander Brown, ed., *The Genesis of the United States . . .*, 4 vols. (Boston, 1890), II, 774–779. See also *The Trade's Increase* (1615), ibid., II, 766; and Wesley Frank Craven, *Dissolution of the Virginia Company: The Failure of a Colonial Experiment* (New York, 1932), 43.

6. Brown, ed., *Genesis*, II, 775. The early history of Bermuda and its relationship with Virginia are discussed in Wesley Frank Craven, "An Introduction to the History of Bermuda," *WMQ*, 2d Ser., XVII (1937), 176–215, 317–362, 437–465, XVIII (1938), 13–63; Craven, *The Southern Colonies in the Seventeenth Century, 1607–1689* (Baton Rouge, La., 1970), chap. 4; Andrews, *Colonial Period*, I, chaps. 5, 11; Edward D. Neill, *History of the Virginia Company of London . . .* (Albany, N.Y., 1869), chap. 3; and John Smith, *The Generall Historie of Virginia, New-England, and the Summer Isles . . .* (London, 1624), in Edward Arber, ed., *Travels and Works of Captain John Smith: President of Virginia and Admiral of New England*, 2 vols., new ed. (Edinburgh, 1910), II, 625–688 (hereafter cited as Smith, *Generall Historie*). The most recent studies are Michael Joseph Jarvis, "'In the Eye of All Trade': Maritime Revolution and the Transformation of Bermudian Society, 1612–1800" (Ph.D. diss., College of William and Mary, 1998); and Virginia Bernhard, *Slaves and Slaveholders in Bermuda, 1616–1782* (Columbia, Mo., 1999), chap. 1.

stock continued to send some laborers and tenants to the public lands to cover the costs of government and defense, most settlers were sent and supplied by individuals or groups of individuals to develop private lands. Settlers became tenants attached to clearly defined tracts of twenty-five acres, paying rents negotiated with absentee landlords. They had the protection of English common law and legal right to any improvements they made, thus encouraging greater commitment to the colony as a permanent home. Shipments of exotic commodities like pearls, coral, ambergris, cedar timber, pineapples, pomegranates, figs, and citrus gave the appearance that the colonists were exploiting a wide variety of natural resources. This presumption made more palatable the reality that by 1616 tobacco, the one new product that James I (although not all his subjects) heartily despised, had become the colony's major export. Private investors reaped the greatest returns, but even the Company magazine yielded a handsome profit in 1618, encouraging the investors "to proceed lustily in their plantation."[7]

By 1618, English adventurers were ready to look again to Virginia. Little Bermuda was well on the way to being filled up, but on the mainland the supply of land seemed unlimited. Although the resources that had been poured into Virginia had as yet yielded no tangible returns and the financially strapped Virginia Company was in no position to replenish the colony's depleted population, still a small toehold had been established, the remaining settlers seemed finally able to feed themselves, and some knowledge of the country had been gained. Moreover, the settlers had concluded a peace with the Powhatans in the spring of 1614 that also encouraged expansion. Heartened by their successes in Bermuda, small groups of investors again became confident that profits could yet be won in Virginia, especially if private developers had a freer rein. As they put it, "divers Lords Knights, gentlemen and Cittizens (grieved to see this great Action fall to nothinge)" offered "to take the matter a new in hand and at their pryvate charges (joyninge themselvs into Societies) to sett upp divers particularr Plantations." By starting anew, the investors would escape the burden of old debts the parent company had incurred in financing past failures and its problems in collecting unpaid subscriptions. They could pool capital to initiate smaller-scale, manageable projects and could spread the burdens of administration among the various shareholders. By selecting their own agents to run what would be almost independent subcolonies, they could implement whatever policies they

7. Smith, *Generall Historie,* II, 661–662; Neil Kennedy, "Imagining Arcadia and Conceiving Bermuda: The Reinvigoration of English Colonization" (paper presented at the conference "The Atlantic World and Virginia, 1550–1624," Williamsburg, Va., Mar. 4–7, 2004).

MAP 3. *Virginia.* By Captain John Smith. Engraved by William Hole, London, 1624 (originally published 1612). Black-and-white line engraving. Courtesy, The Colonial Williamsburg Foundation.
English settlements in the Chesapeake were at first isolated outposts in an overwhelmingly native American world.

decided upon. And, with rights of free trade, they expected to make the most of whatever their plantations might produce or their agents might procure through exchange.[8]

8. Craven, *Southern Colonies,* chap. 4; Craven, "Introduction to the History of Bermuda," *WMQ,* 2d Ser., XVII (1937), 176–215, 317–362, 437–465, XVIII (1938), 13–63; Andrews, *Colonial Period,* I, chap. 5; *RVC,* I, 350; Brenner, *Merchants and Revolution,* chap. 3. The close connection between the experience in Bermuda and new plans for Virginia is evident in a broadside declaration of the Virginia Company's lottery of 1615. The sooner the Company raised money to resupply settlers already in Bermuda, "we should the sooner resolve upon a division of the country by lot, and so lessen the Generall charge, by leaving each several tribe [administrative division in Bermuda roughly equivalent to a county] or family to husband and manure his own" (in Brown, *Genesis,* II, 761–766). The Providence Island Company similarly resorted to opening subsidiary joint stocks for limited ventures in order to raise additional capital when early investors in the

Those adventurers who planned to send large numbers of settlers to Virginia to develop particular plantations led a move to reorganize the Virginia Company. Gentry investors under the leadership of Edwin Sandys seized control of the Company from the great merchants who had previously dominated it. In 1618, new Company rules and bylaws were drafted that clarified the duties of Company officials. In addition, they established fresh policies for the colony (explained in instructions to a new governor, George Yeardley) to set up a uniform land policy, to abolish martial law, and to call a representative assembly in Virginia where both Company officers and representatives of private investors would have a voice in the colony's government. The parent company was to continue to develop the public estate in order to finance the colonial government and military establishment, thus leaving private English investors and Virginia planters free from public taxes. About the same time, parallel policies were effected within the Somers Island Company.[9]

Changes within Virginia also encouraged privately developed settlements. As the few survivors of the first group of settlers became free of their obligations to the Company in 1614, then-governor Dale granted them small individual plots. These, amounting to only three to twelve acres, could provide little more than expanded garden plots. As time passed, however, a small but growing group of "Ancient" Virginia planters joined with English developers in pressing for wider scope for individual initiative within the framework of the Company organization. It was decided in 1618 that all settlers who had arrived in Virginia before the spring of 1616 would be granted one hundred acres for their personal adventure, and those coming thereafter would get fifty acres. Grants of this size would permit not just immediate subsistence but also long-term, small-scale commercial exploitation.[10]

Organizational and policy changes at home and in the colony would have mattered little to the effective development of Virginia, however, had not tobacco emerged at the same time as a viable export commodity. Virginia pitch, tar, clapboards, and potash had produced scant returns to either settlers or investors, and colonial sassafras, sold as a cure for venereal disease, had com-

---

parent company became discouraged by the failure of some adventurers to pay what they had pledged, by slow returns from the settlement, and by the failure of the colony to become self-sustaining (Kupperman, *Providence Island,* chap. 10).

9. Craven, *Southern Colonies,* chap. 4; Craven, *Dissolution of the Virginia Company,* chaps. 2, 3; Andrews, *Colonial Period,* I, chap. 9.

10. Andrews, *Colonial Period,* II, 124–127; Smith, *Generall Historie,* II, 516; Craven, *Dissolution of the Virginia Company,* chap. 3; Edmund S. Morgan, *American Slavery, American Freedom: The Ordeal of Colonial Virginia* (New York, 1975), 81–82; Hecht, "The Virginia Colony," chap. 6; Philip Alexander Bruce, *Economic History of Virginia in the Seventeenth Century . . . ,* 2 vols. (New York, 1896), I, chap. 4; *RVC,* III, 98–109.

pletely glutted the English market. In contrast, tobacco, however frowned upon in official circles since it "served neither for necessity nor for ornament to the life of man," was a commodity in enough demand at home that New World settlers could at least barter it for essential supplies of clothing and tools; the few who could grow the weed in quantity might prosper handsomely.[11]

Spanish colonists and enslaved Africans and free and enslaved Indians in Hispaniola, Cuba, Venezuela, and other Spanish settlements had been growing tobacco since the early sixteenth century. By the 1590s, poor whites and African, mestizo, and mulatto slaves on the northern coast of Hispaniola, Spanish settlers in eastern Venezuela, and Indians on the island of Trinidad were raising large quantities that they sold, despite Spanish prohibitions on trade with foreigners, to English, Dutch, and French ships. In an attempt to curb this illegal trade, in 1605–1606 Spanish authorities prohibited tobacco raising in Hispaniola and Venezuela and forcibly relocated settlers living in areas where contraband trade was most prevalent to more easily policed localities. Their actions served from 1608 to 1612 to divert a growing commerce (involving dozens of English ships alone) further east and south to the island of Trinidad, to the mouth of the Orinoco River, and to Guiana. Dutch, French, and English adventurers also set up small colonies in Guiana, whose settlers traded for and also planted tobacco, learning from Spanish settlers and their enslaved African workers how to cultivate and cure the milder, higher-yielding strains raised in these areas. This trade reached a climax in 1612, prompting the Spanish and eventually the Portuguese in Brazil to more effectively intercept foreign shipping and to expel would-be colonists of rival nations. Curtailment of the Spanish trade, along with newly gained technical expertise, created an opportunity quickly embraced by both farmers in southern and central England and the first settlers on Bermuda, in Virginia, and, soon after, on the Leeward Islands (which Englishmen settled in 1624) and on Providence Island (settled in 1629 off the coast of Central America).[12]

11. *RVC,* II, 36, III, 221; Brown, *Genesis,* II, 789–790; Hecht, "The Virginia Colony," chap. 4; Craven, *Southern Colonies,* chap. 4.

12. Fernadez de Oviedo y Valdés Gonzalo, *La historia general y natural de las Indias . . .* (Seville, 1535), trans. and summarized in Jerome E. Brooks, *Tobacco: Its History Illustrated by the Books, Manuscripts, and Engravings in the Library of George Arents, Jr.,* 5 vols. (New York, 1937), I, 202–209; Girolamo Benzoni, *La historia del Mondo Nuovo* (Venice, 1565), trans. and summarized ibid., I, 224–227; C. T., *An Advice How to Plant Tobacco in England* (London, 1615), ibid., II, 3–7; Andrews, *The Spanish Caribbean,* 213–216, 221–223, 224–236, 243–245; Andrews, *Trade, Plunder, and Settlement,* 285–288, 294–298, 300–303; Jarvis, "'In the Eye of All Trade,'" prologue, chaps. 1, 2; Joyce Lorimer, "The English Contraband Tobacco Trade in Trinidad and Guiana, 1590–1617," in Andrews, Canny, and Hair, eds., *The Westward Enterprise,* 124–150; Marcy Norton

The Virginia Company contingent shipwrecked on Bermuda in 1609 found tobacco of the Oronoco strain, apparently introduced to the island by earlier Spanish castaways, growing there wild. When the first permanent settlers arrived in 1612, they found three Englishmen from the 1609 group who had remained on the island already cultivating it. The following year, the Bermudians sent 170 pounds of tobacco to England and by 1615 had raised a substantial crop. English settlers in the Americas needed about three years to learn how to cultivate the weed in any quantity. Robert Harcourt, who directed a small colony in Guiana from 1609 to 1612, related that when the colonists first arrived "wee altogether wanted the true skill and knowledge how to order it." But they learned its management from Spanish settlers within two or three years. The Bermuda colonists, benefiting from the experience of the three men left on the island in 1610, needed only two additional years to master the basics. Virginia planter John Rolfe, who made the first trial of tobacco on the mainland in 1612, had been among those stranded on Bermuda for ten months in 1609–1610 and probably obtained seeds of the Spanish strain there. Like the Guiana planters, in the first year the Virginians "had not the knowledge to cure, and make up." Yet, by 1614, Ralph Hamor reported that there were "some now resident there, out of the last yeers well observed experience, which both know, and I doubt not, will make, and returne such Tobacco this yeare, that even England shall acknowledge the goodnesse thereof."[13]

Learning how to cure the plant successfully and to preserve and pack it for shipment proved more difficult than learning how to grow it in quantity. Planters in each European colony through trial and error developed unique local methods that shortly became "the custom of the country" for that colony. The evolving customs of particular colonies included blends of expertise and technologies contributed by native Americans (Plate 1), by enslaved West Africans (some of whom had experience in cultivating tobacco both in their homelands and in the Americas), by Spanish, Dutch, French, and English colonists, and by European growers. Newly arrived Bermuda

---

and Daviken Studnicki-Gizbert, "The Multinational Commodification of Tobacco, 1492–1650: An Iberian Perspective," and Philip D. Morgan, "Virginia's Other Prototype: The Caribbean," both in Mancall, ed., *Atlantic World and Virginia,* 251–273, 361–363; Victor Enthoven and Wim Klooster, "'The Dutch Found and Relieved Us': The Shifting Dynamics of Virginia-Dutch Trade in the Long Eighteenth Century," in Douglas Bradburn and John C. Coombs, eds., *Early Modern Virginia: New Essays on the Old Dominion* (Charlottesville, Va., forthcoming).

13. Jarvis, "'In the Eye of All Trade,'" 19–26, 33, 39, 54, 87–88, 101; Robert Harcourt, *A Relation of a Voyage to Guiana . . .* (London, 1613), in Brooks, *Tobacco,* I, 501–503; Ralph Hamor, *A True Discourse of the Present State of Virginia . . . [in 1614]* (1615; rpt. Richmond, Va., 1957), 34–35.

PLATE 1. Detail from *Town of Secota.* Engraved by Theodore de Bry after a drawing by John White, 1588. Courtesy, The Colonial Williamsburg Foundation. *Indians in the Chesapeake were raising an indigenous strain of tobacco when the English arrived. White included only fields of newly planted, green, and ripe corn in his drawing from personal observation. German engraver de Bry added further details, including plots of tobacco, sunflowers, and pumpkins, based on fellow Roanoke expedition member Thomas Harriot's written account,* A Briefe and True Report of the New Found Land of Virginia, *first published in 1588 (Kim Sloan et al.,* A New World: England's First View of America *[Chapel Hill, N.C., 2007], 110–111).*

tenants, for example, were instructed in tobacco cultivation by a Mr. Tickner, "a skilfull planter and curer of tobacco" in England recruited by the Somers Island Company. And they learned both about production in quantity and about a technique developed in Venezuela for processing the leaf into rolls from African slaves who had been captured somewhere in the Spanish Caribbean and brought to Bermuda to instruct English settlers in 1617. Bermuda rolled tobacco found immediate favor with English merchants and smokers, as it replicated the curing and processing techniques used to make the more expensive Spanish tobacco with which English consumers were already familiar. Virginia colonists would struggle some years longer before they discovered curing and packing techniques that would work effectively on the mainland.[14]

Exports of Virginia leaf were negligible in 1613 and 1614, but approximately twelve hundred pounds of the 1615 crop were shipped home. By 1616, twenty-five Company tenants at West and Shirley Hundred were planting the weed to the exclusion of most other activities. Captain John Smith reported that deputy governor Yeardley "applied himselfe for the most part in planting Tobacco, as the most present commoditie they could devise for a present gaine," and both Smith and John Rolfe noted that other colonists were also making "good store of tobacco." That year's crop amounted to at least nine thousand pounds. By the spring of 1617, tobacco mania gripped the Virginia settlement. Despite a continued shortage of cloth and other imports, the colonists' morale improved appreciably. When Governor Samuel Argoll arrived in May, he found "the market-place, and streets, and all other spare places" in Jamestown planted with tobacco, and most of "the Colonie dispersed all about" at other choice spots following the same pursuit. The news that tobacco could be successfully grown in Virginia must have encouraged Englishmen who were making money by raising it in Bermuda or at home in England to try similar ventures on the North American mainland.[15]

The first patents for particular plantations were issued in 1616 and 1617,

14. Virginia Bernhard, "Beyond the Chesapeake: The Contrasting Status of Blacks in Bermuda, 1616–1663," *Jour. So. Hist.*, LIV (1988), 546–548; Jarvis, "'In the Eye of All Trade,'" 95–105, 153–155; J. H. Lefroy, comp., *Memorials of the Discovery and Early Settlement of the Bermudas or Somers Islands, 1515–1685,* 2d ed., rev., 2 vols. (1877–1879; rpt. n.p., 1981), I, 116.

15. Andrews, *Colonial Period,* I, chap. 5; Bruce, *Economic History of Virginia,* I, chap. 4; Craven, *Dissolution of the Virginia Company,* 38–39; Lewis Cecil Gray, *History of Agriculture in the Southern United States to 1860,* 2 vols. (1932; rpt. Gloucester, Mass., 1958), I, 21; [John Rolfe], "Virginia in 1616," *Va. Hist. Reg.,* I (July 1848), 101–113; Smith, *Generall Historie,* II, 527, 535; *RVC,* III, 70–74; Hecht, "The Virginia Colony," chap. 7; Joan Thirsk, "New Crops and Their Diffusion: Tobacco-Growing in Seventeenth-Century England," in C. W. Chalklin and M. A. Havinden, eds., *Rural Change and Urban Growth, 1500–1800: Essays in English Regional History in Honour of W. G. Hoskins* (London, 1974), 88–93; Russell R. Menard, "The Tobacco Industry in the Chesapeake Colonies, 1617–1730: An Interpretation," *Res. Econ. Hist.,* V (1980), 109–177. Recorded imports of colonial tobacco into England do not always distinguish between the quantities brought from Virginia, Bermuda, and other West Indian islands. This chapter employs Menard's estimates of the Chesapeake contribution, which are lower than the figures cited by some other authors. The Jamestown settlers' greater involvement in agriculture after the introduction of tobacco is substantiated by the archaeological record. Only two hoes have been found in deposits at the original fort and its immediate environs dating to 1607–1609, but large numbers have been discovered in deposits left in the 1610s (William M. Kelso with Beverly A. Staube, *Jamestown Rediscovery, 1994–2004* [Jamestown, Va., 2004], 147, 156).

mostly to a few old Virginia hands like John Martin and Samuel Argoll, who had early knowledge of the suddenly more promising state of the settlement, or to groups of Englishmen who had ties to men on the spot, like the Society of Smith's (later Southampton) Hundred that employed George Yeardley, then deputy governor and later governor, as their chief agent. These documents were loosely drawn, either through carelessness or because officers of the nearly prostrate Virginia Company felt compelled to allow extravagant grants of land and jurisdictional rights in order to get some kind of development going. Most early patents were later revoked, to be reissued under the more tightly drawn terms that governed later grants.[16]

Once a number of serious investors began to express interest in Virginia, the Company worked out a clearer policy for particular plantations. These were laid out in Governor Yeardley's instructions of 1618 and were also inserted in subsequent patents. Investors who pooled their resources might obtain large contiguous grants of land. For each Company share bought for £12 10s, the adventurers were to receive one hundred acres of land immediately. They would get another fifty acres for each settler whom they transported to the colony within the next seven years and who remained there for at least three years or died. To prevent subsequent conflicts over boundaries (none of the original grants were surveyed), no particular plantations were to be sited within five miles of Company boroughs or within ten miles of each other. Each plantation was given fifteen hundred acres of land for public use for the support of churches and schools within the plantation. Later the Company also promised one hundred acres of glebe land and agreed to supply three servants for resident ministers (the investors were to send three more) and to loan cattle and corn to the settlers until they were securely established. Once the first grant was sufficiently peopled, a second allotment of equal size would be awarded.[17]

The adventurers got a free hand in administering their enterprises, almost as if they had planted a separate colony. The agent they chose could, "till a forme of Goverment be here settled for them," make "orders ordinances and

16. *RVC,* I, 56–61, 609–610, 614, II, 126–127, III, 98–109; Craven, *Southern Colonies,* chap. 4; Craven, *Dissolution of the Virginia Company,* chap. 3; Andrews, *Colonial Period,* I, chap. 5. For Martin, see James P. C. Southall, "Captain John Martin of Brandon on the James," *VMHB,* LIV (1946), 21–67. The flagging interest in Virginia that forced a switch to private investment also made it difficult for the Virginia Company to regulate particular plantations (Craven, "Introduction to the History of Bermuda," *WMQ,* 2d Ser., XVIII [1938], 40–41).

17. *RVC,* I, 266–269, 314, 539–541, 587, III, 98–109, 130–134, 275–280, 623–628; William Bradford, *History of Plymouth Plantation, 1620–1647,* ed. W. C. Ford, 2 vols. (Boston, 1912), I, 246–251. See also Craven, *Dissolution of the Virginia Company,* chap. 3; and Craven, *Southern Colonies,* chap. 4.

constitutions for the rule governemt orderinge and directinge of all [sorts] persons to be transported and setled" within the grant with the advice of "divers of the gravest and discreetes of their Companies," so long as the regulations were not contrary to the laws of England or to the general orders of the Company. Plantation residents could not be detained for colony service except for defense or to stand trial. Each particular plantation would be a self-sufficient unit of local administration equal with Company boroughs and subordinated only to the colony's central government. The investors could also direct the business of their plantation almost at will. They could send goods and merchandise to England or elsewhere free of any Company imposts. Their agents could trade freely with neighboring Indians and English settlers and could set the rules under which plantation residents and outsiders might carry on trade, or else proscribe such trade entirely.[18]

What did investors in particular plantations expect to achieve? Most were members of the English gentry who had no intention of personally settling in Virginia. Few had prior experience in trade, nor had many joined in other, more strictly commercial joint-stock ventures. From these circumstances, some scholars have surmised that these gentlemen were primarily attracted by the lure of great tracts of land from which they could eventually expect profits from agricultural rents. Certainly some of the gentry adventurers had invested in rental properties in England, and not a few had also acquired Irish tenements. The Virginia Company's announcement of 1616 also suggests that land rents were anticipated as a major source of profit. Absentee investors entitled to dividends of land were expected to send "families to manure it for yearely rent, or for halfe the cleare profits as many others doe." Captain John Smith sensed a great scramble for land; he observed that, in 1621, "there was much suing for Patents for Plantations . . . as though the whole land had beene too little for them."[19]

18. *RVC,* I, 303, III, 130–134; Craven, *Southern Colonies,* 129–131, and chaps. 4, 5; Andrews, *Colonial Period,* I, chap. 5; George Louis Beer, *The Origins of the British Colonial System, 1578–1660* (1908; rpt. Gloucester, Mass., 1959), chap. 8. In 1623, the vice admiral of Southampton Hundred claimed "prerogative before others" in trading with the Indians (*RVC,* IV, 276–271). Similarly, William Claiborne forbade freemen living on the Kent Island post that he commanded from engaging in any trade with the natives ("Claiborne vs. Clobery, et als. in the High Court of Admiralty," *MHM,* XXVII [1932], 340, XXVIII [1933], 28).

19. Rabb, *Enterprise and Empire,* chap. 1; Craven, *Dissolution of the Virginia Company,* 32–33; Brown, *Genesis,* II, 779; Smith, *Generall Historie,* II, 566. That the gentry had a longer-run perspective is reinforced by Francis Bacon's comment that noblemen and gentlemen were the best qualified for establishing successful plantations, compared to merchants who "look ever to the present gain" ("Of Plantations," in Brown, *Genesis,* II, 801). Examples of investments in rental

However, the methods by which contemporary merchants calculated profits from commercial ventures predisposed the gentlemen to expect short-run as well as long-run returns. Overseas trading ventures — whether to proven continental entrepôts or to far-off and risky destinations like the East Indies — were thought of in terms of the returns accruing to individual voyages. Even the most experienced merchants calculated profits only on the basis of net gains acquired from the proceeds of single cargoes. If a shipment of, say, cloth to the continent would not purchase a supply of timber, wine, or hemp that could be resold in England at more than the original cost of the goods ventured plus the expenses of the voyage, that avenue of trade would be quickly discontinued. Consequently, although expectations of rental profits in the distant future might have assuaged some of the anxieties of gentry investors, they continued to evaluate the success or failure of their investments in the same short-run terms as did merchants.[20]

Although hunger for land surely motivated many of the settlers who risked their lives by moving to Virginia, the lure of cheap land had less appeal for the already landed gentry. Motives other than making a profit — glory, converting the Indians, aiding the nation by ridding the commonwealth of excess population, and enhancing the country's prestige over rival European powers — so stressed in the promotional literature of the time, also persuaded gentlemen to put money into risky colonial ventures. About half of the gentry who invested in colonial enterprises by 1621 also sat in Parliament, where they were engaged in discussions of public welfare and were particularly susceptible to propaganda "which shrewdly stressed that the national interest was at stake in overseas ventures." Such nonpecuniary motives, however, easily lost their force when the English domestic economy soured, profits were not forth-

---

properties in England are described in Alan Simpson, *The Wealth of the Gentry, 1540–1660* (Chicago, 1961); and Lawrence Stone, *The Crisis of the Aristocracy, 1558–1641* (New York, 1965), chap. 6. For Irish investments, see *RVC,* I, 446, 476, 554, 626, III, 588.

20. Michael Chatfield, *A History of Accounting Thought,* rev. ed. (Huntington, N.Y., 1977); Peter Ramsey, "Some Tudor Merchants' Accounts," in A. C. Littleton and B. S. Yamey, eds., *Studies in the History of Accounting* (Homewood, Ill., 1956), 185–201; O. ten Have, *The History of Accountancy,* trans. A. van Seventer (Palo Alto, Calif., 1976); James Ole Winjum, *The Role of Accounting in the Economic Development of England, 1500–1750* (Urbana, Ill., 1972); B. W. Higman, *Plantation Jamaica, 1750–1850: Capital and Control in a Colonial Economy* (Kingston, Jamaica, 2005), chap. 4. Irene Hecht estimated that the proprietors of Berkeley Hundred, one of the particular plantations, got back at least 22 percent of the money they had invested in the enterprise from the first shipment of plantation tobacco returned to England in the second supply ship. However, none of the surviving correspondence between the principal investors suggests that they considered the return satisfactory ("The Virginia Colony," chap. 6).

coming, and the attention of those in power turned more to domestic and European issues.[21]

Thoughtful men like Francis Bacon might observe: "Planting of countries is like planting of Woods; for you must make account to lose almost twenty years' profit, and expect your recompense in the end." But even Bacon recognized that investors expected more, and he ended up endorsing a search for commodities like timber, iron, salt, silk, pitch, tar, soap ashes, and sweet woods that would "help to defray" initial charges of settlement. The results of the partly gentry-financed venture in Bermuda could have only reinforced expectations for relatively quick returns from Virginia. Some of the adventurers hoped to recoup part or all of the initial costs of supply from the proceeds of ancillary fishing expeditions, fur-trading ventures, shipments of staple commodities, or profits from carrying passengers and freight. Writing a few years later, William Bullock put his finger on yet another problem. Many of the investors "fully outed themselves at the first, what they intended to adventure." When the initial outlay produced, not "suddaine and certain return," but only demands for additional supplies, they decided to cut their losses by getting out. As the proprietors of Berkeley Hundred put it in 1620 (the first settlers had left England the previous year), the colonists must at the very least get their own subsistence, "for theis great supplyes are insupportable in longer contynuance."[22]

Finally, the investors had one other expectation that went uniformly unrealized. The joint-stock ventures were intended as temporary arrangements, a means to an end, not a permanent form of organization. Once the plantations had been successfully launched by means of cost savings from pooled capital and reduced administrative costs for large-scale settlements, the ad-

21. Rabb, *Enterprise and Empire,* 35–42, 68–69, 92–96.

22. "Of Plantations," in Brown, *Genesis,* II, 799–801; William Bullock, *Virginia Impartially Examined* . . . (London, 1649), 13; *RVC,* III, 399–400. For ancillary ventures, see *CSP,* I, 19; *RVC,* I, 285; and Susan E. Hillier, "The Trade of the Virginia Colony, 1606–1660" (Ph.D. diss., University of Liverpool, 1971), chap. 3. One group of Virginia investors proposed leaving some people to plant tobacco and corn in Virginia while others were to trade with both native Americans and English settlers and to fish off Canada and bring back the catch for resale in the colony. Smith's Hundred also got a license "to go a fishing" in 1619. Similarly, subsidiary joint-stock associations in Bermuda, the Plymouth colony, and the Massachusetts Bay Colony planned projects involving a combination of agriculture, trade, and fishing (Andrews, *Colonial Period,* I, 128), and Cecil Calvert, second Lord Baltimore, organized two associations to manage the fur trade in Maryland (Garry Wheeler Stone, "Society, Housing, and Architecture in Early Maryland: John Lewger's St. John's" [Ph.D. diss., University of Pennsylvania, 1982]). For an example of investors backing out, see *RVC,* I, 577–578, in which the Society for Martin's Hundred took out a new patent in 1622 because Sir John Wolstenholme and some new associates wished to sever themselves from those members who were unwilling to proceed in the venture.

venturers intended to terminate the partnerships and divide the land, goods, servants, and tenants proportionally. Each investor could then run his unit to suit himself. Probably this would have been done once the second dividend of land was acquired. Unfortunately, every project foundered and was abandoned well before it reached a stage at which the assets could be profitably split.[23]

Sponsors of particular plantations were of several sorts. In some cases, "divers of the ancient Adventurers Associating themselves together" (Smith's Hundred and Martin's Hundred are examples) ran their plantations in the same way that the parent company directed the overall operation. Courts of the association held periodic meetings in England and chose officers and standing committees to direct policy while appointing agents in Virginia to manage daily operations. Alternatively, an "ancient Adventurer or Planter" might run the operation with an infusion of capital raised from others (as with Samuel Argoll's Town and Captain John Martin's Brandon), or entirely new groups of adventurers might form subsidiary associations (as the Berkeley Hundred investors). In the latter, four partners jointly purchased forty-five shares in the Company and secured a patent early in 1619. They agreed to split costs and profits equally and to make decisions by a majority vote. Direction of the enterprise was to remain in the hands of the principal investors, although others might buy up some fraction of the original investors' shares. By mid-1620, the first Virginia agent had been fired, and one partner had bowed out, leaving two men to manage the business in England. The third partner, along with a new associate, ended up going in person to manage the venture.[24]

Actual settlement of particular plantations began in 1618 with John Martin's Brandon and Samuel Argoll's Town. Smith's (later Southampton), Flowerdew, and Berkeley Hundreds were started by the end of 1619, as were Captain John Bargrave's, Captain John Ward's, and Captain Christopher Lawne's plantations. Twenty-some other particular plantations were seated over the next three years. By 1623, the Virginia Company had granted and

23. *RVC,* III, 201–207; Craven, *Southern Colonies,* 122–123. For possible subdivision of land among individual investors in the Martin's Hundred settlement by 1622, see Martha W. McCartney, "The History and Historical Context of CG-10, James City County, Virginia," digital report, 11, CWF.

24. *RVC,* I, 577–578, III, 94–98, 103–109 (quotations on 103), 130–134, 201–207, 248–249, 260–261, 271–274, 368–369, 374–375, 376–379, 379–381, IV, 102, 506; Craven, *Southern Colonies,* chap. 4. The Berkeley investors initially hoped that Governor George Yeardley would be a one-fifth partner in the enterprise. He declined for lack of assets, being already heavily committed in the Smith's Hundred venture.

sealed forty-four patents giving a minimum of five thousand acres each to subsidiary groups promising to transport at least one hundred settlers as well as twenty-seven other awards for lesser amounts of land. An additional thirty-six grants were mentioned in the Company records as in some preliminary stage. In all, a total of thirty-some private plantations, some owned by groups of investors and some by individual residents, were actually settled. The Indian attack of 1622, the subsequent year of high mortality, and the crown's dissolution of the Virginia Company in 1624 halted any further attempts to set up new joint-stock estates in Virginia and in most cases ended efforts to effectively resupply those already started. In 1631, however, a Virginian, William Claiborne, along with four English backers, established a similarly joint-stock–financed plantation on Kent Island far up Chesapeake Bay. Aside from some ambiguity about what superior jurisdiction, if any, the settlement should fall under following the founding of the neighboring Maryland colony in 1634, it was organized along the same lines as the earlier particular plantations started under the aegis of the Virginia Company.[25]

Smith's Hundred and Martin's Hundred were numerically the most important settlements. According to Company records, at least 310 settlers had been sent to Smith's Hundred by the spring of 1620 and more than 200 to Martin's Hundred. In March 1620, when Virginia's population stood at about 1,000, at least 338 surviving settlers were living on particular plantations, more than 100 on Smith's Hundred, 77 at Flowerdew, 72 at Martin's Hundred, 37 at Captain Bargrave's, 26 at Captain Ward's, and 21 at Captain Martin's. Little information about the daily life and operation of these settlements has survived. The best-documented are Berkeley Hundred, to which 90-odd settlers were sent between 1619 and 1622, and William Claiborne's Kent Island settlement, which consisted of 18 to 49 workers between 1631 and 1636. The papers of a major investor, John Smyth of North Nibley, describe the English side of the Berkeley Hundred operation and include some correspondence sent back by the settlers. Depositions and documents generated in a series of suits between Claiborne and his English partners supply greater insight into how that particular plantation was operated. The successes and failings of particular plantations will be assessed by examining

25. Hecht, "The Virginia Colony," chaps. 6, 7, and appendix 4C; Charles E. Hatch, Jr., *The First Seventeen Years: Virginia, 1607–1624* (Williamsburg, Va., 1957); Martha W. McCartney, "An Early Virginia Census Reprised," *Arch. Soc. Va. Qtly. Bull.,* LIV (1999), 182–187. Scholars vary slightly in their counts of the number of plantations seated by 1624, because the year in which some were first settled is uncertain. For the most recent discussion of the date of settlement and location of particular plantations, see Martha W. McCartney, *Virginia Immigrants and Adventurers, 1607–1635: A Biographical Dictionary* (Baltimore, 2007).

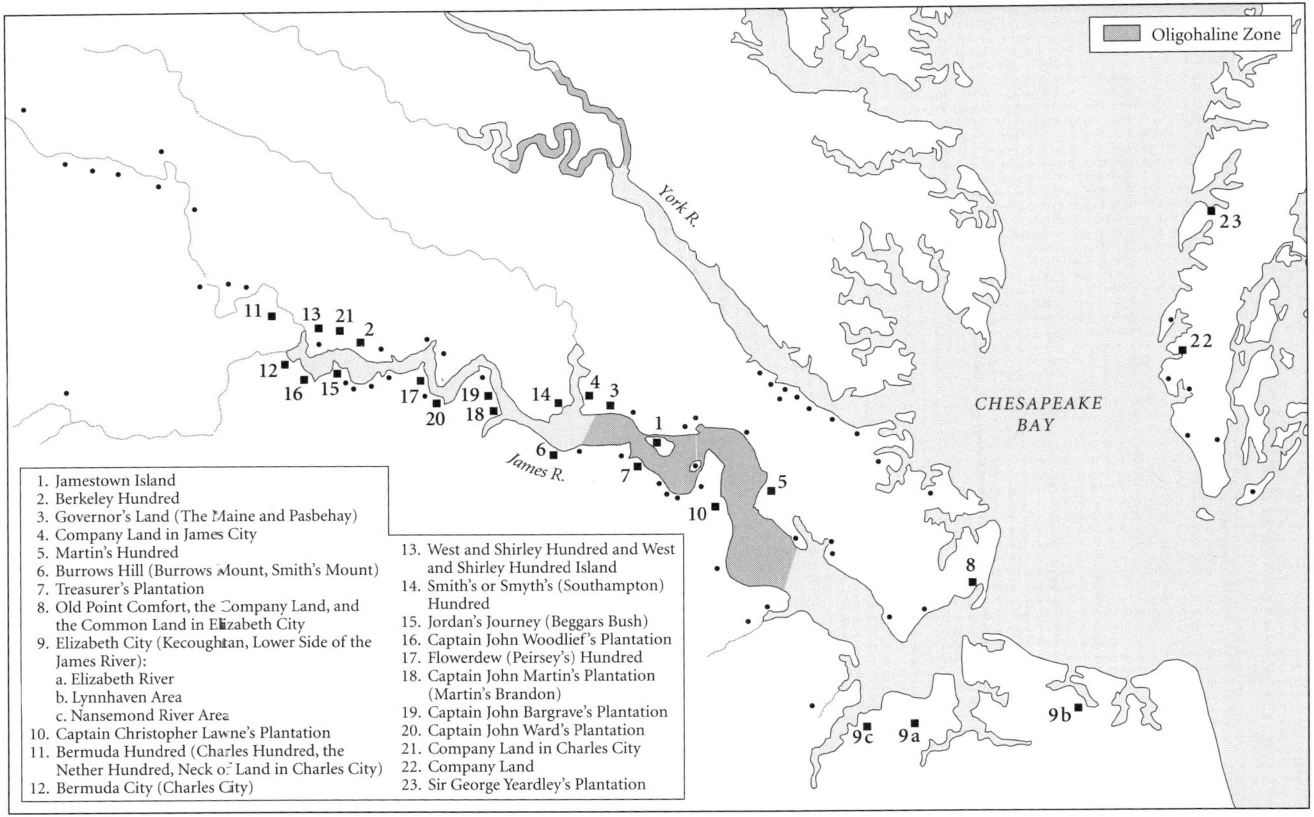

MAP 4. Virginia Company, Particular, and Private Plantations in Virginia Seated by 1635. After Martha W. McCartney, *Virginia Immigrants and Adventurers, 1607–1635: A Biographical Dictionary* (Baltimore, 2007). Dots represent other plantations seated by 1635 that are not discussed in the text.

An oligohaline zone is where fresh and salt water mix, creating maximum concentrations of pathogenic water-borne organisms.

the organization and outcomes of the Berkeley Hundred and Kent Island ventures. In addition, archaeological investigations of Martin's and Flowerdew Hundreds, Jordan's Journey, and of other plantations seated in the 1620s and early 1630s provide some idea of the physical context of early plantation life.[26]

The Problems of Planting

The results of this great surge of private investment in particular plantations between 1618 and 1622 have been variously interpreted. The traditional picture, relying heavily on official Company records, suggests that particular plantations largely replicated the failures of the parent company. Irene Hecht, in contrast, argues that individual or subsidiary joint-stock investors were infinitely more successful than the parent company because they es-

26. *RVC,* I, 349–351; McCartney, "An Early Virginia Census," *Arch. Soc. Va. Qtly. Bull.,* LIV (1999), 182–183. Most of the items in the papers of John Smyth of North Nibley, Gloucestershire, that relate to his involvement in Berkeley Hundred are reprinted in *RVC.* Others are reprinted in *Bulletin of the New York Public Library,* I, III, and XIX (New York, 1897, 1899, 1915), microfilm, CWL. They are described in Anne Barka, "The Smyth of Nibley Papers: The Records of a Particular Plantation," typescript, CWL. For Smyth, see James Horn, *Adapting to a New World: English Society in the Seventeenth-Century Chesapeake* (Chapel Hill, N.C., 1994), 78–81. Development of Martin's Hundred is discussed in McCartney, "History and Historical Context of CG-10." For Kent Island, see "Claiborne vs. Clobery," *MHM,* XXVI (1931), 381–404, XXVII (1932), 17–28, 99–114, 191–214, 337–352, XXVIII (1933), 26–43, 172–195, 257–265.

Archaeological findings are described in Ivor Noël Hume, *Martin's Hundred* (New York, 1982); Ivor Noël Hume and Audrey Noël Hume, *The Archaeology of Martin's Hundred,* pt. I, *Interpretive Studies* (Philadelphia and Williamsburg, Va., 2001); Norman F. Barka, "The Archaeology of Flowerdew Hundred Plantation: The Stone House Foundation Site—an Interim Report," typescript, Southside Historical Sites and Department of Anthropology, College of William and Mary, 1976; and Barka, "Early Seventeenth-Century Architecture at Flowerdew Hundred Plantation: The Enclosed Settlement," typescript, Flowerdew Hundred Foundation, 1981; Alain Charles Outlaw, *Governor's Land: Archaeology of Early Seventeenth-Century Virginia Settlements* (Charlottesville, Va., 1990); Nicholas M. Luccketti, "Archaeology at Arlington: Excavations at the Ancestral Custis Plantation, Northampton County, Virginia," Association for the Preservation of Virginia Antiquities, 1999, http://www.apva.org/resource/other/arlington.pdf; Seth Mallios, "Archaeological Excavations at 44JC568, the Reverend Richard Buck Site," Association for the Preservation of Virginia Antiquities, 1999, http://www.apva.org/resource/other/buckweb.pdf; Mallios, "At the Edge of the Precipice: Frontier Ventures, Jamestown's Hinterland, and the Archaeology of 44JC802," Association for the Preservation of Virginia Antiquities, 2000, http://www.apva.org/resource/sandys.pdf; L. Daniel Mouer et al., "Jordan's Journey: A Preliminary Report on Archaeology at Site 44P.G.302, Prince George County Virginia, 1990–1991," typescript; and Douglas C. McLearen and L. Daniel Mouer, "Jordan's Journey II: A Preliminary Report of the 1992–93 Excavations at Archaeological Site 44P.G.307," typescript, Virginia Commonwealth University Archaeological Research Center prepared for the Virginia Department of Historical Resources, 1994.

chewed costly and unfeasible diversification programs and instead intended from the outset to concentrate on more profitable tobacco. She maintains that it was largely through their efforts that the colony was securely established. More recently, Robert Brenner concludes that neither great London merchants (who initially set the Company's policy) nor gentry investors (who subsequently wrested control of the Company from the merchants and who backed particular ventures) advanced enough capital to set up viable productive plantations. He argues that colonial development instead depended on the combined efforts, following the dissolution of the chartered Company, of more substantial Virginia planters and of a group of less well connected "new" English merchants, most of whom originated in the provinces. These new merchants came to specialize in the colonial trades and were more willing to invest in plantation production as well as in strictly commercial ventures and to accept greater risks and lower profit margins. They retained tight connections to colonial traders and politicians and were willing to advance capital without being granted the monopoly trading privileges that better-established London merchants considered a prerequisite. It was these new merchants who first supplied adequate capital, and, by the later 1620s and 1630s, individual colonists, "who managed their own plantations and did their own shipping, or who worked in collaboration with merchant suppliers and marketers, [who] now took over the primary entrepreneurial tasks."[27]

Other scholars have put even more emphasis on the crucial role of resident planters in effective colonial development. Richard Pares argues that neither the parent nor the subsidiary joint-stock companies provided enough development capital. He stresses instead the slower buildup of capital within colonies, as planters themselves began to finance migration by purchasing servants and to accumulate capital by building farms and by developing a local infrastructure. Frederick Fausz pinpoints the years 1622–1625 as the crucial period when the success of Virginia became assured and identified Virginia leaders as the most dynamic group. It was then that settlements were consolidated, local leaders established some semblance of order, acute crises of mortality and subsistence ceased, population expanded, and economic growth accelerated. Finally, Karen Kupperman concludes that, in all the early colonies, "private property in land, combined with a degree of political devolution, was the key to success." Only the promise of eventually gaining independence through owning land attracted a sufficient number of migrants, overcame problems of low settler morale, and encouraged the repeated experimenta-

27. *RVC,* IV, 130–151; Hecht, "The Virginia Colony," chaps. 3–6; Brenner, *Merchants and Revolution,* chaps. 3, 4 (quotation on 96).

tion essential to developing both viable food and export crops. Economic takeoff could occur only when settlers gained enough knowledge of the new environment to become "active risk-takers" and, through a representative assembly, seized control of their colonies' economic life.[28]

Particular plantations begun under the aegis of the Virginia Company, in the opinion of most scholars, were thus at best a transitional form of plantation organization that largely failed to promote sustained colonial development. The evidence discussed below demonstrates that in some areas, notably attitudes toward settlement and supply, there is little to distinguish the policies of early private investors from those of the parent company. It does appear that, in other areas, such as personnel recruitment and willingness to respond flexibly to economic conditions, private investors had more success than the larger company. Once they landed in the New World, however, all settlers in Virginia faced a number of formidable problems—including poor leadership, low morale, high mortality, and inadequate food, shelter, and defense capability—that largely undermined the efforts of all nonresident investors. New forms of corporate organization and new incentives to invest and to emigrate were not enough to overcome these obstacles. Impatient investors did not take into account the amount of time the settlers would need to learn how to farm in an unfamiliar environment. And they were unwilling to make good the tremendous losses inflicted in the Powhatan uprising of 1622. As Pares, Fausz, and Kupperman observe, only when settlers got clear title to land and gained control of the local economy did sustained development occur.

A comparison of the Berkeley Hundred adventurers' instructions of 1619 for seating their particular plantation with orders issued by the Virginia Company to Sir Thomas Gates ten years earlier demonstrates the similarity of outlook between private investors and the parent company. In 1609, the Company had wanted some new settlements started and advised Gates to choose sites that were dry and wholesome, supplied with fresh water, and that afforded good anchorage for ships. Gates was instructed to build small forts for the protection of the settlers, a secure common storehouse, and a church. Each village was to be built neatly "so you shall prepare for ornament and safety at once." Land was to be set aside for growing corn, and men were to

28. Pares, *Merchants and Planters,* chap. 1; John Frederick Fausz, "The Powhatan Uprising of 1622: A Historical Study of Ethnocentrism and Cultural Conflict" (Ph.D. diss., College of William and Mary, 1977), chaps. 5–7 (quotation on 572); Kupperman, *Providence Island,* 114–116, 126.

be apportioned to tend it; however, it was expected that some maize would also be procured from the Indians. There were to be arrangements for guarding crops and cattle. One man was to be the settlement's commander, aided by lesser appointees. These officers were to supervise the labor of all and to establish a fixed schedule for work, worship, common messes, and military training.[29]

Ten years later, the Berkeley associates still also took for granted the necessity of seating settlers together in units that would afford all the advantages of community life in England. They intended "to erect and build a Towne" with an initial contingent of thirty-two men under the direction of an experienced Virginia planter, Captain John Woodlefe. (A second group of equal size was to be sent the next year under another agent who would have equal powers with Woodlefe.) Woodlefe was to choose a site "healthy for ayre," with fresh water and easy access for ships. The land was to be suitable for farming, pasture, timber, fishing, and hunting. The contingent was to build a framed storehouse for tools, weapons, and food and a hall for "assemblies at time of prayer and time of diet." When it came to other structures, the group was willing to dispense with niceties; additional shelters were "for the better expedicion" to be "built homelike and to be covered wth bordes." Four hundred acres were to be enclosed with a seven-and-a-half-foot-high pale to secure crops and cattle. The "town" the Berkeley settlers eventually built likely resembled contemporary Wolstenholme Town on Martin's Hundred (Plate 2).[30]

Obtaining supplementary maize from the Indians was not mentioned in the instructions, but trade beads and copper figured prominently in the goods supplied, and the adventurers instructed Captain Woodlefe "to make the best speedy use you can, wth the truckinge stuffs nowe sent, of beades, hatchets, Copper, shooes, knives." (Similarly, Smith's Hundred settlers arrived with a supply of copper and beads with which to purchase land and corn from local Indians, persuading them "by composition or other fair means to remove farther off," but apparently anticipating a continued trade in maize.) Five of the Berkeley settlers were named as councillors to the commander. These men, along with others, were assigned various responsibilities — to take charge of arms and tools, to supervise the household and parcel out clothes and bedding, to direct the kitchen, and to serve as usher of the hall and bailiff

29. *RVC,* III, 12–24.

30. Ibid., 130–134, 195–196, 199–201, 207–210. For Woodlefe, see McCartney, *Virginia Immigrants and Adventurers,* 762.

PLATE 2. Wolstenholme Town in Martin's Hundred, circa 1625. Conceptual reconstruction based on extensive archaeological excavations. Permission, Richard Schlecht / National Geographic Image Collection.
This was one of the largest of the particular plantations. The plantation's leader might have lived in the fort (background). The company compound (center) contained a dwelling, an adjoining stable, and a storehouse.

of husbandries. Should the commander die, the five assistants were to take charge of the government.[31]

Establishing and maintaining order was a prime consideration for the Berkeley and Smith's Hundred investors, as it was for the parent company. English projectors decided from their experience with colonization attempts in Ireland that tight control and continuous oversight was essential for transferring social order. In Karen Kupperman's words, "colonists were to be moti-

31. *RVC,* III, 196 (quotation), 207–210, 403; Committee for Smythe's Hundred, Minutes of Meeting, May 8, 1618, Ferrar Papers, 1590–1790, Pepys Library, Magdelene College, Cambridge University, microfilm ed., no. 84, Sir George Yeardley from His Majesty's Council for Virginia, Dec. 2, 1618, no. 92, CWL (numbers cited are those assigned in the microfilm edition).

vated by force at first with incentives to follow." Thus, Captain Woodlefe was instructed to "lyve and converse with and amongst" his colonists, and not elsewhere. He was to cement the five settlers named as assistants to the cause by asking their advice and to reinforce their supervisory authority by eating with them at the same table and by serving them meals equal to his own. Secondary officers like Ferdinando Yate, the "Ancient" (ensign), were also supplied with requisite ceremonial trappings, in Yate's case a "pyke staffe and tassell." Smith's Hundred ancient Ferdinando Shepherd was provided several ells of "Taffety" for the Hundred's standard, and one of the commanders and his son were equipped with "better then ordinary" armor.[32]

Religion as well as rank was seen as a buttress to authority and good order. Woodlefe was to enforce the Sabbath, Anglican holidays, and periods of daily prayer. Initial supplies in 1619 included two Bibles, two common prayer books, and six devotional manuals. The next year saw further support in the form of a minister-surgeon hired on an annual salary. In 1621, four additional servants were sent with another testament, "one of them much desiring the same." Continuing commitment to the support of religion is also evidenced in the Kent Island accounts of a decade later. When a resident minister was available, 7 to 10 percent of local expenditures were devoted to his salary and to Bibles, prayer books, and church furniture. If no regular minister was present, other area preachers were brought to Kent Island whenever possible.[33]

Records of the Berkeley and Smith's Hundred adventurers give some idea of how projectors recruited settlers for particular plantations. At first, the Berkeley investors canvassed their localities for men with desired skills and leadership abilities, looking first to neighbors and acquaintances, negotiating individual and varying agreements with prospective immigrants. The six men who actively invested in the enterprise were all residents of Gloucestershire and were related in differing degrees by marriage. John Smyth of North Nibley, steward of the Manor of Berkeley, probably knew personally many of the people recruited, and others were known by and sometimes related to other of the investors. Apparently, reports of the seven long years of enforced service required of Company settlers had reached the Gloucestershire folk

32. *RVC,* III, 183, 205, 262; Kupperman, *The Jamestown Project,* 209. A silver-tipped staff was also among the ceremonial trappings sent to Providence Island (Kupperman, *Providence Island,* 57–58). For the intended hierarchical organization, see also Alexander B. Haskell, "'The Affections of the People': Ideology and the Politics of State Building in Colonial Virginia, 1607–1754" (Ph.D. diss., Johns Hopkins University, 2005), 16–24.

33. *RVC,* III, 178–189, 207–210, 401–402, 618–619; "Claiborne vs. Clobery," *MHM,* XXVIII (1933), 30–43, 179–189; Haskell, "'Affections of the People,'" 24–25.

among whom the Berkeley investors were recruiting (the Company would raise objections in 1620 to a Plymouth man who was spreading rumors that settlers were "used wth more slavery then if they were under the Turke"), for some drove very astute bargains.[34]

Those named to lead the Berkeley contingent as well as men with special skills such as coopers, carpenters, sawyers, shoemakers, and smiths were to work for the adventurers for only three years and were promised from fifteen to fifty acres of land at the end of their terms of service. As a further inducement to emigration, some of the craftsmen would be allowed to keep a proportion of the proceeds of their labors or were allotted some free time in which they could work for themselves, or they were to be paid wages of seven to forty shillings per year. Men with lesser skills, like cooks, tailors, and gardeners, and men of no listed trade were to serve the investors for longer periods of four to eight years. Some recruits got further concessions. A few were paid part of their wages in advance or were given loans to pay their debts. Others arranged for the proprietors to make annual payments to wives left in England, and in one instance the investors put up money to redeem an emigrant's pawned tools. The Smith's Hundred investors gave several men recruited as lieutenants ten pounds to furnish themselves with supplies in England; the surgeon got six pounds, and the clerk of the store, four pounds. Directors of that settlement were also promised annual salaries of twenty to thirty pounds per annum. Some skilled tradesmen going to Smith's Hundred received money with which to purchase sets of tools in England. Others were promised half of whatever they earned, twenty-five acres of land at the end of five years, and a servant to help them in their trades or in their households. Similarly, six millwrights agreed to go to the Kent Island settlement in the 1630s in return for some combination of wages paid on the spot and annual allowances for their families in England, and the head carpenter contracted to serve a number of years at a fixed wage.[35]

Agricultural tenants recruited for Smith's Hundred in 1618 were promised, in line with Virginia Company policy, a six-month supply of victuals, two sets

34. *RVC,* I, 334–335, III, 137.

35. Ibid., III, 197–199, 210–211, 402–404; Memorandum of Conditions to Be Offered to Men Going to Smythe's Hundred, [May 1618], Ferrar Papers, no. 85, Warrant to Pay Thomas Longe, Jan. 8, 1620/1, no. 214, Warrant to Pay Thomas Bums, Jan. 11, 1620/1, no. 215, Warrant to Pay Thomas Nicolls, Jan. 11, 1620/1, no. 216, CWL; "Claiborne vs. Clobery," *MHM,* XXVIII (1933), 30–43, 172–189. Once private merchants got into the business of transporting servants, various bargains depending on the emigrant's age, training, and desired destination became the rule (David W. Galenson, *White Servitude in Colonial America: An Economic Analysis* [New York, 1981], chap. 7).

of clothing, bedding, household equipment and tools, a loan of cattle to be repaid in later years, and ten shillings in cash to be given to every man upon landing in Virginia. They were to keep half of everything they raised and to be tied to their bargain for no longer than five years. By 1620, the Berkeley proprietors offered still other terms to entire families who wished to emigrate as agricultural tenants. They, too, were to be supplied with a house, ground, household goods and tools, clothing, and such food "as the country can reasonably afford" until the first harvest. Virginia Company tenants had to pay half of everything they produced as rent, but the Berkeley proprietors would be satisfied with half of the increase of the cattle and of export crops like tobacco, but only one-third of whatever maize or wheat they might grow. Tenants recruited for Lord John Zouch's particular plantation were offered two-thirds of the profits of their labors and, when free, were to be charged only "such convenient rent as they shallbe hable to bare."[36]

Because private investors were willing to offer more flexible and more favorable terms to prospective settlers, they were likely able to recruit immigrants of better quality than was the Virginia Company. Possibly, as Irene Hecht has argued, they also recruited personnel more astutely. From the listed occupations of Berkeley settlers, she judged that gentlefolk were not sent in excessive numbers. The craftsmen they selected supplied skills essential to survival—carpenters, coopers, sawyers, and shoemakers. Practitioners of overly exotic trades like the pipemakers, jewelers, refiners, perfumers, and glassmakers recruited by the parent company were notably absent. Half of the initial contingent of settlers were unskilled laborers as were (presumably) even more of those who came with the second supply. These workers could be turned immediately to exploiting the land. Similarly, the Smith's Hundred proprietors initially sent out thirty-five men in 1618 to clear and plant land and erect housing for later emigrants. They recruited four carpenters, two sawyers, and a bricklayer, and the remaining twenty-eight were husbandmen and laborers. Settlers sent over in the next four years included a butcher, porter, miller, millwright, distiller, mason, smith, boatwright, surgeon, nurse, several ironworkers, and more carpenters, bricklayers, and sawyers.[37]

36. *RVC,* III, 77, 393–394; Memorandum of Conditions to Be Offered to Men Going to Smythe's Hundred, [May 1618], Ferrar Papers, no. 85, CWL. For Company tenants, see *RVC,* I, 271, III, 98–109, 307–365.

37. *RVC,* III, 396–397; Committee for Smythe's Hundred, Minutes of Meeting, May 8, 1618, Ferrar Papers, no. 84, List of Passengers for Virginia, [April 1619], no. 107, CWL; Hecht, "The Virginia Colony," chaps. 2, 6. See also Morgan, *American Slavery, American Freedom,* chap. 4. At least some of the settlers in the second supply who had no listed occupation were tenants on half-share arrangements.

The first contingent of settlers the Berkeley Hundred adventurers sent appear to have been thoughtfully selected, although they probably made a serious miscalculation about the proportion of husbandmen and domestics that would be needed. When it came time to assemble a second group of colonists the following year, the supply of willing emigrants in the immediate neighborhood had apparently been exhausted, and the projectors had to turn to other nearby areas, including the Vale of Tewkesbury, a depressed part of Gloucestershire, subsequently described by a local writer as "famous for three things, old clothes, lice, and shitten stiles." Emigrants recruited on the basis of their desperation to get out of an unpromising locale tended to include too few young adult "country laboring men," too many older men who "either live not or doe little service in the County," and too many young boys scarcely able to produce enough to keep themselves alive. Getting ready to depart with the second group in April 1620, agent William Tracy reported that more volunteers had come forward than the ship could carry. He proposed leaving "those that ar of lest imployment tel the next going." Perhaps the proprietors were simply supplying reference books to cover chance contingencies when they sent "Markhams and Goouges books of all kynd of English husbandry and huswifry" with the second group of settlers. Alternatively, there might have been serious gaps in the emigrants' agricultural and domestic expertise.[38]

Perhaps not all private investors were as careful. Those who sent over large numbers of settlers within a short span of time must have had to accept more who lacked the skills most needed in the colony. And even those projectors who recruited primarily men with practical skills failed to send a sufficient proportion of farmers. In 1620, Governor Yeardley (echoing Captain John Smith's plea of 1608 for "Carpenters, husbandmen, gardiners, fisher men, blacksmiths, masons, and diggers up of trees, roots; well provided") was urging the adventurers of Smith's Hundred "to send hither husbandmen truely

38. *Harry Hangman's Honour; or, Gloucestershire Hang-man's Request to the Smokers or Tobacconists in London,* British Library, E 842 (13), quoted in Joan Thirsk, "Projects for Gentlemen, Jobs for the Poor: Mutual Aid in the Vale of Tewkesbury, 1600–1630," in Patrick McGrath and John Cannon, eds., *Essays in Bristol and Gloucestershire History: The Centenary Volume of the Bristol and Gloucestershire Archaeological Society* (Bristol, 1976), 148–149; *RVC,* III, 262–266, 300–306, 397–400. For other decaying areas in Gloucestershire, see Emily Rose, "The Politics of Pathos: Richard Frethorne's Letters Home," in Robert Appelbaum and John Wood Sweet, eds., *Envisioning an English Empire: Jamestown and the Making of the North Atlantic World* (Philadelphia, 2005), 92–108. The books referred to were probably Gervase Markham, *The English Housewife . . .* [London, 1615], and Barnabe Googe, *Foure Bookes of Husbandrie. . . .* At least five editions of the latter, which included directions for husbandry, gardening, grafting, and planting, were published in London between 1577 and 1601.

bred (whereof here is a greate scarcity, or none at all)." What was most needed in a New World settlement was the jack of all trades—a man who could clear land; grow and harvest crops; put up makeshift shelters, storehouses, forts, and fences; tend livestock; start orchards and gardens; hunt and fish; and, in a pinch, cook meals, mend clothes, fashion rudimentary tools, nurse the sick, and fight off intruders. Captain John Smith lauded one Company leader who "plaied the Sawyer, Carpenter, Dauber, Laborer, or any thing" to sustain his settlement. But others might have been so accustomed to a highly segmented and specialized work environment in seventeenth-century England that they would perform only those kinds of labor specified in agreements they had made in the mother country. Some simply might not have known how to do many of the things that would desperately need doing.[39]

Once settlers were recruited, finding and retaining suitable managers remained a major problem for the proprietors of particular plantations, as it did for the parent organization. Many of the agents sent from England were inexperienced and proved unequal, physically or psychologically, to the task of implementing the instructions of their Old World principals. All too often newly arrived leaders quickly succumbed to the hazards of the Virginia environment. William Tracy, for example, one of the Berkeley investors who led the second contingent of settlers, lived only three months after landing in the colony. Upon arriving in Virginia in 1619, George Yeardley found the two principal officers at Smith's Hundred dead, and a Mr. Eps, "a hopefull yong gentellman," in effective command. Lacking any experienced men to direct the plantation, Yeardley left the youthful Eps in charge until Eps killed a ship captain in a drunken brawl and was put in jail. Command then temporarily devolved to one of the ensigns. The more ambitious ironworks and sawmills envisioned for Company lands and for Smith's Hundred foundered when all the principal leaders of these enterprises died.[40]

39. Smith, *Generall Historie,* II, 444, 593; *RVC,* III, 256. See also John Rolfe, quoted in Smith, *Generall Historie,* II, 541, on the absence of men knowledgeable about plowing or training draft animals. Rolfe was disappointed in the quality of the settlers in general: "I never amongst so few, have seene so many falseharted, envious and malicious people (yea amongst some who march in the better ranck)" (*RVC,* III, 247). Similarly, another settler, John Baldwin, complained to a correspondent in Bermuda, "I like yt [the country] well yf the people were good that are in yt: but they are base all over" (Lefroy, comp., *Memorials,* I, 265). William Bradford encountered similar problems with the types of people who emigrated to the Plymouth Colony (*History of Plymouth Plantation,* ed. Ford, I, 325–326). For work conditions in England, see Morgan, *American Slavery, American Freedom,* 64–69.

40. *RVC,* I, 588, III, 118–122, 242, 246, 255–258, 396, IV, 16, 58, 76, 107, 158–161, 171; Bullock, *Virginia Impartially Examined,* 13. For ironworks, see also John Bezís-Selfa, *Forging America: Ironworkers, Adventurers, and the Industrious Revolution* (Ithaca, N.Y., 2004), 45–49.

Investors who employed seasoned Virginia hands were more likely to retain longer-lived agents. Experienced settlers seem to have been almost universally more interested in pursuing private advantage, however, than in carefully managing their employers' affairs. Either they acquired land of their own or pursued trade on their own account and neglected the main enterprise, or they found ways to divert the investors' servants and goods to their own ends. Undoubtedly, the low rewards the projectors offered encouraged such profiteering. The Berkeley Hundred proprietors, for example, paid their managing agent only one-twenty-fifth of the proceeds of goods sold, with a pledge of further benefits when the joint stock should be terminated. Unless the agents were themselves major investors in the enterprise and thus entitled to a further share of the profits from return cargoes, they were probably justified in thinking there were better ways for advancing their fortunes in the colony. The fragility of life in early Virginia and the desire to make a fortune quickly and then perhaps return to England encouraged rapacious behavior among those who survived and rose to positions of power. In a frenetic search for profit, there was little scope for honorable dealings.[41]

Proprietor John Smyth thus might be wrongly incensed at "the foolery of his [agent Woodlefe's] contemplative newe magazen (not to be furnished for 1000 [pounds]) and of the newest fashion, lesse," for Captain Woodlefe had rightly discovered that what would sell best in Virginia was not necessarily what the settlers most needed for sheer survival. On the other hand, Smyth was justified in complaining that Woodlefe had sold the time of some of the settlers transported at the proprietors' expense in order to get funds to buy additional servants for himself, a blatant drawing "to his pryvate . . . our publike." Woodlefe had also failed to observe the adventurers' "orders for dyet and society," presumably by not supplying the settlers with all the provisions promised them and by spending more time on his own plantation across the James River than at the new settlement. Indeed, the adventurers so distrusted Woodlefe that they made agreements with other settlers to send private letters in secret "touching mr Woodleefe and his estate, behaviour and usage of our men." But they received no such communications, perhaps because some of the men from whom they expected reports were already dead, or perhaps because, as the adventurers suspected, Woodlefe intercepted adverse assessments. Similarly, William Claiborne and his partners in the Kent Island venture fell into irreconcilable disputes about what constituted an adequate reward for the managing agent (Claiborne contended he should get a salary of

41. *RVC,* I, 331, III, 201–207, IV, 74, 100–111.

one hundred pounds a year) and also about how to accommodate his simultaneous pursuit of corporate welfare and private profit.[42]

Even with the best of managers, maintaining morale among the colonists was a formidable task. Many immigrants entertained unrealistic expectations about what they would find in Virginia, fed in part by overly optimistic reports circulated by Virginia Company officials and in part by false promises made by recruiting agents. The very different reality they encountered once off the ship produced in many a lethargy stemming from psychological shock that proved as deadly as malaria, dysentery, or typhoid. Like colonist Richard Quailes they found "this place a meere plantacion of sorrowes and Cropp of trobles, haveing been plentifull in nothing but want and wanting nothing but plenty." And some, like Martin's Hundred servant Richard Frethorne, would gladly have sacrificed an arm or a leg if that sacrifice would get them passage back to England.[43]

Diet proved an almost insurmountable problem, both for servants of the parent company and for settlers on particular plantations. Even when there was enough food available in terms of gross caloric intake (and usually this was not the case), many were unable to adjust to a change from traditional, more varied English fare to the monotonous corn (and eventually meat) regimen that would dominate the region's cuisine into the 1800s. Indeed, the rations afforded passengers during the crossing, ideally consisting of biscuit and beer every day; pork and peas on Sunday, Monday, and Thursday; dry fish, oil, and vinegar on Tuesday and Wednesday; and dry fish, butter, and cheese on Friday and Saturday offered much more variety. The frequently hungry,

42. Ibid., III, 292–294; "Claiborne vs. Clobery," *MHM,* XXVI (1931), 383–404, XXVII (1932), 17–28, 191–214, 337–352, XXVIII (1933), 26–43, 172–195, 257–265. One of the Berkeley settlers, Ferdinando Yate, commended "the care and experienc[e] of o[u]r Captain" in establishing the settlement, but the proprietors thought Yate "none of the wiseth" and that he "wryteth too much but all of one syde" (*RVC,* III, 114, 293). A letter of the Virginia Company shows Woodlefe's judgment about the magazine was correct; the Company was sending "necessary and useful" goods instead of "vanities and superfluities, although we find they yield most profit in Virginia" (*RVC,* III, 502–508).

43. *RVC,* II, 112–113, 129–131, III, 262–265 (quotation on 263), 297–300, 534–535, IV, 58–60, 158–161, 468 (quotation); Smith, *Generall Historie,* II, 542–543. An example of the overly optimistic tracts is Alexander Whitaker, *Good Newes from Virginia; Sent to the Counsell and Company of Virginia, Resident in England . . .* (London, 1613). William Bradford similarly reported on the false expectations of some new arrivals in the Plymouth Colony—they expected to find "walled towns, forts, and corne fields, as if townes could have built themselves," and became despondent upon discovering that they could not become "great men and rich, all of a sudaine" (*History of Plymouth Plantation,* ed. Ford, I, 325–326). On Frethorne, see Rose, "The Politics of Pathos," in Appelbaum and Sweet, eds., *Envisioning an English Empire,* 92–108.

ailing, and homesick newcomers longed for not just more "water gruell" but a variety of "English victualls" including peas, oatmeal, biscuit, butter, cheese, malt, cider, salt beef, spices, oil, vinegar, and aquavit. George Yeardley, for example, charged the Virginia Company for peas, oatmeal, and biscuit that he had bought for a contingent of new, sick settlers in June 1620, justifying his action by arguing that meal alone "is very harsh for them to feed upon being new comers" and that a "varyety of victuall to be much to the content and health of the people." George Thorpe, a commander of the Berkeley settlers, was "perswaded that more doe die here of the disease of theire minde then of theire body by havinge this countrey victualls over-praised unto them in England and by not knowinge they shall drinke water here."[44]

In fact, more established settlers confiscated the English provisions intended for new immigrants, giving them corn instead, "the extreame labor of beatinge thereof, being no smale dishearteninge to the new comers; and this suddaine change of diett is affirmed confidently to be the cause of the flix in our men." Moreover, the arrival of ships bringing additional supplies of food as often exacerbated as alleviated shortages, for, when the ships stayed overlong in the colony, their crews ate much of the stock intended for the settlers and commandeered part of what remained as provisions for the return voyage. Psychological and nutritional factors undoubtedly interacted. More than a few settlers found the unaccustomed diet so repugnant and at the same time reacted so negatively to the new environment that they were unable to adjust. The combined effects of malnutrition and mental despair produced a fatal withdrawal from life.[45]

Many proprietors of particular plantations appear to have contributed to the settlers' dilemma by sharing the Virginia Company's lamentable tendency to send settlers without provisions sufficient to last until the next harvest, always vainly trusting that somehow the residue could be gotten in the colony. The Berkeley proprietors, and perhaps others, miscalculated the date

44. *RVC,* III, 262–265, 297–300, 417–418, 455–458, IV, 41–42, 58–62, 231–232, 236, 450–455. For shipboard rations, see Memorandum for Fitting Out Ships [circa 1620], Ferrar Papers, no. 210, CWL. Similarly in Plymouth, William Bradford found some settlers discouraged by "want of those varietis which England affords, from which they are not yet wayned" (*History of Plymouth Plantation,* ed. Ford, I, 326 n. 2). For the nature of the English diet, see Mildred Campbell, *The English Yeoman under Elizabeth and the Early Stuarts* (1942; rpt. New York, 1968), chap. 6; and G. E. Fussell, ed., *Robert Loder's Farm Accounts, 1610–20,* Camden 3d Ser., LIII (London, 1936).

45. *RVC,* III, 495, IV, 175; Virginia Bernhard, "'Men, Women and Children' at Jamestown: Population and Gender in Early Virginia, 1607–1610," *Jour. So. Hist.,* LVIII (1992), 599–618; Karen Ordahl Kupperman, "Apathy and Death in Early Jamestown," *JAH,* LXVI (1979), 24–40; Robert Appelbaum, "Hunger in Early Virginia: Indians and English Facing Off over Excess, Want, and Need," in Appelbaum and Sweet, eds., *Envisioning an English Empire,* 195–216.

of the maize harvest in Virginia, assuming that new corn would be available by Saint Bartholomew's Day (August 24), rather than only toward the end of September or in early October. The Council in Virginia asserted that much of the suffering and starvation that followed the 1622 massacre was a result of private investors inadequately provisioning their tenants. Public letters in England proclaiming abundance in Virginia, Alderman Johnson and others charged, "allured" many to go over meanly provided for. Yet in 1626 when no private adventurer could plead ignorance, Governor Francis Wyatt reported that shiploads of new settlers were still being transported without adequate stocks of food.[46]

Moreover, the settlers arrived in an especially inauspicious moment in the climate history of the Chesapeake region. The winter of 1607–1608 was one of the coldest ever experienced in eastern North America, apparently owing to a strong El Niño appearance. Tree ring reconstruction of growing-season climates indicates that 1606–1612 was the driest seven-year period to occur in 770 years, and drought recurred again in the summers of 1621 and 1622. Mortality among English settlers was significantly higher during dry summers. Then the native Algonquian people were also experiencing shortages of maize and wild fruits and thus had no surplus to trade to the Europeans. English domestic livestock, too, suffered from shortages of drinking water and forage, and reduced freshwater discharge might have contributed to poor water quality in the vicinity of Jamestown. Unusually variable weather also made learning how to grow tobacco more difficult. John Smith noted that a spate of dry weather followed by hail in 1618 spoiled much corn and tobacco. George Yeardley advised in April 1619 that a spring drought would lead to a short crop, but sufficient rain had fallen by September so that both John Rolfe and John Pory reported an abundance of corn and tobacco in what turned out to be the most plentiful year "since the beginning of this Plantacion." The greater "unseasonableness" of the 1621 summer, however, likely contributed to the poor quality of that year's leaf.[47]

Clothing, too, was often in short supply. And if cloth, rather than more ex-

46. *RVC,* III, 275–280, 577–579, IV, 9–17, 65, 109–110, 175, 176, 567–570; [John Harvey], "Harvey's Declaration: A Briefe Declaration of the State of Virginia at My Comminge from Thence in February 1624," *The Aspinwall Papers,* MHS, *Collections,* 4th Ser., XXXIX (Boston, 1871), 60–73; Craven, *Dissolution of the Virginia Company,* chaps. 6, 7; "Letter of Sir Francis Wyatt, Governor of Virginia, 1621–1626," *WMQ,* 2d Ser., VI (1926), 114–121.

47. David W. Stahle et al., "The Lost Colony and Jamestown Droughts," *Science,* CCLXXX (1998), 564–567; Smith, *Generall Historie,* II, 536; *RVC,* III, 244, 257. According to the tree ring reconstruction, 1619 was a year of average rainfall. For winter cold, see Kupperman, *The Jamestown Project,* 168.

pensive ready-made garments, was sent instead, supplies might arrive too late for the few tailors in the colony to make it into warm clothing before winter set in. Other essential manufactured goods were also scarce. Lack of simple equipment like meal sieves could spell the difference between adequate if uninteresting food and indigestible messes that could eventually cause fatal intestinal disorders. Some shortages were also partly the settlers' fault. Hungry, thirsty, and disheartened men might purloin plantation tools and equipment to trade to Indians or to sailors for food, or, more especially, for alcohol.[48]

Despite the intentions of those Company officials and private adventurers who planned early Virginia towns, the colony's crude settlements failed to provide most of the amenities of English community life that the principals envisioned. Overcrowding in tiny, ramshackle houses only exacerbated the situation, contributing both to high death rates from contagious disease and to lowered morale. One Company official found that, when settlers were domiciled six to a small dwelling, bickering and discontent increased while work diminished. Nor was there any place to go to escape other than to religious services. Captain John Smith observed that once the ships departed "there remained neither taverne, beere-house nor place of reli[e]fe but the common kettell." Consequently, when ships did arrive, men desperate for a temporary respite turned the boats into floating taverns, often spending more in drink in one week than they earned in six months.[49]

Communal living was all the more unpleasant because of the scarcity of women. Men who spent their days in unaccustomed toil felling trees or hoeing fields returned from work too tired to cook properly or to wash and mend their clothes, supposing they had the skill, which most probably did not. The few unmarried women sent as wives were "so well sould" that poor men had no hope of obtaining spouses and in fact had to pay exorbitant sums

48. *RVC,* III, 78, 297–300, 448–449, IV, 41–42, 59; John Smith, *A Map of Virginia with a Description of the Countrey . . .* (Oxford, 1612), in Arber, ed., *Travels and Works of Captain John Smith,* I, 127, 155 (hereafter cited as Smith, *Map of Virginia*). Disease rather than the food might have caused the diarrhea; but, if the victims believed the food was the cause, they would have been reluctant to eat even what was available. Union soldiers imprisoned in Andersonville, Georgia, during the Civil War similarly believed they died from eating rations composed only of inadequately milled and sifted corn. Kupperman found that, during the Korean War, Western prisoners believed that the crushed but undercooked soybeans they were given caused their dysentery and so refused to eat them ("Apathy and Death," *JAH,* LXVI [1979], 33).

49. *RVC,* III, 441–443, IV, 59, 176, 271–273, 453; Smith, *Map of Virginia,* I, 95, 103–104, 127. The Virginia Council asserted that merchant Robert Bennet had boasted that the "Sale of fower butts of wyne would be Sufficyent to clere the whole Vioage" (*RVC,* IV, 453). Settlers in Bermuda reacted in much the same way, running to the ships for beer and aquavit, "for which they will give any thing they have" (Smith, *Generall Historie,* II, 682).

for domestic chores such as washing. Men who came as servants resented the women's escaping that condition by marrying and subsequently working only for their own families. This led one bachelor to complain that the women did "nothing but to devoure the food of the land without dooing any dayes deed" for the rest of the men who were dying, lying in their own excrement for want of nursing.[50]

The absence of families or even of much chance for eventually forming them in Virginia also contributed to frenetic profit seeking. Single men "esteeminge Virginia not as a place of Habitacion but onely of a short soiourninge have applyed themselves and their labors wholly to the raisinge of present proffitt and utt[e]rly neglected not only staple Commodities but even the verie necessities of mans life." What the settlers looked for, as had those who temporarily planted tobacco in Guiana a few years earlier, was only "a present Cropp, and theire hastie retourne." Men who believed they had no stake in the continuance of the settlement and who lacked any conception of a common good refused to contribute labor to fortifications and other public works, repining "as much as if all their goods had bene taken from them." Only the presence of women and children would lead men to focus their energies on building for the settlement's future.[51]

Finally, to complete the catalog of problems, high mortality decimated the settlers of particular plantations just as it did those of the parent company. In the best of circumstances, most new settlers suffered an initial bout of seasoning. When large groups of new arrivals were dumped on the colony, often with inadequate food and shelter, some already infected with shipboard diseases, the ensuing mortality was frightful. John Rolfe, for example, reported in January 1620 that the plantations were "much weakened by the greate mortality" the previous year. Smith's Hundred had experienced so much sickness that "no matter of gaine or of greate industry can be expected from them," and at least one settlement, Christopher Lawne's, was entirely abandoned on that account. A letter from the Virginia Company made reference in 1621 to the Society of Martin's Hundred having had "misfortunes as well here as in Virginia" that included heavy loss of life. Groups of individual graves all dug at the same time and also a common burial found on the site of that settlement attest to multiple deaths from the scurvy, bloody flux, and

50. *RVC,* IV, 231–232.

51. Ibid., I, 566, III, 302, IV, 572; Karen Ordahl Kupperman, "The Beehive as a Model for Colonial Design," in Kupperman, ed., *America in European Consciousness, 1493–1750* (Chapel Hill, N.C., 1995), 272–292; Kupperman, *Providence Island,* 158; Kathleen M. Brown, *Good Wives, Nasty Wenches, and Anxious Patriarchs: Gender, Race, and Power in Colonial Virginia* (Chapel Hill, N.C., 1996), chap. 3.

"divers other diseases" mentioned by one Hundred resident. These unnamed diseases probably included the typhoid, dysentery, and salt poisoning from drinking brackish water identified by one modern scholar as major causes of death in early Virginia. Similarly, although the Berkeley proprietors sent over ninety-odd settlers between the fall of 1619 and the spring of 1622, and at least the first group arrived "lusty and in good health," by August 1622 just over twenty remained alive in Virginia. Four had returned to England, and eleven were killed by the Indians. Fifty or more had perished from disease, starvation, or accident within less than three years. Given such an appalling death rate, simply peopling the settlements was a formidable task.[52]

Nonetheless, this catalog of factors contributing to a faltering start was far from unique to the Virginia colony. Many of the settlements the English attempted in the 1600s, from the coast of Maine in the north to Guiana on the northern coast of South America, were abandoned after only a year or so. Plymouth Colony settlers also had insufficient provisions to sustain them until their first harvest, half their number died during their first terrible winter in Massachusetts, and sponsoring English merchants, discouraged by the lack of early returns, stopped sending supplies. The Pilgrims quickly learned to raise New World crops, however, and made some profit for the English adventurers from fishing and trading with the Indians. Stay-at-home investors in the Providence Island Company demanded that the colonists there produce a staggering array of semitropical products, many of which required too much labor and were ill suited to the island's soils and climate. As in Virginia, so many different projects were attempted there that none had a chance of succeeding, and, meanwhile, the investors protested the colonists' reliance on tobacco as their only viable staple. The investors' decision to keep all the Providence Island colonists working as tenants, rather than eventually granting them land, afforded settlers little incentive to develop their holdings or to initiate agricultural experiments. But, however discouraged the Providence Islanders became, and despite hardships brought by drought, failed staple crops, and a diet overheavy on potatoes, there the settlers did not starve.[53]

So the questions remain: Why did Virginia settlers suffer so grievously, and, especially, Why were they repeatedly unable or unwilling to raise enough food to ensure their survival? Karen Kupperman has argued that the high

52. *RVC,* I, 414, III, 197–199, 246–247, 396–397, 502–508, 618–619, 674, IV, 22–26, 58, 232; Noël Hume, *Martin's Hundred,* 34–38, 273–277; Smith, *Generall Historie,* II, 540; Carville V. Earle, "Environment, Disease, and Mortality in Early Virginia," in Thad W. Tate and David L. Ammerman, eds., *The Chesapeake in the Seventeenth Century: Essays on Anglo-American Society* (Chapel Hill, N.C., 1979), 96–125.

53. Andrews, *Colonial Period,* I, 271, 283–285; Kupperman, *Providence Island,* chap. 4.

mortality was caused by a combination of psychological and physical ailments. Protein-calorie malnutrition results in loss of appetite and a reduction in both physical working capacity and in resistance to infection. Nutritional deficiency diseases, including pellagra and scurvy, which were likely present, in their early stages produce symptoms — lethargy and apathy — that appear more psychological than physical. Thus, disease, rather than deficient character, might have been the cause of much of the "Idlenesse and bestiall slouth" that many observers cited as a leading cause of starvation and death. At the same time, a sense of profound isolation and of not knowing when or even if relief would come intensified the symptoms of these diseases. The heightened mortality that followed the massacre of 1622 resulted from the same interaction of famine and disease with despair that reigned during the earlier "starving time" of 1609–1610. As settler William Capps wrote, "The last massacre killed all our Countrie," because "they burst the heart of all the rest." Tenants became so dejected that "most give them selves over, and die of Melanchole." Even those able and willing to work dared not venture far outside fortified compounds for fear of ambush; close confinement and forced idleness also exacerbated the effects of hunger and disease.[54]

Investors' Intentions and New World Realities

What economic regimen did the proprietors, for the most part comfortably ensconced in England, expect their settlers to carry out in these inauspicious circumstances? Virginia Company officials sought to make private investors follow the elaborate schemes for diversified agriculture and staple production pursued by the Sandys administration on the public lands. Patents for particular plantations included a clause that the adventurers would not apply themselves wholly or chiefly to tobacco but also would attend to iron, cordage, pitch, tar, potash, soap ash, timber, silk, vines, salt, and flax. Did the private investors wholeheartedly embrace these objectives, or did they plan to disregard such projects and concentrate on tobacco, the one commodity that had so far produced sure profit? Wesley Craven has argued that adventurers with experience in Bermuda, some of whom had lost heavily in similar diversification schemes there, might have intended to concentrate on tobacco in Virginia. Irene Hecht believes that the Berkeley proprietors' failure to include specifics about economic activity in their agent's instructions

54. Kupperman, "Apathy and Death," *JAH,* LXVI (1979), 24–40; *RVC,* III, 456, IV, 38, 74. See also Bernhard, "'Men, Women, and Children,'" *Jour. So. Hist.,* LVIII (1992), 599–618; and Joanne Bowen and Susan Trevarthen Andrews, "The Starving Time at Jamestown: Faunal Analysis of Pit 1, Pit 3, the Bulwark Ditch, Ditch 6, Ditch 7, and Midden 1" (report submitted to the Association for the Preservation of Virginia Antiquities, October 2000), typescript, CWL.

suggests that they, too, were paying little more than lip service to the Company strictures. On the other hand, there is good evidence that some other private investors, especially those prominent in the parent company but who lacked direct experience of colonial conditions, did intend to pursue at least some of the recommended projects.[55]

First, evidence from practice as well as from stated intention suggests a variety of aims both practical and impractical. The Berkeley proprietors said they hoped eventually to produce iron, cordage, mulberry trees for silkworms, vines, wheat, rice, aniseed, flax, woad, oilseed, and ship timber as well as maize and tobacco. An agreement of 1620 between the proprietors and three Gloucestershire men—a gardener, a glover, and a husbandman—intending to go as agricultural tenants mentioned the planting of orchards, vineyards, tobacco, corn, wheat, woad, silkgrass, flax, hemp, olives, cotton wool, aniseed, and silkworms. To support these projects, the first supply included a variety of "seeds"—aniseeds, cuminseed, fenugreek, wormseed, and almonds—as well as more ordinary unspecified "garden seeds." The settlers did in fact expend effort they could ill afford setting out vines and mulberry trees upon first arrival. In addition, one man was recruited to produce hops and woad and another to do both smithing and to produce fish, pitch, and tar, the latter taking his own fishing gear as well as a seine the investors had purchased. Further supplies sent early in 1620 included forty-one sorts of garden seeds, a half bushel of mustard seed, twenty-eight pounds of aniseed, cotton wool seed, one pound of silkworm eggs, and a bag of plum stones and apple stocks from John Smyth's own nursery. The proprietors also provided two books on silk and silkworms, a pamphlet on the manufacture of artificial wine, and copies of the Virginia Council's order for fishing. Smyth collected additional vines for shipment in 1621 and also advanced a small sum to buy goods for "the trade of Furrs." William Tracy brought additional silkworm eggs in the second expedition and noted that the adventurers had also been discussing the possibility of setting up an ironworks.[56]

55. *RVC,* I, 266–269, III, 130–134; Craven, "Introduction to the History of Bermuda," *WMQ,* 2d Ser., XVIII (1938), 46; Hecht, "The Virginia Colony," chap. 6; Morgan, *American Slavery, American Freedom,* 109. The rationale for the Company's ambitious program was laid out in *Nova Britannia: Offering Most Excellent Fruites by Planting in Virginia . . .* (London, 1609), *Force Tracts,* I, no. 6. See also Craven, *Southern Colonies,* chap. 5.

56. *RVC,* III, 109, 166, 178–189, 197–199, 208, 256, 260–261, 290–291, 389, 393–394, 397–400; Memoranda of Payments Made to the Berkeley Family and of the Personal Expenditure of John Smyth on Visits to London, 1602–1641, John Smyth of Nibley Papers, Gloucestershire Collection 16071, Gloucester City Library (transcripts supplied by James Horn). See also Sir John Berkeley's statement of hopes for ironworks, cotton wool, wine, and silk (*RVC,* I, 623). The colony's first representative assembly similarly enacted laws (clearly never enforced) for planting mulberry

Evidence for other particular plantations is much less complete. We do know that Sir John Zouch and Associates attempted to produce iron. The Society of Southampton Hundred intended to start a saltworks and transported one hundred men under the direction of Benjamin Bluett to establish an ironworks, an investment that came to naught when Bluett and many of his men quickly perished. The remaining settlers were unable to manage the ironworks, "and soe turned good honest Tobaccoe mongers." In 1623, the Southampton Society sought to recoup the costs of resupplying their plantation by sending the ship carrying the goods to trade for furs in the Hudson River and to catch fish off Cape Cod that were then to be brought back to Virginia and sold. The Society of Martin's Hundred provided their settlers with instructions for raising indigo and cassava (written by a minister in Bermuda) as well as for growing tobacco, and the Smith's Hundred investors sent their settlers hemp and flaxseed. Yet, in 1635, the English backers of the Kent Island settlement, primarily a fur-trading venture, would send a contingent of carpenters and millwrights to set up mills on that island and at Kecoughtan in Virginia, apparently envisaging "projects of mills, rape oile [and] iron workes."[57]

Joan Thirsk's studies of gentry investment in similar sorts of projects in England help to explain why enthusiasm for the Sandys program was widespread. It has long been recognized that on the eve of colonization many Englishmen were unemployed or underemployed and that leaders feared the

trees, silkgrass, hemp, flax, and vines. Representatives of particular plantations were already a majority of the body (ibid., III, 153–177). In 1632, a correspondent trying to interest John Smyth in resupplying Berkeley Hundred would use as an argument that other commodities including timber, soap ashes, furs, and wine were being produced so, with new hands, they could "attempt some greater matters" (Smyth of Nibley Papers). French vignerons sent over by the Company also instructed settlers on some particular plantations (*RVC,* III, 166, 256). Mustard balls were a special product of the Vale of Tewkesbury from which the Berkeley Hundred settlers came (Thirsk, "New Crops and Their Diffusion," in Chalklin and Havinden, eds., *Rural Change and Urban Growth,* 88–93).

57. *RVC,* I, 285, IV, 141, 144; *MCGC,* 481; Committee for Smythe's Hundred, Minutes of Meeting, May 8, 1618, Ferrar Papers, no. 84, Memorandum regarding an Iron Furnace [1618], no. 96, Expenses of 44 Persons Sent on Bona Nova, [circa 1619], no. 137, Benjamin Blewitt, Account of Expenses for Southampton Hundred, Jan. 10, 1619/20, no. 142, Michael Lapworth to John Ferrar, Jan. 26, 1621, no. 268, Invoice of Writings Sent to Virginia, [October 1622], no. 428, CWL; *CSP,* I, 19; "Claiborne vs. Clobery," *MHM,* XXVIII (1932), 345. *A Relation of Maryland . . .* (London, 1635) promised potential investors and settlers good prospects for citrus fruit, hemp, flax, pitch, tar, rapeseed, aniseed, woad, madder, saffron, silkworms, potash, wine, iron, lumber, and ship timber in addition to the more realistic possibility of producing grain, meat, and dairy products for export to other colonies (in Clayton Colman Hall, ed., *Narratives of Early Maryland, 1633–1684,* Original Narratives of Early American History [New York, 1910], 82–83).

country was overpopulated. The desire to free England from dangerous dependence on other countries for essentials like ship timber, cordage, and dyestuffs and to improve the nation's balance of payments by halting the outflow of money spent on foreign luxuries like wine, silk, tobacco, and spices is also well understood. Thirsk's contribution has been to document the ways in which private gentry investors acted on these imperatives at the turn of the seventeenth century, putting up capital to initiate a multitude of projects that yielded a profit to the gentleman-undertakers and that put underemployed English men and women to work making a host of previously imported or seldom manufactured products ranging from the essential to the frivolous. Such work included starch, soap, and salt making; the manufacture of pottery and metal wares; brewing and distilling; and the cultivation of rapeseed, flax, hemp, dyestuffs, tobacco, grapes, and mulberries. Once a number of these projects succeeded at home, gentry investors would, not surprisingly, try to set up similar enterprises both on plantations in Ireland and in the New World.[58]

Some of the men who backed particular plantations in Virginia had invested in similar projects at home. It has been suggested that members of the Berkeley family who pursued ironworks in Virginia were previously engaged in iron making in the Cotswolds. Thirsk demonstrates that some of the Berkeley Hundred investors were also involved in flax, rapeseed, and tobacco-growing projects in the Vale of Tewkesbury in Gloucestershire. One of them, William Tracy, became a major stockholder and led the second contingent of settlers to Virginia. Relationships between gentry projectors in England and adventurers in colonial plantations has not been closely studied, but it might have been extensive. It thus becomes more plausible that such men would have supported an extension of activities in the Chesapeake that they were already profitably pursuing at home.[59]

Further evidence that private investors planned to engage in a variety of agricultural projects, with tobacco representing only one of many possibili-

58. Craven, *Southern Colonies,* chap. 4; Hecht, "The Virginia Colony," chap. 4; Beer, *Origins of the British Colonial System,* chap. 3; Joan Thirsk, *Economic Policy and Projects: The Development of a Consumer Society in Early Modern England* (Oxford, 1978). See also Eric Gethyn-Jones, *George Thorpe and the Berkeley Company: A Gloucestershire Enterprise in Virginia* (Gloucester, Eng., 1982).

59. Francis Lewis Berkeley, Jr., "The Berkeleys of Barn Elms, Planters of Colonial Virginia, and a Calendar of the Berkeley Papers, 1653–1767" (master's thesis, University of Virginia, 1940), chap. 1; Thirsk, "Projects for Gentlemen," in McGrath and Cannon, eds., *Essays in Bristol and Gloucestershire History,* 147–169. John Smyth of Nibley, one of the Berkeley Hundred investors, wanted good tobacco seed sent to England by their Virginia agent; presumably he wanted to make a trial of it in Gloucestershire (*RVC,* III, 195).

ties, is seen in the Berkeley proprietors' recruiting not just men to serve for a number of years but also tenant families on shares. The three men and their families sent in 1620 were "our first tenants or undertakers to halfes in that kynd." The proprietors urged their Virginia agent to supply everything that had been promised, hoping that the tenants would send back favorable reports to their friends and neighbors who might then be persuaded to emigrate. The Society of Martin's Hundred eventually adopted the same practice. The agreement between the Berkeley proprietors and their tenants discussed earlier suggests tenants were encouraged not just to plant tobacco but also to practice diversified farming that might yield profits from a variety of staples. In the colonies, however, as we shall see later, half-share tenancy failed to provide returns to investors or incentives to the tenants.[60]

The success of such diverse projects in England depended on some combination of natural resources, the availability of a large, cheap supply of labor, and acquired skills. Mesmerized by the Chesapeake's vast natural resources, English investors seem never to have realized how scarce and dear labor would be there and in what short supply skilled workmen would be. Consequently, they failed to appreciate that, for all labor-intensive commodities except tobacco, the high cost of colonial labor would more than cancel out any advantage afforded by widely available natural resources. Moreover, the more complex projects depended on the survival of particular skilled individuals, and the entire investment was lost when key personnel perished.[61]

Private investors, however, were somewhat more flexible than Virginia Company officers. Gentry projectors in England had been in the habit of undertaking only small trials of several different products, moving quickly from one commodity to another as conditions changed, not throwing all their resources into a single venture. Thus, proprietors of particular plantations probably lost less on diversification projects than did the Virginia Company simply because most of their investment did go to generalized plantation building. If the plantations had been firmly established, they could have afforded to lose on experimental trials of vines, flax, or salt. Finally, unlike Company leaders, plantation commanders seem not to have been tied

60. The agreement is in *RVC,* III, 393–394. One of the men, Richard Smyth, was paid for his garden seeds and another man for carriage of household goods to Virginia. For instructions about the tenants, see ibid., III, 399. For Martin's Hundred, see ibid., IV, 102–103.

61. Hecht, "The Virginia Colony," chap. 4; Beer, *Origins of the British Colonial System,* chap. 4; Sigmund Diamond, "From Organization to Society: Virginia in the Seventeenth Century," *American Journal of Sociology,* LXIII (1958), 457–475. See also *RVC,* III, 455–458, 581–588. The Company envisioned projects that may "set many thousands a worke" (*Nova Britannia, Force Tracts,* I, no. 6, 19–21).

to strict instructions about the projects to be pursued, often being enjoined to employ the servants "to best profiting" and to rent land to tenants on whatever basis proved most practical. They could more quickly abandon diversification schemes that did not succeed and concentrate on the one commodity that produced profits. Rapidly expanding Virginia tobacco exports accompanying the seating of particular plantations suggest that the agents did just that. English imports of Chesapeake tobacco jumped from more than nine thousand pounds in 1617 to nearly twenty-five thousand pounds in 1618 when the first particular plantations were seated, doubled by 1620, and doubled again by 1624.[62]

Whatever their English backers might have envisioned, once the settlers disembarked in the Chesapeake life on particular plantations quickly settled into a routine designed to secure a combination of immediate survival and quick profits. The first steps were to throw up some sort of palisade for protection, to erect a secure storehouse for tools and provisions, and to construct minimal shelters for men and enclosures for livestock. The first group of settlers for Berkeley Hundred left England in mid-September 1619, arrived in Virginia on November 30, and seated their plantation the following month. One of the officers, ensign Ferdinando Yate (who was in charge of armor and tools) reported sometime in the following spring that this initial contingent was well settled on good land, and "our house is bilt with a stoore convenient." In August 1620, the investors stated that the settlers had already erected "dyv[er]s forts and buildings . . . and other enclosures." Perhaps this was an overly optimistic summation of Yate's more modest report, but some Bristol mariners who visited the plantation in 1621 made reference to the "Town of Barkly." Between 1618 and 1622, Martin's Hundred colonists managed to build a paled semirectangular fort, a storehouse, a multiple dwelling unit of fifteen by sixty feet that was likely a barracks, and a church. There were also individual houses, some laid out neatly between the fort and the river in a grid pattern closely paralleling English plantations in Ireland, and others more haphazardly located. Residents of Flowerdew Hundred and Jordan's Journey also built similarly configured settlements.[63]

62. Thirsk, *Economic Policy and Projects;* Thirsk, "Projects for Gentlemen," in McGrath and Cannon, eds., *Essays in Bristol and Gloucestershire History,* 147–169; *RVC,* III, 114, 379–381; Menard, "The Tobacco Industry," *Res. Econ. Hist.,* V (1980), 157. Hecht, "The Virginia Colony," appendix IV-B, reports higher figures that probably include tobacco imported from Bermuda.

63. *RVC,* III, 114, 379–380, IV, 41–42; Gethyn-Jones, *George Thorpe and the Berkeley Company,* appendix 8; Noël Hume, *Martin's Hundred,* chaps. 10, 11, and 12; Noël Hume and Noël Hume, *The Archaeology of Martin's Hundred,* chap. 2; James Deetz, *Flowerdew Hundred: The Archaeology of a Virginia Plantation, 1619–1864* (Charlottesville, Va., 1993), chap. 2; Mouer et al., "Jordan's

The plantation commanders, anxious to reinforce rank with tangible trappings, employed some of their people to construct relatively commodious and reasonably finished dwellings for themselves and their families. Evidence of such houses has been uncovered at Martin's and Flowerdew Hundreds and at Jordan's Journey. Most settlers, however, lived in hastily thrown together hovels. Few houses were framed, most being "punches sett into the Ground And covered wth Boards so as a firebrand is sufficient to consume them all." They were so slightly built that Captain John Smith reported tenant houses required extensive repairs every year. They were also miserably small. Early records specify the construction of a tenant house measuring twelve by fourteen feet and another fifteen by twenty feet. At Martin's Hundred, ordinary domestic units that have been excavated, presumably housing a single family, were also fifteen by twenty feet (Plate 3). In subsequent years, settlers on the more flourishing plantations got around to building tobacco houses, churches, windmills, and additional storehouses. Since there were initially never enough hands to both build and plant, experienced agents like John Martin and William Claiborne wisely hired any available free workmen or loitering sailors to help with construction and cultivation.[64]

Even as the shelters were going up, the laborious clearing and planting had also to be started. Ferdinando Yate reported to his English backers that the

---

Journey"; monthly reports on excavations are posted on the Jamestown Rediscovery website, www.apva.org/jr.html. In January 1620, John Rolfe said the Martin's Hundred settlers had "good and convenyent houses," perhaps the best of any of the particular plantations (*RVC,* III, 247).

64. Noël Hume, *Martin's Hundred,* chaps. 3, 4, and 5, describes a house probably occupied by the commander of Martin's Hundred after the 1622 massacre measuring 40 by 18 feet. Barka, "The Archaeology of Flowerdew Hundred Plantation," reports on a substantial house, possibly a minister's, measuring 41 ½ by 24 ½ feet. See also Deetz, *Flowerdew Hundred,* chap. 2. For hovels, see *RVC,* IV, 259. Captain Nathaniel Butler similarly observed, "Their howses are generally the worste that ever I sawe the meanest Cottages in England beinge every way equall (if not superiour) wth the moste of the best" (*RVC,* II, 374–375; see also I, 567). For more information on impermanent architecture, see Cary Carson et al., "Impermanent Architecture in the Southern American Colonies," *Winterthur Portfolio,* XVI (1981), 135–196. Archaeological excavations at Jamestown have revealed that many of the structures within the fort were lightly framed with clay-covered walls. Archaeology on the various other early plantations has confirmed widespread use of impermanent building techniques. For frequent repairs, see Smith, *Generall Historie,* II, 536; and "The Tragical Relation of the Virginia Assembly, 1624," in Lyon Gardiner Tyler, ed., *Narratives of Early Virginia, 1606–1625,* Original Narratives of Early American History (1907; rpt. New York, 1952), 424. Tenant houses in Bermuda were also generally flimsy and were considered worthless at the end of six years (Lefroy, comp., *Memorials,* I, 417). Sizes of houses are reported in *RVC,* III, 450–451, IV, 288. For other structures and workmen, see Noël Hume, *Martin's Hundred,* 186–213, 239–247, 253–256; Barka, "Archaeology of Flowerdew Hundred Plantation"; Barka, "Early Seventeenth-Century Architecture at Flowerdew Hundred Plantation"; *MCGC,* 167; *RVC,* IV, 510–517; "Claiborne vs. Clobery," *MHM,* XXVIII (1933), 30–43, 179–189.

PLATE 3. Small House in Martin's Hundred. Conceptual reconstruction. Permission, Richard Schlecht / National Geographic Image Collection. *Archaeological excavations reveal that early Virginia settlers lived in crowded, hastily thrown together wattle-and-daub cottages like this one.*

first settlers at Berkeley Hundred were soon all engaged in "daiely husbandtrie sum to clering ground for corne and toback sum to building housses sum to plant vines and mulberie trees." Undoubtedly, gardens and orchards were soon started as well, for, as the Berkeley proprietors reminded their settlers, "a great parte of the food of your family must aryse" from the kitchen garden. Clearing was no small chore and one for which many of the settlers were unprepared. Complaints of working "hard from Sunn rising to Sunn sett at felling of Trees" and "grubbing of woods" were numerous. Some colonists continued to cultivate worn-out fields previously cleared by the Indians because they had not the strength to clear new ground. Perhaps only the later Kent Island settlement, having the fur trade as an alternative source of return, had the luxury of delaying tobacco planting until the third year of settlement. There agent Claiborne reported that during the first two years the settlers built houses, palisaded their fort, cleared and fenced ground for corn and vegetables, raised a herd of swine, secured needed supplies from older settlements on the James, and kept the men "abroad in severall boates a trading, which was our principall worke." Only in the third year did they attempt a small crop of tobacco. The ship carrying the second contingent of settlers to Berkeley Hundred brought back just over five thousand pounds of tobacco early in 1621, an apparently handsome return. But it is unlikely that much had been produced on the new plantation. One Company employee, joiner Richard Godfrey, shipped two hundred pounds on his own account, and some of that sent by George Thorpe might have been produced by Berkeley Hundred settlers. At least half of the tobacco, however, had been raised by other, more established planters.[65]

Since future supplies depended primarily on the settlers' getting in a good crop of tobacco, every energy went into cultivating the weed, often to the neglect of prudent steps to guarantee survival for all until the next growing season. Without some return, investors were unwilling to advance additional stocks of food, clothing, and tools, and without tobacco the settlers could not purchase such necessities independently. Take the basics of shelter and food. Immigrants who arrived too late in the year to plant tobacco could expect little help in constructing shelters from those engaged in the crop. The only available building material was unseasoned green timber. Skilled workmen could not be hired to help except at exorbitant wages since they were loathe to quit their tobacco fields. Newcomers who became too ill to put up their own huts might well perish from cold and exposure in the ensuing winter. Settlers who had somehow managed to build houses were understandably reluctant to take new immigrants into their already cramped dwellings. Those at Martin's Hundred apparently proved so recalcitrant in 1621 that the Virginia Company ordered the colony's governor to forcibly billet the Society's new settlers on the old Wolstenholme Town tenants if they refused to take them in voluntarily. At the same time, newcomers who were not provided their own housing grew anxious from "long expecting of what shall become of them."[66]

Food was often similarly neglected. Although the more seasoned colonists had learned the rudiments of maize culture from the Indians, many failed to plant enough to feed themselves, and certainly not enough to carry over additional ill-supplied settlers who arrived late in the season. If corn was to be the

65. *RVC,* III, 109–114, 380, IV, 228, 580. For orchards, see also [Rolfe], "Virginia in 1616," *Va. Hist. Reg.,* I (July 1848), 101–113; and *RVC,* IV, 228. The orchards described by Lady Wyatt were subsequently neglected, however, and in "their diligence about Tobacco left them to be spoiled by the cattell" (John Smith, *A Continuation of the Generall Historie of Virginia, New England, and the Summer Islands, 1624–1629* [London, 1630], in Arber, ed., *Travels and Works of Smith,* II, 887 [hereafter cited as Smith, *Continuation of the Generall Historie*]). For clearing, see *RVC,* III, 557, IV, 236. For Kent Island, see "Claiborne vs. Clobery," *MHM,* XXVIII (1933), 181–187. For tobacco shipments, see *RVC,* III, 435–436; Hecht, "The Virginia Colony," chap. 6.

66. *RVC,* I, 589, III, 226, 300–306, 441–443, 455–458, 502–508, IV, 158–161. Robert Bennet wrote from Bennet's Welcome in 1623 that, although "vytellse" of all sorts were scarce and dear, he was concentrating on the tobacco and corn crops (IV, 220–222).

major element of the diet, it required more acreage to grow than the settlers could readily clear. Captain Thomas Nuce reported of Company settlers in 1621: "The labor is infynite that they are here putto for Corne yeerly. In so much that it takes up att the least three parts of our hands." George Sandys similarly reported, "*It is not* a small proporcion of Corne that will feed a man, when that is his onelie sustenance." The cash crop, however, took even more time. John Pory estimated that tobacco, "in sowing plantinge, weedinge, worminge, gatheringe, Cureinge, and making up, it Consumes ten monethes at least, if not eleven." With labor in short supply, there were not enough hands to do both.[67]

Tending two acres in corn, the minimum needed for subsistence, by the late seventeenth century required a man to work only about thirty-five days a year, whereas cultivating three acres in tobacco (then the usual amount tended by one laborer) took up approximately 126 days. Labor constraints were particularly marked in April and May, when a sufficient number of hills had to be prepared for both corn and tobacco before the planting season became too advanced, and again from mid-June to mid-July, when both crops required repeated weeding. Less experienced planters might not have been able to make hills at the customary late-seventeenth-century rate of 320 hills per man per day and also might have needed more time for weeding if they had not burned the fields beforehand to destroy weed seed. Settlers in Bermuda, who managed to avoid repeated subsistence crises, did not have to work under such severe seasonal constraints. On that more temperate island they could prepare ground for tobacco in August, sow the seed in mid-October for a late April to mid-May harvest, and then plant summer crops of corn and potatoes while the tobacco was drying. Virginia officials rightly warned the parent company that new settlers who arrived in May or June could neither grow food for themselves nor be supplied by more settled residents. If ground had not been cleared and a sufficient number of hills prepared by the end of May, no additional corn could be planted that season.[68]

Incentives for working hard also might have been lacking on particular plantations, just as on Company lands. The initial contingents of settlers almost certainly planted the first year's crops as a group and then put the harvest into a common storehouse. Although individual private plots were undoubtedly subsequently allotted, the settlers still might have been expected

67. Ibid., I, 423, III, 257, 455–458, IV, 65, 239.

68. Lois Green Carr and Russell R. Menard, "Land, Labor, and Economies of Scale in Early Maryland: Some Limits to Growth in the Chesapeake System of Husbandry," *JEH,* XLIX (1989), 413–414; Jarvis, "'In the Eye of All Trade,'" 95–97; *RVC,* III, 78, 297–306.

to contribute to a continuing common stock. Certainly, the English investors favored communal storehouses in order to reduce the cost of provisioning subsequent groups of new settlers. To the extent that common food stores were continued, the difficulties of communal food production so well documented for Company settlements also might have hampered the private enterprises. Ralph Hamor explained in his account of the colony published in 1615, "When our people were fedde out of the common store and laboured jointly in the manuring of the ground, and planting corne, glad was that man that could slippe from his labour, nay the most honest of them in a generall businesse, would not take so much faithfull and true paines, in a weeke, as now [after allocation of individual plots] he will doe in a day." On Berkeley Hundred, the initially optimistic Ferdinando Yate admitted tellingly that the laborers had to be "seen unto"; otherwise "they will not follow their busines." Hamor noted that provision from a common stock afforded a further disincentive to hard work; "neither cared they for the increase, presuming that howsoever their harvest prospered, the generall store must maintain them, by which meanes we reaped not so much corne from the labours of 30 men, as three men have done for themselves."[69]

Since they could not do both, colonists faced a choice between either growing enough corn to feed everyone in the colony and going without all other necessities or concentrating on planting tobacco with which they might purchase both food and other needed commodities. They almost invariably opted to plant tobacco. Official orders to the contrary, settlers long neglected to plant adequate amounts of corn, trusting that they could get what they needed from the Indians, either through trade or by force if that proved more expeditious. As ex-governor Yeardley allegedly told some of his tenants, he "should provide them Corne if they would provide Tobacco." The situation likely persisted in part, too, because recurring maize shortages benefited some leading Virginians. Starving settlers were forced to sell everything they had and then to pledge future tobacco crops in order to buy corn from the privileged few who had the boats and hands needed to secure grain from the Indians. As colonist Peter Arundel protested, there was no corn "to bee had but with great men which endanger mee and myne to sterve beefore harvest." Between 1622 and 1624, there were "notable examples of food profiteers who

69. *RVC,* III, 109; Hamor, *True Discourse,* 17. Francis Bacon explained English thinking about common stocks: "Let the main part of the ground employed to gardens or corn, be to a common stock; and to be laid in, and stored up, and then delivered out in proportion; besides some spots of ground that any particular person will manure for his own private use" ("Of Plantations," in Brown, *Genesis,* II, 800). The prominence of company storehouses on the particular plantations also hints at continued farming in common.

contributed to the misery and high mortality of their countrymen," including, not surprisingly, the colony's councillors and military commanders.[70]

Corn shortages were all the more critical because meat, too, was always in short supply. Initially, the settlers lacked the skill to hunt game, and meat was not among the items local Indians were willing to trade. As with planting corn, so, too, was hunting time consuming. It seemed much simpler to hire Indians to do it instead. By the time some of the colonists had learned to hunt for themselves, the massacre occurred, making hunting in any but large parties too dangerous to attempt. As one colonist lamented in 1623, despite "all the Deere, Fish and Fowle . . . so talked of in England," he had none since his arrival, not "so much as the s[c]ent." In Jamestown during the "Starving Time" of 1609–1610, those who survived did so only by eating horses, cats, dogs, snakes, musk turtles, rats, mice, and, eventually, human corpses. Some of those who perished were perhaps unwilling to purchase survival at such a price.[71]

The lack of game could not be made up with domestic livestock because that, too, was scarce. The Virginia Company sent a few cattle for its tenants, but, according to a recently discovered census of the colony, in March 1620 only forty-seven bulls and oxen, apparently intended to be used as draft animals, were counted on Company land. Many more cattle were sent to particular plantations—six, for example, to the Berkeley settlement in its second year. Perhaps as well residents of particular plantations managed to convert some Company cattle to their own use, for, by March 1620, 238 "horned cattle," including all the female breeding stock, were in the hands of private investors. Since it cost but six pounds to transport a man or woman to Virginia and ten pounds to ship a cow, however, human emigrants greatly outnumbered bovines, whose survival was as problematic as that of the settlers.

70. *RVC,* IV, 89–90, 186 (see also II, 374–376, IV, 217, 229–230, 275–276, 450–455); Smith, *Generall Historie,* II, 586, 595, 600; Morgan, *American Slavery, American Freedom,* chaps. 5, 6. Whether Berkeley Hundred settlers bought corn from the Indians is not known, but they did borrow some from more established settlers (*MCGC,* 11, 17, 36). Still, in 1635 Virginians needed to get corn from the Indians in order to feed newcomers (Captain Samuel Mathews, "The Mutiny in Virginia, 1635," *VMHB,* I [1893–1894], 416–424). For food profiteers, see Fausz, "The Powhatan Uprising of 1622," 475.

71. *RVC,* IV, 58–62, 167–168, 228, 231–232 (quotation), 232–233, 235; Willie Graham et al., "Adaptation and Innovation: Archaeological and Architectural Perspectives on the Seventeenth-Century Chesapeake," *WMQ,* 3d Ser., LXIV (2007), 470–475. Captain John Smith was amazed in 1621 to learn that the Indians were being employed to hunt and fowl and were supplied with guns for that purpose, while the English were "rooting in the ground about Tobacco like Swine" (*Generall Historie,* II, 563). Bowen and Andrews, "Starving Time at Jamestown," reports that faunal remains at the fort are dominated by small-bodied species, not deer. Post–Company-era sites dating to the late 1620s and 1630s, when constraints on colonists' movements decreased, contain higher proportions of deer remains.

It was intended that herds should expand through natural increase. To this end, severe penalties were established for killing breeding stock for food. Their numbers, though, grew slowly. Animals accustomed to a milder climate and to supplementary food in winter had, like the humans, to adapt to a new, hazard-filled environment. Settlers who could not manage to grow enough food for themselves clearly could not produce fodder for overwintering livestock. Consequently, the cattle had to fend for themselves, sometimes falling prey to wolves or hungry native or English hunters, or drowned in swamps into which they ventured for early spring grasses. Some of the settlements did attempt to fence in the cattle and to provide shelter for them at night. But, with no proper pastures, no adequate source of winter food, and sometimes no nearby source of drinking water, the cattle had to range dangerously far during the day. Not surprisingly, a number of the valuable animals perished through "want of good lookeinge to." Other animals were killed in the 1622 attack or perished in the chaos that followed.[72]

Swine, on the other hand, adopted to the New World environment much more readily than did cattle, sometimes only too well from the settlers' perspective. In March 1620, 308 tame swine were counted in the colony, but "the Swine of the Forest belonginge to the publique" were "not to be numberd." Feral hogs soon bred prolifically in the swamps and woods, but hungry colonists who could not catch and kill them discovered they were of no more benefit than the wild deer, being beasts "belonginge to noe man" that "serve to feed the Savages or the Maryners that live idlie there, rather then the people of the Colony, who have starved in the middest of this Imaginary plenty."[73]

The adventurers' demands for quick returns on their investments, coupled with their failure to provide adequate supplies, left most Virginia settlers pre-

72. *RVC,* I, 501–502, 525, 535, 618, 626, III, 118–122, IV, 137–138; McCartney, "An Early Virginia Census," *Arch. Soc. Va. Qtly. Bull.,* LIV (1999), 178–181 (table 2, 181, gives in error the number of privately owned oxen in Charles Hundred as eighty; the figure on the manuscript is instead eight). For fencing at Martin's Hundred, see Noël Hume, *Martin's Hundred,* 236–240. That cattle normally ranged widely is indicated by a 1626 Virginia Council directive that, when settlers erected pales across necks for protection from the Indians, they were not to fence off the cattle from a great part of their food (*MCGC,* 120).

73. McCartney, "An Early Virginia Census," *Arch. Soc. Va. Qtly Bull.,* LIV (1999), 178; *RVC,* I, 501–502, II, 40, III, 118, 589; Smith, *Generall Historie,* II, 471. Since swine become sexually mature earlier than cattle, have a shorter gestation period and larger litter sizes, and can bear two litters per year, a sow and her offspring might produce more than one hundred female piglets in the time required for a cow to produce just two female calves (Alan L. Olmstead and Paul W. Rhode, *Creating Abundance: Biological Innovation and American Agricultural Development* [New York, 2008], 266–267).

cariously dependent for survival on a risky combination of the good will of the Indians, the arrival of uncertain shipments from England, and the sometimes questionable mercies of a few highly placed leaders. Doubtless, few in England ever appreciated the real conflict that the colonists encountered between producing the cash crop and raising enough food to stay alive.

Successes and Failures on Kent Island

The Kent Island settlement was not established until 1631, after most particular plantations in Virginia had been abandoned. This settlement was atypical not only in its late appearance but also in that it concentrated on the fur trade rather than on growing tobacco. Moreover, unlike the Virginia settlers, the Kent Islanders managed to maintain good relations with the native American groups with whom they traded. However, the plantation's form of organization did parallel that of earlier enterprises along the James River. Financial records and depositions from this venture supply the only detailed information so far discovered about the operation of a particular plantation over a number of years. These materials aid understanding of how plantation managers might reap handsome profits while the joint-stock organization in England foundered in debt and of how much Virginia Company officials underestimated the time and effort required to carry out traditional women's work essential to successful, lasting New World settlements. Finally, the Kent Island experience demonstrates the great difference made by better access to imported goods and to free laborers available for hire.

Four English investors, three of them well-connected London merchants with extensive interests in the colonial fur and tobacco trades and in trade with Africa, financed the transportation of the first contingent of settlers and one additional group along with trade goods and provisions at a cost of about £3,660. Two of the investors became quickly discouraged by the lack of quick returns and sold their shares to others. The accounts of the resident manager and one-sixth owner of the project, William Claiborne, show that he spent an additional £5,200 running the plantation between 1631 and 1637, either out of his own pocket or from goods sold in the country. Claiborne, who brought experience in local trade and agriculture and influential local connections, was a younger son of a Norfolk trading family and came to Virginia in 1621 as the colony's first surveyor. Within three years, he became a member of the Virginia Council. High office enabled Claiborne to garner special privileges for his particular plantation.[74]

74. "Claiborne vs. Clobery," *MHM,* XXVI (1931), 381–404, XXVII (1932), 17–28, 99–114, 191–214, 337–352, XXVIII (1933), 26–43, 172–195. The other partners refused to accept Clai-

As had been the case in the 1620s, labor proved the dearest commodity on Kent Island. Between twenty-two and forty-four colonists worked for the joint-stock association each year. The investors transported most of them, but labor shortages led Claiborne to rent between seven and eleven additional servants every year from other established planters. Since the initial contingent included few skilled workmen, Claiborne had as well to hire a further six to fourteen freemen for all or part of each year. These included an interpreter and assorted coopers, carpenters, boatwrights, tailors, surgeons, smiths, mariners, boat masters, and one "negro" whose duties Claiborne did not specify. Wages due to Company employees and to skilled freemen and the costs of hiring servants together constituted the largest local expense, a quarter of Claiborne's annual disbursements.[75]

Administration, food, clothing, and trade goods were the enterprise's other major costs (Table 2). Four officers ran the plantation, Captain Claiborne being assisted by a lieutenant, an overseer, and a storekeeper. The one-hundred-pound annual salary Claiborne paid to himself over and above wages paid to the other officers made administration the second-greatest local expense—18 percent of disbursements. Food and apparel each accounted for 12 percent of expenditures. The Company employees grew garden crops and corn, but not in sufficient quantity. Over seven years, nearly one thousand bushels of corn were acquired from area Indians for the plantation's use. Otherwise, Claiborne bought about thirty-five-pounds' worth of English victuals a year—sugar, spice, oil, cheese, liquor, and the like, partly to treat those who fell sick. The main difference from earlier settlements was that meat was no longer scarce. Two dozen hogs acquired in the first year grew into a herd large enough to feed the settlers and occasionally to allow sales of surplus meat. During the first three years, a huntsman was also employed to bring in game. In addition, Claiborne supplied the employees with milk from his private stock of cattle. Consequently, the diet was better and costs were undoubtedly lower than earlier.

A good portion of the initial stock of trading goods was destroyed when

borne's accounts, asserting they were "not merchantlike." They were clearly drawn up in a manner designed to favor his cause, maximizing expenditures he had made and minimizing total returns. The English partners knew what they had spent but seem to have had only the vaguest notions about costs and wage and price levels in the colonies. The Kent Island enterprise is analyzed in J. Frederick Fausz, "Merging and Emerging Worlds: Anglo-Indian Interest Groups and the Development of the Seventeenth-Century Chesapeake," in Lois Green Carr, Philip D. Morgan, and Jean B. Russo, eds., *Colonial Chesapeake Society* (Chapel Hill, N.C., 1988), 47–98; and Brenner, *Merchants and Revolution*, 120–124.

75. "Claiborne vs. Clobery," *MHM*, XXVIII (1933), 30–43, 172–189.

TABLE 2. Expenditures on Kent Island, 1631–1637

Category	1631 (%)	1632 (%)	1633 (%)	1634 (%)	1635 (%)	1636 (%)	1637 (%)	Mean (%)
Wages	12.0	23.9	25.2	20.2	21.8	32.4	42.8	25.5
Administration	25.3	19.7	20.2	17.0	10.5	15.6	14.4	17.5
Clothing	13.7	14.4	11.9	9.5	12.4	20.3	4.3	12.4
Food	17.1	13.0	13.0	9.2	11.5	13.4	8.8	12.3
Trade goods	0.2	4.4	6.1	20.0	22.0	0.0	0.0	7.5
Boatbuilding and repair	4.1	4.9	6.0	5.9	9.9	6.1	0.2	5.3
Religion	0.0	10.0	8.7	6.5	0.5	4.8	0.0	4.4
Supplies from England	3.2	0.5	0.0	0.8	0.0	0.0	14.0	2.6
Itinerant charges	3.4	0.0	1.1	1.8	1.3	1.5	8.6	2.5
Weapons	3.2	2.6	4.6	3.5	1.1	0.0	0.0	2.1
Construction of mills	0.0	0.0	0.0	0.0	7.4	3.1	0.0	1.5
Furniture and utensils	4.0	1.8	1.0	1.1	0.5	0.3	0.3	1.3
Livestock	7.3	1.4	0.0	0.0	0.0	0.0	0.0	1.2
Legal expenses	0.0	0.0	0.0	1.2	0.0	0.0	6.5	1.1
Construction of buildings	4.1	1.0	0.6	1.1	0.1	0.0	0.0	1.0
Tools and casks	0.7	1.2	0.2	0.4	0.6	0.7	0.0	0.5
Miscellaneous	1.7	1.2	1.4	1.8	0.4	1.8	0.1	1.2
Total	100.0	100.0	100.0	100.0	100.0	100.0	100.0	99.9
Expenditures in sterling	£411	£661	£732	£949	£1,108	£641	£696	£743

Note: Figures for 1637 do not cover the whole year. Mean total does not add up to one hundred owing to rounding.

Source: "Claiborne vs. Clobery et als. in the High Court of Admiralty," *MHM,* XXVIII (1933), 30–43, 172–189.

a storehouse burned, and, since the English backers sent additional supplies only once in the settlement's first five years, most clothing had to be purchased in Virginia at high rates—five pounds or more per man per year. Local purchases of goods for the Indian trade—cloth, hoes, axes, knives, combs, and roanoke (wampum)—accounted for 8 percent of expenditures, and building and maintaining boats took up another 5 percent. The minister's salary averaged 4 percent of disbursements, although in one year it amounted to as much as 10 percent. Weapons and ammunition averaged 2 percent annually. Disbursements for other domestic items such as building materials, tools, household goods, and livestock were minimal.

In recompense for a total outlay of about £8,800, Claiborne claimed to have gained for the investors £5,200 from trade. The greatest returns came from 5,000 pounds of beaver skins sold in England for £4,500. Company servants, Claiborne stated, also produced 6,348 pounds of tobacco between 1633 and 1636, an item he seems not to have accounted for. Other receipts came from 1,853 bushels of corn secured from the Indians and sold to other settlers for £582. Claiborne also marketed £100 of trade goods and £26 of pork to English colonists. Altogether local trade yielded about £700 income, or about 10 percent of total receipts. The investment in the millwrights, of which Claiborne disapproved, seems to have come to naught. Claiborne had advised his partners not "to undertake projects of Mills[,] rape oile[,] iron workes or the like, because he had formerlie experience in Virginia that Marchants are usually in the like kind defective in sending all necessaries." The English partners, Claiborne alleged, sent defective millstones and failed to pay the millwrights wages or allowances promised to their wives in England, leaving the artisans discontented and unwilling to work. On the face of it, the joint-stock venturers appear to have failed to recoup their initial outlay, much less to have made any profit.[76]

The accounts suggest, however, that some of the adventurers made further gains from the enterprise outside the framework of the joint investment. One profitable sideline arose from Claiborne's purchasing a dozen cattle from the James River settlements and transporting them to the island, where they multiplied prolifically, safe from predators and fed by the island's rich natural grasslands, reaching a herd of around 150 by 1637. (Testimony differed as to the extent to which Company funds had been used to pay for the animals.) With this stock, Claiborne ran a lucrative dairy service, renting cows to freemen living in his settlement at one hundred pounds of tobacco per year for their milk. He also employed servants of his own not financed by the joint stock who probably grew tobacco. From these and other ventures not so clearly described, Claiborne accumulated thirteen thousand pounds of tobacco that he sold in Ireland on his own account. Part of the misunderstandings between Claiborne and his fellow investors arose because Claiborne paid for only a one-sixth share of the first supply, promising to allow the other investors satisfaction for further sums due to the group from the profits of commodities sold in England. He maintained, however, that he had promised to report to his partners only accounts "soe farr as concerned the trade with the Indians, the transportation of corne and of all the profits and benefittes on the wages made by the joint stock and not otherwise" and steadfastly refused to give any accounting of his own particular estate or dealings. At least two more of the English investors also traded with area settlers on the side, sending car-

76. Ibid., XXVII (1932), 99–114, XXVIII (1933), 30–43, 345–346.

goes on their personal account, probably to secure a profit that could not be taken to cover the joint venture's outstanding debts.[77]

Intercolonial and international politics contributed to the eventual failure of Kent Island as a commercial venture. William Claiborne was among those elite Virginians who vigorously opposed Charles I's grant of the northern Chesapeake to the Calvert family. Claiborne's Kent Island enterprise threatened Maryland colonists' access to the lucrative northern Chesapeake fur trade, and the political maneuvering of Claiborne and his backers in Virginia and in England threatened both the proprietary charter and Cecil Calvert, second Lord Baltimore's control over the Maryland colony. Kent Islanders attacked vessels of Maryland settlers attempting trade with local Indians in 1635, and three years later the Marylanders invaded Kent Island and Claiborne's private trading base on Palmer's Island at the mouth of the Susquehanna River, seizing or destroying livestock, tools, and trade goods and appropriating four of Claiborne's servants. Claiborne's successes in the Indian trade encouraged competitors to undermine the Kent Island enterprise, and his reckless actions alienated key supporters in England.[78]

Annual lists of the number of Company servants and their principal tasks for the year show what proportion of the workforce was needed to carry out various tasks on the plantation (Table 3). The four administrators constituted between 8 and 13 percent of the total workforce. Initially, only one full-time carpenter was attached to the Company, so additional free workers had to be hired periodically to assist him. Once the investors sent a contingent of carpenters and millwrights who were supposed to set up a series of commercial mills, the number of builders rose to seven and a half in 1634, 15 percent of the workforce. Other craftsmen such as tailors and smiths varied in number from three to seven between 1634 and 1636. Since experience was essential to successfully trading with the Indians, only between one and seven of the men sent by the Company were assigned to this major plantation activity. Hired

77. For cattle, see ibid., XXVII (1932), 26, 108, 342, XXVIII (1933), 193, 257–258; and Lois Green Carr, "Livestock Increase and Its Implications for Diet in the Seventeenth century Chesapeake" (typescript report prepared for Historic Saint Mary's City, 1982). Claiborne's and other of the investors' private trading are documented in "Claiborne vs. Clobery," *MHM,* XXVI (1931), 394–399, XXVII (1932), 100 (quotation), 109, 338–340, XXVIII (1933), 28. Cf. *RVC,* II, 82, 322–323, 525. The Virginia Company stated that most of the tobacco imported into England from Virginia belonged either to particular plantations or to private men "acquired by purchase or personal adventure." Much of the latter was probably gained by such side trading.

78. William Hand Browne et al., eds., *Archives of Maryland,* 72 vols. (Baltimore, 1883–1972), I, 23–24, III, 24–39, 64–67, 71–73, 76–77; Fausz, "Merging and Emerging Worlds," in Carr, Morgan, and Russo, eds., *Colonial Chesapeake Society,* 47–98.

TABLE 3. Occupational Distribution of Labor Forces, Berkeley Hundred, 1619, and Kent Island, 1631–1636

	Berkeley Hundred	Kent Island					
Category	1619	1631	1632	1633	1634	1635	1636
Administration							
No. of workers	4	2–3	3–4	2.5–3.5	3–4	2	3
% of total	10	11–12	12–13	10–11	7–8	5	8–11
Housekeeping							
No. of workers	1	5	6	5	5	7	4
% of total	3	20–28	20–24	15–23	10–12	17–19	11–15
Craft							
No. of workers	10.5	1–5	1–2	1–4	11.5–13.5	9	4–7
% of total	27	6–20	4–7	5–12	27	21–24	15–19
Agriculture							
No. of workers	3.5	1	2	3	3	3	1
% of total	9	4–6	7–8	9–14	6–7	7–8	3–4
Trading							
No. of workers	0	1–2	4–7	5–10	6–10	4–9	7–13
% of total	0	6–8	16–23	23–29	14–20	11–21	26–36
Unspecified labor							
No. of workers	20	7–8	8	5–8	13	12	8
% of total	51	32–39	27–32	23–24	26–31	29–32	22–30
Minister							
No. of workers	0	1	1	1	1	0	0
% of total	0	4–6	3–4	3–5	2	0	0
Total no. of workers	39	18–25	25–30	22.5–34.5	42.5–49.5	37–42	27–36

Note: The higher numbers reported in each category include hired laborers and other freemen identified in Claiborne's accounts as working at least part-time for the plantation but who are not included in the official lists of personnel. Claiborne's accounts usually specified the services that the hired or paid workers provided, allowing most to be assigned to an occupational category. The first figure in each of the percentage rows is the percentage of workers in the occupational category of the total officially listed workforce. The second figure in each of the percentage rows is the percentage in the occupational category of both officially listed and hired or paid workers.

1619: includes the captain who was present but not included in the official lists of personnel.

1631: excludes 1 man hired out and 6 of unspecified occupation who died shortly after arrival.

1634: excludes 2 men who died shortly after arrival.

1635: excludes 1 man hired out and includes the captain not listed in 1635 but who was present.

1636: excludes 3 men hired out and includes the captain not listed in 1636 but who was present.

Sources: For Berkeley Hundred, see *RVC*, III, 197–199. For Kent Island, see "Claiborne vs. Clobery et al. in the High Court of Admiralty," *MHM*, XXVIII (1933), 180–187.

workers raised the total number of traders to between seven and thirteen from 1632 on, or about a quarter of all employees.

Laborers or men with no listed trade varied from seven to thirteen (or about another quarter of the workforce) and were presumably engaged in agriculture along with one or two others specifically designated as gardeners. One or two rangers sufficed to look after the cattle and hogs. A major unanticipated finding is the number of people required for "dressing victuals," washing, and running the dairy. The joint stock supplied one maidservant for these tasks between 1631 and 1634 and three women servants from 1635. In addition, between one and five men were also "imployed in the Ketchin to beate corne and dresse Victualls . . . and other worke in the howse," making a total domestic workforce of four to seven, or 11 to 28 percent of the workers. Although some of the kitchen hands were boys or ailing or injured men, a surprisingly large portion of the total available workers were employed in "huswifery."[79]

A comparison of the Kent Island workforce with the passenger list for the first (and similarly sized) contingent of Berkeley Hundred settlers in 1619 is instructive. The first document describes actual practice, the second anticipated allocations. The proportion of settlers assigned to the tasks of administration, building, other crafts, and gardening was similar in both cases. Among the Berkeley settlers, just over half had no specified skills and were presumably to work mainly at husbandry. The proprietors, though, seriously underestimated the number of men needed to "dresse Victualls" and perform other domestic chores. The sergeant of the company was also to be clerk of the kitchen and, in turn, was to be assisted by "two or more (as Caters or cuntry Purvires)." Two men were specifically to work as both cooks and gardeners. Altogether it seems that no more than four full-time domestics were intended and perhaps fewer. This workforce would probably suffice for either a ship's crew or a company in a European army backed by victuallers or quartermasters who could buy or commandeer provisions from a resident settled population.[80]

If, as on Kent Island, however, a quarter of the hands were at first needed just to beat corn and prepare meals, then up to nine of the intended Berkeley laborers would have had to be diverted to the kitchen. If the men recruited

79. "Claiborne vs. Clobery," *MHM,* XXVIII (1933), 180–187. When Claiborne first arrived at the settlement, he found the men sent by the company so sick and weak they were "scarce able to dress their owne victualls," much less do anything else (194).

80. *RVC,* III, 209–210. Cf. Francis Bacon, "Of Plantations," in Brown, *Genesis,* II, 799: "The people wherewith you plant ought to be gardeners, ploughmen, laborers, smiths, carpenters, joiners, fishermen, fowlers, with *some few* apothecaries, surgeons, cooks and bakers" (emphasis added).

as craftsmen did not also help with the planting, there would in fact have been only eleven or twelve men available both to grow corn to feed the entire group of thirty-nine and to produce the tobacco crop. Although John Rolfe said that one man could tend enough corn to feed five and tobacco to clothe two, his estimates were clearly based on the most favorable of circumstances, hardly allowing for either the toll of seasoning or the labor of first clearing the land. Considered in this light, Captain Nuce's statement that it took three-quarters of his hands to get corn seems entirely reasonable—half the men to grow it and another quarter to beat and cook it.[81]

Given prevailing attitudes in England toward women's work, that investors might underestimate the time that would be required for essential domestic chores is not surprising. The tasks were, after all, regarded as essentially "unprofitable." William Tracy chafed at the ship room taken by four maidservants and three married women in the second group of settlers; he could take fewer men "being the mor excelent and yousefull Cretuers." Or, when the stay-at-home Berkshire farmer Robert Loder considered possible ways to cut farm costs around 1615, dispensing with the maid usually came first to mind. She did not produce enough malt to cover her board and wages; her time was instead taken up in "the doing of the thinges, that must indeed be donne." If other groups planning particular plantations failed to allow for sufficient hands, male or female, to perform frontier housewifery, it becomes quickly apparent why healthy settlers, much less ailing ones, might have had difficulty building shelters, clearing land, raising tobacco, and feeding themselves.[82]

The End of Corporate Plantations

Particular plantations marked a transitional phase in Chesapeake agricultural history between the first communal Company-sponsored settlements and the spread of individual private plantations worked by resident planters and their servants. At the beginning of 1617, Virginia's future was yet by no means assured. The colony's population remained small, its reputation unattractive, and its returns to investors negligible. Subsidiary joint-stock companies operating between 1617 and 1623 revived hopes for the colony and raised in six years more than twice the amount of capital that the parent company had in the previous nine. These resources were used more efficiently than were those of the Company; private groups managed to establish more than

81. [Rolfe], "Virginia in 1616," *Va. Hist. Reg.,* I (July 1848), 101–113.

82. *RVC,* III, 395; Fussell, ed., *Robert Loder's Farm Accounts,* 56, 71, 155. On the ways the English housewife contributed to the operation of a farm, see Campbell, *The English Yeoman,* chap. 6.

thirty new settlements in the colony between 1617 and 1623, whereas the Company set up only four. This rapid territorial expansion assured Virginia's continued existence despite the fearful setback of 1622–1623. In addition, expanding tobacco exports from private plantations at last supplied the colony with a viable economic base.[83]

Despite these successes, the joint-stock associations that sponsored particular plantations did not long survive the 1622 massacre and the subsequent dissolution of the parent organization. Returns had been too disappointing or in some cases the amount of destruction too devastating. The commander of Martin's Hundred, for example, estimated "that 3000 pounds will not make good our Plantacion againe." The "extreame penury of monney" and "generall sickness of mynde . . . weake hands and feeble knees" that nearly paralyzed the parent company by December 1622 similarly prevailed among the subsidiary joint-stock investors. The legal situation after 1624 also might have been too uncertain to interest absentee owners in further attempts to develop holdings to which their title was no longer clear.[84]

In the aftermath of the 1622 attack, "every man now of meaner sort, who before lived well . . . upon their owne land [were forced] . . . to forsake their houses . . . and to joyne themselves to some great mans plantation." The palisaded dwellings of more affluent settlers afforded some measure of protection against continued Indian raids. Abandoning indefensible and thinly peopled outlying plantations, survivors relocated in more clustered communities around Jamestown and Elizabeth City, which had emerged from the attack virtually unscathed, or farther up the James near Charles City at the confluence of the Appomattox and James Rivers. Newly patented lands remained concentrated in these areas until the mid- to late 1630s.[85]

Settlers attached to particular plantations who survived the Powhatan uprising and subsequent famine found themselves largely abandoned by adven-

83. Hecht, "The Virginia Colony," chap. 7. How many settlers the private investors sent as opposed to the parent company is uncertain. Although Company records of 1621 stated that nine hundred of one thousand settlers were sent by private men's charges (*RVC,* I, 538), Hecht's analysis of the provenance of all settlers and ships shows that the Company and private investors each supplied about half of those that could be classified. However, nearly one-third could not be classified one way or the other.

84. *RVC,* IV, 41–42, 228–229; *MCGC,* 43, 55, 167; John Ferrar to Sir Francis Wyatt and Mr. George Sandys, Dec. 19, 1622, Ferrar Papers, no. 437, CWL. The Berkeley proprietors were concerned, for example, about a proposal to challenge all cattle belonging to nonresidents for the king (Smyth of Nibley Papers). See also Craven, *Southern Colonies,* chap. 5.

85. Fausz, "The Powhatan Uprising of 1622," chaps. 6, 7; Robert C. Johnson, ed., "The Indian Massacre of 1622: Some Correspondence of the Reverend Joseph Mead," *VMHB,* LXXI (1963), 410 (quotation).

turers in England and exploited by local leaders who were blatantly converting the adventurers' remaining assets to private ends. Three men at Martin's Hundred, for example, Robert Addams, a planter who came from Jamestown Island to assist the settlers of the Society of Martin's Hundred, Thomas Ward, a potter, and John Jackson, a bricklayer, demanded that the adventurers do something to assist them in their efforts to rebuild. Plantation commander William Harwood allotted Addams, his wife, and a servant the remains of a house and some ground previously tended by a family who had been killed by the Indians. The Addamses repaired the house and "fell to setting and planting," with Addams's wife joining him and the male servant in the fields. Harwood, however, gave them little assistance, refusing to stir out of a fortified storehouse to help them when Indians returned to kill a woman servant, wound Addams and pursue his wife (who survived the attack), and cut down their corn. Addams stayed that year "despite discouragement," left with two other Society servants in 1623, but was persuaded by Ellis Emmerson, who had been sent by the adventurers in late 1622, to return to Martin's Hundred the following season. Emmerson was earnestly recruiting freemen like Addams since, of some thirty new tenants who had arrived with him around Christmas 1622, twenty were dead by the following March.[86]

Addams then built yet another house that he attempted to better protect with a palisade. Harwood continued to provide "little comfort," refused to give the Company tenants powder and shot from fresh supplies the adventurers had sent, was lending or selling goods from the store to other planters rather than to the Society's settlers, and was charging exorbitant prices. Ward and Jackson reported in May 1625 that, although only one of them could work at a time as the other had to stand guard, Harwood continued to demand half their crops as rent and to charge unreasonable prices for the little corn and meat he was willing to sell. Nor did Harwood supply any of the tools, clothing, or loan of cattle the adventurers had promised, instead forcing them to forgo new clothing and to go into debt to buy essential tools. "We will not indure this kynd of liveing any long time," Ward and Jackson warned. Addams, at least, did remain at Martin's Hundred, but he died four years later in 1629. By that time, Harwood had secured a minister to serve the settlement, but his plans for enlarging the parish by annexing some adjoining private plantations had fallen through. Consequently, the Council ordered Harwood himself to pay one-third of the annual salary he had promised the

86. Robert Addams to Nicholas and John Ferrar, June 16, 1625, Ferrar Papers, no. 572, CWL; Fausz, "The Powhatan Uprising of 1622," 468; *RVC,* IV, 58–60. See also McCartney, "History and Historical Context of CG-10," 13–26.

minister—two thousand pounds of tobacco and enough corn for food—since "the plantacin is very small in soe much that it is likely to ly very heavie on the parishioners."[87]

Once the terms of their servants and tenants expired, the adventurers in England were left with only the land and a few cattle. (Tenants that had belonged to the defunct parent company were divided among the governor and his Council in 1627.) Although the Society of Martin's Hundred, for example, again attempted to resupply that settlement in 1632, it failed to grow appreciably. The Society's claim to vast, uncultivated acres became a public nuisance, blocking fortification and development of a critical location. Many of the Berkeley Hundred survivors removed to better-defended plantations, such as Jordan's Journey, taking with them the remaining cattle, and ended up working for whomever could provide protection and advance essential supplies. The Berkeley proprietors apparently failed to repeople their foundering enterprise, rejecting a proposal for reviving the project on a much reduced scale under the direction of one or two of their former servants, and instead by 1636 sold their interests to a group of London merchants. By 1635, the Society of Southampton Hundred had "nothing left but a stock of cattle" to show for an investment of more than six thousand pounds. Resident merchant-planters and groups of London merchants acquired the lands allotted to the corporate plantations through purchase or seizure. All Virginia particular plantations had apparently been absorbed into that colony's regular land system by 1641, and the later Kent Island settlement was taken over by Maryland's proprietor. Only some of the smaller particular plantations granted to investors who lived in Virginia stayed in the hands of their owners or their descendants who eventually either enlarged or sold them.[88]

87. Thomas Ward and John Jackson to Nicholas Ferrar, May 20, 1625, Ferrar Papers, no. 569, Addams to Nicholas and John Ferrar, June 16, 1625, no. 572, CWL; *MCGC,* 196. Addams's unnamed widow was granted administration of his estate in 1629. It is not known whether she was the woman who survived the 1622 Indian attack or a subsequent wife (*MCGC,* 181).

88. *RVC,* II, 486–487, III, 683–690, IV, 16–17, 109–110, 245–246; "Letter of Sir Francis Wyatt," *WMQ,* 2d Ser., VI (1926), 114–121; Governor Wyatt and Council of Virginia to the Privy Council, May 17, 1626, in "Sir Thomas Wyatt, Governor: Documents, 1624–1626," ibid., VIII (1928), 163–167; *MCGC,* 136–137, 196; McCartney, "History and Historical Context of CG-10," 19–32; Craven, *Southern Colonies,* chap. 5; Smyth of Nibley Papers; *VMHB,* V (1899), 185, VIII (1900–1901), 401–402; Brenner, *Merchants and Revolution,* 145–146; Anthony S. Parent, Jr., *Foul Means: The Formation of a Slave Society in Virginia, 1660–1740* (Chapel Hill, N.C., 2003), 26–27. In 1625, for example, fifteen men belonging to the Berkeley proprietors were seated at Shirley Hundred under the direction of Sergeant Gabriel Holland (*MCGC,* 42–43, 55). In contrast to other historians, Gethyn-Jones believes that John Smyth might have sent over some additional settlers in 1623 (*George Thorpe and the Berkeley Company,* 204)..

But, despite continuing problems, some individual freemen managed to establish viable plantations, and more were able to survive long enough to reap some of the benefits of farm building. John Smith noted that "being thus in a manner left to themselves, since then within these foure yeeres [1625–1629], you shall see how wonderfully they have increased beyond expectation." By 1638, an English placeman observed, "That slaughter [1622] much abated Adventurers for a while, but afterwards coveting quick retornes for profitt, they proceeded to new Expedicion without, att all, any regard of Meanes [or] any Redresse of ould defaults."[89]

Indeed, already by the spring of 1625 the Virginia colony was "becoming a secure and prosperous homeland for many colonists." The subsistence crises had ended, and, with power concentrated in a few hands, political stability increased. Members of Virginia's Council, in their capacity as military leaders, assembled private armies of colonists that they employed to pursue revenge against Indian foes and to trade for corn with friendly, more distant Indian allies. Instead of attempting to wipe out hostile groups of native Americans entirely, they conducted more modest raids at harvest time to appropriate the Indians' maize. These "feedfights" did not overly interfere with tobacco production and helped to justify continued exploitation of English servants laboring under the leaders' protection. With the Indians successfully pushed back or sufficiently harried that they did not often pose a significant threat to the English, ordinary planters could now concentrate their energies on building farms and growing tobacco. By 1632, when Cecil Calvert, second Lord Baltimore, established his Maryland colony in the northern Chesapeake, these English settlers experienced no initial starving time. Settlement quickly spread from the first fortified landing at Saint Mary's City. Early schemes for an Old World society of landlords and tenants were soon abandoned, and, after about a decade, small independent planters, most of them former servants, began growing tobacco on their own farms.[90]

Warfare with the Powhatans, according to Frederick Fausz, provided the emerging Virginia elite "with the monopoly on servants, ships, and political authority that launched their careers as merchant-planters." The councillors quickly established "an effective, if unlikely, oligarchy based on their common acquisitive ambitions" that "successfully merged public service with private

89. Smith, *Continuation of the Generall Historie,* II, 884; T. H. Breen, "George Donne's 'Virginia Reviewed': A 1638 Plan to Reform Colonial Society," *WMQ,* 3d Ser., XXX (1973), 466.

90. Fausz, "The Powhatan Uprising of 1622," chap. 7 (quotation on 548); Fausz, "Merging and Emerging Worlds," in Carr, Morgan, and Russo, eds., *Colonial Chesapeake Society,* 52–56; Russell Robert Menard, "Economy and Society in Early Colonial Maryland" (Ph.D. diss., University of Iowa, 1975).

gain." Consequently, their interests sometimes coincided with those of ordinary planters, but at other times the councillors' interests were separate from and even opposed to the interests of most colonists. The councillors' desire to escape the control of meddling English officials encouraged economic expansion, for example, but their support for various schemes to grant monopolies for selling tobacco in England went against the interests of small and middling planters. Warren Billings similarly characterizes these early leaders as "a rapacious lot at first who sought to exchange the intricate legalities of England for the opportunities to stalk personal gain in nearly untrammeled freedom." In contrast, Alexander Haskell argues that the main concern of provincial leaders was the "fragility of authority," and in the 1620s they began to realize that the continued failure of the government to fulfill settlers' basic expectations for order and justice was itself an obstacle to establishing order. Virginia councillors thus tried to "buffer the colony from the disruptive influence of poorly conceived metropolitan policies" that they believed were undermining their efforts to secure the colonists' obedience and to protect "the liberties that settlers expected of an English commonwealth." Jon Kukla points out that, whatever the character of the first councillors, by the early 1640s civilian offices did supplant military offices as the primary locus of power, resulting in a transformation from a "control regime" into "an enduring civilian polity."[91]

Councillors, numbering between ten and eighteen, were, like the governor, from 1625 on appointed by the crown, and they served in office indefinitely. They were, indeed, powerful men, exercising wide-ranging executive, legislative, and judicial roles. In their executive capacity, they advised the governor and, sitting as the Council of State, constituted the chief administrative agency of the colony. The governor was bound to respect the wishes of a majority of the councillors on all important decisions, and he did not have independent power to remove advisers who opposed him. Should the governor's chair become vacant, the councillors had the authority to select a temporary replacement until the crown named a new administrator. Councillors were also members of the General Assembly that enacted laws to govern the colony. Finally, they, along with the governor, served as Virginia's highest regular court, with original jurisdiction over capital crimes and civil suits

91. Fausz, "Merging and Emerging Worlds," in Carr, Morgan, and Russo, eds., *Colonial Chesapeake Society,* 53–56; Warren M. Billings, *A Little Parliament: The Virginia General Assembly in the Seventeenth Century* (Richmond, Va., 2004), foreword, chap. 5 (quotation on xviii); Haskell, "'Affections of the People,'" introduction, chaps. 2, 3 (quotations on 7, 103, 134); Jon Kukla, "Order and Chaos in Early America: Political and Social Stability in Pre-Restoration Virginia," *AHR,* XC (1985), 275–298 (quotation on 282).

involving large sums, and had appellate jurisdiction over decisions of lower courts.[92]

There were no set requirements for appointment to the Council, although wealth, residence near Jamestown, experience in Virginia politics, and personal or kinship ties to important people in England all mattered. Initially, most were of middling origins and had mercantile backgrounds or connections. They were also political novices; only one councillor had ever served in Parliament, and none had held an influential local office in England. Councillors were not paid a salary, nor were they reimbursed for their expenses, although they and up to ten members of their families were exempted from paying taxes except those due to the established church. Their position at the apex of power, however, afforded them ample opportunities for profit. They monopolized all lucrative provincial offices that awarded fees for collecting colonial revenues and appointed kinsmen to lesser government posts. Their legislative and judicial functions afforded them opportunities for legitimating and enforcing the privileges and perquisites they awarded themselves in their capacity as administrators. And, since Charles I and his ministers failed to exercise the same oversight over the councillors' actions as had officers of the Virginia Company, there were few effective restraints on their activities. In 1635, for example, the councillors got away with deposing a royal governor whose conservative land policies threatened their interests. High office also gave councillors an entrée into strategic social and mercantile networks in the colony and abroad, grants of free land, a monopoly of trade with the Indians, effective first choice of new bound laborers, and enhanced possibilities for strategic marriages that further consolidated their political and economic power.[93]

In addition to consolidating the power of the elite, a growing population and spreading settlement led to other changes in political organization that did afford ordinary planters some control over the laws that governed their society and over the administration of justice and the settling of disputes. So long as settlers were concentrated in or near Jamestown Island, the Virginia governor and his council had sole authority to make and to enforce laws. The creation in 1619 of a representative assembly, initially seen as no more than an appendage of the Virginia Company and an adjunct to the Council of State, gave freeholders and some tenants a say in lawmaking and in hearing the

92. Billings, *A Little Parliament,* chap. 5; Craven, *Southern Colonies,* 154–158.

93. Bernard Bailyn, "Politics and Social Structure in Virginia," in James Morton Smith, ed., *Seventeenth-Century America: Essays in Colonial History* (Chapel Hill, N.C., 1959), 90–98; Morgan, *American Slavery, American Freedom,* 123–126, 230; Billings, *A Little Parliament,* chap. 5; Haskell, "'Affections of the People,'" chap. 3.

inhabitants' grievances. After the dissolution of the Virginia Company, the crown made no explicit provision for an assembly (burgesses being the only elective officers in the colony) to continue. Governors continued to summon assemblies throughout the 1620s and 1630s, however, and in 1639 Charles I confirmed the General Assembly's legality. Moreover, in the 1620s, some judicial duties, originally the prerogative of top officials, were assigned to the proprietors or to delegated commanders of particular plantations and to judges of several monthly courts created by the Assembly in 1624. This practice led at first to an irregular and ineffective patchwork of jurisdictions.[94]

Subsequently, in 1634, the Assembly divided Virginia into eight counties, each governed by a county court composed of justices of the peace, a sheriff, a clerk, and lesser officials. These justices had functions similar to local officers in England. Like the councillors, magistrates received no salary, but they were reimbursed for any expenses they incurred in the performance of their duties. As individuals they could settle minor disputes and collectively they handled petty criminal offences and civil cases that did not involve large sums. Creation of county governments was the first step in transferring control of local matters to resident planters rather than a limited number of officials selected in England, often recent arrivals who had scant knowledge of actual conditions.

Officials of the Virginia Company had assumed that the leaders of the colony would be gentlemen whose power would be enhanced by recognized status, influential connections, formal learning, and wealth. In the aftermath of the 1622 Indian attack and the dissolution of the Company, which had successfully recruited leaders drawn from the higher ranks of English society, such men were in short supply and were far too few to fill the growing number of local offices. A motley group of hardened survivors of varied backgrounds who had managed to wrest some profit in extremely difficult times began to share effective power with governors and councillors chosen by the king. Local officers were appointed by the governor with the advice of the councillors who were most familiar with the residents of particular areas. A similar system of county courts was soon established in Maryland. Despite the undistinguished and often ill-educated status of many of the magistrates, county courts provided an efficient system of administration by officers living in the midst of their constituents, and the number of courts could easily be expanded as settlement spread. These local governments quickly proved

94. Billings, *A Little Parliament,* foreword, chap. 1; Warren M. Billings, "The Growth of Political Institutions in Virginia, 1634 to 1676," *WMQ,* 3d Ser., XXXI (1974), 226; Brenner, *Merchants and Revolution,* chap. 4; Craven, *Southern Colonies,* 159–160, 165–169.

effective in maintaining order in both colonies, in part because they gave free settlers (if not bound servants) a sense that they had the same legal protections that they had enjoyed in England. Moreover, ordinary planters, now working land of their own, also began to take responsibility for keeping the peace and helping to recreate, as far as possible, the sorts of "friendly neighborhoods" they had been accustomed to in the Old World.[95]

Advances in Agriculture and Changes in the Labor Force

Although particular plantations fell dismally short of investors' expectations, the men and women who settled and directed them did gain some knowledge of the sorts of crops, cultivation techniques, and methods of managing labor that were appropriate to the region. This hard-bought expertise enabled surviving colonists to recover surprisingly quickly from the enormous setbacks of 1622–1623. In some cases, residents of particular plantations might have been more willing to try some of the indigenous labor-saving methods, novel to Europeans, identified by more pragmatic Virginia Company leaders such as Captain John Smith.

Some settlers began to learn how to better select and manage land. From the Indians the English learned to choose the most fertile soils, at first by taking over old fields the Indians had cleared, farmed, and later abandoned. (Comparisons of the sites of native settlements with modern soil surveys show that the Indians had selected the choicest tracts in the region.) Without using fertilizers, however, even the best of coastal Chesapeake land had to be rested in long fallow after only a few years of cultivation. Old fields needed to remain untilled long enough for nitrogen-bearing organic matter, removed from plots when crops were harvested, to reaccumulate in the upper layers of the soil. Coastal soils are also generally acidic and deficient in phosphorous and base elements such as potassium, calcium, and magnesium necessary for plant growth. The burned limbs and, in subsequent years, fallen rotten tree trunks replenished organic matter and supplied other nutrients drawn up from far underground and stored in the wood. Shallow-rooted food crops could then flourish for a year or so from the nutrients contained in the soil.[96]

95. Billings, "The Growth of Political Institutions," *WMQ*, 3d Ser., XXXI (1974), 225–242; Lois Green Carr, "The Foundations of Social Order: Local Government in Colonial Maryland," and Robert Wheeler, "The County Court in Colonial Virginia," both in Bruce C. Daniels, ed., *Town and County: Essays on the Structure of Local Government in the American Colonies* (Middletown, Conn., 1978), 72–110, 111–133.

96. Timothy Silver, *A New Face on the Countryside: Indians, Colonists, and Slaves in South Atlantic Forests, 1500–1800* (Cambridge, 1990), 46–50. For descriptions of contact period Virginia

Unable to manage the hard work of clearing the forest, Company settlers clung to Indian old fields even after the soil was thoroughly depleted, claiming either that cutting down trees was too arduous or that they lacked the proper tools. But the lament of a Virginia Company official that planters chose new sites for good tobacco soil, and not for a healthy situation, or the statement of an unknown author quoted by Captain John Smith that they planted wherever "a choice veyne of rich ground invited them" indicates that some had learned how to judge land quality for themselves. George Sandys's statement that the Company tenants were farming the barrenest places in the country because of the Company's policy of placing them on previously cleared ground without regard for the fertility of the plots implies that other settlers on some of the private plantations were managing to clear virgin sites and were reaping the benefits of fresh soils. Indeed, in 1619 John Rolfe reported there were markedly higher corn yields on "new ground."[97]

Moreover, since some of the particular plantations were located higher up the James River beyond the deadly oligohaline zone around Jamestown Island (where the mixing of salt and fresh water in tidal rivers created an environment in which disease organisms such as typhoid and dysentery became more concentrated in warm weather than in either fresh water upstream or saltier water downstream), private settlers on these plantations had somewhat better chances for avoiding some of the diseases that repeatedly decimated the colonists at Jamestown (Map 4). After 1624, settlers were free to move into healthier places, and a more dispersed pattern of settlement also mitigated the spread of epidemic diseases. Although more than half the Vir-

Indian agriculture, see Helen C. Rountree, *The Powhatan Indians of Virginia: Their Traditional Culture* (Norman, Okla., 1989), 32, 44–49; and Rountree and E. Randolph Turner III, *Before and After Jamestown: Virginia's Powhatans and Their Predecessors* (Gainesville, Fla., 2002), 85, 103–104. Some archaeologists argue that native Americans cropped fields in nutrient-poor coastal areas for just two to three years before allowing a fallow period of between twenty-one and forty-two years (Turner, "An Archaeological and Ethnohistorical Study on the Evolution of Rank Societies in the Virginia Coastal Plain" [Ph.D. diss., Pennsylvania State University, 1976], 192–195). Others argue that fields were cropped for longer periods or that enough arable land was available so that, although individual households might shift garden plots and dwelling locations, entire villages seldom had to move (Stephen R. Potter, *Commoners, Tribute, and Chiefs: The Development of Algonquian Culture in the Potomac Valley* [Charlottesville, Va., 1993]), 95–97. For southern coastal plain soil characteristics and shifting cultivation, see Douglas Helms, "Soil and Southern History," *Ag. Hist.,* LXXIV (2000), 723–730.

97. *RVC,* III, 73–74, 79, 92, 219–221, IV, 179 (quotation); Smith, *Map of Virginia,* I, 54; Smith, *Generall Historie,* II, 573. Some colonists were pleased to have the 1622 uprising as an excuse to seize the Indians' clearings: "Now their cleared grounds in all their villages (which are situate in the fruitfullest places of the land) shall be inhabited by us, whereas heretofore the grubbing of the woods was the greatest labour" (*RVC,* III, 557).

ginia population died between 1625 and 1632, annual mortality was nonetheless cut in half.[98]

Some possibly also mastered more efficient clearing methods. By 1612, both Captain John Smith and William Strachey had recognized that the Indian style of girdling trees was the easiest way to clear fields for planting. However, all references to early English practice describe the more arduous method of cutting down all the timber, undoubtedly in part because most new settlers arrived too late in the season to effectively girdle trees before the first year's planting (Smith noted that the Indians barked trees one year before putting in a crop). Moreover, ethnocentric settlers were often unwilling to embrace indigenous methods that they considered inferior to more labor-intensive European farming practices. Still, some settlers on particular plantations might have adopted the Indians' method. Smith's Hundred, for example, was reported to have a "great store of good Cleared grounds" by 1619. On the whole, though, land clearing ceased to be a major obstacle only when individuals acquired an assured interest in the ground they worked. John Rolfe noted in 1620 that freemen "knowing their owne lands" were prepared to build houses and to clear, "w[hi]ch giveth the greate incouragemt, and the greatest hope to make the Colony florrish that ever yet happened to them." Martin's Hundred tenant Robert Addams, who in 1625 commanded only his own labor and that of his wife and one servant, instead of complaining about the difficulty of clearing, requested a lease for "a reasonable term" to ensure that he would enjoy the fruits of his labor, explicitly requesting "liberty to clear more into the wood."[99]

98. McCartney, "An Early Virginia Census," *Arch. Soc. Va. Qtly. Bull.,* LIV (1999), 181; Earle, "Environment, Disease, and Mortality," in Tate and Ammerman, eds., *Chesapeake in the Seventeenth Century,* 96–125.

99. Smith, *Map of Virginia,* I, 61–62; Strachey, *Historie of Travell into Virginia Britania,* ed. Wright, 118; Bruce, *Economic History of Virginia,* I, 196–197; *RVC,* III, 120, 245 (quotations), IV, 236; Lefroy, comp., *Memorials,* I, 265; Addams to Nicholas and John Ferrar, June 16, 1625, Ferrar Papers, no. 572, CWL; Henry Spelman, *Relation of Virginia* (1609), in Arber, ed., *Travels and Works of Captain John Smith,* cxi. Whether native Americans used different methods for clearing virgin forest and for clearing old gardens grown up in brush is uncertain, as the only evidence comes from three brief accounts written by Europeans (Rountree, *Powhatan Indians,* 46–47). Settlers who could not manage the heavy labor of clearing would likely have had to spend more time weeding than did those who adopted the Indians' methods. Agricultural experiment stations in twentieth-century British Africa found that clearing well-established forests followed by burning is the only system of shifting cultivation that does not require subsequent weeding. Intense burning kills not only weeds but also harmful insects and fungi and diminishes their impact for several years. When old fields are recycled after a fallow of only a few years, the effort of clearing brush is greater than that required for clearing forest, the fields must be hoed to remove roots of grasses and other vegetation that remain intact after burning, and more effort must

Some of the most critical advances were made in tobacco culture, and most occurred after particular plantations had been started. In all the New World European colonies, settlers were experimenting with different methods for cultivating, harvesting, curing, storing, and packing the weed. By midcentury, growers in all the mainland and island colonies had developed similar methods, but ones that also included variations that made the most of differing environments and natural resources. The evolving customs of particular colonies included blends of knowledge and technologies contributed by native Americans, by enslaved West Africans, by English and Dutch tobacco growers, and by English, Dutch, and French colonists. There was doubtless considerable exchange of ideas and experiments in this formative period, although little documentary evidence of such exchanges have survived. As Captain John Smith underscored, "No discoveries of any thing more than the curing of Tobacco, by which hitherto, being so present a commodity of gaine . . . hath brought them to this abundance." Settlers might have learned how to plant tobacco in hoed up hills and to remove secondary shoots (termed "succors") in order to promote the growth of prime leaves by observing native American practices. They also might have benefited from the knowledge of slightly more experienced Bermuda growers. In the 1630s, perhaps six hundred to seven hundred Bermudians emigrated to Virginia, lured by the promise of obtaining land on the mainland and bringing with them new ideas about cultivating, curing, and processing the weed.[100]

Awareness of tobacco's growing cycle sharpened, encouraging planting

---

be put into controlling weeds during the growing season (Erik Green, "Diversification or De-Agrarianization? Income Diversification, Labor, and Processes of Agrarian Change in Southern and Northern Malawi, Mid-1930s to Mid-1950s," *Ag. Hist.*, LXXXII [2008], 169; Helms, "Soil and Southern History," ibid., LXXIV [2000], 727).

100. See, for example, the detailed account of tobacco cultivation in the English and French West Indies in Charles de Rochefort, *Histoire . . . des Ils Antilles . . .* (Rotterdam, 1658), trans. in Brooks, *Tobacco,* II, 329–335. For the Caribbean in general, see Norton and Studnicki-Gizbert, "Multinational Commodification of Tobacco," in Mancall, ed., *Atlantic World and Virginia,* 253–263. For Dutch colonies, see Enthoven and Klooster, "'The Dutch Found and Relieved Us,'" in Bradburn and Coombs, eds., *Early Modern Virginia.* For Bermuda, see Jarvis, "'In the Eye of All Trade,'" 92–105, 194–195. For Providence Island, see Kupperman, *Providence Island,* 87, 91–92. For Virginia, see Smith, *Continuation of the Generall Historie,* II, 888 (quotation); and Daniel Tucker to John Ferrar, Apr. 6, 1623, Ferrar Papers, no. 465, David Lewis to John Ferrar, Apr. 7, 1623, no. 466, Miles Kendall to John Ferrar, Apr. 15, 1623, no. 467, ? to John Ferrar, [1623], no. 518, CWL. By 1705, many Virginia Indians had ceased growing tobacco, instead obtaining smoking tobacco from the English, and few settlers knew how native Americans had "order'd" their crops before the arrival of the English. Historian Robert Beverley was, however, informed by older settlers that the Indians had practiced succouring (Beverley, *The History and Present State of Virginia* [1705], ed. Louis B. Wright [Chapel Hill, N.C., 1947], 145).

early enough in the spring to avoid losses to fall frosts. Initially, the labor required for curing the harvested leaf was a major bottleneck. John Rolfe reported in 1618 that two men were required to house and cure what one worker could grow. A Mr. Lambert discovered around 1617 that the weed cured better hung on lines rather than piled into heaps and sweated with hay; in 1618, Governor Argyll forbade the latter practice in order both to improve the quality of the cured weed and to reserve scarce hay for cattle. Settlers on particular plantations apparently quickly adopted line curing and asked their English backers to send supplies of pack thread to use for lines. George Thorpe reported in May 1621 that lines must be sent immediately "because the Crope of this yeare will as it were utterly perish wth out that supplie," and in 1623 an overseer at Flowerdew Hundred was reported hanging tobacco to dry on lines. The less costly and labor-saving method eventually adopted in the Chesapeake and in the West Indian islands was air drying of the harvested leaf on wooden poles hung in well-ventilated earthfast "tobacco houses." Exactly when Virginia colonists began using poles for drying is uncertain. Most likely it was adopted as the technology of tobacco house construction evolved.[101]

The first structures used for housing tobacco in Virginia and in Bermuda were storehouses probably first erected for the safe storing of supplies that drew on ancient European building antecedents. These storehouses had no exterior walls; instead, their A-shaped roofs reached to the ground, creating an interior storage space that was lined with plank. One such building has been excavated at Martin's Hundred, and a similar vault, expressly designed for storing tobacco, was built in Bermuda by 1620. When the Berkeley Hundred adventurers instructed Captain Woodlefe to erect a framed storehouse in 1619 for the safekeeping of tools, munitions, armor, and food, "strongly planted on the inside," they likely intended a structure similar to the early storehouse excavated at Martin's Hundred that might later have been used for storing tobacco (Plate 4). Indians burned at least one outbuilding that John Smith explicitly termed a "Tobacco-house" in 1622, and the muster of 1625 noted four "tobacco houses" on Flowerdew Hundred, one at Burrows Hill in James City, and three on Mr. Blaney's plantation in James City. Raftered structures without walls, sometimes termed "roof huts," continued to be used for curing tobacco into the 1660s. A York County, Virginia, will of 1658 mentioned a sixty-foot tobacco house "with rafters upon the ground," and a tentlike tobacco house appears on a 1664 plat of Jamestown Island.

101. *RVC,* III, 78, 92–93, 183, 186, 257, 448–449, 581–588; *MCGC,* 27; Bruce, *Economic History of Virginia,* I, chap. 4.

PLATE 4. Storehouse at Martin's Hundred. Drawn by Cary Carson.
Such a storehouse might have been used for keeping tobacco.

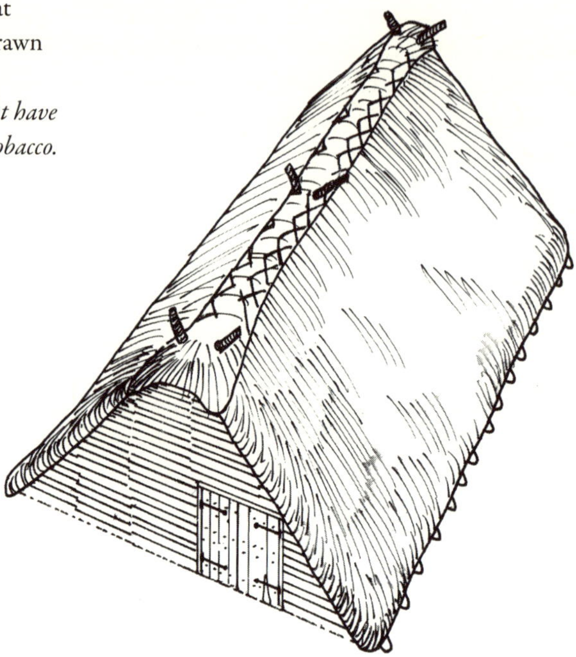

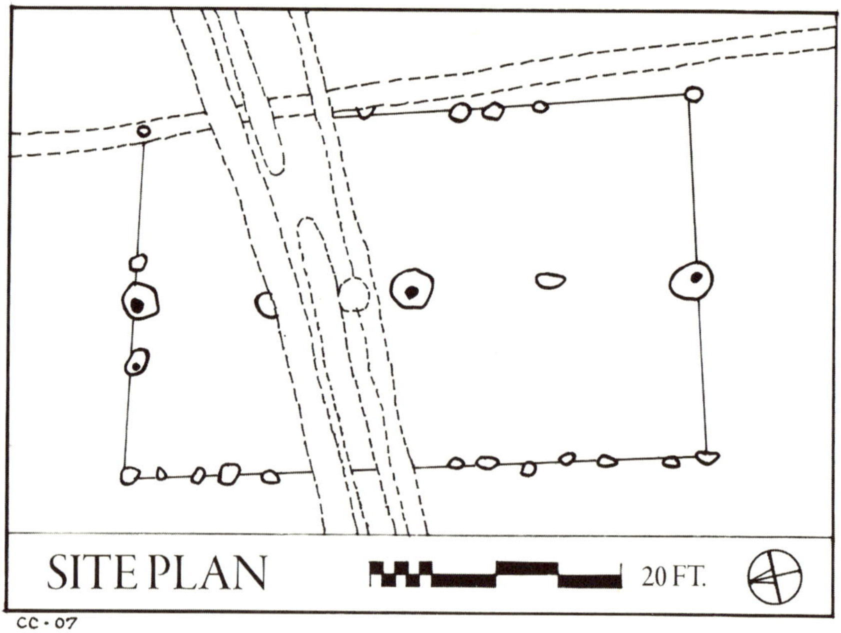

These secure buildings would have served to protect the crop from pilfering but would likely not have afforded optimal ventilation for drying the leaf. Other "tobacco houses" referred to in the later 1620s were perhaps earthfast storing and curing houses covered with nailed planks. Settlers at Jordan's Journey might have experimented with the use of small fires inside a barn for curing the crop. An earthfast structure on that plantation dating to approximately 1620–1635 might have been a tobacco house that incorporated a small internal heating flue.[102]

In addition to trying different methods of housing and curing the crop, the Virginia colonists also experimented with various processing techniques. Some learned to pack their tobacco in twisted rolls, the method employed by Spanish growers in Trinidad and the Orinoco basin and taught to Bermudians by African slaves. (This technique was also adopted by English tobacco growers on Barbados and French planters on Martinique [Plate 5].) And they learned as well the fine art of concealing trash tobacco inside the rolls. As a consequence, in 1621 the Virginia Company demanded that the settlers ship their tobacco in the form of leaves laid straight rather than rolled. Packing in leaf soon became the usual mode of preparation, but for some years thereafter a few planters continued to make up rolls. Laws enacted in 1633 allowed only the present year's crop to be processed into rolls and limited the time in which they could be made up to the immediate post-harvest season, thus seeking to prevent the concealing of old tobacco. To reduce the expenses of freight and of customs duties that were assessed by weight, growers were instructed to ship "noe more stalke then needs must." By 1627, large debts in tobacco were being paid "according to the custome of this Cuntry," half in leaf with the stalk half stripped out and half with no allowance for the stalk. Over the next decade, there were experiments with further processing of some of the leaf, including "stripping and smoothing."[103]

By 1621, some planters had begun to send their tobacco overseas in sturdy hogsheads, a large wooden cask used in Europe for holding wine, beer, fish,

102. Noël Hume and Noël Hume, *Archaeology of Martin's Hundred,* 131–133; Carson et al., "Impermanent Architecture," *Winterthur Portfolio,* XVI (1981), 154; *RVC,* III, 209; Smith, *Generall Historie,* II, 575–576; *MCGC,* 142, 159, 186; plat of 119 acres surveyed for John Knowles, Aug. 6, 1664, Ambler Manuscripts, 1638–1809, LC, microfilm, M-104, CWL. For fire curing in Hispaniola, see Benzoni, *La Historia del Mondo Nuovo,* in Brooks, *Tobacco,* I, 224–226; and, in Virginia, see Mouer et al., "Jordan's Journey," 34–35.

103. *RVC,* III, 492–498; C. T., *An Advice How to Plant Tobacco in England,* in Brooks, *Tobacco,* II, 3–7; *MCGC,* 27, 128–129; York County Deeds, Orders, and Wills, 2: 105, microfilm, VSL; William Waller Hening, ed., *The Statutes at Large: Being a Collection of All the Laws of Virginia . . .* (1809–1823; rpt. Richmond, Va., 1969), I, 205, 224–225; [Richard] Caswell to George Menify and Edmund Pickard, Sept. 19, 1624, Ferrar Papers, no. 554, CWL.

PLATE 5. *Ménagerie.* From Jean Baptiste du Tertre, *Histoire generale des Antilles . . .*, II (Paris, 1667). Courtesy, John Carter Brown Library at Brown University, Providence, R.I.

In the West Indies and in Spanish colonies on the South American mainland, enslaved Africans prepared tobacco for shipment to Europe by twisting it into long rolls, as the men in the tobacco house on the right are doing.

and other liquid or semiliquid products. The hogshead eventually also became the standard shipping container for tobacco produced in most of the English colonies. Still, in 1628, however, Virginia tobacco was being shipped, not just in hogsheads, but in whatever miscellaneous containers the colonists could lay their hands on. A few casks might have been newly made in the colonies, but most were surely recycled containers in which goods had been shipped from England, as evidenced by tobacco arriving in London and the outports packed in a bewildering assortment of barrels, pipes, puncheons, butts, and chests in addition to hogsheads. And sometimes the leaf was apparently just stowed loose in the hold. Transatlantic freight charges were customarily assessed by the ton, a measure of volume rather than of weight (a practice apparently carried over from the medieval wine trade) in which a freight ton was considered equivalent to four hogsheads. In the 1620s, Virginia growers made limited progress in saving on transport charges by pressing tobacco more tightly into casks. In 1628, the average Virginia hogshead contained only 140 pounds of tobacco, whereas those shipped from Bermuda weighed twice as much, a result of the islanders' practice of more highly processing the cured leaves. By the late 1630s, however, Chesapeake planters had learned how to pack 250 pounds into their casks.[104]

Over time, settlers did become more skilled in judging the various grades of tobacco. They began to take some steps, however ineffective, to deal with the problem of bad tobacco and on the whole improved the quality of what they grew. Observers noted the "amendment of it in the quality, learned by time and experience," and Bermuda growers began to encounter stiff competition from the mainland, having reluctantly to concede that Virginia tobacco was "farre better then it was wont to be." And, as colonists and importing merchants gained experience in the timing of the harvesting and subsequent processing cycle, attempts were made to coordinate the coming of the fleet with the time the crop was ready to ship so that the cargo stood a better chance of arriving at market in good condition. As mariners gained more experience with the several possible routes ships might take across the Atlantic, chances improved that the ships would actually arrive in the right place at the right time.[105]

Perhaps the major advance in tobacco culture was a pronounced increase in the volume of tobacco one laborer could produce. The reasons for the shift and its timing are obscure. Observers seemed to agree that, around 1620, one thousand plants were the maximum one man could tend. It was variously estimated that this number of plants would yield from 100 to 250 pounds of cured tobacco. Scattered observations of crops per laborer for the 1620s range from 100 to 700 pounds per hand, the mean being 400 pounds (Table 4). By 1629, as tobacco prices fell, the Virginia Assembly ordered that a worker could tend no more than three thousand plants, a number further reduced to two thousand in 1630 and to fifteen hundred in 1633. Thus, by the end of

104. *RVC,* III, 509–510; *CSP,* I, 36; *MCGC,* 19; Ronald Edward Zupko, *A Dictionary of Weights and Measures for the British Isles: The Middle Ages to the Twentieth Century* (Philadelphia, 1985), 186–189; Neville Williams, "England's Tobacco Trade in the Reign of Charles I," *VMHB,* LXV (1957), 403–449; John J. McCusker, "The Tonnage of Ships Engaged in British Colonial Trade during the Eighteenth Century," in McCusker, *Essays in the Economic History of the Atlantic World* (London, 1997), 61–65; Menard, "Economy and Society in Early Colonial Maryland," 455. In 1630, John Pory deposed that a hogshead of leaf tobacco weighed about two hundred pounds and a hogshead of rolled tobacco about three hundred pounds ("Interrogatories on Behalf of Edmund Rossingham, plt v. Ralph Yardley," Mar. 18, 1629/30–November 1630, C24/560, pt. 2, 84, transcription in Special Collections, CWL).

105. *RVC,* III, 228, 546, 581–88, IV, 271–273; *MCGC,* 128–129; Lefroy, comp., *Memorials,* I, 397–399; Hillier, "Trade of the Virginia Colony," chap 3.

TABLE 4. Tobacco Crops per Laborer, Virginia, 1623–1626

Year	No. of laborers	No. of full hands	Size of crop	Crop share per hand	Source
1623	4	4	2,800 lbs.	700 lbs.	*RVC,* IV, 524
1623	3	1.5	3,000 lbs.	2,000 lbs.[a]	*RVC,* IV, 524
1623	1	1	2,000 plants	500 lbs.	*RVC,* IV, 38
1623	16	8[b]	1,000 lbs.	125 lbs.	*RVC,* IV, 104–106
1623	10	10	3,000 lbs.	300 lbs.	*MCGC,* 139
1624	16–31	16–31[c]	9,000 lbs.	290–563 lbs.[d]	*MCGC,* 27
1625	1	1	1,500 plants	150–375 lbs.[e]	*MCGC,* 80
1626	6–7	6–7	3,000–4,000 lbs.	429–667 lbs.	*MCGC,* 99
1626	1	1	100 lbs.	100 lbs.	*MCGC,* 99

Note:

[a] This observation is an obvious exaggeration. The same estimate reports an equally impossible corn crop of 367 bushels per laborer.

[b] The 16 laborers were half sharers.

[c] This estimate assumes that the 31 men present on the plantation in the Virginia census of 1625 were present the previous year.

[d] The high productivity estimate assumes one-half of them planted tobacco; the low productivity estimate assumes that all did.

[e] The low estimate assumes that 1,000 tobacco plants equals 100 pounds cured tobacco; the high estimate assumes that 1,000 plants equals 250 pounds.

the 1620s, the number of plants an experienced worker could handle in one season had apparently trebled over the number tended in the 1610s. Subsequently, some combination of better plant quality and better cultivation techniques enabled a laborer to produce 800 to 1,000 pounds a year by 1634, and perhaps more by the 1640s.[106]

To the extent that these early stinting regulations were enforced (and the level of enforcement is questionable, since those who enacted these laws had

106. Smith, *Generall Historie,* II, 541, 564–565; *RVC,* III, 581–588, IV, 37–39; *A Relation of Maryland,* in Hall, ed., *Narratives of Early Maryland,* 96; Morgan, *American Slavery, American Freedom,* 142–143; Stone, "Society, Housing, and Architecture in Early Maryland," chap. 2. In the 1630s on Providence Island, one hundred pounds of tobacco was considered the usual annual crop of one servant, and two hundred pounds was the maximum that could be raised in the most propitious circumstances (Kupperman, *Providence Island,* 79). One to two hundred pounds was also the crop typically made in Bermuda in the 1610s; however, captured Iberian slaves who were experienced in growing tobacco made nearly seven hundred pounds per laborer (Jarvis, "'In the Eye of All Trade,'" 155).

the most to gain by disregarding them), some planters were likely stimulated to experiment with new, higher-yielding strains, with different spacing between plants, and with the optimal number of leaves allowed to grow to maturity. (Others, however, simply tried to make the most from the plants they were legally allowed to tend, curing inferior leaves growing from the base of the plants and others that sprouted after the first cutting.) Laws of the 1630s enjoining the planting of new varieties ("the long sorts"), wider spacing between plants (four and one-half feet apart), and a reduction in the number of leaves to be tended and harvested (to increase the size of the leaves allowed to remain) suggest the areas in which experimentation was occurring. Whether any of these innovations can be traced to experience gained on particular plantations is uncertain. Those who grew the most tobacco should have been the first to profit from learning by doing. On the other hand, most of the gains seem to have occurred only after the corporate plantations had disappeared and were replaced with individually held farms.[107]

These increases in productivity were, however, not an unmixed blessing for Virginia planters. As more and more tobacco poured into England from Virginia and Bermuda, the high prices it initially commanded began to fall precipitously. Moreover, the establishment between 1624 and 1629 of four additional English colonies in the Caribbean—Saint Christopher's, Barbados, Nevis, and Providence Island—whose settlers also hoped to profit from growing tobacco, promised ever greater increases in the supply of the weed. By the mid-1620s, Virginians experienced the first of many cyclical depressions that, in the words of Russell R. Menard, "reverberated through the entire economy and affected the volume and pace of immigration, the growth of population, the spread of settlement, the extent of opportunity, government policy, experimentation with other resource-intensive staples, the rise of a manufacturing sector, and the level of material well-being in the colonies." At first, Virginia tobacco sold in the colony for as much as 3s a pound. But, by 1623, Virginia Company officials complained that it was worth only half the legal rate of 3s for the best grades and 1s 6d for poorer sorts, and merchants began to increase the prices of merchandise to reflect tobacco's diminishing value. The price might have fallen to as low as 2d a pound by 1629 when the first cyclical trough was reached, and the economy remained severely depressed until mid-1633. Shipping was scarce, freight charges high,

107. Beer, *Origins of the British Colonial System,* 96. The Virginia laws of the 1630s were similar to laws enacted in Bermuda in 1621. See [Nathaniel Butler], *The Historye of the Bermudaes or Summer Islands . . .* , ed. J. Henry Lefroy, Hakluyt Society, 1st Ser., LXV (London, 1882), 235–236; Jarvis, "'In the Eye of All Trade,'" 96.

and English manufactured goods expensive and in short supply. Prices again rose beginning in 1634, and immigration increased. Thousands of acres were patented by new settlers, old planters, and London merchants who obtained rights to land for financing the transportation of indentured servants. The renewed investment in land and labor led again to a significant growth in output, leading to a second depression in the late 1630s that persisted for some years.[108]

Marketing arrangements also changed to the advantage of both individual planters and individual merchants. In general, Charles I retained his father's antipathy toward tobacco, but his need, occasioned by his frequent disputes with Parliament, to increase customs revenues led to inconsistent policies toward the tobacco trade. Charles wanted to curtail tobacco consumption in England but also wanted to benefit from the further revenues he hoped to obtain by granting various favored courtiers and merchants monopolies for selling tobacco. In 1625, the Privy Council, responding to complaints about the inhibiting effects of a succession of monopoly contracts for the purchase and distribution of tobacco granted to various subsidiary groups of officials of the Virginia Company, decreed that all parties "shall have free trade and liberty to make the best of their own labours." Tobacco growers were then free to make the best deals they could with private merchants, and the merchants were able to sell freely in England after paying impost duties on the weed. Together the merchants and planters were able to defeat attempts by Charles I and assorted groups of would-be projectors to create a new royal tobacco monopoly in the late 1620s and again in the 1630s. By the early 1630s, Charles did encourage the expansion of England's reexport trade to continental markets by instituting a drawback system under which half or more of duties on tobacco shipped from England were refunded.[109]

At this early stage in the tobacco trade, however, there was little continuity among English merchants, since most employed a "single adventure" system. A small group of traders, almost all of them living in London, chartered a ship to carry a cargo of European goods to the Chesapeake and to bring back

108. Menard, "The Tobacco Industry," *Res. Econ. Hist.,* V (1980), 109–177 (quotation on 123); John R. Pagan, "Growth of the Tobacco Trade between London and Virginia, 1614–40," *Guildhall Studies in London History,* III (1979), 248–262; Robert C. Batie, "Why Sugar? Economic Cycles and the Changing of Staples on the English and French Antilles, 1624–54," *Journal of Caribbean History,* VIII–IX (1976), 1–41.

109. Brenner, *Merchants and Revolution,* chaps. 3, 4; Beer, *Origins of the British Colonial System,* chaps. 6–8; Hillier, "Trade of the Virginia Colony," chaps. 5–6; Pagan, "Growth of the Tobacco Trade," *Guildhall Studies in London History,* III (1979), 255; Thomas Cogswell, "'In the Power of the State': Charles I, Virginia, and Mr. Anys's Project," manuscript in the possession of the author.

tobacco. The vessel often carried outward and inward cargos for fifteen to twenty others; thus, the "adventurers" earned a profit from freight and could more quickly procure return cargos. Initially, the ship captain or a supercargo bartered the goods for tobacco, but by the 1630s some merchants were employing commission merchants (called "factors") who resided in the colony to handle sales and purchases. When the ship returned, the adventurers divided up the tobacco among themselves. Most did not otherwise trade as partners, and tobacco was only one of many commodities in which they dealt. Between 1616 (when groups of private individuals were given permission to send out vessels) and 1660, more than one thousand English men and women are known to have participated in the Virginia tobacco trade, and most invested in only one to three voyages. Consequently, planters did not usually form relationships with those who sold their tobacco in England, unless the merchants also happened to be relatives.[110]

In addition to the marked advances in tobacco culture and more limited advances in marketing, as they learned more about the local environment Chesapeake settlers increased agricultural productivity by further modifying Old World agricultural practices to suit New World conditions. For example, they made some improvements in the way they cared for their livestock, especially cattle. The first step was to fence in ranges on necks of land for their better protection and control. The first fences were made with boards inserted into trenches in the ground, a technique that was somewhat labor intensive but that did not require the use of scarce nails. Phillip Bruce maintained that by 1621 some planters had given up paling and had learned to put up more easily constructed split-rail fences, thus making it easier both to contain cattle and to protect crops. But, as with tobacco growing and curing, major improvements came only after 1625. In that year, 369 cattle were counted in the colony census, evidence that by the mid-1620s cattle were reproducing well enough to replace those lost in the previous difficult two years. Indeed, Governor John Harvey estimated that the true size of the Virginia herd was by then already around 500. Shipments of additional cattle to supplement herds ceased after 1624, but by 1629 the animals had multiplied rapidly to between 2,000 and 5,000. Consequently, colonists were able to meet more of their meat requirements from domestic livestock rather than from hunting and fishing. Settlers became more willing to risk waiting for the delayed returns afforded by raising large animals to maturity rather than opting, be-

110. Jacob M. Price and Paul G. E. Clemens, "A Revolution of Scale in Overseas Trade: British Firms in the Chesapeake Trade, 1675–1775," *JEH,* XLVII (1987), 2–6; Hillier, "Trade of the Virginia Colony," chaps. 1, 4, and 6.

cause of uncertainty about the future, for immediate returns from hunting. Archaeologists have found that by the 1630s up to 70 percent of the meat colonists consumed came from domestic livestock, with beef dominating, followed by pork, mutton, wildlife, and domestic fowl.[111]

Through the early 1640s, however, cattle remained relatively scarce, and most of them were still owned only by wealthier householders. Local customs did develop for better allocating cattle and cattle byproducts—annual rental of cows to small planters for their milk, for example, or killing the excess bull calves and very old cows for meat while preserving the young cows for future increase. Some planters also became more adept at hunting wild swine and began by design to include some swampland in their farms to supply forage for semidomesticated hogs.[112]

Improvements in grain culture, on the other hand, seem to have been minimal. One positive (if apparently retrogressive) step was abandoning most attempts to grow wheat and other English grains. These grains produced lower yields per acre than did corn, required thoroughly cleared ground, and did well only if plowed, which proved impossible in stump- and root-filled fields. Settlers thus planted wheat and other small grains only on relatively clear

111. For fencing, see *RVC,* I, 508, III, 441–443; Smith, *Generall Historie,* II, 509–510; Bruce, *Economic History of Virginia,* I, chap. 5. On the importance of cattle to the agricultural regimen, see Morgan, *American Slavery, American Freedom,* 136–140. Archaeologists have found the remains of such cattle palisades on numerous early Virginia sites. For numbers, see Annie Lash Jester and Martha Woodruff Hiden, *Adventurers of Purse and Person* (Richmond, Va., 1964), 5–69; [Harvey], "A Briefe Declaration," *Aspinwall Papers,* MHS, *Colls.,* 4th Ser., IX (Boston, 1871), 60–73. Various observers quoted by Captain John Smith estimated two thousand in 1627 and two thousand to five thousand in 1629 (*Continuation of the Generall Historie,* II, 885). In 1630, five deponents with experience in the colony stated that "allowing some loss for casualties according to the usual increase of cattle in Virginia," four young cows could be expected to produce a herd of twenty to twenty-five cattle in seven years ("Interrogatories on Behalf on Edmund Rossingham, plt. v. Ralph Yardley," Mar. 18, 1629/30–November 1630, C24/560, pt. 2, 84, transcription in Special Collections, CWL). See also Cary Carson et al., "New World, Real World: Improvising English Culture in Seventeenth-Century Virginia," *Jour. So. Hist.,* LXXIV (2008), 31–88; and Graham et al., "Adaptation and Innovation," *WMQ,* 3d Ser., LXI (2007), 470–475.

112. Stone, "Society, Housing, and Architecture in Early Maryland," chap. 2; "Claiborne vs. Clobery," *MHM,* XXVII (1932), 342; *RVC,* II, 215–216, III, 171, IV, 137–138; Smith, *Continuation of the Generall Historie,* II, 885–886; [Harvey], "A Briefe Declaration," *Aspinwall Papers,* MHS, *Colls.,* 4th Ser., IX (Boston, 1871), 60–73. William Claiborne stated, "The like bargaine of hireing landing and restoring of cattle is usuall in Virginia." An observer in 1632 reported to Berkeley proprietor John Smyth that the settlers generally had sufficient butter and cheese (Smyth of Nibley Papers); similarly, another traveler stated in 1634 that "ordinary planters houses of the better sort" had enough dairy products and meat from swine, goats, and poultry (Extract from a Letter of Captain Thomas Yong to Sir Toby Matthew, 1634, in Hall, ed., *Narratives of Early Maryland,* 60).

fields whose fertility had already been sapped by preceding crops of tobacco and corn. By 1629, ordinary colonists seem to have recognized that corn was far more suited to a labor-short economy and left more complicated English-style arable husbandry to the few large planters who had enough hands and draft animals to pursue it. Learning how to effectively combine corn and tobacco culture so that the work requirements of the two crops did not compete with one another apparently took more time. Corn shortages continued for a number of years, and legislation enjoining the planting of at least two acres per hand continued to be enacted in Maryland up to 1654 and in Virginia through 1662.[113]

Advances in frontier housewifery awaited the arrival of more women. Evidence from Claiborne's Kent Island settlement suggests something of the difference that the presence of even a few women could make in mitigating the sparseness of domestic life. During the first three years on the island, beating corn, "dressing vicutals," and "other worke in the howse" occupied not only the one woman in the settlement but also four to six males, a quarter of the workforce. The boys and sometimes injured or ailing men assigned to unaccustomed domestic tasks likely carried them out with considerable reluctance. During the settlement's final three years, when there were three women servants on the domestic staff, the percentage of laborers so engaged dropped to between 11 and 19. In addition to fetching water, preparing meals, and washing clothes, the women also took up dairying, something that had not been attempted when only men were present. On private plantations, women typically cared for poultry and cattle, milked cows, tended kitchen gardens, and gathered wild foods. These important additions to an otherwise numbingly monotonous diet of "water gruell" alleviated many of the complaints about victuals so prevalent in earlier years. More largely, the presence of women increasingly contributed to the stability of the colony as men who were able to marry came to see Virginia as their home and to work diligently to advance the fortunes of their families.[114]

113. *RVC,* I, 349–351, III, 73–74, 455–458, 581–588; Smith, *Generall Historie,* II, 538, 541; Smith, *Continuation of the Generall Historie,* II, 885; Bruce, *Economic History of Virginia,* I, 198–199; "Wyatt Manuscripts," *WMQ,* 2d Ser., VIII (1928), 48–57; Gray, *History of Agriculture,* I, 171–172; Morgan, *American Slavery, American Freedom,* 136. In New England, one Indian woman could raise between twenty-five and sixty bushels of corn on one or two acres. These yields supplied half the annual caloric requirements for a family of five (Cronon, *Changes in the Land,* 44). Captain John Smith reported that by 1629 the Virginia settlers "finde the *Indian* corne so much better than ours, they beginne to leave sowing it *[i.e., wheat].*" Planters not only fed their servants hominy but themselves relished it (Smith, *Continuation of the Generall Historie,* II, 886–889).

114. Brown, *Good Wives,* 84–85; "Claiborne vs. Clobery," *MHM,* XXVIII (1933), 180–189. See also Kupperman, *Providence Island,* 158–161.

The period also saw changes in labor management. When planning their settlement in 1619, the Berkeley proprietors devised a cumbersome division of subordinate administrative duties between a bailiff of husbandries, a clerk of the kitchen, a steward of the household, a clerk of the store, a sergeant of the company, an usher of the hall, and a keeper of armor and tools. These functionaries, customary officers on many an English manor, were apparently expected to carry out their duties for the honor conferred by the position and an eventual reward of extra land when their terms of service were up. This system probably broke down quickly once the settlers left ship. In 1621, the Virginia Company first described a productive unit in the colony as consisting of one thousand acres worked by five persons and an overseer. By 1625, when colonial records become more plentiful, most of the administrative duties associated with plantations had been consolidated under one officer, the overseer. Taking direction from the plantation's commander, the overseer supervised daily routines and managed the enterprise when superior officers were absent. In return, he got some form of immediate compensation for his efforts—either an annual wage or a share of the crops that he was responsible for producing. These arrangements were part of rapidly evolving local colonial custom for which Old World traditions provided inadequate guidelines, and they would become an enduring feature on Chesapeake plantations.[115]

The severe shortage of labor was also leading to changes in the position of bound workers. In the Company era, servitude, whether for the parent company or for proprietors of particular plantations, was of three sorts. Tenancy was the form most advantageous to the emigrant and least advantageous to the labor owner, as these bondpeople were entitled to keep some portion of what they earned from the outset. Those who contracted to go as ordinary servants were entitled only to food, clothing, and shelter during their terms of service, unless before emigrating they negotiated for additional benefits such as wages, time to work for themselves, or payments to wives remaining in England. Otherwise, the Company or the owners of particular plantations who recruited them kept whatever they produced. The work these servants did most closely resembled that of servants in husbandry in England, but the contracts they signed were derived from urban artisan apprenticeship contracts, whereby the apprentice promised to serve faithfully for a term and the master promised food, clothing, shelter, training in some skill, and perhaps freedom dues at the end of the contract. What was lacking in Virginia were the laws, customs, and oversight by local rulers that provided both con-

115. *RVC,* I, 539–541, III, 207–210, 241; *MCGC,* 43, 47–48, 96, 124, 159; Lefroy, comp., *Memorials,* I, 265.

tracted apprentices and rural servants with important protections. The third and least favorable form of servitude the Company employed was derived from pauper apprenticeship in England, whereby poor children were bound to service by court order to relieve the community of the cost of their support. With the rationalization of ridding the country of idle poor, in 1618 and 1619 the Company persuaded London officials to round up indigent children (often termed "Duty Boys" after the ship that carried some of them) whom it then transported to Virginia to be bound as servants for seven years to anyone willing to pay ten pounds for them. Leading officials, in fact, made universal the practice of buying and selling the unexpired terms of hapless servants and tenants of all sorts. As early as 1619, John Rolfe reported that "buying and selling men and boies," which "was held in England a thing most intolerable," had become common practice in Virginia. By 1625, as more independent merchants and ship captains entered the Virginia trade, it had become common practice for servants to be "sold heere upp and downe like horses."[116]

Tenancy was not surprisingly the form of servitude that Virginia officials wanted to discontinue, in part because it had proved impractical and in part because it was the form least favorable to labor owners. In 1619, the Virginia Council wrote a partially self-serving letter to the parent company justifying renting out some of the newly arrived Company tenants to old planters. There was some truth to their assertions that new settlers fared better if split up among experienced planters where they might be adequately housed, nursed through seasoning, if lucky, and "trayned up . . . in all the usuall workes of the Country." Greed for labor, though, was a more important motive. By 1623, the Council urged complete abandonment of the half-share tenant system, asserting it had proved disastrous both for the Company and for the owners of particular plantations where it had been employed.[117]

Share tenancy was one form of short-term leasing that was used in England in areas where arable farming predominated and occasionally in dairy-

116. Morgan, *American Slavery, American Freedom,* 116–129; Andrews, *Colonial Period,* I, 134–136; Smith, *Generall Historie,* II, 542; *MCGC,* 82; Kupperman, *The Jamestown Project,* 286. Billings, *A Little Parliament,* argues that, although artisan apprenticeship was the model for indentured servitude, the institution "met peculiar labor conditions in Virginia in a peculiar way" and did not take "much substance from precedents drawn out of English history or from common law" (204). This argument is further elaborated in Billings, "The Law of Servants and Slaves in Seventeenth-Century Virginia," *VMHB,* XCIX (1991), 45–62. For the work, contracting procedures, and legal and customary protections of servants in husbandry, see Ann Kussmaul, *Servants in Husbandry in Early Modern England* (Cambridge, 1981), chap. 2.

117. *RVC,* III, 226, IV, 13, 16, 623; for the conditions offered half-share tenants, see I, 271, III, 98–109, 307–365, 534–535.

ing areas as well. Such arrangements gave the tenant no interest in the land and hence were often only oral agreements. Usually the persons so engaged were undertenants on lands held by a variety of tenures. Most frequently, the landlord and tenant divided the crop equally, although "sowing to thirds" might also be negotiated. For example, in Gloucestershire undertenants grew tobacco (illegally after 1619) on a half-share basis. The system was generally suited to highly developed farms. Landlords customarily supplied land and housing and sometimes seed as well. English backers saw tenancy at halves as a reasonable way of sustaining a colonizing enterprise and for meeting common costs, and it was adopted in Bermuda, Virginia, Plymouth, and Providence Island. Perhaps it was also viewed as an arrangement for extending credit to farmers with few resources of their own. If so, the system was doomed to failure when absentee adventurers could offer only uncleared land and when they proved unwilling, after financing the tenants' not inconsequential cost of transportation across the Atlantic, to advance adequate supplies of food, clothing, and tools.[118]

A remonstrance of 1625 of the Bermuda Assembly (where half-share tenancy was the norm) helps to explain why the arrangement did not work well there, and not at all in Virginia. There was, the Bermudians asserted, "as much difference betweene a husbandman's sowing of wheate to halves in England, and planting Tobacco at halves in Somer Islands, as is betwixt black and white." English tenants farmed cleared land with adequate houses and barns; they had draft animals and plows; they grew a crop that required little labor, apparently affording them time for by-employments; and, once harvested, grain had only to be housed and stored for sale with no further processing. In contrast, tobacco farmers had to clear land, build houses, cultivate their crops entirely by hand, work at them "daylie and hourely," allowing no other employments, and yet, when the crop was harvested, still risk losing all if they botched the curing process. Bermuda tenants were unhappy with their situation, but at least they were not starving. In Virginia, Company tenants on half shares were not able to feed and clothe themselves and their families on 50 percent of what they produced, and the Council asserted that was also

118. Eric Kerridge, *Agrarian Problems in the Sixteenth Century and After* (New York, 1969), 52; C. W. Chalklin, *Seventeenth-Century Kent: A Social and Economic History* (London, 1965), 63; Joan Thirsk, ed., *The Agrarian History of England and Wales,* IV, *1500–1640,* in H. P. R. Finberg, gen. ed., *The Agrarian History of England and Wales* (Cambridge, 1967–2000), 689; Thirsk, "New Crops and Their Diffusion," in Chalklin and Havinden, eds., *Rural Change and Urban Growth,* 88–93. For Gloucestershire, see Carew Reynell, *The True English Interest . . .* (London, 1674), in Brooks, *Tobacco,* II, 428–430; Kupperman, *Providence Island,* 21, 33–34, 38, 122–126.

the case with tenants on particular plantations. The Company demanded an annual rent of sixty pounds of tobacco, twenty bushels of corn, one pound of silk, and six days of labor for a fifty-acre farm. If tenancy was to sustain the tenants and yield some reliable return, the Council recommended shifting to individually negotiated fixed annual rents on less confiscatory terms. Subsequent Council records suggest that this was precisely the system adopted both by private Virginia landowners and the few English investors who retained some interest in their colonial properties after 1625. Tracts of fifty acres began to be let for annual rents that amounted to only seven and a half bushels of corn, or fifty pounds of tobacco, or fifty shillings in cash.[119]

Once investors stopped recruiting English men and women to go to Virginia in service as tenants, tenancy began to evolve into a status that ex-servants moved into after their terms expired. Now trained "in all the usuall workes of the Country," they were well equipped to pursue Chesapeake-style farming. Tenancy, albeit on terms more suitable to the region, did not disappear with the fall of the Company and the decline of particular plantations. Rental arrangements figured prominently in Virginia's land records between 1626 and 1634. In Maryland in 1642, more than half of the colony's freemen were tenants. Although these arrangements did not evolve into the permanent lifelong positions that some landlords initially envisaged, tenancy would continue as a vital intermediate stage in a man's progression from servant to freeholder. Tenants benefited from the protection their landlords' fortified dwellings offered, from supplemental supplies of corn the landlord might sell them, from rent or sale of domestic livestock, and from credit extended to buy essential imports. Most of these were advantages that English investors had promised their settlers but that they had been unable or un-

119. Lefroy, comp., *Memorials,* I, 432–434 (quotation). On common share tenancy in Bermuda, see ibid., 165–166, 181–182, 203, 397–399, 400–401. W. F. Craven noted that, although the small-scale, diversified farming originally envisaged by Bermuda adventurers called for responsible tenants, the shift to consolidated holdings and one-crop plantation agriculture encouraged the use of unskilled servants and slaves ("Introduction to the History of Bermuda," *WMQ,* 2d Ser., XVII [1937], 357–358). For Virginia, see *RVC,* III, 666–673, IV, 65–74, 107, 145, 175, 177; Hecht, "The Virginia Colony," chap. 5; *MCGC,* 124, 129, 201. Virginia Company tenants were paying two years' worth of work for a twelve-month supply of food and, with high corn prices and a stint on tobacco imposed by the Company, had nothing left with which to buy clothing and tools. Even in the mid-eighteenth century when land had become scarce in older areas, rents on single family farms usually took half the tenant's tobacco crop but only a small part or none of his corn or livestock. William Bradford commented on the failure of similar schemes introduced by proprietors of particular establishments associated with the Plymouth Colony. There tenants were to pay into a common store one-half of their surplus crops, what they actually ate being exempted. Settlers who tried farming on these terms soon wearied of the attempt (Bradford, *History,* I, 352–353).

willing to supply. At the same time, resident landowners benefited from the cleared land, housing, fencing, and orchards that the tenants produced.[120]

What all Virginia settlers most wanted, however, once they had experienced colonial conditions firsthand, was, not inexperienced, poorly supplied, and poorly motivated tenants, but servants bound for a term of years who could be employed in whatever work the manager decided upon, including, on the larger enterprises, gang labor for clearing and planting. Both the parent company and subsidiary joint-stock investors were reluctant to supply any but officers with the extra hands needed to produce enough tobacco to procure anything beyond a miserable subsistence. This reluctance perhaps came in part from Old World notions about what was appropriate to various ranks and statuses. A Virginia tenant, bound to an obligatory term of service, was viewed as having similar status as the English husbandman, "who followeth in the next place [above landless laborers] though hee thriveth ordinarily well, yett he laboreth much." No one questioned the suitability of Berkeley Hundred commander William Tracy, for example, taking with him not only his wife and children but also six men and four maidservants. This gentry family had likely employed a workforce of similar size at their Gloucestershire home. On the other hand, investor John Smyth was disturbed when five of the Berkeley tenants requested two servants apiece for their private use—this "desire seemed somewhat strange." The only servants any of them obtained were two young "Duty boys" transported by the parent company.[121]

Once English joint-stock associations ceased to finance migration of organized groups to the Chesapeake, private merchants quickly took over, recruiting and transporting the kinds of laborers that the settlers wanted. "Virginia differed from later American boom towns in that success depended not on acquiring the right piece of land but on acquiring men," in the words of Edmund Morgan. With some planters willing and able to pay the costs of transporting and clothing new immigrants and able to grow or to trade for enough provisions to feed extra hands, the need for intermediate groups in

120. Hecht, "The Virginia Colony," chap. 8; Menard, "Economy and Society in Early Colonial Maryland," chap. 2; Stone, "Society, Housing, and Architecture in Early Maryland," chap. 2. John Zouch apparently expected former servants on his plantation to become tenants for life. The Berkeley proprietors granted leases to ex-servants for a term of three lives (*RVC,* III, 77, 210–211).

121. *RVC,* I, 391, III, 313, 367–368, 397–400, 443–444, 460–461; Governor Wyatt and Council of Virginia to the Privy Council, May 17, 1626, in "Sir Thomas Wyatt, Governor," *WMQ,* 2d Ser., VIII (1928), 163–167; Morgan, *American Slavery, American Freedom,* chaps. 5, 6; Campbell, *The English Yeoman,* 31; *MCGC,* 36.

England to raise money to finance migration quickly disappeared. Although servants arriving in Virginia after 1624 did not face threats of imminent starvation, more than a few would encounter exploitation, material deprivation, and physical brutality during their terms of service, and they had fewer avenues for relief than did servants in England. "The Conductors of them are drivers of the Trade, vallueing whome they carry at no other rate then theire passage-mony," observed one royal official in 1638.[122]

The practice of tying acquisition of land to importation of people to work it originated from grants of fifty acres of land awarded to any settler who financed his or her own passage or who paid for the transportation of another immigrant. (These grants were termed "headrights" in Virginia and "land rights" in Maryland.) The headright system ensured that henceforth the taking up of land would be related to investments in settlers who could develop the new holdings. Devised by the Virginia Company in 1618 as a means for encouraging investors to develop the land awarded them as a dividend, the practice was continued by the Virginia Council in the 1620s and ultimately approved by the Lords Commissioners of Foreign Plantations in 1634. It became the principal method of patenting land in early Virginia. Until later in the century, when headrights diminished as servant immigration declined and Virginia provincial elites devised other means for amassing vast tracts of undeveloped land beyond the fringes of settlement for speculative purposes, land acquired through headrights was usually developed into working farms within a reasonable time. The land-granting system thus promoted rather than hindered settlement and development. The promise of fifty acres of land to every immigrant able to pay his or her own passage attracted new waves of free migrants, settlers who were in a position to contribute quickly to the carving of new farms out of the wilderness. The Lords Baltimore adopted similar measures for rewarding those who financed their own passage or that of others, and between 1649 and 1684 also granted a fifty-acre land right to servants who completed their terms as well as to the masters or ship captains who imported them. In addition to encouraging immigration, however, the headright system also served to strengthen the power and control of mer-

122. Edmund S. Morgan, "The First American Boom: Virginia 1618 to 1630," *WMQ,* 3d Ser., XXVIII (1971), 183; Breen, "George Donne's 'Virginia Reviewed,'" ibid., XXX (1973), 457 (quotations); R. C. Nash, "The Organization of Trade and Finance in the British Atlantic Economy, 1600–1830," in Peter A. Coclanis, ed., *The Atlantic Economy during the Seventeenth and Eighteenth Centuries: Organization, Operation, Practice, and Personnel* (Columbia, S.C., 2005), 96–97. For transportation of servants, see Galenson, *White Servitude in America,* chap. 1; Craven, *Southern Colonies,* 219–220; Pares, *Merchants and Planters,* 7.

chants and of elite planters who acquired vast tracts from transporting servants and, beginning in 1635, from transporting enslaved Africans as well.[123]

Before the late 1630s or early 1640s, there is no evidence that Chesapeake settlers envisioned African slaves as a possible solution to the scarcity of laborers that was hindering economic development. This is surprising, since Englishmen who attempted to seat colonies in more southerly areas almost universally incorporated enslaved Africans into their workforces. From the 1550s, when English mariners began to venture into the wider Atlantic, they found black men, women, and children being sold in West Africa as commodities by other Africans, and most blacks they encountered in the Caribbean were slaves. Sailors and merchants, April Hatfield argues, "readily accepted enslaved Africans' status in the Atlantic world as trade goods and the degradation inflicted upon enslaved Africans by African and Portuguese enslavers and by Spanish and Portuguese masters." Entering the area "after Iberian-African-American contacts had already shaped many of its contours," English mariners viewed enslaved blacks bound for the Americas or laboring in the Caribbean as exploitable commodities the English might seize and turn into "very good marchandise," in Richard Hakluyt's words. Capture of slave ships became more attractive to northern European privateers after the creation of fledgling English and Dutch colonies, since they could sell slaves there with much less risk than attended selling them as contraband in Spanish colonies. As early as 1586, Sir Francis Drake attempted to disembark up to 250 Africans, whom he picked up in raids on Spanish settlements, at the first faltering Roanoke colony, apparently intending to reinforce the fragile outpost with agricultural workers. Short-lived English colonies located at the mouth of the Amazon and in Trinidad in the 1600s, 1610s, and 1620s quickly acquired African slaves. Bermudians began importing Africans in 1616, and Saint Christopher, the first successful English colony in the Caribbean, seated in 1624, had a workforce of about 60 captured slaves by 1626. In the next few years, planters in new English colonies on Barbados, Nevis, and Providence Island almost immediately began buying African slaves taken by privateers.[124]

123. Wesley Frank Craven, *White, Red, and Black: The Seventeenth-Century Virginian* (Charlottesville, Va., 1971), 9–10, 82–85; Parent, *Foul Means,* chap. 1; Brenner, *Merchants and Revolution,* 140–148; Craven, *Southern Colonies,* 127–128, 175–176; *RVC,* III, 100–101, 107. According to Warren Billings, freedom dues in Virginia were not specifically defined in statutes until 1705; consequently, in the seventeenth century, whether or not a freed servant was entitled to a tract of land usually depended on the inclusion of such a provision in his or her indenture ("Law of Servants and Slaves," *VMHB,* XCIX [1991], 51–52).

124. Linda M. Heywood and John K. Thornton, *Central Africans, Atlantic Creoles, and the*

The first step toward the development of a slave society in the colonial Chesapeake came in August 1619, when an English privateer operating under a letter of marque from Flushing, Holland, the *White Lion,* carried in "20. And odd Negroes" whom Captain John Colyn Jope, in concert with another English privateer, the *Treasurer,* under Captain Daniel Elfrith, had seized from the *São João Bautista,* a Portuguese slaver en route from Loango in Angola to Vera Cruz in Spanish America. These captives Jope bartered to Virginia officials in exchange for provisions. Elfrith arrived four days later carrying another 30 or so of the captured Angolans. He was unable to sell them, however, since his letter of marque from the duke of Savoy was no longer valid, and he had to beat a hasty retreat in order to avoid arrest as a pirate. This chance arrival might have afforded English colonists who came from rural areas in Britain their first introduction to Africans, although some had surely already encountered blacks—some enslaved and some free—in one or more English towns. Perhaps the mariners relayed more information about the race-based slavery prevailing in the Iberian colonies as well as news of how useful Bermuda settlers were finding Africans in that island colony. More educated settlers, however, would have already read about African slavery in the Spanish colonies in travel literature, and some might have gotten firsthand accounts of how Africans were being treated as commodities from Virginia Company leaders with long experience in the West Indies like Captain John Martin of Brandon (who had been involved in Sir Francis Drake's 1585–1586 raiding expedition), Captain Christopher Newport (who had captured 300 Africans from a Portuguese ship in the West Indies in 1591 and sold them in Puerto Rico), and Captain William Parker (who captured

---

Foundation of the Americas, 1585–1660 (Cambridge, 2007), chap. 1; April Lee Hatfield, "A 'Very Wary People in Their Bargaining' or 'Very Good Marchandise': English Traders' Views of Free and Enslaved Africans, 1550–1650," *Slavery and Abolition,* XXV (2004), 1–17; "The First Voyage of the Right Worshipfull and Valiant Knight Sir John Hawkins . . ." (1562), in Richard Hakluyt, *The Principal Navigations, Voyages, Traffiques and Discoveries of the English Nation . . . ,* 12 vols. (New York, 1969), X, 7; Morgan, "Virginia's Other Prototype," in Mancall, ed., *Atlantic World and Virginia,* 371–378; Susan Dwyer Amussen, *Caribbean Exchanges: Slavery and the Transformation of English Society, 1640–1700* (Chapel Hill, N.C., 2007), 68–69, 91–94. Drake might have promised at least some of the blacks he captured or who joined his fleet in 1585–1586 their freedom. Many might have perished in the hurricane that struck Drake's ships at Roanoke, and others might have been left in Carolina when the English settlers decided to abandon the colony (David Beers Quinn, ed., *The Roanoke Voyages, 1584–1590: Documents to Illustrate the English Voyages to North America under the Patent Granted to Walter Raleigh in 1584,* 2 vols., Hakluyt Society, 2d Ser., CIV [London, 1955], I, 251–255; Quinn, "Turks, Moors, Blacks, and Others in Drake's West Indian Voyage," *Terrae Incognitae,* XIV [1982], 93–100; Karen Ordahl Kupperman, *Roanoke: The Abandoned Colony* [New York, 1984], 88, 92).

370 slaves off a ship from Angola bound for Cartagena in 1601 and sold them in Panama).[125]

Another dozen or so Africans were brought into Virginia in 1619 or early 1620, since by March 1620 there were a total of thirty-two unnamed "negroes"—seventeen women and fifteen men—listed in that year's muster as "not Christians in the Service of the English." In the next extant census of February 1623/4, twenty-two blacks survived, and another was listed as having died since the previous April. A year later, twenty-three were enumerated in the 1624/5 census. Most of those present in 1624 and 1625 were survivors from the *São João Bautista,* either members of the group disembarked in 1619 or carried back into Virginia later. Captain Elfrith took his share of the Angolans to Bermuda where some were sold and others were put to work on behalf of the Somers Island Company. The woman Angelo (or Angela), named in the 1624/5 muster, and a half dozen or so others returned to Virginia in the *Treasurer* in February 1620. John Pedro, Antonio, and Mary, noted as having arrived in English ships between 1619 and 1623, all of whom became free, were also members of the 1619 group transported from Bermuda to England and subsequently back to Virginia by the Rich family, who had extensive holdings in Virginia, Bermuda, and the recently chartered Council of New England and were patrons of Elfrith. Also, by early 1625, two of the African women had by then borne a child, beginning a new generation of African Virginians whose status as free or enslaved had yet to be resolved. Not surprisingly, these exotic but desirable workers had been snapped up by eight members of the colony's labor-hungry elite—sometime governor Sir George Yeardley and Cape Merchant Abraham Peirsey (who together purchased the twenty-odd from Captain Jope in 1619), militia captains Samuel Matthews, William Tucker, William Pierce, and Francis West, prominent planter Richard Kingsmill, and planter-merchant Edward Bennett.[126]

125. Engel Sluiter, "New Light on the '20. and Odd Negroes' Arriving in Virginia, August 1619," *WMQ,* 3d Ser., LVI (1997), 395–398; John Thornton, "The African Experience of the '20. and Odd Negroes' Arriving in Virginia in 1619," ibid., LX (1998), 421–434; Bernhard, "Beyond the Chesapeake," *Jour. So. Hist.,* LIV (1988), 545–564; Tim Hashaw, *The Birth of Black America: The First African Americans and the Pursuit of Freedom at Jamestown* (New York, 2007), 70–94; Heywood and Thornton, *Central Africans,* 5–9, 17–19; Morgan, "Virginia's Other Prototype," in Mancall, ed., *Atlantic World and Virginia,* 356; Southall, "Captain John Martin," *VMHB,* LIV (1946), 21–67; Samuel M. Bemiss, "John Martin: Ancient Adventurer," *VMHB,* LXV (1957), 209–221. For Virginians' knowledge of the slave systems of other colonies, see April Lee Hatfield, *Atlantic Virginia: Intercolonial Relations in the Seventeenth Century* (Philadelphia, 2004), 137–142. Because Captain Jope, who was an English subject, had a letter of marque from Dutch authorities, Virginia colonists described the ship as Dutch or Flemish rather than English.

126. The recently discovered 1620 census is in the Ferrar Papers, nos. 138, 139, 159, 178 (or,

As long as blacks were acquired as booty from raids on Iberian ships or settlements, rather than through an established trade in slaves between African and English merchants in West Africa, the prices freebooters asked for their captives remained low. So the elite Virginians who bought the blacks brought to the labor-short colony in 1619 would not have been overly concerned about the absence of any law upholding their property rights in non-Christian laborers who might, simply by following pan-Atlantic custom, be held to extraordinary terms of service. Tellingly, by 1620 "negroes" were considered a special category of labor explicitly set apart from "Christian" servants. These categories were borrowed from Iberian, not English, precedents and replicated the opposition of "Christian" and "black" *(negro)*

according to an earlier numbering system, nos. 1597 A, B, C, D), CWL. The census was first described in William Thorndale, "The Virginia Census of 1619," *Magazine of Virginia Genealogy,* XXXIII (1995), 155–170. Thorndale ascribed the document (ambiguously dated so that it could have been generated in either March 1619 or March 1620) to March 1619, New Style. Martha McCartney has since demonstrated, from a close analysis of all four of the documents in the Ferrar Papers relating to the census, some of which mention events that occurred after March 1618/19, that the date of the census is March 1620, New Style (McCartney, "An Early Virginia Census," *Arch. Soc. Va. Qtly. Bull.,* LIV [1999], 178–181). The 1624/5 census, which provides information on who held the blacks and on the ships on which four of them arrived, is transcribed in full in Jester and Hiden, eds., *Adventurers of Purse and Person,* 5–69, and in part in John Camden Hotten, ed., *The Original Lists of Persons of Quality . . . and Others Who Went from Great Britain to the American Plantations, 1600–1700* (1874; rpt. Baltimore, 1962), 201–265. The 1623/4 census is reproduced ibid., 169–196. The death of one "negar" at West and Shirley Hundred in 1623 is mentioned ibid., 190. The 1625 census is analyzed in Irene W. D. Hecht, "The Virginia Muster of 1624/5 as a Source for Demographic History," *WMQ,* 3d Ser., XXX (1973), 65–92. For Angelo, see Heywood and Thornton, *Central Africans,* 7–9. John Pedro's subsequent freedom is well documented (Heywood and Thornton, *Central Africans,* 283–284; Hashaw, *Birth of Black America,* 127–151). Scholars disagree about which of the several Anthonys or Antonios listed on the 1625 census became the freedman Anthony Johnson as well as about the identity of Anthony's wife Mary. That the Johnsons (who stated in 1653 that they had been inhabitants of Virginia for more than thirty years) were the Antonio listed as arriving on the *James* in 1621 and Mary, the woman who came from England a year later and who lived on the same plantation as Antonio, seems most plausible. For the Johnsons, see Heywood and Thornton, *Central Africans,* 282; Hashaw, *Birth of Black America,* 212–217; Joseph Douglas Deal III, "Race and Class in Colonial Virginia: Indians, Englishmen, and Africans on the Eastern Shore during the Seventeenth Century" (Ph.D. diss., University of Rochester, 1981), 196–201; and T. H. Breen and Stephen Innes, *"Myne Owne Ground": Race and Freedom on Virginia's Eastern Shore, 1640–1676* (New York, 1980), 8–10. See also Michael Joseph Guasco, "'Encounters, Identities, and Human Bondage': The Foundations of Racial Slavery in the English Anglo-Atlantic World" (Ph.D. diss., College of William and Mary, 2000), chap. 8; Alden T. Vaughan, "Blacks in Virginia: Evidence from the First Decade," in Vaughan, *Roots of American Racism: Essays on the Colonial Experience* (New York, 1995), 128–135; and Robert McColley, "Slavery in Virginia, 1619–1660: A Reexamination," in Robert H. Abzug and Stephen E. Maizlish, eds., *New Perspectives on Race and Slavery in America: Essays in Honor of Kenneth M. Stampp* (Lexington, Ky., 1986), 11–24.

that underpinned racial slavery in Spain, Portugal, and their colonies. In the early sixteenth century, the Portuguese had begun using *negro* not only as a generic term for Africans, as did all Europeans, but also as a synonym for *slave.* The English did not necessarily view slavery in exactly the same way as the Spanish and Portuguese, and before the 1650s Africans living in English colonies were almost always referred to as "negroes" rather than explicitly termed slaves. But, given the well-established commodification of captured Africans throughout the Atlantic world, the English regarded any Africans taken from Portuguese slave ships or, in later years, purchased in Africa, as slaves. English colonists, however, as Linda Heywood and John Thornton demonstrate, had not yet defined slavery as lifelong, inheritable servitude but rather as indefinite service. If Virginia colonists did not purchase the *São João Bautista* survivors with the intention of holding them for life, they might have been held, as became practice in Bermuda by the 1630s, as servants for a term of ninety-nine years, slaves for life in all but name.[127]

That a few Africans did come to the Chesapeake colonies as indentured servants has for too long confused the issue of the fate of the great majority. The only direct evidence about the status of the earliest blacks is that, unlike almost all of the European servants, most were either not recorded by any

127. On low prices of slaves, see Kupperman, *Providence Island,* 170–173, 176, who found that in the 1630s the Providence Island settlers could buy slaves at much lower prices than they could English servants. Bernhard, "Beyond the Chesapeake," *Jour. So. Hist.,* LIV (1988), 545–564, found that Bermuda settlers initially acquired slaves captured on the high seas from ship captains for little or nothing but that subsequently Africans were exchanged between planters on the island at much higher prices. For ninety-nine-year terms of service, see ibid., 556–557; and A. C. Hollis Hallett, ed. and comp., *Bermuda under the Sommer Islands Company, 1612–1684: Civil Records,* 3 vols. (Bermuda, 2005), III, 260–291. For discussion of the use and meaning of *negro, servant,* and *slave,* see Heywood and Thornton, *Central Africans,* chap. 6; McColley, "Slavery in Virginia," in Abzug and Maizlish, eds., *New Perspectives,* 12–14; and Alden T. Vaughan, "The Origins Debate: Slavery and Racism in Seventeenth Century Virginia," in Vaughan, *Roots of American Racism,* 158–159, 172. For use of similar terms in the Caribbean, see Richard S. Dunn, *Sugar and Slaves: The Rise of the Planter Class in the English West Indies, 1624–1713* (Chapel Hill, N.C., 1972), 224–229; and Amussen, *Caribbean Exchanges,* 91–94. For the development of racial categories in the early British Empire, see also Carl H. Nightingale, "Before Race Mattered: Geographies of the Color Line in Early Colonial Madras and New York," *AHR,* CXIII (2008), 48–71. For the absence of law, see Billings, *A Little Parliament,* 206–207; Jonathan A. Bush, "Free to Enslave: The Foundations of Colonial American Slave Law," *Yale Journal of Law and the Humanities,* V (1993), 417–470; Bush, "The British Constitution and the Creation of American Slavery," in Paul Finkelman, ed., *Slavery and the Law* (Madison, Wis., 1997), 379–418; Thomas D. Morris, *Southern Slavery and the Law, 1619–1860* (Chapel Hill, N.C., 1996), 39–43, who argues that blacks were viewed as slaves and hence as property, so English colonials "would apply English notions and rules of property law to slaves" (42); and Guasco, "'Encounters, Identities, and Human Bondage,'" chap. 8, who argues for "the presence of a functional and highly adaptive Iberian model" of slavery (466).

name in the 1624 and 1625 colonywide censuses or were accorded only a given name. In addition, none had their ages recorded, and all but four lacked any information about their date of arrival—both age and arrival date being crucial information for establishing the years of service owed by servants who were either indentured for a specified time or to be freed once they reached a specified age. The few blacks who claimed indentured status in Virginia or Maryland courts, along with others who appear to have gained their freedom after a term of service, came into the Chesapeake from England, elsewhere in Europe, or from other colonies where they were able to prove that they had been free. A handful arrived with written contracts, and others added baptism or conversion to Christianity to bolster their claims of difference from the majority of forced African migrants. These blacks were exceptions whose challenges to enslaved status demonstrate that the rule, spelled out by the Barbados Assembly in 1636, that "negroes and Indians that came here to be sold, should serve for life unless a contract was before made to the contrary" was also the practice in Virginia.[128]

Indirect evidence about the fate of the *São João Bautista* survivors suggests that many remained enslaved for life, as probably did their descendants. One of the original purchasers, Governor George Yeardley, directed in his will of October 1627, eight years after he acquired the Africans, that all his goods, including servants and "negars," be sold. His widow remarried but died before settling the estate. Subsequently, Yeardley's brother Ralph, who lived in London, retained control of the property in his capacity as guardian to the children, and still in 1630 that property included "negroes." In late 1637 or early 1638, the couple's eldest son, Argoll, returned from England (where he and two younger siblings traveled to join their uncle and guardian after their mother's death) at the age of eighteen to take charge of his inheritance.

128. I find no evidence to support the still widely held contention, first advanced by James Curtis Ballagh, *A History of Slavery in Virginia* (Baltimore, 1902), and reiterated by Oscar Handlin and Mary F. Handlin in 1950, that these and other early blacks were held as indentured servants rather than as slaves ("Origins of the Southern Labor System," *WMQ*, 3d Ser., VII [1950], 199–222). Vaughan, "Blacks in Virginia," in Vaughan, *Roots of American Racism*, 128–135, provides a more updated and accurate summary of the sparse available evidence. The racial categories employed in the recently discovered 1620 census add further proof to Vaughan's contention that lifetime servitude was the condition of almost all blacks from the outset. Deal, whose study is widely cited for the relatively favorable legal position some Eastern Shore blacks enjoyed in early years, states that he had found no evidence for indentures and that from the outset Africans on the Eastern Shore were normally held as slaves for life ("Race and Class in Colonial Virginia," 196–201). The most comprehensive discussion is Heywood and Thornton, *Central Africans*, chap. 6 (quotation on 1636 law from [William Duke], *Some Memoirs of the First Settlement of the Island of Barbados and Other Caribbee Islands* . . . [London, 1743], 20, from now lost records).

Argoll claimed land for having transported his wife Frances, six servants, and two "negroes," Andolo and Maria. These blacks were likely two of the Angolans who arrived in Virginia in 1619 and who later accompanied the Yeardley children to England, since it seems unlikely that Argoll would have acquired Africans other than through inheritance before his coming of legal age. Cape merchant Abraham Peirsey (who bought Flowerdew Hundred and possibly some Africans living on it from George Yeardley in 1624) left similar directions for sale of all his goods in his will of March 1627. (In it he referred to his bound workers simply as "servaunts.") Peirsey's widow Frances married again to Captain Samuel Matthews, who, the husband of Peirsey's daughter Mary charged in 1633, embezzled Peirsey's estate after Frances's death. In 1639, Matthews still had failed to account for the estate and was known to have retained some of it "in kynd." Some of the forty "Negro servants" reportedly working on Matthews's highly diversified plantation in 1648, whom he "brings . . . up to Trades in his house," were likely children of the *São João Bautista* captives. The language implies that at least some of Matthews's "servants" were youths born in Virginia of parents who arrived no later than the early 1630s. It is unlikely that Matthews, who made his fortune by marrying Frances Peirsey, could have acquired what was for the time an extraordinarily large number of blacks other than through natural increase of Africans enslaved for life.[129]

129. For George Yeardley, see "Flowerdew Hundred and Sir George Yardley," *Tyler's Quarterly Historical and Genealogical Magazine,* II (1920–1921), 115–129; McCartney, *Virginia Immigrants and Adventurers,* 771–774; and Hashaw, *Birth of Black America,* 103–106. His will is transcribed in Nora Miller Turman, *George Yeardley: Governor of Virginia and Organizer of the General Assembly in 1619* (Richmond, Va., 1959), 183–185. For Argoll Yeardley, see Nell Marion Nugent, *Cavaliers and Pioneers: Abstracts of Virginia Land Patents and Grants, 1623–1800,* 7 vols. (Richmond, Va., 1934–1999), I, 81; McCartney, *Virginia Immigrants and Adventurers,* 548–549; and Nora Miller Turman and Mark C. Lewis, eds., "Inventory of the Estate of Argoll Yeardley of Northampton County, Virginia, in 1655," *VMHB,* LXX (1962), 410–419. The seven unnamed "Negro servants," spanning two generations, listed in Argoll's estate inventory cannot be definitely linked to the Angolans purchased by his father in 1619. For Abraham Peirsey, see "Letter from the Governor and Council of Virginia concerning Captain Mathews," *VMHB,* XI (1903–1904), 174–182 (quotation on 180). A transcription of Abraham Peirsey's will is in Edward D. Neill, *Virginia Carolorum: The Colony under the Rule of Charles the First and Second, A.D. 1625–A.D. 1685, Based upon Manuscripts and Documents of the Period* (Albany, N.Y., 1886), 404. For Matthews, see "Letter from the Governor," *VMHB,* XI (1903–1904), 174–182; "Governor and Council to the Privy Council," ibid., 285–287; and McCartney, *Virginia Immigrants and Adventurers,* 483–485. For Matthews's 1648 workforce, see *A Perfect Description of Virginia . . .* (1649), *Force Tracts,* II, no. 8, 14–15, which is based on letters John Ferrar received in England in March 1648/9, so presumably written in Virginia in 1648. Some of the conclusions advanced in Hashaw, *Birth of Black America,* go beyond what the sparse evidence will support. Nonetheless, Hashaw's contention that the

In the 1620s, the handful of men who owned blacks in Virginia as yet had no need to enact laws "to restrayne the insolencies of the Negroes" from carrying weapons and stealing food, as did officials in the Bermuda colony who had been acquiring blacks from the West Indies since 1616. The harshness and brutality that Virginia masters were able, more or less with impunity, to exercise over European servants afforded them more than sufficient license with which to control the twenty to thirty blacks living in the colony in the early to mid-1620s. And, in those precarious years when early death remained the fate of most colonists, bound or free, slaves or servants for life might afford their buyers only a limited advantage over other workers clearly bound for shorter terms. The magnates who bought them, however, were apparently already concerned about establishing a right to the indefinite labor of any Africans who might live longer than the four to seven years of service required of most European servants.[130]

That African laborers "in the service of sev[er]all planters" were already in 1620 differentiated from English servants in terms both of their ethnicity and of their presumed non-Christian status was ominous. Repeated references to "negroes" instead of "slaves" collapsed status with color and "served to foreground the workers' origins, rather than their legal status." By 1628, the number of Africans in Virginia might have reached more than one hundred, as a Massachusetts Bay ship, the *Fortune,* reportedly took "an Angola man with many negroes, which the captain bartered in Virginia for tobacco." Unfortunately, another plantation-by-plantation census "of all the inhabitants men woemen and Children as well Englishe as Negroes" that the Virginia Council called for in early 1629 either was not taken or has not survived. Whatever their numbers, these disadvantaged settlers who had not freely chosen to move to Virginia in the first place would have limited, if any, chances for profiting personally from their own hard labor. Major markets for Africans quickly shifted away from the North American mainland, preventing an early conversion to slave labor in the Chesapeake. Attitudes and mechanisms favoring such a conversion, however, were already evident during the Company era.[131]

---

history of Virginia's first African Americans can be expanded and enhanced by tracing the chain of their transmission (or appropriation) as human property is perceptive and promising.

130. Bernhard, "Beyond the Chesapeake," *Jour. So. Hist.,* LIV (1988), 545–564; Amussen, *Caribbean Exchanges,* 91, 134.

131. Elizabeth Donnan, ed., *Documents Illustrative of the History of the Slave Trade to America,* Carnegie Institution of Washington Publication no. 409, 4 vols. (Washington, D.C., 1930–1935), IV, 49, 118; *MCGC,* 196.

Firmly establishing viable settlements in the Chesapeake proved extraordinarily difficult and was a long, drawn-out process during which the enterprises might have foundered for a multitude of reasons. Developments that can be seen in retrospect as crucial to the colonies' eventual success were not necessarily perceived as marked turning points by the colonists themselves. Moreover, even the most positive changes all had distinctly negative aspects. Finding a viable export was essential for eliciting and maintaining adequate levels of investment for getting people and supplies to the region and for enabling the settlers to acquire the means for survival. By 1617, tobacco had already provided a solution to the lack of export products, but adoption of this lucrative crop did not bring about the levels of political and social stability requisite for sustaining a viable colony. Moreover, near total reliance on a single luxury product subject to declining returns once production expanded and also subject to cycles of boom and bust meant that Chesapeake planters would remain vulnerable to periodic bouts of hard times well into the eighteenth century. Nonetheless, tobacco did prove the colony's salvation since it remained sufficiently profitable to induce a host of private merchants to step in to provide supplies and capital after gentry promoters pulled out. The colony's poor reputation in England, however, continued to deter emigration and to divert investment to more easily settled islands in the Caribbean.

The shift from a corporate form of organization to one based on private property occurring between 1616 and 1625 was critical for providing adequate incentives for settlers to commit to remaining in Virginia, to working hard, and to improving their estates. But unregulated private enterprise also created conditions in which profiteering and exploitation of bound laborers ran rampant. A nascent gentry class did emerge in the aftermath of the Indian attack that proved capable of defeating the native Americans and of promoting the interests of more privileged settlers. Most of these leaders lacked the social attributes traditionally expected of a ruling class, though, and began building a regime rooted in greed, favoritism, and arbitrary rule. Revocation of the Virginia Company's charter replaced an ineffective and stifling monopoly with free trade, but schemes hatched in England for reinstituting some other form of trading monopoly continued to threaten settlers' interests into the 1630s.

After the 1622 attack, defeat of the Powhatan empire became essential to Virginia's survival. A policy of harrying and pushing back rather than annihilating the Indians saved the English colonists precious time and resources, but it meant that those living on the fringes of settlement remained vulnerable to further attacks. Abandoning towns as the focus of settlement was an important step toward adopting systems of landholding and farming suitable to the region and encouraged settlers to move from some of the most unhealthy areas, but it also promoted dispersed settlement patterns that led to poor communication and weak social organization. Giving up attempts to replicate English agriculture freed time and resources farmers needed to devote to the demanding staple but also meant that settlers had to accept radical changes to their accustomed diets and to adapt to landscapes that appeared uncivilized to European eyes. The representative assembly established in 1619 was an important step in giving settlers some control over their internal governance, but its very continuance remained in doubt up to 1639. The creation in 1634 of counties and of governing institutions capable of supplying essential services on a neighborhood level and of maintaining local order was another important step, yet it took time for inexperienced magistrates, many with limited social standing, to learn how to carry out their administrative and judicial duties effectively. Even a well-managed and profitable venture based on peaceful rather than hostile relations with the Indians, like the Kent Island settlement, failed once entangled in jurisdictional disputes. Advances in settlement, agriculture, and social organization were thus slow and incremental, and the ultimate outcome far from assured until the early 1630s.

In contrast to the middling and most especially the elite Virginia planters who began to realize substantial profits by the late 1620s, absentee English investors reaped little benefit from the advances in land selection and management, tobacco culture, livestock care, colonial housewifery, or labor management and recruitment discussed above. By and large, neither did most of the men and women they sent to develop their particular holdings; a little knowledge had been gained at the cost of many lives. The English settlers who survived, however, and others who followed made rapid progress in the immediate post-Company years. The promise of wealth that had briefly sustained the subsidiary joint-stock associations now motivated thousands more English men and women to try their fortunes in the Chesapeake. A number of them soon afterward began to realize that promise.

CHAPTER 2

The Age of the Small Planter

1640–1679

As England descended into civil war in 1642, new groups of Puritan gentry and merchants rose to power who had quite different conceptions of how trade should be regulated and of the role of overseas colonies. The Commonwealth government wanted to mold the disparate colonies it had seized upon the execution of Charles I into a centrally governed, commercially integrated empire. Merchants in Parliament and on the Council of State intended to funnel all colonial trade through English merchants' hands. Opposing the monopolistic approach to trade of the early Stuart kings, the Commonwealth granted equal opportunity to all English traders, but it also sought to deny European rivals access to colonial Atlantic trade. The first step was the Navigation Act of 1651, which provided that all goods traded between England and its colonies and between foreign countries and English colonies had to be carried in English or colonial ships of which the master and a majority of the seamen were English. The law was aimed primarily at the Dutch, who had been busy establishing New World colonies and building an extensive international carrying trade, including one with the English colonies. Two years of inconclusive naval war with the Dutch followed, and scarcely had that conflict ended when Lord Protector Oliver Cromwell, who came to power in 1653, launched an ambitious military offensive against the Spanish West Indies. That effort resulted in the capture of Jamaica but otherwise had scant success. In reality, during the chaotic Interregnum years the Commonwealth government lacked the capacity to enforce such unprecedented interventions in colonial affairs. Shipping between England and the colonies was severely disrupted, and settlers traded with shippers of all nations, especially the Dutch, a practice planters in the Chesapeake and West Indies found greatly to their advantage and that they were loath to give up.[1]

1. Carla Gardina Pestana, *The English Atlantic in an Age of Revolution, 1640–1661* (Cambridge, Mass., 2004), chaps. 1–5; Wesley Frank Craven, *The Southern Colonies in the Seventeenth Century,*

When Charles II was restored to power in 1660, Restoration authorities accepted many of the changes in trade and colonial policies instituted by the Commonwealth and Protectorate governments. The policy of requiring that all overseas trade be funneled through England continued, and a new Navigation Act limiting shipping to English vessels was enacted in the first year of Charles II's reign. Among its other provisions, that act stipulated that all tobacco, along with other enumerated commodities such as sugar, cotton, and dyewood, be transported only to England, Ireland, or some other English province. Following mercantilist policy, valuable raw materials were to be kept within the Empire. Moreover, the buildup of English naval power begun under the Protectorate (216 ships were added to the navy between 1649 and 1660) allowed the national government to assume a more direct role in regulating trade and in protecting civilian shipping. So also did reforms of the customs service, as a significantly weakened crown needed above all to maximize the customs revenues that the Restoration Parliament had confirmed to the king.[2]

Although Charles I had repeatedly castigated the Chesapeake and West Indian colonies for overdependence on tobacco, for having economies "wholly built upon smoke," royal enthusiasm for agricultural diversification was much less pronounced in the Restoration monarchy. Charles II lent Virginia governor William Berkeley only limited support for a diversification policy reminiscent of the one promoted by the defunct Virginia Company that Berkeley went to England in 1661 to promote. Charles committed no resources to carry out the program, merely empowering Berkeley and the Virginia General Assembly to impose taxes on the colonists to implement it. The king allowed that some restraint might be put on the planting of tobacco if the government of Maryland agreed (which it did not). Charles II's main aim was to stop Virginians from trading with the Dutch, however, not to encourage them to diminish royal customs revenues by ceasing to grow tobacco.[3]

1607–1689 (Baton Rouge, La., 1970), chaps. 7–8; Robert Brenner, *Merchants and Revolution: Commercial Change, Political Conflict, and London's Overseas Traders, 1550–1653* (Princeton, N.J., 1993), chaps. 7–13; J. E. Farnell, "The Navigation Act of 1651, the First Dutch War, and the London Merchant Community," *Economic History Review*, 2d Ser., XVI (1964), 439–454; Victor Enthoven and Wim Klooster, "'The Dutch Found and Relieved Us': The Shifting Dynamics of Virginia-Dutch Trade in the Long Seventeenth Century," in Douglas Bradburn and John C. Coombs, eds., *Early Modern Virginia: New Essays on the Old Dominion* (Charlottesville, Va., forthcoming).

2. Michael J. Braddick, "The English Government, War, Trade, and Settlement, 1625–1688," in Nicholas Canny, ed., *The Origins of Empire: British Overseas Enterprise to the Close of the Seventeenth Century*, vol. I of Wm. Roger Louis, ed., *The Oxford History of the British Empire* (Oxford, 1998), 286–308.

3. Warren M. Billings, *Sir William Berkeley and the Forging of Colonial Virginia* (Baton Rouge,

By the 1660s and 1670s, issues of trade and empire were becoming equally or more important than issues of religion in struggles for power in Europe. Charles II's marriage to Catherine de Braganza of Portugal in 1661 brought these two colonial powers into alliance and gave England possession of Tangier (critical to control of the Straits of Gibraltar) and of an outpost at Bombay, and hence a toehold in Asia. Charles's chartering of the Royal African Company in 1660 heightened Anglo-Dutch rivalry over the gold and slave trades in West Africa as well as over islands in the Caribbean. Fearing that the Spanish, French, and Dutch might combine to curb its overseas ambitions, England launched two more wars against its Protestant Dutch rivals between 1664 and 1667 and again from 1672 to 1674. Catholic Spain supported the Dutch to prevent its former subjects there from being overwhelmed by the English, who now posed the most significant threat to Spain's New World colonies, to counter French designs on the Spanish Netherlands, and to regain control of Portugal's overseas possessions (which the Dutch, too, coveted). France, also wishing to curb growing English imperial power, supported the Dutch in the second Anglo-Dutch conflict. France joined with England to try to dismember the Dutch republic in the third Anglo-Dutch war, however, hoping to decrease Dutch penetration of the French domestic market and to seize Dutch colonies in Asia. Naval warfare erupted on a global scale with fighting in northern European waters, the Mediterranean, the Caribbean, West Africa, and Asia.[4]

Order and Disorder in the Chesapeake

During these years of international conflict and changing conceptions in England over the role its empire should play, the Chesapeake colonies were themselves plagued by a series of internal political disruptions. The bitter political and religious quarrels that precipitated the English Civil War and raged during the Interregnum period inevitably crossed the Atlantic, contributing to instability at the colony level in both Virginia and Maryland. Maryland's proprietary government was particularly fragile owing to the Calvert family's adherence to Roman Catholicism at a time when Puritans were in power in England. In 1645, a Protestant ship captain, Richard Ingle, subdued the settlers around the capital of Saint Mary's City, demanded that all Mary-

---

La., 2004), chaps. 8–9; Billings, "Sir William Berkeley and the Diversification of the Virginia Economy," *VMHB,* CIV (1996), 433–454; George Louis Beer, *The Origins of the British Colonial System, 1578–1660* (1908; rpt. Gloucester, Mass., 1959), chap. 4 (quotation on 91).

4. Braddick, "English Government, War, Trade, and Settlement," and Jonathan I. Israel, "The Emerging Empire: The Continental Perspective, 1650–1713," both in Canny, ed., *Origins of Empire,* 286–308, 423–444.

landers take an oath supporting the Puritan Parliament, and plundered the estates of those who refused. At the same time, Virginian William Claiborne temporarily reseized Kent Island, which Maryland forces had taken from him in 1638. Not until late 1646 did Ingle leave and proprietary governor Leonard Calvert assemble enough men to regain control of the government. A number of Maryland settlers fled the colony for Virginia during this "time of troubles."[5]

Next the Virginia government was disrupted owing to the settlers' overwhelming support for Charles I. Virginia escaped most of the sectarian disorders of the 1640s in large measure because of Governor William Berkeley's policy of upholding the established Anglican church by expelling troublesome Puritans and Quakers from the colony. Berkeley and the Virginians could not prevail, however, against a Commonwealth government determined to retain control of England's colonies. In 1649, the Virginia Assembly denounced Parliament's execution of the king, Berkeley proclaimed his exiled son Charles II the rightful ruler, and he and the Assembly enacted penalties against anyone who supported the Commonwealth. The next year, a parliamentary commission, backed by a fleet of ships, persuaded Virginia leaders to surrender to the English state. Berkeley was replaced as governor by one of the commissioners, but the lower house of assembly in practice held the upper hand. The parliamentary commissioners then proceeded to demand the surrender of the Maryland government, although the Commonwealth did not consider the colony of Cecil Calvert, second Lord Baltimore, to be in rebellion. Puritan commissioners ousted the proprietary governor in 1654, and a Protestant assembly outlawed the practice of Catholicism and disenfranchised Catholics. When proprietary governor William Stone tried to reimpose his authority the next year, a group of Virginia Puritans, who had at the governor's invitation established a settlement at Providence on the Severn River in 1648, defeated the proprietary forces in a pitched battle. Two years later, the parliamentary commissioners agreed to return the province to Calvert, the Puritans received amnesty, and the policy of religious toleration for Christians was restored. Virginia reverted quietly to royal rule upon Charles II's restoration in 1660, and William Berkeley was reappointed royal

5. Robert J. Brugger, *Maryland: A Middle Temperament, 1634–1980* (Baltimore, 1988), 18–36; J. Frederick Fausz, "Merging and Emerging Worlds: Anglo-Indian Interest Groups and the Development of the Seventeenth Century Chesapeake," in Lois Green Carr, Philip D. Morgan, and Jean B. Russo, eds., *Colonial Chesapeake Society* (Chapel Hill, N.C., 1988), 47–98; James Horn, *Adapting to a New World: English Society in the Seventeenth-Century Chesapeake* (Chapel Hill, N.C., 1994), chap. 8; Charles M. Andrews, *The Colonial Period of American History,* II (New Haven, Conn., 1936), chap. 8; Craven, *Southern Colonies,* chaps. 7–8.

governor. In Maryland, then proprietary governor Josias Fendall, a Protestant, joined the Maryland lower house in repudiating Lord Baltimore's authority. Fendall's revolt ended a few months later when Charles II restored the province to the Calverts.[6]

Finally, in 1676, Virginia was convulsed by Bacon's Rebellion, and Maryland was briefly disturbed by a group of Protestant insurgents who met in Calvert County to protest against what they perceived as corruption, nepotism, arbitrary taxation, and mismanagement in Charles Calvert, third Lord Baltimore's regime in the same year. Historians have since strenuously debated the causes for the conflict in Virginia. Most agree that high taxes (and the uses to which these taxes were put) was a widespread grievance and that the precipitating factor was a series of raids committed by nonresident Indians on Virginia settlers (resident Indians were only rumored to have taken part), followed by counterraids by the settlers upon the Indians. Virginia planters, especially those living on the frontier, then wanted to declare a general war on all Indians, both for security and because they coveted the Indians' land. They revolted under the leadership of Councillor Nathaniel Bacon when Governor Berkeley failed to aggressively pursue the war himself or to permit Bacon and an army of volunteers to do so.

Modern-day historians have emphasized internal political and social factors such as Berkeley's autocratic rule, monopolization of offices by a small circle of the governor's favorites, oppression of ordinary settlers by a grasping ruling clique, the elites' failure to marshal support for an overly ambitious diversification plan, unrest among indentured servants and enslaved Africans, rebellious frontiersmen, and disputed conceptions of masculine honor. Virginia's first historian Robert Beverley (himself a participant in the conflict, so hardly an unbiased observer), however, listed recent changes in English imperial policy, along with the "Disturbance offer'd by the *Indians,*" as "the main Ingredients towards this intestine Commotion." The problems Beverley cited included low tobacco prices that the Virginia government could do nothing to remedy because of the failure of Marylanders to cooperate in stinting schemes, Charles II's granting of all or parts of Virginia to proprietors (threatening land titles and future expansion), Parliament's prohibition of trade with the Dutch, and activities of the English (who had conquered Dutch New Netherland in 1664) that disrupted the intercolonial

6. Alexander B. Haskell, "'The Affections of the People': Ideology and the Politics of State Building in Colonial Virginia, 1607–1754" (Ph.D. diss., Johns Hopkins University, 2005), chap. 4; Brugger, *Maryland,* 18–36; Craven, *Southern Colonies,* chaps. 7–8; Horn, *Adapting to a New World,* chap. 8.

Indian trade and angered the Indians. Although ordinary Virginia settlers listed internal problems of governance as their main grievance, the economic hardships wrought by new imperial policies and by commercial disruptions caused by international warfare in the mid-1660s and again in the mid-1670s surely contributed to widespread popular discontent. Given the crucial importance of Virginia tobacco to English government revenues, the home government reacted quickly by sending a fleet and more than one thousand soldiers to restore order and to conclude peace with the Indians. The colony by and large returned to a prerebellion status quo, although some of Berkeley's archsupporters were replaced with new leaders, including a number who had supported Nathaniel Bacon. Councillors resumed their usual leading role, although they more often shared some power with representatives in the lower house. The royal commissioners and the royal governors who followed them had little success in limiting the power of the Virginia Assembly, and the Virginia governors and legislature were, at the same time, unable to effect changes in imperial policy.[7]

Instability at the provincial level, however, had limited effect on the lives of ordinary settlers who by and large continued to go about their business building farms and growing tobacco. Who controlled the colony was vitally important to men who held or who aspired to hold important provincial offices, as it was to religious sectarians who believed that salvation in the next world might well be dependent on establishing a godly government on earth. Most of the contests for power at the top were conducted with little or no loss of life, nor were there widespread reprisals against political foes, except those wrought by Governor Berkeley after the collapse of Bacon's Rebellion. The "reduction" of Virginia and Maryland to parliamentary rule during the Interregnum, for example, was achieved without military intervention, and the only consequence of Fendall's "rebellion" is that he and one of his prime supporters were proscribed from holding political office in Maryland. New administrations routinely confirmed land titles granted under previous

7. Edmund S. Morgan, *American Slavery, American Freedom: The Ordeal of Colonial Virginia* (New York, 1975), chaps. 12–14; Wilcomb E. Washburn, *The Governor and the Rebel: A History of Bacon's Rebellion in Virginia* (Chapel Hill, N.C., 1957); David S. Lovejoy, *The Glorious Revolution in America* (New York, 1972), chap. 3; John C. Rainbolt, *From Prescription to Persuasion: Manipulation of Eighteenth [Seventeenth] Century Virginia Economy* (Port Washington, N.Y., 1974), chap. 4; Stephen Saunders Webb, *1676: The End of American Independence* (Cambridge, Mass., 1985), book 1; Kathleen M. Brown, *Good Wives, Nasty Wenches, and Anxious Patriarchs: Gender, Race, and Power in Colonial Virginia* (Chapel Hill, N.C., 1996), chap. 5; Brugger, *Maryland,* 36; Billings, *Sir William Berkeley,* chaps. 13–14; Horn, *Adapting to a New World,* chap. 8; Haskell, "'The Affections of the People,'" chap. 5; Robert Beverley, *The History and Present State of Virginia* (1705), ed. Louis B. Wright (Chapel Hill, N.C., 1947), 74–78.

ones, and in most cases the losing factions were granted blanket political amnesty.[8]

More crucially, despite repeated turmoil at the top, county courts and justices of the peace proved markedly effective in maintaining local order. Between 1640 and 1679, the Virginia Assembly and Maryland's proprietor delegated more powers to county officials, a group that dramatically increased as additional counties were created in the wake of expanding settlement. According to Warren Billings, the county courts "emerged by 1662 as the units of government whose wide authority touched the colonists' lives far more immediately than did that of the Assembly or the governor." Once settlers could conduct most of their business and settle most of their disputes in county courts, they had fewer occasions to travel to Jamestown or Saint Mary's City to meet with provincial officials.[9]

Whoever claimed control of the provincial government, the local magistrates conducted business as usual in the county courts. Land transactions continued to be recorded, claims to headrights and to livestock marks registered, and civil cases litigated. Justices listened to grievances of servants against masters and of masters against servants, determined the number of years indentured servants had to serve, and set punishments for servants and slaves who ran away. Wills continued to be probated, decedents' estates administered, and orphans cared for. Justices of the peace took bonds of those who threatened the peace and punished those who refused to obey court orders or to assist in arrests. Grand juries decided who should be prosecuted for crimes and for offences against public morality, constables or sheriffs brought malefactors to the bar, and punishments for noncapital transgressions were meted out. Ferries were established, and roads and bridges were maintained. Ordinary keepers were licensed, and the prices they could charge customers for drink, food, and accommodations were regulated. Taxes were assessed to cover the expenses of county government, and poor relief was administered. In addition to serving as officers in the county militia, the justices were charged with securing sufficient arms and ammunition for defense, and they judged cases involving violations of maritime law. They also conducted elections for representatives to the lower house of assembly. In Vir-

8. Jon Kukla, "Order and Chaos in Early America: Political and Social Stability in Pre-Restoration Virginia," *AHR,* XC (1985), 275–298; Horn, *Adapting to a New World,* chap. 8.

9. Lois Green Carr, "The Foundations of Social Order: Local Government in Colonial Maryland," in Bruce Daniels, ed., *Town and County: Essays on the Structure of Local Government in the American Colonies* (Middletown, Conn., 1978), 72–110; Warren M. Billings, "The Growth of Political Institutions in Virginia, 1634 to 1676," *WMQ,* 3d Ser., XXXI (1974), 225–242 (quotation on 228).

ginia, parish vestries oversaw the building of churches, hiring of ministers, and processioning of land boundaries. Familiar and effective local institutions provided a sense of order and stability, despite turmoil at the top and high population turnover, and reinforced the efforts of household heads and neighborhood leaders to maintain local peace.[10]

In newer areas, these officials were simply the most prosperous planters in their neighborhoods. In places where recognized gentlemen were few, county benches might include former servants, men who might have been able to read but could not sign their names, and respectable planters of middling economic status. Such men sometimes had difficulty commanding respect from neighbors of similar background since they lacked the assets of great wealth, gentle birth, and influential kin or mercantile connections that settlers associated with a ruling class. Even magistrates of undistinguished social and economic status took their duties seriously, however, and most performed adequately. Some rose to positions of honor and influence they could not hope to achieve in the home country. As more younger sons of English gentlemen and merchants with capital and influential connections immigrated to the Chesapeake, where they managed to amass relative fortunes in land and labor, a nascent gentry class gained a growing proportion of local government appointments. Most Chesapeake settlers accepted this strengthening of traditional social hierarchy and gentry rule, so long as the gentlemen tempered their authority with respect for English law and for the rights of individual heads of households.[11]

Ordinary settlers' worlds were intensely local, confined more to neighborhoods—organized around nodes of activity such as a county officer's home, a planter-merchant's landing, or a church—than opened to the larger county, much less the province. Scattered, low-density settlements and the absence of any towns (aside from the capitals of Jamestown and Saint Mary's City) limited the number of other households that one family could physically reach. Moreover, a landscape bisected by numerous rivers and creeks, with

10. Carr, "The Foundations of Social Order," and Robert Wheeler, "The County Court in Colonial Virginia," both in Daniels, ed., *Town and County,* 72–110, 111–133; Billings, "The Growth of Political Institutions in Virginia," *WMQ,* 3d Ser., XXXI (1974), 225–242; James R. Perry, *The Formation of a Society on Virginia's Eastern Shore, 1615–1655* (Chapel Hill, N.C., 1990), chap. 8; Craven, *Southern Colonies,* 269–289, 302–309; Horn, *Adapting to a New World,* chap. 8; Kukla, "Order and Chaos," *AHR,* XC (1985), 275–298.

11. Martin H. Quitt, "Immigrant Origins of the Virginia Gentry: A Study of Cultural Transmission and Innovation," *WMQ,* 3d Ser., XLV (1988), 629–655; Carr, "Foundations of Social Order," in Daniels, ed., *Town and County,* 72–110; Billings, "Growth of Political Institutions," *WMQ,* 3d Ser., XXXI (1974), 225–242; Horn, *Adapting to a New World,* chap. 8.

few roads, ferries, or bridges, and, before the 1670s, few horses, tended to limit movement to foot traffic whenever boats were unavailable. Repeated and ordinary contacts between tenant farmers and small landowners, who could not leave crops unattended for long, involved households lying within an approximate five-mile radius of one another, a journey of an hour or two. The most frequent contacts between such men took place within the smaller scope of two to three miles of the home, a radius that also encompassed most married women's social contacts.

Middling planters, who could leave their servants to tend the farm from time to time, were more likely to have wider connections with households between five and ten miles around their farms, as well as those nearby, and they also had more occasion to travel to meetings of the county court and to militia musters as well as to buy or sell produce. Even much county business, however, was conducted on a neighborhood basis. Most counties had yet to build centrally located courthouses. County court meetings were instead held in ordinaries or in the houses of one of the justices, and the locations were rotated among different parts of the jurisdiction. Many settlers likely deferred recording deeds or initiating suits until a meeting was held nearby. All freemen between the ages of sixteen and sixty were supposed to attend a general muster of the county militia once a year, but otherwise even the militia trained in neighborhood troops or companies so that men did not have to travel far from their farms. Only county and provincial officers and planter-merchants, who had both the means and occasion to travel more widely, enjoyed significantly expanded communication networks, encompassing at least a twenty-five-mile radius from their homes and involving several adjoining counties.

Analysis of settlers' association patterns based on a broad range of sources demonstrates that the often spatially constricted activities reflected in planters' account books and private correspondence are not just artifacts of these particular source materials; they are a reality of the generally small physical worlds within which all but provincial elites lived. Difficulties of moving around within the region meant that contacts between disparate parts of the colonies were often few and far between. It did not mean, however, that ordinary settlers were similarly isolated from the wider Atlantic world. The annual arrival of the tobacco fleet afforded those who could write a chance to exchange letters with relatives, patrons, and business contacts in Britain. And those living in areas engaged in intercolonial provisioning trades had similar opportunities for contact with New England, New Amsterdam, and the Caribbean. Moreover, given the dispersed and time-consuming nature of the tobacco trade (it could take as long as 210 days for a ship to load a full cargo of tobacco), most settlers could learn what was occurring outside the colony from the ship captains and mariners plying Chesapeake waters.[12]

Immigration, Settlement, and the Labor Problem

Despite the political turmoil in both England and the colonies, the late 1640s, 1650s, 1660s, and early 1670s marked the Golden Age of the small planter in the colonial Chesapeake. Thousands of English men and women were not deterred from pursuing the opportunities they continued to believe the Chesapeake offered, and even the disruptions of Bacon's Rebellion failed to interrupt a steady rise in the quantities of tobacco pouring into England. A boom in tobacco prices that began in the middle 1640s encouraged immigration. According to Edward Hyde, first earl of Clarendon, the Chesapeake region "was more improved in people and stock" during the late 1640s and early 1650s "than it had been from the beginning to that time." Tobacco production increased more than tenfold between 1640 and the late 1660s. The number of men and women who worked the fields (termed "taxables" or "tithables" because they were liable to a poll tax) in Virginia and Maryland more than tripled between 1645 and 1665, and the total population grew even faster, from about eight thousand in 1645 to nearly thirty-four thousand in 1665. Thousands of acres of new land were surveyed and brought into production, and settlers felt secure enough from the threats posed by local Indians, after Virginia colonists decisively defeated the Pamunkeys following their second major attack in 1644, to begin farming all along the edges of the James, York, Rappahannock, Potomac, and Patuxent rivers, along the upper shores of Chesapeake Bay, and in the Albemarle region of Carolina.[13]

12. Lorena S. Walsh, "Community Networks in the Early Chesapeake," in Carr, Morgan, and Russo, eds., *Colonial Chesapeake Society,* 200–241; Darrett B. Rutman and Anita H. Rutman, *A Place in Time: Middlesex County, Virginia, 1650–1750* (New York, 1984), chap. 4; April Lee Hatfield, *Atlantic Virginia: Intercolonial Relations in the Seventeenth Century* (Philadelphia, 2004), chaps. 2–3; Perry, *Formation of a Society,* chaps. 4–5, 7; Horn, *Adapting to a New World,* 237–239; Beverley, *History and Present State of Virginia,* ed. Wright, 268–270.

13. Edward, earl of Clarendon, *The History of the Rebellion and Civil Wars in England: Begun in the Year 1641,* ed. W. Dunn Macray, 6 vols. (Oxford, 1888), V, 263. An overview of social, political, and economic developments in the colonies in this period is Pestana, *The English Atlantic,* introduction, prologue, and epilogue. The history of settlement patterns and population growth are described in Russell R. Menard, "The Tobacco Industry in the Chesapeake Colonies, 1617–1730: An Interpretation," in *Res. Econ. Hist.,* V (1980), 109–177; and Morgan, *American Slavery, American Freedom,* appendix. A good contemporary description is R. G., *Virginia's Cure; or, An Advisive Narrative concerning Virginia . . .* (1662), *Force Tracts,* III, no. 15, 3–4. For inland settlement on the James-York peninsula, see Philip Levy, "A New Look at an Old Wall: Indians, Englishmen, Landscape, and the 1634 Palisade at Middle Plantation," *VMHB,* CXII (2004), 226–266.

Generally, every male between the ages of sixteen and sixty was liable for a poll tax in both

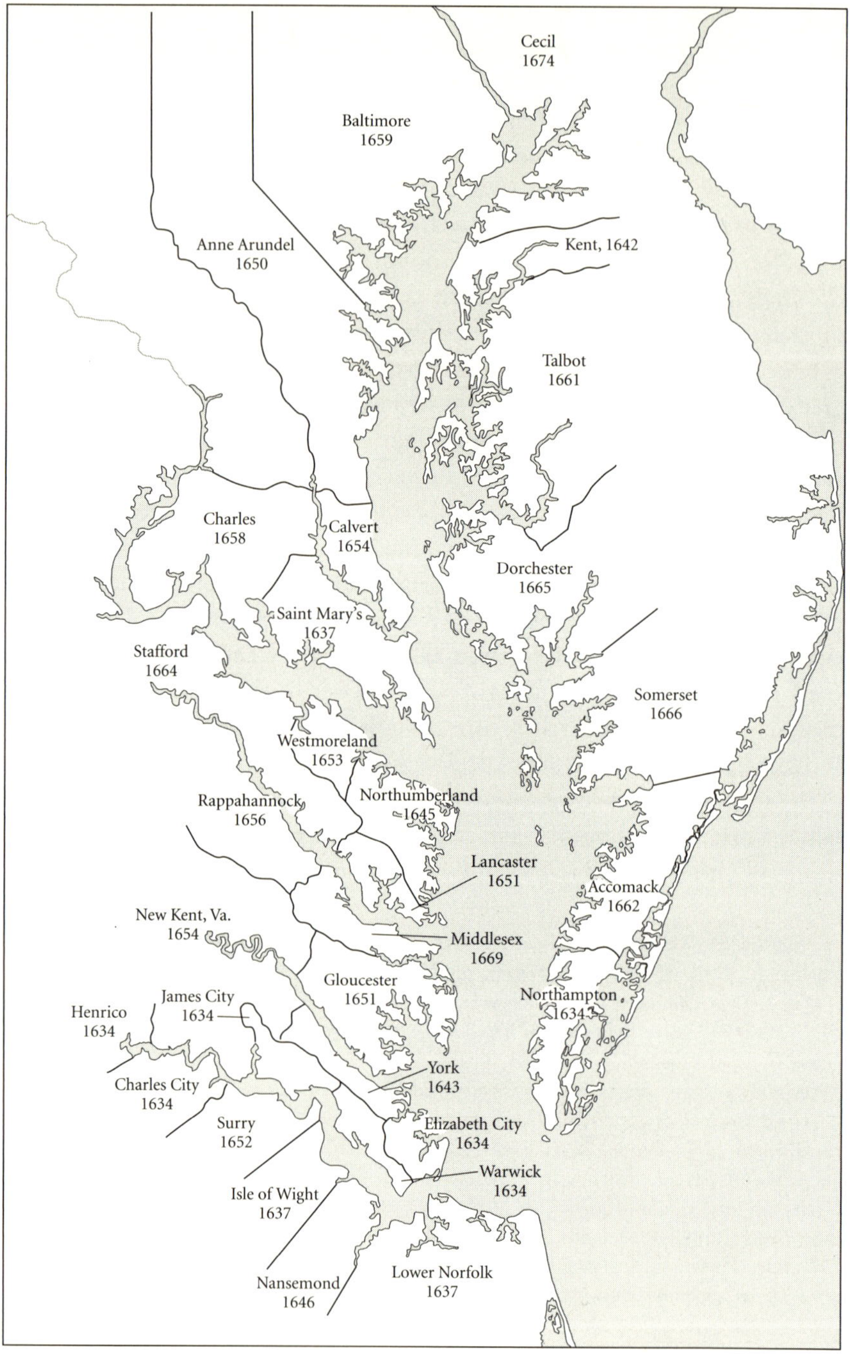

MAP 5. Chesapeake Counties Established through 1679

Some wealthy men carved out vast landed estates for themselves and sought, as had proprietors of particular plantations in earlier years, to make their fortunes at least in part as rentiers. But with seemingly limitless land available almost (if not quite) for the taking, would-be colonial landlords were nearly universally disappointed in their expectations. Immigrants with at least enough capital to pay their passage to the New World and, increasingly, large numbers of ex-servants as well acquired freehold land and became planters in their own right. In Virginia, most good land in older settled areas had already been taken up, forcing new immigrants to move to less developed parts of the tobacco coast in order to acquire fresh land with access to water. In Maryland, the proportion of both very large and very small landholdings declined, indicating less inequality in landownership than at the colony's beginning, and those with middling estates increased. In the third quarter of the century, most householders who survived in the colony more than a few years were able to establish themselves as planters on land of their own. Given the scarcity of men born to or claiming gentle status in the newer areas, a number of middling planters assumed roles of leadership in their communities, serving as jurors, justices of the peace, and sometimes as delegates to the lower house of the colony's legislature.[14]

---

colonies, as were bound women who worked in the fields. The legal definition of who was taxable, however, varied before 1705. For Virginia, see William Waller Hening, ed., *The Statues at Large: Being a Collection of All the Laws of Virginia* . . . (Richmond, Va., 1809–1823; rpt. Richmond, Va., 1969), I, 128, 144, 242, 292, 361, 454, II, 170, 267, 296, 479–480, 492, III, 258–259; Morgan, *American Slavery, American Freedom,* 400–401; and Brown, *Good Wives,* 108, 116–125. For Maryland, see William Hand Browne et al., eds., *Archives of Maryland,* 72 vols. (Baltimore, 1883–1972), I, 527, II, 399, 538–539; and Arthur E. Karinen, "Maryland Population, 1631–1730: Numerical and Distributional Aspects," *MHM,* LIV (1959), 365–367.

14. Russell Robert Menard, "Economy and Society in Early Colonial Maryland" (Ph.D. diss., University of Iowa, 1975), 178–238; Lois Green Carr, Russell R. Menard, and Lorena S. Walsh, *Robert Cole's World: Agriculture and Society in Early Maryland* (Chapel Hill, N.C., 1991), chaps. 1, 5; Walsh, "Land, Landlord, and Leaseholder: Estate Management and Tenant Fortunes in Southern Maryland, 1642–1820," *Ag. Hist.,* LIX (1985), 373–396. In parts of Virginia where the amount of land was limited, especially in the Eastern Shore counties of Accomack and Northampton, early landlords might have done somewhat better (Perry, *Formation of a Society,* chap. 2). For the settlement patterns and social structure of individual Chesapeake counties, see Paul G. E. Clemens, *The Atlantic Economy and Colonial Maryland's Eastern Shore: From Tobacco to Grain* (Ithaca, N.Y., 1980), chaps. 1–3; Robert Anthony Wheeler, "Lancaster County, Virginia, 1650–1750: The Evolution of a Southern Tidewater County" (Ph.D. diss., Brown University, 1972); Lorena S. Walsh, "Charles County, Maryland, 1658–1705: A Study of Chesapeake Social and Political Structure" (Ph.D. diss., Michigan State University, 1977), chap. 5; Walsh, "Land Use, Settlement Patterns, and the Impact of European Agriculture, 1620–1820," in Philip D. Curtin, Grace S. Brush, and George W. Fisher, eds., *Discovering the Chesapeake: The History of an Ecosystem* (Baltimore, 2001), 220–248; and Horn, *Adapting to a New World,* chap. 4.

Starting out, although easier than in earlier years, did remain fraught with difficulties. Most settlers arrived in the fall and early winter when the ships came from Europe to load tobacco, and, weakened by their passage on often-crowded vessels, some were ill prepared to survive the cold. Whether they began farming land of their own or worked for or rented land from an established planter in the ensuing growing season, by midsummer almost all fell ill with "seasoning." A good number died shortly of malaria or gastrointestinal diseases. As one settler explained in the early 1650s, "Here is many that lives very plentifully and gett good estates, but they are such whose long experience hath dearely bought it, and like wise with unwearyed labour hath undergon both the parchinge heate of the summer and the Freizinge of the winter regardinge neither."[15]

Those settlers who survived learned from the mistakes of the first colonists. Most immigrant European planters (unlike promoters of colonial enterprises in England) soon gave up any notions they might have entertained about producing a wide variety of semitropical products. In the Chesapeake, tobacco reigned, the only commodity residents could raise that was in sufficient demand in Europe to overcome the disadvantages of high regional labor costs and long-distance transportation charges. Moreover, tobacco was the only crop for which there was a reliable marketing network extending from the Chesapeake Bay to England and, in the 1640s and early 1650s, to Holland as well. If colonists were going to purchase manufactures and thus achieve some semblance of what was by their standards the good life, they had to make tobacco. Even though prices for the weed dropped steadily as production increased (falling about 4 percent a year from the mid-1620s to the early 1660s and about 2 percent annually from 1660 into the 1680s), these diminishing returns were still sufficient not only to sustain planting families but also to underwrite respectable improvements in wealth.[16]

15. Russell R. Menard and Lorena S. Walsh, "Death in the Chesapeake: Two Life Tables for Men in Early Colonial Maryland," *MHM,* LXIX (1974), 211–227; Susan E. Hillier, "The Trade of the Virginia Colony, 1606–1660" (Ph.D. diss., University of Liverpool, 1971), chap. 3; Beauchamp Plantagenet [Edmund Plowden], *A Description of the Province of New Albion* . . . (1648), *Force Tracts,* II, no. 7, 4–5; William Bullock, *Virginia Impartially Examined* . . . (London, 1649), 4, 11, 47–48; David Peterson de Vries, *Voyages from Holland to America A.D. 1632 to 1644* (1655), trans. Henry C. Murphy, New-York Historical Society, *Collections,* 2d Ser., III, pt. 1 (New York, 1853), 37; Jasper Danckaerts, *Journal of Jasper Danckaerts, 1679–1680,* ed. Bartlett Burleigh James and J. Franklin Jameson, Original Narratives of Early American History (New York, 1913), 151–152; Lawrence Ward to John Ferrar, July 12, 1653, Ferrar Papers, 1590–1790, Pepys Library, Magdelene College, Cambridge University, microfilm ed., no. 1231, CWL.

16. Garry Wheeler Stone, "Society, Housing, and Architecture in Early Maryland: John Lewger's St. John's" (Ph.D. diss., University of Pennsylvania, 1982); Russell R. Menard, "A Note

In the 1640s, 1650s, and 1660s, the innovations and adaptations that the hundreds of small planters made shaped the agricultural regimen more profoundly than in any other period in the region's history. For about fifty years, planters experimented widely with different strains of tobacco, with different methods of cultivation, curing, and housing, and with improved ways for packing their crops. They also learned how to mesh the seasonal growing cycles of tobacco and maize, assuring their families an adequate supply of food at minimal cost in time and labor. By the mid-1670s, the agricultural techniques had been established that would prevail throughout the colonial period. Small tobacco growers on Bermuda, Barbados, Montserrat, and Martinique as well as in England, the Netherlands, France, and Germany were engaging in similar experiments that soon resulted in regional variations that reflected differences in climate, natural resources, and the available labor force. The basic tasks were everywhere the same, although the methods of raising the seedlings, the types of fencing, and the materials and architecture of barns differed widely (see Plate 6).

The principal shortage new Chesapeake planters faced remained labor, just as it had been earlier for the proprietors of particular plantations. Attempts to recruit mature tenant farmers and their families had been largely abandoned in Virginia with the fall of the Virginia Company, and in the 1630s Maryland manor lords had similarly limited success. Instead, young men and women were recruited as servants under fairly uniform contractual conditions—those spelled out in indentures drawn up by merchants in England or, for the growing numbers that arrived without specific prior agreements, by "the custom of the country" as defined by colonial legislatures and courts. Although kidnapping or "spiriting" of unwilling or unsuspecting persons continued in England, as well as recruiting prospective emigrant servants with unrealistic promises and fraudulent contracts, Chesapeake courts did remedy some of the most blatant injustices. Leading planters found it in

---

on Chesapeake Tobacco Prices, 1618–1660," *VMHB,* LXXXIV (1976), 401–410; Menard, "Farm Prices of Maryland Tobacco, 1659–1710," *MHM,* LXVIII (1973), 80–85; Menard, "The Tobacco Industry," *Res. Econ. Hist.,* V (1980), 109–177; *A Perfect Description of Virginia . . .* (1649), *Force Tracts,* II, no. 8, 6. For the few big planters who did try to produce other staple commodities, see ibid., 13–15. By 1635, Robert Wintour, an Englishman well informed about the new Maryland colony, advised that mining ought not to be undertaken until the country was better established—"to undertake it sooner would both prejudice the plantation, and not answere in such a measure the adventure employed upon it as other commodities"—and that wine might be made, but it would not be profitable to export it to Britain; see John D. Krugler, ed., *To Live Like Princes: A Short Treatise Sett Downe in a Letter Written by R. W. to His Worthy Friend C. J. R. concerning the Plantation Now Erecting under the Right Ho[nora]ble Lord Baltemore in Maryland . . .* (Baltimore, 1976), 36.

PLATE 6. Tobacco Plantation in Holland. From an agricultural manual published in the Hague in 1692. Permission, George Arents Collection, The New York Public Library, Astor, Lenox and Tilden Foundations.
Here the seedlings were raised in glass-covered hotbeds, hedges were used as fencing, and the tobacco house in the foreground has large shuttered openings for ventilation and a thatched gable roof decorated with elaborate finials on the pediments.

their interest to promote regularity and consistency in the servant trade in order to encourage emigration.[17]

17. The standard work on servants is Abbot Emerson Smith, *Colonists in Bondage: White Servitude and Convict Labor in America, 1607–1776* (Chapel Hill, N.C., 1947). See also David W. Galenson, *White Servitude in Colonial America: An Economic Analysis* (New York, 1981); Richard S. Dunn, "Servants and Slaves: The Recruitment and Employment of Labor," in Jack P. Greene and J. R. Pole, eds., *Colonial British America: Essays in the New History of the Early Modern Era* (Baltimore, 1984), 157–194; Menard, "Economy and Society"; and Lorena S. Walsh, "Servitude and Opportunity in Charles County, Maryland," in Aubrey C. Land, Lois Green Carr, and Edward C. Papenfuse, eds., *Law, Society, and Politics in Early Maryland* (Baltimore, 1977), 111–133. David W. Galenson, "British Servants and the Colonial Indenture System in the Eighteenth Century," *Jour. So. Hist.*, XLIV (1978), 41–66, concluded that, when age is taken into account, servants without indentures serving according to the custom of the country were not

Although not every servant who came to the Chesapeake was satisfied with his or her lot there, and, indeed, some suffered grievous abuse and profound hardship, nonetheless good reports—filtering into villages from the port towns, included in letters sent back home, or published in more formal tracts—came to outweigh the bad. Enough young English men and women were persuaded to immigrate that through the 1660s even very ordinary Chesapeake planters had a reasonable chance to acquire an extra hand or two.[18]

Once their time had been purchased, these men, women, boys, and girls could be organized into work groups to undertake primitive hand cultivation of food and cash crops, livestock care, and farm building. Since most planters were either unmarried or very recently married, few had children of age to help with farm work. Although most plantations were now single family enterprises, they remained organized around bound servant, not family, labor. For those able to buy bound workers, life spans remained too short and the profits a man and his wife could wrest from the land by their own labor too small to afford the luxury of waiting for their own children to grow into effective farm hands.

On the plantations of elite officeholders, a very different form of labor organization appeared by the mid-seventeenth century. In the 1630s and 1640s, the richest Virginia planters, along with a few wealthier Marylanders, started to buy slaves and to do so as a matter of deliberate choice rather than of chance, as had been the case with the first Africans carried to Virginia in 1619. These planters decided early on that higher profits could be obtained from enslaved workers. Dutch traders, who were by then carrying slaves from West Africa to the Caribbean, supplied most of the laborers. The Chesapeake elite developed close connections with these merchants, and some had direct family ties. Virginia planters engaged in the provisioning trade to the islands or enmeshed in intercolonial trading networks with the Dutch acquired most of the Africans brought to the Chesapeake in the 1630s and 1640s. In the absence of any basis in law, European merchants determined who was

---

unduly exploited by being assigned terms appreciably longer than those who negotiated indenture contracts.

18. John D. Krugler, *English and Catholic: The Lords Baltimore in the Seventeenth Century* (Baltimore, 2004), chap. 6, provides a discussion of the early recruitment efforts for Maryland, which were based primarily on "word of mouth and a limited campaign using printed material and public statements" (133). See also Menard, "Economy and Society," chap. 5; and Wheeler, "Lancaster County, Virginia," chaps. 1, 2. An example of a promotional tract for recruiting settlers is George Alsop, *A Character of the Province of Maryland [1666],* in Clayton Colman Hall, ed., *Narratives of Early Maryland, 1633–1684,* Original Narratives of Early American History (New York, 1910), 335–387.

or was not a slave, and the merchants almost invariably sold any Africans they carried into the colonies, regardless of how they might have obtained them, as slaves. Between three and five hundred blacks were reported to be in Virginia by 1647. How they were employed and disciplined was left largely to the discretion of individual masters but was occasionally adjudicated in county courts. A middling planter living in York County, Virginia, reported in the same year that "Negroes remaine in particular mens hand and are soe dispersed that I can make noe narration of them."[19]

During the 1630s and 1640s, in their capacity as private individuals, elite Chesapeake slaveholders made a number of critical decisions that established African slavery as a form of bondage different from all other kinds of unfree labor. Then, in their capacity as councillors, burgesses, and magistrates, the slaveowning elite upheld and enforced these private acts. By the late 1640s, bills of sale demonstrate that Africans were clearly being held according to common custom, among English settlers in both the mainland and West Indian colonies, that "negroes" brought for sale from Africa should serve for life, unless they arrived with a contract specifying a lesser term of service. Moreover, slaveowners quickly decided that enslaved status was not just lifelong but also hereditary. Contrary to English custom, by which a child's status was determined by that of the father, slaveowners decided that the status of children born to an enslaved woman would be determined by the condition of the mother. Although ambiguities about whether children of

19. For the absence of law, see Jonathan A. Bush, "Free to Enslave: The Foundations of Colonial American Slave Law," *Yale Journal of Law and the Humanities,* V (1993), 417–470; Bush, "The British Constitution and the Creation of American Slavery," in Paul Finkelman, ed., *Slavery and the Law* (Madison, Wis., 1997), 379–418; Warren M. Billings, "The Law of Servants and Slaves in Seventeenth-Century Virginia," *VMHB,* XCXI (1991), 51–52. For knowledge of Iberian customs, see Michael Joseph Guasco, "Encounters, Identities, and Human Bondage: The Foundations of Racial Slavery in the Anglo-Atlantic World" (Ph.D. diss., College of William and Mary, 2000), chap. 8. John Ferrar dispatched a long list of queries about natural resources and the state of population, agriculture, and trade to several Virginia planters in 1646, including Captain Barnard, Captain [Argoll?] Yeardley, and Richard Collett (Questionnaire, Aug. 30, 1646, Ferrar Papers, no. 1106, CWL). The answers formed the basis of the tract, *A Perfect Description of Virginia, Force Tracts,* II, no. 8. Ferrar's 1646 query list includes annotations he made later from the answers he received; some of the letters on which he based his notations may no longer be present in the Ferrar Papers. *A Perfect Description of Virginia* states the number of "negroes" in Virginia as three hundred. Ferrar's annotations give the number as five hundred. The quotation is from Richard Collett to John Ferrar, 1647, Ferrar Papers, no. 1121, CWL. For the modes of acquisition and documentation on the status of blacks, see John C. Coombs, "Building 'The Machine': The Development of Slavery and Slave Society in Early Colonial Virginia" (Ph.D. diss., College of William and Mary, 2004), chaps. 2, 3; and Coombs, "Beyond the 'Origins Debate': Rethinking the Rise of Virginia Slavery," in Bradburn and Coombs, eds., *Early Modern Virginia* (forthcoming).

mixed race should be enslaved or free persisted into the 1690s, by the mid-1640s boys and girls born to a white father and an enslaved African mother were being sold as lifelong slaves. Other official actions imposed discriminatory measures against African men and women. In 1640, the Virginia legislature decreed that, unlike European men, "negroes" were not to be provided with arms and ammunition for defense of the colony, and the General Court authorized an armed militia expedition to pursue and capture "certain runaway negroes" at public expense. The practice of putting African women to heavy field labor equal to that of men also began in the 1640s, if not earlier. Their status as permanent agricultural workers was confirmed by a law classifying all "negro women" age sixteen and above as tithables in 1643. Gender, as Jennifer Morgan argues, "furnished one of the crucial axes around which the organization of enslavement and slave labor in the Americas took place." Turning "women" into "workers" was a critical step in expropriating both the productive and reproductive labor of black women. Thus, many of the conditions that came to define African slavery—lifelong service, heritability, descent through the mother, public support for policing private property, relegation of women to heavy field labor, and restrictions on men's access to weapons—had already been established in practice, if not in law, within the first three decades after Africans arrived in the region.[20]

A regular trade in African slaves developed by the 1650s and 1660s, as English merchants entered the lucrative direct trade with Africa previously monopolized by the Portuguese and the Dutch. This trade, however, became

20. For service for life as the usual condition of imported blacks in Barbados, see Winthrop D. Jordan, *White over Black: American Attitudes toward the Negro, 1550–1812* (Chapel Hill, N.C., 1968), 63–65; and Susan Dwyer Amussen, *Caribbean Exchanges: Slavery and the Transformation of English Society, 1640–1700* (Chapel Hill, N.C., 2007), 69–71, 90–101. Providence Island settlers were informed by company officials in 1638 that blacks could be "kept as perpetuall servants." Similarly, the Rhode Island Assembly, when it prohibited slavery in that colony in 1652, noted the "common course practised amongst English men to buy negers, to that end they may have them for service or slaves forever" (Jordan, *White over Black,* 64, 70). For sales of hereditary slaves in Virginia, including those of mixed race, see Carl N. Degler, "Slavery and the Genesis of American Race Prejudice," *Comparative Studies in Society and History,* II (1959), 49–66; Winthrop D. Jordan, "Modern Tensions and the Origins of American Slavery," *Jour. So. Hist.,* XVII (1962), 18–30; Linda M. Heywood and John K. Thornton, *Central Africans, Atlantic Creoles, and the Foundation of the Americas, 1585–1660* (Cambridge, 2007), chap. 6; and Coombs, "Building 'The Machine,'" chap. 5. For the work of enslaved women, see Jennifer L. Morgan, *Laboring Women: Reproduction and Gender in New World Slavery* (Philadelphia, 2004), 69–106 (quotation on 69); and Brown, *Good Wives,* chap. 4. For arms, women as tithables, and public policing, see William Waller Hening, ed., *The Statutes at Large: Being a Collection of All the Laws of Virginia . . .* (1809–1823; rpt. Richmond, Va., 1969), I, 226, 242, 292, 454–455; Browne et al., eds., *Archives of Maryland,* I, 342; *MCGC,* 468.

firmly focused on the rapidly expanding Caribbean sugar industry, which offered a much larger market for slaves than did the Chesapeake. Meanwhile, a succession of monopolies of the English trade with Africa that Charles II granted to favored groups in England severely restricted the supply of slaves, especially to minor markets. Moreover, in contrast to the 1620s and 1630s when the prices of slaves captured from Spanish and Portuguese ships or settlements were low, in the ensuing decades their prices rose as English planters on Barbados first bought slaves to grow tobacco and cotton, and, after they turned to sugar, demanded ever-increasing numbers of Africans to grow it. Nonetheless, laborers bound for life, who had little or no protection from the courts, and who included "outlandish" women who could be exploited in the fields as completely as men, proved attractive to Virginia and Maryland elites wealthy and well connected enough to buy slaves and to bear the risk of early loss of expensive workers. By the 1650s, blacks were commonly present on large plantations in all areas of Virginia, generally working in concert with indentured servants. They were most numerous along the lower James, the area most involved in the West Indian provisioning trade, and along the York River, where planter wealth was most concentrated. By the mid-1660s, blacks were more than 15 percent of the population of York County.[21]

When the English Parliament outlawed trade with the Dutch in 1660, elite Chesapeake planters then turned to West Indian merchants and planters, primarily Barbadians, to obtain enslaved laborers. English merchants, unable to meet the labor demands of West Indian sugar growers, sent only two or three shiploads of Africans to Virginia in the 1660s. Maryland's proprietary rulers sought in 1664 to negotiate the shipment of between one and two hundred Africans to that colony every year from the Company of Royal Adventurers into Africa but discovered that their cost was more than Maryland planters could afford. Only one ship bringing a cargo of slaves directly from Africa is known to have come to Maryland in the 1660s. But, however they were acquired, the number of blacks laboring in the region continued to grow across the 1650s and 1660s. Governor Berkeley estimated that there were two thousand Africans in Virginia by 1671, about 5 percent of the colony's total population. This number was sufficient to pose problems of social control that extended beyond the boundaries of private plantations. It also by then

21. Coombs, "Building 'The Machine,'" chaps. 2, 3; David W. Galenson, "Economic Aspects of the Growth of Slavery in the Seventeenth-Century Chesapeake," in Barbara L. Solow, ed., *Slavery and the Rise of the Atlantic System* (Cambridge, 1991), 265–292; David Eltis et al., *The Trans-Atlantic Slave Trade: A Database on CD-ROM* (Cambridge, 1999), voyage no. 25149; Russell R. Menard, *Sweet Negotiations: Sugar, Slavery, and Plantation Agriculture in Early Barbados* (Charlottesville, Va., 2006), chap. 2; Pestana, *The English Atlantic,* chap. 6.

represented a substantial capital investment, one that was unprotected by colonial law. Consequently, during the 1660s, when the number of burgesses with a personal interest in protecting slavery was growing every year, the Virginia and Maryland assemblies enacted the region's first laws defining the status of most blacks as slaves and closing off legal loopholes some enslaved blacks had been using to sue for freedom.[22]

Shortages of indentured servants indeed occurred during cyclical downturns in the tobacco economy. Such shortages, however, could not have forced elite planters to turn to slaves had they preferred indentured servants and been willing to pay higher prices for them, for it was the elite who had first choice of any laborers offered for sale in the colony. Moreover, they had close relationships and frequent communication with European merchants who could have supplied them with whatever sort of laborers they requested as well as access to better and cheaper credit and more information about the productivity of African slaves in the West Indies.[23]

Those at the pinnacle of wealth and power, the councillors, took the lead in buying slaves, followed by burgesses and other county-level officeholders. From 1635, the Virginia Council awarded headrights to those who imported Africans as well as Europeans, and it was members of the council who regis-

22. Coombs, "Building 'The Machine,'" chaps. 2, 3, 5; Galenson, "Economic Aspects of the Growth of Slavery," in Solow, ed., *Slavery and the Rise of the Atlantic System,* 265–292; Anthony S. Parent, Jr., *Foul Means: The Formation of a Slave Society in Virginia, 1660–1740* (Chapel Hill, N.C., 2003), chap. 4; David Eltis, *The Rise of African Slavery in the Americas* (Cambridge, 2000), 208; Eltis et al., *The Trans-Atlantic Slave Trade,* voyage no. 21502. The estimate of two thousand blacks and the information about direct shipments is from Gov. William Berkeley to the Board of Trade, 1671, in Hening, ed., *Statues at Large,* II, 515. For the organization of the early English slave trade, see David Eltis, "The British Transatlantic Slave Trade before 1714: Annual Estimates of Volume and Direction," in Robert L. Paquette and Stanley L. Engerman, eds., *The Lesser Antilles in the Age of European Expansion* (Gainesville, Fla., 1996), 182–205; Herbert Klein, "The Atlantic Slave Trade to 1650," in Stuart B. Schwartz, ed., *Tropical Babylons: Sugar and the Making of the Atlantic World, 1450–1680* (Chapel Hill, N.C., 2004), 201–236; and Stephanie Smallwood, *Saltwater Slavery: A Middle Passage from Africa to American Diaspora* (Cambridge, Mass., 2007), chaps. 2, 6, 7. For the West Indian trade, see Gregory E. O'Malley, "Beyond the Middle Passage: Slave Migration from the Caribbean to North America, 1619–1807," *WMQ,* 3d Ser., LXVI (2009), 125–168.

23. The argument that planters embraced slaves only when they could no longer get enough servants was developed by Russell Menard, "From Servants to Slaves: The Transformation of the Chesapeake Labor System," *So. Stud.,* XVI (1977), 362–390. Dunn, "Servants and Slaves," in Greene and Pole, eds., *Colonial British America,* 167, and Parent, *Foul Means,* chap. 2, have also accepted Menard's account of the timing and reasons for the transition. For the credit and information advantages of elites, see Galenson, "Economic Aspects of the Growth of Slavery," in Solow, ed., *Slavery and the Rise of the Atlantic System,* 265–292. For the legalization of slavery as a conscious decision, see Billings, "Law of Servants and Slaves," *VMHB,* XCIX (1991), 54–62. For legalization in the Caribbean, see Amussen, *Caribbean Exchanges,* 107–114, 121–139.

tered three-quarters of the headrights claimed for importing blacks ("negro rights," in period terminology) that were used to obtain certificates ordering the surveying of a parcel of unclaimed land, the first step in acquiring acreage granted for importing settlers. Councillors also registered more than half of black headrights claimed in land patents, the final step in getting a title. (Maryland slave buyers got no additional land for importing Africans, since the Lords Baltimore granted land rights only for transporting persons of British or Irish descent.) Between 1640 and 1669, more than half of the inventoried estates of Virginia provincial and county-level officeholders included one or more slaves, as did half of the estates of Maryland legislators. Overall, officeholders controlled between two-thirds and three-quarters of enslaved bondpeople. Most councillors built up holdings of at least ten slaves, and some owned two to four times that number. After 1670, when direct shipments from Africa became more common, the planter elite had less difficulty buying slaves, although demand still outran the supply. At least nine vessels, most from the port of London, came to the Chesapeake in the 1670s, carrying perhaps two thousand more Africans. Others were brought by migrating Barbadian planters squeezed off that island in the course of the sugar boom.[24]

Changes in sources of supply led to greater diversity among Africans carried into the Chesapeake. Most Africans arriving between 1619 and 1659 were from West Central Africa, a region where coastal peoples had more than a century of contact with Portuguese traders, missionaries, and colonizers and where an Atlantic Creole culture had developed. The early captives' familiarity with European languages and customs, and with Christianity, which some had embraced in their homeland, made them appear less like outsiders than later arrivals and equipped them to make the most of loopholes in the slave system. In the 1660s and 1670s, direct shipments of slaves supplied by English traders came primarily from Senegambia, the Gold Coast, and the Bight of Biafra, whereas most of those acquired from Barbados likely came from the Gold Coast or the Bight of Benin. These less homogeneous forced migrants, most of whom arrived with little knowledge of European culture

24. Coombs, "Building 'The Machine,'" chap. 3; Coombs, "Beyond the 'Origins Debate,'" in Bradburn and Coombs, eds., *Early Modern Virginia* (forthcoming); David W. Galenson, "Labor Market Behavior in Colonial America: Servitude, Slavery, and Free Labor," in Galenson, ed., *Markets in History: Economic Studies of the Past* (Cambridge, 1989), 52–96; Hatfield, *Atlantic Virginia,* chap. 6; Eltis et al., *The Trans-Atlantic Slave Trade,* voyage nos. 9948, 15064, 15065, 15174, 20837, 20908, 20914, 21245, 21246. For an explanation of the use of headrights to obtain land in Virginia, see Edmund S. Morgan, "Headrights and Head Counts: A Review Article," *VMHB,* LXXX (1972), 361–371.

or religion, might have posed greater problems of control, thus prompting laws sanctioning violent punishments and restricting slaves' movement.[25]

By the late 1670s, the share of inventoried officeholders who owned slaves rose to two-thirds of all county officers in Virginia, and a similar proportion prevailed in estates of Maryland legislators worth £150 or more. By then, the region's elites were so committed to slavery that, on many of their plantations, there were more slaves than servants, and some were relying almost solely on black agricultural labor, even in poorer or more recently settled areas like the lower Eastern Shore and the Potomac River basin.[26]

The majority of Chesapeake tobacco growers were small to middling planters, however, who lacked both the wealth and the mercantile connections necessary for acquiring slaves. The price of a prime slave, about twenty-three pounds sterling in the 1670s, was well beyond their means. Moreover, they could not afford the risk involved in borrowing money to buy one or two high-priced, permanent bound laborers who might die within the year. Before 1670, slaves appeared in only 6 percent of the estates of Virginians who did not hold important offices, a proportion that would change little until the very end of the century. Because the only substantive plantation records so far discovered for the mid-seventeenth century are those for the estate of Robert Cole, a middling planter in Maryland who farmed solely with indentured servants, discussion of enslaved workers and management of slaves on Chesapeake plantations will be deferred to subsequent chapters.

25. The concept of the "Charter Generation" was first developed in Ira Berlin, "From Creoles to African: Atlantic Creoles and the Origins of African-American Society in Mainland North America," *WMQ,* 3d Ser., LII (1996), 251–288. Heywood and Thornton restrict the members of the Charter Generation to West Central Africans (*Central Africans,* 236–248, 268–293). For ethnic groups in the West Indies, see Eltis, *Rise of African Slavery,* 244–247. Ethnic groups in the Chesapeake are from Lorena S. Walsh Database of Slave Trading Voyages to Virginia and Maryland (November 2008). This database was compiled to supplement Eltis et al., *The Trans-Atlantic Slave Trade,* by adding other voyages recorded in Walter Minchinton, Celia King, and Peter Waite, eds., *Virginia Slave-Trade Statistics, 1698–1775* (Richmond, Va., 1984), in Maryland naval office records not included in Elizabeth Donnan, ed., *Documents Illustrative of the History of the Slave Trade to America,* 4 vols. (Washington, D.C., 1900–1935), IV, in newspaper advertisements in the *Maryland Gazette,* and in other miscellaneous documentary sources containing information on slaving voyages that I identified. I have updated the database to incorporate additional voyages and revised information in the online edition of *The Trans-Atlantic Slave Trade* (David Eltis et al., *Voyages: The Trans-Atlantic Slave Trade Database* [Atlanta, Ga., 2008], http://www.slavevoyages.org), which David Eltis kindly made available before the public launch. Most (but not all) of the information in the Walsh Database appears in the 2008 online edition. Voyage identification numbers are identical in the 1999 and 2008 editions, but the 2008 edition (which will be regularly updated every three years) may contain revised or additional information.

26. Coombs, "Building 'The Machine,'" chap. 3; Galenson, "Labor Market Behavior," in Galenson, ed., *Markets in History,* 79–82.

The labor regime employed on Cole's well-documented plantation reflects the one prevailing on the vast majority of tobacco farms in the 1640s, 1650s, and 1660s.[27]

Agricultural Techniques and Marketing Arrangements

Recognizing that the long-fallow, shifting cultivation techniques of the local Indians produced the largest returns for the least inputs of labor, by the 1640s all but the richest planters had abandoned more intensive English-style agriculture. "This truely is the great labour in Virginia, to fell Trees, and to get up the Roots, and so make Cleare ground for the Plow," stated the author of *A Perfect Description of Virginia* in 1649. But most planters had entirely given over any attempt to clear land for plowing in order to raise European grains. Plowing proved extremely difficult in stump-filled fields; there were few oxen trained for draft; and many planters could not spare enough tobacco to buy plow gear, which was both scarce and expensive in the colonies. More typically, as the author of *Virginia . . . Truly Valued* noted, if the planter "but unbarke the Trees one foot round after the Indian mode to prevent the shade occasioned by the leaves, which such unbarking quite destroyes, the Corne (set betwixt those Trees) will thrive and prosper exceedingly, and their Ground thus prepared will last seven or more yeares successively, and this worke cannot last him above five dayes at the longest" — if he were lucky enough to have a wife and children to help (which was not always the case).[28]

By using this method, Chesapeake planters were able to produce adequate supplies of food. Grain (almost exclusively maize), legumes, garden

27. Coombs, "Building 'The Machine,'" chap. 3. For prices, see Menard, "From Servants to Slaves," *So. Stud.*, XVI (1977), 372.

28. Ester Boserup, *Population and Technological Change: A Study of Long Term Trends* (Chicago, 1981), chap. 15, discusses the relationship between population density, agricultural techniques, and labor supply. See also Lewis Cecil Gray, *History of Agriculture in the Southern United States to 1860*, 2 vols. (1933; rpt. Gloucester, Mass., 1958), chap. 19; Collett to Ferrar, 1647, Ferrar Papers, no. 1121, CWL; *A Perfect Description of Virginia, Force Tracts*, II, no. 8, 10; Edward Williams, *Virginia: More Especially the South Part Thereof, Richly and Truly Valued* . . . (1650), *Force Tracts*, III, no. 11, 48. For Indians' clearing methods, see also Richmond Croom Beatty and William J. Mulloy, eds. and trans., *William Byrd's Natural History of Virginia; or, The Newly Discovered Eden* (Richmond, Va., 1940), 92–93. Martin L. Primack estimated that in the nineteenth-century South it cost 40 percent less to girdle trees on uncleared land than it did to chop them down. In 1895, a worker in North Carolina could girdle an acre in thirteen and a half days but required seventeen to twenty days to chop down the trees. Ultimately, the dead trees would have to be burned, but this effort could be deferred to later years ("Land Clearing under Nineteenth-Century Techniques: Some Preliminary Calculations," *JEH*, XXII [1962], 484–497).

vegetables, and meat (domestic livestock supplemented seasonally with fish, shellfish, and game) could be produced with minimal effort by skimming the resources of rich virgin tracts and by capitalizing on usually abundant rainfall and a long growing season. Croplands were cleared of timber only to the extent necessary to get in a crop of maize around the stumps, and cattle and hogs were left to forage for themselves in woods, swamps, natural meadows, and abandoned old fields. A shift from a diet based on wheat and other small grains to one based almost exclusively on maize was one of the more important adaptations Chesapeake colonists had to make. Raising English grains fertilized with animal manure and cultivated with draft animals and plows required more time and labor than planters could spare from the staple. To make enough money from tobacco to purchase the imports they deemed necessities, colonists became willing to sacrifice the accustomed, more varied, Old World diet.[29]

A second major adaptation was in livestock husbandry. Instead of raising animals for draft and fertilizer as well as for food, Chesapeake planters raised them only for meat and milk and as a store of wealth, and did so by giving the creatures minimal attention and almost no supplemental feed. Cattle and hogs were allowed to forage freely, but planters retained some measure of control over them, as these social animals tend to live in herds composed of females and their young and do not often stray far from their home ranges. Although some few cattle and hogs did become truly feral, when it came

29. Contemporary descriptions of the composition of the diet are found in *A Relation of Maryland* . . . (London, 1635), in Hall, ed., *Narratives of Early Maryland,* 82; *A Perfect Description of Virginia, Force Tracts,* II, no. 8; John Hammond, *Leah and Rachael; or, The Two Fruitful Sisters Virginia and Mary-Land* (1656), in Hall, ed., *Narratives of Early Maryland,* 277–308; Nathaniel Shrigley, *A True Relation of Virginia and Mary-Land* . . . (1669), *Force Tracts,* III, no. 7; Joseph Ewan and Nesta Ewan, eds., *John Bannister and His Natural History of Virginia, 1678–1692* (Urbana, Ill., 1970), 360–361; and Collett to Ferrar, 1647, Ferrar Papers, no. 1121, Michael Upchurch to John Ferrar, 1650, no. 1182, CWL. The most complete contemporary description of period agriculture and diet is Thomas Glover, *An Account of Virginia, Its Scituation, Temperature, Productions, Inhabitants, and Their Manner of Planting and Ordering Tobacco etc.* . . . (1676; rpt. Oxford, 1904). See also Henry M. Miller, "An Archaeological Perspective on the Evolution of Diet in the Colonial Chesapeake, 1620–1745," in Carr, Morgan, and Russo, eds., *Colonial Chesapeake Society,* 176–199; and Carr, Menard, and Walsh, *Robert Cole's World,* chap. 4. Jasper Danckaerts, traveling on Maryland's upper Eastern Shore in 1679, noted the relationship between a diet limited in variety and the staple crop. On tobacco plantations, servants, slaves, and the planters themselves often had nothing to eat but maize, bread, or hominy, the bread tasting "miserable" to Danckaerts's European palate. He observed that "the servants and negroes after they have worn themselves down the whole day, and come home to rest, have yet to grind and pound the grain, which is generally maize, for their masters and all their families as well as themselves, and all the negroes, to eat" (*Journal,* ed. James and Jameson, 111, 117, 129, 130, 133, 134).

time to mark or brand the young or to pen mature animals for a brief period of fattening before slaughter, most of them could be found relatively easily in places where their favorite foods grew abundantly. Handling the young, penning calves so that the cows would return each evening to suckle them, and occasionally offering animals a little corn served to keep these creatures to some degree domesticated. One result of such minimal management was that losses of livestock in winter and early spring when forage was scarce were numerous, but rates of reproduction were high enough that planters could tolerate the losses and still have an adequate stock to afford their families relatively abundant meat and milk.[30]

Analysis of the remains of cattle excavated on Chesapeake sites demonstrates that the colonists' choice to save labor by limiting the care they gave to livestock was rational. The mean size of cattle increased in the second half of the seventeenth century, indicating that the animals were able to find adequate nutrition from foraging. Indeed, cleared forests and abandoned fields were ideal feeding grounds for cattle. "What they cann gett in the woods," the faunal remains suggest, for some time did afford "very fatt [and large] beefe." And when forage alone no longer sufficed to maintain the animals, planters did begin to provide some supplemental feed. Not until the early to mid-1700s do faunal remains show a decline in the size of cattle, indicating the onset of overgrazing and shortages of woodland and old field pastures. Hogs also fed in the woods and swamps, with the numbers allowed to grow to maturity dependent on the annual supply of mast. Planters at first also raised goats, but by the late 1640s most stopped keeping these more troublesome creatures once cattle became plentiful. Among familiar Old World animals, only sheep were in short supply owing to a lack of suitable pastures and predation by wolves. In the second half of the seventeenth century, the time saved in producing grain and meat over that commonly required in Europe by using extensive rather than intensive farming methods and according livestock minimal care could be devoted to the labor-intensive staple.[31]

30. Shrigley, *True Relation, Force Tracts,* III, no. 7, 5; Colonel Norwood, *A Voyage to Virginia* (1649), ibid., III, no. 10, 7, 47; Williams, *Virginia . . . Truly Valued,* ibid., III, no. 11, 6, 48; Bullock, *Virginia Impartially Examined,* 7–8; Glover, *An Account of Virginia,* 18–19; Danckaerts, *Journal,* ed. James and Jameson, 134; Upchurch to Ferrar, 1650, Ferrar Papers, no. 1182, CWL; Edmund Berkeley and Dorothy Smith Berkeley, eds., *The Reverend John Clayton: A Parson with a Scientific Mind: His Scientific Writings and Other Related Papers* (Charlottesville, Va., 1965), 54–55, 79–80, 87–89; Joanne Bowen, "Historical Ecology and the British Landscape" (paper presented at the annual meeting of the Society for Historical Archaeology, Mobile, Ala., 2002).

31. Bowen, "Historical Ecology and the British Landscape"; Virginia DeJohn Anderson, "Animals into the Wilderness: The Development of Livestock Husbandry in the Seventeenth-Century Chesapeake," *WMQ,* 3d Ser., LIX (2002), 377–408; Carville V. Earle, *The Evolution of*

Nicotiana tabacum is an adaptable plant that can thrive in a wide variety of environments. Chesapeake planters soon learned that different soils yielded quite different finished products. As naturalist John Clayton stated, "The same sort of Seed in different Earths" produced tobaccos "much different, as to goodness." With the discovery in the 1640s of a new strain of tobacco, sweet-scented, that would grow only in the rich bottomlands of lower tidewater Virginia, the fortunes of tobacco growers living on the middle and lower Peninsula began to differ from the prospects of those farming other parts of the Tobacco Coast.[32]

By 1647, some planters living along the York River had begun growing this new variety, which was soon favored by English consumers and so commanded high prices on the London market. In that year, for example, planter Richard Evens of York County agreed to pay a debt in the "best sort of sweete scented leafe tobacco." Experiments made with the strain in the early 1650s throughout the Chesapeake proved that it could be successfully grown only along the York and Rappahannock rivers. Elsewhere in Virginia and throughout Maryland, nothing but the lower-valued oronoco, in demand in northwestern Europe, would thrive. By the late 1650s, most York County planters had taken to growing the more highly valued strain, and promissory notes made in the lower Virginia tidewater began to specify that payment should be made only in sweet-scented leaf.[33]

The leaves of sweet-scented tobacco were more rounded than those of oronoco, had finer fibers, were cured a darker brown, and had a milder taste. The leaves were also denser so that a hogshead of sweet-scented tobacco weighed more than a hogshead of oronoco, saving on shipping costs and export duties, which were charged by the hogshead rather than by weight. Sweet-scented

---

a Tidewater Settlement System: All Hallow's Parish, Maryland, 1650–1783 (Chicago, 1975), 24–30; Philip Alexander Bruce, *Economic History of Virginia in the Seventeenth Century* . . . , 2 vols. (New York, 1896), I, 322; Avery Odelle Craven, *Soil Exhaustion as a Factor in the Agricultural History of Virginia and Maryland, 1606–1860* (1926; rpt. Gloucester, Mass., 1965); Cary Carson et al., "New World, Real World: Improvising English Culture in Seventeenth-Century Virginia," *Jour. So. Hist.*, LXXIV (2008), 36–49.

32. Berkeley and Berkeley, eds., *The Reverend John Clayton*, 60. The fullest description of the culture and geographic distribution of sweet-scented tobacco is David Scott Hardin, "'Alterations They Have Made at This Day': Environment, Agriculture, and Landscape Change in Essex County, Virginia, 1600–1782" (Ph.D. diss., University of Maryland, 1995), chap. 4; and Hardin, "'The Same Sort of Seed in Different Earths': Tobacco Types and Their Regional Variation in Colonial Virginia," *Historical Geography*, XXXIV (2006), 137–158.

33. York County, Virginia, Deeds, Orders, and Wills, 2: 319, microfilm, VSL. Promissory notes in tobacco are recorded throughout this volume. See also Brown et al., eds., *Archives of Maryland*, XLI, 93, 425, LIV, 221, 559–561.

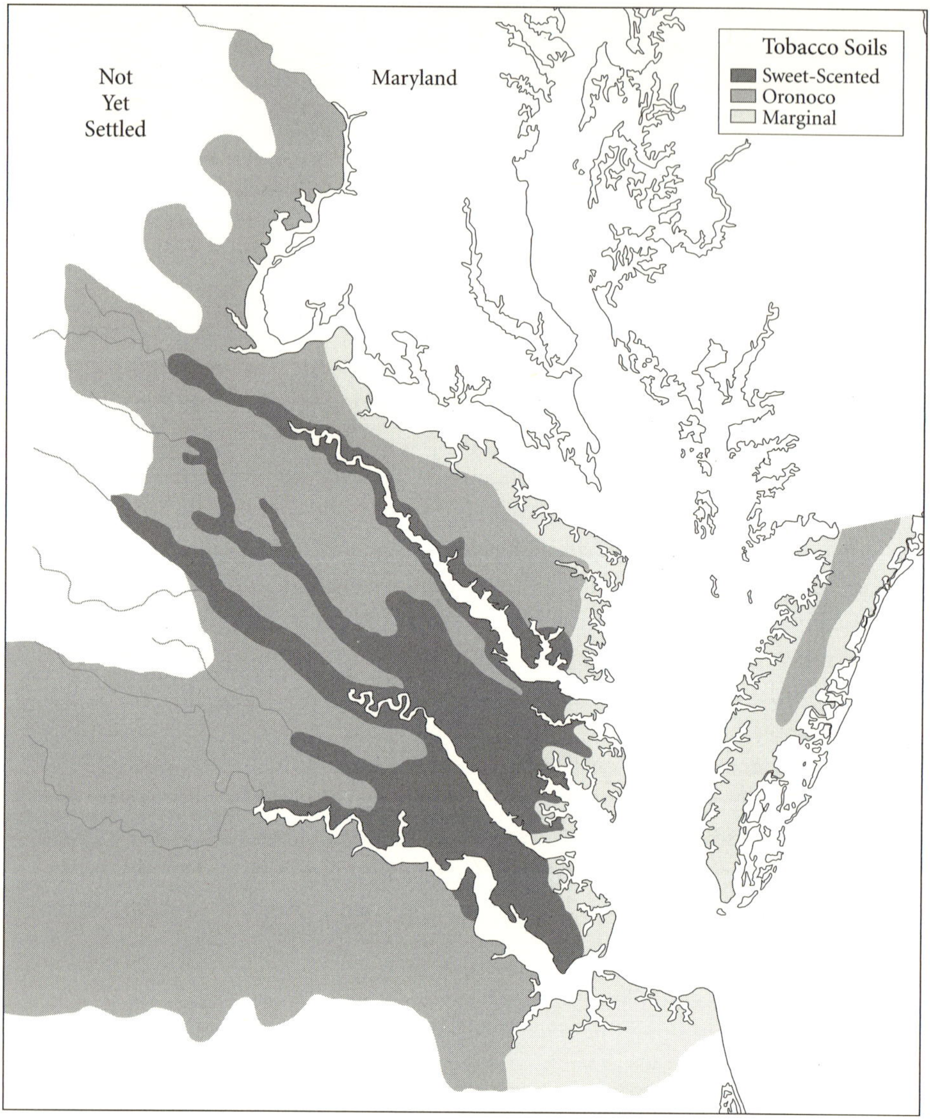

MAP 6. Tobacco Soils in Tidewater Virginia. After David S. Hardin, "'The Same Sort of Seed in Different Earths': Tobacco Types and Their Regional Variation in Colonial Virginia," *Historical Geography*, XXXIV (2006), 137–158.
Within areas settled in Virginia by 1724, soils suitable for sweet-scented comprised 1,524 square miles, soils suitable for oronoco 6,516 square miles, and marginal soils 1,652 square miles.

tobacco was topped lower than oronoco, allowing only six to eight leaves to grow to maturity, rather than the ten to sixteen typically allowed for oronoco. The bottomland soils on which sweet-scented was grown were among the best in the region, and with low topping the plants absorbed fewer nutrients each season, leaving more for subsequent crops. Sweet-scented could apparently be cultivated for as long as six to eight years on the same plot, whereas the sandier, less fertile Chesapeake soils on which oronoco was grown could produce reasonable crops for only three seasons. Consequently, planters of sweet-scented also saved by having to clear fresh land less often than did oronoco growers.[34]

Soils suitable for producing the superior strain were in short supply, however, limited to low river terraces at the ends of the middle and lower Peninsula and to narrow tracts along the shores of the Rappahannock, Mattaponi, Pamunkey, Chickahominy, and James rivers. Consequently, only a small number of planters could benefit from the discovery of the new sort. Soils capable of sustaining oronoco were much more widely distributed, and, as settlement spread, the oronoco regions expanded. Through the 1670s, the fortunes of tobacco grown in sweet-scented and oronoco areas traced a parallel course, with the prices of sweet-scented and lesser-valued oronoco rising and falling in tandem. Thereafter, the returns to the two varieties began to follow different trajectories, and the courses of subregional economies increasingly diverged.[35]

The character and conduct of the tobacco trade also changed over this period as the commodity gained rapidly in importance. English tobacco imports increased nearly fiftyfold between 1628–1631, when an average of 365,000 pounds were brought in each year, and 1672, when 17.6 million pounds were imported. After 1660, English merchants more frequently owned shares in ships plying the trade as well as in outbound cargoes. Some tobacco traders abandoned barter and let their factors in the Chesapeake sell some goods on credit. Shipownership and credit sales required that the merchants shift from adventuring single cargoes to a system of sending out continuing adventures, thus bringing more continuity to the trade and enabling relationships to develop between large planters and leading merchants. Moreover, planters were somewhat com-

34. Hardin, "'Alterations They Have Made,'" chap. 4. Barbara Hahn, "Paradox of Precision: Bright Tobacco as Technology Transfer, 1880–1937," *Ag. Hist.,* LXXXII (2008), 222, states that tobacco varieties "demand" different cultivation techniques, because they are not genetically distinct and do not reliably reproduce from seed generation after generation.

35. Hardin, "'Alterations They Have Made,'" chap. 4; Lorena S. Walsh, "Summing the Parts: Implications for Estimating Chesapeake Output and Income Subregionally," *WMQ,* 3d Ser., LVI (1999), 53–94.

pensated for the closing of commerce with the Dutch by the vigorous entry of Bristol merchants into the tobacco market, ending London merchants' virtual monopoly of tobacco sales. London's share of the trade dropped to about 60 percent in the 1660s and 1670s. Expanding imports (from less than 2 million pounds per year in the late 1630s) to more than 11 million pounds in 1676, however, still attracted ever more city dealers to the trade. The number of London importers rose from 264 in 1627–1628 to more than 300 in 1640 and reached about 575 by 1676. One-third of London's trade was in the hands of just 9 large firms, the next 40 percent was controlled by 51 substantial firms, and the remaining one-third was handled by more than 500 small firms and individuals, including tobacconists, retailers and wholesalers in other trades, merchants trading primarily in other goods, ship captains, and sailors. By 1672, the London dealers were augmented by 467 separate traders operating from Bristol. When it came time to sell their tobacco, all planters benefited from increased competition between the growing number of merchants as well as from competition between London and outport traders. Moreover, the larger firms were in a better position to extend credit to favored planters.[36]

A Middling Planter's Profits

A set of accounts for Robert Cole's farm in Saint Mary's County, Maryland, covering the years 1662–1673, supply the most detailed information about the operation of characteristic small to middling mid-seventeenth century plantations. The accounts also provide insights into the development of plantations in oronoco-growing areas and hence are typical of the kinds of farms that would proliferate across the Chesapeake landscape for the remainder of the colonial era.

Robert Cole (circa 1628–1662) was descended from an English farming family from Heston Manor, Middlesex County, about fifteen miles from London. A map of Heston and its surroundings drawn by an estate agent in 1635 identifies the house lots of six yeomen with the surname Coale living in Heston and the neighboring village of Suttone (Plate 7). Cole's parents and other relatives likely owned or held a lease for one or more lives that included a house lot, a one-to-four-acre pasture, and from five to nineteen acres of arable fields on which they grew wheat and other small grains. They might also have had the right to pasture livestock on town commons and to till strips in Heston's three common fields. Cole immigrated to Maryland in 1652

36. Jacob M. Price and Paul G. E. Clemens, "A Revolution of Scale in Overseas Trade: British Firms in the Chesapeake Trade, 1675–1775," *JEH,* XLVII (1987), 1–43; Hillier, "Trade of the Virginia Colony," chap. 6.

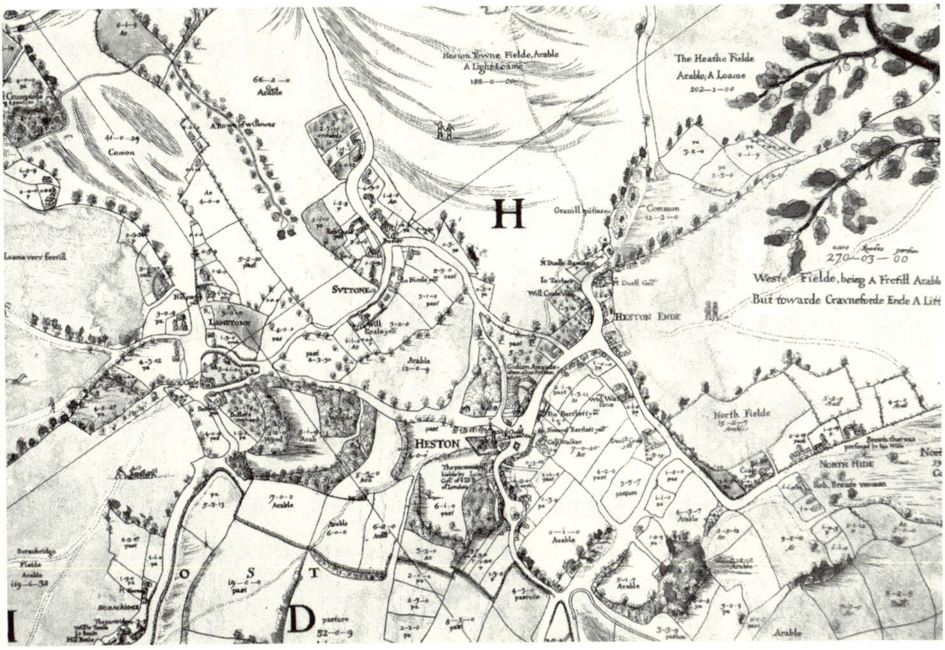

PLATE 7. Detail from *Map of the Hundred of Isleworth and the Manor Syon.* By Moses Glover, 1635. Permission, His Grace The Duke of Northumberland, Syon Park, Middlesex, UK.
The plate shows Heston Manor from which Robert Cole emigrated in the early 1650s. Members of his family had farmed here for generations.

or 1653 when he was in his early twenties with his wife, two male servants, and four young children (including at least two stepchildren). He arrived with some capital (probably about fifty pounds obtained by selling land or heritable leases on the manor), for he quickly acquired a three-hundred-acre farm on a tributary of the Potomac. A decade later, he embarked on a trip to England from which he never returned, meanwhile leaving a family of seven now motherless children and four servants in the care of two neighboring planters. One of the neighbors, Luke Gardiner, assumed responsibility for the estate and in 1673 supplied an account of his management of the farm over the preceding eleven and a half years.[37]

37. The documents are in Robert Cole Plantation Accounts, Testamentary Proceedings, 6: 118–147, MSA. They are discussed and analyzed in Russell R. Menard, Lois Green Carr, and Lorena S. Walsh, "A Small Planter's Profits: The Cole Estate and the Growth of the Early Chesapeake Economy," *WMQ*, 3d Ser., XL (1983), 171–196, and transcribed in Carr, Menard, and Walsh, *Robert Cole's World*, appendix 1; for biographical information, see appendix 4.

Cole began as a small planter with his three-hundred-acre tract (perhaps as much as ten times the acreage he might have owned in Heston) and two servants. The costs of transporting his family, feeding them for the first year, buying land, and stocking his plantation required an initial outlay of roughly £160. He left Maryland ten years later with an established farm, four servants, and a larger estate than he began with. At his departure in 1662, Cole was better-off than the majority of small planters, but still below the level of the county gentry. His family lived comfortably, but was not rich, and not so much better-off that their quality of life was sharply different from the majority of Chesapeake planters. Cole had all the land he needed to produce tobacco and food crops with the labor at hand, but his three-hundred-acre farm was not far above the median landholding for the county, which was two hundred acres in 1660. The total value of Cole's land and movable goods was about £200 in 1662, twice the median but just about the mean value of other landowning households in the county.[38]

The accounts suggest that Robert Cole worked within an English agricultural tradition modified by the demands of tobacco production in New World conditions. When it came time to write his will, Cole chose to call himself a "yeoman," suggesting that he continued to think of himself as a freeholding farmer of respectable, but not elevated, status. Certainly, when he began farming in Maryland, Cole was no careless husbandman who devoted all his energies to the staple, neglected his livestock, and depleted his land only in pursuit of quick profits. He chose his land carefully, selecting a site with good access to waterways, several springs for drinking water, and soils that were well suited to diversified agriculture. In addition to the inevitable tobacco, Cole raised grain and garden crops and established an orchard. Cole also built a dairy, purchased nails for a hog pen just before he left, and his administrator later built a henhouse on the plantation, all suggesting concern for the condition of his livestock. In his will, Cole stipulated that the herd of cattle should not be allowed to exceed fifty; thus, he was well aware of the carrying capacity of his farm. Indeed, Cole's animals were sufficiently well managed that livestock products accounted for a high proportion of farm income.[39]

The care with which Cole inventoried his possessions and instructed his overseers in preparation for his voyage to England also suggests a careful,

38. Menard, Carr, and Walsh, "A Small Planter's Profits," *WMQ*, 3d Ser., XL (1983), 171–196.

39. Cole almost certainly had already adopted the common practice of letting his cattle feed in the woods and marshes of lands adjoining his farm. One cow would have needed twenty to thirty acres of browse in order to stay healthy. Fifty head of cattle could not have been supported for long on his three-hundred-acre tract (Bowen, "Historical Ecology and the British Landscape").

intelligent farmer. Neither Cole nor the great majority of other small planters who left inventories in Maryland and Virginia during the period were what would later be known as "improving farmers." They did not own plows or harrows, nor did they manure their fields or rotate crops by any scheme other than the primitive long-fallow method. Cole was perhaps more careful and certainly better educated than the average immigrant planter in the mid-seventeenth-century Chesapeake, but he probably more closely represents reality than the old stereotype of the slovenly Chesapeake planter who depended "altogether upon the Liberality of Nature, without endeavouring to improve its Gifts, by Art or Industry."[40]

There is no doubt that tobacco was the major source of market income on Cole's plantation. Production averaged about 5,800 pounds a year and 1,667 pounds per worker. Output was fairly steady, although there were three short crops and one unusually large one, reflecting both fluctuations in the number of servants and the vagaries of climate. Tobacco accounted for two-thirds of market revenues over the eleven years, 1662–1673, never less than 45 percent and sometimes as much as 80 percent. Revenues, like output, were relatively stable, falling between thirty and forty pounds in most years (Table 5).

The state of the tobacco market affected not only foreign-sector income but also what might be gained locally. During a depression in the export trade between 1665 and 1667 and again in the early 1670s, the farm's income fell, and the share that tobacco contributed to total income rose. In the boom year of 1668, total farm income increased, but mainly from produce other than tobacco. Apparently, prosperity in the staple trade encouraged expansion and created markets for livestock and food products, whereas depressed tobacco markets discouraged immigration and led to a decrease in all areas of economic activity. A more diversified agriculture could thus provide only a slight cushion against the uncertainties of foreign trade.[41]

The local market supplied a far from negligible income to the Cole plantation, however, and in many years one that made the difference between profit and loss. As John Hammond put it in 1656, "He is a very poor man that hath not sometimes provision to put off." Livestock and livestock products—meat, tallow, hides, butter, and cheese—contributed about 20 percent of the Cole farm's income. Because cattle and swine increased with only minimal care and herd size soon outstripped the family's needs, income could be gained from this source with little effort. As Nathaniel Shrigley noted in

40. Beverley, *History and Present State of Virginia,* ed. Wright, 319.

41. For the timing of cycles in the tobacco trade, see Menard, "The Tobacco Industry," *Res. Econ. Hist.,* V (1980), 109–177.

TABLE 5. Revenues, Cole Plantation, Saint Mary's County, Maryland, 1662–1672

Year	Tobacco	Livestock	Dairy	Grains	Orchard	Misc.	Total revenue
1662	30.62	0	0	0	0	8.93	39.55
% total revenue	*77.4*					*22.6*	
1663	27.77	17.29	3.47	0	0	6.84	55.37
% total revenue	*50.2*	*31.2*	*6.3*			*12.4*	
1664	32.43	6.99	0	0	0	2.76	42.18
% total revenue	*76.9*	*16.6*				*6.5*	
1665	31.24	4.94	0	0	0	0.94	37.12
% total revenue	*84.2*	*13.3*				*2.5*	
1666	13.97	0.51	0.32	0	1.09	0	15.89
% total revenue	*87.9*	*3.2*	*2.0*		*6.9*		
1667	20.14	4.67	0	0	0	0.78	25.59
% total revenue	*78.7*	*18.2*				*3.0*	
1668	37.58	32.53	0	0	3.22	9.54	82.87
% total revenue	*45.3*	*39.3*			*3.9*	*11.5*	
1669	17.37	0	0	0	0	14.82	32.19
% total revenue	*54.0*					*46.0*	
1670	19.28	8.23	0	0	0	0	27.51
% total revenue	*70.1*	*29.9*					
1671	23.77	6.51	0	1.09	0	1.15	32.52
% total revenue	*73.1*	*20.0*		*3.4*		*3.5*	
1672	22.48	6.15	0	1.03	0	1.09	30.75
% total revenue	*73.1*	*20.0*		*3.3*		*3.5*	
Total	276.65	87.82	3.79	2.12	4.31	46.85	421.54
Mean	25.15	7.98	0.34	0.19	0.39	4.26	38.32
% total revenue	*65.6*	*20.8*	*0.9*	*0.5*	*1.0*	*11.1*	

Note: Currency values are in constant pounds. The Cole estate and other period documents report nearly all values in pounds of tobacco, the money of account in Maryland before the early 1680s. These were translated into local currency via a tobacco price series published in Russell R. Menard, "Farm Prices of Maryland Tobacco, 1659–1710," *MHM*, LXVIII (1973), 80–85. The figures in local currency were then reduced to constant value by a commodity price index discussed in Menard, Lois Green Carr, and Lorena S. Walsh, "A Small Planter's Profits: The Cole Estate and the Growth of the Chesapeake Economy," *WMQ*, 3d Ser., XL (1983), 171–196. There is no direct information on the sterling value of local currency, but most scholars agree that in the seventeenth century the value of Maryland local currency was usually the same as sterling (John J. McCusker, *Money and Exchange in Europe and America, 1600–1775: A Handbook* [Chapel Hill, N.C., 1978], 189–190; P. M. G. Harris, "Inflation and Deflation in Early America, 1634–1860," *Social Science History*, XX [1996], 479).

Source: Cole Plantation Accounts, Testamentary Proceedings, 6: 118–147, MSA.

1669, "The planters feed not their Swine nor Cattle, but kill them fat out of the Woods."[42]

The situation of the Cole plantation is probably typical of most farms of the period. Planters just starting out could spare little effort beyond clearing land, planting enough food crops to feed family and servants, and producing tobacco. On better established farms, such as Cole's would have been had he remained at home and had his wife survived, more effort could be spared to raise produce that both improved the family's diet and could be sold on the local market. As it was, income from sales of corn and cider were inconsequential, although perhaps understated, since some of the small bills paid to the estate by local residents for unspecified goods or services might well have been for agricultural products.

Most effort on mid-seventeenth-century plantations was put into producing a large crop of tobacco along with enough corn to meet the requirements of self-sufficiency. Accounts sent by several Virginia planters to the Ferrar family in England and the published accounts of George Alsop for Maryland in 1666 and of Thomas Glover for Virginia in 1671 are in general agreement on the agricultural calendar in the Chesapeake. All underscore the high labor requirements of the staple, the complementary requirements of other food crops, and the minimal effort devoted to ancillary activities.[43]

In late winter or early spring, a planter prepared a small, well-cultivated seedbed of fine mold and sowed tobacco seed saved from the previous year. He then covered the seedbed with leaves or boughs to protect the tender seedlings from frost. When the plants were "grown to the bredth of a shilling," they were drawn out and transplanted to hills in the main fields, carefully chosen to include the planter's best soils. These hills, prepared in advance while the seedlings were maturing, were spaced four to four and a half feet apart, "somewhat after the Manner of our Hop-yards," Glover noted (Plate 8). Replanting took place in moist weather, "a season," as the planters

42. Hammond, *Leah and Rachael,* in Hall, ed., *Narratives of Early Maryland,* 299; Shrigley, *True Relation, Force Tracts,* III, no. 7, 5. See also Upchurch to Ferrar, 1650, Ferrar Papers, no. 1182, CWL.

43. Period accounts of tobacco culture include Alsop, *A Character of the Province of Maryland,* in Hall, ed., *Narratives of Early Maryland,* 363; Glover, *An Account of Virginia,* 28–30; Bannister, *Natural History of Virginia,* 360–361; and Questionnaire, Aug. 30, 1646, Ferrar Papers, no. 1106, Collett to Ferrar, 1647, no. 1121, John Stirrup to John Ferrar, Jan. 26, 1649/50, no. 1152, Answer to Queries concerning Virginia from ? to [John Ferrar], Aug. 1, 1650, no. 1175, Upchurch to Ferrar, 1650, no. 1182, CWL. Gloria L. Main, *Tobacco Colony: Life in Early Maryland, 1650–1720* (Princeton, N.J., 1982), chap. 1, provides a good synthesis of these and later published accounts of tobacco culture. See also Carr, Menard, and Walsh, *Robert Cole's World,* chap. 3. In the present chapter, I have confined the text to descriptions written between the 1640s and early 1680s.

PLATE 8. Detail of "Everyday's Observation . . . Montgomery County." From Lewis Miller, Sketchbook of Landscapes in the State of Virginia, circa 1853–1867. Watercolor and ink on paper. Courtesy, Abby Aldrich Rockefeller Folk Art Museum, The Colonial Williamsburg Foundation. Gift of Dr. and Mrs. Richard M. Kain in memory of George Hay Kain.
Detail shows laborers hoeing in a recently cleared tobacco field. Techniques of land clearing and hand cultivation had not changed from the mid-seventeenth century to the mid-nineteenth century when this sketch was made.

termed it, usually in May in Virginia and June in Maryland. Once set out, the hills might have to be replanted two or three times, as groundworms and budworms often destroyed the freshly transplanted seedlings.[44]

Once the tobacco was sown and the main fields prepared, the planter then concerned himself with his next year's bread. Corn was planted between March and June, depending on the weather and other labor requirements. It, too, was grown in previously prepared hills spaced, Indian fashion, three to five feet apart, according to one observer in order to leave room to weed it thoroughly. Three to five grains were planted in each hill, and in favorable

44. Alsop, *A Character of the Province of Maryland,* in Hall, ed., *Narratives of Early Maryland,* 363; Glover, *An Account of Virginia,* 28–30.

years one hill yielded six to eight "faire and large" ears of corn. Many planters, again in the Indian fashion, also planted beans in the corn hills and allowed the legumes to run up the cornstalks. Most of the labor went into preparing the corn ground and keeping it weeded. Actual planting took little time, although the field might have to be gone over several times to replace the seeds that failed to germinate or that were consumed by birds.[45]

The tobacco and cornfields were kept clear throughout the summer growing season "with diligent weedings." Since the fields had been prepared at slightly different times, the hands could alternately hoe first the tobacco and then the corn. Fortunately, maize required no other efforts aside from weeding. After two weedings, the corn was usually sufficiently tall and robust to survive without further attention. Additional earth was thrown around the stalks, and it was then left to mature, or "laid by," in period parlance. Such was not the case with tobacco. When each plant had "put out so many Leaves as the ground will nourish to a substance and largeness that will render them Merchantable," the plant was topped to prevent flowering and to produce greater substance in the remaining leaves. Thereafter, the plant grew no higher, "but afterwards it will put out Suckers between their leaves, which they pluck away once a week, till the plant comes to perfection." In addition, voracious tobacco worms had to be regularly picked off the now rapidly growing plants.[46]

The tobacco became ripe in August in Virginia, in mid-September in Maryland. "Then in dry weather," according to Glover, "when there is a little breez of wind, they cut down what is ripe, letting it lie about four hours on the ground, till such time as the leaves, that stood strutting out, fall down to the stalk, then they carry it on their shoulders into their Tobacco-houses, where other Servants taking of it, drive into the stalk of each plant a peg, and as fast they are pegg'd they hang them up by the pegs on Tobacco-sticks, so nigh each other that they just touch, much after the manner they hang Herrings in *Yarmouth*." The trick was to harvest the tobacco only when it was fully ripe but to get it in before first frost.[47]

The family began to eat green corn as soon as it ripened—from June to

45. De Vries, *Voyages,* trans. Murphy, NYHS, *Colls.,* 2d Ser. III (1853), pt. 1, 107; *A Perfect Description of Virginia, Force Tracts,* II, no. 8, 4; Bullock, *Virginia Impartially Examined,* 9, 34–35; Williams, *Virginia . . . Truly Valued, Force Tracts,* III, no. 11, 12–13, 47–48; Shrigley, *True Relation,* ibid., III, no. 7, 5; Danckaerts, *Journal,* ed. James and Jameson, 134; Collett to Ferrar, Mar. 18, 1646/7, Ferrar Papers, no. 1111, Collett to Ferrar, 1647, no. 1121, CWL.

46. Glover, *An Account of Virginia,* 28–30.

47. Ibid. John Bannister stated that sweet-scented was hung by individual leaves, whereas stalks of oronoco were pegged (*Natural History of Virginia,* 360).

August, depending on weather and when the corn had been planted—either boiling it or roasting the ears in the shuck. Mature corn was gathered from the fields once the tobacco had been harvested, usually in October. Here the timing was much less critical, since the corn would "keep" on the stalk well into the winter. Eventually, all the ears were gathered and "lofted" in sheds or in the main dwelling to be husked and further processed as needed.

Glover's 1671 account picks up the tobacco cycle again in October: "Thus they let [the pegged plants] . . . hang five or six weeks, till such time as the stem in the middle of the leaf will snap in the bending of it; then, when the Air hath so moistened the leaf as that it may be handled without breaking, they strike it down, strip it off the stalk, bind it up in bundles, and pack it into Hogsheads for use." These later stages in the process required both good judgment and the cooperation of the weather. If the leaves were struck while too green or damp, the packed tobacco would not keep. On the other hand, if the weather remained too dry, the cured leaves would crumble with handling. The planter thus had to delay striking until there was moist weather, again "a season." He could not afford to delay packing too long, however, for the tobacco ships would arrive between November and January to carry off his finished product.[48]

The planter's wife followed a rather different schedule, albeit one that differed less from housewifery in Britain than did the fieldwork of the men. Evidence from later periods suggests that she probably helped prepare the tobacco seedbeds and aided in the transplanting process. Otherwise, she devoted her energies in the spring to planting the vegetables and tending to the livestock. On a large plantation, a mid-seventeenth-century garden might occupy as much as two acres. On smaller farms, where there were no spare laborers to help prepare the ground, the size of the garden also would likely have been smaller. Among the foodstuffs cultivated during this period were peas, cabbage, cauliflowers, potatoes, yams, asparagus, carrots, turnips, parsnips, onions, artichokes, pumpkins, cucumbers, coleworts, muskmelons, watermelons, various herbs, and the more vaguely identified "salats." The Dutch traveler Jasper Danckaerts wrote that Chesapeake vegetable gardens were "of the coarsest kinds" and were "cultivated in the coarsest manner." Still, they supplied many fresh foods for the table in a season when corn and meat might be in short supply.[49]

48. Glover, *An Account of Virginia,* 28–30; Alsop, *A Character of the Province of Maryland,* in Hall, ed., *Narratives of Early Maryland,* 363; Bannister, *Natural History of Virginia,* 360–361.

49. De Vries, *Voyages,* trans. Murphy, NYHS, *Colls.,* 2d Ser., III (1853), pt. 1, 34; Danckerts, *Journal,* ed. James and Jameson, 134–135. See also Shrigley, *True Relation, Force Tracts,* III, no. 7, 15; Bullock, *Virginia Impartially Examined,* 7, 16.

Chesapeake women doubtless also became quite efficient gatherers of wild foods. We know that they collected a variety of edible greens from the woods. Travelers' accounts note a proliferation of wild fruits—strawberries, gooseberries, raspberries, plums, cherries, and grapes—as well as a variety of nuts. So long as these were abundantly available in the wild simply for the picking, few probably bothered to spend time cultivating them in regular gardens.[50]

Once the cows began to drop their calves in the spring, women busied themselves as well with daily milking and butter making. Butter was commonly made in May until the middle of June "for men's winter store," uncured cheese from mid-June through August, and then butter again in the early fall. Spring, too, was the time to set poultry and to gather eggs. Since every household kept chickens, and some raised turkeys and geese, there was no market for fowl or eggs. Although little has been recorded about women's work routines during the months when the men were fully occupied with tobacco and corn, their efforts likely kept family and laborers adequately fed from one maize harvest to the next and provided supplemental proteins during the hot summer months when the butchering of large animals was impractical. Most women were so occupied with these tasks that they had no time for brewing beer, tending flax, or for spinning or weaving.[51]

At the onset of winter, the pace of work slackened for all. Once the tobacco was packed and the corn stored, there remained little to do but slaughter hogs and a steer or two and salt them down, hunt for game, cut firewood, and clear new land for the next year's planting.

On Robert Cole's plantation, the administrator, Luke Gardiner, seems to have made tight calculations about the amount of food crops to be raised, refusing to commit the available workforce to producing more secondary crops than were absolutely necessary for domestic consumption. Four times in the eleven and a half years covered by the account, the corn crop fell short, and it was necessary to buy more. (Two of these purchases might have been required, not for home consumption, but to pay rents in kind. There was no reason to pay local debts in exportable tobacco if they could be discharged in less valuable corn.) Only once was there a surplus of corn to be marketed.[52]

50. Bullock, *Virginia Impartially Examined,* 15–18; *A Perfect Description of Virginia, Force Tracts,* II, no. 8, 6; ? to [John Ferrar], Aug. 1, 1650, Ferrar Papers, no. 1175, CWL.

51. Collett to Ferrar, 1647, Ferrar Papers, no. 1121, Upchurch to Ferrar, 1650, no. 1182, Answer to Queries by [Michael Upchurch], Mar. 27, 1651, no. 1204, Ward to Ferrar, July 12, 1653, no. 1231, CWL.

52. Cole owned a freehold tract on Saint Clement's Manor, the property of manor lord Thomas Gerrard. Consequently, he paid rents to Gerrard rather than to the proprietor. These amounts were slightly more than the equivalent quitrents charged on proprietary land. The mar-

Given corn's relatively high yields, and a seasonal schedule that meshed into rather than conflicted with the demands of the tobacco crop, more might have been grown. Indeed, all area planters could not have been calculating corn production as closely as Gardiner did on the Cole plantation, for, if they had, periodic shortages would surely have been the result. No scarce years were reported in this period, however, and in fact nominal corn prices in all parts of the Chesapeake had begun to fall by the 1660s even in newly settled areas. The author of *A Perfect Description of Virginia* noted as early as 1649 that "a man can plant two Thousand waight a yeare of it [tobacco], and also sufficient Corne and Rootes, and other provisions for himselfe." In the next year, another observer estimated that one man could prepare ground and set enough corn for a year's worth of food in forty-eight hours. Similarly, in 1669, Shrigley thought one "laborious man" could plant, tend, and get in fifty barrels of corn without the help either of additional laborers or of horses or oxen.[53]

The demands of tobacco and corn imposed marked seasonal labor constraints that help explain much about Chesapeake husbandry. The months of April and May were frantic. On the Cole plantation, each worker had to make nearly twelve thousand hills for tobacco and corn. They had as well to keep the tobacco beds watered and weeded and begin transplanting tobacco seedlings as soon as they were ready. During these months, the workers could do nothing else if the tobacco and corn crops were to be planted in time to

---

ginally higher cost of manor rents was probably offset by privileges of pasturing livestock, cutting timber, and hunting on ungranted portions of the manor (Carr, Menard, and Walsh, *Robert Cole's World,* chap. 5).

53. *A Perfect Description of Virginia, Force Tracts,* II, no. 8, 4; *Virginia . . . Truly Valued,* ibid., III, no. 11, 12; Shrigley, *True Relation,* ibid., III, no. 7. The price series was compiled by Historic Saint Mary's City, MSA. Some Maryland commodity prices can be found in Menard, "Economy and Society," appendix V. On the suitability of corn as a complementary crop with cotton, another staple whose peak demand for labor was during the harvesting season, see Robert E. Gallman, "Self-Sufficiency in the Cotton Economy of the Antebellum South," and Raymond C. Battalio and John Kagel, "The Structure of Antebellum Southern Agriculture: South Carolina, a Case Study," both in William N. Parker, ed., *The Structure of the Cotton Economy of the Antebellum South* (Washington, D.C., 1970), 23, 33–34. These authors point out that, if a planter had a labor force sufficient to meet the harvest requirements of the staple, in all other seasons he would have a labor surplus available for other pursuits "at little or no opportunity cost" with respect to production of the staple. On methods of corn cultivation in the colonial Chesapeake, see Harold B. Gill, Jr., "Cereal Grains in Colonial Virginia" (Colonial Williamsburg Foundation Research Report Series, October 1974), typescript, CWL; David O. Percy, "Corn: The Production of a Subsistence Crop on the Colonial Potomac," National Colonial Farm Research Report no. 2 (Accokeek, Md., 1977); and Collett to Ferrar, 1647, Ferrar Papers, no. 1121, Upchurch to Ferrar, 1650, no. 1182, CWL.

mature before frost. In these circumstances, it is no surprise that planters did not raise additional corn for their livestock or produce much extra for sale. The hilling and weeding required did not easily fit into the months in which such work was best done.[54]

Other evidence suggests that Cole's administrator considered production for the local market either too unimportant or too troublesome to be worth the effort. In all but one year, the produce of the Cole orchard was entirely consumed by family and servants. If Cole had started the orchard by the mid-1650s, it should have been fully productive by the early 1660s. ("Once planted," according to one account, orchards "grow while you sleepe.") Indeed, the orchard was probably neglected in these years when there was no resident owner to look after it. Certainly, other planters in Cole's neighborhood were producing and selling large quantities of cider on the local market. Some Virginia planters had done so by the late 1640s. Any planter who went to the trouble to protect his trees from damage by livestock and to make and store the beverage could augment family income through sales to thirsty, but less industrious or established, neighbors.[55]

Similarly, with only one adult woman on hand (and by 1670 only a ten-year-old girl) dairying was also neglected. In earlier years, Rebecca Cole had procured both butter and cheese making equipment, and she usually had a maidservant to help with dairying. After Rebecca died, cheese making was tried in only one year, and surplus butter was made in just two years. Because of labor shortages and the absence of both the planter and his wife, the Cole estate was earning minimal income from local sales.[56]

54. Lois Green Carr and Russell R. Menard, "Land, Labor, and Economies of Scale in Early Maryland: Some Limits to Growth in the Chesapeake System of Husbandry," *JEH,* XLIX (1989), 407–418.

55. Williams, *Virginia . . . Truly Valued, Force Tracts,* III, no. 11, 49. For orchards, see *A Perfect Description of Virginia,* 13–15; Shrigley, *True Relation, Force Tracts,* III, no. 7, 5; Bullock, *Virginia Impartially Examined,* 7, 13–14; Glover, *An Account of Virginia,* 13–14; and Danckaerts, *Journal,* ed. James and Jameson, 134–135. According to Robert Beverley, an orchard would be bearing fully six to seven years after it was started (*History and Present State of Virginia,* ed. Wright, 314). In 1664, Thomas Gerrard, Cole's landlord, made at least 984 gallons of cider on his Saint Clement's Manor plantation, which was valued at £24.47. A lease for nearby Westwood Manor in 1664 anticipated that cider production there might exceed 1,100 gallons per year (Browne et al., eds., *Archives of Maryland,* XLIX, 517, 520, LVII, 32–36). For women's role in cider making, see Sarah Hand Meacham, "'They Will Be Adjudged by Their Drink, What Kinde of Housewives They Are': Gender, Technology, and Household Cidering in England and the Chesapeake, 1690–1760," *VMHB,* CXI (2003), 117–150.

56. Danckaerts associated failure to care for orchards once they were planted, neglecting to cask and store cider properly so that it would keep, indifferent truck gardening, and failure to do dairying even in summer with the heavy labor demands of the staple. Tobacco planters, he found,

The farm income reported in Gardiner's account, however, grossly understates total farm production, for it does not take into consideration the goods and services produced and consumed on the spot that never entered the market. Such understatements are characteristic of most of the plantation accounts examined. Planters almost invariably reckoned only market income and failed to account for the value of subsistence production. The Cole plantation like most others was nearly self-sufficient in foods that would otherwise have to be purchased. The workforce was provided shelter, fuel, and washing, and the small children were supervised and nurtured. From references to the costs of room and board that appear occasionally in other records, one can make rough estimates of the annual value of subsistence production on the Cole plantation. The value of goods and services produced on the farm proved substantial—nearly 45 percent of estimated total returns, even more than that earned by tobacco and twice the share earned from local exchange. This high level of self-sufficiency even on small plantations helped to stabilize a family's standard of living, at least enough to weather short-term fluctuations in external markets (Table 6).[57]

A decision to pursue a combined strategy of self-sufficiency in foodstuffs with maximum tobacco production for foreign exchange made sense given the state of the local marketing networks in the mid-seventeenth century. In years of heavy immigration, a number of new households were formed, all requiring corn and livestock to get started. In the absence of towns, however, there were no established local marketplaces (except in Virginia's faltering capital at Jamestown) where produce could be readily traded. Buyers usually found sellers simply through word of mouth, learning who had a cow or hogs or corn to spare through the neighborhood grapevine. Because almost all families aimed at self-sufficiency, perishables like butter, eggs, and poultry scarcely had a market value, and since there were no professional butchers and little meat exchanged between households, even fresh meat lacked an established price. And, should the local market fail, there were no assured overseas outlets. English merchants did not want to deal in small parcels of unrelated products, and ship captains did not want to take on bulky and irregular cargoes of grain or lumber so long as they could get a more uniform cargo of hogsheads of tobacco. Moreover, since the arrival of ships was concentrated in the fall and winter months, the provisioning of ships afforded

---

"bestow all their time and care in producing tobacco" (*Journal,* ed. James and Jameson, 135). On lack of equipment for cheese making and dairying, see Wesley Newton Laing, "Cattle in Early Virginia" (Ph.D. diss., University of Virginia, 1952), chap. 3; and Brown, *Good Wives,* chap. 3.

57. For methods of estimation, see Menard, Carr, and Walsh, "A Small Planter's Profits," *WMQ,* 3d Ser., XL (1983), 171–196.

TABLE 6. Sources of Returns, Cole Plantation, Saint Mary's County, Maryland, 1662–1672

Year	Estimated value of subsistence production	Export earnings	Local earnings	Total returns
1662	32.24	30.62	8.93	71.79
1663	40.76	27.77	27.60	96.13
1664	42.93	32.43	9.75	85.11
1665	40.30	31.24	5.88	77.42
1666	29.85	13.97	1.92	45.74
1667	29.19	20.14	5.45	54.78
1668	28.17	37.58	45.29	111.04
1669	27.70	17.37	14.82	59.89
1670	25.88	19.28	8.23	53.39
1671	25.12	23.77	8.75	57.64
1672	21.02	22.48	8.27	51.77
Mean	31.20 (44.9%)	25.15 (36.2%)	13.17 (18.9%)	69.52 (100%)

Note: Currency values are in constant pounds.
Sources: Cole Plantation Accounts, Testamentary Proceedings, 6: 118–147, MSA; Lois Green Carr, Russell R. Menard, and Lorena S. Walsh, *Robert Cole's World: Agriculture and Society in Early Maryland* (Chapel Hill, N.C., 1991), 79–81.

limited seasonal opportunities. "Where money and Trade is not quicke," William Bullock observed, "diet is cheap." Conversely, the rewards for producing foodstuffs for market were not great.[58]

Neither was there a stable group of local producers and another of consumers; buyers and sellers were seldom the same from one year to the next. New householders soon achieved self-sufficiency and no longer needed to purchase foodstuffs and livestock. Many established planters who could have produced commodities for the local market seem not to have done so intentionally. Individual plantations, like that of the Cole family, were too isolated

58. Collett to Ferrar, 1647, Ferrar Papers, no. 1121, Upchurch to Ferrar, 1650, no. 1182, Answer to Queries by [Michael Upchurch], Mar. 27, 1651, no. 1204, CWL; Bullock, *Virginia Impartially Examined,* 11–37 (quotation on 37); Sister Joan de Lourdes Leonard, "Operation Checkmate: The Birth and Death of a Virginia Blueprint for Progress, 1660–1767," *WMQ,* 3d Ser., XXIV (1967), 44–74, esp. 61.

and nearby households too few to make producing for neighbors worthwhile (Plate 9). Rather, they simply sold off accidental surpluses, for example, when there was extra grain in favorable years, or livestock herds grew large enough to press on farm resources, or orchards planted on tracts intended for a child's inheritance began to bear.[59]

In 1643, the Dutch merchant David de Vries advised ship captains to buy their provisions in New Netherlands, for they could not be gotten in Virginia "because every one there only produced for himself." In contrast, five years later another visitor found Virginians willing to victual ships and cited Captain Matthews as an example of a planter who was raising cattle for the express purpose of supplying ships. Similarly, another writer stated, "There is no man will ever be denyed the loane of Corne for his house-spending, and seed till the Harvest." Clearly, much depended on how good one's local contacts were.[60]

The limits to the local market can be clearly seen where estimates of population are available. For example, Charles County, sited on the Potomac contiguous to Saint Mary's, was settled about ten years later than the section of Saint Mary's County where the Cole farm lay. In 1660, Charles County had a total population of about 460. By the 1680s, there were more than 1,300 residents, a rate of growth in excess of 5 percent per year. Even growth as rapid as this, however, happened in very small increments. In the 1660s and 1670s, only five to seven new households were started each year, and in the 1680s households increased by only six to seven a year. Perhaps one or two more families began housekeeping every year in older Saint Mary's County, but no more. If we assume that these families regularly bought food or livestock for the first two years, a more established planter in the county might have found between ten and fourteen good customers within the jurisdiction. However, an ordinary planter's spatial contacts were usually restricted to a much smaller area than an entire county. It thus seems unlikely that there were more than half a dozen families who were buying provisions and livestock in any quantity within the normal reach of most planters and probably very often no more than three or four such families. Even in rapidly growing areas, the domestic market offered a planter limited opportunities indeed.[61]

59. Walsh, "Charles County, Maryland," chap. 5.

60. De Vries, *Voyages*, trans. Murphy, NYHS, *Colls.*, 2d Ser., III (1853), pt. 1, 120; *A Perfect Description of Virginia, Force Tracts*, II, no. 8, 13–15; *Virginia . . . Truly Valued*, ibid., III, no. 11, 47.

61. Lorena S. Walsh, "Estimated Charles County, Maryland, Population, 1642–1755" (paper in possession of author). For surviving census totals, see Russell R. Menard, "Five Maryland Censuses, 1700 to 1712: A Note on the Quality of the Quantities," *WMQ*, 3d Ser., XXXVII (1980), 616–626. On planter networks, see Perry, *Formation of a Society*, chaps. 4, 5; and Walsh, "Commu-

PLATE 9. Plat of the Cole Plantation, circa 1724. *Guibert v McWilliams,* Provincial Court (Ejectment Papers), MSA S549. Courtesy, Maryland State Archives, Annapolis. *Robert Cole acquired ten times more acreage in Maryland than he might have owned had he stayed in England, but neighbors were few and far between.*

TABLE 7. Expenditures, Cole Plantation, Saint Mary's County, Maryland, 1662–1672

Category	1662	1663	1664	1665	1666	1667
Clothing	13.41	13.82	12.58	16.75	7.02	9.99
Tailor's wages	1.89	1.44	1.69	0.52	1.58	3.12
Indentured servants	9.67	10.21	0	0	0	5.54
Maid's wages	0	3.33	4.67	4.29	2.72	0
Provisions	1.79	0.47	0.90	1.72	0.95	2.42
Tools	1.03	1.31	1.71	0.13	0	0
Casks	0	2.28	2.10	2.68	0.91	1.38
Rent	0	2.42	0	0	0	0
Taxes	0.90	1.59	1.87	2.00	1.39	1.41
Building repairs	0	1.88	0	4.94	0	0
Medical, education	0.05	1.40	0.93	5.19	3.81	6.18
Misc. local goods and services	2.75	6.00	12.17	0.77	0.16	4.57
Misc. imported goods and services	3.34	0.39	0.27	0.24	0.50	0.99
Unidentified	11.20	4.30	0	0	0	0.11
Total	46.03	50.84	38.89	39.23	19.04	35.71

Note: Currency values are in constant pounds. The expenditure listed for indentured servants in 1670 is for capturing a runaway servant.

[a] £8.01 in local provisions and £16.01 in imported provisions.

[b] £12.76 in taxes paid the government and £4.86 in donations to the church.

What were the major costs on a mid-seventeenth-century plantation? On the Cole plantation, clothing for the children and servants was the biggest expense. Between 1662 and 1673, expenditures for cloth, thread, pins, needles, buttons, shoes, stockings, and ready-made clothes—all English imports—were 34 percent of total costs. Labor required the second largest outlay. Cole had instructed that his workforce be maintained at its current size with new servants to be purchased when those present were freed. Gardiner apparently tried to maintain a workforce of four to five men, including Cole's maturing sons. New servants were thus purchased in five of the eleven years, on average 13 percent of total expenditures (Table 7).

nity Networks in the Early Chesapeake," in Carr, Morgan, and Russo, eds., *Colonial Chesapeake Society,* 200–241.

1668	1669	1670	1671	1672	Total	Mean
14.22	10.97	16.14	12.38	11.71	138.99	12.64
1.17	0.74	1.54	1.21	1.14	16.04	1.46
7.27	18.70	9.17	0	0	60.56	5.51
0	0	0	0	0	15.01	1.36
1.41	3.16	4.66	3.36	3.18	24.02[a]	2.18
0	0	0	0	0	4.18	0.38
2.72	1.81	0.46	0.78	0.74	15.86	1.44
0	2.04	2.06	0.64	0.60	7.76	0.71
3.86	1.42	0.83	1.21	1.14	17.62[b]	1.60
1.82	0	0.55	0	0	9.19	0.84
4.54	0	1.83	0.86	0.82	25.61[c]	2.33
5.33	2.87	7.50	1.15	1.09	44.36[d]	4.03
0.85	2.79	2.12	1.80	1.70	14.99	1.36
0	0	0	1.82	1.72	19.15	1.74
43.19	44.50	46.86	25.21	23.84	413.34	37.58

[c] £3.09 in medical expenses; £22.52 for education.

[d] Includes £16.84 in administrative expenses.

Source: Cole Plantation Accounts, Testamentary Proceedings, 6: 118–147, MSA.

Purchases of goods not produced locally—sugar, spices, salt, and soap, for example—were minimal. Except among the very rich, it seems to have been accepted practice that a family learned to do without amenities that could not be grown or raised locally. Little, too, was needed in the way of tools and household goods on this established plantation—less than 4 percent of all expenditures. Clearly, these goods would have taken up a larger share of expenditures on farms just starting up. Even so, the axes, hoes, and knives needed for Chesapeake agriculture were never a major expense. Cooking utensils, bedding, tableware, and linens cost much more and would necessarily have consumed more of the resources of new households. Most Chesapeake families clearly had an amazing capacity to get along without all but a mattress and a pot, if need be. Some perhaps had grown up in English households that were blessed with precious few comforts and had no great expectations for their own homes. Others had known better in the Old World. These families were apparently willing to sacrifice much in the way of everyday comfort for a chance to acquire and improve their own land.[62]

Comfortable housing was one of those amenities that was usually forgone. Archaeological investigations confirm contemporary comments on the limited character of period Chesapeake dwellings. Many new planters began by throwing together improvised cabins, of course, often a light frame encased in clay, to shelter themselves and their families for the first year or so until they had time to build something better. Almost universally that "something better" was a small, impermanent, and unlovely wooden house, usually of two rooms, raised upon an earthfast frame. Just as farming techniques extracted the greatest possible returns with the least amount of work, so the planters' building methods supplied shelter quickly, cheaply, and with a bare minimum of labor.[63]

The first area of economy was size; the houses of poorer planters might measure no more than twenty feet long by sixteen or twenty feet wide, whereas the homes of more prosperous farmers seldom exceeded fifty by twenty feet. The second sacrifice was permanence. House frames set in the ground (as opposed to fully framed buildings set on waterproof foundations) lasted only a decade or so without repair. On the other hand, such framing minimized both the amount of timber that had to be cut and shaped and the number and kinds of joints required to hold it together. Earthfast houses could be repaired several times or abandoned and replaced with a similar structure with much less labor than it would take to build one fully framed house with a brick or stone foundation. Other common concessions to economy included the use of wooden chimneys, which saved labor by eliminating the need to make brick, and earthen floors, which eliminated sawing floor planks or constructing sills.[64]

62. For consumer patterns, see Lois Green Carr and Lorena S. Walsh, "Changing Life Styles in Colonial St. Mary's County," *WPREHRC,* I, no. 3 (1978), 73–118; Carr and Walsh, "Changing Lifestyles and Consumer Behavior in the Colonial Chesapeake," in Cary Carson, Ronald Hoffman, and Peter J. Albert, eds., *Of Consuming Interests: The Style of Life in the Eighteenth Century* (Charlottesville, Va., 1994), 59–166; Main, *Tobacco Colony,* chaps. 5–7; and Clemens, *The Atlantic Economy,* chap. 3. For comparisons with English standards of living, see James Horn, "Adapting to a New World: A Comparative Study of Local Society in England and Maryland, 1650–1700," in Carr, Morgan, and Russo, eds., *Colonial Chesapeake Society,* 133–175; and Horn, *Adapting to a New World,* chap. 7.

63. Garry Wheeler Stone, "The Roof Leaked, but the Price Was Right: The Virginia House Reconsidered," *MHM,* XCIX (2004), 313–328.

64. Cary Carson et al., "Impermanent Architecture in the Southern American Colonies," *Winterthur Portfolio,* XVI (1981), 135–196. See also Willie Graham et al., "Adaptation and Inno-

We cannot know what expedients Robert Cole adopted in housing his family, for the probable site of the farmstead has since slid down the banks of now-eroded Tomakonkin Creek. He might have chosen a plan from a variety of vernacular English traditions familiar to Chesapeake settlers. He might also have built an early version of a house type peculiar to the Chesapeake, the "Virginia house." These structures would remain an enduring feature in the region into the nineteenth century. Beginning with a post-in-the-ground frame, such dwellings were further simplified from heavier English models in that the roof frame was constructed independently of the wall frame (eliminating much complicated joinery) and was then covered with short riven clapboards that held the roof together with only an underpinning of light rafters. The outside was then covered with riven clapboards, and the inside, in better houses, lined with more riven boards or daubed with white-limed mortar. Since Cole mentioned no other housing on his farm, the indentured servants lived and ate with the family, likely sharing sleeping space with the older Cole children in a loft or on the hall floor.[65]

Having provided shelter for himself and his dependents, the mid-seventeenth-century planter invested little more in farm buildings, with the exception of putting up one or more impermanent tobacco houses that typically equaled or exceeded human dwellings in size (Plate 10). Planters continued to experiment to find the most efficient technical solutions for providing a sheltered space for curing and securing tobacco and with the most economic forms of construction. In Virginia, raftered structures without walls (like those described in Chapter 1, above) remained in use into the 1660s. In Maryland into the 1680s, tobacco houses were often roofed with thatch, though as early as 1648 there is mention of one wall-plated, weatherboarded tobacco house similar in construction to the labor saving "Virginia house." By the mid-1670s, most newly built tobacco houses were probably of the sort later described by Bannister: "usually built of clapboard about 20 or 25 foot wide; at every 5 foot their goes a range or rail quite cross the house, the Joists are ranges themselves, and there are usually three tire [tier] above and one below. These are to bear up the sticks on which the Tobacco is hung like

vation: Archaeological and Architectural Perspectives on the Seventeenth-Century Chesapeake," *WMQ,* 3d Ser., LXIV (2007), 458–470; and Frazer D. Neiman, *The "Manner House" before Stratford (Discovering the Clifts Plantation)* (Stratford, Va., 1980), 5–21.

65. Carson et al., "Impermanent Architecture," *Winterthur Portfolio,* XVI (1981), 135–196. Collett to Ferrar, 1647, Ferrar Papers, no. 1121, Upchurch to Ferrar, 1650, no. 1182, CWL. Cary Carson dated the appearance of the Virginia house to the third quarter of the century ("The 'Virginia House' in Maryland," *MHM,* LXIX [1974], 185–196). Graham et al., "Adaptation and Innovation," *WMQ,* 3d Ser., LXIV (2007), 65–70, have subsequently found examples dating to the 1640s.

PLATE 10. Reconstructed Mid-Seventeenth-Century Plantation at Historic Saint Mary's City. Photograph courtesy, The Colonial Williamsburg Foundation. *The tobacco barn (foreground) dwarfs the middling planter's dwelling.*

shingles on a lath." Otherwise, planters economized by fencing in the smaller acreage devoted to crops rather than the more extensive ranges reserved for livestock. They did without most other sorts of farm outbuildings, especially shelters for livestock, or put up small, highly impermanent ones.[66]

The spending patterns found in the Cole account are clear evidence of the colonial structure of the Chesapeake economy. Foreign sales accounted for the bulk of farm revenues (aside from those goods and services produced and consumed on the farm without entering the market). The major part of the income that was above what was required to keep body and soul together was exported to the metropolis to purchase essential manufactured goods and services and some desirable nonessentials like hard liquor. Because of the commercial policies of England's Commonwealth and Restoration governments that funneled all colonial trade into the hands of English merchants, the planters' product enriched the customs revenues of the English king,

66. Carson et al., "Impermanent Architecture," *Winterthur Portfolio,* XVI (1981), 154; Browne et al., eds., *Archives of Maryland,* IV, 455–456, X, 278, 561, LI, 290, 330; Charles County, Maryland, Court Records, H1: 70, 139, I1: 259, MSA; Bannister, *Natural History of Virginia,* 360.

filled the coffers of English merchants, and swelled the pocketbooks of English craftsmen. Small wonder, then, that Chesapeake planters, who would have preferred an open trade with all nations, often felt they were getting all too little back in return.[67]

About 40 percent of expenditures on the Cole farm, however, were local ones, and county craftsmen were the main beneficiaries. If many plantations were nearly self-sufficient in food, they were ill supplied with essential services, especially craft skills, probably owing to the settlers' prior experience. In some parts of England at least, the occupational structure was highly segmented. Many immigrants had not learned how to construct casks, build houses and barns, tan leather, or make shoes, for example, because their parents or masters had hired specialists to do these things. If his servants were no more proficient in these tasks than the planter himself, it was much more efficient to keep the servants in the fields tending tobacco while hiring proven local workmen to perform jobs that required special skills. Providentially, such a local workforce was generally available. Although a dearth of craftsmen had been an acute problem in the Chesapeake in the 1610s and 1620s, by the 1660s and 1670s usually enough free immigrants had arrived with only their craft expertise, or skilled ex-servant coopers, carpenters, shoemakers, and tailors who had not yet formed households or acquired land were eager to work for wages.[68]

On the Cole plantation, there was a clear division between owned unskilled servant labor and purchased craft services. All the casks for packing tobacco were purchased from local coopers, even though coopering tools were available on the farm. A local carpenter was employed for building repair as well as for new construction. No attempt was made to process hides from slaughtered animals; they were sold in the raw state, and, when finished leather was needed for patching shoes, it was purchased. A tailor made all the clothes for both children and servants. Even the mending of old clothes was hired out. The purchase of such services accounted for about 14 percent of total expenditures and 35 percent of the portion spent locally.

Aside from the purchase of the services and products of craftsmen, other

67. The literature on export-led growth in early America is surveyed by David W. Galenson and Russell R. Menard, "Approaches to the Analysis of Economic Growth in Colonial British America," *Historical Methods,* XIII (1980), 316; and John J. McCusker and Russell R. Menard, *The Economy of British America, 1607–1789* (Chapel Hill, N.C., 1985). See also Pestana, *The English Atlantic,* chap. 5.

68. On the early Chesapeake occupational structure, see Walsh, "Charles County, Maryland," chap. 5; and Jean B. Russo, *Free Workers in a Plantation Economy: Talbot County, Maryland, 1690–1759* (New York, 1989).

local outlays were small. Gardiner did occasionally pay a physician or local healer to treat the children and servants, but medical bills were minimal. Only 6 percent of total expenditures went for medical care and the education of the Cole children. Even this amount was unusual for the time and place. Cole had directed that his sons be taught reading, writing, and arithmetic, an advantage that by no means all boys had, and that his daughters learn reading and sewing, a fairly unusual provision for girls at this time. The sums spent on the children's education amounted to 13 percent of local expenditures. Administration accounts for other estates suggest that usually only the elite invested comparable amounts for schooling their children.[69]

Rents and taxes constituted a hardly noticeable drain on the estate—1.9 and 4.4 percent, respectively, the latter figure including contributions to the church as well as county and province taxes. Without question, rent bore much more heavily on nonlandowners. For those who had their own land, however, rent and taxes cut into income hardly at all.[70]

A comparison of annual market income and outlays on the Cole farm shows that expenditures were closely related to income. In good years, additional goods were purchased, and spending was curtailed when earnings were low. There was much less variation in expenditures, however, than in income. The resources of the estate allowed the executor to operate in the red for a year or two, either by drawing on the earnings of earlier years or by buying on credit, which served to smooth the effects of fluctuations in income, allowing the family to maintain fairly even standards of consumption (Table 8). John Hammond noted this strategy of curtailing family consumption in 1656. "By their labours is produced corne and Tobacco, and all other growing provisions, and this Tobacco however now low-rated, yet a good maintenance may be had out of it, (for they have nothing of necessity but cloathing to purchasse)."[71]

Planters who could not draw on the resources of an established farm nor persuade merchants or large local landowners to extend them credit must have felt the effects of falling tobacco prices or a poor crop most acutely. The Cole family provides a clear illustration. Although they were better-off than most, they were certainly not living lavishly. The one new suit of clothes a year plus a hat, stockings, and shoes that Gardiner allotted the servants was

69. Lorena S. Walsh, "Child Custody in the Early Colonial Chesapeake" (paper presented at the Berkshire Conference on the History of Women, June 1981).

70. *A Relation of Maryland,* in Hall, ed., *Narratives of Early Maryland,* 350; Hammond, *Leah and Rachel,* ibid., 298.

71. Hammond, *Leah and Rachel,* in Hall, ed., *Narratives of Early Maryland,* 298.

TABLE 8. Summary of Revenues and Expenditures, Cole Plantation, Saint Mary's County, Maryland, 1662–1672

Year	Revenues	Expenditures	Balance
1662	39.55	46.03	−6.48
1663	55.37	50.84	4.53
1664	42.18	38.89	3.29
1665	37.12	39.23	−2.11
1666	15.89	19.04	−3.15
1667	25.59	35.71	−10.12
1668	82.87	43.19	39.68
1669	32.19	44.50	−12.31
1670	27.51	46.86	−19.35
1671	32.52	25.21	7.31
1672	30.75	23.84	6.91
Mean	38.32	37.58	0.75

Note: Currency values are in constant pounds.
Source: Cole Plantation Accounts, Testamentary Proceedings, 6: 118–147, MSA.

probably the standard issue of the time, the minimum required to keep them decently covered and warm enough for outdoor work in winter. Although heirs to a reasonable estate, the children fared little or no better than the laborers. They, too, could expect no more than one new set of clothes, shoes, and stockings a year, although the younger children's wardrobes might have benefited somewhat from the outgrown garments of their older siblings. Possibly some of the children's clothing was of better quality than that of the servants, but where comparisons could be made—for example, shoes and stockings—the children and servants got articles of identical value.[72]

The reason for such a cautious and frugal expenditure pattern becomes clear when one calculates the farm's net market income. In most cases, the account specifies whether expenditures were made for maintaining the workforce and the plantation or were made for the support of the family. Some entries, however, merely note purchases of food, cloth, or unspecified goods. These costs were allocated between the expenses of maintaining the workforce and net income to the family according to the ratio of servants to all members of the household. Once the operating expenses are subtracted, aver-

72. Walsh, "Charles County, Maryland," 112–116.

TABLE 9. Net Revenues, Cole Plantation, Saint Mary's County, Maryland, 1662–1672

Year	Market revenues	Workforce/ plantation expenses	Net market revenues	Net subsistence	Total net
1662	39.55	25.31	14.24	14.92	29.16
1663	55.37	37.42	17.95	15.92	33.87
1664	42.18	32.59	9.59	15.92	25.51
1665	37.12	25.98	11.14	15.16	26.30
1666	15.89	10.41	5.48	13.62	19.10
1667	25.59	18.36	7.23	12.88	20.11
1668	82.87	33.61	49.26	9.79	59.05
1669	32.19	38.76	−6.57	10.73	4.16
1670	27.51	31.87	−4.36	7.82	3.46
1671	32.52	18.26	14.26	4.89	19.15
1672	30.75	19.67	11.08	0.00	11.08
Mean	38.32	26.57	11.75	11.06	22.81

Note: Currency values are in constant pounds. Market revenues are export earnings plus local earnings (Table 6). Net market revenues are market revenues minus workforce and plantation expenses. Expenditures for the whole household (Tables 7, 8) differ from expenditures for the workforce only shown here. Net subsistence is the estimated value of subsistence production (Table 6), minus the estimated value of subsistence production for adults and servants age twelve to fifteen. Total net is net market revenues plus net subsistence.
Sources: Cole Plantation Accounts, Testamentary Proceedings, 6: 118–147, MSA; Lois Green Carr, Russell R. Menard, and Lorena S. Walsh, *Robert Cole's World: Agriculture and Society in Early Maryland* (Chapel Hill, N.C., 1991), chap. 4.

age annual net income from market transactions amounted to only £11.75 between 1662 and 1672, hardly a princely sum (Table 9).

The family, however, was doing much more than just scraping by. A comparison of the inventory Robert Cole made in 1662 with one Gardiner presented in 1673 reveals a dramatic rise in the value of the estate—from just over £208 to nearly £360 (Table 10). The value of labor, tools, and consumer goods fell, because Gardiner failed to replace many of the articles that a housewife might have insisted be renewed. The value of livestock, however, had increased fourfold, and the value of the home farm and its improvements rose by half. In addition, Gardiner had acquired more than five hundred acres of unimproved land for Cole's three sons. Altogether the estate had grown

TABLE 10. Value of Cole Plantation, Saint Mary's County, Maryland, 1662 and 1673

Category	1662	1673
MOVABLES		
Labor	27.14	15.70
Livestock	56.79	217.07
Other capital	6.50	0.28
Consumer goods	63.48	11.71
Total	153.91	244.76
LAND		
Home plantation	54.29	72.38
Other land		41.42
Total	54.29	113.80
Overall	208.20	358.56

Note: Currency values are in constant pounds.
Sources: Cole Plantation Accounts, Testamentary Proceedings, 6: 118–147, MSA; Lois Green Carr, Russell R. Menard, and Lorena S. Walsh, *Robert Cole's World: Agriculture and Society in Early Maryland* (Chapel Hill, N.C., 1991), chap. 1 n. 76, chap. 4 n. 8.

at an annual rate of 5 percent. The rate of return on the initial investment in land and labor was at least 20 percent per annum.[73]

From other available alternatives, Cole had chosen the most viable route for increasing his family's wealth. The results appear extremely modest by present-day standards but were impressive by the standards of the seven-

73. Menard, Carr, and Walsh, "A Small Planter's Profits," *WMQ*, 3d Ser., XL (1983), 171–196. Some of the cattle and swine were distributed to the older children before the 1673 inventory was taken. If we include that livestock in the estate and make a reasonable allowance for its increase, the value of the livestock would rise to roughly £265, the estate to more than £400, and the 1662–1673 growth rate to more than 6 percent per year. The documents convey the impression that Gardiner was less careful in constructing the inventory of the plantation in 1673 than Cole had been in 1662. If so, some of the categories in the column for 1673 in Table 10 may be undervalued and hence also the rate of growth in total wealth. The calculation of the internal rate of return, a measure of the rate at which invested capital is growing, and a figure comparable to the interest rate on a modern savings account, is explained ibid., 190. For a discussion of this measure, see Robert C. Higgins, *Financial Management: Theory and Applications* (Chicago, 1977), 47–51, 95–108. Useful comparative material is found in Robert William Fogel and Stanley L. Engerman, "The Economics of Slavery," in Fogel and Engerman, eds., *The Reinterpretation of American Economic History* (New York, 1971), 321–328, 338–339.

teenth century. Lending money at interest, investing in unimproved land for appreciation on resale, or renting land (assuming tenants could be found) would all have been less profitable. Only by buying retail goods for local sale might he have done better. The capital required to begin trading on a profitable scale, however, was more than middling planters like Cole could command. Indeed, Cole's rate of profit was comparable to the 5 to 6 percent per year earned by sugar planters in Barbados in the late seventeenth and early eighteenth centuries and half the usual 10 percent rate prevailing in more recently settled Jamaica. The small scale at which Cole and most other Chesapeake tobacco planters were forced to operate, not a markedly lower rate of return, was what distinguished them from West Indian sugar magnates.[74]

74. Economic historians are reluctant to compare rates of return because investment opportunities and risks were not comparable in England and in the colonies. Investments in land (which included an element of nonmonetary prestige value) were not necessarily substitutable for investments in loans that might have negative social value. Moreover, rates of return varied between short- and long-term loans. In the seventeenth and early eighteenth centuries, even English merchants, presumably the most knowledgeable investors, attempted no more than to compare rates of return between real estate and usury.

In England, the statutory maximum interest rate on loans was 10 percent during 1571–1623, 8 percent during 1624–1651, and 6 percent during 1651–1691. By the 1670s in England, returns on safe loans (secured by bonds or mortgages) fell below the statutory maximum rate. Speculative loans (like most made in the colonies) were made at higher rates through concealed interest in discounts or exchange manipulations. Nominal interest rates cited in early Chesapeake sources follow the English statutory limits, but the real rate was likely higher owing to various hidden charges (P. G. M. Dickson, *The Financial Revolution in England: A Study in the Development of Public Credit, 1688–1756* [London, 1967], chaps. 2, 19).

Richard Grasby estimates that the normal profit of an experienced and fortunate English merchant before 1650 ranged from 10 to 15 percent, though more profitable business ventures yielded rates up to 25 percent. Between 1651 and 1691, returns to joint-stock ventures fell, with rates of return on trade and manufacturing yielding between 6 and 12 percent (Grasby, "The Rate of Profit in Seventeenth-Century England," *English Historical Review,* LXXXIV [1969], 721–751). Gross returns on land are estimated at 5 percent during most of the seventeenth century, and net yields, at 3 to 3.5 percent. Urban property returned roughly 8 percent per year. In the first half of the eighteenth century, returns to money laid out in agricultural improvements in England were between 6 and 12 percent (H. J. Habahkuk, "The Long-Term Rate of Interest and the Price of Land in the Seventeenth Century," *Economic History Review,* 2d Ser., V [1952], 26–45; Christopher Clay, "Landlords and Estate Management in England," in Joan Thirsk, ed., *The Agrarian History of England and Wales,* V, part 2, *1640–1750: Agrarian Change,* in H. P. R. Finberg, gen. ed., *The Agrarian History of England and Wales* [Cambridge, 1985], 245–251). For rates of return on shipping, see Ralph Davis, *The Rise of the English Shipping Industry in the Seventeenth and Eighteenth Centuries* (London, 1962), chap. 17. For colonial comparisons, see J. R. Ward, "The Profitability of Sugar Planting in the British West Indies, 1650–1834," *Economic History Review,* 2d Ser., XXXI (1978), 197–213; and Clemens, *The Atlantic Economy,* 150–161.

The Robert Cole Plantation Compared

Lacking comparable accounts for other farms operating in the third quarter of the seventeenth century, it is difficult to determine how representative the Cole plantation was. We can begin with some speculations based on the farm's own peculiar situation. Earnings might well have been below average, for the individual with the most powerful incentives for hard work, savings, and investment, Cole himself, was absent. The farm was probably as well managed as any detached quarter manned exclusively by servants. (Gardiner's home was but two farms removed from the Cole plantation, as close or closer than most of the outlying units run by other area landowners.) On the other hand, Gardiner collected annual fees for his management that cut into net profits, and still the laborers doubtless worked less intensively than they would have had the owner been working alongside them every day.

Second, Gardiner, as an administrator, might have acted more conservatively than Cole would have done as the owner of the estate. Had the value of the property declined while Gardiner was responsible for it, he might have found himself liable for at least part of the loss. On the other hand, had he taken greater risks in an attempt to increase either the present income or future value of the estate, he stood to reap little benefit. Perhaps in consequence, Gardiner invested in livestock rather than increasing the labor force (and thus the size of the tobacco crop). Livestock needed little supervision and could be expected to increase of its own accord, and some losses among the stock would prove much less catastrophic than the death of one or more servants.[75]

Even here, Gardiner was fairly innovative. Rather than raising more cattle and hogs, he invested in horses. In 1660, horses were still scarce in the Chesapeake, and usually only the rich could afford them. (In 1647, a Virginia planter had reported that at that time most horses were imported from New England, and at a cost of fifteen or twenty pounds each they were "soe deare but little benefitt comes on them.") By the late 1660s, however, imported horses had adjusted to the new environment and had begun to breed as prolifically as the cattle and hogs introduced earlier. A mare that Gardiner purchased in 1664 along with her offspring produced thirty head of horses by 1673, besides one horse already sold. Horses would find a ready local market

75. For a discussion of the importance of livestock in the estates of orphans, see Morgan, *American Slavery, American Freedom,* 136–140, 168–170.

in the 1670s and 1680s, as planting families sought to acquire one or more for transportation and even nonhouseholding freedmen began saving to acquire mounts that would enable them to travel widely in search of work.[76]

Special circumstances, though, tended to counterbalance the administrator's conservatism. The ratio of dependents to workers on the Cole plantation declined sharply over the life of the accounts. Had Robert Cole stayed home, he would almost certainly have remarried and continued to father children who would have had to be supported from the efforts of the adult workforce. Moreover, the plantation was exceptionally lucky with both its laborers and its livestock. None of the servants died, and only one ran off and had to be captured and returned. In addition, the livestock herds grew rapidly without suffering major losses to disease, miring, or predators. One would like to argue that these several considerations balanced each other and that the returns to the Cole plantation were close to the average. To do so, however, would be to build upon a much too flimsy base.[77]

Comparison with travelers' accounts and with the numerical data on plantation performance and wealth levels that can be collected for the period helps further to put the Cole account in perspective. In 1652, Robert Cole spent about £60 transporting himself and a labor force of two servants to the Chesapeake. This was the cost estimated by the writer of *A Relation of Maryland* (1635), by William Bullock in *Virginia Impartially Examined* (1649), and by Edward Williams in *Virginia . . . Truly Valued* (1650), and something less than the £100 cited by Edmund Plowden in 1648. Once arrived, though, Cole's income doubtless fell far short of the £170 per year (or about £57 per hand) anticipated by Bullock, or the £50 to £70 per laborer expected by Plowden, or even the £49 or £50 cited in *A Relation of Maryland.*[78]

By the 1660s, Cole's servants were generating an income ranging from £4 to £9 per hand per year from tobacco, averaging £6.5 over the eleven-year period for which records are available, an annual market income from all sources ranging from just less than £5 up to £20 per laborer (averaging £9.75), and total income (including subsistence production) of approximately £18 per

76. On horses, see Lois Green Carr and Russell R. Menard, "Immigration and Opportunity: The Freedman in Early Colonial Maryland," in Thad W. Tate and David L. Ammerman, eds., *The Chesapeake in the Seventeenth Century: Essays on Anglo-American Society* (Chapel Hill, N.C., 1979), 217–218; and Main, *Tobacco Colony,* chap. 1. Thomas Glover's narrative of 1676 was the first contemporary account to note an abundance of horses (*An Account of Virginia,* 19).

77. On livestock increase, see Carr, Menard, and Walsh, *Robert Cole's World,* appendix 3.

78. *A Relation of Maryland,* in Hall, ed., *Narratives of Early Maryland,* 93–96; Bullock, *Virginia Impartially Examined,* 35–36; Williams, *Virginia . . . Truly Valued, Force Tracts,* III, no. 11; Beauchamp Plantagenet [Plowden], *Description of the Province of New Albion,* ibid., II, no. 7, 32–34.

hand each year. Although falling well below the most optimistic contemporary estimates, these figures approached the range of incomes that three other writers cited as average returns for workers engaged in tobacco production. Father Andrew White estimated a yearly income of about £25 per laborer in 1639; the author of *A Perfect Description of Virginia* estimated a yearly income of £20 per laborer circa 1649; and Samuel Hartlib cited an average annual market income from tobacco of £14 per hand in 1655. These on-the-ground estimations from the only well-documented mid-seventeenth-century plantation are significant, for, as we will see in subsequent chapters, returns from this plantation staffed only with indentured servants proved equal to or better than returns obtained from poorly managed plantations worked primarily or exclusively with enslaved Africans a few decades later, but they were well below the annual revenues that the most efficient early-eighteenth-century slaveowners were able to extract from their bondpeople.[79]

Similarly, although the value of the Cole plantation did indeed grow, Cole's total wealth rose at a much slower rate than some of the promotional tracts suggested. Bullock claimed that an outlay of £60 in the early 1650s would yield an estate worth £120 at the end of the first year. Through judicious reinvestment, he estimated, the same planter might be worth £600 in two to three years. Instead, Robert Cole's investment of £60 in the early 1650s produced an estate of about £200 in 1662 and roughly £360 ten years later.[80]

Other evidence suggests that such respectable, if more modest, returns were not atypical. Menard has estimated that real, nonhuman, physical wealth per capita grew at about 2.7 percent per annum on Maryland's lower western shore from the 1650s to the 1680s. Period inventories from a number of Chesapeake counties demonstrate that, among probated estates, mean movable wealth increased at even higher rates, and the average worth of decedent householders grew faster still. In some of the earlier-settled Virginia counties, aspiring planters with few resources were probably already discouraged by a scarcity of good, undeveloped land. There were, however, numerous and not far distant places they could go where they might succeed in duplicating Robert Cole's achievements.[81]

79. Father Andrew White to Lord Baltimore, February 1638/9, *The Calvert Papers,* 3 vols., Maryland Historical Society Fund Publication, nos. 28, 34, 35 (Baltimore, 1889–1899), I, 206–207; *A Perfect Description of Virginia, Force Tracts,* II, no. 8, 6; Samuel Hartlib, *The Reformed Virginian Silk-Worm . . .* (London, 1655), ibid., III, no. 13; Krugler, ed., *To Live Like Princes,* 35.

80. Bullock, *Virginia Impartially Examined,* 37–44.

81. Russell R. Menard, "Comment on Paper by Ball and Walton [D. E. Ball and G. M. Walton, "Agricultural Productivity Change in Eighteenth-Century Pennsylvania," 102–117]," *JEH,* XXXVI (1976), 124–125; Menard, P. M. G. Harris, and Lois Green Carr, "Opportunity and

The evidence from the Cole account shows that most promotional writers exaggerated the profits that could be earned, failed to take into account many of the continuing costs of maintaining the labor force, and ignored altogether the sharp reduction in income per head that was inevitable once a planter married and had young, nonproductive children to support. On the other hand, the Cole account also suggests that both contemporary complaints and some scholarly studies have painted an excessively pessimistic picture of a small planter's prospects in the middle of the seventeenth century. Although he did not become a rich man, after ten years of farm building Robert Cole did enjoy all the benefits of independent landownership that had induced Robert Wintour to emigrate in the 1630s.[82]

Cole also enjoyed benefits other than profit. He had the pleasure of the company of other like-minded Catholics living on and around Saint Clement's manor and joined with them to build a Roman Catholic chapel on nearby Newtown Neck, to which he paid voluntary taxes. Honor came from the opportunity to practice his faith freely and openly and from the ability to fulfill family obligations that the "worldly Estate as the lord in mercy hath lent" permitted. These obligations included providing his widowed mother in Heston with a yearly income should she "bee necessitated and empoverished," arranging for his children to "bee brought and taught in the Roman Catholique Religion," and providing each of his sons with the means to become independent farmers when they came of age and his daughters with household goods and education so that they might marry well. Cole never identified himself in a record as "mister" or "gentleman," but he could have chosen to so style himself. Although his role in local government was limited to frequent jury service and an ensign's post in the militia, his wealth equaled and even surpassed that of many men with seats on the county bench and in the provincial assembly. Beginning in 1658, he was occasionally called "mister" or "gentleman" in public records, a clerical inconsistency reflecting the transitional stage he had reached in his career. Had he lived longer, he might have had the honor of being appointed to the county bench and assuming the designation of gentleman.[83]

Inequality: The Distribution of Wealth on the Lower Western Shore of Maryland, 1638–1705," *MHM,* LXIX (1974), 169–183; and Harris, "Integrating Interpretations of Local and Regionwide Change in the Study of Economic Development and Demographic Growth in the Colonial Chesapeake, 1630–1775," *WPREHRC,* I, no. 3 (1978), 35–72.

82. For example, Morgan, *American Slavery, American Freedom,* 180–195. For an assessment of the costs and benefits of emigration, see Lois Green Carr, "Emigration and the Standard of Living: The Seventeenth Century Chesapeake," *JEH,* LII (1992), 271–291.

83. *Robert Cole's World,* 28–29, 126. Quotations are from Cole's will transcribed on 170–174.

Farm Building and the Tobacco Economy

Statements that excessive concentration on tobacco reduced planters to an untenable condition are not hard to find in either official pronouncements or scattered private correspondence. As Virginia's governor Berkeley put it in 1663, "They can neither handsomely subsist with it, nor without it." Contemporary complaints about low prices, overproduction, glutted markets, shortages of manufactured goods, scarce and dear shipping, and planter indebtedness were not, however, constant across the period. As Menard has demonstrated, expressions of distress and legislative action to limit tobacco production and to promote a more diversified economy tended to cluster during cyclical downturns in the tobacco trade—in the mid-1620s and early 1630s, the late 1630s and early 1640s, and the mid-1660s. In addition to low tobacco prices, the second Anglo-Dutch war and the resulting exclusion of the Dutch from the Chesapeake trade contributed to shortages of imported goods and raised return freight rates. But, in contrast, both complaints and legal restrictions were rare or entirely absent during upturns in the trade—the mid-1630s, late 1640s and early 1650s, and the late 1670s. The period, then, was not uniformly bad. In addition, many of the people who complained the loudest had special reasons for doing so. Virginia's governor Berkeley, for example, wanted to demonstrate the disastrous consequences of the Navigation Acts in order to effect their repeal and also hoped to secure crown backing for an ambitious program to diversify Virginia's economy that in many ways resembled the fondest schemes of the old Virginia Company.[84]

There is no doubt that the farm price of tobacco declined sharply across the period, from two to three shillings a pound in the early 1620s to only a penny a pound by the 1670s. That farm income did not drop nearly so precipitously is also clear, for the costs of production also fell. We have seen that, as the number of farms increased, as the Indian threat diminished, and as livestock herds multiplied, the food supply became assured and the prices of grain and meat declined. Similarly, as the Chesapeake trade became better organized, the number of ships (primarily from London, Bristol, and Holland) arriving annually increased from about thirty in the late 1640s to fifty in the early 1650s. As a consequence of larger supplies and more competition

84. Sir William Berkeley, *A Discourse and View of Virginia* (1663; rpt. Norwalk, Conn., 1914), 6; Menard, "The Tobacco Industry," *Res. Econ. Hist.*, V (1980), 109–177; Morgan, *American Slavery, American Freedom*, 186–195; Leonard, "Operation Checkmate," *WMQ*, 3d Ser., XXIV (1967), 44–74; Rainbolt, *From Prescription to Persuasion*, chaps. 2–4; Billings, "Sir William Berkeley," *VMHB*, CIV (1996), 433–454; Billings, *Sir William Berkeley*, 66–74, 114–116, 130–155, 195–199.

between merchants, the prices of clothing and tools fell, and the cost of credit must have also fallen as risks diminished and merchants became satisfied with lower returns for the goods they ventured. At the same time, the cost of indentured servants was at worst stable from the 1630s to the 1660s and then clearly dropped into the early 1670s. Although mortality levels remained high, nonetheless the chances for a new laborer's surviving seasoning did increase substantially across the middle decades of the century. Consequently, for the planter, the annual cost of indentured workers fell.[85]

Nor were planters content simply to reap the benefits that obtained from the regularization of the tobacco and servant trades. In addition, they took steps to increase the productivity of individual laborers and to reduce further their costs of production. Selective breeding produced better and more diverse varieties of the weed, enabling planters to choose the strains most suited to the particular soils they farmed. Improvements in methods of transplanting and cutting tobacco greatly increased the number of plants a worker could tend. In the late 1620s, a hand could grow between one thousand and three thousand plants. By the end of the century (and probably by the 1670s), ten thousand plants were considered the norm. Yields per hand consequently rose from a mean of four hundred pounds in the 1620s to about fifteen hundred pounds per worker in oronoco-growing areas by the 1650s, and to about thirteen hundred pounds in the 1670s in Virginia counties that concentrated on the more valuable sweet-scented strain.[86]

85. Menard, "A Note on Chesapeake Tobacco Prices," *VMHB,* LXXXIV (1976), 401–410; Menard, "Farm Prices of Maryland Tobacco, 1659–1710," *MHM,* LXVIII (1973), 80–85; Menard, "The Tobacco Industry," *Res. Econ. Hist.,* V (1980), 109–177; Russell R. Menard, "Plantation Empire: How Sugar and Tobacco Planters Built Their Industries and Raised an Empire," *Ag. Hist.,* LXXXI (2007), 309–332; James F. Shepherd and Gary M. Walton, *Shipping, Maritime Trade, and the Economic Development of Colonial North America* (New York, 1972), chaps. 1, 2, 5. On the mortality decline, see Morgan, *American Slavery, American Freedom,* 180–185; Beauchamp Plantagenet [Plowden], *Description of the Province of New Albion, Force Tracts,* II, no. 7, 45; *A Perfect Description of Virginia,* ibid., II, no. 8, 6–7; and Hammond, *Leah and Rachel,* in Hall, ed., *Narratives of Early Maryland,* 285.

86. A Virginia statute of 1633 enjoined planting of only the "large sortes" (Hening, ed., *Statutes at Large,* I, 204–205). A letter of Reverend Moray in 1665/6 mentions a "new sort of sweet sented Tobacco, which yet I have not had time to have emproved" ("Letters Written by Mr. Moray, a Minister to Sr. R. Moray, from Ware Diver in Mock-Jack Bay, Virginia, Feb. 1, 1665," *WMQ,* 2d Ser., II [1922], 158). Maryland Provincial Court records of the 1650s show widespread experimentation with sweet-scented varieties along the Potomac. The strain did not prosper there, however, and was soon abandoned in favor of oronoco varieties better suited to the locality (Browne et al., eds., *Archives of Maryland,* XLI, 93, 289–290, 425; *Calvert Papers,* I, 263). For outputs per laborer in the 1620s, see Chapter 1, Table 4; Morgan, *American Slavery, American Freedom,* 177–178; Menard, "The Tobacco Industry," *Res. Econ. Hist.,* V (1980), 145–146; Browne et al., eds., *Archives of Maryland,* XXXVI, 267; and Main, *Tobacco Colony,* 39.

TABLE 11. Tobacco Crops per Laborer, Virginia and Maryland, 1635–1679

	Virginia			Maryland		
Years	*N*	Mean crop per laborer	% change	*N*	Mean crop per laborer	% change
1635–1649	15	1,073		9	739	
1650–1659	25	1,025	−5	4	1,475	50
1660–1669	18	1,091	6	18	1,532	4
1670–1679	20	1,356	20	14	1,559	2

Source: Appendix 1.

Increased yields were probably achieved almost entirely by an increase in the number of plants each worker could manage. Some writers have speculated that growers learned how to increase the weight each plant produced in response to laws limiting the number of plants that one worker could tend. Weights per plant probably did not rise, though, once planters introduced further measures of quality control—topping, suckering, the destruction of ground leaves, and more selective cultivation of second growth crops. A pattern of increase in crop output per worker can be seen in both Virginia and Maryland, although the data is based on a limited number of observations (Table 11).[87]

Restrictions imposed by the Virginia legislature on the amount of tobacco that planters might cultivate appears to have had some effect on the size of crops that Virginia planters pitched in the 1630s, early 1640s, and again in the 1660s. The effect of the legislation was probably temporary, but it at times gave Maryland planters, who were encumbered by no restrictions, a temporary advantage over Virginia growers. Nor is it certain how well the rules for stinting or for a system of inspection to destroy bad tobacco were enforced. Part of Cecil Calvert, second Lord Baltimore's opposition to intercolonial

87. Menard, "The Tobacco Industry," *Res. Econ. Hist.,* V (1980), 146. William Capps stated that Virginia planters were getting a quarter of a pound of tobacco per plant in 1623 (*RVC,* II, 524). In 1634, David de Vries expected settlers in Guiana to get half a pound of tobacco per plant (*Voyages,* trans. Murphy, 62). In 1648, Edmund Plowden said Virginians got a pound of cured tobacco per plant in fresh, prime land, but only one-fifth to one-sixth of a pound from plants growing in worn land (*Description of the Province of New Albion, Force Tracts,* II, no. 7, 25–26). But, in 1697 and again in 1724, plants were estimated to yield only one-eighth of a pound (Arthur Pierce Middleton, *Tobacco Coast: A Maritime History of Chesapeake Bay in the Colonial Era* [Newport News, Va., 1953], 378–379).

stint plans of the 1660s in fact turned on the issue of enforcement. A system that relied on neighbors informing on one another was, he believed, both unworkable and likely to increase social tensions. "There wilbe no means of convicting transgressors without either encouraging servants to inform against their Masters, or next Neighbours one against another, both which wilbe odious and dangerous for perjury and setting families in combustion both within themselves, and one against another." Otherwise, many officials would have to be appointed to visit every tobacco farm soon after the day appointed to cease planting, "which in so many Plantations so farr distant one from another . . . wilbe a matter of great charge and trouble and no officers wilbe found willing to undertake it without considerable salary for their paines, and no way to raise those salarys." Planters also feared that there "wilbe that negligence, corruption or partiality in it, as will render it ineffectuall to the end intended as was found by experience in a like Act in Virginia for burning Tobacco, which by this meanes came to nothing." In addition, too many planters conceived it "their Birthright as Englishmen of planting what quantity they please of that Commodity." Indeed, Virginia governor Berkeley noted that the Virginia legislature's imposed limits on the amount of tobacco that planters could raise encouraged some to move to Maryland.[88]

By the mid-1650s, measures enacted by the Virginia and Maryland assemblies were much less severe, prohibiting the cultivation of ground leaves and of "seconds," rather than placing limits on the total amount of tobacco that might be cultivated. In 1659, Virginia finally suppressed entirely the cultivation or sale of ground leaves, which might have made up as much as a quarter of the tobacco previously sold, but partially compensated planters in 1663 by again allowing them to tend seconds at their discretion. Whether these measures were effectively enforced, such legislation probably did tend, in the

88. In the 1630s, English officials on the Antilles and in Virginia formed a cartel aimed at reducing the supply of tobacco grown on both the mainland and in the Caribbean. In 1639, the English on Saint Christopher even forced French settlers there to support the scheme (Robert C. Batie, "Why Sugar? Economic Cycles and the Changing of Staples on the English and French Antilles, 1624–54," *Journal of Caribbean History,* VIII–IX [1976], 212). Virginia and Maryland laws regarding tobacco are thoroughly surveyed in Vertrees J. Wyckoff, *Tobacco Regulation in Colonial Maryland* (Baltimore, 1936). Some assessment of the degree of enforcement requires careful study of the county court records that have survived for the period, a task that remains to be done for most jurisdictions. In remote areas such as the lower Eastern Shore, the acts regarding tobacco might have been little enforced. The Virginia Assembly explained in 1628 that the planters' indebtedness had caused them not to stint as much as the law demanded. Debt no doubt continued to be an incentive for evasion ("The Randolph Manuscript," *VMHB,* XVII [1909], 3–4). For Baltimore, see Browne et al., eds., *Archives of Maryland,* III, 509. For Virginia governor Berkeley's aims, see Billings, *Sir William Berkeley,* 150–152, 188–199.

long run, to encourage voluntary adoption of some quality controls. By the 1670s, conventional wisdom dictated, for example, limiting the number of leaves allowed to grow to maturity on each plant and more careful topping, suckering, and worming of the crop.[89]

Indeed, merchants might have been more successful than lawmakers in encouraging planters to tend and cull their crops more carefully. As the price of the staple fell, traders became more selective in the types of tobacco they would accept in payment of debts. Promissory notes as well as legislation increasingly included definitions of "merchantable" tobacco, which excluded not only stalks, stones, and dirt but also ground leaves and seconds. The quality-control efforts of both merchants and lawmakers, however, continued to be undercut to some extent by a clandestine trade between planters and sailors in legally unmerchantable grades of tobacco.[90]

Better curing and packing of the harvested crop was more a matter of art than legislation, but here, too, growers made improvements. Planters were able to reduce substantially the costs of transporting the staple through better packaging. Nominal freight rates fell somewhat from the 1630s to the 1660s, although they increased sharply in wartime. Planters' efforts to improve packing, however, effected much more substantial reductions. By the late 1630s, most Virginia tobacco was being shipped in casks of more uniform shape and size, a move that made possible fuller use of available shipping space. In 1640, Virginia legislators defined a ton of freight as equivalent to four hogsheads of tobacco, regardless of weight. At first, Virginia planters managed to pack only about 150 pounds into each hogshead, but by the 1640s growers had learned

89. Hening, ed., *Statutes at Large,* I, 487–488, 524, II, 119, 190–191; Berkeley, *Discourse and View of Virginia,* 6; Browne et al., eds., *Archives of Maryland,* I, 360, 372. See also Beverley, *History and Present State of Virginia,* ed. Wright, 59, 70–72.

90. A Virginia law of 1632, for example, prohibited tending more than fourteen leaves or gathering more than nine and directed that the crop be struck before the end of November. As early as 1628, the Virginia Assembly advised tending only twelve leaves to a plant instead of the twenty-five or thirty that had previously been allowed to grow ("Randolph Manuscript," *VMHB,* XVII [1909], 3–4). By 1676, Thomas Glover noted that planters had accepted these practices. They normally tended twelve to fourteen leaves per plant in rich lands and only nine to ten in meaner ground (Hening, ed., *Statutes at Large,* I, 164–165; Glover, *An Account of Virginia,* 28–30). On the other hand, some other practices seem to have been difficult to root out. The 1632 act also forbade the cultivation of slips of old stalks or second crops. More than a century later, the Virginia legislature was still trying to devise effective methods to curtail the culture of second crops. In 1720, for example, lawmakers found it necessary to spell out that the forbidden "cultivating" included weeding, topping, hilling, succoring, housing, curing, stripping, or packing seconds, suckers, or slips. In the 1730s, they directed constables to view all tobacco patches in their precincts in late summer to further discourage what apparently continued to be a common practice (Hening, ed., *Statutes at Large,* I, 164–165, 189, II, 119, III, 34, 435–440, IV, 87–89, 241–244, 507–508).

TABLE 12. Weights of Tobacco Hogsheads, Virginia and Maryland, 1627–1679

	Virginia				Maryland	
Year	Hogsheads (*N*)	Mean pounds	Low pounds	High pounds	Hogsheads (*N*)	Mean pounds
1627–1639	268	144	141	300	—	—
1640–1649	179	324	220	419	92	312
1650–1659	677	398	225	624	388	381
1660–1669	761	463	332	670	238	391
1670–1679	222	506	358	701	167	397

Sources: For Virginia, the data are compiled from Neville Williams, "England's Tobacco Trade in the Reign of Charles I," *VMHB,* LXV (1957), 421–449; and York County Deeds, Orders, and Wills, 1–6, microfilm, VSL. For Maryland, they are adapted from Russell Robert Menard, "Economy and Society in Early Colonial Maryland" (Ph.D. diss., University of Iowa, 1975), 455.

how to double that amount (Table 12). In the 1650s, Chesapeake residents attempted to effect further savings by making ever-larger containers. English merchants' complaints, though, forced the Virginia and Maryland legislatures to specify the legal size for hogsheads in 1658.[91]

By the 1670s, hogsheads packed in Maryland averaged four hundred pounds each, and Virginia hogsheads at least five hundred pounds apiece. Some elite Virginia planters managed to press up to seven hundred pounds of tobacco into each container. Some of this increase might have been achieved by more carefully stripping the leaves from the stalks, by smoothing them, and by packing different sizes and grades of leaf separately. Separation by grade became feasible once more planters began producing several hogsheads per year. Virginia planters who were growing sweet-scented tobacco could effect more substantial reductions in freight charges since they were able to pack more of this variety into a standard cask. By the 1650s, hogsheads from sweet-scented counties consistently weighed more than those coming from oronoco areas. Probably as early as the 1660s, this differential had stabilized to an additional one hundred pounds per hogshead.[92]

91. For period freight rates and hogshead weights, see Menard, "The Tobacco Industry," *Res. Econ. Hist.,* V (1980), 146–147; Menard, "Economy and Society," appendixes A10, A11; and Shepherd and Walton, *Shipping,* 49–90, 191–192. For laws regulating the size of hogsheads, see Hening, ed., *Statutes at Large,* I, 225, 456; Browne et al., eds., *Archives of Maryland,* I, 371–372.

92. Merchants encouraged planters to pack their tobacco in small bundles made up of leaves of the same size (Capt. Richard Longman to Mr. Richard Jones, June 15, 1659, in *Tyler's Quarterly Historical and Genealogical Magazine,* I [1919–1920], 272). Additional processing of the harvested

So long as cost reductions could be made in the production of the staple, the great majority of planters spent their energies to this end. Both literary evidence and period probate inventories show that efforts to cultivate the land more intensively with plows, to manure it, to grow low-yielding English grains, or to improve livestock through selective breeding, penning, or winter feeding were few and far between. Even the most ardent advocates of more diversified enterprises recognized that the certain annual income that could be gotten from tobacco put attempts to produce other staples "out of the heads and hands of the Common people." It was "the wealthier sort of men [who] must begin and give the example, and make the gain of other Commodities as apparent to them, by the effecting them to perfection, or it will not (as it hath not hither unto) go forward."[93]

Many Virginia leaders nevertheless continued to advocate the kinds of ambitious projects so often urged by the old Virginia Company—producing a variety of commodities that England could not itself produce. And they sought to achieve these objectives by similar methods—mainly exhortation, restraints, and bounties that benefited only the elite. Their actions seem to have stemmed from a naive belief that planters could shift overnight to other economic activities if they only would and from an exaggerated faith in the natural potential of the country. Meanwhile, in practice large planters strove to produce ever-greater quantities of tobacco on their own plantations, thus contributing significantly to the very problem their public policies sought to cure.[94]

Some of the contemporary proponents of diversification recognized the economic disincentives that doomed these projects. In 1628, it was stated that pitch and tar and pipe staves could not be produced profitably because

---

leaves must have contributed to the increase in hogshead weights. In 1628, it was asserted that stripping would spoil the tobacco and hinder both other work and expeditions against the Indians ("Randolph Manuscript," *VMHB*, XVII [1909], 3–4). A Virginia law of 1640, however, mentioned stripping and smoothing as standard practice (Hening, ed., *Statutes at Large*, I, 224–225). In Maryland, various attempts were made to force planters to pack ground leaves separately from better grades of the weed (for example, Browne et al., eds., *Archives of Maryland*, X, 117, 380), and period promissory notes generally specified hogsheads free from ground leaves (X, 361, 497, 510, XLI, 192, 280, 537, 541).

93. See, for example, Lois Green Carr, "Diversification in the Colonial Chesapeake: Somerset County, Maryland, in Comparative Perspective," in Carr, Morgan, and Russo, eds., *Colonial Chesapeake Society*, 342–388. Contemporary sources make the same point; for example, *A Perfect Description, Force Tracts*, II, no. 8, 6 (quotation); Berkeley, *Discourse and View of Virginia*, 4–8; Bullock, *Virginia Impartially Examined*, 32–33; Glover, *An Account of Virginia*, 12.

94. Rainbolt, *From Prescription to Persuasion*, chap. 4; Leonard, "Operation Checkmate," *WMQ*, 3d Ser., XXIV (1967), 44–74; Beer, *Origins of the British Colonial System*, II, chap. 8.

of high labor costs and freight rates, that the settlers lacked the expertise to make wine or potash, and that iron manufacture required too much capital. The author of *A Perfect Description of Virginia* noted in 1649 that hemp, flax, indigo, pitch, tar, potash, and soap ash had all been tried with some success; mulberries for silk and grapes for wine were widely available, but "onely hands are wanting to this and other workes." In the same year, Bullock stated that the West Indies already had a head start in sugar and indigo, so these crops were not going forward in Virginia. Wines, dyes, drugs, silk, potash, and mines, he declared, all required too much experience, time, and capital. Pitch, tar, and timber products were bulky and thus too expensive to transport unless desperate captains were willing to carry them cheaply for lack of other freight. Governor Berkeley said he expected to produce silk, flax, hemp, pitch, tar, iron, ship's timber, and potash within a few years but admitted a want of skillful men for all these enterprises and insufficient capital to pursue iron. Similarly, in 1665/6 Reverend Moray, a recent settler on the Ware River, reported he was experimenting with silk, barley, rice, and distilling. None of these projects had yet produced any salable products, however, and he admitted he had lost more than twenty thousand pounds of tobacco in the attempt.[95]

In contrast to the earlier Company period, only a handful of leading planters even considered anything other than producing the maximum amount of tobacco possible. The few highly publicized attempts to produce silk, rice, potash, salt, hemp, and flax failed for lack of skilled workmen and lack of capital, just as they had in the initial years of settlement. Most attempts at diversification seem to have been abandoned by the 1670s. Glover found in 1676 that there were no vineyards, that Virginians considered themselves too poor to attempt ironworks, and that mulberry trees "were planted at first to feed Silkworms, but that design failing, they are now of little use amongst them." Planters great and small now concentrated all their efforts on the staple. As an Anglican minister in James City County lamented, "For as the old planters began with Smoke soe they will dye in Smoke I think for never were men soe bewiched with soe base a Commodity in a Country which will yield better things."[96]

95. "Randolph Manuscript," *VMHB,* XVII (1909), 3–4; *A Perfect Description of Virginia, Force Tracts,* II, no. 8; Bullock, *Virginia Impartially Examined,* 32–34; Berkeley, *Discourse and View of Virginia,* 5; "Letters Written by Mr. Moray," *WMQ,* 2d Ser., II (1922), 157–161.

96. Glover, *An Account of Virginia,* 12–16; Edward Johnson to Ferrar, Mar. 25, 1650, Ferrar Papers, no. 1160, CWL (quotation). Various attempts at diversification in these years are discussed in Bruce, *Economic History of Virginia,* I, chap. 6; Gray, *History of Agriculture,* I, chap. 8; Theodore

The Cole account supplies some insight into why small-scale farms were the dominant feature of the Chesapeake in the third quarter of the century. Initially, most settlers began with little. As planters cleared land, erected buildings and fences, built up livestock herds, planted orchards, and cultivated kitchen gardens, all while getting out a tobacco crop, both income and wealth increased. Once the farm was well established, however, growth in income soon failed to keep pace with growth in wealth.

On the Cole plantation, income from tobacco did not grow; instead, it declined slightly, which was probably the case on most plantations. The amount of tobacco exported from the Chesapeake increased enormously in this period, but primarily because of the addition of new farms and new workers. Through the 1660s, planters were able to maintain relatively stable incomes by taking advantage of falling costs and by increasing output per worker. By the 1670s, costs had ceased falling, and possibilities for dramatically increasing output per hand were diminishing. Sales of tobacco placed a floor under Chesapeake farm incomes, but limits to expansion were quickly reached, and there were few opportunities for further growth.[97]

Similarly, the local market did not offer many opportunities for expansion. On the Cole plantation, earnings from local exchange also failed to increase. This is hardly surprising given that all planters strove to become self-sufficient in foodstuffs and that only a handful of new households were likely to be formed in a county in any given year. A family dramatically raised their income when they were able to grow their own food rather than buying it from others. Thereafter, there was little incentive in prime tobacco-growing areas to produce more. The main exception was the lower Eastern Shore and the lower James River basin. There, by the 1640s, some residents (later joined by migrant merchants from Barbados) had begun building coasting vessels and commenced trade in provisions and naval stores with New England, Dutch New Amsterdam, the West Indies, and, from the 1670s, the Carolinas.[98]

Initially, a planter increased his wealth in a newly settled area through

Saloutos, "Efforts at Crop Control in Seventeenth Century America," *Jour. So. Hist.,* XII (1946), 45–66; and Beverley, *History and Present State of Virginia,* ed. Wright, 66–68. Attempts to manufacture salt on the Eastern Shore in the 1660s were also ineffective ("Randolph Manuscript," *VMHB,* XVII [1909], 240). George Melvin Herndon, "The Story of Hemp in Colonial Virginia" (Ph.D. diss., University of Virginia, 1959), chap. 1, discusses the failure of that fiber. In the 1650s, returns anticipated from hemp were less than half of what could be expected from tobacco. Laws requiring the sowing of hemp and flax in the 1660s were suspended for lack of seed.

97. Menard, Carr, and Walsh, "A Small Planter's Profits," *WMQ,* 3d Ser., XL (1983), 171–196.

98. Ibid.; Hatfield, *Atlantic Virginia,* chaps. 2, 4.

farm building. The labor expended in creating a working farm from a wilderness yielded substantial returns. Such improvement came partly from limiting current consumption—forgoing the purchase of nonessential household goods or curtailing meat consumption until livestock herds were built up, for example. Most savings were achieved by sacrificing leisure time. Tobacco growing absorbed a good portion of the year, yet there remained time during the winter and early spring in which one could clear additional land, erect fences, or build more buildings without cutting back on the size of the crop and thus sacrificing export earnings. In a newly settled area, neighbors were few and opportunities for more organized diversions almost nonexistent, making steady devotion to work easier. In addition, once a planter acquired one or more servants, he could curtail their leisure rather than that of family members.[99]

Contemporary observers did suggest that, despite the very demanding requirements of the staple, a great deal of leisure time remained. Even servants, burdened as they were with legal disadvantages, on the whole succeeded in retaining customary rights to a substantial amount of free time. Certainly, some masters brutally exploited their servants. Many others, however, seem to have demanded or got little more from their hands than more labor rich English employers might have required. John Hammond observed that no work was done in Virginia in winter but dressing victuals and making fires. Even during the growing season, he asserted, laborers had five hours rest in the heat of the day, Saturday afternoon off, and the "old Holidayes" were observed. Sir John Colleton noted that laborers in Barbados worked the year round, but the Virginians worked diligently "but halfe the year." George Alsop, himself a former servant, stated that servants in the Chesapeake worked only five and a half days a week, with three hours rest in the summer heat. During the

99. Menard, Carr, and Walsh, "A Small Planter's Profits," *WMQ*, 3d Ser., XL (1983), 171–196. The concept of farm building as capital accumulation is discussed in Percy Wells Bidwell and John I. Falconer, *History of Agriculture in the Northern United States, 1620–1860* (Washington, D.C., 1925), 82. Several contemporary writers commented on this farm building and savings process. Bullock found that the proceeds of the first year's crop would pay for a "pretty house" and then thought (rather too optimistically) that by the second year a Virginia planter would—by virtue of his estate—become a gentleman (*Virginia Impartially Examined*, 35). Edward Williams advised a new settler to save carefully for three years; thereafter, he could afford "large latitude of expences" (*Virginia Truly Valued, Force Tracts*, III, no. 11, 48). William Berkeley advised that, "if we will be provident and industrious for a year of two, we may provide for our Posterity for many Ages; the manifest knowledge of this makes men industrious and vigilent with us" (*Discourse and View of Virginia*, 4). Even in the depths of the 1665–1667 depression, Lord Baltimore stated that industrious planters "live in much greater plenty, and gaine estates much sooner than those of their quality in England" (Browne et al., eds., *Archives of Maryland*, V, 16).

three winter months, they generally did little but cut firewood and hunt. Although such promotional literature might well be suspect, much supporting evidence can be found. For example, the 1657 ruling of Maryland's provincial court that a master could not keep his servants in the fields so long that they must beat corn for their victuals in the night "but at a seasonable time in the yeare or in case of necessity."[100]

Servants would naturally prefer not working over working, whatever the circumstances. But planters, too, had particular reasons for valuing leisure above work beyond a certain point. Many did not expect to engage in the sorts of gentlemanly pursuits that Robert Wintour had envisioned, but, nonetheless, most free families valued the ordinary pleasures that more populous Chesapeake neighborhoods were beginning to afford. Since opportunities for increasing either export or local earnings on a seventeenth-century Chesapeake plantation were in most cases quite limited, the majority of planters soon reached a level at which additional effort yielded diminishing returns. In the early 1660s, the Cole estate generated £1 in income for each £2.4 in wealth. By the early 1670s, it took £4.3 of wealth to produce £1 in income.[101]

Growing wealth did, of course, have an effect on a family's welfare. Parents could be assured that their estates would provide their offspring a reasonable start in life. Meanwhile, ownership of an estate growing in value made it more likely that credit could be obtained when needed to weather cyclical downturns. The rising ratio of wealth to income does suggest, however, that there were definite limits to early growth. In the absence of dependable markets for any products other than tobacco, diminishing returns soon lowered incentives for intensified farm-building activities. Only if they sold their farms could planters realize profits from farm building, and selling an estab-

100. Hammond, *Leah and Rachel,* in Hall, ed., *Narratives of Early Maryland,* 290; Sir John Colleton to Peter Carteret, Sept. 9, 1665, in William S. Powell, ed., *Ye Countie of Albemarle in Carolina: A Collection of Documents, 1664–1675* (Raleigh, N.C., 1958), 8; Browne et al., eds., *Archives of Maryland,* X, 521. See also R. G., *Virginia's Cure, Force Tracts,* III, no. 15; and Lois Green Carr and Lorena S. Walsh, "Economic Diversification and Labor Organization in the Chesapeake, 1650–1820," in Stephen Innes, ed., *Work and Labor in Early America* (Chapel Hill, N.C., 1988), 144–188. The "old Holidayes" Colleton referred to were approximately forty-nine saints' and other holy days that were supposedly abolished during the Protestant Reformation, but some continued to be observed in various parts of England and the Netherlands. At the beginning of the sixteenth century, the number of days officially available for work ranged from 250 to 260. Abolishing holy days allowed expansion of the work year by more than 20 percent to a maximum of 307 days (Jan de Vries, *The Industrious Revolution: Consumer Behavior and the Household Economy, 1650 to the Present* [Cambridge, 2008], 87–90).

101. Menard, Carr, and Walsh, "A Small Planter's Profits," *WMQ,* 3d Ser., XL (1983), 195.

lished, favorably situated plantation away from their heirs was the last thing most Chesapeake settlers wished to do. The few privileged farmers who could command more land and especially more labor than families like the Coles generally turned to building additional new farms, while modestly expanding and improving old ones.

In subsequent decades, the early growth process ended, and many Chesapeake planters encountered a spate of hard times. As immigration from Europe dropped off, most small planters could no longer find any indentured servants offered for sale and thus were unable to increase their fortunes by growing more tobacco. We now know from the benefit of hindsight that increasing reliance on enslaved Africans among those who could procure and afford them was beginning to transform Chesapeake society in tragic ways. But that middling planters like Robert Cole, much less his poorer neighbors, were entirely cognizant of the direction their society was heading is unlikely.

When Cole left for England in 1662, he had probably encountered no more than one to two dozen enslaved Africans anywhere on Saint Clement's Manor, and likely even fewer. Thomas Gerrard, the lord of the manor, had become a slaveowner by the early 1670s, and he might have bought one or two bondpeople before Cole's departure. Only Robert Slye, Gerrard's son-in-law and one of the richest planter-merchants in Maryland, is definitely known to have acquired several slaves before Cole left Maryland. All other planters on the manor and across Saint Clement's Bay on Newtown Neck, which Cole frequently visited, farmed only with the labor of family members or with that of indentured servants. And, although Cole was doubtless aware of the Africans' peculiar status as lifetime chattels, the conditions under which they lived were similar to those of European servants. Cole probably knew Cloise, an enslaved man who was the foreman of one of Slye's quarters, who was paid in 1663 for doctoring one of Cole's children. A few years later, when Thomas Gerrard deeded an enslaved couple and their two children "with their increase and posterity forever" to his stepchildren, he nonetheless referred to the blacks somewhat ambiguously as "slaves servants" and accorded the woman, Joane, the distinctly English-sounding title of "now wife" to Frank. Although we cannot know what Cole thought of these Africans as individuals or as a group, he probably could not have envisioned that colonial legislatures and courts would soon systematically reduce them to the status of chattels entirely without rights, nor how their involuntary presence would shortly transform his children's social and economic world. Cole almost certainly did not think about risking buying a slave himself and likely regarded slaveholding as the limited and peculiar practice of a few economic and social superiors.[102]

Cole surely also could not have anticipated that a rapid decline in the opportunities that had prompted him to emigrate would shortly deter most English men and women from financing a move to the Chesapeake by selling their labor for a term of years. Nor could he have imagined that, as a result, his and most of his neighbors' children might not be able to replicate his own rise in fortune. Between Cole's departure in 1662 and the end of the next decade, every one of the county officers who were mentioned in the account of his estate purchased one or more slaves, and several by then owned a dozen or more. Cole could hardly have foreseen how quickly these more privileged men would advance their fortunes, nor that many of his middling-planter neighbors who survived into the 1670s would no longer be able to buy indentured servants.[103]

Moreover, Cole could not have imagined that his oldest son Robert would fail to prosper, despite the superior education and practical training in planting his father provided along with a legacy of land and livestock, and instead die in 1693 at age forty with personal property worth only a third of what Cole had accumulated, working his tobacco fields without any assistance, except perhaps that of his wife. More unlikely is that Cole could have foreseen that his youngest son Edward, who lived to the advanced age of sixty, would, in contrast, flourish, accumulate assets worth more than twice those of his parents, and end up at his death in 1717 able to trade toiling in the fields for supervising a workforce of seven enslaved Africans. The subsequent fortunes of the Cole children, however, are typical of the range of experiences Chesapeake settlers encountered in ensuing troubled decades.

102. The labor forces and estate values of individuals mentioned in the Cole account are summarized in Carr, Menard, and Walsh, *Robert Cole's World,* appendix 4. Robert Slye's slaves are listed in Testamentary Proceedings, 5: 152–190, MSA. Gerrard's deed of gift is in Westmoreland County Wills and Deeds, 1665–1677: 142–143, microfilm, VSL.

103. Carr, Menard, and Walsh, *Robert Cole's World,* appendix 4.

CHAPTER 3

An Era of Hard Times

VIRGINIA 1680-1729

Historians conventionally subdivide the time periods they study by events such as the span of a monarch's reign or the beginning or ending of wars or revolutions. Economic historians are as likely to subdivide time periods by trends in prices of critical commodities or by structural changes in market organization. For 1680–1729, Chesapeake historians have taken their cue from economic historians, defining these years as a distinct era characterized by a prolonged stagnation in the prices planters obtained for their tobacco, by a failure to increase the total amount of tobacco they produced, and by declining per capita incomes in the region overall. This ending of the initial period of expansive growth had profound impacts, both on the Chesapeake economy and on its society. The term "stagnation," of course, implies a time of little change. But once one takes into account diverging trends within three different economic subregions of the Tobacco Coast, it becomes clear that stasis is not an accurate description. The period was in fact one of marked adaptation and change as planters struggled to overcome the problems they encountered in making and selling tobacco and as they worked to develop a variety of alternative exports in order to take advantage of new markets throughout the Atlantic world.[1]

1. Russell R. Menard, "The Tobacco Industry in the Chesapeake Colonies, 1617–1730: An Interpretation," *Res. Econ. Hist.,* V (1980), 109–177; John J. McCusker and Russell R. Menard, *The Economy of British America, 1607–1789* (Chapel Hill, N.C., 1985), chap. 6. Charles Wetherell, from an analysis of price and production data, disputes the presence of cycles in the tobacco economy. Although he raises some important questions about the measurement and interpretation of preindustrial cycles, among other things, his decision to use only English imports as a measure of production (excluding the increasing volume of Chesapeake tobacco shipped to Scotland) raises doubts about the merits of his interpretation ("'Boom and Bust' in the Colonial Chesapeake Economy," *Journal of Interdisciplinary History,* XV [1984–1985], 185–210). The most cogent critique is Douglas M. Bradburn and John C. Coombs, "Smoke and Mirrors: Reinterpreting the Society and Economy of the Seventeenth-Century Chesapeake," *Atlantic Studies,* III (2006), 131–157.

Nor was the Chesapeake untouched during these years by significant changes in the structure of the British government and by international contests for empire between Britain and France. Charles II died in 1685. The troubled reign of his successor, James II, who wished to re-Catholicize the Anglican church, to strengthen royal prerogatives, and to centralize administration of English colonies, lasted only four years. William, Prince of Orange and husband of James's sister Mary, at the invitation of leading men who wanted to remove James, invaded England in 1688. The next year, a Convention Parliament offered the British crown to William and Mary jointly. As a result, Roman Catholics were excluded from succession to the throne, Trinitarian Protestants were granted freedom of worship, and the powers of Parliament and of the peers and gentry significantly increased. Cabinet ministers answerable to the majority in the Commons emerged as the real executive government, with regular elections being mandated, the king's right to suspend legislation abolished, and nonparliamentary taxation made illegal, as was the maintenance of a standing army in peacetime and arbitrary royal fines and forfeitures. Henceforth, the crown was entirely dependent on Parliament for all public expenditures. This "Glorious Revolution" ended royal succession crises. Upon William III's death in 1702, Mary's sister Anne became queen, and, when she died without direct heirs in 1714, a transfer of the crown to the Protestant House of Hanover was effected with only minor disturbances.[2]

Most settlers in the English colonies in America enthusiastically supported the Glorious Revolution, hardly anticipating that as a result they would become involved in more than twenty years of global wars between 1689 and 1713. Partly as a result of mutual animosity toward the Netherlands and sympathy for the Catholic religion, Charles II and James II had pursued generally pro-French policies. William III, in contrast, implacably opposed French king Louis XIV's expansionist plans. Consequently, William led Britain into an alliance with the Netherlands against France, now indisputably Britain's major rival for empire in Europe, North America, and Asia. In addition to protracted warfare in Europe, fighting erupted on the North American continent between English and French colonists and groups of native Americans allied to one or the other of the rival European powers as well as between English and French in the Caribbean. King William's War (1689–1697) ended

2. A useful summary is Ian K. Steele, "The Anointed, the Appointed, and the Elected: Governance of the British Empire, 1689–1784," in P. J. Marshall, ed., *The Eighteenth Century,* vol. II of Wm. Roger Louis, ed., *The Oxford History of the British Empire* (Oxford, 1998), 105–127. See also David S. Lovejoy, *The Glorious Revolution in America* (New York, 1972); and Linda Colley, *Britons: Forging the Nation, 1707–1837* (New Haven, Conn., 1992), 46–52.

inconclusively; the Peace of Rijswijk returned captured territories to their prewar status. Subsequently, England, the Netherlands, and Austria declared war on France in 1702 to prevent Louis XIV from uniting the thrones of France and Spain, thus also potentially gaining control of Spain's New World colonies in addition to new settlements the French were planting in Louisiana and in the area west of the Appalachian Mountains as a barrier against Anglo-American expansion. This European conflict over the succession to the Spanish throne quickly spread to North America as Queen Anne's War (1702–1713), with fighting for control of Canada, Spanish Florida, and the West Indies. The resulting peace gave England possession of Newfoundland, Nova Scotia, the region around Hudson's Bay, Saint Christopher, Gibraltar, and Minorca as well as a lucrative contract (the Asiento) to supply African slaves to Spanish colonies in North America. The main effects of these wars for Chesapeake colonists were losses of shipping to French privateers, disruption of trade to and from England and with other English colonies, and a drop in the number of slaves carried to the Tobacco Coast from Africa.[3]

Another result of the Glorious Revolution was a temporary end to proprietary rule in Maryland. Charles Calvert, third Lord Baltimore, then in England defending his charter, sent orders to proclaim allegiance to William and Mary, but the bearer of the message died en route, and the Catholic president of the Maryland Council failed to acknowledge the new rulers. A group of Protestant gentry (some of whom had already attempted a revolt in 1681) took over the province in a bloodless coup. Discontent with high taxes, with Baltimore's disallowing laws that did not please him, and especially with his policy of appointing primarily Roman Catholics and close relatives to provincial offices contributed to settlers' acquiescence in the takeover. The colony was controlled by a committee of Protestant Associators until 1692, when William III stripped Lord Baltimore of his political authority and sent a royal governor to rule Maryland in concert with the assembly. The Anglican church was established, and Catholics and Quakers, who could not or would not take the oaths now required to hold office or serve on juries, were barred from public life. Again, however, county courts succeeded in maintaining local order. Although conditions for poorer planters continued to worsen,

3. Jonathan I. Israel, "The Emerging Empire: The Continental Perspective, 1650–1713," and Richard S. Dunn, "The Glorious Revolution and America," both in Nicholas Canny, ed., *The Origins of Empire: British Overseas Enterprise to the Close of the Seventeenth Century,* vol. I of Louis, ed., *Oxford History of the British Empire,* 423–444, 445–466; Bruce P. Lenman, "Colonial Wars and Imperial Instability, 1688–1793," and N. A. M. Rodger, "Sea-Power and Empire, 1688–1793," both in Marshall, ed., *The Eighteenth Century,* vol. II, ibid., 151–168; Jeremy Black, *Trade, Empire, and British Foreign Policy, 1689–1815: The Politics of a Commercial State* (London, 2007), chap. 5.

lower-class discontent was not a factor in this colonial "revolution of government." In fact, proprietary rule was restored in 1715, when the fourth and fifth Lords Baltimore, who had converted to the Anglican faith, succeeded their father and grandfather. Virginia, in contrast, experienced no major disruptions during this era. After 1676, politics at the provincial level were characterized by contests for power between royal governors, the councillors, and the lower house of assembly. Colonial assemblies were bent on claiming local powers similar to those gained by Parliament, leading inevitably to conflict with royal governors enjoined with enforcing metropolitan policies. Among ordinary settlers, the troubled tobacco economy remained their major concern along with a desire to retain their rights as Englishmen and to avoid arbitrary governments.[4]

Through the Glorious Revolution, Parliament won not only a larger role in governing England but also the right to intervene more deeply in colonial affairs. It resolved to broaden its mercantilist policy by more rigorously enforcing the Navigation Acts, by more consistently disallowing colonial laws in conflict with those of the mother country, and, in order to protect English manufacturers, by taking steps to stifle colonial manufacturing. Vice-admiralty courts were established in the colonies, the powers of customs and naval officers were enhanced and their numbers increased, and more effective registration of ships trading to the colonies was implemented. During the early years of William III's reign, the Privy Council and other royal administrative agencies neglected colonial matters. Parliament, responding to complaints from merchants about heavy losses of shipping in the war with France and to a perceived escalation of illegal trade, planned itself to establish and nominate members to a council for trade and plantations. To forestall

4. Lois Green Carr and David William Jordan, *Maryland's Revolution of Government, 1689–1692* (Ithaca, N.Y., 1974); Charles M. Andrews, *The Colonial Period of American History* (New Haven, Conn., 1936), II, chap. 9; Robert J. Brugger, *Maryland: A Middle Temperament, 1634–1980* (Baltimore, 1988), 31–40; Jack P. Greene, *The Quest for Power: The Lower Houses of Assembly in the Southern Royal Colonies, 1689–1776* (Chapel Hill, N.C., 1963), 22–31; David Alan Williams, "Political Alignments in Colonial Virginia Politics, 1698–1750" (Ph.D. diss., Northwestern University, 1959), chaps. 1–6; Edmund S. Morgan, *American Slavery, American Freedom: The Ordeal of Colonial Virginia* (New York, 1975), 346–362; Carter Lee Hudgins, "Patrician Culture, Public Ritual, and Political Authority in Virginia, 1680–1740" (Ph.D. diss., College of William and Mary, 1984), chap. 4; Warren M. Billings, "The Glorious Revolution in Virginia: The Revolution That Wasn't" (paper presented at "The Glorious Revolution in America—Three Hundred Years After," College Park, Md., 1988); Alexander B. Haskell, "'The Affections of the People': Ideology and the Politics of State Building in Colonial Virginia, 1607–1754" (Ph.D. diss., Johns Hopkins University, 2005), chaps. 6–7; Dunn, "The Glorious Revolution and America," in Marshall, ed., *The Eighteenth Century*, vol. II of Louis, ed., *Oxford History of the British Empire*, 456–457; Lovejoy, *Glorious Revolution in America*, 257–270.

such a diminution of royal prerogative, William created a new administrative agency, the Lords Commissioners of Trade and Plantations, or Board of Trade, in 1696. Composed of the president of the Privy Council, executive officers, and representatives of mercantile and industrial interests, the board was an advisory body that collected information, conducted investigations, and advised the crown (and Parliament upon request) about all aspects of colonial commerce, drafted instructions to colonial governors, and reviewed all colonial laws. They could not make decisions, but their recommendations were usually accepted and implemented by orders in council or acts of Parliament. According to Charles M. Andrews, "no part of the English executive was more consistently mercantilist in its relations to trade and the colonies or more conservative in its defense of the king's prerogative in America" than the Board of Trade. Unfortunately for Chesapeake planters, the board was especially receptive to the views of English tobacco merchants, who were often opposed to regulations designed to improve the quality and hence raise the price of tobacco (which colonists favored), and in favor of trade restrictions that injured tobacco growers.[5]

Stricter enforcement of the Navigation Acts, restrictions on various forms of intercolonial commerce, royal disallowance of laws passed by colonial legislatures to regulate their internal economies, and Britain's discouragement of colonial manufacturing all made it more difficult for Chesapeake colonists to cope with the severe economic problems that plagued the tobacco trade during the fifty years between 1680 and 1729. Although the period saw three or four short booms in the tobacco market, there were still almost four full decades when trade was severely depressed. Everywhere stagnant tobacco prices and adverse terms of trade and credit curtailed opportunities for small planters. The supply of indentured servants contracted drastically, especially in wartime, and minor operators remained unable to take advantage of more available, but still costly, slave labor to realize the sort of modest prosperity enjoyed by farmers of limited means in the mid-seventeenth century. Unable to obtain indentured servants and too poor to afford or too ill connected to procure slaves, small and middling planters had limited chances for improving their fortunes by growing more tobacco. Such men continued to make up the majority of planters and to produce more than half of the region's tobacco, but they no longer played a dynamic role in advances in agricul-

5. Andrews, *Colonial Period,* IV, chaps. 6–9 (quotation on 291); Ian K. Steele, *The English Atlantic, 1675–1740: An Exploration of Communication and Community* (New York, 1986), chap. 12; Black, *Trade, Empire, and British Foreign Policy,* chap. 6.

tural productivity. As the price of tobacco fell, poorer folk even had difficulty finding the resources with which to buy the imported cloth, shoes, and tools on which they were dependent. Meanwhile, the favored few—large planters with access to credit in Britain, more land than they could develop, African slaves, and often nonagricultural sources of capital as well—fared better. Everywhere the rich garnered an increasing share of available wealth. Above all, the shift to African slave labor enhanced the fortunes of the white elite and fixed upon the region a subordinate caste of permanently unfree and disadvantaged blacks with tragic and far-reaching consequences for Chesapeake society for many generations to come.[6]

As returns from the staple—and with it the opportunity for poor immigrants to achieve at least a modest competence—diminished, European immigration dropped sharply. Wages had been rising in England, and the prices of necessities had been falling for some time. Potential emigrants thus had increasing incentives for remaining at home, while, at the same time, chances for freed men and women to advance their status in the Chesapeake were declining. King William's War and Queen Anne's War might have tipped the balance. Transatlantic migration became more risky, while military employment expanded and other jobs abandoned by those who joined armies and navies opened. English real wages surged upward during the wars—and between 1688 and 1713 there were only five years of peace. At the same time, the planters' need for new field laborers continued to grow, as more new households were formed each year. The price of servants rose, especially in oronoco-growing areas where planters were heavily dependent on servant labor, even as the numbers arriving declined. Small and middling planters, who occupied positions similar to that of Robert Cole, could no longer af-

6. For accounts of the Chesapeake economy in this period, see Menard, "The Tobacco Industry," *Res. Econ. Hist.*, V (1980), 109–177; McCusker and Menard, *Economy of British America*, chap. 6; George Louis Beer, *The Old Colonial System, 1660–1754*, 2 vols. (1912; rpt. New York, 1933), chap. 8; Margaret Shove Morriss, *Colonial Trade of Maryland, 1689–1715*, Johns Hopkins University Studies in Historical and Political Science, XXXII, no. 3 (Baltimore, 1914); Charles George Gordon Moss, "The Virginia Plantation System: A Study of Economic Conditions in That Colony for the Years 1700 to 1750" (Ph.D. diss., Yale University, 1932); Jacob M. Price, "The Economic Growth of the Chesapeake and the European Market, 1697–1775," *JEH*, XXIV (1964), 496–511; John Mickle Hemphill II, "Virginia and the English Commercial System, 1689–1733: Studies in the Development and Fluctuations of a Colonial Economy under Imperial Control" (Ph.D. diss., Princeton University, 1964); Peter V. Bergstrom, *Markets and Merchants: Economic Diversification in Colonial Virginia, 1700–1775* (New York, 1985); Paul G. E. Clemens, *The Atlantic Economy and Colonial Maryland's Eastern Shore: From Tobacco to Grain* (Ithaca, N.Y., 1980); and Gloria L. Main, *Tobacco Colony: Life in Early Maryland, 1650–1720* (Princeton, N.J., 1982).

ford to buy the additional laborers that were critical to any marked improvement in family fortunes.[7]

Large planters were also caught short of cheap white labor, but, as mentioned previously, many of them had already switched to slaves between 1640 and the mid-1670s. The timing of their decision is critical, since most scholars argue that slavery was a labor system forced upon elites by shortages of indentured servants in the 1680s and 1690s, accompanied by a need to make tobacco more cheaply in depressed markets after 1680. But, in fact, by the late 1670s the region's provincial elites were already thoroughly committed to slavery, and about two-thirds of all Virginia county officers had purchased at least one slave. With black slaves already outnumbering white servants on many elite Virginians' plantations, the only additional European laborers many of them needed were a few servants with special skills, and these they continued to obtain through contacts in the British merchant community. The embrace of slave labor by the region's wealthiest, best-connected, and most-powerful planters who had the widest possible range of choices for solv-

7. Russell Menard, "From Servants to Slaves: The Transformation of the Chesapeake Labor System," *So. Stud.*, XVI (1977), 355–390; McCusker and Menard, *Economy of British America,* chap. 6; James Horn, "Servant Emigration to the Chesapeake in the Seventeenth Century," in Thad W. Tate and David L. Ammerman, eds., *The Chesapeake in the Seventeenth Century: Essays on Anglo-American Society* (Chapel Hill, N.C., 1979), 51–95; Horn, "British Diaspora: Emigration from Britain, 1680–1815," in Marshall, ed., *The Eighteenth Century,* vol. II of Louis ed., *Oxford History of the British Empire,* 28–52; Henry A. Gemery, "Markets for Migrants: English Indentured Servitude and Emigration in the Seventeenth and Eighteenth Centuries," in P. C. Emmer, ed., *Colonialism and Migration: Indentured Labour before and after Slavery* (Dordrecht, 1986), 33–54; Nicholas Canny, "English Migration into and across the Atlantic during the Seventeenth and Eighteenth Centuries," in Canny, ed., *Europeans on the Move: Studies on European Migration, 1500–1800* (Oxford, 1994), 39–75; Farley Grubb and Tony Stitt, "The Liverpool Emigrant Servant Trade and the Transition to Slave Labor in the Chesapeake, 1697–1707: Market Adjustments to War," *Exp. Econ. Hist.,* XXXI (1994), 376–405; David W. Galenson, "The Settlement and Growth of the Colonies: Population, Labor, and Economic Development," in Stanley L. Engerman and Robert E. Gallman, eds., *The Cambridge Economic History of the United States,* 3 vols. (New York, 1996), I, 135–208; Alison F. Games, "Migration," in David Armitage and Michael J. Braddick, eds., *The British Atlantic World, 1500–1800* (New York, 2002), 31–50; Lorena S. Walsh, "The Differential Cultural Impact of Free and Coerced Migration to Colonial America," in David Eltis, ed., *Coerced and Free Migration: Global Perspectives* (Stanford, Calif., 2002), 117–151; Christopher Tomlins, "Indentured Servitude in Perspective: European Migration into North America and the Composition of the Early American Labor Force, 1600–1775," in Cathy Matson, ed., *The Economy of Early America: Historical Perspectives and New Directions* (University Park, Pa., 2006), 146–182. The earlier literature posited that the new colonies of Pennsylvania and Carolina diverted servants away from the Chesapeake after 1680. The evidence on servant destinations shows, however, that less than 4 percent of servants went to these destinations in the 1680s and less than 2 percent between 1697 and 1707 (Grubb and Stitt, "Liverpool Servant Trade," *Exp. Econ. Hist.,* XXXI [1994], 378).

ing their labor problems can only have been a "thinking decision" initiated by elites before either a shortage of servants or falling tobacco prices might have forced such a course upon them.[8]

As more slaves were brought into the Chesapeake, less wealthy county officeholders, who had still relied primarily on servant labor through the 1670s, also converted to enslaved Africans in the 1680s and 1690s. By the late 1690s, all the region's large planters had chosen slavery irrevocably. It would take another thirty-odd years, however, before their thirst for ever more enslaved Africans was largely slaked. The transition was fueled in part by a dramatic increase in the supply of slaves. Beginning in the mid-1680s, a chronic depression in the West Indian sugar islands led to a drastic reduction in the price of new Africans there, from thirty pounds sterling for a prime hand in the mid-1650s to about fifteen pounds in the 1680s. This decline prompted English slave traders to explore alternative markets for their human wares. Barbadian merchants might also have started shipping more slaves to the mainland in the 1680s, but this traffic became too risky during King William's War.[9]

The vigorous entry of private merchants into the African slave trade after the abolition of the Royal African Company's monopoly in 1698, a change effected by a lobby of separate slave traders closely connected to the tobacco trade, produced a further marked upsurge in the numbers of blacks transported to the Chesapeake. From 1680 to 1697, roughly three thousand new Africans were brought into the region. Between 1698 and 1703, at least an additional forty-six hundred arrived. The numbers jumped to more than fourteen thousand new slaves brought in between 1704 and 1718, followed by another nineteen thousand between 1719 and 1729. Bristol merchants, who could deliver slaves to the Americas more efficiently than metropolitan

8. John C. Coombs, "Building 'The Machine': The Development of Slavery and Slave Society in Early Colonial Virginia" (Ph.D. diss., College of William and Mary, 2003), chaps. 2, 3; Lorena S. Walsh, "Thinking Decisions? Migration, Servitude, and Slavery in the British North American Mainland Colonies" (paper presented at a seminar of the McNeil Center for Early American Studies, 2006).

9. Coombs, "Building 'The Machine,'" chaps. 2, 3; John C. Coombs, "Beyond the 'Origins Debate': Rethinking the Rise of Virginia Slavery," in Douglas Bradburn and Coombs, eds., *Early Modern Virginia: New Essays on the Old Dominion* (Charlottesville, Va., forthcoming); Main, *Tobacco Colony,* chaps. 2, 3; Grubb and Stitt, "Liverpool Servant Trade," *Exp. Econ. Hist.,* XXXI (1994), 397–398; David Eltis, *The Rise of African Slavery in the Americas* (Cambridge, 2000), 49–54, 151–153; Lorena S. Walsh, "Mercantile Strategies, Credit Networks, and Labor Supply in the Colonial Chesapeake in Trans-Atlantic Perspective," in David Eltis, Frank D. Lewis, and Kenneth L. Sokoloff, eds., *Slavery in the Development of the Americas* (Cambridge, 2004), 92–98.

slavers, began aggressively contesting London's dominant role in the African trade. By the 1720s, Bristol became for a time Britain's leading slaving port as well as the leading supplier of Africans to the Chesapeake. As a result, the ethnic origins of new Africans changed yet again, since Bristol merchants were then buying most of their human cargoes in the Bight of Biafra, a region where the price of slaves was lower than on the Gold Coast and Slave Coast and European competitors fewer. More than half of the slaves brought into the Chesapeake between 1698 and 1729 were from this region, of whom the majority was probably Igbo, who spoke a common language and shared a common culture. In the early 1700s, slave buyers thus had to learn to manage an influx of bondpeople from an unfamiliar ethnic group that was numerous enough to retain some elements of homeland culture and so was perhaps slower to adopt English language and customs than the more diverse groups arriving earlier.[10]

Chesapeake planters obviously benefited from increased competition among slave traders as well as from another decline in West Indian sugar prices in the 1720s. As the volume of slave imports rose, the price of a prime male hand fell from a range of twenty-seven pounds to thirty pounds sterling in the 1710s to a range of eighteen pounds to twenty pounds in the 1720s. At this rate, slave buyers fortunate enough to lose few hands to seasoning might recoup their initial outlay with the proceeds of two or at most three tobacco crops. Once a planter bought one or two slaves, it became increasingly easy for him to buy more.[11]

The transition to slave labor was swift and dramatic throughout the re-

10. Lorena S. Walsh Database of Slave Trading Voyages to Virginia and Maryland (November 2008); preliminary results are published in Lorena S. Walsh, "The Chesapeake Slave Trade: Regional Patterns, African Origins, and Some Implications," *WMQ*, 3d Ser., LVIII (2001), 139–170. See also P. M. G. Harris, *The History of Human Populations,* II, *Migration, Urbanization, and Structural Change* (Westport, Conn., 2003), 313–318. For the campaign of separate traders, see William A. Pettigrew, "Free to Enslave: Politics and the Escalation of Britain's Transatlantic Slave Trade, 1688–1714," *WMQ*, 3d Ser., LXIV (2007), 3–38. For Bristol's greater efficiency, see David Eltis and David Richardson, "Productivity in the Slave Trade," *Exp. Econ. Hist.,* XXXII (1995), 465–484; and Eltis, *Rise of African Slavery,* chap. 5. For Igbo in Virginia, see Lorena S. Walsh, *From Calabar to Carter's Grove: The History of a Virginia Slave Community* (Charlottesville, Va., 1997); and Douglas Brent Chambers, "'He Gwine Sing He Country': Africans, Afro-Virginians, and the Development of Slave Culture in Virginia, 1690–1810" (Ph.D. diss., University of Virginia, 1996).

11. Chesapeake slave prices were compiled from period planter and merchant correspondence. For trends in the prices of African slaves, see also David Eltis and David Richardson, "Prices of African Slaves Newly Arrived in the Americas, 1673–1865: New Evidence on Long-Run Trends and Regional Differences," in Eltis, Lewis, and Sokoloff, eds., *Slavery in the Development of the Americas,* 181–218.

gion. Between the mid-1680s and mid-1690s, the aggregate unfree labor force in both Virginia and Maryland changed from primarily indentured servants to primarily slaves. In wealthy Virginia counties like York, white servants virtually disappeared from the workforce. Between 1701 and 1730 in Virginia as a whole, in 86 percent of inventoried estates with just one bound laborer that worker was a slave, and every estate with five or more laborers included at least one slave. Since very few slaves ever gained freedom, and since some of them also eventually produced children whose fate was the same perpetual bondage that their parents endured, the size of labor forces on large plantations increased dramatically. Shortly after the turn of the century, some of the richest planters accumulated enslaved labor forces of one hundred or more. No seventeenth-century planter ever acquired an indentured servant workforce even approaching this size. Although enslaved workers were divided among numerous quarters, on the bigger plantations planters had enough adult field hands to organize them into labor gangs. The extent to which the use of gang labor might have increased labor productivity on large plantations remains a subject for debate among economic historians. The early Chesapeake elite, like their sugar-growing counterparts in the Carribean, apparently considered gang labor advantageous.[12]

Between 1680 and 1720, the enslaved population in the Chesapeake increased at almost twice the rate of the white population. The proportion of blacks rose from a mere 7 percent in 1680 to around 15 percent in 1690 and reached about 25 percent by 1720. Slaveowning elites had ever-greater need to protect their substantial investments in human property and to enact laws to control a growing segment of the population who had little or no stake in the society to which they had been forcibly brought. Individual plantation discipline proved inadequate to police growing numbers of blacks who were sometimes able to run away and remain at large for months, killing livestock and stealing from neighboring plantations, and occasionally violently resisting whites who attempted to capture them. Moreover, fears increased about slave marronage and the possibility of slaves' combining and organizing to foment insurrections. At the turn of the century, the Virginia and Maryland assemblies enacted fully articulated slave codes that denied blacks the civil

12. Menard, "From Servants to Slaves," *So. Stud.*, XVI (1977), 354–390; Coombs, "Building 'The Machine,'" chap. 7. For shifts in the size of work groups, see Lois Green Carr and Lorena S. Walsh, "Economic Diversification and Labor Organization in the Chesapeake, 1650–1820," in Stephen Innes, ed., *Work and Labor in Early America* (Chapel Hill, N.C., 1988), 144–188; Allan Kulikoff, *Tobacco and Slaves: The Development of Southern Cultures in the Chesapeake, 1680–1800* (Chapel Hill, N.C., 1986), 330–331; and Philip D. Morgan, *Slave Counterpoint: Black Culture in the Eighteenth-Century Chesapeake and Lowcountry* (Chapel Hill, N.C., 1998), 41–42.

rights enjoyed by whites and that prescribed extraordinary measures for the policing of slaves. The early 1730s, in the wake of a series of threatened slave revolts, saw the passage of additional laws for suppressing and controlling slaves. Planters were accorded a free hand in doing whatever they deemed necessary to control their human property and to force them to work. Moreover, the authority of the slaveholder was by then effectively backed by a legal infrastructure of state-sponsored racism.[13]

Racism developed and spread throughout all levels of society as colonists of European descent began to think and speak of themselves first and foremost as "white," rather than as European or Christian, and to draw pride from the superior status that whiteness conferred. Already in the 1660s, censuses in the sugar colonies of the British West Indies universally used the dichotomy of "white" and "black," and slave laws outlined a system of governing blacks based entirely on coercion. Racial consciousness developed more slowly in the Chesapeake; "white" began to replace "Christian" or "free" in official documents only near the end of the century. In the 1680s and 1690s, for example, Maryland and Virginia laws regulating interracial sex were directed to "freeborne Englishe or other white women"; by 1699, "white" sufficed to define the identity of persons of European ancestry. Similarly, "white" first appeared as a category in a Maryland census of 1710, replacing the earlier category of "free," a shift that also appeared in more northerly mainland colo-

13. Coombs, "Building 'The Machine,'" chap. 5; Anthony S. Parent, Jr., *Foul Means: The Formation of a Slave Society in Virginia, 1660–1740* (Chapel Hill, N.C., 2003), chap. 4; Ira Berlin, *Many Thousands Gone: The First Two Centuries of Slavery in North America* (Cambridge, Mass., 1998), chap. 5. Estimates of the black population of the Chesapeake in the early eighteenth century are based on scanty information and vary widely. Parent gives the number of blacks in Virginia in 1698 as 5,000, almost certainly an underestimate (*Foul Means,* 74). Peter H. Wood provides a similarly low estimate of 5,500 blacks in Virginia in 1700 but offers a high figure of 49,700 in 1730 ("The Changing Population of the Colonial South: An Overview by Race and Region, 1685–1790," in Wood, Gregory A. Waselkov, and M. Thomas Hatley, eds., *Powhatan's Mantle: Indians in the Colonial Southeast* [Lincoln, Nebr., 1989], 38). McCusker and Menard, *Economy of British America,* 136, estimates the numbers in both Virginia and Maryland as about 13,000 in 1700 and about 53,000 in 1730. Philip Morgan estimates there were 13,000 blacks in Virginia alone in 1700 and 40,000 in that colony in 1730 (*Slave Counterpoint,* 61). Allan Kulikoff's estimate for the Virginia black population in 1730 is 36,600 ("A 'Prolifick' People: Black Population Growth in the Chesapeake Colonies, 1700–1790," *So. Stud.,* XVI [1977], 415). Berlin presents a higher estimate for 1700—16,000 in Virginia and an additional 3,000 in Maryland; he has no estimate for 1730 (*Many Thousands Gone,* 369). A frequently cited estimate is 19,600 blacks in Virginia and Maryland in 1700 and 53,800 in 1730 (John J. McCusker, "Colonial Statistics," in Susan B. Carter et al., eds., *Historical Statistics of the United States: Earliest Times to the Present,* millennial ed., 5 vols. [New York, 2006], V, 561–562). Since more recent research on the transatlantic slave trade suggests that larger numbers of Africans were imported at the turn of the century than previously thought, the higher estimates for 1700 are probably closer to the mark.

nies a few years later. In the 1680s, 1690s, and 1700s, Chesapeake lawmakers made whiteness a legal category that conferred race-specific political privileges and denied to people of mixed ancestry the rights accorded to those of "all-white" ancestry. By the second quarter of the century, when the few white servants who remained were normally assigned only to preferential craft or supervisory positions and the social and economic status of poor free whites ceased deteriorating, while the status of despised and disprivileged blacks continued to fall, elites could be confident of enlisting the help of nonslaveholding whites in supporting the slave regime.[14]

A final characteristic of the period was irregular but continued growth of the white population as a result of natural increase. Total numbers of whites rose from about 57,000 in 1680 to 69,000 in 1700 and reached around 158,000 in 1730. As immigration fell off, the proportion of native-born colonists grew, and, by the 1690s, creoles became a majority among adults. Owing to the pronounced decline in immigration and consequent slowing down in white population growth, no new counties were formed in either Virginia or Maryland from the early 1670s until the early 1690s. Indeed, the most recent estimates of Virginia's white population imply there was almost no increase in the number of white residents between 1680 and 1700. Then, as births among native whites surged and forced migration of Africans accelerated at the turn of the century, wide new areas were settled.[15]

Three differences between European immigrants and white Chesapeake natives had important implications for the region's agricultural development. First, creole sons and daughters owed no period of service beyond the age of majority, and, unlike indentured servants, most could expect to begin adult life with some inherited assets. Given the region's high mortality rates, few parents lived to see their children through adolescence. As a result, most youngsters got control over whatever assets they were to inherit as soon as they came of age. Nor did native-born men have the same problems finding mates as did many immigrant men. Sex ratios among creoles were nearly even; in addition, native-born girls often seem to have preferred native-

14. Kathleen M. Brown, *Good Wives, Nasty Wenches, and Anxious Patriarchs: Gender, Race, and Power in Colonial Virginia* (Chapel Hill, N.C., 1996), 187–244; Morgan, *American Slavery, American Freedom,* 327–346; Susan Dwyer Amussen, *Caribbean Exchanges: Slavery and the Transformation of English Society, 1640–1700* (Chapel Hill, N.C., 2007), 121–123, 129–144; Carl H. Nightingale, "Before Race Mattered: Geographies of the Color Line in Early Colonial Madras and New York," *AHR,* CXIII (2008), 62–63; William Hand Browne et al., eds., *Archives of Maryland,* 72 vols. (Baltimore, 1883–1972), VII, 204, XIII, 547, XXII, 552, XXV, 256, 258–259; William Waller Hening, ed., *The Statutes at Large: Being a Collection of All the Laws of Virginia . . .* (1809–1823; rpt. Richmond, Va., 1969), III, 86–87; Coombs, "'Building the Machine,'" 245–250.

15. McCusker, "Colonial Statistics," in Carter et al., eds., *Historical Statistics,* V, 652.

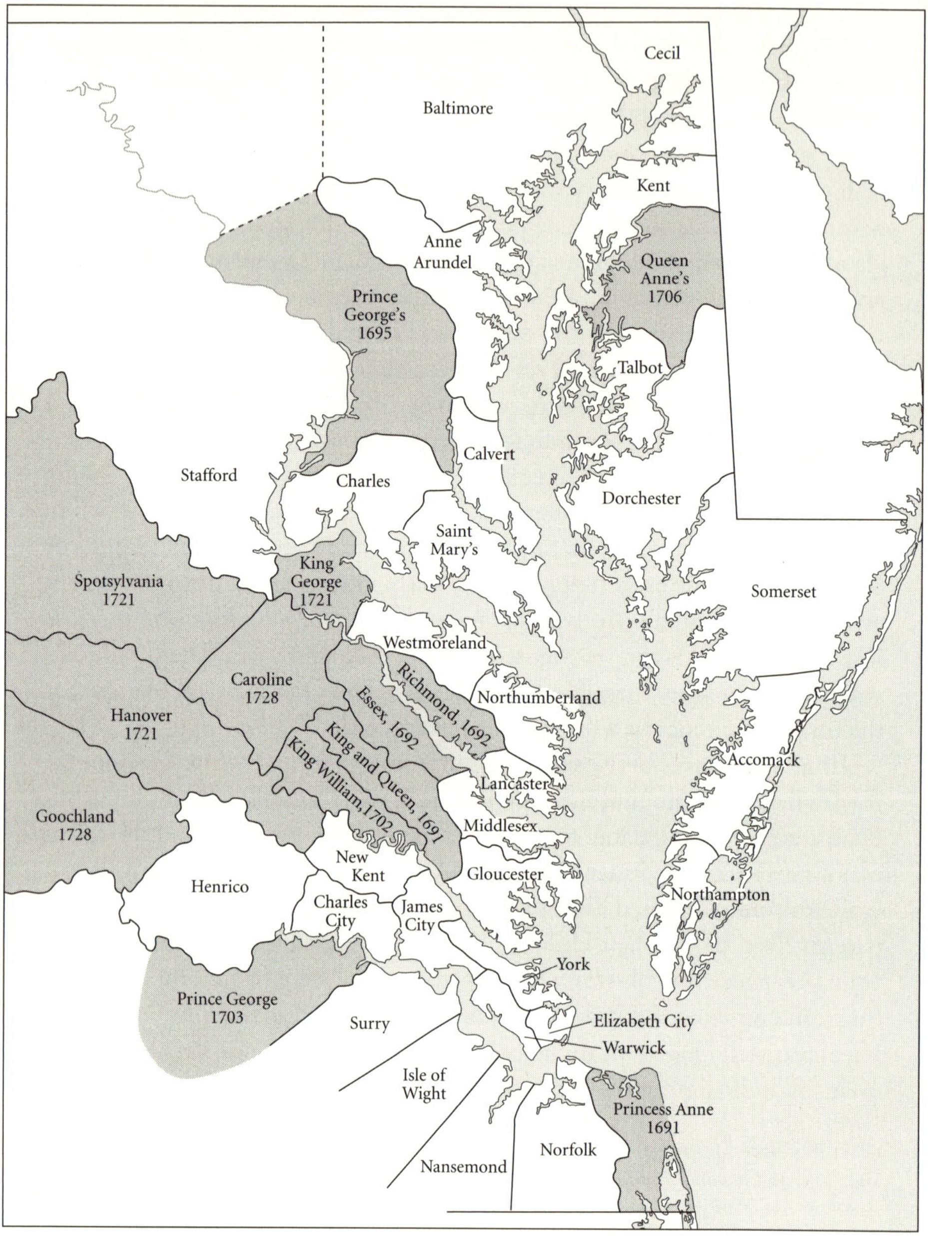

MAP 7. Chesapeake Counties Established through 1729

born men for husbands. Consequently, creoles had little difficulty setting up households. Most prospective native-born couples could, with the help of parents, guardians, or other relatives, piece together the cow, sow, three barrels of corn, bed, pot, hoe, and ax that were the prerequisites for starting housekeeping in the Chesapeake. As Robert Beverley put it at the turn of the century, when native-born sons and daughters came of age, "they who have taken any care to improve themselves, generally get well Married, and live in Plenty, tho they had not a farthing of paternal Estate." Consequently, once native-born young adults grew to a majority in a local population, the area's economy would become somewhat less closely tied to the vicissitudes of the tobacco trade and rates of household formation among whites would become less closely related to spurts in immigration.[16]

The ordinary creole's initial advantage in getting started in planting was soon offset, however, by a growing number of young dependents. Since natives, and especially native women, married at much younger ages than did immigrants, they had more children. The amount of corn and tobacco that a couple could produce by themselves declined once the wife was encumbered with repeated pregnancies and the need to care for young children. A constant or even falling farm income had thus to support more children too small to make any significant work contribution. Since most adults still died young, only the very lucky or very healthy would benefit from the labor of maturing sons and daughters. Men born in the colony did often enjoy slightly longer life spans than had their immigrant fathers, but, so long as settlement remained largely confined to unhealthy tidewater areas, new, more lethal strains of malaria, introduced by forced African migrants, continued to claim a heavy toll, especially among children and pregnant women. A few years earlier, the rising fortunes of ordinary planters had been built in large part on the chance to acquire highly productive single white youths and adult indentured laborers. The shift to a predominately native-born population produced a permanent increase in the proportion of young dependents as contrasted to prime-age laborers. To meet these new local demographic

16. In this period, parents who lived to see one or more children reach majority usually turned over land to sons at age twenty-one, and sometimes before. Daughters received their dowries at marriage. Sons who lost their fathers in their minority got their inheritances at age twenty-one, and sometimes they were allowed to manage their property from age seventeen or eighteen. Orphaned daughters received their portions at age sixteen in Maryland and eighteen in Virginia, or earlier if they married young. See Lorena S. Walsh, "'Till Death Us Do Part': Marriage and Family in Seventeenth-Century Maryland," in Thad W. Tate and David L. Ammerman, eds., *The Chesapeake in the Seventeenth Century: Essays on Anglo-American Society* (Chapel Hill, N.C., 1979), 126–152; Robert Beverley, *The History and Present State of Virginia* (1705), ed. Louis B. Wright (Chapel Hill, N.C., 1947), 260.

demands, planters shifted more of their energies into producing goods and services for a growing resident population. This was a one-time change that inevitably brought a drop in per capita income earned from exports. Falling tobacco prices and a shortage of cheap labor made the transition infinitely more difficult for the ordinary Chesapeake planter.[17]

A third difference between immigrants and natives was that creoles were bound—both economically and socially—to the area of their birth in a way few immigrants ever were. Within a locality, at least the first and perhaps the second native-born generation found the obligations that kin ties imposed or the help that kin networks could supply a sufficient inducement to remain in the area where they had been raised. In the later seventeenth and early eighteenth centuries, immigrants who failed to acquire land in one place were likely to move to another where opportunity appeared greater. Natives were much more likely to stay put, either from an obligation to help younger brothers, sisters, and cousins or from an expectation of aid and support from other kin. Given their higher level of commitment to a locality, creoles might have been more willing during bad times to devise ways to get by in the neighborhood where they had been born rather than to throw in the towel and move on.[18]

Sometime later than among whites, similar demographic changes appeared in the black population. Although in the region overall enslaved blacks were not yet fully reproducing, in the first three decades of the eighteenth century there were marked differences between reproductive outcomes on estates composed almost entirely of newly acquired slaves and on estates whose owners had earlier embraced slavery. Planters who obtained bondpeople through the provisioning trade with the West Indies in the 1660s and 1670s

17. Russell R. Menard, "Immigrants and Their Increase: The Process of Population Growth in Early Colonial Maryland," in Aubrey C. Land, Lois Green Carr, and Edward C. Papenfuse, eds., *Law, Society, and Politics in Early Maryland* (Baltimore, 1977), 88–110; Menard, "British Migration to the Chesapeake Colonies in the Seventeenth Century," in Lois Green Carr, Philip D. Morgan, and Jean B. Russo, eds., *Colonial Chesapeake Society* (Chapel Hill, N.C., 1988), 99–132; Menard, "The Tobacco Industry," *Res. Econ. Hist.*, V (1980), 109–177; Menard and Lorena S. Walsh, *The Demography of Somerset County, Maryland: A Progress Report,* Newberry Papers in Family and Community History, paper 81-2 (Chicago, 1981); Lorena S. Walsh, "Summing the Parts: Implications for Estimating Chesapeake Output and Income Subregionally," *WMQ,* 3d Ser., LVI (1999), 53–94; Walsh, "Differential Cultural Impact of Free and Coerced Migration," in Eltis, ed., *Coerced and Free Migration,* 117–151; James Horn, *Adapting to a New World: English Society in the Seventeenth-Century Chesapeake* (Chapel Hill, N.C., 1994), chaps. 1, 3; Darrett B. Rutman and Anita H. Rutman, "Of Agues and Fevers: Malaria in the Early Chesapeake," *WMQ,* 3d Ser., XXXIII (1976), 31–60.

18. Lorena S. Walsh, "Staying Put or Getting Out: Findings for Charles County, Maryland, 1650–1720," *WMQ,* 3d Ser., XLIV (1987), 89–103. See also Clemens, *Atlantic Economy,* chap. 5.

bought, as was the common practice on Barbados, relatively even numbers of men and women. Such balanced sex ratios enabled most enslaved adults living on these plantations to find partners and to begin having children. By the beginning of the eighteenth century, these early slaveowners' sons and daughters were inheriting enslaved workforces that were primarily native born. On the whole, second- and third-generation elite creole planters inherited labor forces that were growing through births at a much higher rate than those of planters who had more recently acquired their first permanently bound human property. The costs for further increasing their workforces were thus lower than among families who started acquiring slaves at a later date.[19]

By the 1720s, although Africans still predominated among enslaved adults in the two colonies overall and sex ratios still favored males (who might have been brought in slightly higher proportions in cargoes of slaves coming directly from Africa), the proportions of black women and children were rising, especially in long-settled counties. In a given area, the proportion of dependent slaves could rise dramatically over the course of a few years, should the number of new imports decline. The Rutmans have estimated, for example, that in Middlesex County, Virginia, in 1710 about 70 percent of the black labor force were prime adults but that, by the 1720s, the proportion of prime workers had plummeted to near 40 percent. Similarly, in parts of Maryland in the first quarter of the eighteenth century nearly 70 percent of enslaved workers were productive adults, but in subsequent decades this proportion declined to less than 60 percent. These demographic changes had an impact on plantation costs, organization of work, and product per hand. At the same time, however, native-born slaves' ability to more than reproduce themselves eliminated the recurring capital expenditures planters who purchased indentured servants were required to make, and which West Indian slaveowners continued to incur owing to the lethal disease environment of the Caribbean and the harsh work routines associated with the cultivation and processing of sugar. The costs of rearing enslaved children to maturity were spread over a number of years, and, unlike the purchase of expensive Africans, could be paid for in large part by diverting a bit more labor into raising extra corn. Cloth was the only item that had to be purchased for young slaves from the proceeds of the tobacco crop, and, since planters allotted children only a bare minimum, this added outlay was hardly significant. Moreover, planters passed some of the costs of child rearing onto enslaved parents. Enslaved

19. Coombs, "Building 'The Machine,'" chap. 7. For slave-buying patterns in Barbados, see Jennifer L. Morgan, *Laboring Women: Reproduction and Gender in New World Slavery* (Philadelphia, 2004), 84–87.

mothers were required to make clothing for their younger children, and both parents often devoted much of their scant spare time to raising or catching food to improve the family's diet. As in the white population, an increased dependency ratio was a onetime transition from an age-imbalanced immigrant workforce with extremely high productivity to a more demographically normal but less productive one overall.[20]

Developing Regional Economies

The prolonged slump in tobacco prices between 1680 and 1729 prompted planters to adopt new strategies for coping with hard times. The strategy varied depending on the strain of tobacco grown and on the sorts of soils available. Beginning in 1681, prices fell and remained depressed (except for a brief recovery around 1700) until the 1710s, and then they improved modestly. The price of oronoco tobacco, the sort grown on Maryland's lower Western and upper Eastern shores, the Virginia side of the Potomac, and the north side of the Rappahannock, slumped most precipitously. It fell under a penny a pound in 1681 and remained stubbornly below this dismal level until 1713. The price of the sweet-scented strain, the variety usually cultivated between the north side of the James River and the south side of the Rappahannock, and the sort most favored by English consumers, also dropped to little more than a penny a pound in the early 1680s. Sweet-scented growers enjoyed a brief boom in the middle 1700s and again between 1715 and 1720 but suffered another severe slump in the 1720s (Figure 1).[21]

Stagnant European demand, warfare from 1689 to 1697 and again from 1702 to 1713 that disrupted markets and sent shipping costs skyrocketing, and additional English taxes and marketing charges all contributed to these woes. Until the late 1680s, planters and merchants continually improved productivity in cultivating and processing the leaf, in packing, and in shipping, thus lowering costs. In these circumstances, Chesapeake planters could still maintain a modest prosperity, while ever more European consumers could

20. Russell R. Menard, "The Maryland Slave Population, 1658–1730: A Demographic Profile of Blacks in Four Counties," *WMQ*, 3d Ser., XXXII (1975), 29–54; Allan Kulikoff, "The Beginnings of the Afro-American Family in Maryland," in Land, Carr, and Papenfuse, eds., *Law, Society, and Politics in Early Maryland*, 171–196; Kulikoff, *Tobacco and Slaves*, chap. 8; Kulikoff, "A 'Prolifick' People," *So. Stud.*, XVI (1977), 391–428; Morgan, *Slave Counterpoint*, 58–101; Darrett B. Rutman and Anita H. Rutman, *A Place in Time: Middlesex County, Virginia, 1650–1750* (New York, 1984), 187–188; Rutman and Rutman, *A Place in Time: Explicatus* (New York, 1984), chap. 12; Trevor Burnard, *Creole Gentlemen: The Maryland Elite, 1691–1776* (New York, 2002), 38–41; Coombs, "Building 'The Machine,'" chap. 7; Harris, *History of Human Populations*, II, *Migration, Urbanization, and Structural Change*, 340–342, 346–350, 400–401.

21. Walsh, "Summing the Parts," *WMQ*, 3d Ser., LVI (1999), 53–94.

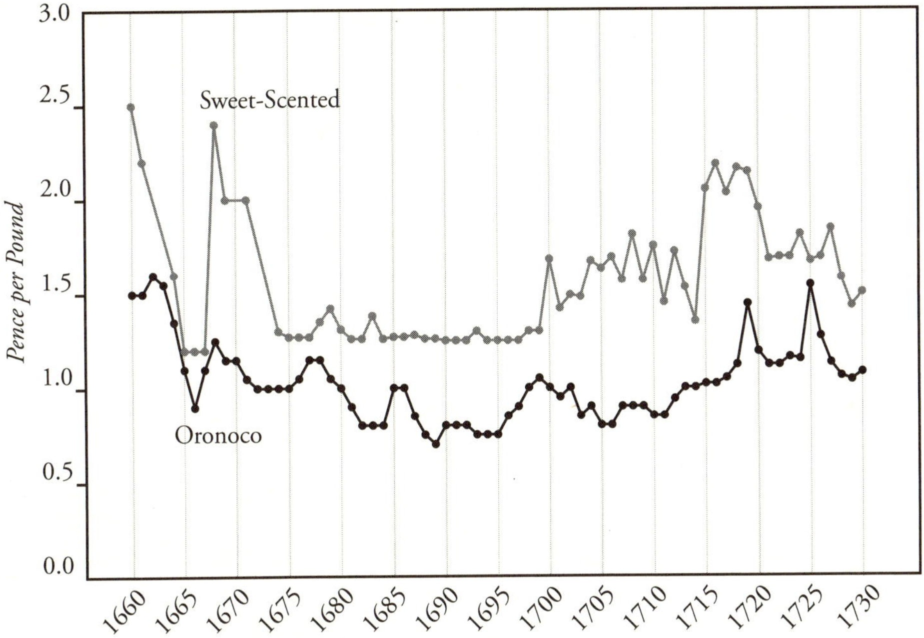

FIGURE 1. Oronoco and Sweet-Scented Tobacco Prices, 1660–1730. Drawn by Kimberly Foley. Menard, "The Tobacco Industry in the Chesapeake Colonies, 1617–1730: An Interpretation," *Res. Econ. Hist.,* V (1980), 157–166; John J. McCusker, "Colonial Statistics," in Susan B. Carter et al., eds., *Historical Statistics of the United States: Earliest Times to the Present,* millennial ed., 5 vols. (New York, 2006), V, 681–683.

afford to buy the alluring weed. By the early 1690s, most established planters had achieved the highest output per worker that could be obtained with the existing technologies, and, indeed, these production levels were never again equaled. Rising prices of land and labor meant that the average planter could cut his costs no further. And, with the price of manufactured tobacco no longer falling, few additional Europeans took up smoking. Chesapeake planters—in the aggregate—ceased expanding production. Consequently, income from tobacco, both per taxable and per capita, declined to a low in 1705 and exhibited only a modest recovery by the 1720s.[22]

Regionwide estimates of tobacco production and income, however, ob-

22. See sources cited in note 6, above, and R. C. Nash, "The Organization of Trade and Finance in the British Atlantic Economy, 1600–1830," in Peter A. Coclanis, ed., *The Atlantic Economy during the Seventeenth and Eighteenth Centuries: Organization, Operation, Practice, and Personnel* (Columbia, S.C., 2005), 111, 126.

scure the markedly different experiences of planters farming in sweet-scented, oronoco, and peripheral areas. Through 1730, residents of the peripheral tobacco areas (the lower James River basin and the lower Delmarva Peninsula), where only low-quality tobacco could be grown on the small proportion of this region's land capable of supporting any sort of agriculture, were about 20 percent of the total Chesapeake population. Few soils in these places were suitable for tobacco, and most planters dropped the staple crop entirely when tobacco prices were low overall, and especially abysmal for the inferior grades that they could raise. These planters turned instead to the production of naval stores, timber, cider, small grains, corn, and livestock as well as to subsistence farming. This shift was well under way by the mid-1670s, when tobacco from the periphery constituted only about 10 percent of the region's total exports and export earnings had already dropped to about two pounds per taxable, less than half that obtained elsewhere in the Chesapeake.

Between the mid-1670s and 1702, residents of the periphery suffered a further halving in export income, and yearly returns per taxable remained under one pound thereafter. Falling tobacco prices after 1680 served to reinforce, rather than to initiate, changes in the way Euro-American settlers on the periphery made their living. By about 1700, in the periphery earnings from the staple had become a minor factor in the calculations of all but a few fortunate planters who had access to exceptionally fertile soils capable of producing crops that could compete with those of the Western Shore. By 1710, planters in these areas produced only 5 percent of the region's tobacco exports. Most local entrepreneurs who were seeking greater fortunes looked primarily to some combination of local and intercontinental markets, unrelated to the tobacco trade, for new opportunities.[23]

When the peripheral tobacco-growing areas are excluded, revised output and income estimates suggest that planters in prime tobacco country realized higher earnings from exports across the seventeenth century than previously thought—about £5 per taxable rather than about £3.5, a sufficient inducement to persuade most to continue growing the weed. Similarly, income per capita hovered around £2 rather than £1. Moreover, after 1710, residents of oronoco areas were earning about £1 more apiece each year from the staple than were residents of sweet-scented areas (Figures 2, 3). The oronoco-growing areas accounted for about a third of the region's total population, but it produced half of the tobacco exported in the last quarter of the seventeenth century. Between 1710 and 1730, this subregion produced two-thirds of Chesapeake tobacco exports. Oronoco growers countered the lower prices their more

23. Walsh, "Summing the Parts," *WMQ*, 3d Ser., LVI (1999), 53–94.

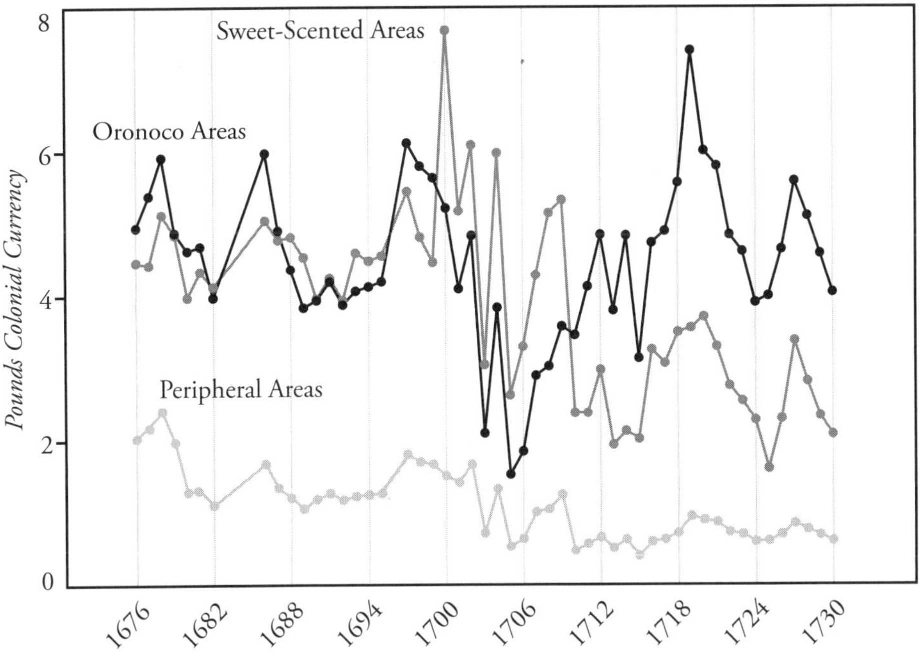

FIGURE 2. Tobacco Income per Taxable, 1676–1730. Drawn by Kimberly Foley. Lorena S. Walsh, "Summing the Parts: Implications for Estimating Chesapeake Output and Income Subregionally," *WMQ*, 3d Ser., LVI (1999), 88–94.

ordinary leaf commanded by making as much tobacco as they could. This strain was marketed in Britain and, from the turn of the eighteenth century, also increasingly reexported to new continental European consumers.

Sweet-scented growers, whose family members and large workforces accounted for nearly half of the total population, in contrast, attempted to capitalize on the price differential between the two strains by improving the quality rather than increasing the quantity of their output. Since the soils where sweet-scented could be grown were limited to the lower peninsula of Virginia and parts of the lower Rappahannock basin (see Map 6), this strain accounted for a steadily decreasing proportion of Chesapeake tobacco exports. Between 1675 and 1705, over 40 percent of exports came from the sweet-scented subregion, but as settlement expanded into the hinterlands, where only oronoco would grow, the share dropped to just over a quarter.

Up to the decade of the 1700s, the oronoco and sweet-scented areas remained in a relatively integrated market. Prices for the two strains moved in tandem, and levels and trends in income per taxable and per capita were quite similar. Neither subregion emerged with a clear advantage; what sweet-

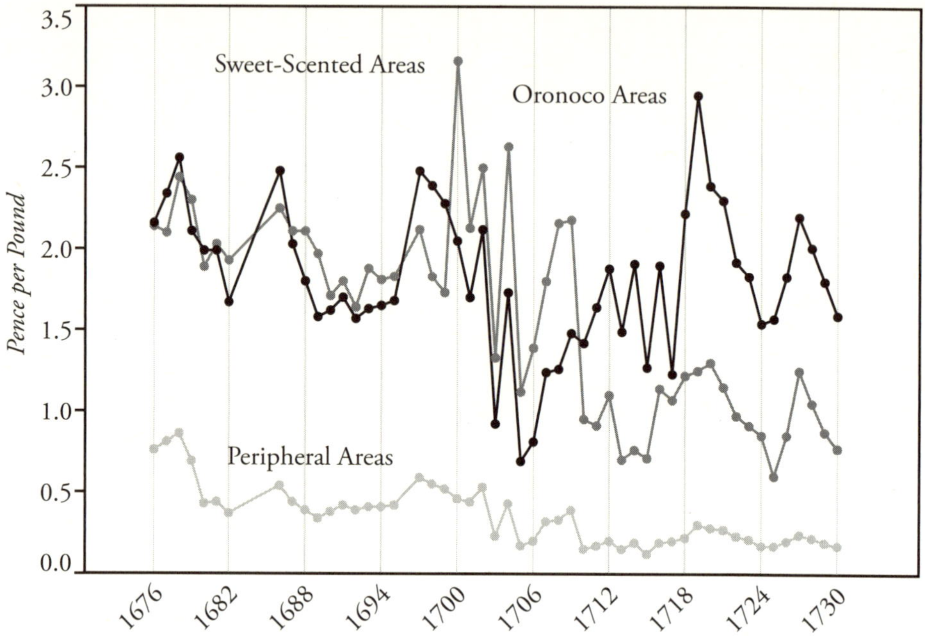

FIGURE 3. Tobacco Income per Capita, 1676–1730. Drawn by Kimberly Foley. Lorena S. Walsh, "Summing the Parts: Implications for Estimating Chesapeake Output and Income Subregionally," *WMQ*, 3d Ser., LVI (1999), 88–94.

scented growers lost in diminished output was offset by higher prices for what they did produce. Production per taxable dropped drastically in both oronoco and sweet-scented areas after 1700, but with differing results in terms of income. In the first decade of the eighteenth century, with the price of sweet-scented tobacco rising, producers in this subregion fared much better than did those who grew the more common oronoco strain. For a short time, the shift in emphasis from quantity to quality paid off. Higher profits underwrote new investments in African slaves, sustained the political power of the tidewater Virginia elite, and helped to finance the construction of most of the early, well-built public buildings and the initial development of viable urban settlements in the lower tidewater. Meanwhile, in the oronoco areas output and earnings dropped to lower levels than planters had experienced up to this time. Many of the first generation of native-born whites coming of age around the turn of the eighteenth century in the oronoco-growing region fared less well economically than had their immigrant parents, as, for example, did Robert Cole, Jr.[24]

24. Ibid.

But the sweet-scented growers' advantage proved extremely short-lived. Robert "King" Carter's pessimistic prediction in 1702, "He that measures his Expenses by his Seven Last years gettings in Virginia, may Seven years hence probably have Little to spend," proved all too true. In contrast to the 1700s when output per taxable slumped in oronoco areas, in the 1710s and 1720s, as prices rose, oronoco growers again raised levels of output to around one thousand pounds per worker. They concentrated solely on quantity; almost none invested any labor in further processing their product. English tobacco manufacturers were willing to pay more for processed leaf, but continental manufacturers, who were the ultimate buyers of much of the oronoco, were not. Although production per worker remained lower than the twelve hundred to sixteen hundred pounds achieved in the 1680s and 1690s, this level of output was enough to enable growers to capture the benefits of expanding continental markets for cheaper grades of the weed. After about 1710, incomes per taxable and per capita in the oronoco areas were consistently higher than in sweet-scented counties. Since sweet-scented growers invested more labor into further processing of their crops, small and middling growers might have been especially hard-hit by a shortage of bound workers.[25]

Political developments in the two colonies reflect the differing production strategies that emerged in the two prime tobacco-producing subregions. In periods when glutted markets were a problem, Virginia governors and legislators, the majority of whom came from sweet-scented areas, had long tended to favor a variety of regulations to limit crop size and improve quality. Tobacco-cutting "riots" that occurred in sweet-scented-growing Gloucester, Middlesex, and New Kent counties in 1682 suggest that some small Virginia planters (and their wives who participated in the destruction) also supported cessation or stinting. It seems likely, however, that some political leaders who were frustrated in their plans for enacting a cessation law (including members of the lower house and county magistrates) had a hand in orchestrating what were well-coordinated, nonviolent crowd actions in order to force the absent royal governor to return to Virginia to address economic grievances. Other gentry instigators were alleged to have hoped to profit from large quantities of tobacco they had engrossed in anticipation of a proposed cessation that they had expected would lead to rising prices. The deputy governor had to call out the militia to suppress the rioters, almost all of whom were subsequently

25. Ibid.; Carter to Matthews and Goodwin, July 15, 1702, Robert Carter Letterbook, 1701–1710, Wormely Estate Papers, 1701–1716, in the Christ Church Parish Records, VSL, microfilm, M-1173.1, CWL.

pardoned. An upturn in tobacco prices beginning the next year brought a temporary halt to agitation for a cessation.[26]

In 1713, in response to another price slump, the Virginia governor and legislature tried a new tact, enacting a tobacco-inspection system that included most of the features of later, successful acts. This law foundered in 1717, though, more from problems with political patronage than from planter opposition to regulation per se. Governor Alexander Spotswood wanted to reward burgesses who supported his policies with tobacco inspectorships. Middling and small planters feared that such politically and economically powerful men who were both growers and buyers of tobacco would act partially to advance their own interests, so they refused to obey a law that was shortly disallowed by the king at the behest of British tobacco merchants. From 1724 on, limitations on the number of plants each laborer might tend were in effect. Although this later stinting might have forced some planters to cut back on their crops, in general it seems simply to have made into law what was already widespread practice. Those most affected by the new limits were probably small and middling growers who had not voluntarily switched to producing higher-quality tobacco, given the extra labor required for intermediate processing, and planters whose soils were incapable of yielding high-quality crops.[27]

In contrast, Maryland legislators, governors, and its proprietor were much less enamored of laws that would seriously curtail production. Repeatedly, they cited small-planter opposition as one reason for their failure to support either cessations, stints, or inspections. Periodic acts forbidding the tending of seconds (probably widely ignored) and the inclusion of trash tobacco or other foreign matter in hogsheads (much better enforced through individual information) was as far as that colony's officials were willing to go until 1728.

26. John C. Rainbolt, *From Prescription to Persuasion: Manipulation of Eighteenth [Seventeenth] Century Virginia Economy* (Port Washington, N.Y., 1974), chap. 5; Vertrees J. Wyckoff, *Tobacco Regulation in Colonial Maryland* (Baltimore, 1936), chaps. 3, 4, 5; Warren M. Billings, John E. Selby, and Thad W. Tate, *Colonial Virginia: A History* (White Plains, N.Y., 1986), chap. 10. For the 1682 riots, see *EJCCV,* I, 17–37, 48–49, 55; and Haskell, "'Affections of the People,'" 304–313.

27. Leonidas Dodson, *Alexander Spotswood, Governor of Colonial Virginia, 1710–1722* (Philadelphia, 1932), chap. 4; Billings, Selby, and Tate, *Colonial Virginia,* chap. 10; Moss, "The Virginia Plantation System," chaps. 4, 5; Hemphill, "Virginia and the English Commercial System," chaps. 1–3; Calvin Brewster Coulter, Jr., "The Virginia Merchant" (Ph.D. diss., Princeton University, 1944), chap. 1; David Alan Williams, "Political Alignments in Colonial Virginia Politics, 1698–1750" (Ph.D. diss., Northwestern University, 1959), chaps 4–6; Beverley, *History and Present State of Virginia,* ed. Wright, 70–72, 74–75; Lewis Cecil Gray, *History of Agriculture in the Southern United States to 1860,* 2 vols. (1932; rpt. Gloucester, Mass., 1958), I, chap. 10.

In that year, they did pass a stint act (set at a higher limit than Virginia's), but the law seems to have been observed mostly in the breach and was soon revoked by the proprietor. Some of the representations might have been in part self-serving, but there seems no good reason to doubt Lord Baltimore's earlier assertion that most Maryland planters continued to regard the liberty of planting as much tobacco as they wished "their Birthright as Englishmen." The Maryland planters' intransigence was rooted in a strategy of maximum production that usually produced good results for oronoco growers. It was thus surely no accident that in 1732 it was poor oronoco planters in Northern Neck and Eastern Shore counties (who shared the Marylanders' interest in maximum production) who violently resisted the regulations regarding quality imposed by Virginia's new tobacco inspection act.[28]

The Rise of the Commission System

As increases in the volume of tobacco imports slowed between 1676 and 1700, and growth stopped completely over the first quarter of the new century, institutional arrangements in the tobacco trade became increasingly complex. In Britain, the small, temporary "adventure" form of enterprise that had prevailed at midcentury was supplanted by more complex and credit-dependent forms. Changes in the level of customs duties and in marketing arrangements encouraged concentration of the trade in the hands of a smaller number of large, heavily capitalized firms that specialized exclusively in Chesapeake commerce. Low prices and high duties hurt not only small planters in the colonies but also petty retailers and minor merchants in Britain.[29]

Beginning in the 1660s, large Barbadian sugar planters had responded to a similar problem—falling sugar prices on that island—by assuming the risk of shipping their crops to England, where they could obtain much higher prices than they could get by selling to merchants on the island. All staple planters who sold their crops in the colonies had the advantage of getting their proceeds immediately without having to wait until the crop was sold in Europe, and they did not have to hazard the risks of the ship's sinking or being captured, the crop being damaged in transit, or prices in Europe sud-

28. Wyckoff, *Tobacco Regulation,* chaps. 6, 7; Answer of Lord Baltimore to the Privy Council, Nov. 19, 1664, in William Hand Browne et al., eds., *Archives of Maryland,* 72 vols. (Baltimore, 1883–) III, 506–510; Billings, Selby, and Tate, *Colonial Virginia,* 240. Planter Thomas Cable reported that one Northampton County justice of the peace flagrantly violated the law by tending seconds and encouraged other planters to follow his example. See Cable to John Custis, Jan. 19, 1728, Thomas Cable Letterbook, 1723–1733, MS 171, MHS.

29. Jacob M. Price and Paul G. E. Clemens, "A Revolution of Scale in Overseas Trade: British Firms in the Chesapeake Trade, 1675–1775," *JEH,* XLVII (1987), 1–43.

denly collapsing. On the other hand, they gave up any opportunity to take advantage of the much higher prices offered in England from which planters who could wait longer for their pay might benefit substantially. Moreover, the Royal African Company, which had a monopoly of the trade in Africans, preferred to sell slaves for bills of exchange or for ready cash rather than for produce, and planters who consigned could draw such bills on their London commission agents (or "factors"). By the 1670s, the consignment system had become dominant in Barbados and was spreading to Jamaica and the Leeward Islands.[30]

As larger Chesapeake planters increasingly felt the pinch of falling tobacco prices in the 1680s and 1690s, they began to follow the West Indian example. In effect, the planters became entrepreneurs, shipping their crops on their own account and at their own risk to a commission agent in London or Bristol. Commission agents arranged for the unloading and storing of the cargo, paid the charges incurred for transport, freight, insurance, customs, and warehousing, sold the crop, and applied the profits according to the planter's instructions in return for a commission of 2.5 percent calculated on the gross value of all the transactions handled. During the war years of 1689 to 1697 and 1702 to 1713, the commission trade was suspended in the West Indies owing to heavy losses of sugar-carrying ships at sea. It expanded in the Chesapeake, however, where high wartime transport costs lowered tobacco prices in the country and raised them in Europe, but where shipping losses were not so severe. The large difference between prices offered in the country and those offered in Europe made elite planters more willing to hazard the greater risks the new system entailed. Moreover, an expanding marine insurance industry began to reduce the risks for those who were willing to pay the premiums. Consigning was most profitable during the peace years from 1714 until 1725, when tobacco prices in Europe began to fall. By 1730, as much as 40 percent of Chesapeake tobacco was shipped on consignment, either by planters or by local merchants. This tobacco was principally the produce of large planters who could afford to wait a year for their pay in order to get European prices for their weed, and who, as shipments of slaves direct from Africa became more common, had increasing need to draw bills of exchange to pay for them. The more regular correspondence between merchants and planters that the commission system necessitated also improved market in-

30. K. G. Davies, "The Origins of the Commission System in the West India Trade: The Alexander Prize Essay," Royal Historical Society, *Transactions*, 5th Ser., II (1952), 89–107; Nash, "Organization of Trade and Finance," in Coclanis, ed., *Atlantic Economy*, 95–151; Russell R. Menard, *Sweet Negotiations: Sugar, Slavery, and Plantation Agriculture in Early Barbados* (Charlottesville, Va., 2006), 2, 5–6, 91, 102–103, 121.

formation on both sides of the Atlantic. Middling and small planters, however, who still produced more than half of the region's tobacco, continued to sell their crops in the country to British factors or to local merchants.[31]

The consignment system promoted concentration of firms in England for several reasons. First, there were increased shipping risks. Merchants had to buy or charter ships with a large return capacity, and they would lose money if they failed to attract enough consignments to fill the vessels they sent out. Second, at the accession of James II, Parliament raised the duty on tobacco consumed in Britain from two pence to five pence per pound, more than 100 percent of the pre-duty price prevailing in England. Merchants opted either to pay the new impost immediately in order to obtain a 10 percent discount or to give bond for it. These financial responsibilities discouraged marginal importers, since only affluent merchants could find adequate securities to countersign their bonds. In addition, only larger firms could command the confidence of elite planters and could provide the range of commercial and quasi-banking services that these planters came to expect. Moreover, larger firms might have enjoyed economies of scale in wholesaling tobacco and in purchasing goods directly from manufacturers. Between 1676 and 1686, the number of Londoners who imported Chesapeake tobacco was halved from 573 to 292, and the average annual volume of tobacco these importers handled rose from 22,000 pounds to 59,000 pounds. Bankruptcies among merchants between 1710 and 1716, occasioned in part by more stringent collection of overdue customs bonds, further reduced the number of London tobacco importers to only 117 in 1719. These importers now handled an average of 193,000 pounds a year. No really large tobacco merchants operated in Bristol, but there, too, the number of importing firms declined after 1680, while the average poundage the surviving firms handled also rose.[32]

During King William's and Queen Anne's wars, trade with the outports expanded at the expense of London and Bristol because vessels sailing from

31. Price and Clemens, "Revolution of Scale," *JEH,* XLVII (1987), 6, 16–21; Nash, "Organization of Trade," in Coclanis, ed., *Atlantic Economy,* 98–112; Bergstrom, *Markets and Merchants,* chap. 6. In contrast to sugar and tobacco growers, South Carolina rice planters seldom used the consignment system. There were few quality gradations in rice and, hence, few opportunities to earn higher prices by consigning better grades to British markets. Rice planters did, however, later use the consignment system for indigo, a crop also characterized by wide variations in quality (Nash, "Organization of Trade," in Coclanis, ed., *Atlantic Economy,* 107–110). For general increases in the speed and extent of transatlantic communications, see Steele, *English Atlantic,* chaps. 7–9, 11.

32. Jacob M. Price, *Perry of London: A Family and a Firm on the Seaborne Frontier, 1615–1753* (Cambridge, Mass., 1992), 95–100; Price and Clemens, "Revolution of Scale," *JEH,* XLVII (1987), 9–22.

ports such as Liverpool, Whitehaven, and Glasgow on Britain's west coast were in less danger from capture by French privateers than were vessels whose home ports were on the south coast within easier reach of the enemy. Starting in the mid-1680s, Liverpool merchants began competing for a share of the region's tobacco, but initially they had limited success. In the early 1700s, some Liverpool traders still found themselves shuttling ships from one Chesapeake river basin to another in order to obtain a full cargo of tobacco, while others sent ships to the lower Eastern Shore, where they encountered few competitors for that region's poor-quality crops. Early Liverpool tobacco traders seem to have found ingenious ways to defraud royal customs agents by claiming to have imported large quantities of damaged, and thus duty-free, tobacco and by finding ways to understate the amount of tobacco they actually imported. Buying tobacco with indentured servants lured from Lancashire and Ireland at a time when European laborers were in short supply also afforded some aspiring Liverpool tobacco dealers a foothold in the staple trade. After the union with Scotland in 1707, Glasgow merchants, who enjoyed both a quicker and safer shipping route and lower operating costs than the English, also formally entered a trade in which they had previously participated only as interlopers. By the 1720s, the outports and Scotland together handled about 40 percent of Chesapeake tobacco.[33]

London, outport, and Scottish merchants catered to different sets of British and of European consumers with varying tastes. Sweet-scented growers enjoyed an advantage over oronoco growers in the home market at the beginning of the period, but by the 1720s their position was less favorable. Bristol merchants, who continued to deal primarily in finer tobacco for home consumption, did compete aggressively against London factors for a larger share of the crops from the York and Rappahannock river basins. Given the high duties imposed on tobacco sold domestically, however, the number of British consumers, the major buyers of sweet-scented, was not expanding much, and they were also finding less expensive, good-grade oronoco increasingly attractive. London merchants Micajah and Richard Perry, for example, informed Virginia planter John Custis in 1716: "You would hardly believe if we should tell you what a run there hath been in the home consumption of smaller sort of Browne aranoka tobacco which hath ben sold from 8 ½ to 9d p[er] pound and is so to this day. Which wonderfully lengthens out

33. Price and Clemens, "Revolution of Scale," *JEH,* XLVII (1987), 24–37; Lorena S. Walsh, "Liverpool's Slave Trade to the Colonial Chesapeake: Slaving on the Periphery," in David Richardson, Suzanne Schwarz, and Anthony Tibbles, eds., *Liverpool and the Transatlantic Slave Trade* (Liverpool, 2007), 98–137.

[sales of] sweets[cente]d and show us that cheapnesse hath an Argument above quality." Consequently, the main result of competition between merchants serving the domestic market was to force small-scale and less efficient English traders out of business, rather than to raise the prices that sweet-scented growers received. Neither was sweet-scented much in demand in Holland and Germany, markets that reopened with the resumption of peace in Europe; instead, these markets favored the bright grades of oronoco grown primarily along the Patuxent River and on Maryland's upper Western Shore. Although the oronoco tobacco market was probably less well differentiated than that for sweet-scented, there are nonetheless hints that bright-leaf growers also encountered somewhat better prospects in the 1690s and 1700s than did planters in dull-leaf counties. Nonetheless, the main effect of the Navigation Acts and of wartime disruptions, both of which raised the price of Chesapeake tobacco on the Continent, was to encourage domestic tobacco production in the Dutch Republic, which in turn reduced demand for colonial leaf.[34]

During the same years, France surpassed Germany to become the second most important market, after the Dutch, for British tobacco. The French trade was a state monopoly farmed out to private buyers who were not bound by conventional mercantilist ideas about national self-sufficiency. The entry of the French "created a new force in the English tobacco market, a single large buyer." The power of the French monopolists was enhanced by their employing a limited number of agents—and sometimes only one—thus driving down prices much more readily than less organized, competing British merchants could keep them up. Moreover, the French agents paid either in ready cash or in "tobacco notes" redeemable in thirty to sixty days (which could alternatively be sold immediately at a discount). The French agents also often advanced money to tobacco importers before purchase, which allowed them to pay customs duties promptly. Cash-strapped British merchants were thus strongly tempted to sell at least a portion of their tobacco low to the French rather than in the domestic, Irish, Dutch, or German markets, where they might have to wait months before receiving the proceeds.[35]

34. Jacob M. Price, *France and the Chesapeake: A History of the French Tobacco Monopoly, 1674–1791, and of Its Relationship to the British and American Tobacco Trades,* 2 vols. (Ann Arbor, Mich., 1973), I, chap. 25; Micajah and Richard Perry to John Custis, 1716, Custis Family Papers, 1638–1858, MSS1 C9698, VHS, microfilm, M-58.1, CWL. For the Dutch market, see Victor Enthoven and Wim Klooster, "'The Dutch Found and Relieved Us': The Shifting Dynamics of Virginia-Dutch Trade in the Long Seventeenth Century," in Bradburn and Coombs, eds., *Early Modern Virginia.*

35. Price, *France and the Chesapeake,* I, chaps. 20, 25 (quotation on 531); Jacob M. Price, *The*

After 1713, merchants from the English outports and shortly thereafter from Scotland began to compete vigorously with London factors for the lower-grade, dull oronoco leaf grown along the James, the Potomac, and on the Eastern Shore that was suited to the French market. Outport merchants bought tobacco in the country at increasingly attractive prices, and, in order to secure their cargoes, either sold their merchandise more cheaply or extended planters more credit—or both. Growers of sweet-scented tobacco realized fewer benefits, since, with the partial exception of Bristol, most outport merchants had little interest in handling the more expensive strain and most did not care to trouble themselves with buying on consignment. In addition, the practice of stemming sweet-scented tobacco, which was a major means for cutting the cost of freight, limited possible alternative markets, since stemmed tobacco was a commodity the Scotch "did not care to be concerned in." The entry of Scottish and outport merchants thus benefited primarily small and middling oronoco growers. Large planters, who continued to consign their own crops to England, realized gains from increased competition among outport merchants only if they sold ordinary oronoco leaf received for land rents, official salaries, or professional fees in the country.[36]

Pragmatic Adaptations

The Reverend Hugh Jones, an Anglican minister who taught natural philosophy and mathematics at the College of William and Mary between 1717 and 1721, concluded an account of Virginia that he published in England in 1724 with a long appendix in which he laid out a scheme for improving

Tobacco Adventure to Russia: Enterprise, Politics, and Diplomacy in the Quest for a Northern Market for English Colonial Tobacco, 1676–1722, American Philosophical Society, *Transactions,* n.s., LI, pt. 1 (Philadelphia, 1961); Price, "The Economic Growth of the Chesapeake," *JEH,* XXIV (1964), 496–511.

36. Robert Carter to William Dawkins, Aug. 1, 1723, Robert Carter Letterbook, 1723–1724, UVa (quotation); Jacob M. Price, "The Rise of Glasgow in the Chesapeake Tobacco Trade, 1701–1775," *WMQ,* 3d Ser., XI (1954), 179–199; Paul G. E. Clemens, "The Rise of Liverpool, 1665–1750," *Economic History Review,* 2d Ser., XXIX (1976), 211–225; Clemens, *Atlantic Economy,* chaps. 1, 4; Hemphill, "Virginia and the English Commercial System," chaps. 1, 2. For the Dutch market, see David Ormrod, *The Rise of Commercial Empires: England and the Netherlands in the Age of Mercantilism, 1650–1770* (Cambridge, 2003), 198–201. John Oldmixon, *The British Empire in America* (1741), 2 vols. (New York, 1969), 340, noted an increase in demand for oronoco in Continental markets. See also William Woodford to Charles Chiswell, July 31, 1725, Letterbook of William Woodford, 1723–1737, University of Kentucky, Lexington, microfilm, M-1060-1, CWL; John Tayloe's letters for 1716–1719, in Stephen Lyde Account Book, 1708–1711, and Accounts and Letters of John Tayloe I and John Tayloe II, 1714–1778, Tayloe Family Papers, MSS1T2118, b1, VHS; *Byrd Correspondence,* I, 376–377; and Higginson and Bird Letterbook, Galloway / Maxcy / Markoe Family Papers, 1654–1888, LC.

the colony by promoting "projects, inventions, and manufactures." Taking a page from the defunct Virginia Company, he urged the current British government to provide funds for transporting poor and idle people from the mother country to Virginia as indentured servants. There, Jones maintained, they could live better than at home and, at the same time, enrich the empire by producing essential commodities that Britain had to import from other nations. The projects Jones proposed included "all sorts of English husbandry," raising surplus cattle and hogs for victualing ships and for supplying food-short West Indian islands, rearing sheep to lessen imports of wool from Spain, raising hemp and flax, manufacturing cloth, building ships, making wine and cider, collecting medicinal plants, sawing timber, manufacturing paper, fishing, creating water meadows, tanning leather, and making iron. Rather than interfering with the interests of the mother country, as was then commonly perceived in Britain, Jones argued that such enterprises would strengthen the empire without lessening the planting of tobacco and, hence, crown revenues. The enslaved Africans who now raised most of the tobacco could continue to be so employed, while the new European immigrants would be assigned to these other "useful" enterprises.[37]

If the British government failed to execute such schemes, Jones added, "it might not be cost and labour thrown away if the Virginia gentlemen, traders and planters attempted at some, if not all the things mentioned." He believed that such trials would not entail great expense, and some of the designs "might answer expectation." In reality, by the 1720s a number of Virginia and Maryland planters had already embraced many of the projects Jones enumerated and, in so doing, had begun the transformation of the Chesapeake from a place dependent on the fortunes of a single staple into a more diversified plantation region drawing income from a variety of trades pursued throughout the Atlantic world.[38]

The list of projects Jones proposed included many that English rulers and colonial promoters had urged on Chesapeake colonists in earlier years. In the last quarter of the seventeenth century and in the turbulent years of war in the early eighteenth, the new British rulers were distracted by political problems at home and conflicts in Europe. They had little interest in the colonies and no inclination for prescribing alternative sorts of agriculture in which they thought colonists ought to engage or alternative projects

37. Hugh Jones, *The Present State of Virginia, from Whence Is Inferred a Short View of Maryland and North Carolina* (1724), ed. Richard L. Morton (Chapel Hill, N.C., 1956), 75–154 (quotation on 137).

38. Ibid., 146.

that they should undertake. The emphasis had instead shifted to more closely regulating existing trades, maximizing customs revenues, and discouraging manufacturing to protect industries at home. For example, legislation of 1691 mandating the building of towns — a step British rulers had been urging on the Chesapeake colonists from the 1620s until 1680 — was opposed by crown officials and English merchants and manufacturers who now considered colonial cloth making and other industries that towns might encourage a threat to the economy of the home country. From 1713 on, Virginia governors and councillors routinely defended proposed tobacco inspection or stint laws by assuring British officials that, without the improvement in tobacco prices that the legislation was designed to effect, the people "will soon fall into Manufactures of their own." Beginning in the 1680s, then, initiatives for change came, not from the metropolis, but from the periphery. Chesapeake colonists adopted new expedients more in contradiction to metropolitan expectations than with their blessing.[39]

Among the new strategies many large and some middling and small planters adopted was the introduction of more elements of traditional European husbandry such as plowing, ditching, wheat culture, sheep rearing, and improved pastures. Clearly, however, the activities and publications of period English agricultural improvers had virtually no influence on colonial adaptations. In Britain, prices of grain fell in the late seventeenth and early eighteenth centuries, encouraging a shift in emphasis and resources toward livestock rearing. The attention of agricultural improvers turned primarily to techniques for carrying more livestock on a given acre of land. This was a problem most Chesapeake colonists did not have, and those living in crowded older areas who were encountering difficulties maintaining the size and quality of their livestock could find cheaper and less labor-intensive ways for getting supplemental feed than creating water meadows or instituting elaborate crop rotations or raising problematic new fodder crops.[40]

Although some elite planters had editions of one or two books on English husbandry published in the late sixteenth or early seventeenth century in their libraries (texts similar to those investors had sent with the first settlers

39. John W. Reps, *Tidewater Towns: City Planning in Colonial Virginia and Maryland* (Williamsburg, Va., 1972), chap. 4; *EJCCV,* IV, 46; Rainbolt, *From Prescription to Persuasion,* chaps. 6, 7; Steele, "The Anointed, the Appointed, and the Elected," in Marshall, ed., *The Eighteenth Century,* vol. II of Louis, ed., *Oxford History of the British Empire,* 105–119; Lovejoy, *Glorious Revolution in America,* chap. 9; Andrews, *Colonial Period,* IV, chaps. 10, 11.

40. E. L. Jones, *Agriculture and the Industrial Revolution* (Oxford, 1974), chaps. 1–3; Mark Overton, *Agricultural Revolution in England: The Transformation of the Agrarian Economy, 1500–1850* (Cambridge, 1996), chap. 3.

at Jamestown), more recent publications on husbandry were almost entirely absent. Among the eleven planters of the era discussed below, information is available for four of their libraries. Virginia planter William Fitzhugh never mentioned having read any agricultural literature in the numerous letters he wrote between 1679 and 1701, and, although he ordered many books from England, none had anything to do with farming. William Byrd II had perhaps the largest library in Virginia, and, although he owned scores of books in several languages on the gentlemanly pursuits of gardening and horticulture and about thirty volumes on architecture, only six titles in a collection of more than 3,500 books deal extensively with the culture of field crops. Robert "King" Carter had a library of 263 volumes, but he owned not one book on husbandry. Maryland planter James Carroll's more modest collection of 63 books included only Gervase Markham's *Whole Art of Husbandry,* first published in 1631. There is no reason to suppose that the libraries of other elite planters assembled between the 1670s and 1730 were dissimilar to those of these four men. Most apparently considered contemporary European agricultural literature irrelevant to Chesapeake plantation management.[41]

Neither were larger planters attempting a wide range of diversification schemes to set an example for others, as had Virginia governor Sir William

41. Louis B. Wright, *The First Gentlemen of Virginia: Intellectual Qualities of the Early Colonial Ruling Class* (San Marino, Calif., 1940), esp. chaps. 6 (Fitzhugh), 9 (Carter), and 11 (Byrd). Carter's older brother John did own two books on husbandry, one by Gervase Markham, who died in 1637, and one by Walter Blith on English husbandry, probably *The English Improver Improved . . . ,* published in various editions between 1649 and 1653 ("Libraries in Colonial Virginia," *WMQ,* 1st Ser., VIII [1899–1900], 18–22). For Carter, see also George K. Smart, "Private Libraries in Colonial Virginia," *American Literature,* X (1938–1939), 24–52. For Byrd, see also the catalog of his library in John Spencer Bassett, ed., *The Writings of "Colonel William Byrd of Westover in Virginia Esqr"* (New York, 1901), 413–443; Margaret Beck Pritchard and Virginia Lascara Sites, *William Byrd II and His Lost History: Engravings of the Americas* (Williamsburg, Va., 1993), chap. 5; and Mark R. Wenger, "Westover: William Byrd's Mansion Reconsidered" (master's thesis, University of Virginia, 1980). The books on field agriculture in Byrd's library included John Mortimer's *Whole Art of Husbandry . . .* (1707), Richard Bradley's *General Treatise of Husbandry and Gardening . . .* (1721), and John Worlidge's *Systema agriculturae . . .* and *Dictionarium rusticum . . . ,* both published in multiple editions between 1669 and 1716. According to the long title of *Systema agriculturae,* it dealt with *New and Most Advantagious Ways of Tilling, Planting, Sowing, Manuring, Ordering, Improving of All Sorts of Gardens, Orchards, Meadows, Pastures, Cornlands, Woods, and Coppices* and the culture of fruits, grain, livestock, and poultry. The relatively extensive library of elite Middlesex County planter Edmund Berkeley (1665–1718) contained no titles on agriculture ("Virginia Council Journals, 1726–1753," *VMHB,* XXXV [1927], 36–37). For Carroll's library, see Charles M. Flanagan, "The Sweets of Independence: A Reading of the 'James Carroll Daybook, 1714–21'" (Ph.D. diss., University of Maryland, College Park, 2005), 156–168. For Marylanders generally, see Carl E. Garrigus, Jr., "The Reading Habits of Maryland's Planter Gentry, 1718–1747," *MHM,* XCII (1997), 37–53.

Berkeley between 1642 and 1677. As historian Robert Beverley explained, Berkeley "was full of Projects, so he was always very fickle, and set them on Foot, only to shew us what might be done, and not out of Hopes of any Gain to himself; so never minded to bring them to Perfection." Part of the problem with Berkeley's efforts had been his continued promotion of the quest to produce semitropical products unsuited to the region. In the judgment of John Rainbolt, "unwittingly the government sought to force the planters into economic endeavors with little or no potential," but offered few incentives for pursuing the provisioning trade with the West Indies, "where progress was possible." From the 1680s onward, gain, not glory, became the planters' principal, if not sole, motive for trying alternative crops. William Byrd II, for example, when embarking on an experimental shift in crop mix, did advise English correspondents that he intended to "set my countrymen an example" by sowing hemp. Byrd's main reason for trying the fiber crop, though, was that he was "out of humour with tobacco." Once Byrd discovered that hemp culture was unprofitable in Virginia, he quickly dropped it and again turned most of his energies to growing tobacco. None of the other planters, even the region's premier agricultural producer, Robert "King" Carter, claimed to be an "improving" farmer.[42]

Indeed, reintroducing European-style agriculture proved a much harder feat than it would have been in the 1660s and 1670s. In earlier years, many immigrants were present, like Robert Cole, who had grown up in rural areas in Britain and who had practical experience in growing wheat and other small grains and in plowing with draft animals. Many of the immigrants who arrived later, like most of those whose management practices will be shortly examined in greater detail, were sons of merchants or artisans and had grown up in towns or cities, and hence might not have known much about farming. A few sons of the planter elite did go to England for schooling in this period, but, since they spent most of their time in boarding schools or counting-houses—and their school vacations sampling urban pleasures—it is unlikely that these adolescents learned much about European farming methods while living abroad. Those who did travel in the countryside, like William Byrd II, seem to have paid much more attention to the architecture of manor houses and layouts of formal gardens than to the culture of field crops. When they returned to the Chesapeake and began managing plantations, these men learned Chesapeake husbandry from their fathers or from other successful large planters. Sons of small planters whom they hired as overseers or as farm

42. Beverley, *History and Present State of Virginia,* ed. Wright, 135; Rainbolt, *From Prescription to Persuasion,* chap. 3 (quotation on 88); *Byrd Correspondence,* I, 364, 379, 397, 410.

workers had little or no formal education and probably knew nothing about how to plow with draft animals or to grow grain. Many had perhaps never so much as seen a plow or a wheat field. Nor did enslaved Africans have any experience with such crops, unless they had been brought from the area of Senegambia where millet and other small grains were the staple starch. Consequently, after the turn of the century most planters who wanted to diversify their crop mixes by pursuing British husbandry had to import experienced indentured English, Irish, or Scottish plowmen to train both workers and supervisors in the proper techniques.

Throughout the British Empire, this era was one of internally generated transformations and adaptations. In all the staple-producing colonies, as settlement expanded and output rose the prices of their principal products fell. Planters had either to learn how to increase production and lower costs, to increase the value of their product by further processing it, or to find alternative sources of income. Barbados sugar growers had faced a similar decline in the price of their staple beginning in the 1660s and had responded by creating plantations that integrated all phases of production, by moving toward the gang system of labor, by shifting from less efficient, animal-driven mills to windmills, by adding steps to the refining process to produce purer grades of sugar, and by beginning to make rum from sugar by-products. Settlers on Bermuda, having exhausted their limited arable land with tobacco, abruptly abandoned the crop in 1691 and restructured their island economy around ship building and the provisioning and carrying trades. New England colonists, who never found a profitable staple, similarly expanded their maritime activities — fishing, ship building, and the intercolonial carrying trade — as well as manufactured at home some of the textiles that they lacked export earnings to buy from Britain. South Carolina settlers, who introduced rice as the third major colonial staple around 1700, encountered the same problem of expanding output and falling prices in the 1740s. There planters shifted the locus of production from moist uplands to inland swamps, improved irrigation systems, and added indigo as a second staple crop. A study of South Carolina planters concludes, "Homegrown experience, not imported models for emulation, yielded the critical innovations that made planting possible, profitable, and secure." Such was also the case with Chesapeake tobacco growers between 1680 and 1729.[43]

43. Menard, *Sweet Negotiations,* chaps. 4, 5; Michael J. Jarvis, *In the Eye of All Trade: Bermuda, Bermudians, and the Maritime Atlantic World, 1680–1783* (Chapel Hill, N.C., forthcoming), chap. 2; McCusker and Menard, *Economy of British America,* 97–101; S. Max Edelson, *Plantation Enterprise in Colonial South Carolina* (Cambridge, Mass., 2006), chaps. 3, 4 (quotation on 166).

Accounting for Success

We now turn to examining the management strategies of individual planters. Plantation records for this era are limited to a few scattered letterbooks, diaries, and account books left by a handful of very wealthy men. There are no comparable materials that would enable us to assess how middling and small planters of the period managed their farms. Some of the account books that are mentioned in elite planters' diaries and letters have since been lost; later generations might have valued literary documents more than arcane ledgers. It is also likely that some of the sorts of financial records we might expect to find for larger, more complex operations were in fact never kept. No consensus had yet been reached in the Chesapeake, or indeed in contemporary Britain, about what record-keeping and accounting practices were best suited for managing day-to-day agricultural operations or for devising longer-run financial strategies for landed estates. Nor was there agreement about whether it was necessary or desirable for gentlemen possessed of private fortunes who were not engaged in commerce to keep financial records aside from loose papers documenting individual dealings on credit and perhaps notations of cash receipts and expenditures.

In an economy like that of the Chesapeake that was overwhelmingly dependent on credit, keeping written records (or, among those who could not write, corroborating oral testimony) of transactions between individuals was essential. In addition, executors and administrators of estates were legally obligated to present an annual written account of their management to county or provincial probate courts, as were guardians of orphans with property. Despite widespread understanding among literate Chesapeake planters about how to record and balance ordinary one-on-one business transactions, and among traders and storekeepers about how to keep mercantile records, in the first third of the eighteenth century the practice of keeping discrete accounts for farms operated by a living owner was rare. Had elite Virginia and Maryland landowners followed the example of estate owners in Britain, their managers would have employed the medieval "charge and discharge" method, still widely used in the mother country until the end of the eighteenth century, that typically reported only what the estate official received and how he disposed of any assets under his charge. This aristocratic Old World accounting system proved irrelevant for colonial stewards whose primary task was to oversee overseers of slaves living on scattered quarters—not tenant farmers renting land on long-established manors—and who usually had no responsibility for marketing the major cash crops. The only examples of charge and discharge accounts so far discovered for the Chesapeake are for a manor in Maryland operated by Jesuits. All late-seventeenth- and early-eighteenth-century elite Chesapeake planters kept some form of accounts with individuals, but they employed no generally accepted method of plantation bookkeeping, if indeed they kept formal account books.[44]

Those who did keep account books derived their methods from, and, indeed, had often been trained in, contemporary counting houses. Men who did not serve a mercantile apprenticeship could turn to a growing number of "how-to" manuals on double-entry bookkeeping. One or more such treatises was published in London in the decades of the 1650s, 1660s, 1680s, and 1690s. Starting in the 1710s, new texts or revised editions of earlier ones began to appear more regularly in London and Edinburgh, generated by and often tailored to meet the needs of expanding Atlantic commerce. Of the six early-eighteenth-century planters for whom account books relating to plantation management survive, two, Robert Goldsborough and Stephen Loyd (who both came from mercantile families), learned standard double-entry accounting in England in the 1670s and 1690s, respectively. Two others, James Carroll and Edmund Bagge, who might have become estate managers had they not immigrated to the Chesapeake, got their training in bookkeeping in Ireland in the 1690s and 1700s. The remaining two, native-born planters Richard Tilghman, Sr., and John Tayloe, found means for learning similarly up-to-date bookkeeping practices in Maryland and Virginia by the mid-1680s.

Early eighteenth-century bookkeeping manuals recommended that merchants record business transactions in two or more separate account books. The first and most fundamental book was the *waste-book,* which began with an inventory of assets and of debts owed and continued with a record of all

44. For estate accounting in Britain, see Christopher Clay, "Landlords and Estate Management in England," in Joan Thirsk, ed., *The Agrarian History of England and Wales,* V, part 2, *1640–1750: Agrarian Change,* in H. P. R. Finberg, gen. ed., *The Agrarian History of England and Wales* [Cambridge, 1985], 245–251; Christopher J. Napier, "Aristocratic Accounting: The Bute Estate in Glamorgan, 1814–1880," *Accounting and Business Research,* XXI (1991), 164–165; Rob Bryer, "The Roots of Modern Capitalism: A Marxist Accounting History of the Origins and Consequences of Capitalist Landlords in England," *Accounting Historians Journal,* XXXI (2004), 1–56; and B. W. Higman, *Plantation Jamaica, 1750–1850: Capital and Control in a Colonial Economy* (Kingston, Jamaica, 2005), 95–96. For seventeenth-century colonial accounting, see Phyllis Whitman Hunter, "Containing the Marvellous: Instructions to Buyers and Sellers," in Natasha Glaisyer and Sara Pennell, eds., *Didactic Literature in England, 1500–1800: Expertise Constructed* (Aldershot, Eng., 2003), 169–185; and Donna Merwick, "A Genre of Their Own: Kiliaen van Rensselaer as Guide to the Reading and Writing Practices of Early Modern Businessmen," *WMQ,* 3d Ser., LXV (2008), 669–712. The Maryland manor records are in Maryland Provincial Archives, St. Thomas Day Book, 1741–1743, Special Collections, Georgetown University Library, Washington, D.C.

subsequent transactions written down as they occurred, "narrated in a plain, simple style, and in the exact order of time as they succeed one another." The waste-book was "written up in the time of business, and commonly too by different hands," and thus was often not "fair and uniform, nor very accurate." The second essential book was the *ledger,* the "chief or principal book of accounts," in which "the articles of each particular accompt, that lie scattered in the other books according to their dates, are collected and placed together" in double-entry format, with debits to an individual or to a category of goods appearing on the left page and credits appearing directly opposite on the right page. Some instructors also recommended keeping a third, intermediate book, the *journal* or *daybook.* This book was a waste-book "in another Stile" that retained the sequential ordering of transactions in single-entry format but that also "resolves every Thing into its proper *Debtors* and *Creditors"* in an "artificial" style that required some level of abstract reasoning to execute properly. Most early-eighteenth-century gentry planters kept only one kind of account book—either a sequentially ordered waste-book or daybook (which occasionally morphed into double-entry ledger style) or they constructed a double-entry ledger at the outset from informal memorandum books or from loose accounts supplied by individuals or firms with whom they dealt. Planters' ledgers primarily recorded transactions with individuals residing in the colonies, where an account book could be introduced as evidence in a local court in the event of a dispute, and not infrequently omitted altogether any record of transactions with overseas factors.[45]

Since diaries, letterbooks, and account books supply different kinds of information about plantation agriculture, and no one source touches on all aspects of farming, the topics vary with each of the individuals discussed. Only rarely do both outgoing business correspondence, which explains what planters hoped to do, and account or crop books, which reveal what they actually

45. Quotations are from Alexander Malcolm, *A New Treatise of Arithmetick and Book-Keeping*... (Edinburgh, 1718), 126; and *Encyclopedia Britannica,* 1st ed., s.v. "Bookkeeping" (which provides clearer descriptions than many of the earlier texts). Representative manuals can be found in Early English Books Online and Eighteenth-Century Collections Online. For Scottish treatises, see M. J. Mepham, "The Scottish Enlightenment and the Development of Accounting," in R. H. Parker and B. S. Yamey, eds., *Accounting History: Some British Contributions* (Oxford, 1994), 268–293. For contemporary accounting practices in the West Indies, see Higman, *Plantation Jamaica,* chap. 4; and J. R. Ward, "The Profitability of Sugar Planting in the British West Indies, 1650–1834," *Economic History Review,* 2d Ser., XXXI (1978), 197–213. The main record series for sugar plantations were similarly accounts of sales of sugar (or of the weights shipped) net of marketing expenses, accounts of local sales and expenditures, and invoices for European goods for the plantation.

did, survive for one individual. What these materials left by the planter elite do demonstrate is a wide variety of approaches to plantation management and to estate building between 1680 and 1729. There was no one optimal strategy that a majority of planters chose; instead, there were several, depending on topography, soil quality, the strain of tobacco raised, the supply of imported laborers, the practices of British merchants who traded to the area, and the presence or absence of alternative mercantile networks linked to other products and to nonmetropolitan regions. That these homegrown indigenous adaptations took many forms should not be surprising.[46]

The surviving account books also fail to provide a complete record of individual farm revenues and expenses, as the Robert Cole estate account did, making the relative profitability of individual enterprises impossible to assess accurately. The one measure that can be calculated from many of the accounts is that of gross returns per laborer (reduced to a constant value) for the major field crops of tobacco and corn, and, less often, for other crops such as wheat. These returns are the standard measure employed throughout the remainder of this study for assessing the agricultural success or failure of individual planters. This measure incorporates elements of both a planter's ability to raise and process crops and of his ability to manage his workforce. Returns to workers on individual plantations can also be compared with the estimated subregional incomes per taxable discussed above. Such comparison permits a rough assessment of which management strategies yielded above average returns and which produced ordinary or below average results. Where available, values of probate inventories taken upon a planter's death (expressed in constant pounds sterling) are also presented as a further means for comparing his relative success or failure in building an estate equal in value to the estates of other large planters, and equal to or larger than the one he inherited or purchased at the beginning of his career.

Only rarely did planters themselves provide estimates of the value of their estates or of their total annual incomes. Few gentlemen saw the advantage, put forward by numerous authors of bookkeeping manuals such as John Mair, of being able "at any time to compute, by the help of his books, what he is worth to a farthing." Perhaps large planters kept an idea of their total net worth always in the back of their minds, but generally they set such calculations down on paper only during marriage proposals and dowry negotiations,

46. Main, *Tobacco Colony,* 57–96, similarly found that during this era planters in the six Maryland counties she studied adapted at least three distinct coping strategies, depending on the mix of local resources.

when asking to borrow money from a stranger, or when explaining to their heirs the value of the estate they would inherit.[47]

Planters did use outputs of tobacco and corn per laborer when comparing the success of their management strategies with those of their neighbors. They did not normally measure the financial status of their enterprises by the value of these outputs, however, as is done in this study. All large planters calculated outputs of major crops per worker, along with the prices at which they sold, because they needed these figures to determine what to pay their overseers as their shares of each year's crops. They also used the measure to evaluate the overseers' effectiveness in managing bound workforces. But planters reckoned annual plantation income primarily as the net proceeds of the tobacco crop. The balance their English factors reported each year told them what they could spend on clothing and tools for their workforces and on consumer goods for their families; it also also told them what debts remained to be paid or what savings they had that they could use to buy more land or slaves, lend out at interest, or invest in other enterprises. As Reverend Andrew Burnaby rather condescendingly explained, when large Chesapeake planters had enough tobacco "to pay their merchants in London, and to provide for their pleasures; they are satisfied, and desire nothing more."[48]

Because many still aimed to produce only enough corn for plantation needs, maize was thought of primarily as a local subsistence crop rather than as a staple cash crop. This mindset is reflected in the way yields, uses, and sales of grains were recorded. All planters instructed overseers or estate managers to carefully measure the corn crop and to record the number of barrels produced no later than December of each year so that they would know whether there was enough to supply workers and livestock until the next harvest as well as how many barrels to pay the overseer as his share. Those who found they had a surplus to sell put corn and other grains into a different conceptual category than tobacco. The sale of grains might offset local expenses such as goods or services purchased from artisans, medical bills, taxes and legal fees, and goods bought from local storekeepers, especially salt, sugar, and rum from the West Indies. Grain sales were recorded more haphazardly and inconsistently than were sales of tobacco, in part because overseers or agents, not the planter, did the selling. An overseer, for example, might simply be credited in an account book for "sundries sold from the plantation," an entry that could

47. John Mair, *Book-Keeping Methodiz'd; or, A Methodical Treatise of Merchant-Accompts, According to the Italian Form* . . . (Edinburgh, 1736), 6.

48. Rev. Andrew Burnaby, *Travels through the Middle Settlements in North-America, in the Years 1759 and 1760; with Observations upon the State of the Colonies,* 2d ed. (1775) (Ithaca, N.Y., 1960), 27–28.

include livestock, butter, cider, firewood, and other miscellaneous products in addition to grain. Not until planters began deliberately and consistently raising grains as cash crops around midcentury did they begin to keep more complete records of how and on what terms grain surpluses were disposed of. Even then, most planters continued to think of plantation revenues as either local revenues or metropolitan ones, rather than as unitary returns to an integrated enterprise.[49]

Agriculture was the base on which all large Chesapeake estates were built in this era, and the workers who produced the crops were almost invariably the most valuable asset that planters possessed. Land speculation, land leasing, slave dealing, moneylending, mining, or trade did not offer returns as high as those that could be obtained through farming, but these activities could provide important supplements to agricultural revenues, particularly when a planter engaged in several of them. In general, the wealthier the individual, the more likely he was to invest in nonagricultural pursuits. As Gloria Main put it, "A solid foundation in agriculture—investment in livestock and bound labor—took clear priority over alternative uses for the capital, but once this was established, the state of the economy (or their expectations concerning it) promoted or deflected interests in mercantile ventures." Building up their estates to enhance their present wealth and social standing and to provide suitably for their heirs were among the principal goals of large planters. In the discussions of individual planters here, revenue-producing activities other than agriculture are mentioned, but the primary emphasis is on their farming and labor-management strategies. The extent to which planters borrowed money and for what is discussed as well as whether and how quickly they repaid their debts. Prudent use of credit was critical for building up labor forces and for expanding estates and is evidence of entrepreneurial behavior. Failure to pay one's debts and a penchant for borrowing to finance the purchase of nonessential consumer goods or extravagant houses were then, as now, considered poor business practices.[50]

Any evidence of negative personal behaviors, such as excessive drinking, gambling, or financial irresponsibility, that might have had an adverse im-

49. Similarly, in eighteenth-century Jamaica, plantation owners normally sold their main staple, sugar, themselves but assigned to estate managers the responsibility for selling some or all of the less valuable rum and molasses locally. It was hoped that the proceeds from rum and molasses, like those from corn and livestock in the Chesapeake, also usually sold by plantation managers, would suffice to cover local expenditures. This strategy allowed the entire sugar or tobacco crop to be exported. See Higman, *Plantation Jamaica,* 108; Ward, "Profitability of Sugar Planting," *Economic History Review,* 2d Ser., XXXI (1978), 199.

50. Main, *Tobacco Colony,* 87–91 (quotation on 90).

pact on the solvency of estates, or of sexual abuse of dependents that might have impaired a planter's ability to manage his labor force, are also noted. These behaviors do not appear to have been prominent failings among late-seventeenth- and early-eighteenth-century planters, as they would sometimes be for planters from the 1770s onward. Drinking and gaming were nearly ubiquitous recreations in the Chesapeake from the 1610s, however, and sexual temptations, including easy opportunities for abuse of indentured and enslaved women, were surely always present. Hence, evidence for negative behaviors such as chronic alcoholism and sexual predation are simply more likely to surface in the fuller collections of family papers that survive for later planter generations.[51]

On the other hand, evidence that compulsive gambling and overspending on consumer goods did not threaten the solvency of estates in this era is much stronger. Bookkeeping manuals advised those who kept accounts to record "money gained or lost on wagers," since such "private expenses" affected a man's net worth. Several planters did consistently record their winnings or losses on card games and horse races either as part of their cash account or in accounts with individuals. The wagers were almost always small, only a few shillings, and sums won or lost stayed in rough balance. For the individuals considered here, gambling evidently remained a recreation, not an addiction.[52]

Under the commission system, large planters ordered most of their manufactured goods from England, so the issue of levels of spending on consumer goods is most appropriately addressed by determining, from account book entries or from annual statements sent by factors, whether the value of the goods and services elite planters ordered was less than, equal to, or exceeded the value of the tobacco and bills of exchange they remitted. In al-

51. Robert "King" Carter's and William Byrd II's diaries do address these issues, particularly Byrd's, since he kept his diaries in a shorthand code to prevent others from reading it. Both men regularly drank alcohol, and earlier excesses might have contributed to the gout that so afflicted Carter toward the end of his life. Byrd did at times become "merry" when drinking in company but attempted to practice moderation and, given the detail in which he recorded his day-to-day physical condition, would likely have noted—and castigated himself—had he regularly become inebriated. Carter moderated his drinking as his health deteriorated and avoided alcohol when his gout flared up. In both cases, one does not have a sense that either man was frequently impaired by drink. Byrd II fondled white women of all stations and had sex with prostitutes in England and with white maidservants on both sides of the Atlantic. According to Kenneth A. Lockridge, Byrd "long avoided sexual play with his female slaves as a matter of pride" and began having sex with an enslaved woman only in his late sixties (*The Diary, and Life, of William Byrd II of Virginia, 1674–1744* [Chapel Hill, N.C., 1987], 54, 101, 104–105, 148).

52. John Mair, *Book-Keeping Methodiz'd*, 6.

most every case for which this information exists, the late-seventeenth- and early-eighteenth-century planters discussed here strove to balance the two. Of course, large planters did frequently borrow to finance the purchase of land and, especially, of slaves, so it could be argued that those who neglected to cut consumer spending to the bone until these debts were repaid were financially imprudent. Such an argument, however, would require a stricter definition of financial irresponsibility than their contemporaries normally employed or that is employed today. Merchants or farmers who borrowed money to increase their productive capacity were not obliged to reduce whatever standard of living they and their families enjoyed before taking out the loan, although such curtailment might become appropriate should an unanticipated economic downturn occur.

Moreover, colonial planters could only estimate what the proceeds of their tobacco would be when they made out their orders for goods, the prices of which they had also to estimate. As Sir William Keith explained, a planter had to exercise "a careful Occonomy to inlarge or contract his yearly Commissions to Britain for Goods, according to his crop, and the current Value of Tobacco." Consigning planters could control expenditures only by specifying the highest cost per unit they were willing to pay for a good, the maximum amount they were willing to spend for certain items, or direct the factor to decide which goods to omit from their order if the total cost exceeded remittances. Should tobacco prices fall below the planters' expectations and the cost of the goods they ordered exceed returns or, as was more often the case, they reserved insufficient funds both to pay for new goods and to repay borrowed money, the only remedy was to curtail the next year's invoice if the factor was generous—or foolish—enough to ship all that had been ordered. There were certainly levels of family comfort that elite planters were unwilling to sacrifice, but most did give repayment of debts a high priority and did moderate consumer spending until that goal was achieved.[53]

53. Sir William Keith, *The History of the British Plantations in America* . . . (London, 1738), 185–186 (quotation). Scholars have often castigated elite colonial Chesapeake planters for "overspending" on nonessential consumer goods or for "excessive" consumption, but these are terms lacking in analytical rigor. "Overspending" or "excessive" spending would imply either that they allocated a higher proportion of their incomes to consumer goods than was usual among families of comparable economic and social status or that annual expenditures regularly exceeded annual revenues. The proportion of income or of assets allocated to consumer goods that might be considered "normal" and the proportion judged "excessive" are not defined, much less measured. Analyses of the proportions of income-producing or of consumption goods found in probate inventories decisively answer in the negative the question of misallocation of *stocks* (the value of assets at a given date that have been accumulated over time by inflows or depleted by outflows of revenue). Similarly, the evidence from mercantile and probate accounts from this period demon-

Another strategy some larger planters employed around the turn of the century was to reduce the amount of money they lent in the colony to more quickly pay off debts they had incurred to English merchants for buying slaves. This course, although prudent for those who were indebted to English lenders, meant that fewer funds were available to middling and small planters, who sold their tobacco in the country and who depended on local credit. Elite planter indebtedness was thus a hot political issue. It was important, not because ordinary planters cared deeply about the fortunes of their betters, but rather because elite borrowing abroad reduced opportunities for the majority to borrow locally.[54]

A second means for assessing planter indebtedness is an analysis of the accounts a planter's executors or administrators made of his estate after death. Such accounts survive more frequently for Maryland estates, where they were required by law, than for Virginia, where appraised inventories were not required for estates deemed solvent by local authorities and formal accounts were presented and recorded only when required by county magistrates. Among the twelve planters discussed here, the estates of the eight who enjoyed relatively long lives were free from significant debt by the time the men died. It was planters who died young shortly after borrowing to buy land or slaves whose estates were often encumbered with large debts and mortgages. Moreover, studies of probate inventories consistently show that large planters routinely allocated three-quarters or more of their movable assets to productive goods — such as labor, livestock, and agricultural and craft tools — and only a quarter to a fifth to consumption items. These findings underscore planters' generally conservative, production-oriented spending priorities. The only economic group that consistently allocated a third or more of movable assets to consumer goods had small estates worth fifty pounds or less constant value.[55]

---

strate balanced spending *flows* (the total value of income or expenditures over a given period of time). Despite such overwhelming quantitative evidence (more than eleven thousand Virginia and Maryland probate inventories were analyzed in the studies of Gloria Main [*Tobacco Colony*] and of Lois Green Carr and Lorena S. Walsh ["Changing Lifestyles and Consumer Behavior in the Colonial Chesapeake," in Cary Carson, Ronald Hoffman, and Peter J. Albert, eds., *Of Consuming Interests: The Style of Life in the Eighteenth Century* (Charlottesville, Va., 1994), 59–166]), a number of Chesapeake scholars continue to accept impressionistic literary accounts of planter extravagance or an increase in the amount of consumer goods imported as firm evidence for "excessive" consumer spending.

54. Main, *Tobacco Colony,* 81–82.

55. For Virginia probate practices, see Harold B. Gill, Jr., and George M. Curtis III, "Virginia's Colonial Probate Policies and the Preconditions for Economic History," *VMHB,* LXXXVII

An approach that emphasizes quantitative measures of production and performance and that defines success as the accumulation of large estates, on the one hand, makes more precise our understanding of the occasionally extraordinary opportunities the Chesapeake plantation economy offered and of when and how income-advancing increases in production were achieved. On the other hand, such a detached approach inevitably glosses over the unconscionable human costs entailed by entrepreneurial successes or even entrepreneurial failures in a society where "success" by any definition required the enslavement, domination, and exploitation of other human beings and their future progeny. To question the morality of privileged planters who were engrossing thousands of acres of land when about a third of the region's free white householders had none is easy enough. Efficient accumulation of commodified human beings, defined by the color of their skin as culturally—and eventually biologically—inferior in order not only to exploit their labor but also to exploit their reproductive capabilities, is impossible to justify today. Scholars have yet to find a compelling phrase to characterize these elite colonial planters that expresses a combination of fascination with their extraordinary achievements and revulsion for the means by which they made their fortunes in the way the term "robber baron" does for later-nineteenth-century capitalists.

During the 1680s and 1690s, the wealth of large planters first surged dramatically upward, while the value of estates of poor and middling planters ceased growing, and the widening gap between the living standards of rich and poor became more evident. As Gloria Main states in her study of probate inventories in six Maryland counties, "The bottom 90 percent of estates was perhaps a little better off, in terms of material goods, at the close of the period [1660 to 1720] than they had been earlier, but the rich were far richer, no matter which measure we use." The growing disparity in movable assets can be attributed in large measure to the rising value of bound laborers, as planters bought ever more African slaves and also began to benefit from the slaves' natural increase. The shift to slave labor did not, however, entail marked changes in the way large planters allocated their resources. Planters continued, as they had in earlier years, to devote about three-quarters of their movable wealth to capital assets, but rising incomes allowed them to

---

(1979), 68–73; and Hening, ed., *Statutes at Large,* IV, 12–25, V, 454–467. For allocation of assets, see Carr and Walsh, "Changing Lifestyles and Consumer Behavior," in Carson, Hoffman, and Albert, eds., *Of Consuming Interests,* 114–115; Burnard, *Creole Gentlemen,* 48; Main, *Tobacco Colony,* 60–62; Hudgins, "Patrician Culture," 210–261.

live more comfortably on the quarter or so of their assets they reserved for durable consumer goods.[56]

Although large Chesapeake planters might have ignored trends in British agriculture in the late seventeenth and early eighteenth centuries, the same cannot be said about metropolitan fashions in consumer goods and in housing. New kinds of household furnishings that facilitated social rituals associated with dining, drinking alcoholic beverages, consuming new and popular imported groceries such as tea, coffee, and chocolate, and dressing activities and toilet preparations were gaining popularity among more affluent consumers in Britain. Consumer goods were becoming "tools for actively cultivating a distinctive, genteel style of life that set off 'polite society' from the 'meaner sort.'" Social status was coming to be gauged not just by large holdings of land and labor but also by elegance in dress and in household arrangements and by increased emphasis on polished manners and on knowing how to use consumer goods appropriately when participating in elaborate social ceremonies. These new ways for using material possessions appealed to status-conscious, wealthy Chesapeake planters whose forebears' social origins might not be commensurate with the elevated status to which they felt their landed wealth and high political offices entitled them. In addition, members of the creole gentry who operated in high colonial social circles often thought themselves slighted by emigrant English officials, or, among the more select few who actually visited the mother country, unwarrantably devalued by the British gentry, who derided their inferior colonial origins, education, and manners.[57]

The shift to the commission system starting around 1680 afforded such planters not only increased revenues from their tobacco but also unprecedented access to stylish metropolitan goods chosen personally by their London factors or by the factors' wives. Probate inventories demonstrate that by the mid-1680s the wealthiest planters had acquired not only long-available comforts like chairs, feather beds, and linens but also one or two luxuries like pictures, watches and clocks, large looking glasses, or richly wrought bed curtains. A few showed signs of taking up newer fashions, such as the use of table knives and forks and the acquisition of a few pieces of imported fine furniture and, most consistently, increasingly larger quantities of elaborately wrought silver plate. Thereafter, the pace with which the planter elite took up

56. Main, *Tobacco Colony,* 53–57, 80–81 (quotation on 53).

57. Kevin M. Sweeney, "High-Style Vernacular: Lifestyles of the Colonial Elite," and Cary Carson, "The Consumer Revolution in Colonial British America: Why Demand?" both in Carson, Hoffman, and Albert, eds., *Of Consuming Interests,* 1–24 (quotation on 6), 483–697; Hudgins, "Patrician Culture," chap. 4.

the new fashions accelerated, and by 1725 dining tables and matching chairs, chests of drawers, drinking glasses, tea services, and sets of fine dining wares could be found in most elite planter homes. These families were increasingly convinced that a combination of fashionable props and proper manners would demonstrate their cultural equality with Europeans.[58]

Large planters could now also afford to spend more on their houses as well as on the goods with which they furnished them. Bigger and more fashionable houses became a new symbol of a planter's success along with the large landholdings, herds of livestock, and bound labor forces that had been the primary indicators of status from the 1620s into the 1670s. Beginning in the 1690s, the built environment began more clearly to reflect the increasing wealth and rising social pretensions of the gentry. Houses of two rooms rather than one, with a raised wood floor rather than one of packed bare earth, and with one or two brick chimneys rather than a comparable number of ramshackle flues of wattle and daub had sufficed to set apart more affluent planters from their neighbors in the 1660s and 1670s. Near the end of the century, such indigenous status markers began to be replaced by more metropolitan markers of gentility.

Members of the provincial elite increasingly chose to build with brick rather than wood and to incorporate some elements of metropolitan architecture, such as a full two stories, decorative brickwork, diagonally set chimney stacks, tile roofs, stone-paved floors, casement windows, and curvilinear Dutch gables. Many added specialized spaces, such as dining rooms, studies, and individual bedrooms, and laid out elaborate symmetrical formal gardens adjoining their homes. In the late 1690s, Anglican commissary James Blair described a house proper for a Virginia clergyman as "a framed house" of at least one story. It would require "brick Chimneys and glass windows with Casements, that the walls within be plaistered and the roof shingled . . . with two rooms and a large Closet, besides Cellars and Garrets," along with a detached kitchen "and whatever other outhouses shall be judged necessary." Although Virginia ministers were seldom so well housed until later in the century, the Chesapeake planter elite did strive to provide their families with comparable accommodations. By 1705, historian Robert Beverley could note that private houses in Virginia had "of late very much improved," as several gentlemen had built "large Brick Houses of many Rooms on a Floor, and several Stories high. . . . and their Windows large, and sasht with Cristal Glass."[59]

58. Carr and Walsh, "Changing Lifestyles and Consumer Behavior," in Carson, Hoffman, and Albert, eds., *Of Consuming Interests,* 59–166.

59. Samuel Clyde McCulloch, "James Blair's Plan of 1699 to Reform the Clergy of Virginia,"

Modeled on sizable merchant or minor gentry houses in provincial British towns, turn-of-the-century gentry houses were "planned in response to local notions of what constituted a proper genteel environment." As Cary Carson explains, "standardized architectural spaces equipped with fashionable furnishings" were considered increasingly essential as "universally recognized settings for social performances that were governed by internationally accepted rules of etiquette." One example is Fairfield, built about 1694 by wealthy Gloucester County planter Lewis Burwell, a justice of the peace, burgess, and nominee to the council (Plates 11a–b). This great brick, hip-roofed house, built atop a tall English basement, contained on its front range a large, high-ceilinged hall (or drawing room), said to have been finished with paneled walls, a carved chimneypiece and a marble mantel, and an adjoining room perhaps used as a dining parlor. The two-story wing at the rear might have included a fashionable best bedchamber. With most of the ground floor devoted to public spaces, this plantation house was "more showplace than dwelling." This new house demonstrated Burwell's "command of capital and his cultivated taste" through the money and attention devoted to elaborate decoration of public rooms. The simpler pleasures colonists like Robert Wintour and Robert Cole enjoyed earlier in the century no longer sufficed for the provincial elite.[60]

WMQ, 3d Ser., IV (1947), 79–80 (quotation); Beverley, *History and Present State of Virginia,* ed. Wright, 289 (quotation); Bradburn and Coombs, "Smoke and Mirrors," *Atlantic Studies,* III (2006), 131–157; Ann B. Markell, "Solid Statements: Architecture, Manufacturing, and Social Change in Seventeenth-Century Virginia," in Paul A. Shackel and Barbara J. Little, eds., *Historical Archaeology of the Chesapeake* (Washington, D.C., 1994), 51–64. For earlier housing styles, see Cary Carson et al., "Impermanent Architecture in the Southern American Colonies," *Winterthur Portfolio,* XVI (1981), 135–196; Willie Graham et al., "Adaptation and Innovation: Archaeological and Architectural Perspectives on the Seventeenth-Century Chesapeake," *WMQ,* 3d Ser., LXIV (2007), 451–522; and Hudgins, "Patrician Culture," chap. 2.

60. Edward Chappell, "Looking at Buildings," *Fresh Advices: A Research Supplement* (November 1984), i–vi; Sweeney, "High-Style Vernacular," Chappell, "Housing a Nation: The Transformation of Living Standards in Early America," and Carson, "The Consumer Revolution," all in Carson, Hoffman, and Albert, eds., *Of Consuming Interests,* 1–24, 167–232 (quotation on 179–180), 523–524 (quotation on 524), 620–638; Graham et al., "Adaptation and Innovation," *WMQ,* 3d Ser., LXIV (2007), 451–522. For Fairfield, see Thomas Tileston Waterman and John A. Barrows, *Domestic Colonial Architecture of Tidewater Virginia* (1932; rpt. New York, 1968), 31–35; Cary Carson, "Plantation Housing: Seventeenth Century," in Carson and Carl Lounsbury, eds., *The Chesapeake House: The Practice of Architectural Investigation by Colonial Williamsburg* (Chapel Hill, N.C., forthcoming); and *Quarterly Bulletin of the Archaeological Society of Virginia,* LXII, no. 3 (2007), which is devoted to architectural and archaeological research at Fairfield Plantation. For other mansions built in this period, see Carter Hudgins, "The 'King's' Realm: An Archaeological and Historical Analysis of Robert Carter's Corotoman," research report, 1981, CWL; Rachel Most, *Discovering Rosewell: An Historical, Architectural, and Archaeological Over-*

PLATES 11A–B. Fairfield, Gloucester County, Virginia. Conjectural elevations based on archaeology and historical photographs. Drawn by Richard Anderson. Courtesy, the Fairfield Foundation.

Fairfield was built by Lewis Burwell II about 1694.

By the 1720s, members of the county elite joined in, either building new, larger, more specialized frame houses (albeit on a more modest scale than the provincial gentry), or adding rooms or new half stories onto older, more vernacular houses. As one scholar puts it, more of the gentry were coming to believe that they could reinforce their social dominance "as much by their cultural hegemony as by the power their positions in the government gave them." An example of a house remodeled to fulfill such aspirations is Holly Hill in Anne Arundel County, Maryland, which began as a small, unremarkable one-story, clapboard dwelling built on an outlying quarter by a successful immigrant planter in 1698 (Plate 12). In the 1720s, the son who inherited the tract, Samuel Harrison, enclosed the former cottage with brick walls and built a new brick addition, the entrance graced by a piazza fronting on a fenced, formal garden. The first floor of the new section contained all the requisite rooms for genteel entertaining—a large parlor, a dining room with a built-in "bowfat" to display china, and a generous stair hall—and the upper floor accommodated two finely finished chambers. In the painting Harrison commissioned to hang above his parlor fireplace, he chose to display both his improved house and a plat of the inherited plantations whose revenues made such refined pleasures possible.[61]

In contrast, ordinary Chesapeake landholders lacked the time, resources, and social pretensions that might have encouraged them to build more permanent or fashionable houses. Even middling planters continued to build relatively small, wood-framed and covered vernacular structures little changed from the sort of house Robert Cole put up in the 1650s. A reconstruction of a twenty-seven-by-sixteen-foot house, based on archaeological excavations of a post-in-the-ground dwelling occupied by middling planter Thomas Atkinson in the 1680s, 1690s, and early 1700s, is probably typical of the "pretty

view (Gloucester, Va., 1994); Nicholas M. Luccketti, "Archaeology at Arlington: Excavations at the Ancestral Custis Plantation, Northampton County, Virginia," Virginia Company Foundation and the Association for the Preservation of Virginia Antiquities, 1999, http://www.apva.org/resource/other/arlington.pdf; and Timothy B. Riordan, "The Formal Frontier: Transition from Impermanence to Formal Architecture in Colonial Maryland, 1660–1700" (paper presented at the annual meeting of the Society for Historical Archaeology, 2007). John Harris, *The Artist and the Country House: From the Fifteenth Century to the Present Day* . . . (London, 1995), provides illustrations of period houses in England.

61. Hudgins, "'King's' Realm," 10 (quotation); Carson et al., "Impermanent Architecture," *Winterthur Portfolio,* XVI (1981), 176–178; "Holly Hill," National Register Listings in Maryland, http://www.marylandhistoricaltrust.net (accessed May 20, 2007). For other surviving early-eighteenth-century houses, see Herman J. Herkkenen and Mark R. Edwards, "The Key-Year Dendrochronology Technique and Its Application in Dating Historic Structures in Maryland," *Bulletin of the Association for Preservation Technology,* XV, no. 3 (1983), 2–25.

PLATE 12. Samuel Harrison's Plantation, Holly Hill, Anne Arundel County, Maryland. Before 1733. Framed overmantel painting, oil on wood. Courtesy, Brice M. Clagett. Photograph, The Colonial Williamsburg Foundation. *The inset shows Harrison's 1698 house recently remodeled.*

timber houses" Reverend Hugh Jones observed in Virginia in the 1720s (Plate 13). Atkinson did not hold any major county offices and owned only a sixty-four-acre tract along the James River that was once part of Martin's Hundred, one of the particular plantations. Despite the small size of his farm, however, Atkinson was wealthy enough to build a relatively sizable house of two to three rooms and to have acquired several indentured servants or slaves. For most families less well off than Atkinson, one-room dwellings continued to suffice.[62]

What did change on most plantations with bound laborers, however, was

62. Graham et al., "Adaptation and Innovation," *WMQ*, 3d Ser., LXIV (2007), 495, 505–506, 516–518; Carson et al., "Impermanent Architecture," *Winterthur Portfolio*, XVI (1981), 135–196; Jones, *Present State of Virginia*, ed. Morton, 74; Steve Archer, "Atkinson Site," http://research.history.org/Archaeological_Research/Research_Articles/ThemeAfrAmer/Atkinson.cfm (accessed Aug. 1, 2007).

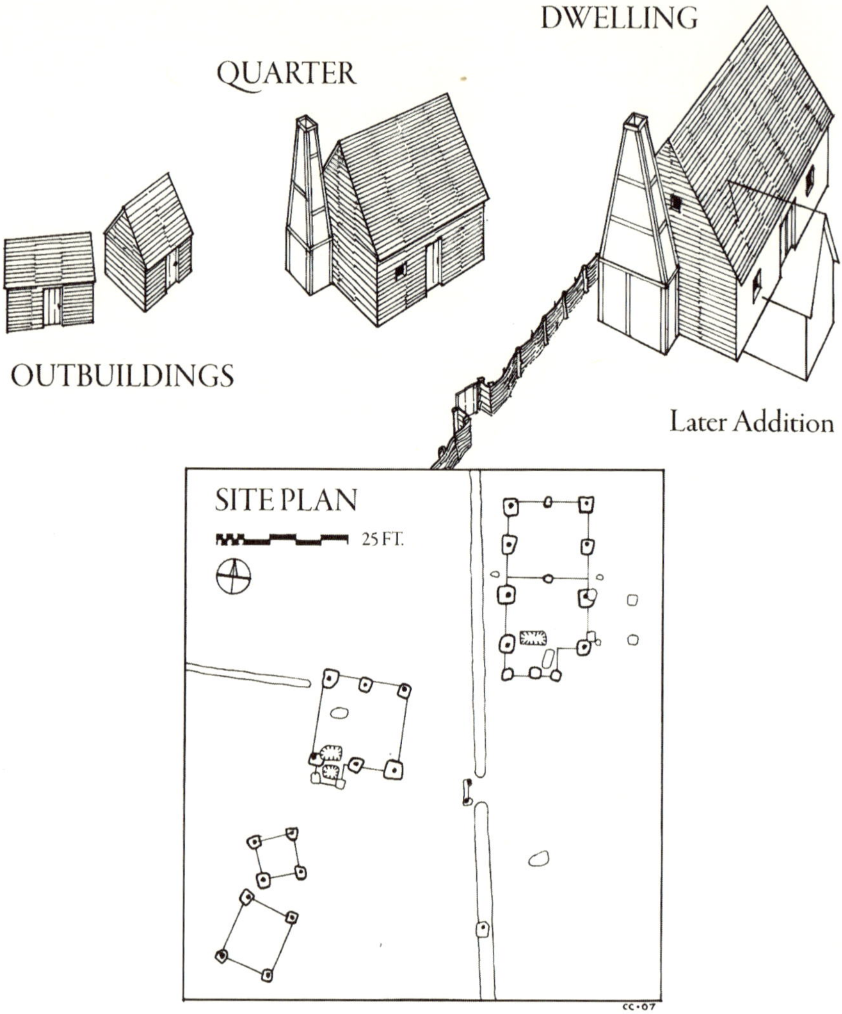

PLATE 13. Atkinson Site, James City County, Virginia, 1680/90–1711.
Drawn by Cary Carson.
Conjectural reconstruction of a middling planter's dwelling based on archaeological excavations.

where the servants and especially where the slaves worked, ate, and slept. By the third quarter of the seventeenth century, planters were experimenting with a variety of layouts that divided even two- or three-room houses along social lines by separating the space family members used from the one in which servants lived and worked and that changed the location of dwelling entrances to increase family privacy. By the 1680s and 1690s, new houses were being built or old ones remodeled so that they accommodated only the white family. Kitchen functions were moved into adjacent workhouses, and cooks were moved to separate quarters. Fenced yards adjoining the principal dwelling limited access to formally laid out gardens, which were more ornamental than functional. With these changes, two-room houses equipped with a hall and the best bedchamber on the ground floor were all planters of middling means needed to afford the white family increased privacy, and, in their eyes, enhanced status. Atkinson again serves as an example; he built a separate sixteen-foot-square post-in-the-ground laborer's quarter near his main dwelling but separated the work area from his house with a fence.[63]

By the early 1700s, those few white servants who remained, although now somewhat elevated in social status, were nonetheless more often housed in lofts of outbuildings or in separate quartering houses. Most of the slaves (and almost invariably all enslaved field hands) lived and ate in separate flimsy, wooden quarters, often constructed at a greater distance from the owners' dwelling than the separate structures built a decade or so earlier to house white servants. As a foreign traveler observed in 1686, "a separate house for the Christian slaves, [and another] one for negro slaves" were becoming the norm. With the exception of the separate kitchens, the house plans were borrowed from British prototypes, where a desire for greater privacy and rooms reserved for special purposes had nothing to do with racial separation. The switch among the Chesapeake elite to enslaved labor, however, surely made architectural arrangements that physically segregated domestic workers from the planter's family and guests another reason for remodeling older dwellings or for building new ones that employed architecture to enforce social hierarchy.[64]

63. Carson et al., "Impermanent Architecture," *Winterthur Portfolio,* XVI (1981), 135–196; Graham et al., "Adaptation and Innovation," *WMQ,* 3d Ser., LXIV (2007), 506–510; Cary Carson and Lorena S. Walsh, "The Material Life of the Early American Housewife" (paper presented at the Conference on Women in Early America, Williamsburg, Va., 1981), typescript, CWL; Archer, "Atkinson Site."

64. Coombs, "Building 'The Machine,'" chap. 6; Fraser D. Neiman, "Domestic Architecture at the Clifts Plantation: The Social Context of Early Virginia Building," *Northern Neck Hist. Mag.,* XXVIII (1978), 3096–4128; Hudgins, "Patrician Culture," chaps. 2, 6; [Durand, of Dauphiné],

Two Planters of the 1680s and 1690s

Minor adjustments of crop mix, attempts to replace imports during economic downturns, new marketing strategies, rapid expansion of ancillary productive units, investments in nonagricultural enterprises, and new methods for managing enslaved labor forces all contributed to rising great-planter wealth in the last decades of the seventeenth century. The experiences of two of the best-documented period planters, William Fitzhugh and William Byrd I, illustrate the estate-building strategies that ambitious immigrants adopted to provide for their posterity. These two men, and others like them who immigrated around 1670, passed on an economic, political, and social status higher than they could have achieved in England.

William Fitzhugh (1651–1701), the youngest son of a Bedford, England, woolen draper, immigrated to Virginia in the early 1670s with few resources of his own, but he managed to marry an eleven-year-old Virginia heiress in 1674 (Plate 14). He acquired some land through his wife and soon purchased other tracts. By 1680, Fitzhugh was operating a one-thousand-acre home plantation on the Potomac River in Stafford County, Virginia. He began farming with indentured servants and at least two slaves who were part of his wife's dowry. Fitzhugh was one of the few planters living outside the lower Virginia tidewater who were able to convert to farming with large numbers of enslaved workers before the turn of the century. In the 1680s, he purchased a few Africans from ships arriving in the York River, the only place where slavers regularly called, using bills of exchange, and he bought at least two others from lower tidewater planters. He continued to employ indentured servants—there were seven working on Fitzhugh's farms when he made his will in 1700—but the majority of his workforce were slaves. By 1686, they numbered twenty-nine and, by the time his estate inventory was taken in December 1701, that number had increased to fifty-one.[65]

Convinced that slaves represented the key to success, Fitzhugh sought to become a slave dealer himself. In 1683, he offered to sell new Africans for

A Huguenot Exile in Virginia; or, Voyages of a Frenchman Exiled for His Religion; with a Description of Virginia and Maryland, from the Hague edition of 1687, ed. Gilbert Chinard (New York, 1934), 119–120 (quotation). Donald W. Linebaugh, in contrast, emphasizes warm summer temperatures, high humidity, and the presence of vermin as the main reasons for removing services to outbuildings in "'All the Annoyances and Inconveniences of the Country': Environmental Factors in the Development of Outbuildings in the Colonial Chesapeake," *Winterthur Portfolio,* XXIX (1994), 1–17.

65. Richard Beale Davis, ed., *William Fitzhugh and His Chesapeake World, 1676–1701* (Chapel Hill, N.C., 1963). See also Parent, *Foul Means,* 30–31; and Coombs, "Building 'The Machine,'" chap. 7.

PLATE 14. *William Fitzhugh* (1651–1701). By an unidentified artist, copied by John Hesselius. Courtesy, Virginia Historical Society, Richmond

tobacco payable a year later in partnership with a New Hampshire mariner. This scheme came to naught, probably because of the New England projector's inability to get a contract from the Royal African Company. Although Fitzhugh was fortunate enough to purchase slaves through his connections with gentlemen in the lower tidewater, given his remote location he had to accept more women than men. Even though Fitzhugh eventually scattered these bondpeople among four separate quarters, they soon overcame their isolation and formed families. The slaves apparently enjoyed relatively good health; at least Fitzhugh never complained about excessive losses. The women proved exceptionally prolific; for example, between April 1700 and December 1701, the eleven adult women on Fitzhugh's plantations bore seven children. Fitzhugh's workforce grew more through natural increase than additional purchases; by 1686, he noted that the laborers were already "most country born."[66]

Tobacco growing was the heart of Fitzhugh's agricultural operation. At first he raised only oronoco tobacco, but by 1682 he was experimenting with the sweet-scented strain. Fitzhugh had greater success with sweet-scented than did most other planters along the Potomac, who had no luck trying to grow it; sometimes he was able to sell his crop for prices equal to York River tobacco. Fitzhugh explained to his London factor that he had made the change to sweet-scented, "not that I expect that will bear a much better price, but for that it will weigh heavyer, and therefore better answer freight and keep better and so not so great danger or loss, if want of freight or a Market make one keep it by him." By the early 1690s, copying the practice of lower tidewater planters, Fitzhugh had the slaves stem the tobacco before packing it. This further increased the weight of each hogshead and so saved more on freight, an especially heavy change during wartime.[67]

In addition to cutting costs at home, Fitzhugh was assiduous in seeking out the best markets in England. In most years, he consigned the bulk of his tobacco to two or more London factors, primarily to John Cooper but also to Thomas Starke, John Taylor, Roger Jones, and Henry Hartwell. But he also sent part of each crop to one or more Bristol firms, including Stephen Watts, Cornelius Serjeant, George Mason, William Dains, and John Pemberton. In addition, he made a few trial consignments to Liverpool and to Plymouth, but in 1686, 1687, and 1688, when prices were exceptionally low and freight scarce, he instead sold his crop in the country for a mix of cash and imported manufactured goods. During King William's War, Fitzhugh again decided

66. Davis, ed., *William Fitzhugh,* 175.
67. Ibid., 257.

to risk consigning his tobacco, but he lost two of three crops to accidents or seizure at sea. "It is more uncertain for a Planter to get money by consigned Tobo. than to get a prize in a lottery," he complained. By 1694, he was shipping oronoco tobacco that he bought from neighboring planters on consignment as well on the assumption that the market for oronoco was rising in England.[68]

Although tobacco was Fitzhugh's major cash crop, local sales of surplus corn, livestock, and cider provided steady if secondary revenues. Stocks of cattle, hogs, and sheep, a dovecote, and a henhouse supplied the Fitzhughs' table with an abundance of meat and milk, supplemented with produce from an orchard of twenty-five hundred fruit trees and a one-hundred-square-foot kitchen garden. Although Fitzhugh found the tobacco market sometimes disappointing, he seldom diverted labor to other pursuits. He did occasionally send shipments of plank and staves to England or to the West Indies, but with poor results. Minor experiments with flax, hemp, hay, olives, and iron ore came to nothing. When the market was at its worst, Fitzhugh's workers began to spin a little yarn and to tan leather for shoes. Although, as he wrote in 1681, necessity was the mother of invention and the nurse of such import replacement schemes, any rise in the price of tobacco gave a clear signal to abandon them. Most Chesapeake planters followed similar strategies, diverting labor to make goods normally imported when tobacco prices were especially low but shifting most of that labor back into tobacco when prices rose.

In the late 1680s, Fitzhugh's annual income from all sources was about thirteen hundred pounds. Although he considered himself above all a planter, land rents, tolls from a gristmill, income from his legal profession and various government offices, and an extensive mercantile business might have brought in more revenues than did farming. This handsome combined income enabled the Fitzhughs to build a thirteen-room frame dwelling with brick chimneys and a variety of outbuildings, which in Fitzhugh's estimation afforded "a comfortable and gentile living." They were also able to set up three additional agricultural quarters, each with "a choice crew of Negros," on tracts intended for the Fitzhughs' sons. By 1688, Fitzhugh judged his fortune sufficient to permit him to indulge in the purchase of a carriage and "an handsom Cupboard of [silver] plate," an investment both "well politic as reputable . . . which gives my self the present use and Credit, is a sure friend at a dead lift, without much loss, or is a certain portion for a Child after my decease." Otherwise, Fitzhugh believed that land and slaves were the safest

68. Ibid., 256.

investments a Chesapeake planter could make. He bought up land with a passion or granted it to himself as agent for the Northern Neck proprietary. At his death, Fitzhugh held more than fifty thousand acres scattered throughout the Northern Neck, with many of his farms in already-settled areas leased to tenants. Although he occasionally borrowed money to purchase land, he quickly repaid these debts and always refused to purchase goods from England on credit.[69]

Fitzhugh's contemporary, William Byrd I (1652–1704), the son of a London goldsmith, immigrated sometime before 1670, while still in his teens, to join his childless maternal uncle Thomas Stegge in trading with the Indians. As a consequence, Byrd traveled widely in southern Virginia and developed a keen eye for choice tracts of land. Over a span of thirty years, he amassed almost thirty thousand acres. He acquired more than half of this land using headrights, with a third of that coming from importing African slaves. Byrd's surviving correspondence does not reveal much about how he managed his plantations; it is known, however, that he, like Fitzhugh, started growing tobacco and corn with a combination of slaves and servants but that for field hands he switched entirely to slaves in the mid-1680s. Byrd also operated two gristmills and dealt in tobacco that he purchased with manufactured goods. A combination of a large inheritance from his uncle Stegge, a better location for trade than Fitzhugh enjoyed (first at the falls of the James and later nearer the mouth of that river at Westover), better international connections with leading London merchants and English officials, and more lucrative provincial offices—deputy auditor and receiver general in addition to being a member of the council—afforded Byrd more options for building a fortune. Byrd consigned his tobacco primarily to the large-scale and politically influential London firms of Micajah Perry and Thomas Lane and of Arthur North and was a part-owner of at least three ships engaged in the tobacco trade. In addition to planting, Byrd pursued the tobacco and Indian trades, exchanged goods with New England merchants, together with three partners imported at least one shipload of slaves brought directly from Africa, and sent corn, flour, and staves to the West Indies to exchange for sugar, rum, and smaller lots of slaves. Byrd's endeavors built a handsome estate on which his better-known son would expand.[70]

69. Ibid., 175–176, 246.

70. *Byrd Correspondence,* I. For brief biographies of Byrd I, see Martin H. Quitt, "Immigrant Origins of the Virginia Gentry: A Study of Cultural Transmission and Innovation," *WMQ,* 3d Ser., XLV (1988), 649–650; Parent, *Foul Means,* 31–32; and "William Byrd (ca. 1652–4 December 1704)," in Sara B. Bearss et al., eds., *Dictionary of Virginia Biography,* 3 vols. (Richmond, Va., 1998–

Early-Eighteenth-Century Plantation Empires

In the first quarter of the eighteenth century, large Virginia planters were able to further increase their scale of operations through the continued acquisition of ever more land and slaves. The high prices the coveted sweet-scented strain of tobacco fetched at the turn of the century, coupled with the larger amounts of credit competing British merchants were willing to extend, facilitated the purchase of productive resources that could be passed on to the next generation. Only in the 1720s did a slump in the sweet-scented market lower export earnings, force planters to cut back on discretionary spending, shake their confidence in their future prospects, and encourage efforts to diversify their enterprises to try to maintain the levels of income to which they had become accustomed. Like the large sugar growers in Barbados, most elite Virginia planters responded to falling prices by trying to improve the quality of their premier product, to raise its value by further processing, and to expand output by increasing their labor forces. A few native-born planters who inherited already large estates went on to build veritable plantation empires in the first third of the eighteenth century. European immigrants who began their planting careers a few years later, however, were more likely to realize tobacco revenues per laborer no higher than those estimated for the region overall—in other words, an "ordinary" performance.

Rather than emulating the island planters by concentrating all their activities on one large, integrated plantation, however, Virginia planters set up multiple ancillary quarters on which the workers concentrated primarily on raising staple crops and livestock. They could then farm the best tobacco soils they owned, maintain large herds of livestock, and begin developing additional plantations that they could eventually give to their children. Placing Africans on outlying quarters also served to keep them isolated from whites other than an overseer and to keep most enslaved field workers at a distance from the master's dwelling. However, elite planters did reserve for their home plantations most processing, craft, and manufacturing activities—which expanded during economic downturns—as well as the culture of crops (like small grains) and the raising of livestock (like sheep), to which overseers and workers were unaccustomed.

Soon after his father's death in 1704, William Byrd II (1674–1744) returned from an extended mercantile and legal education on the Continent and in England to manage the large Virginia estate that he inherited (Plate

---

2006), II, 463–466. For the ranking of London merchants of the period, see Price and Clemens, "Revolution of Scale," *JEH,* XLVII (1987), 10–15.

15). Byrd enjoyed the sophisticated London life and would have preferred to remain in England. He soon married a Virginia heiress, Lucy Parke, however, and, between visits to England in 1715–1720 and 1724–1726, set about learning how to handle tobacco from older planters, buying additional land and new African slaves, and expanding his farming and trading operations. Byrd's philosophy of plantation management was to "take care to keep all my people to their duty, to set all the springs in motion, and to make every one draw his equal share to carry the machine forward." Evidence about Byrd's planting activities are fullest for his earlier career, and the discussion that follows concentrates on the years between 1704 and the early 1730s.[71]

Tobacco provided the greatest share of Byrd's agricultural income. Unlike most other large planters who dealt with several commission agents, for many years he consigned his crops almost exclusively to one London factor, Micajah Perry, perhaps a strategy occasioned by his frequent absences from Virginia. When in the colony, however, Byrd also pursued a great variety of other activities on his home plantation and on a number of quarters widely scattered along the James. "Like one of the patriarchs, I have my flocks and my herds, my bond-men, and bond-women, and every soart of trade amongst my own servants, so that I live in a kind of independence on every one, but Providence," he declared in 1726.[72]

His estate, as he described it in 1718, consisted of forty-three thousand acres of land and about 220 slaves. Some of the land was leased to tenants, but most was cultivated by enslaved laborers and a few indentured servants, and some distant tracts were reserved for future development. Outlying farms, managed by overseers, produced not only tobacco but also surplus cattle, hogs, barreled pork, hides, wheat, and lumber as well as wild game for the Byrds' table. Slaves at a quarter on the Chickahominy made shingles. At Falling Creek, Byrd established a coal pit, a tannery, a copper mine, and a sawmill that turned out boat timber, garden pales, planks, and staves. An enslaved shoemaker and a joiner also worked there. A warehouse at Shockoe brought in fifty pounds in rental income per year. From a store at Appomattox, Byrd continued the Indian trade his great uncle had started, exchanging salt and

71. *Byrd Correspondence,* I, 355; Louis B. Wright and Marion Tinling, eds., *The Secret Diary of William Byrd of Westover, 1709–1712* (Richmond, Va., 1941); Maude H. Woodfin, ed., *Another Secret Diary of William Byrd of Westover, 1739–1741,* trans. Marion Tinling (Richmond, Va., 1942). Most of the evidence on plantation operations is drawn from *The Secret Diary.* For biographies of Byrd II, see "William Byrd (28 March 1674–26 August 1744)," in Bearss et al., eds., *Dictionary of Virginia Biography,* II, 466–470; and Pierre Marambaud, *William Byrd of Westover, 1674–1744* (Charlottesville, Va., 1971).

72. *Byrd Correspondence,* I, 355.

PLATE 15. *Portrait of William Byrd II* (1674–1744). By Studio of Sir Godfrey Kneller, London, 1700–1704. Oil on canvas. Courtesy, The Colonial Williamsburg Foundation

manufactured goods for skins with the native Americans and for pork and tar with North Carolina whites. A hatter and a smith plied their trades there as well. The produce garnered from all these activities was consumed or sold locally or sent to England, the West Indies, or the wine islands.

On the home plantation at Westover, activities were even more diverse. Orchards and gardens provided a wide variety of fruits and vegetables for the Byrds and their guests. Sheep were raised in fenced pastures for meat and for wool, which a group of enslaved spinners and a servant weaver processed. Slaves cultivated wheat for sale and oats for home consumption and mowed marsh hay to feed the livestock. They also pursued cidering, wine making, dairying, fishing, beekeeping, brickmaking, and the building of small boats. Tobacco dealing, land rental, sloop hire, and shares in ships in the tobacco trade provided Byrd with further revenues. These enterprises yielded him a clear income of eighteen hundred pounds sterling between 1715 and 1717, or a net profit of very roughly ten to twelve pounds per adult slave. This figure is more than five times the estimated *gross* annual tobacco income per tithable for the sweet-scented region overall. Either Byrd exaggerated his income (the estimate was prepared for a man whose daughter he hoped to marry) or his plantations were exceptionally well managed and the additional enterprises in which he was engaged generated additional revenues nearly equaling those derived from tobacco.[73]

After his return to Virginia from England in 1726 with a second wife, Maria Taylor, Byrd began expanding and improving the garden his father had begun laying out. Seeking to demonstrate his membership in an international genteel culture, as well as his position of power within Virginia, Byrd set about replicating the formal structures and landscapes that he had so admired on the estates of English gentlemen and nobles that he had visited in England. Between 1731 and 1735, Byrd built a handsome brick great house to replace his father's more modest wooden dwelling. The house employed geometric proportions, exterior details, and interior arrangements borrowed from English sources. Byrd undoubtedly hoped to use the fortunes his wives ostensibly brought to their marriages to further augment his estate. Unfortunately, the inheritance of the first was hopelessly encumbered with debts, and that of the second was tied up in the courts for more than two decades. And, in contrast to most other large planters who received income from lucrative provincial offices, Byrd sold the office of receiver general he had inherited from his father because he hoped to secure the higher position of lieutenant governor. When this effort failed, Byrd retained only his seat on the colony's

73. Ibid., 311–312.

council, which provided no direct annual salary but did offer avenues for advancing his private interests. Still, Byrd had more success building his fortune in Virginia than he would have had in England; in the colony, Byrd could "live in great luxury, without being any thing out of pocket" and could "save great part of what my effects produce."[74]

Hoping in 1728 to decrease his dependence on tobacco, Byrd put a number of his hands to "hemp and husbandry." He marketed three tons of hemp in 1729 and had ninety bushels of seed sown for the next year's product, but shortly after he abandoned this crop, as the necessity of processing it by hand so increased labor costs that Virginia fiber was not competitive in English markets. That Byrd and his overseers were "unacquainted with dressing it" also contributed to the unsatisfactory returns he received. Byrd's efforts to institute "husbandry"—sowing of small grains, use of plows, and production of forage crops—on the other hand, marked an early attempt to achieve a more diversified agriculture in which revenues from grains might eventually lessen dependence on the tobacco market. Byrd also aggressively explored a variety of markets for the diverse local products that he either traded for or had made on his multiple quarters.[75]

One senses that Byrd might have made more from his expansive estate had he paid more attention to it. His diaries reflect no extensive interest in farming and no sustained involvement in day-to-day management of farming operations. Toward the end of his life, he scarcely even mentioned plantation affairs in his diary. True, Byrd made a bad bargain by assuming to pay the debts of his first wife's profligate father, Daniel Parke, in order to get a share of Parke's Virginia land and slaves, rather than accepting the promise of a one-thousand-pound marriage portion that might never be paid. But Byrd also spent ten years in England between 1714 and 1726 obsessively seeking to get himself appointed governor of some English colony, pursuing a political vendetta against then Virginia governor Alexander Spotswood, and, after the

74. Ibid., 358. For the Westover house, see Wenger, "Westover." Architectural historians are no longer certain what the house looked like when it was first built, as they have discovered that many of the exterior and interior decorative elements long considered to have been William Byrd II's creation were in fact added by his son William Byrd III in the 1750s and 1760s.

75. George Melvin Herndon, "The Story of Hemp in Colonial Virginia" (Ph.D. diss., University of Virginia, 1959), chap. 2, discusses Byrd's hemp experiment. Byrd's inexperience in dressing hemp is documented in Fayrer Hall, *The Importance of the British Plantations in America to This Kingdom; with the State of Their Trade, and Methods for Improving It; as Also a Description of the Several Colonies There* (London, 1731), 75–76. For another large estate on the south bank of the James whose owner was pursuing mercantile and self-sufficient activities similar to those of Bryd, see "An Inventory of the Estate of Nathaniel Harrison Esquire, Decd., July 15th, 1728," *VMHB*, XXXI (1923), 361–380.

death of his first wife, seeking frantically to find a wealthy English spouse. These extended stays in England consumed income he could have more purposefully spent on further developing his vast holdings and "laying up" something for his children, as did his contemporary Robert "King" Carter. Ultimately, Byrd's difficulties in paying Parke's debts stemmed, not so much from his choosing to "live in great luxury" in Virginia (although his style of life contributed to the slowness with which he made headway repaying the debt), but rather the years he spent in England attempting to live like and to be accepted as an English gentleman.[76]

Robert "King" Carter (1663–1732) was the son of John Carter (circa 1613–1669), a royalist sympathizer who moved to Virginia in the 1640s (Plate 16). Owing to his influential mercantile and kin connections in England, the elder Carter quickly rose to county-level political offices and eventually became a member of the council. He amassed several thousand acres by patenting land due for transporting servants and slaves or by purchase, and at his death in 1670 he owned an exceptionally large labor force composed of thirty-four indentured servants and forty-three slaves. Like William Byrd II, Carter enjoyed a long classical and mercantile education in England and began his career as a planter and businessman in the early 1680s with the advantages of inherited land, labor, and political connections. As a second son, Carter initially received only one thousand acres of land and one-third of his father's bound laborers. Upon the death of his elder brother in 1690, however, Carter inherited the entirety of the family lands. Over the ensuing forty-two years, he accumulated the largest estate ever amassed in Virginia through a combination of agriculture, trade, and the fruits of officeholding. As a justice of the peace, militia commander, naval officer, burgess, from 1700 a councillor, and in 1726–1727 the interim governor, Carter was active in shaping the legal system under which all Virginia planters operated. He gained income from his office as treasurer of the colony between 1699 and 1705. And, as agent for the Northern Neck proprietary between 1703 and 1712 and again from 1722 to 1732, Carter granted himself, his children, and grandchildren nearly two

76. Lockridge, *Diary, and Life, of William Byrd II,* 69–70, 73–100, 117–123; *Byrd Correspondence,* I, 358. When Byrd finally decided to pay the Parke debt of approximately three thousand pounds in 1736 (having lost a battle to prevent the passage of the Colonial Debts Act of 1732 that made all land and slaves in the colonies liable to seizure for debts owed in Britain), he was able to remit enough tobacco and bills of exchange to continue paying the interest and to discharge five hundred pounds of the principal in that year and in the next. Byrd eventually had to sell some land and slaves to complete repayment (ibid., II, 499, 522; Lockridge, *Diary, and Life, of William Byrd II,* 126–127, 149).

PLATE 16. *Robert "King" Carter* (1663–1732). By an unidentified artist.
Courtesy, Shirley Plantation in Charles City, Virginia

hundred thousand acres of undeveloped land in the hinterlands of the Rappahannock.[77]

Carter clearly sought to build a family dynasty and aggressively bought up scores of slaves in order to establish working plantations that he intended as inheritances for his children and for many of his grandchildren. In the judgment of one biographer, "No other Virginian of his generation was so successful in his political career, in the marriages made by his children, and so ruthless in building his estate for the benefit of those children." Carter himself viewed his estate building more as a divine duty: "We are but stewards of God's building: the more he lends us the larger accounts he expects from us, and happy they that make a right use of their Master's talents." Taking advantage of a Virginia law that, like that of Barbados, permitted the annexing of slaves to landed estates, he tried to ensure that an enslaved workforce as well as land would descend from generation to generation in the Carter family. By his death in 1732, Carter owned 333,000 acres of land and 829 slaves who worked on the home plantation, Corotoman, in Lancaster County, and at least forty-seven ancillary quarters scattered in Lancaster, Richmond, Northumberland, Westmoreland, Caroline, King George, Stafford, Spotsylvania, and Prince William counties. Moreover, before he died Carter had already transferred some land and numerous slaves to children and grandchildren. Carter operated his vast estate with a cadre of subordinate managers, including head overseers for clusters of the more distant plantations and ordinary overseers

77. Manuscript sources for Robert Carter include Robert Carter Diary, 1722–1727, and Robert Carter Letterbooks, 1723–1724, 1727–1728, 1728–1731, 1731–1732, all in the Robert ("King") Carter Papers, MS 3807, UVa; Robert Carter Inventory, 1733, Robert Carter Will, 1733, and Robert Carter Letterbooks, 1723–1724, 1727–1728, 1728–1730, all in VHS; Robert Carter Letterbook, 1701–1710, in the Wormely Estate Papers, 1701–1716, VSL; Louis B. Wright, ed., *Letters of Robert Carter, 1720–1727: The Commercial Interests of a Virginia Gentleman* (San Marino, Calif., 1940). There are four nonduplicate Letterbooks, two dated 1723–1724 and two dated 1727–1728; one Letterbook of each of the two dates are in both UVa and VHS. All extant Robert Carter texts are being transcribed in Edmund Berkeley, Jr., ed., "The Diary, Correspondence, and Papers of Robert 'King' Carter of Virginia, 1701–1732," http://etext.virginia.edu/users/berkeley. Brief biographical sketches of Robert Carter are in Wright, ed., introduction, *Letters of Robert Carter;* Louis Morton, *Robert Carter of Nomini Hall: A Virginia Tobacco Planter of the Eighteenth Century* (Williamsburg, Va., 1941), chap. 1; Bearss et al., eds., *Dictionary of Virginia Biography,* III, 84–86; and Edmund Berkeley, "A Brief Life of Robert Carter," in Berkeley, ed., "Diary, Correspondence, and Papers," 5–8. The will of John Carter, Robert's father, are in Lancaster County Wills, etc., 10, part 2, 1709–1727: 416, microfilm, VSL; the inventory is in Lancaster County Loose Papers, VSL; and a brief biography is in Bearss, et al., eds., *Dictionary of Virginia Biography,* III, 72–73. The will and inventory of Robert's brother John is in Lancaster County Wills, etc., 8, 1690–1709: 3, 22, microfilm, VSL. See also Edmund Berkeley, Jr., ed., "Robert Carter as Agricultural Administrator: His Letters to Robert Jones, 1727–1729," *VMHB,* CI (1993), 273–295.

for each of the quarters, who in turn managed the slaves with the assistance of an enslaved foreman. Most of Carter's surviving papers date from the 1720s, when he had nearly forty years of experience managing plantations, so they reflect the operation of a well-established enterprise.[78]

Carter bought dozens of new African slaves nearly every year. He took first choice of the shiploads of slaves that he himself sold and, by offering to pay ready cash for large lots that he bought from others, was able to persuade the dealers to allow him to make his purchases before the captives were exposed to public sale. Carter's correspondence offers unique insights into how the new Africans were broken in. Within a few days of their purchase, Carter assigned them English names, and he and his sons "repeated them so often to them that every one knew their names and would readily answer to them." Carter instructed overseers to "take care that the negros both men and women . . . go by the names we gave them" and to not allow them to use their African names. The purpose of the naming policy, according to Carter, was so that "we can always know what sizes [of clothing] they are of," but, of course, the practice also served to strip the captives of an important element of their African identity.[79]

Weakened and traumatized by the Middle Passage and unaccustomed to the Chesapeake disease environment, many of the new slaves immediately fell sick or never recovered their health. Some died within a few days. Many new hands who arrived in the summer also fell sick by early fall from malaria, which was endemic in the region. In October 1729, for example, Carter reported: "My new Hands are all down in their seasoning Except one. I must wait with Patience till they recover their strength before I can send them [to a distant quarter]." Carter hoped that adequate food and warm clothes and bedding would speed their recovery. Unfortunately, many of those who survived the fall did not live through their first winter in a colder climate, a season in which both whites and blacks often succumbed to respiratory diseases. Carter blamed "the carelessness and cruelty of the overseers in turning the people out in hard and bitter weather." The winter of 1726–1727 was especially deadly: "I have had prodigeous loss in my Slaves this winter and have bin forced to recruit by purchasing a large number of new Negroes," Carter

78. Berkeley, "Brief Life of Robert Carter," in Berkeley, ed., "Diary, Correspondence, and Papers"; Carter to Dawkins, Feb. 23, 1721, in Wright, ed., *Letters of Robert Carter,* 81. Carter's plantation operations are also discussed in Hudgins, "Patrician Culture," 290–305.

79. Carter to Robert Jones, Oct. 10, 1727, Carter Letterbook, 1727–1728, Carter Papers, fol. 72, UVa; Carter to Jones, Oct. 24, 1729, Carter Letterbook, 1728–1730, fol. 41, VHS. The following segment on Carter's management of slaves summarizes an extended discussion in Walsh, *From Calabar to Carter's Grove,* 81–94.

wrote one correspondent. At least seventy adult slaves perished that winter, and in April 1727 Carter bought eighty more new Africans.[80]

Some of the strongest Africans immediately tried to escape their captor, as did seven men who took off from Corotoman in a slave's canoe five days after their ship arrived and managed to elude recapture for a week. Others, once they had regained some strength and found an opportunity, took off into the woods. By the 1720s, Carter had enough experience with newly arrived Africans to expect these attempted escapes, and for the first months he ordered his overseers "to be kind to all the negros but especially to the new ones." Of one new slave woman who had been hiding in the neighborhood of one of the quarters, Carter observed, "Now she hath tasted of the hardship of the woods she will go near to stay at home where she can have her belly full." Carter undoubtedly hoped that "kind" treatment would minimize early deaths occasioned by "seasoning" or sheer despair.[81]

Eventually, however, Carter ran out of patience with Africans who refused to accept their new situation. He turned to brutal measures. When, for example, in 1727 Carter decided that chronic runaway Ballazore was an "incorrigeble rogue," he ordered the overseer to have the man's toes cut off; "nothing less than dismembring will reclaim him." This was not an isolated act of brutality, for Carter asserted, "I have cured many a negro of running away by this means." Lancaster County court records for 1710 and a private diary entry for 1722 show at least three other instances of a similar "cure." Other slaves who resisted in unspecified ways were whipped and branded.[82]

One of Carter's strategies for forcing new Africans to become productive workers and reconciling them to bondage was to encourage them to form families as soon as possible. Married slaves were much less likely to try to run away than were unattached individuals. Like most other period slaveholders, Carter was in too much of a hurry to sacrifice short-term returns from expanded tobacco crops for the possibility of longer-term returns in the form of slave children. He generally bought nearly twice as many men as women, clearly a deliberate policy since he had first choice among captives offered for

80. Carter Diary, Nov. 8, 1722, Carter Papers, UVa; Carter to ?, partial letter of 1727 or 1728, Carter Letterbook, 1727–1728, Carter Papers, fol. 83, UVa, Carter to John Stark, May 19, 1727, fol. 22, Carter to John Pemberton, June 3, 1727, fol. 25.

81. Carter Diary, July 12, 17, 25, 1727, Carter Papers, UVa; Carter to Robert Jones, Oct. 10, 1727, Carter Letterbook, 1727–1728, Carter Papers, fol. 72, UVa.

82. Carter to Jones, Oct. 10, 1717, Carter Letterbook, 1727–1728, Carter Papers, fol. 72, UVa; Carter Diary, Oct. 26, 1722, Carter Papers, UVa. See also Philip J. Schwarz, *Twice Condemned: Slaves and the Criminal Laws of Virginia, 1705–1865* (Baton Rouge, La., 1988), 80–81; Terrence W. Epperson, "Race and the Disciplines of the Plantation," *Historical Archaeology,* XXIV, no. 4 (1990), 29–36.

sale and could have bought men and women in equal numbers. But he did encourage those men who could find mates to enter into regular unions. By 1733, almost no adult women on Carter quarters lived by themselves; most either were married according to whatever understanding they and their husbands had negotiated between themselves and with Carter or were part of a household consisting of two or three other apparently unrelated men and women. Wherever possible, Carter kept husbands and wives together on the same plantation and allotted them a house to themselves. When spouses lived on another quarter or had died, mothers and their children or, less often, fathers and their children lived together. Unattached and probably recently arrived men usually shared a cabin with one to three other males. Because no more than fifteen to twenty-five slaves resided at any individual quarter, there were no large concentrations of men living in barracks for any length of time. An additional incentive to marry for both men and women was the chance to acquire "securable and separate domestic environments," a privilege usually reserved only for married couples or for women with one or more children. The greater privacy a house of one's own afforded also offered greater control over some aspects of private life.[83]

Another of Carter's strategies for extracting work from the slaves was regimented gang labor. Rather than assigning slaves individual tasks, overseers put all the laborers to work on one job at a time, keeping watch over all and forcing all to work at a similar pace. Carter's letters include repeated references to slave gangs. Whether it was planting, weeding, or harvesting corn or tobacco, gathering corn fodder, or digging a millrace, his slaves usually worked as a single unit. And, indeed, there was just one unit, for the total labor force on any one of Carter's quarters averaged only nine able-bodied men, women, and older children. On the quarters inventoried in 1733, the range was from two to twenty-three laborers; however, there were more than fifteen hands of laboring age on only three quarters. The overseers were "always to be with the people to keep them to their work." Very likely Carter thought regimented labor was the best way to control the many new African slaves he added to his workforce, most of whom understood little or no English and whose main goal was to escape their captors. Slaves too young, old, or sick to keep up with the other laborers were assigned light tasks with little or no supervision.[84]

Carter usually mixed more acclimated slaves with the "new Negroes" on the various quarters, probably hoping to speed assimilation but also aiming

83. Robert Carter Inventory, VHS.

84. Carter to ?, [1728] [torn], Carter to ?, [1729] [torn], Carter to ? [torn], Aug. 27, 1729, Carter Letterbook, 1728–1730, fols. 1, 33–34, 36, VHS.

to gain more control over the recent arrivals through the mediation of the more acculturated slaves. With the exception of skilled artisans on his home plantation, by 1733 Carter owned few white indentured servants. In earlier years, such servants might have served as an intermediating group; now their presence, on outlying quarters at least, would likely have complicated the imposition of radically un-English work regimens.

Carter's third strategy for extracting work from new arrivals was to appoint a slave man on each quarter as foreman. Regular use of slave foremen—working as drivers under the supervision of a white overseer—was something of an innovation for the region. In the early eighteenth century, quarters were not uncommonly run by trusted slaves with no overseer present, despite laws that made owners responsible for any thefts or crimes slaves committed on quarters where there was not a resident white supervisor. Designated drivers, on the other hand, common on larger West Indian plantations, were much less frequently part of the formal management structure on Chesapeake plantations. These men must have demonstrated superior skills in raising tobacco and corn and managing livestock and probably were among the most influential quarter residents. Carter advanced each foreman certain privileges, likely including the "provisions Equall to an overseer (vizt) halfe a middleing beefe[,] one barrow Hogg for bacon and one small hogg for Pork and other things as [are] . . . reasonable for such a trusty negro" that were enumerated by a contemporary Virginia planter. Carter provided each slave supervisor a house to himself, even if he was single; on one quarter, one of the foremen had two wives, a common custom in Africa but unusual in the Chesapeake. In return, Carter presumably made the foreman responsible for keeping the other slaves in line and working at an acceptable pace. Some of the foremen had been enslaved for some time (several are described as "old"), and a few were creoles. There were not enough such men to go around, however, and Carter selected a number of more recently arrived Africans as foremen including Ebo George at Totuskey, another Ebo George at Head of the River, and likely King Tom at Hamstead and George at Forrest with the two wives.[85]

Almost all the work the slaves were forced to do was physically demanding and entirely manual. By 1729, Carter was encouraging some of his overseers to train oxen to help with heavy hauling. In the 1733 inventory, however, evidence shows working oxen on only six of the forty-seven quarters. By Carters' own account, most of the slaves had to carry all fencing materials

85. Will of Edwin Thacker, May 1, 1704, Middlesex County Will Book A, 1698–1713: 161, microfilm, VSL; Robert Carter Inventory, VHS.

from the woods to the fields, tote firewood from wherever they cut it to their cabins, and haul corn and tobacco from field to barn either on their backs or, African-style, on their heads. Robert's son Landon recollected in 1776 that his father had used no plows and had no more than twelve draft oxen on all the quarters. On quarters where Carter raised wheat, the slaves leveled the ground and chopped in the seed with hoes. They also rolled tobacco hogsheads to landings and loaded ships with thousands of bushels of wheat and corn entirely by hand. The agricultural methods Carter employed might, in fact, have made optimal use of the skills of his labor force, rather than being the result of a short-sighted, slovenly, and stingy nature. The overseers presumably had some knowledge of how to train oxen to pull carts, even if most failed to follow through, but they might not have had any experience with plowing. The cultivation techniques Carter and his son described were similar to those that African grain farmers used. It is uncertain whether Carter was making conscious or unconscious use of some of the enslaved Africans' expertise—most of the Africans whose region of origin is known came from the Bight of Biafra rather than Senegambia—but he was at least using hand cultivation techniques with which all would have been familiar.[86]

Carter did not assign many of the slaves to work as craftsmen or domestics. He preferred to use European indentured servants as artisans. In 1733, white servants, most living on the home plantation, followed the trades of carpenter, glazier, bricklayer, tailor, gardener, blacksmith, miller, sailor, and cook. Most of the privileged slaves were either foremen or sailors, and many of these were Africans. The scattering of enslaved artisans were mostly carpenters and coopers, many of whom were native born. A few others were assigned to domestic trades such as butcher, tailor, and cook.

Production of high-quality tobacco commanded a great deal of Carter's attention. Like William Fitzhugh, by the 1690s Carter had begun having his tobacco stemmed and tightly packed to save on freight charges and to cater to the English domestic market. In 1720, following complaints about the quality of his consignments, he advised London merchant Richard Perry: "I

86. Carter to Robert Jones, Oct. 24, 1729, Carter Letterbook, 1728–1730, fol. 41, VHS; Jack P. Greene, ed., *The Diary of Colonel Landon Carter of Sabine Hall, 1752–1778* (Charlottesville, Va., 1965), 1038–1039. At least as late as the 1770s, slaves on some of the Carter plantations still often carried harvested crops from field to barn on their heads. In 1770, Landon Carter noted that a slave man needed to make at least twenty-four "head baskets" for the coming corn harvest; a few months later, he commented sourly that, when "pregnant wenches" were "indulged in not carrying weights on head or shoulders," they did little other work (Greene, ed., *Diary of Carter,* 523, 554). For lack of plows, see also Gregory A. Stiverson and Patrick H. Butler III, eds., "Virginia in 1732: The Travel Journal of William Hugh Grove," *VMHB,* LXXXV (1977), 44.

take up resolutions to be nicer in the handling of mine than ever I was yet. It troubles me not a little that Mr. Pratt should get more for his miscellaneous stuff that comes from all parts of their county, low lands as well as high, than I do for mine that is made with my own people and upon as good, proper [land] as anybody can boast of." Carter considered himself "as much master of the planting trade as any one you know." He monitored the packing of the tobacco on the quarters nearest Corotoman and wanted accounts kept for each of the many plantations he managed in order to determine which sorts of tobacco sold best in which markets and to identify which overseers made good tobacco and which inferior.[87]

Carter kept testing different markets and factors, sending the bulk of his tobacco (including most of the sweet-scented) to London but also shipping lesser amounts to Bristol, Liverpool, Whitehaven, Weymouth, Biddeford, and Lyme Regis. As he explained to Edmund Jenings in 1702, he sent the "most to them who makes the best Returns and when there is money before hand, to Lodge in the ablest hands where there is least appearance of failure." "I have no friends or Relations there [in England] further then they Rend[e]r themselves by their good Managem[en]t in my favour." In the 1720s, he dealt with at least thirteen factors, and by 1730, with nineteen. Each merchant had his own ideas about how tobacco should be processed and packed, and their directions were often contradictory. Carter found himself engaged in continuous struggles, trying to make so many different overseers and laborers comply not only with the merchants' recommendations but also with changing laws in the 1720s and early 1730s regarding the amount of tobacco that might be tended and how it should be processed.[88]

Like William Byrd II, Carter made some efforts to diversify his operations. The average price of sweet-scented dropped below two pence a pound in 1721, and Carter soon complained that "these hard times will Drive us into All the ways of husbandry we are capable of." By the later 1720s, overseers on Carter's numerous quarters made annual returns not only of tobacco but also of surplus corn, cattle, pork, wheat, butter, apples, and cider. Some of these returns were for consumption at the great house, but not even the Carter establishment could have consumed all that was produced. At Corotoman, beans, clover, and oats were added to the crop mix, and the gathering of cherries and walnuts, cidering, distilling, wine making, ale brewing, and wool

87. Carter to Perry, July 7, 1720, in Wright, ed., *Letters of Robert Carter,* 2; Carter to Dawkins, July 3, 1723, Carter Letterbook, 1723–1724, Carter Papers, fols. 11–12, UVa.

88. Answer to Jenings's proposals, Apr. 6, 1702, Carter Letterbook, 1701–1710, Carter Papers, UVa, letter to ? [torn], Aug. 20, 1706; Wright, ed., *Letters of Robert Carter,* 2.

processing had become regular features of the plantation calendar. But, as we have seen above in the example of draft animals and plows, Carter made little effort to adopt European methods of husbandry.[89]

Agriculture remained the base of Carter's empire. Like most of the other large planters, however, additional income from land speculation, rents, shares in transatlantic ships, tobacco trading, slave dealing, moneylending, and provincial officeholding was also important to Carter's virtually unparalleled prosperity. In 1728, he and other family members attempted to diversify even further, starting a copper mine on the Frying Pan tract near the present boundary of Fairfax and Loudoun counties. This investment produced poor results owing to constant disputes about food and work requirements with English miners whom the Carters had imported from Falmouth and to the low quality of the ore.

Despite his great wealth and a fondness for some forms of social display, such as the fine clothing evident in his portrait, silver plate, and carriages, Carter abhorred spending more for consumer goods than he considered necessary. He lived for thirty-five years in an unprepossessing twenty-four-by-fifty-two-foot, three-room brick house that he had built shortly after his return to Virginia in 1685. Only near the end of his life did Carter seek to demonstrate his high status not just by accumulating land and labor but also by building around 1720 one of the largest brick houses in the Chesapeake. Corotoman measured an impressive ninety feet by twenty-five feet, not including an arcaded piazza running the length of the house, with a brick cellar underneath and some sort of second story. Even this impressive dwelling, however, reflected Carter's practicality and relative frugality. The house was but a single room wide, and the ground floor consisted of only a central passage flanked by a pair of single rooms with oversized closets. Lacking was any suite of specialized rooms designed primarily for public entertainments like those at Lewis Burwell's Fairfield. And when his fine new house burned down in 1729, Carter apparently decided that he was too old to bother rebuilding and instead moved back into his original modest three-room dwelling.[90]

Toward the end of his life, Carter admitted, "The Irons I have already in the fire are so many one half of them burn for want of a rigorous Application." After several years of bad growing seasons and an unexpected drop in prices for sweet-scented tobacco at the close of the 1720s, Carter was not get-

89. Carter to [Thomas] Evans, Aug. 3, 1723, Carter Letterbook, 1723–1724, Carter Papers, fol. 24, UVa.

90. Hudgins, "Patrician Culture," 248–261; Hudgins, "'King's' Realm."

ting as much revenue from his plantations as he had anticipated. The downturn in the tobacco market forced Carter to work ever harder at developing alternative sources of revenue if he was going to provide as generously for his progeny as he intended. "There seems to be a necessity upon us to find some other ways to imploy our people or else in a few years to become beggars," he complained. Still, Carter's position at the apex of the Virginia plantation economy was only slightly impaired by a spate of hard times.[91]

Gentlemen possessed of a plentiful fortune derived from a landed estate, bookkeeping instructor Alexander Malcolm urged in 1718, ought "to know readily the distinct State and Posture of their Affairs," including "what Part of his Fortune yields best, and what Part of his Expenses is greatest," in order to "make the best he can of his Fortune, and the best Use of what he receives." Being able to ascertain from properly kept accounts "what is the Occasion of his Loss, what Branch has failed," and consequently to "foresee the Way to help such Misfortune for the Time to come," held out "prospect both of Pleasure and Profit from his Labour." Nonetheless, Malcolm knew that many British landed gentry, "who visibly have so great Use for it," neglected the keeping of systematic accounts that might enable them to make better-informed decisions for maintaining or augmenting their estates. Although in their youths William Byrd II and Robert "King" Carter were schooled in commercial methods as well as in the classics in England, both men appear to have neglected to keep systematic accounts of their far-flung enterprises. In their diaries, both Byrd and Carter frequently mentioned *settling* (that is, reconciling) accounts with individuals in person and, less regularly, *settling* accounts received by post from overseas correspondents. Carter used his diary for noting down travel expenses and miscellaneous business transactions, but neither he nor Byrd mentioned regularly *recording* transactions in an account book or *consulting* journals or ledgers kept by subordinates. Carter admitted to Micajah Perry in 1729 that he could not answer many of the merchant's questions about revenues and expenditures for an estate that Carter had helped administer in the 1710s because he no longer had the original "vouchers" (loose paper accounts), and the estate accounts that Carter had taken the trouble to transcribe were, "as I find them in my books," far from complete. "You cannot expect wee are as regular in our books as you merchts. are whose business it is to transact Affairs for other men," Carter explained. Thus, it is possible that two of the region's largest landowners and labor owners made

91. Carter to Col. John Tayloe, Jan. 29, 1729, Carter Letterbook, 1728–1731, Carter Papers, fol. 16, UVa.

decisions about managing their vast estates without benefit of any systematic form of accounting.[92]

"Ordinary Success" in the 1710s and 1720s

By the early 1700s, most ordinary Englishmen no longer believed that the Chesapeake afforded them economic opportunities. A trickle of young men from solid middling backgrounds, however — merchants, ministers, lawyers, royal officers, and the like — continued to try to improve their fortunes in the region around the turn of the century. This small but continuing migration stream ensured that there were fresh infusions of knowledge about current English practices in farming, building, and social organization. Records for three other tidewater Virginia plantations begun in the 1710s and 1720s reveal how much more difficult estate building had become for men with fewer economic resources and less managerial ability than a Carter or a Byrd. Initial start-up costs were high, and planters who continued to rely for market income almost exclusively on tobacco raised by slave labor seldom realized the level of profits they hoped for. It was not that the costs of raising tobacco consistently exceeded returns from it. Had this been the case, Western Shore tidewater planters would have ceased growing it, as did most of their contemporaries living on the Eastern Shore and along the lower James River. It was rather that the returns often failed to meet expectations based on earlier, more prosperous years or based on the example of highly skilled planters like Robert "King" Carter. In addition, those who did try to reduce their dependence on tobacco discovered that implementing a more diversified agriculture and finding alternative markets for other products was far from easy, especially when they lived far from foci of intercolonial trade. Identifying the problems that absentee owners and less skilled managers encountered adds to our understanding of why regional tobacco production failed to keep pace with increases in the labor force in the first quarter of the eighteenth century.

92. Malcolm, *New Treatise of Arithmetick and Book-Keeping,* appendix, 34–35; Carter to Alderman [Micajah] Perry, Apr. 10, 1729, Carter Letterbook, 1728–1730, fol. 57, VHS. Roger North, *The Gentleman Accomptant . . .* (London, 1714) was one of the first double-entry bookkeeping manuals directed especially to gentlemen of private fortune rather than to merchants and tradesmen. The method was little used on English estates, however, until about 1750 (Bryer, "Roots of Modern Capitalism," *Accounting Historians Journal,* XXXI [2004], 1–56). Edmund Berkeley, "The Extant Texts," in Berkeley, ed., "Diary, Correspondence, and Papers," found reference in Carter's diary to only one "other book," now lost, that contained financial notations but was apparently not a journal or ledger.

Stephen Loyd (Loyde, Lyde, 1681–1711), the first example, was a tobacco merchant from Bristol who immigrated around 1705. He married Elizabeth Gwyn, a Richmond County heiress, whose planter father had close connections with merchants in Stephen's home city, in December 1706. Through this fortunate alliance, Stephen, who had as yet received no portion from his father, acquired five hundred acres of "choice," "well-seated" land on the Rappahannock River, five slaves, an unspecified number of indentured servants, livestock, household furnishings, store goods, and more than two thousand pounds sterling. Stephen and one of his brothers, James, decided that high freight rates during Queen Anne's War justified setting up a "factory" for stemming oronoco tobacco that they would buy from area planters; after processing the leaf, they would reprize it into heavier hogsheads of nearly double the usual weights, thus reducing shipping costs by almost half. In 1707, James Loyd and Isaac Elton, another Bristol merchant who had joined the partnership, purchased Chatterton, a 550-acre, partially developed plantation on the Potomac River in Stafford County, adjoining William Fitzhugh's lands, from the widow of a Bristol mariner. Since the title could not be conveyed until two young daughters who would inherit the land came of age, the partners agreed to pay rent of £7 per year (less the cost of the quitrents) for nineteen years, and £140 sterling when they finally got a clear title.[93]

Stephen and James agreed that they would buy equal numbers of slaves for the enterprise and jointly bear most costs of maintaining the workforce. James would remain in Bristol and handle the English side of the business. Stephen was to build convenient houses on his own plantation for stripping, prizing, and warehousing tobacco and to supply the slaves with provisions. It was intended that the slaves would work for Stephen during the winter and early spring, collecting tobacco, stemming it, and then repacking it into heavier hogsheads. In the summer, between the time the tobacco fleet left the Chesapeake and late winter when the new crop was ready for collection, James would employ the slaves on his own account to put in a crop of tobacco on the Potomac tract. Initially, Stephen hired John Peters to oversee the partner's plantation and sent some of his own slaves there to work under

93. The accounts are in the Stephen Lyde Account Book, 1708–1711, and Accounts and Letters of John Tayloe I and John Tayloe II, 1714–1778, Tayloe Family Papers, MSS1T2118, b1, VHS; and Loyde Account Book, 1708–1710, 38–62, UVa. For Elizabeth Gwyn Loyd Tayloe, see *VMHB,* XXXVIII (1930), 383–384; and *WMQ,* 1st Ser., XVIII (1909–1910), 62. For Stephen Loyd, see *VMHB,* XXXVII (1929), 23. For the land transaction, see Stafford County Will Book Z, 1699–1709: 410–412, 420–422, microfilm, VSL. William Fitzhugh had wanted to buy this tract, but the owners in England would not sell to him (Davis, ed., *William Fitzhugh,* 205–208, 209n, 221–223, 277n, 308–309).

Peters. He also rented part of the tract to a tenant, who helped with putting up buildings and fences and with rounding up hogs.[94]

Despite his acquisition of land and slaves (he owned eight adults and five children when his estate was inventoried in 1712), Stephen Loyd remained more merchant than planter. In 1708, he was busy disposing of a cargo of English goods in exchange for tobacco and was running the tobacco-repacking operation. In July 1710, he and another brother, Lyonel, sold a shipment of 164 slaves brought on the *Leopard Galley* from Calabar for Bristol slave trader Laurence Hollister. Buyers came from Essex, Richmond, Lancaster, Westmoreland, Stafford, Gloucester, King and Queen, King William, New Kent, and Henrico counties. The brothers were able to sell the choicer captives for bills of exchange (the Loyds had to make good any bills that were protested) but could sell the poorer ones only for tobacco, some payable a year later, which proved troublesome to collect. Stephen explained that "the many maugre ones did very much prejudice the sale of them being at least 60 or 70 of which some dyed in the time of sale and many are dead since." That same year, Stephen also sold a cargo of English goods for his brother James and a consignment of rum and molasses from the West Indies, and he arranged shipment of a load of tar for a North Carolina planter. The next year, he was again buying up large quantities of tobacco for several Bristol merchants along with enough timber to ensure that the ships left fully laden.[95]

Loyd started keeping a separate account for the plantation when serious development began in 1708. In 1690, William Fitzhugh of Stafford County had estimated the cost of setting up a plantation in the Northern Neck at £350 to £400 sterling—between £150 and £200 for a tract of improved land with water access and an equal amount to buy seven to ten slaves costing from £16 to £20 apiece. Horses, cattle, and hogs, Fitzhugh advised, could be easily purchased in the colony "and continually raised for a future support." In the first decade of the 1700s, however, the cost to the partners of setting up their plantation was more than £550 sterling constant value. Fitzhugh had accurately estimated what the land would eventually cost, although at first the partners paid only rent and taxes. They had to pay about £100 for buildings and livestock, however, and they spent, not the £200 Fitzhugh estimated, but more than £450 for slaves.[96]

94. Stephen Lyde Account Book, 1708–1711, VHS.

95. Loyd to Gentlemen, Oct. 10, 1710, Stephen Lyde Account Book, 1708–1711, VHS. The sales of slaves from the *Leopard Galley* are recorded in the Loyde Account Book, 1708–1710, UVa. For the voyage, see David Eltis et al., *Voyages: The Trans-Atlantic Slave Trade Database* (Atlanta, Ga., 2008), http://www.slavevoyages.org, voyage no. 16052.

96. Davis, ed., *William Fitzhugh*, 279–280.

The high costs associated with building that Fitzhugh had enumerated in 1687 still prevailed. In the environs of Stafford County, as Fitzhugh explained to an English correspondent,

> Labour is so intolerably dear, and workmen so idle and negligent that the building of a good house, to you there will seem insupportable . . . notwithstanding we have timber for nothing, but felling and getting in place, the frame of my house stood me in more money . . . than a frame of the same Dimensions would cost in London, by a third at least where every thing is bought, and near three times as long preparing.

Although Stephen had no need to build an English framed house on the partners' plantation, between 1708 and 1710 he did pay carpenters fifty-six pounds sterling constant value to build a storehouse, tobacco house, henhouse, sheep pen, dairy, and kitchen, to repair an existing dwelling and build two welsh (wattle-and-daub) chimneys, to pale in a vegetable garden, and to make a cart and a plow. In addition, Loyd had to buy hogs, cattle, and several horses, including two cart animals (and, at some point, sheep as well) for nearly fifty pounds. That Stephen had a dairy and a sheep pen built, had two plows made, and asked his brother to send some "English hands" for the plantation shows that he also planned to pursue some elements of English husbandry. No indentured servants were in fact sent, however, and there is no evidence that anyone ever raised any wheat or pursued dairying.[97]

In 1708, Stephen purchased four slaves for the farm, eight more in 1709, and seven in 1710, six of them captives from Calabar, part of the *Leopard Galley* shipment. The total of nineteen slaves included nine men, four women, two boys, two girls, and two young children. The partners paid twenty-three or twenty-four pounds apiece for the adults, but less for the children. The slaves were disembarked too late in the season in 1708 to plant any crops, so Loyd had to buy corn in that year and the next for their provisions. They planted no tobacco in 1709 either. Those brought in 1708 must have planted some corn the next year, but otherwise the hands might have been fully occupied in clearing land. In 1710, Stephen employed Thomas Leftwick as overseer. By 1711, the slaves were growing corn (although not quite enough to feed themselves) and potatoes and got in a crop of tobacco. Yield per hand was unimpressive, about 490 pounds each. Meanwhile, outlays continued to mount as new clothing and tools had to be provided every year and bedding occasionally.

Stephen Loyd died in October 1711 at age thirty, already possessed of a

97. Ibid., 202–203.

personal estate of £720 sterling constant value. Eighty percent of his movable assets were invested in capital goods—primarily slaves, livestock, and trade goods. Stephen and Elizabeth had not stinted, however, when it came to furnishing their five- or six-room home, one of them a recently added "new room." They had already acquired several richly curtained beds and a variety of chairs and tables and had imported a fashionable cane couch, china dishes, tea and gaming tables, books, a gold watch, and cutlery. Their home plantation, with its separate dairy, kitchen, storehouse, and tobacco prizing house showed every evidence of Stephen's rising fortunes. Stephen's estate was indebted £325 sterling to James for money or goods advanced, and some of the debts owed to the estate for merchandise Stephen had sold on credit proved uncollectible. James and the executor seem eventually to have reached a settlement over Stephen's accounts, although, after the couple's only surviving child died without issue in 1723, Stephen's relatives in Bristol pressed claims (which were ultimately unsuccessful) for the child's share of Stephen's estate, which they maintained should revert to them. Meanwhile, the partners' Chatterton plantation suffered misfortune and neglect after Stephen died. When appraisers came to value the assets in June 1714, only nine of the nineteen Africans imported between 1708 and 1710 survived, along with three young children likely born in Virginia. These slaves were valued at £318, more than one hundred pounds less than the amount Loyd had invested in bondpeople, and the nineteen cattle and one horse remaining on the plantation were worth only about half of what had been paid for livestock.[98]

A neighbor, John Tayloe (1687–1747), took charge of the partners' plantation in return for a fifteen-pound-per-year salary. (He initially asked for twenty pounds but later settled for fifteen, although partner Isaac Elton considered ten pounds more than enough recompense.) Tayloe had only recently come of age but was already becoming one of the richest men in Virginia. His father, William, an immigrant from London, had done well, and at his death in 1710 John inherited a half share of more than three thousand acres and twenty-one slaves that William owned. Since 1708, Tayloe had worked for Stephen Loyd as a tobacco receiver and was thus well-acquainted with the partners' affairs. He did not remain a mere manager for long. In 1713, he married Stephen's widow and through her acquired the use of much of Loyd's property. The couple set up housekeeping on Tayloe's home planta-

98. Inventories of the two plantations and a preliminary account with James Loyd are in Essex County Deeds and Wills, 14, 1711–1716: 395–402, microfilm, VSL. For the dispute over Elizabeth's inheritance, see "State of the Case inter Tayloe and the Lydes of Bristoll," Tayloe Family Papers, MSS1T2118, d121–122, VHS.

tion on Rappahannock (later Cat Point) Creek in Richmond County. Tayloe continued to manage the Potomac farm through at least 1723, when the record book ends. Tayloe, like Loyd, began dealing in slaves, commissions gained through the influence of Stephen's brothers James and Lyonel. In 1717, Tayloe joined with John Baylor in selling a consignment of 156 slaves and the next year sold a cargo of 72 slaves for Bristol slave trader Samuel Jacob, both groups arriving on the *Charfield* from the River Gambia. In 1723, Tayloe joined with another slave dealer, Augustine Moore, to sell 172 Africans carried on the *Greyhound,* and he sold 300 carried on the *Shirley* for Bristol partners Isaac Hobhouse, William Challoner, and Lyonel Loyd. He also exchanged imported goods that would not sell along the Rappahannock with lower James River merchants for tar, shingles, plank, and tallow and brought shipments of corn, wheat, pork, bacon, and myrtle wax from the Eastern Shore.[99]

Between 1714 and 1723, from seven to eleven adult slaves worked the Potomac plantation. Supervision posed a perpetual problem, for the quarter lay far from both Loyd's and Tayloe's main base of operation — the Rappahannock. Overseers must have been in short supply in Stafford County in 1710, for Thomas Leftwick secured an extraordinary salary of three and a half shares of corn and tobacco for managing fewer than a dozen hands. Tayloe discharged Leftwick in 1715, alleging he had "dealt unfairly with us," but admitted he did not have enough proof of wrongdoing to prosecute Leftwick. The next overseer had to be content with the more usual recompense of a share and a half of the major crops. In 1718, Tayloe wrote James and Lyonel Loyd that he thought he had hired a good overseer — but, to be certain, he was paying half of his fifteen-pound manager's salary to William Fitzhugh II's general manager, James Green, who lived near the plantation, "to have a constant Ey Over the overseers management so that I hope we might do well if we had a Larger Gang." In 1722, the overseer ran off in the middle of the growing season, making it difficult to get the tobacco harvested and packed.[100]

99. For Tayloe, see *VMHB,* XVII (1909), 369n–375n; Richard S. Dunn, "A Tale of Two Plantations: Slave Life at Mesopotamia in Jamaica and Mount Airy in Virginia, 1799 to 1828," *WMQ,* 3d Ser., XXXIV (1977), 32–65; Walter E. Minchinton, ed., "The Virginia Letters of Isaac Hobhouse, Merchant of Bristol," *VMHB,* LXVI (1958), 278–301; and Laura Croghan Kamoie, *Irons in the Fire: The Business History of the Tayloe Family and Virginia's Gentry, 1700–1860* (Charlottesville, Va., 2007), chap. 1. At his death in 1747, Tayloe's personal estate was worth more than £10,500 sterling constant value, including nearly three hundred slaves (Richmond County Will Book, 5, 1725–1753: 547–553, microfilm, VSL). William Tayloe's inventory is in Richmond County Wills and Inventories, 1709–1717: 41–44, microfilm, VSL.

100. Tayloe to James Lyde, May 13, 1715, Tayloe to James and Lyonel Lyde, July 12, 1718, Accounts and Letters of John Tayloe I and John Tayloe II, 1714–1778, Tayloe Family Papers, VHS.

On the Loyd-Tayloe plantation, even more than on Robert Cole's plantation forty years earlier, tobacco was the main source of income (Table 13). The weed accounted for more than three-quarters of farm revenues. Annual production was usually seven thousand to eight thousand pounds. Only 1715 and 1720 saw unusually large crops of more than twelve thousand pounds. Tobacco revenues fluctuated widely, ranging from £33 to £123 a year, depending on prices and the size of the crop. Export earnings accounted for no less than 57 percent of farm revenues in any season and were usually 80 percent or more of total revenues.[101]

As on the earlier Cole plantation, livestock was the second most important source of local income, on average 9 percent of annual revenues. Between 1714 and 1723, there were surplus cattle (primarily steers) or hogs in eight of nine years. Corn, the only other important plantation product, also accounted for 9 percent of sales. Tayloe's strategy was likely to have the slaves grow only enough corn to feed themselves and the livestock. They must have been successful, for Tayloe did not have to buy any additional grain. Corn was sold in five of nine years, but these seem to have been accidental surpluses made in especially favorable seasons rather than intentional production for market. Sales in 1716 might have resulted from a surplus carried over from the previous year when the tobacco crop was much larger than usual, an indi-

101. Tables throughout were prepared by reassembling the accounts (usually found in the form of credits and debits posted to individuals over a number of years) into annual totals of sales and purchases. Revenues and expenditures are reported in pounds sterling constant value to facilitate comparisons between different times and between different colonies. Prices recorded in pounds of tobacco were converted either at the rates stated in the individual accounts or, where no rates were given, by the most appropriate available price series. For the upper Eastern Shore, prices are from Clemens, *Atlantic Economy,* tables A2 and A3; for the rest of Maryland, from Menard, "The Chesapeake Tobacco Industry," *Res. Econ. Hist.,* V (1980), 157–161. For sweet-scented areas in Virginia, a series compiled by the author was used. Through 1730, these are reported in Walsh, "Summing the Parts," *WMQ,* 3d Ser., LVI (1999), 92–93. Maryland and Virginia currencies were converted to sterling using annual exchange rates in John J. McCusker, *Money and Exchange in Europe and America, 1600–1775: A Handbook* (Chapel Hill, N.C., 1978). Finally, the resulting sterling figures were reduced to a constant value using a Chesapeake-wide sterling deflator prepared by P. M. G. Harris for Historic Saint Mary's City with base years 1700–1709. For construction of the deflator, see Harris, "Inflation and Deflation in Early America, 1634–1860," *Social Science History,* XX (1996), 469–505.

For the Loyd-Tayloe plantation, tobacco income was figured using prices from the accounts where available and otherwise those in Rutman and Rutman, *A Place in Time: Explicatus,* chap. 1. Although Stafford County produced primarily oronoco, sweet-scented was also grown on the nearby Fitzhugh estate. The available prices seemed to better fit the Middlesex levels than did Menard's oronoco series, and the results obtained using the Middlesex series corresponded much more closely with Tayloe's observations about costs and profits on the plantation. If in fact oronoco was being grown, the plantation would have just broken even.

TABLE 13. Revenues, Expenditures, and Investments on the Loyd-Tayloe Plantation, Stafford County, Virginia, 1707–1723

Category	1707	1708	1709	1710	1711	1713	1714
REVENUES							
Tobacco	18.75	0	0	0	33.45	—	47.61
Corn	0	0	0	5.57	0	—	0
Livestock	0	0.34	0	1.52	0	—	0
Rent	0	0	13.44	0	0	—	0
Slaves	0	0	19.20	0	0	—	0
Misc.	0	0	13.26	0	0	—	4.50
Total	18.75	0.34	45.90	7.09	33.45	—	52.11
EXPENDITURES							
Clothes	0	0	0	6.78	0	—	0.43
Tailor	0	0	0	0	0	—	2.96
Cooper	0	0	0	0	0	—	2.45
Blacksmith	0	0	0	0	0	—	—
Building repair	0	0	0	0	0	—	1.53
Provisions	0	0	9.60	5.19	4.00	—	—
Tools	0	0.15	0	3.04	1.25	0.78	—
Marketing	0	0	0	0	0	—	—
Medical	0	0	0	0	0	—	0.51
Supervision	5.38	17.83	0	14.55	19.98	—	27.74
Taxes and rent	0	7.26	7.26	7.69	7.69	7.69	6.36
Misc. local goods and services	0	4.04	0	2.04	1.75	—	0.50
Misc. imported goods	0	0	7.38	0	0	—	—
Total	5.38	29.28	24.24	39.29	34.67	8.47	42.48
INVESTMENTS							
Slaves	0	98.55	190.87	163.74	0	—	—
Livestock	0	5.78	9.60	22.28	3.00	—	—
Buildings	0	28.75	5.62	21.57	0	—	—
Total	0	133.08	206.09	207.59	3.00	—	—

Note: Currency values are in constant pounds sterling. For methods of calculation, see note 101. Accounts are missing for 1712 and 1718; they are incomplete for 1713 and 1714.

Sources: Stephen Lyde Account Book, 1708–1711, and Accounts and Letters of John Tayloe I and John Tayloe II, 1714–1778, Tayloe Family Papers, VHS; Loyde Account Book, 1708–1710, UVa.

1715	1716	1717	1719	1720	1721	1722	1723	Total
122.95	50.86	53.52	53.59	90.09	53.68	54.46	38.91	617.87
15.37	0	27.05	0	12.04	0	2.00	11.93	73.96
4.64	4.01	12.87	7.86	10.89	18.13	2.50	12.79	75.55
0	0	0	0	0	0	0	0	13.44
0	4.69	0	0	0	0	0	0	23.89
12.11	0	0	0	0	0	0	0	29.87
155.07	59.56	93.44	61.45	113.02	71.81	58.96	63.63	834.58
0	0	0.29	5.31	9.32	0	0	0	22.13
2.08	2.03	2.08	3.11	2.24	2.54	0	2.79	19.83
4.18	3.72	1.60	3.66	1.68	3.70	0.86	1.90	23.75
0	0	0	1.26	0	0.62	0	0	1.88
0	0	0	0.42	0.84	0	0	0	2.79
0	0	0	2.06	0.33	0	0	0	21.18
0	0	1.65	0.76	0	0	0	4.08	11.71
0	3.00	0	0	2.73	0.45	0.45	1.31	7.94
0.87	1.95	2.40	6.20	4.27	0	0	1.70	17.90
67.14	25.90	27.49	25.21	30.03	23.06	23.18	21.00	328.49
27.12	8.07	10.21	6.92	4.97	4.79	6.42	6.22	118.67
0	0	0	9.25	12.75	0	0	0	30.33
1.19	0.23	0	0	0	0	1.20	0	10.00
102.58	44.90	45.72	64.16	69.16	35.16	32.11	39.00	616.60
0	0	0	0	0	0	0	0	453.16
0	0	7.11	0	0	0	0	0	47.77
0	0	0	0	0	0	0	0	55.94
0	0	7.11	0	0	0	0	0	556.87

cation of good weather. For both corn and tobacco, 1720 was a bumper year. Even if surpluses were not planned, however, maize proved especially useful in the local market. Tayloe was sometimes able to barter grain for goods and services, thus avoiding the need to pay tobacco or coin.[102]

Supervision of the labor force was by far the largest single expense, accounting for 53 percent of recorded outlays. Overseer's wages plus the cost of overseeing the overseers ran high. Clothing for the slaves—yard goods, shoes, stockings, caps, and the cost of employing a tailor to make up the clothes—was the second major outlay. The expenditures reported in Table 13 (7 percent of total outlays) in most years reflect only Tayloe's local purchases. The other partners supplied the bulk of the clothing and bedding needed in annual shipments that represented part of their share of plantation costs, the value of which Tayloe neglected to record in the plantation accounts. Only in 1719 and 1720, when no shipments appear to have been sent, did Tayloe purchase yard goods, thread, and shoes locally. In these two years, clothing accounted for 13 to 17 percent of total expenditures, a more accurate reflection of the proportion required to clothe the workers. Other expenses—cask, tools, household goods, medicine, and the like—were minimal, reflecting both that little equipment was needed for staple agriculture and that planters could hold the living standards of their slaves to a bare minimum. Rents and taxes, higher than on the Cole plantation, constituted the third largest expense, 19 percent of costs. Stephen Loyd failed to record what he paid in county and parish taxes and had avoided paying quitrents altogether. In 1715, Tayloe had to compound with the Northern Neck proprietary for twenty-five years of back taxes, a painful outlay. In addition, county and parish levies amounted to between five and ten pounds per year, enough to make Tayloe grumble. He wrote James Loyd in 1715, "Levies run so prodigious high it sweeps away a great deal."[103]

A comparison of plantation revenues and expenditures shows that beginning in 1714 revenues exceeded expenditures, but the operation was far from a spectacular success (Table 14). The farm ran two seasons in the red under Tayloe's management, and in only three years were there substantial profits. Tayloe made gross revenues from tobacco of six and a half pounds a year and from all sources of about nine pounds a year per hand, comparable to the revenues that indentured servants produced on the Cole Plantation

102. Similarly, in 1720 Robert Carter noted that he tried to defray most "contingent charges" of an estate he was administering by the sale of corn and provisions (Wright, ed., *Letters of Robert Carter,* 21–22).

103. Tayloe to James Lyde, May 8, 1716, Accounts and Letters of John Tayloe I and John Tayloe II, 1714–1778, VHS.

TABLE 14. Summary of Revenues and Expenditures on the Loyd-Tayloe Plantation, Stafford County, Virginia, 1707–1723

Year	Revenues	Expenditures	Balance
1707	18.75	5.38	13.37
1708	0.34	29.28	−28.94
1709	45.90	24.24	21.66
1710	7.09	39.29	−32.20
1711	33.45	34.67	−1.22
1714	52.11	42.48	9.63
1715	155.07	102.58	52.49
1716	59.56	44.90	14.66
1717	93.44	45.72	47.72
1719	61.45	64.16	−2.71
1720	113.02	69.16	43.86
1721	71.81	35.16	36.65
1722	58.96	32.11	26.85
1723	63.63	39.00	24.63
Mean	59.61	43.44	16.18

Note: Currency values are in constant pounds sterling. Accounts are missing for 1712 and 1718; they are incomplete for 1713 and 1714.
Sources: Stephen Lyde Account Book, 1708–1711, and Accounts and Letters of John Tayloe I and John Tayloe II, 1714–1778, Tayloe Family Papers, VHS; Loyde Account Book, 1708–1710, UVa.

forty years earlier. Total net annual market income, however, averaged a most unimpressive twenty-eight pounds a year, or only about four pounds per enslaved worker. Expenditures were more constant than revenues, for little could be cut. Rent, taxes, and overseers had to be paid, and the slaves required tools and clothing if they were to work. A poor yeoman family might do without new clothes in a bad year, making do with rags and hand-me-downs. Slaveowners were likely afraid to risk the reaction of their bondpeople, already allowed so little, should they fail to come through with the customary annual issue of new clothing.[104]

104. Planter John Custis protested to Glasgow merchant John Stark in 1729 that the plaid (a coarse woolen cloth) sent for his slaves' winter clothing was rotten and moth-eaten. Custis complained about the labor required to darn the cloth but also explained that he "expected to give them better than this." See Custis to John Stark, [1729], John Custis Letterbook, 1717–1741, Library of Congress, microfilm, M-8, CWL.

The plantation was probably on a somewhat firmer footing than Table 14 suggests. The slaves were producing enough food for their own subsistence, a form of farm income not considered in the accounts. They were also helping to increase the value of the farm through land clearing, fencing, and other unrecorded improvements. When Stephen Loyd ran the plantation, the wintertime tobacco stripping and reprizing operation generated additional, nonagricultural earnings that were not credited to the plantation. Tayloe, too, reprized tobacco he bought from small planters. He might have employed his own slaves in repacking rather than those from the Potomac plantation, but some of them might have helped to collect tobacco along the Potomac and take it around to the Rappahannock. Still, one suspects that the English partners were less than fully satisfied with the returns to their investment.

Spates of low tobacco production were part of the problem (Table 15). Initially, the newly enslaved Africans produced very little—less than five hundred pounds each, a mediocre record to which several seasons of below-average rainfall might have contributed. When Tayloe first assumed management of the farm in 1715 (a year of exceptionally good weather), he was able to effect a jump in production to about 1,100 pounds a hand, an increase maintained through 1720. Tayloe was apparently satisfied; in 1718, he advised James and Lyonel Loyd, "We have had ordinary Success wth that Plantation." After 1720, production declined again to about 650 pounds per laborer. The weather was bad in 1721, but the next two years saw little improvement in yields. Falling prices might have discouraged both owner and overseer. Tayloe was either unwilling or unable to begin production of alternative cash crops to maintain or augment revenues on this ancillary farm, however, although he probably pursued more diversified activities on his home plantation.[105]

Sickness and deaths among the slaves surely help to explain their relatively low productivity. One enslaved woman bought in 1708 proved "foolish," lead-

105. Tayloe to James and Lyonel Lyde, July 12, 1718, Accounts and Letters of John Tayloe I and John Tayloe II, 1714–1778, VHS. Throughout, a crop share is taken to be what planters designated as a full crop share. Surviving accounts clearly indicate that in this period planters counted all healthy adult slaves, male and female, as full hands. (William Tatham's description in 1800 of shares differentiated by sex simply does not apply here.) Only the very old or younger adolescents were rated as less than full hands. For listings of sharers, see Edmund Bagge Account Book, 1726–1733, CWL; Robert Gouldesborough Memorandum Book, 1696–1771, privately owned, microfilm, MSA; List of Negroes in York and King William counties belonging to Edmund Jenings, 1712 [and 1713], MS, Duke University, microfilm, M-36-3, CWL. From the 1710s, women made up about 40 percent of field hands where sharers were listed by name. To convert share observations for this period presented in the appendixes into prime adult male crop equivalents, the figures should be increased 20 percent.

TABLE 15. Tobacco and Corn Crops and Revenues per Laborer, Loyd-Tayloe and Bagge Plantations, Stafford, Essex, and Spotsylvania Counties, Virginia, 1707–1723

Year	Tobacco (lbs.)	Tobacco revenue (£)	Corn (barrels)	Corn revenue (£)
LOYD-TAYLOE PLANTATION				
1707	661	4.23		
1714	492	3.55		
1715	1,129	9.84		
1716	1,008	7.88		
1717	726–812	6.01		
1719	1,004	7.66		
1720	1,429	10.01		
1721	740	5.18		
1722	683	4.87		
1723	520	3.54		
Mean	844	6.28		
BAGGE PLANTATIONS				
1726	556	4.74	8.3	3.17
1728	314	2.50	15.9	7.23
1729	334	2.58	11.6	4.58
1730	538	3.34	6.9	2.41
1731	325	2.05	—	—
1732	531	4.30	8.3	3.23
1733	623	5.70	—	—
Mean	460	3.60	10.2	4.12

Note: Currency values are in constant pounds sterling.
Sources: Stephen Lyde Account Book, 1708–1711, and Accounts and Letters of John Tayloe I and John Tayloe II, 1714–1778, Tayloe Family Papers, VHS; Loyde Account Book, 1708–1710, UVa.

ing Loyd to sell her along with her child the next year for less than he had paid for them. Another woman proved too sick to do any work aside from hand weeding a vegetable garden, and Tayloe eventually sold her for five pounds, after paying a doctor two and a half pounds in a vain attempt to cure her. A third woman died in or soon after childbirth in 1715. Two men were sick for an extended period in 1716, another man was incapacitated by frostbite in 1719, and still others were ailing in 1720. The Africans who survived did find partners, though, and natural increase did begin. Between 1708 and 1723, eight children were born to enslaved mothers, but it is uncertain how many survived. Tayloe did not spend much on feeding and clothing these children, but frequent pregnancies must have lowered the mothers' productivity, and hence lowered the investors' short-run profits.[106]

High start-up costs, Loyd's frequent absences while trading for tobacco coupled with his death at a young age, unsatisfactory overseers, problems encountered in managing a distant plantation, and frequent sickness and death among the newly enslaved all contributed to the less-than-optimal revenues this frontier plantation produced. Under Tayloe's more experienced management, however, gross revenues per laborer from tobacco were in most years similar to or slightly higher than the annual tobacco income per tithable estimated for the oronoco-growing subregion overall. Tayloe thus aptly characterized the plantation's success as "ordinary."[107]

The second set of planters who experienced difficulties managing plantations in this period were Robert Bristow (1643–1707), an absentee Virginia landowner who, after suffering losses in Bacon's Rebellion, had gone back to England in 1677 to become a London tobacco merchant, and his grandson Robert (1688–1737), who inherited the farms upon the death of his father in 1706. Bristow's letters and those of his grandson, written between 1705 and 1719, describe an operation similar to that of Stephen Loyd and his partners. The elder Bristow had begun buying land in the 1660s, and by 1705 he owned 2,000-acre Fleets Bay plantation in Lancaster County, three farms totaling 2,242 acres in Gloucester County, and an undeveloped 7,500-acre tract in Prince William County. The size of the labor force at the turn of the century is unknown, but there were probably at least thirty adult slaves. Some of them might have been brought from the Bight of Biafra on the *Calabar Merchant* in 1702, a slaving voyage in which Bristow was an investor. The problems

106. Stephen Lyde Account Book, 1708–1711, and Accounts and Letters of John Tayloe I and John Tayloe II, 1714–1718, VHS.

107. Tayloe to James and Lyonel Lyde, July 12, 1718, Accounts and Letters of John Tayloe I and John Tayloe II, 1714–1718, VHS.

the Bristows encountered were precisely those that plagued Loyd and many other tidewater Virginia planters—high supervisory and marketing costs, small, poorly handled crops, unreliable overseers, and slaves who were unwilling or unable to work hard for their owners.[108]

A general manager, Thomas Booth, supervised the Gloucester County plantations himself and was responsible for hiring an overseer for the Lancaster County farm. Booth had many concerns of his own, and the Bristows at times suspected he was giving much more attention to advancing his own fortunes than to watching out for theirs. Little was made on any of the farms besides tobacco and enough corn and cider for plantation use. The Bristows expected some profit from the sale of surplus horses and cattle, but they were disappointed; the overseers either neglected the stock or converted any increase to their own use.

Consequently, all depended on producing top-quality tobacco for the London market, but the Bristows enjoyed few good cropping years. The Gloucester County plantations on Mobjack and Purton's bays lay in the heart of the sweet-scented region and should have produced high-quality leaf, although the soils on the Lancaster quarter at Fleet's Bay might have been less suitable for the more valued strain. In 1705, the Lancaster County crop was small, badly cured, and badly packed. The elder Bristow wrote the overseer: "If you make little or no crop nor anything of my stock I had better give away my Servants and Land than keep it." Instead, he willed them to his grandson, who took over in 1707 with no more satisfactory results. The previous year's crop was packed when the tobacco was too damp ("in too high case," in period parlance), and the hogsheads were light. The 1707 crop was "poor funky stuff," not worth consigning. That of 1709 was better cured, but some of it had turned black from overprizing. Bristow advised the overseer to learn the art of packing "weighty" hogsheads without damaging the tobacco in the press. It was the same story in 1714, when that year's overly dark tobacco could be sold only in Spain. Booth made a different error in 1716, stemming some of the crop in order to get a higher price but then packing stemmed indiscriminately with unstemmed so that the whole was counted only as less valuable leaf tobacco. In 1718, on Booth's advice, Bristow switched his Lancaster County overseer's thirty-pound annual salary to a one-twelfth share of the crop. The share amounted to a healthy six hogsheads—but they were of such poor quality that all six netted less than four pounds. The next

108. Robert Bristow Account and Letterbook, 1688–1750, MS, VSL, microfilm, M-1164, CWL; Robert Bristow Letterbook, PRO/A.O. 13/84, microfilm, M-494, CWL; Eltis et al., *Trans-Atlantic Slave Trade Database,* http://www.slavevoyages.org, voyage no. 21258.

year, a disgusted Bristow was looking for a new overseer whom he intended to pay in crop shares, not an annual wage.[109]

The distant Lancaster County plantation was clearly the most difficult to manage. In 1717, Bristow calculated that it was operating at a net loss. He claimed that at some point "in his grandfathers time" the farm had cleared between one hundred and two hundred pounds sterling per year. Now he was getting only eight to nine hogsheads of tobacco that netted between seventy and eighty pounds. Charges on the plantation for new African slaves, clothing, and tools were running to sixty pounds a year, not including the overseer's salary. It was necessary to keep buying slaves just to keep up the size of the labor force. "Whatt care you take of my negroes I know nott but I find noe increase of them but all loss," the elder Bristow had written in 1707; "I fear you take no care of the young children that are born to bread them up." High adult and infant mortality decimated the workforce and demoralized those who survived, while poorly supervised overseers took insufficient pains to properly cure the crops and paid more attention to furthering their own concerns rather than those of the owners. These plantations eventually again became profitable, but only after Bristow changed the crop mix and devised harsher methods for forcing the slaves to work harder.[110]

Finally, an account book dealing with several farms in Essex and Spotsylvania counties in the late 1720s underscores the extreme difficulties immigrant and young creole planters faced when sweet-scented tobacco prices fell to an unprecedented low in the last half of the decade. The planters, John and later Edmund Bagge, like Robert Bristow, pinned their hopes on making high-quality tobacco for the English domestic market. But both had limited managerial skills, and their efforts met with little success. Like Stephen Loyd and John Tayloe, John Bagge made few efforts to supplement disappointing returns from tobacco by pursuing a more diversified agriculture. Edmund Bagge went to the other extreme, attempting to earn extra income by trying such a wide range of new activities that none stood much chance of success.

John Bagge (circa 1682–1726), of County Waterford, Ireland, first came to Virginia in 1709 as a deacon of the Episcopal Church. Bagge was not an ordained minister, but, given the shortage of clergy in Virginia, he was allowed to take charge of Saint Anne's Parish in Essex County, and he also preached in adjacent Sittinbourn Parish. He married and acquired land near the Sittinbourn Parish church; he stopped preaching there in 1716 after a

109. Bristow to John Grason, Jan. 23, 1705/6, Bristow Account and Letterbook, 1688–1750, M-1164, CWL.

110. Ibid., Sept. 15, 1707.

quarrel with the vestry over his salary. In 1717, Bagge went back to England to take priests' orders, returned to Virginia, resumed the cure of Saint Anne's Parish (after a dispute over salary with that vestry), and died in Essex County in 1726. Bagge had got some training in mercantile accounting practices in Ireland and is known to have kept two double-entry ledgers. On the other hand, he might have arrived with little knowledge of farming or of how to run a plantation and manage enslaved workers. But, by virtue of his appointment to Saint Anne's Parish, whose minister had the use of a glebe farm, and through marriage to the daughter of a Virginia planter who brought slaves to the union, he nonetheless became a planter and a slaveowner. Bagge's main goal likely continued to be augmenting his minister's salary through farm revenues; he seems to have remained contented to leave day-to-day management of his farms to overseers. A fondness for drinking and socializing also might have contributed to Bagge's inattention to farming. John Mercer of Stafford County noted in 1724 that planter Harry Beverley owed Mercer for "2 quarts rum with Parson Bagge."[111]

Bagge ended up a wealthy man, although not on the scale of Byrd, Carter, or Tayloe. His personal estate, inventoried in 1726, was worth approximately £1,074 sterling constant value. It included lavish household furnishings and a carriage, evidence of a tendency for display. Most of his capital, about £810, was tied up in labor—twenty-six slaves and two white servants. Some of the slaves Bagge had purchased, but others almost certainly came as part of his wife's dowry. Bagge's only other important productive asset was livestock—sixty-nine assorted cattle, four draft oxen, and five horses worth just over £110. (Presumably he also kept hogs, but they were not inventoried.) Eight slaves and two white servants worked on the Saint Anne's Parish glebe. An additional five slaves lived on a second quarter, the Range. Each of these quarters was stocked with some cattle and a horse or two and little else besides the slaves' working tools. Thirteen slaves lived on Bagge's home plantation—five men, six women or girls (one of them, Cate, was a "house wench"), and two boys. Here there was an orchard where the slaves made cider and distilled brandy. Mary Bagge ran a dairy, and the presence of smith's tools suggests that at one time there had been a slave or servant blacksmith. The Bagges

111. Essex County Wills, 4, part 2, 1722–1730: 138, 184, 189, microfilm, reel 40, VSL; William Stevens Perry, ed., *Historical Collections Relating to the American Colonial Church,* 3 vols. (Hartford, Conn., 1870), I, 313–318; George Maclaren Brydon, *Virginia's Mother Church and the Political Conditions under Which It Grew* (Richmond, Va., 1947), 347–348; John K. Nelson, *A Blessed Company: Parishes, Parsons, and Parishioners in Anglican Virginia, 1690–1776* (Chapel Hill, N.C., 2002), 304; John Mercer Account Book, 1725–1732, fol. 29, Bucks County Historical Society, microfilm, M-164, CWL.

also used a small boat and a little cart on the home plantation. There were no other indications of diversified agriculture — no plows, no sheep, no cards or spinning wheels, and no sign of any grains other than maize. Production of tobacco for export was clearly the plantation's primary activity.

John Bagge left thirteen of the slaves and the dwelling plantation to Mary during her lifetime. Since they had no children, after Mary's death her dower land and slaves, along with other tracts and laborers, were to go to two orphaned nephews, sons of Bagge's deceased brother, Leonard, who lived on the Eastern Shore. One of the heirs, William Bagge, along with his older brother, Edmund, were to serve as executors. Mary Bagge died within a few months of her husband, and she made nephew Edmund her sole executor.

Edmund Bagge (?–1733) was by then living somewhere in the Northern Neck, where he worked as an agent for Colonel Alexander Spotswood's iron mines. Upon the deaths of John and Mary Bagge, Edmund assumed management of their Essex County estate, married an heiress, and fathered a child. He, in turn, died young in 1733, leaving his estate to his infant son. Edmund directed the two friends he made his executors to buy nearby Nehocking plantation, which he had been farming as a tenant but had not yet purchased. He had already bought other land and by 1731 had set up a second quarter, Indian Fields, on a 2,284-acre tract "at the Mountain" in Spotsylvania County. Although his career was cut short by early death, Edmund, too, died a relatively rich man with a personal estate worth about £912. Like his uncle, Edmund had almost all his capital, nearly £695, in slaves. Most of his twenty-seven blacks were acquired through his wife, although he had bought four new Africans — two men and two women — in 1731 for £80 sterling. Otherwise, he had about £120 worth of livestock: eighty-two cattle, two horses, thirty-six sheep, and doubtless unappraised hogs.[112]

Edmund Bagge kept an account of his management of John and Mary Bagge's property and of his own plantations from 1726 up to his death in 1733. The records proved too confused, too incomplete, and too badly preserved to permit a reconstruction of revenues and expenditures for the various farms. Nonetheless, they reveal a great deal about Edmund's management decisions. Bagge likely learned bookkeeping in Ireland, as had his uncle John. Edmund's "Journal-Book" (Plate 17) is a neatly kept example of

112. Essex County Wills, 5, 1730–1735: 242, 354, microfilm, reel 41, VSL. Bagge was living somewhere in the Northern Neck by 1725, since in that year John Mercer debited him for a small sum lost at a horse race and credited him the next year for six and a half shillings Bagge won at a card game at Colonel Mason's. During these years, Bagge sold Mercer three iron firebacks, a product of Spotswood's ironwork, and arranged for the making of two irons for prizing tobacco at the mines (Mercer Account Book, 1725–1732, fol. 20, M-164, CWL).

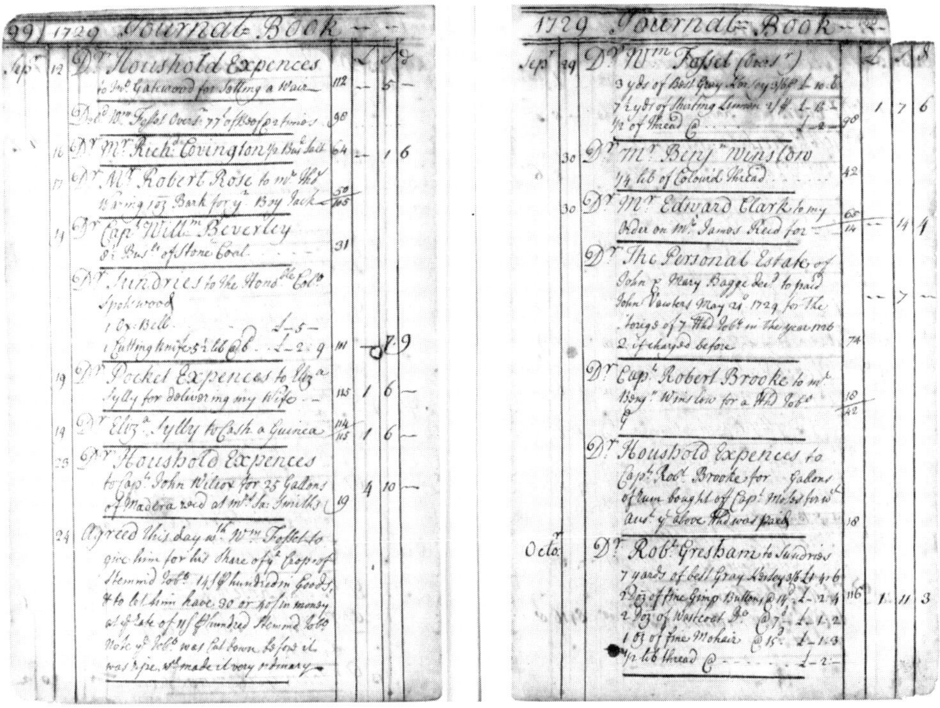

PLATE 17. Page from Edmund Bagge's Account Book, Virginia, 1726–1733. Entries for September and October 1729. Courtesy, Special Collections, John D. Rockefeller, Jr., Library, The Colonial Williamsburg Foundation

the intermediate record book that accounting instructors recommended be transcribed from a waste-book before constructing a double-entry ledger. He might have studied Alexander Malcolm's *New Treatise of Arithmetick and Book-Keeping* or Alexander Macghie's *Principles of Book-Keeping Explain'd,* both published in Edinburgh in 1718, or have been taught by someone familiar with these treatises. Both texts provide extended printed examples of a "Journal-Book" similar to Bagge's, and Malcolm's treatise also advocated using highly detailed impersonal categories for classifying transactions like those Bagge adopted. Bagge recorded transactions in his journal chronologically in single-entry form along with miscellaneous notations of things to which he might wish to refer later, such as agreements with overseers and other hired employees, lists of kinds and amounts of clothing and bedding distributed to slaves, of the weights of hogs killed and of meat delivered to overseers, and of the size of tobacco and corn crops. In the journal, Edmund rendered everything he received or paid as a debit to something else, following the instruction, "In the *Journal,* the Debtors and Creditors are both distinctly marked, tho' we find only the Word *Debtor* mentioned." Unfortunately for later readers, Edmund's application of bookkeeping conventions tended to the abstract rather than to the practical. Some of the impersonal categories Bagge used were ones other period planters and merchants commonly employed, such as "cash," "sundries," and "expenses of housekeeping," but others are more abstract categories, such as "John Bagge's Freehold," "The Personal Estate of John and Mary Bagge," "Account of Servants," "Account of Slaves" (costs associated with their purchase), "Account of Negroes" (expenses incurred for slaves after their purchase), "Account of Buildings," "Shoemaker's Work," "Wheelwright's Account," "Iron Acccount," and the like, which most Chesapeake bookkeepers would have simply recorded as a credit or a debit to an individual. References in one of the columns (given in the form of a fraction in which the numerator is the folio in the ledger on which a debit was recorded and the denominator the folio on which an offsetting credit was entered) indicate that Bagge later posted the transactions to a double-entry ledger that has not survived. How much Edmund's rather bookish and overly complex accounting practices helped him assess the day-to-day financial status of his various farming operations is problematic.[113]

Running his uncle's property was probably not an enviable responsibility, in any event. John Bagge had farmed three separate tracts—his freehold of 419 acres, the Range quarter (probably a 200-acre purchase), and the parish glebe. Bagge had mortgaged the home plantation and twenty slaves to Robert "King" Carter to secure a loan of two hundred pounds sterling plus interest that John had presumably borrowed from London tobacco merchant Robert Cary to finance the purchase of some of the slaves and had failed to pay on time. There are hints that John had not been a particularly skilled planter. In 1717, for example, John Tayloe, acting as agent for a British merchant, bought a shipload of tobacco from a number of planters living along the Rappahannock. High freight charges had led many to take special care to pack their hogsheads tightly in order to cut transport costs. Of the eighty-six casks Tayloe purchased, only nine weighed less than five hundred pounds gross, and one of these light hogsheads came from Bagge. In 1718, a London firm, Higginson and Bird, refused Bagge a new supply of goods in light of his unpaid balance of sixty-three pounds. John did not fare much better with local Virginia merchants. In 1723, William Woodford, a Caroline County trader, wrote Bristol merchant George Mason that he had had to abate money on some tobacco Bagge had sold him and was glad to be rid of it at any price;

113. Bagge Account Book, 1726–1733, CWL; Malcolm, *New Treatise of Arithmetick and Book-Keeping*, 129.

"such Tobos . . . is not only a los to you but a discredit to those that dispose of them." The next year, Woodford found some of John's hogsheads "much funkt." Edmund then seems to have inherited a load of debt, an operation with an unenviable reputation, and quite possibly severe labor problems as well.[114]

After Mary Bagge's death, Edmund moved onto the Bagges' freehold and ran it as his home farm. The glebe reverted to Saint Anne's Parish once the 1726 crop was finished, and Edmund shuffled the slaves who had worked there between the home farm and the Range quarter. By 1728, Edmund had married. He then used his wife's dower slaves to set up a quarter of his own on 470-acre Nehocking plantation, which he rented, and three years later he moved some of them to his own new quarter in Spotsylvania County. The account thus deals with four farms—Bagge's freehold (home farm) and the Range quarter for which Edmund was administrator, Nehocking plantation that he was developing for his children, and Indian Fields, the newly established quarter in Spotsylvania.

Labor supervision seems to have posed even more of a problem for Edmund Bagge than it did for John Tayloe. He normally employed three different overseers in a given year, one for the Range, one for the home plantation who usually also oversaw the Nehocking farm, and one for Indian Fields in Spotsylvania. No overseer lasted more than two years on any one plantation, nor for more than a total of three years anywhere in Bagge's employ. In 1728, the home house overseer ran off in the middle of the growing season. That year and the next, Thomas Greenwood supervised the Nehocking farm. Greenwood gave up his customary right to provisions of corn and meat in exchange for the privilege of keeping his own cattle and hogs on Bagge's land and for a right to ground on which Greenwood's children might tend corn and tobacco. Bagge paid him only a single share of the crop; in 1728, this share amounted to a mere 314 pounds, since several of the hogsheads he produced were rejected as worthless (Table 15). Bagge noted that even a single share was "a great deal more than the idle dog deserved."[115]

In 1728, William Fossett, overseer on the home plantation, cut the tobacco before it was ripe, "w[hi]ch made it very ordinary." Bagge's efforts to encour-

114. Carter Diary, Aug. 7, 1723, Carter Papers, UVa; Stephen Lyde Account Book, 1708–1711, and Accounts and Letters of John Tayloe I and John Tayloe II, 1714–1778, Tayloe Family Papers, MSS1T2118, b1, VHS; Gilbert Higginson and Robert Bird to John Bagge, Dec. 15, 1718, Higginson and Bird Letterbook, Galloway / Maxcy / Markoe Family Papers, LC; William Woodford to George Mason, Nov. 25, 1723, Woodford to John Bagge, Nov. 7, 1724, Letterbook of William Woodford, CWL.

115. Bagge Account Book, 1726–1733, CWL.

age Fossett to make the slaves work harder were to no avail; in 1729, he promised Fossett a share and a half of the Nehocking tobacco crop if he could produce 900 pounds or more per hand. In fact, Fossett managed to make only 321 pounds per worker, thus no more than matching the poor record of that "idle dog," Thomas Greenwood. Bagge fired Fossett in the summer of 1730; his "Idle and unjust behavior obliged me to turn him off." In 1731, Bagge shifted Joseph George, Fossett's replacement, to the Range. For the previous two years, one of the slaves, Cornwall, had been in charge of that plantation, delivering modest tobacco crops of 346 to 538 pounds a hand. George's record for the next two years almost exactly duplicated Cornwall's achievement, despite Bagge's promise to George of a larger share of some of the crops should he manage to increase production. Tobacco revenues averaged only about three and a half pounds sterling constant value per hand, about half of those obtained on the Loyd-Tayloe farm. Although this earnings record at first glance appears to be evidence of truly abysmal management, a comparison with tobacco income per taxable estimated for the entire sweet-scented region in the late 1720s reveals that earnings on the Bagge plantation exceeded the regional average by one pound per annum. Bagge and his overseers had then achieved more than "ordinary" success, but at a level that left a great deal of room for improvement.[116]

Perhaps part of the problem was that the enslaved workers had developed effective means for resisting imposed labor. They might have succeeded in establishing the tending of modest amounts of tobacco as the custom of the plantation but strongly resisted any attempts by supervisors to increase their workload. They were apparently able to undermine any white overseer that Bagge employed and could not be made to take much care in curing and packing the crops. None of the white overseers were able to force the slaves to make more tobacco than they did when working at a pace more of their own choosing under Cornwall. Perhaps as well some were either too sick to work hard or were forced at times to work so hard that they became sick. Between 1730 and 1732, white midwives attended births of ten enslaved women on various of Bagge's quarters. Six of the ten infants were either born dead or died within a few days of birth, tragic evidence of unhealthy living conditions and perhaps of pregnant women's being forced to labor excessively hard.

Certainly, Edmund Bagge found himself in a difficult situation. On the whole, he clung to traditional solutions, trying, with little success, to produce high-quality tobacco with alienated or sick workers and incompetent or lazy overseers. Edmund consigned his crops to a number of London, Bristol,

116. Ibid.

Liverpool, and Glasgow merchants, casting about for factors who could produce favorable returns. In one year, Edmund's consignment sold for twenty shillings a hundredweight, the going direct purchase price for sweet-scented, but in other years he got only nine to fourteen shillings per hundredweight, returns more characteristic for lower-grade oronoco. Edmund did try to improve the quality of the crop—for example, in 1729 he paid a laborer to cut up tobacco suckers (an action required by law), and in 1731 he had a prize made to better pack his hogsheads. Unfortunately, such efforts were no substitute for careful day-to-day supervision of the overseers. From 1724 on, Virginia laws also set a limit on the number of plants that individual workers could tend. These laws probably had no effect on Edmund's plans, for it appears that on none of Bagge's operations did the workers tend anything near the maximum number of plants the law allowed. For the benefit of his younger brothers, Edmund did manage to pay his uncle's debt to Robert Carter, and thus got a clear title to the land and slaves that they were to inherit. To provide for his own children, however, Edmund had to borrow money to buy more slaves and new land farther west. Since he died young, the estate Edmund bequeathed his infant son was heavily indebted.

As on the Loyd-Tayloe quarter, corn might have provided some additional revenue. Table 15 in fact suggests that some of the time not spent on tobacco was put into maize. Corn yields were usually between eight and twelve barrels a hand, more than was required for plantation use. One year, Edmund employed two plowmen to help the field hands expand maize production at Nehocking, resulting in a large crop of nearly sixteen barrels per worker. However, Edmund seems not to have continued the practice. Perhaps the extra plowing proved too much work for the slaves, or the overseer neglected unfamiliar farming practices once the hired plowmen left. Like occasional surplus livestock, corn was usually sold throughout the year in small amounts. Indeed, there seems to have been no organized market for grain along the Rappahannock. This doubtless discouraged Bagge in his efforts to increase production. Still, corn yielded about four pounds per hand per year. Annual net revenues from maize, though, after deducting that part of the crop required for plantation needs, would have amounted to no more than two pounds per hand.[117]

117. Contemporary comments on lack of markets, especially for grain, were numerous. Francis Michel observed that poor planters sometimes had better food on their tables than the rich because they had few opportunities to sell eatables except to ship captains and innkeepers (William J. Hinke, ed. and trans., "Report of the Journey of Francis Louis Michel, from Berne, Switzerland, to Virginia, October 2, 1701–December 1, 1702," *VMHB,* XXIV [1916], 140). Bergstrom, *Markets and Merchants,* table 5.4, shows some export of staves from the Rappahannock and South Poto-

By 1731, Bagge was also trying to grow wheat and oats. In that year, he promised one overseer a share of any small grain made except barley. Not enough was made for home use much less for sale, it seems, for no division of the crop with the overseer appeared. Bagge continued to buy small amounts of wheat and barley from local planters and to get shipments of oats from a relative living on the Eastern Shore. An attempt to capitalize on most planters' never-ending taste for alcohol probably had more potential. Edmund maintained and expanded John Bagge's orchards and continued cidering operations. He also traded in hops, grew barley, and constructed a malt kiln, perhaps hoping to start a beer-brewing business. He was able to market some malt and hops to a Fredericksburg innkeeper, but the success of this enterprise was limited.

Bagge's main innovation was an attempt to make cloth. With his slaves growing so little tobacco, he had few assets with which to buy English manufactures, so even small savings were important. Having lived on the Eastern Shore, he likely had some experience with its more diversified farming practices. In 1727, he transported a flock of sheep from Accomack County and acquired wool cards and spinning wheels. He then hired local white women to knit stockings, spin yarn and candlewick, and weave cloth. By 1731, when he had brought an additional forty sheep from the Eastern Shore, he employed a neighbor woman to make yarn and to instruct one of his female slaves in spinning, and another local housewife wove him sixty-three yards of cloth. With time, such efforts would make indigenous cloth production an enduring part of the local economy. Since labor costs remained high, however, homespun was not a particularly cheap cloth. Nor did planters usually consider spinning, knitting, and weaving sufficiently worthwhile to divert able-bodied adult slaves from the fields. What such import replacement activities, carried out largely by white women, did achieve was a saving of sterling balances in England.

Bagge tried fishing as a second experiment for cutting costs. In 1731, he had a man set a weir and purchased a large seine. The results must have been disappointing, for Edmund subsequently sold the net to Eastern Shore relatives. The Bagges also began making their own candles, using homespun wick and bringing myrtle wax and tallow from Accomack County, and they hired a local man to make soap. On Edmund and Katherine Bagge's plantation, then,

---

mac Naval districts in 1727, but very little corn compared to other parts of Virginia. Across the two colonies, the amount of corn raised per hand showed no increase from the early seventeenth century except on the upper Eastern Shore. For corn crop shares, see Appendix 2.

necessity led to a heroic if not always successful increase in self-sufficient activities.

But had the Bagges had to live on farm income alone, they would have been in trouble. For both John and Edmund, other earnings probably served to keep their operations afloat. John estimated that his annual salary of sixteen thousand pounds of tobacco had brought in about forty-eight pounds a year during 1709–1714 and 1721–1724, and better than eighty pounds a year between 1715 and 1720. Edmund, in turn, tried his hand at a variety of money-making schemes, including the malt experiment, marketing other planters' tobacco, purchasing lots of manufactured goods for retail in the piedmont, going into partnership to buy a cargo of West Indian rum, sugar, and molasses for local resale, and acting as an agent for Alexander Spotswood's ironwork at Germanna. In the latter capacity, Edmund purchased slaves for Spotswood, arranged for shipments of foodstuffs and manufactures to be sent to the settlement and for return loads of iron to be put on ships, and even peddled small consignments of cast-iron firebacks. Had Bagge lived longer, these additional sources of income would have helped pay back the money he had borrowed to buy land and slaves, and with time he might have learned how to better select overseers and manage slaves. As it was, Bagge simply had too many irons in the fire at one time.[118]

Comparison of the Loyd-Tayloe, Bristow, and Bagge plantations reveals many similarities. All were begun with considerable capital, either on hand or borrowed. Costs of starting up quickly—the immediate purchase of both land and especially slaves—were high. And high mortality among recently arrived captives necessitated further outlays just to keep up the size of the workforce. Returns to these investments were not great. Tobacco was the only significant cash crop produced, but output per hand was often low. Neither was the quality always suitable for the English domestic market. Much of the tobacco was of lower grades normally reexported to the Continent, and Loyd, Tayloe, and the Bagges dealt as often with Bristol, Liverpool, and Glasgow merchants as with London factors in a struggle to get the best prices for their crops. Supervision of a reluctant labor force posed continual problems. Good overseers were hard to find, and the owners themselves were either too busy with other activities or lived too far away to keep a close eye on daily operations. Absentee owners like the Bristows found that overseers and even estate managers spent a great deal of time pursuing their own advantage,

118. Perry, ed., *Historical Collections,* I, 316–317; Bagge Account Book, 1726–1733, CWL; Mercer Account Book, 1725–1732, fol. 20, M-164, CWL.

leaving the slaves to work more or less as they would. Income from office or mercantile pursuits was often required to permit the owners' families to live in a style they considered suitable to their station. One can appreciate why planters like these were ready to consider more drastic measures for improving the price of Virginia's staple in the early 1730s.

For those who had the resources of a Byrd or a Carter, a continued concentration on accumulating land and slaves and on producing ever more tobacco clearly constituted a viable agricultural strategy. That these men built two of the largest personal estates ever amassed in the region throughout the colonial era is powerful evidence that accumulation and production on a grand scale could pay handsomely. Most tidewater Virginia planters did not command comparable resources or managerial skills and often succeeded in extracting only average returns from their farms and labor forces. Could they have done better had they put less emphasis on tobacco?

CHAPTER 4

Strategies of Adaptation and Change

MARYLAND, THE PERIPHERY, AND REGIONAL DIVERGENCE

1680–1729

When trends in Chesapeake tobacco prices, production, and revenues are broken down into three agricultural subregions characterized by the kind of tobacco produced—sweet-scented, oronoco, and peripheral—the fortunes of full-time growers of either of the two major strains were less dismal during the 1680s, 1690s, and early 1700s than commonly portrayed. Export earnings per taxable on Maryland's Western and Upper Eastern shores and on Virginia's Northern Neck averaged £4.66 per year between 1680 and 1702, fell precipitously to only a little more than £3 during the next decade, but at the end of Queen Anne's War rose again to an average of £5 a year. After 1710, for a time oronoco growers in both colonies realized higher average earnings per worker than did growers of the more prestigious sweet-scented strain. Although all Maryland planters confronted similar challenges posed by international wars and stagnant European demand, the strategies they adopted to cope with these problems began to vary, depending on the nature of the soils they farmed and the presence or absence of outlets for alternative products. Both contemporaries and present-day scholars often further subdivide the colony into four distinct geographic areas. Between 1680 and 1729 in these localities the typical crop mix and the time and resources allotted to commerce and to import replacement activities diverged. Three of the four areas—the lower and upper Western Shore and the upper Eastern Shore—fall within the oronoco subregion. The fourth area, the lower Eastern Shore, was a part of the periphery where the contribution of tobacco to the local economy atrophied during the era of hard times. In the three oronoco-growing areas, planters either retained earlier practices or adopted new strategies, depending on a combination of comparative advantage in producing different crops and the availability of alternative markets.[1]

1. Lorena S. Walsh, "Summing the Parts: Implications for Estimating Chesapeake Output and Income Subregionally," *WMQ*, 3d Ser., LVI (1999), 53–94; Gloria L. Main, *Tobacco Colony: Life*

No accounts survive for these years for large plantations on Maryland's lower Western Shore. Probate inventory studies have amply demonstrated, however, that there planters continued to concentrate most of their energies on growing tobacco in quantity and made only minimal shifts into import replacement activities or English-style husbandry, just as did Northern Neck Virginia planters like William Fitzhugh and the managers of Chatterton plantation. This allegedly backward-looking strategy in fact made optimal use of lower Western Shore land, assuming that soil qualities then were similar to or better than those found in the late twentieth century. Almost 60 percent of lower Western Shore soils remain suitable for growing tobacco, with between a third to a half of this acreage capable of producing either high-yield or high-quality crops. Moreover, on about a quarter of the productive soils, tobacco growers can make larger than average crops that are also of high quality. Lack of markets along the Potomac discouraged area planters from raising surplus grains. But equally or more critical to their decision to put minimal effort into diversifying their crop mix was the unsuitablility of a quarter to a third of the soils for raising either corn or wheat. Even on soils suitable for these grains, most yielded crops of low or, at best, medium size. Consequently, it remained most advantageous for lower Western Shore planters to concentrate on the crop for which the soils were most appropriate.[2]

In the 1680s and 1690s, oronoco growers on Maryland's upper Western and upper Eastern shores pursued a similarly tobacco-centered agriculture, as did those residing in the lower Potomac basin. But after the turn of the century, there and in areas to the south where tobacco was peripheral management strategies were often more innovative and dynamic than in the Virginia tidewater and on Maryland's lower Western Shore. Counties at the head of Chesapeake Bay and on the lower Delmarva peninsula were hardest hit by low tobacco prices and scarce shipping during wartime as well as by continuing shortages of indentured servants and, to a lesser extent, of enslaved laborers. In addition, some Eastern Shore planters were already encountering a shortage of good, arable land by the early 1700s. Unable to expand their operations by adding more workers, land-short managers had instead either to exploit other natural resources and alternative trading networks or to find ways to increase revenues per laborer by adding new crops and by inducing

in Early Maryland, 1650–1720 (Princeton, N.J., 1982), chap. 2; Lois Green Carr, "Diversification in the Colonial Chesapeake: Somerset County, Maryland, in Comparative Perspective," in Carr, Philip D. Morgan, and Jean B. Russo, eds., *Colonial Chesapeake Society* (Chapel Hill, N.C., 1988), 342–388.

2. Carr, "Diversification in the Colonial Chesapeake," in Carr, Morgan, and Russo, eds., *Colonial Chesapeake Society,* 348–349.

their laborers to work more intensely. These planters *had* to make more pronounced changes in their strategies for generating income than did richer and better-situated planters in the Virginia tidewater who faced no constraints to expansion. Adjustments in crop mix, in levels of self-sufficient activities, and in participation in local and intercolonial markets first initiated on Maryland's upper Eastern Shore and at the head of Chesapeake Bay would in later years become common practice on most large tobacco plantations throughout the region. Initially, however, marked changes in crop mix were the exception rather than the rule along the Potomac, Patuxent, Rappahannock, and York rivers.

Three Innovative Maryland Planters

By the 1720s, agricultural diversification and intensifying local exchange were beginning to afford gentry planters living at the head of Chesapeake Bay new avenues for building estates. Double-entry account books kept by two upper Eastern Shore planters along with one from the upper Western Shore demonstrate how such changes were effected. No longer did planters have few options for making money aside from growing tobacco or buying it from others and only limited chances for developing surplus land. And, as large planters diversified their agricultural operations, they began to play a more dynamic role in stimulating intensive rather than extensive development in surrounding communities by providing employment for artisans and free laborers and by exchanging more and more diverse goods and services with their neighbors. This situation was especially true in areas where local and intercolonial trades were expanding and that were served by British tobacco merchants who were entering the rapidly growing reexport trade to continental Europe. Rising demand among the French for lower-quality oronoco and among the Dutch and English for higher-quality, bright-leaf oronoco (a variety that, like sweet-scented, grew only in particular kinds of soils) offered new opportunities to those who grew either of these strains. Planters on Maryland's upper Eastern Shore, an area from which tobacco commanded lower than prime prices but that had good access to expanding grain markets, were among the first to adopt innovative agricultural strategies that combined a small sacrifice in tobacco revenues (by reducing the amount they planted by about one-third) with a more-than-compensating increase in revenues from expanded grain farming—a productivity improvement accomplished, moreover, without increasing the size of bound labor forces.[3]

3. Paul G. E. Clemens, *The Atlantic Economy and Colonial Maryland's Eastern Shore: From Tobacco to Grain* (Ithaca, N.Y., 1980); Jean B. Russo, "Self-Sufficiency and Local Exchange: Free

Geography profoundly influenced the new directions taken by upper Eastern Shore Maryland planters. The amount of land available for agricultural expansion on the Delmarva peninsula was finite. Acreage suitable for growing tobacco was even more limited, and most of it had been taken up by the 1690s. Travel and communication between the Eastern and Western shores was so slow and unpredictable as to deter planters from trying to farm land on both sides of Chesapeake Bay. Few Marylanders owned land in both places, and those who did had almost invariably acquired the tracts on the opposite shore through inheritance, marriage, defaulted mortgages, or other chance occurrences, not by patent or purchase as part of a deliberate estate-building strategy. In addition to problems inherent in managing distant plantations separated by an extensive body of water, by about 1690 all desirable waterfront land on the Western Shore had also been taken up. Consequently, in order to expand westward, Eastern Shore planters had to move beyond established Western Shore settlements to look for undeveloped inland tracts in competition with better-informed locals and with Western Shore land speculators. Moreover, after 1681, expansion northward and eastward was blocked by the newly established Pennsylvania proprietary and the formerly Swedish "Lower Counties" (which later became Delaware) annexed to Pennsylvania. Since it was often impossible to acquire any fresh land, upper Eastern Shore planters were under more pressure than those on the tidewater mainland to shift to more intensive farming methods.

The settlement of Pennsylvania, although barring Marylanders from territorial expansion, also offered upper Eastern Shore planters exceptional new opportunities (Map 8). Philadelphia soon emerged as a major center of intercolonial and international trade, specializing in the export of grains, and was within easy reach of upper Eastern Shore residents either overland or by water. (From 1730, wheat prices in Philadelphia and on the upper Eastern Shore followed parallel trends, an indication of an integrated market.) In addition, the fertile areas of Pennsylvania adjoining Maryland's northernmost counties were rapidly settled by German, English, Welsh, and Scots-Irish immigrants who practiced European-style general mixed farming with great success. They raised a wide variety of small grains, but wheat quickly became their key crop. Hence, upper Eastern Shore planters had greater access to information about international grain markets through Pennsylva-

Craftsmen in the Rural Chesapeake Economy," in Carr, Morgan, and Russo, eds., *Colonial Chesapeake Society,* 389–432; Jacob M. Price, *France and the Chesapeake: A History of the French Tobacco Monopoly, 1674–1791, and of Its Relationship to the British and American Tobacco Trades,* 2 vols. (Ann Arbor, Mich., 1973), I, chap. 25.

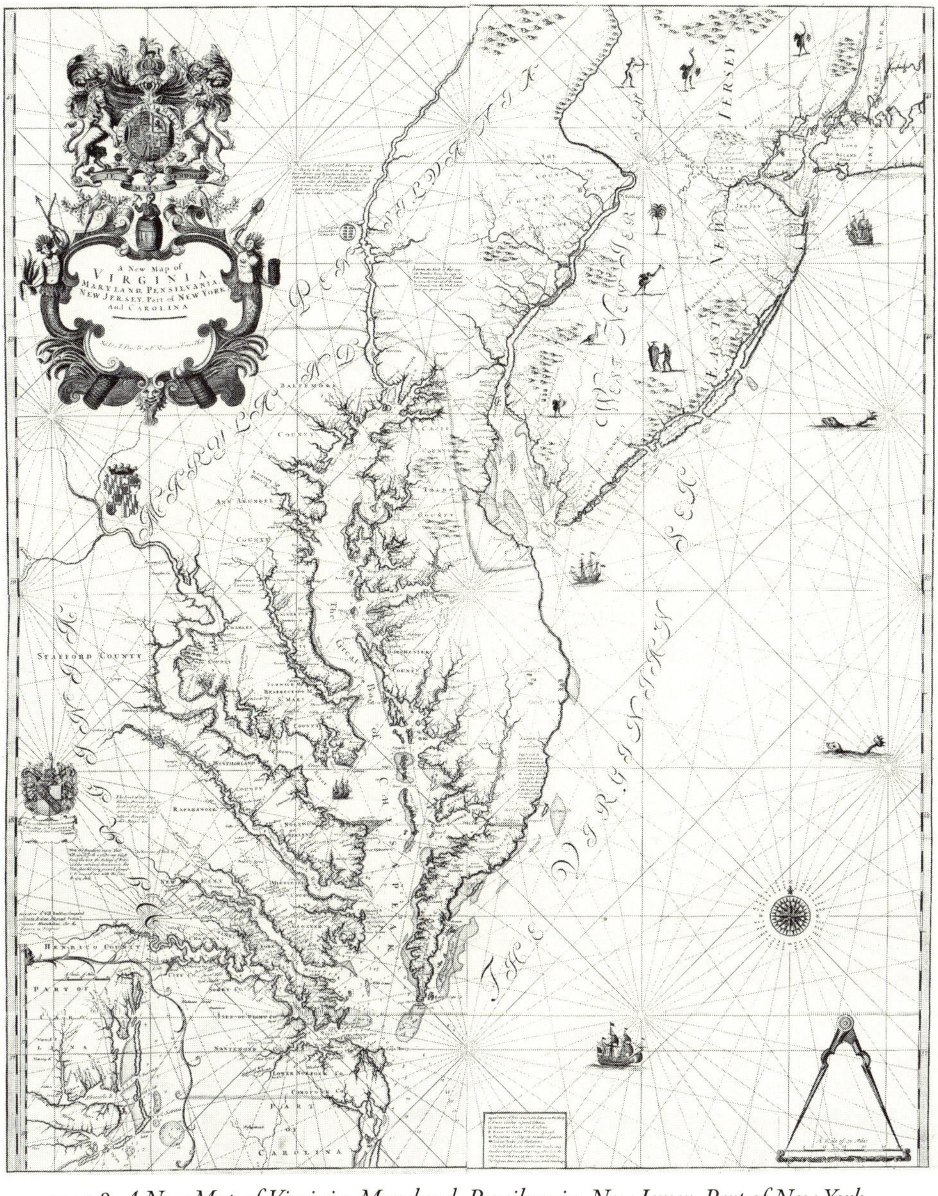

MAP 8. *A New Map of Virginia, Maryland, Pensilvania, New Jersey, Part of New York, and Carolina.* By John Thornton, London, circa 1728–1738 (originally published circa 1701). Black-and-white line engraving with period outline color. Courtesy, The Colonial Williamsburg Foundation.

The recently established adjoining colonies of Pennsylvania and the Carolinas afforded Chesapeake residents new trading opportunities but also limited the areas into which tobacco planters could expand.

nia traders and to farmers and laborers experienced in plowing, in raising and harvesting small grains, and in cultivating English forage grasses than did planters elsewhere in the Chesapeake. Finally, the upper Eastern Shore, unlike areas across the Chesapeake Bay, was also uniquely suited for raising grains. Not all upper Eastern Shore land is good for growing tobacco, but as much as 90 percent of its nearly uniformly level soils are highly suited to grains, with almost all producing medium to high yields. These geographical constraints and opportunities must have influenced the new courses Robert Goldsborough, Richard Tilghman, and other elite Eastern Shore planters embarked on between 1680 and 1729.[4]

Robert Goldsborough (1660–1746), of Talbot County, was born in Dorsetshire, England, the oldest son of merchant Nicholas Goldsborough. His father had tried his fortune in the colonial trades in Barbados, New England, and, finally, in Maryland, where he died in 1670. After finishing a legal education in England, Robert moved to Maryland in 1677. He settled first in swampy Dorchester County but moved to Talbot by the late 1680s, where he acquired a farm located on a neck of land endowed with fertile soil. He was admitted to the bar of Talbot County in 1687 and to the bar of the colony's Provincial Court in 1693. By 1696, he had been appointed a justice of the peace, and in 1704 he was elected to the Maryland assembly. Goldsborough apparently began with only a modest inheritance aside from his education, and his economic rise was relatively slow. Marriage in 1697 to Elizabeth Greenbery, daughter of a well-off and politically powerful Anne Arundel County planter, doubtless improved his prospects. By 1704 at age forty-four, Goldsborough had acquired more than nine hundred acres, a large estate by ordinary Chesapeake standards but a much less spectacular landed empire than that inherited by many native-born gentry sons. By 1735 (at the advanced age of seventy-five), he had patented or purchased more than two thousand acres and had acquired as many as thirty adult slaves. Goldsborough became and remained an able and energetic lawyer, and income from his profession gave a substantial boost to his rising fortunes. At the same time, his agricultural operations contributed heavily to his overall success.[5]

4. Arthur L. Jensen, *The Maritime Commerce of Colonial Philadelphia* (Madison, Wis., 1963), 1–10, 77–79; James T. Lemon, *The Best Poor Man's Country: A Geographical Study of Eastern Pennsylvania* (Baltimore, 1972), chaps. 2, 6; Clemens, *Atlantic Economy,* 178–180; Carr, "Diversification in the Colonial Chesapeake," in Carr, Morgan, and Russo, eds., *Colonial Chesapeake Society,* 348–349. Clemens, *Atlantic Economy,* 44–47, characterizes some of the bayside soils as less fertile than does Carr but is in general agreement about the high quality of most upper Eastern Shore soils.

5. Robert Gouldesborough Memorandum Book, 1696–1771, Robert Goldsborough Ledgers,

Goldsborough began keeping two ledgers in 1687, one recording transactions related to his legal practice, and the second contained all other business transactions. The general business ledger reveals that he began farming in Talbot County on a modest scale with a handful of white indentured servants. Already an established lawyer, Goldsborough likely never worked the soil alongside his bound laborers. He probably did spend time in the fields directing their work, as it would take him another eighteen years before he had enough workers to merit hiring an overseer. Goldsborough had connections with several Eastern Shore merchants, but he was seldom inclined to risk his own assets in trade. At one point in the 1690s, Goldsborough tried to cash in on the West Indian market. He bought a sloop and sent a cargo of Eastern Shore cider, meat, and staves to Barbados. Unfortunately, the ship foundered on her return voyage. This loss seems to have ended Robert's mercantile ambitions. Thereafter, he concentrated on planting, and not until 1729 did he again participate in another trading venture to the islands.

Goldsborough, like Robert Cole some years earlier, at first relied heavily on tobacco. His decision was not misplaced, for, one way or another, despite low prices, losses of part of the 1703 and 1705 crops to the French, and the disastrous West Indian venture, he must have done well overall. Farm profits, along with legal fees, enabled him to make extensive improvements to his land through a series of generally bad years for the Chesapeake economy. In 1692, he built four tobacco houses and a twenty-five-foot dwelling house. By 1700, he added two more tobacco houses and a shed. The next year, he built another tobacco barn and a tenant house or quarter. In 1703, he enlarged a kitchen; built three more tobacco houses, a corn house, and a henhouse; erected a more substantial forty-five-foot dwelling; and paled in his vegetable garden. The next year, he repaired some existing buildings and fenced in his hen yard and a bee house. In 1705, he built three new tobacco houses, added another corn house and additional henhouses, and built a sheepfold. The next year, he added two laborers' quarters. By 1708, he needed yet another tobacco

1687–1719, 1718–1742, 1751–1755, Robert Goldsborough Ledger of Legal Fees, 1687–1698, all privately owned, microfilm, MSA; see also Wills, 24: 521, MSA. For a biography, see Edward C. Papenfuse et al., *A Biographical Dictionary of the Maryland Legislature, 1635–1789* (Baltimore, 1979), I, 361–362. I did not try to estimate Goldsborough's legal income, since his fee book covers only the earlier years of his legal career, and one cannot be certain what proportion of fees charged were actually collected. For another discussion of the Goldsborough farm, see Clemens, *Atlantic Economy,* 186–188. The analysis presented here is more complete, since Clemens did not have access to all the Goldsborough materials. For John Gresham, another upper Eastern Shore planter who followed a strategy similar to Goldsborough, see Christine Daniels, "Gresham's Laws: Labor Management on an Early-Eighteenth-Century Plantation," *Jour. So. Hist.,* LXII (1996), 205–238.

house and began assembling timber and bricks for a new frame dwelling for his growing family; the dwelling was built in 1709 at a cost of sixty pounds.

In 1703, Goldsborough began keeping more detailed records of his farming operations, starting a separate crop book, a type of record that became fairly common among large planters by the middle of the century. In it he listed the size of each year's tobacco crop, the number of hands who produced it, the overseer's share, and the names of the merchants to whom the tobacco was consigned and of the vessels that carried it. (Subsequently, Goldsborough also included information on the size of annual grain crops.) By 1703, he was farming the home plantation and had also set up a new quarter on his outlying three-hundred-acre Four Square plantation. There, in the first year, two slaves, Mercury and Isabell, working together without much supervision, grew a spectacular crop of more than three thousand pounds apiece. In addition to enlarging the labor force on the home plantation, Goldsborough later added more workers to the Four Square quarter. By the 1720s, groups of eight or nine enslaved adults toiled on both the home plantation and the Square. In addition, there were two smaller quarters on the main plantation managed by two enslaved men, Tom and Sam, who seem not to have grown much tobacco, but who did plant corn.

Exactly when Goldsborough decided to begin buying slaves is uncertain, since he did not record purchases of laborers in his account book. His wife's dowry apparently included at most one slave, but the Goldsboroughs' 1697 marriage might have been the occasion that prompted him to change the size and composition of his workforce. Clearly, he made no abrupt switch, since he bought at least two English servant boys in the 1690s whose ages were judged and terms of service were set by the county court. Both were only eleven years old, an indication of the scarcity of servants offered for sale in this area. It is also possible that Goldsborough decided to switch from servants to slaves at the outbreak of Queen Anne's War in 1702, when servants were in even shorter supply, slaves were more available, and adequate returns from a short-term investment in servants were made more uncertain by the conflict. Goldsborough could have traveled to Annapolis, Maryland's major slaving port, to purchase new Africans from British slave ships, or he could have bought them from local Eastern Shore dealers who purchased small lots of newly arrived Africans in Annapolis and carried them across the Chesapeake Bay for resale. We know that in 1703 he owned at least two slaves and at least two indentured servants. The backgrounds of the remaining four or five hands who worked his fields between 1704 and 1716 is uncertain, since up to 1716 Goldsborough listed only the name of the overseer and the number of "servants" the overseer supervised in his crop book. Beginning in 1717, he recorded the names of the field hands, and by then all were African slaves.[6]

Goldsborough's decision to acquire a larger and more permanent workforce made it necessary to use subordinate managers, rather than directing all himself. He first employed an overseer at the Square in 1705, when an expanding legal practice required that he be often absent from his Eastern Shore plantations. Most of the overseers were young local white men who served for only a year or so, although Denis Larey, who supervised four workers at the Square in 1709 and 1710, was a twenty-two-year-old freedman who had just finished an eleven-year term of service with Goldsborough. Beginning in 1711, when Robert Lowder settled in for a five-year stint, Goldsborough changed overseers less frequently. In 1716, Lowder was succeeded by Peter Rusham, who in turn managed the Square for seven years. Rusham then left, and, for at least a year or so, Goldsborough put an enslaved man, Jack, who had some prior experience overseeing the laborers on the home plantation, in charge of the Square. By 1727, he had shifted Edward Rusham (perhaps a younger brother of Peter) from the main plantation to the Square, leaving the home farm under the direction of another slave, Anthony. Rusham would remain at the Square for a minimum of five years, through at least 1731.

Goldsborough's accounts are largely silent about how he managed a workforce that initially included both indentured servants and slaves who worked the fields together. The relatively long tenures of Goldsborough's later overseers, coupled with the workers' extraordinarily high production, suggest fewer tensions and less day-to-day resistance to work requirements than was evident on many other plantations of the period. Perhaps Goldsborough made more use of incentives than did many planters; he certainly entrusted a number of the male slaves with more responsibilities than did most slaveowners. At least five of the men, Jack, Anthony, Tom, Sam, and Billy, oversaw one of the quarters for a year or more. Another, Mercury Barrett (almost certainly the Mercury who produced extraordinary crops with little supervision), eventually purchased or was granted his freedom and went on to become a small planter in Talbot County.

Between 1703 and 1720, the enslaved workers on the outlying Square quar-

6. Clemens, *Atlantic Economy,* 61, suggests that Talbot County landowners who could afford to do so began investing more systematically in slaves during Queen Anne's War owing to an increased difficulty in obtaining adequate returns from a short-term investment in servants during years of low wartime tobacco prices. See also Farley Grubb and Tony Stitt, "The Liverpool Emigrant Servant Trade and the Transition to Slave Labor in the Chesapeake, 1697–1707: Market Adjustments to War," *Exp. Econ. Hist.,* XXXI (1994), 376–405.

ter produced on average 1,724 pounds of tobacco per hand, ranging from a high of 3,003 pounds in 1711 to a low of 635 pounds in 1714, a bad growing year throughout the Chesapeake. On the home farm, the average share per hand was lower, 1,365 pounds, or about three-quarters of the usual crop produced at the outlying Square (Tables 16 and 17). Probably this lower share reflects the pressure of other duties that home house laborers were assigned—prizing tobacco, caring for riding horses and sheep, raising bees and poultry, and domestic work for the Goldsborough family.

In the 1690s, when he was still farming with a small number of indentured servants, Goldsborough sold his tobacco in the country to local Eastern Shore merchants. By 1705, when his expanding enslaved workforce was producing larger crops, he chose instead, like elite Virginia planters, to send his crops on consignment to London merchants from whom he got annual shipments of imported goods in return. At first he tried one or more different merchants each year, including Jonathan Scarth, Thomas Perrine, Thomas Bond, and Robert Dunckley, but by 1712 had settled on dealing with just two firms, those of Benjamin Hatley and Thomas Bond. After Hatley's death, Goldsborough turned to two other prominent London traders, John Hanbury and John Hyde. These influential mercantile connections might have facilitated purchases of additional slaves.

In 1721, Goldsborough began to grow less tobacco and to produce more grain. Probably he hoped to lessen his dependence on the traditional staple at a time when future prospects for the weed appeared especially uncertain. Local tobacco prices had been rising since 1714, but the 1720 crop was a poor one. Normally, prices would then have risen; unfortunately, the panic in Europe that followed the bursting of the South Sea bubble seriously depressed the market. As London merchant John Hyde explained to a Maryland correspondent in 1722: "The greatest part of the yeare 1720, no business at all was done but South Sea, and then Towards the close of that yeare, mankinde was generally Struck, with Such a Consternation, that 'till about May last none would Turn their thoughts toward their former business, so that the Greatest part of Two Cropps was on hand, and a third on the Ground."[7]

Production records for 1721–1731 indicate that Goldsborough made a conscious choice to reduce the amount of tobacco he expected the slaves to raise. Between 1721 and 1725, average outputs of tobacco per hand declined from seventeen hundred pounds on the Square quarter, which was devoted primarily to earning export income, and from thirteen hundred pounds on

7. John Hyde to Clement Hill, Feb. 19, 1721/2, Clement Hill Letterbook, 1683–1732, Clement Hill Papers, MS 446, MHS.

TABLE 16. Tobacco and Corn Crops and Estimated Revenues per Laborer, Robert Goldsborough's Home Plantation, Talbot County, Maryland, 1703–1730

Year	Tobacco (lbs.)	Tobacco revenue (£)	Corn (barrels)	Corn revenue (£)	Total revenue (£)
1703	980	3.44	5.00	1.96	5.40
1705	—	—	5.00	1.94	—
1706	—	—	5.00	1.98	—
1707	—	—	5.00	1.88	—
1708	—	—	5.00	1.95	—
1709	—	—	5.00	1.88	—
1710	—	—	5.00	2.03	—
1711	1,220	5.20	5.00	1.82	7.02
1712	1,221	4.66	5.00	2.09	6.75
1713	1,306	4.84	5.00	1.95	6.79
1714	—	—	5.00	2.16	—
1715	1,383	6.08	5.00	1.88	7.96
1716	1,253	6.06	5.00	1.96	8.02
1717	1,853	9.91	5.00	2.37	12.28
1718	1,038	6.61	5.00	2.49	9.10
1719	2,033	10.05	5.00	2.20	12.25
1720	1,364	4.92	5.00	1.66	6.58
1721	390	1.55	7.00	2.92	4.47
1722	774	2.93	—	—	[2.93]
1723	1,095	4.29	—	—	[4.29]
1725	1,278	8.07	—	—	[8.07]
1726	1,256	5.14	11.58	4.05	9.19
1727	870	3.56	—	—	—
1728	941	3.73	12.00	4.58	8.31
1729	636	2.45	13.00	4.61	7.06
1730	861	3.36	—	—	—
MEAN					
1703–1720	1,365	6.18	5.00	2.01	8.22
1721–1725	884	4.21	7.00	2.92	[4.94]
1726–1730	913	3.65	12.19	4.41	8.19

Note: Currency values are in constant pounds sterling. Corn output before 1721 was estimated at five barrels per laborer. Tobacco and corn revenues were calculated by multiplying outputs by tobacco and corn prices for Talbot County in Paul G. E. Clemens, *The Atlantic Economy and Colonial Maryland's Eastern Shore: From Tobacco to Grain* (Ithaca, N.Y., 1980), 226. Brackets indicate the sum is incomplete. The mean total revenue for 1721–1725 is incomplete. It includes corn revenue for only one year.
Source: Robert Gouldesborough Memorandum Book, 1696–1771, privately owned, microfilm, MSA.

TABLE 17. Tobacco and Corn Crops and Estimated Revenues per Laborer, Robert Goldsborough's Square Quarter, Talbot County, Maryland, 1703–1731

Year	Tobacco (lbs.)	Tobacco revenue (£)	Corn (barrels)	Corn revenue (£)	Total revenue (£)
1703	2,880	10.10	5.00	1.96	12.06
1705	759	2.50	5.00	1.94	4.44
1706	1,522	5.34	5.00	1.98	7.32
1707	1,529	4.55	5.00	1.88	6.43
1708	1,568	5.48	5.00	1.95	7.43
1709	1,390	4.57	5.00	1.88	6.45
1710	1,044	3.04	5.00	2.03	5.07
1711	3,003	12.83	5.00	1.82	14.65
1712	2,656	10.15	5.00	2.09	12.24
1713	2,144	7.95	5.00	1.95	9.90
1714	635	2.50	5.00	2.16	4.66
1715	2,615	11.51	5.00	1.88	13.39
1716	1,936	9.36	5.00	1.96	11.32
1717	1,392	7.44	5.00	2.37	9.81
1718	1,226	7.81	5.00	2.49	10.30
1719	2,101	10.39	5.00	2.20	12.59
1720	905	3.27	5.00	1.66	4.93
1721	1,158	4.62	9.00	3.75	8.37
1722	768	2.90	17.69	6.61	9.51
1723	996	3.91	6.27	2.44	6.35
1725	—	—	—	—	—
1726	1,605	6.56	—	—	—
1727	1,757	7.18	15.28	5.35	12.53
1728	1,868	7.40	20.00	7.63	15.03
1729	885	3.42	23.00	7.52	10.94
1730	1,101	4.29	22.40	8.38	15.12
1731	1,094	4.05	23.00	7.52	11.57
MEAN					
1703–1720	1,724	6.99	5.00	2.01	9.00
1721–1725	974	3.81	10.99	4.27	8.08
1726–1731	1,385	5.48	20.74	7.28	13.04

Note: Currency values are in constant pounds sterling. Corn output before 1721 was estimated at five barrels per laborer. Tobacco and corn revenues were calculated by multiplying outputs by tobacco and corn prices for Talbot County, Maryland, in Paul G. E. Clemens, *The Atlantic Economy and Colonial Maryland's Eastern Shore: From Tobacco to Grain* (Ithaca, N.Y., 1980), 226. The total revenue for 1730 and the mean total revenue for 1726–1731 includes 2.45 pounds from 19.10 bushels of wheat.

Source: Robert Gouldesborough Memorandum Book, 1696–1771, privately owned, microfilm, MSA.

the home plantation to about nine hundred pounds per worker on both farms, a drop of more than a third from prior production levels. In subsequent years, Goldsborough continued to raise reduced amounts of tobacco on the home plantation but did return to planting somewhat larger crops at the Square quarter. There were no corresponding increases in the market prices of either corn or wheat that Goldsborough might have taken as a clear signal to shift his crop mix. Rather, he, along with a number of other similarly situated upper Eastern Shore large planters, apparently decided that the best way to increase agricultural income was to increase grain production with his existing workforce with some sacrifice in tobacco revenues. Given the relative shortage of good tobacco land on the upper Eastern Shore and the high proportion of lands well suited to growing grains, a tilt toward grain was a sensible strategy for the long run.

Goldsborough's enslaved workers had long been growing their own food. Not since 1701 had Goldsborough had to buy corn to feed the workforce. When he began keeping records of the size of his corn crops in 1721, the workers were harvesting between six and seven barrels a hand. Once the needs of the white family, plantation livestock, slave children, and nonfield laborers were met, little was left. Within a few years, Goldsborough had the workers double or treble the amount of corn they tended. Between 1728 and 1731, slaves on the home plantation produced twelve to thirteen barrels a hand, and workers on the Square made twenty-two to twenty-three barrels each. Clearly, Goldsborough was growing substantial amounts of corn for the market. Which market, local or West Indian, is uncertain, since he failed to record any grain sales (or to specify that sums received were the proceeds of such sales) in his ledger. No crop records survive for subsequent years.

By 1730, when he recorded a crop of 190 bushels, Goldsborough had made wheat a third market product. His decision to try a third staple, like the earlier expansion into corn, seems to have been in response to a slump in tobacco prices. As early as 1705, Goldsborough had sown small amounts of wheat for his own table. Doubtless he continued to have the workers grow small grain for home use, and he might even have been gradually increasing outputs. His decision in 1730 to start keeping records of the size of the wheat crop, however, suggests that this year saw a real jump into commercial production.[8]

By expanding grain production, Goldsborough nearly doubled gross revenues per laborer. Between 1703 and 1720, tobacco revenues per hand averaged six pounds on the home plantation and seven pounds at the Square quarter.

8. The price of wheat relative to tobacco was high in the late 1720s and mid-1730s; see Clemens, *Atlantic Economy,* chap. 6.

If the workers on both plantations produced the five barrels of corn apiece per year recommended for subsistence, then maize would have added about two pounds additional gross revenues per hand per year. Thus, in the decades during which he concentrated on tobacco, revenues from field crops were on average about eight pounds per hand on the home plantation and nine pounds on the ancillary quarter, where workers did little else besides tending crops. Goldsborough's income did decline for the first two or three years after he decided to expand grain production, a period in which he would have had to put more labor into clearing land and training draft animals to assist in plowing. Incomplete crop records between 1722 and 1725 suggest a drop of at least one pound gross revenues per worker per year. Between 1726 and 1731, however, when Goldsborough had fully effected a transition to increased grain production, total revenues per hand rose to thirteen pounds at the quarter, and probably also increased on the home plantation if, as seems likely, he was growing wheat as well as corn on both farms. If Goldsborough continued grain production at the 1730 level, wheat would have further increased revenues at the quarter to at least fifteen pounds per hand. This was a handsome return that only the most skilled growers of sweet-scented tobacco were able to achieve in Virginia in the 1750s and early 1760s. Contemporary oronoco growers who did not also raise grains as market crops could not match Goldsborough's returns.

Up to 1720, the field hands were already tending the maximum amount of tobacco any worker could handle in a season. The crops of twenty-six hundred to three thousand pounds that some made in the more favorable years were among the highest tobacco outputs per worker ever recorded in the Chesapeake, at least before the use of artificial fertilizers. Goldsborough thus could have increased his tobacco revenues only by buying more slaves and expanding the number of acres in tobacco, and he did not have enough land to make this a viable long-run strategy. Consequently, Goldsborough chose the only means he had for increasing revenues—using the resources he had more intensively. The increased outputs required that the enslaved laborers work harder and sacrifice leisure they might have previously enjoyed in the off-season, since they would have to process grain and prepare more acres for spring planting during the late fall and winter months.

Richard Tilghman (1673–1739) of Queen Anne's County, the second Maryland planter for whom we have a detailed farming record, was the only surviving son of a former London surgeon who had immigrated to Maryland with his wife and first-born children in 1657. From his father (who died in 1676, three years after Richard was born), he inherited all the family lands including the home farm, Hermitage plantation on the Chester River, and two outlying quarters, Forlorn and Dugoes, and shared with his mother and sister a £675 personal estate when he came of age in 1693. Young Tilghman rose rapidly in Maryland political circles. He served his county in the 1690s as a justice of the peace, sheriff, militia leader, and vestryman and eventually served the colony as a provincial court justice, chancellor, or councillor for nearly forty years. By the late 1690s, when Tilghman was only in his mid-twenties, he held more than eighty-five hundred acres, and he would amass more than fifteen thousand acres before he died in 1738.[9]

Tilghman's father had farmed almost exclusively with indentured servants; at his death in 1676, he owned thirteen white servant men and women and only one old enslaved man. As late as the 1720s, the younger Tilghman continued to employ two or three servants and one or two hired free white workers, most of whom were likely skilled tradesmen rather than farm hands. We know that one man was a tailor and another a joiner because Tilghman sold their services to neighbors when these men were not fully employed working for the family. Other servants likely included a shoemaker and a cooper, since Tilghman seldom had to pay free artisans to make shoes or casks.

By 1720, however, Tilghman had turned exclusively to slaves for bound agricultural labor. In 1724, he owned between fifteen and eighteen adult workers, and by 1729 he probably held altogether between thirty and forty slaves, young and old. The blacks on these upper Eastern Shore plantations enjoyed somewhat better health than did many living in tidewater Virginia. On both the Goldsborough and Tilghman plantations, losses to seasoning appear to have been minimal, and already by the 1720s the numbers of the enslaved on both plantations were expanding through natural increase. Indeed, the only bound laborers that Tilghman purchased during the 1720s were indentured servants, not more slaves. Tilghman also divided his workforce between two plantations under the direction of two overseers, and Tilghman's white supervisors, like Goldsborough's, had relatively long tenures. James Backhurst acted as main overseer for ten years (1721–1731). Edward Wright lasted only four years as a quarter manager (1723–1726), but his successor, one of Tilghman's former tenants, Michael Bateman, remained for eight years (1729–1736). The family papers reveal almost nothing about labor relations on the plantation except that in 1726 two of the slaves ran away; they were

9. Ledger A, 1718–1741, and Loose Papers, folder 6, Tilghman Family Papers, 1712–1843, Special Collections, Georgetown University Library, Washington, D.C.; Wills, 5: 120, 22: 35, Inventories and Accounts, 4: 29–36, MSA. For the Tilghman family, see Papenfuse et al., *Biographical Dictionary,* II, 830–831; Christopher Johnston, "Tilghman Family," *MHM,* I (1906), 181–184, 280–284, 369–376.

subsequently recaptured. Perhaps the slaves' flight contributed to Wright's being dismissed at the end of that year.

Tilghman's ledger covers the years 1719 to 1738, but here we will follow the farm only through 1729. Richard had clearly gotten some practical training in bookkeeping as a youth and kept his accounts in standard double-entry form (Plate 18a–b). His use of the heading "Laus Deo" on every page, however, was somewhat exceptional. Such pious expressions were common in sixteenth-century English account books, but they are not ordinarily found in late-seventeenth- or early-eighteenth-century records. Whether Tilghman was simply repeating a formulaic term or whether this practice reflected some deeper sense of being a steward of God's gifts, as Robert "King" Carter professed, is uncertain.

Unfortunately, Tilghman's accounts provide a much less complete record of farm management than does Robert Goldsborough's crop book. They are almost solely a record of local transactions, not of all commercial exchanges. Tilghman, for example, noted the sale of his tobacco crop in only one year, although his accounts with overseers for their shares of the crop make it clear that he always grew large amounts of the weed. Since the Tilghman Family Loose Papers include a number of annual statements of account sent by the English factors to whom the Tilghmans consigned their crops (see, for example, Plates 19a–b), one suspects that these loose accounts might have been the only record of tobacco crops and their proceeds that the Tilghmans kept. In the case of grain, we know that Richard was growing wheat for market only because in five years he listed payments to hired laborers for threshing his crop; otherwise, there is no record of either production or sales. Similarly, corn crops were never accounted for; however, shares of twelve barrels a hand paid to overseers recorded in two years indicate some market production of maize.

Like Goldsborough, Tilghman must initially have concentrated on tobacco. He recorded sales only in 1725, when twenty-nine hogsheads netted £138; this amount included tobacco earned from other sources, primarily rents, in addition to the plantation crop. Tilghman, also like Goldsborough and other large planters, consigned his tobacco to London merchants. During the 1710s and 1720s, his primary factors were Samuel and John Hyde, who became major suppliers to the newly formed French tobacco monopoly in the 1720s. Tobacco crop shares paid to overseers between 1723 and 1729 ranged from 476 to 2,766 pounds; the mean was 1,498 pounds. The account does not make clear whether these were single shares or a share and a half. If they were single shares, Tilghman's slaves produced about 1,500 pounds each; if the overseers got a share and a half, output per slave was about 1,000 pounds, approximately the same size crop as laborers on the Goldsborough farms were making.[10]

By 1722, Tilghman had turned to wheat as a second staple. In that year, his workers grew at least 116 bushels; by 1725, they produced 269 bushels. Tilghman apparently thought he needed a man experienced in raising grain to help further expand production and perhaps to train the workers. In 1727, he hired a white thresher-plowman to work on his farms for one year at the substantial annual salary of ten pounds. Corn was a third market crop, although Tilghman seems to have produced much less surplus than did Goldsborough, perhaps because wheat could be more readily marketed from Tilghman's home on the Chester River, closer to Pennsylvania shipping at the head of the Chesapeake Bay than from the Goldsborough plantations in more southerly Talbot County. Across the 1720s, crops of 36 to 107 bushels of oats are also mentioned, grain that was probably used to feed draft or carriage horses. Although he did not record his grain sales, Tilghman was almost certainly shipping some of his surplus corn to the West Indies; the accounts reveal annual local sales of rum, sugar, and molasses that must represent the returns from regular West Indian ventures.

Tilghman's recorded farm revenues, expenditures, and investments for 1720 though 1729 are shown in Table 18. If 1725, the only year for which tobacco revenues are known, was a typical year, Tilghman earned about one hundred pounds per annum from his tobacco crops, excluding rent tobacco. Tobacco thus probably provided about half of Tilghman's returns. Second in importance were wheat, cider, and rents; each contributed around 15 percent of revenues. In reality, of course, grains made up an even larger share of farm income and tobacco a lesser share, since some corn must have been marketed. Tilghman had extensive orchards that provided a handsome return. He sold between 140 and 1,415 gallons of surplus cider a year. Some went to area farmers, but most he sold to a tenant who in turn retailed the beverage from an ordinary in Queenstown. Occasionally, there were also surplus livestock and dairy products, although the bulk was consumed at home. On average, livestock products supplied about 2 percent of Tilghman's farm returns. Finally, sale of the services of a servant tailor and of furniture made by a servant joiner added an additional 3 percent. Other minor additions came from periodic flatboat hire and from the sale of plank, timber, and shingles.

Even though Tilghman had acquired a relatively large bound labor force,

10. For the Hydes, see Price, *France and the Chesapeake,* 651–654.

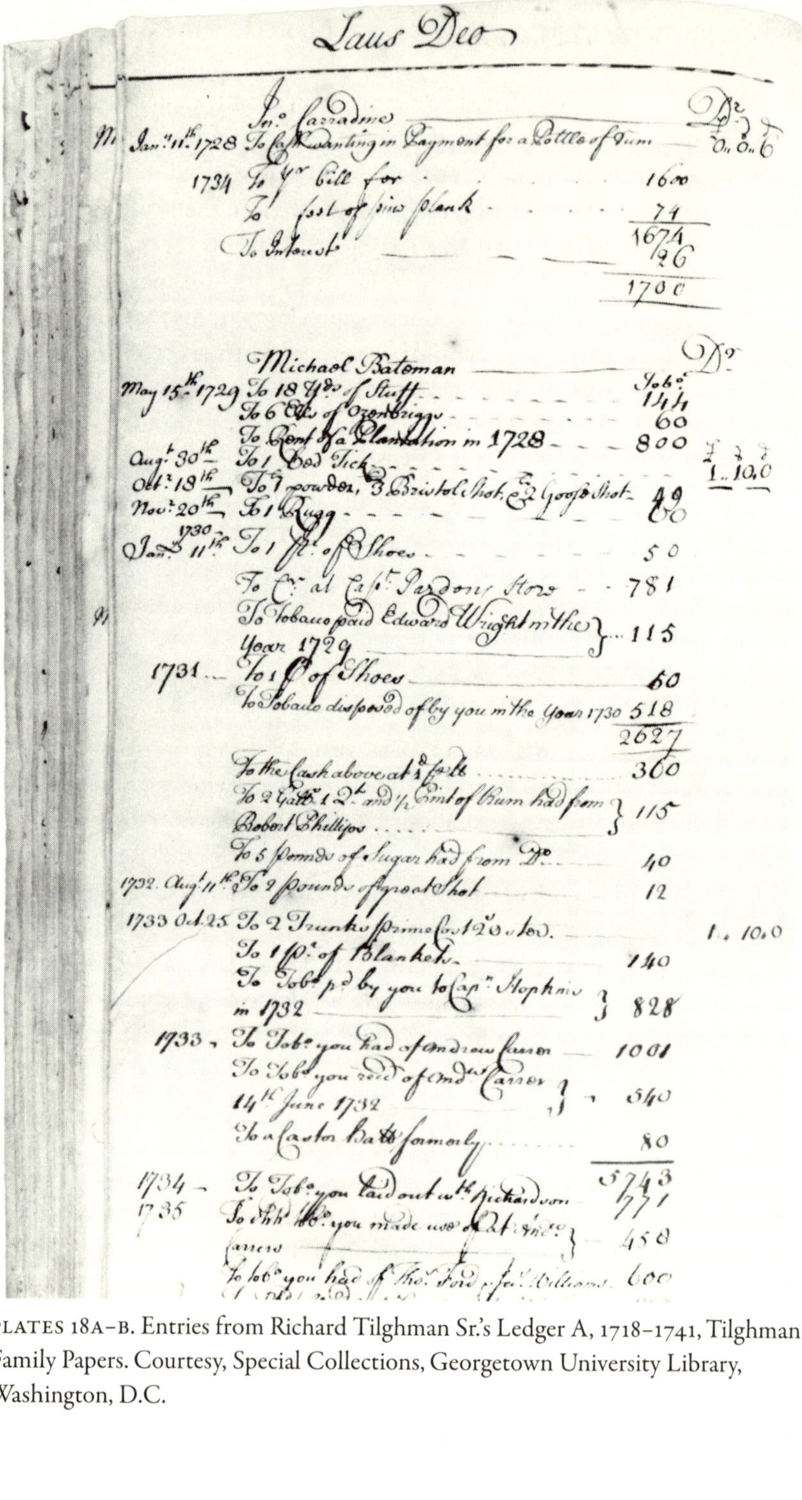

PLATES 18A–B. Entries from Richard Tilghman Sr.'s Ledger A, 1718–1741, Tilghman Family Papers. Courtesy, Special Collections, Georgetown University Library, Washington, D.C.

Dr. Contra Cr.

	By Tob°. paid my Son Richard	1214	
1737 April 18th	By Tobacco at Mr. Jacobs	451	
		1665	
1740	By Richd Tilghman	35	
		1700	

Dr. Contra Cr.

			£
	By 1 Hhd. Tobacco of Weat	852	
1729	By your Share of the Crop this Year	908	
1730	By your Share of the Crop this Year	1270	
		3030	
	By the Cash as p. Cont charged in Tobacco		1..10..0
1731	By your Share of good tobacco this Year	1116	
	By your Share of the ground Leaves	207	
1732	By yr. Share of the good Tob°. this Yr.	981	
	By yr. Share of the ground Leaves	282	
1733.	By yr. Share of the Crop this Year	1347	
		6963	
1734	By yr. Share of this Year's Crop	1538	
1735	By Work done by Your wife formerly		"19"00
	By work done by do. this Year		8"06
	By yr. Share of a Hogd. Tob°. made formerly	113.	
	By yr. Share of the Crop	766	
		9380	
1736 Novr. 1st.	By Work done by yr. Wife to this Day		..8..9
			3..6..3

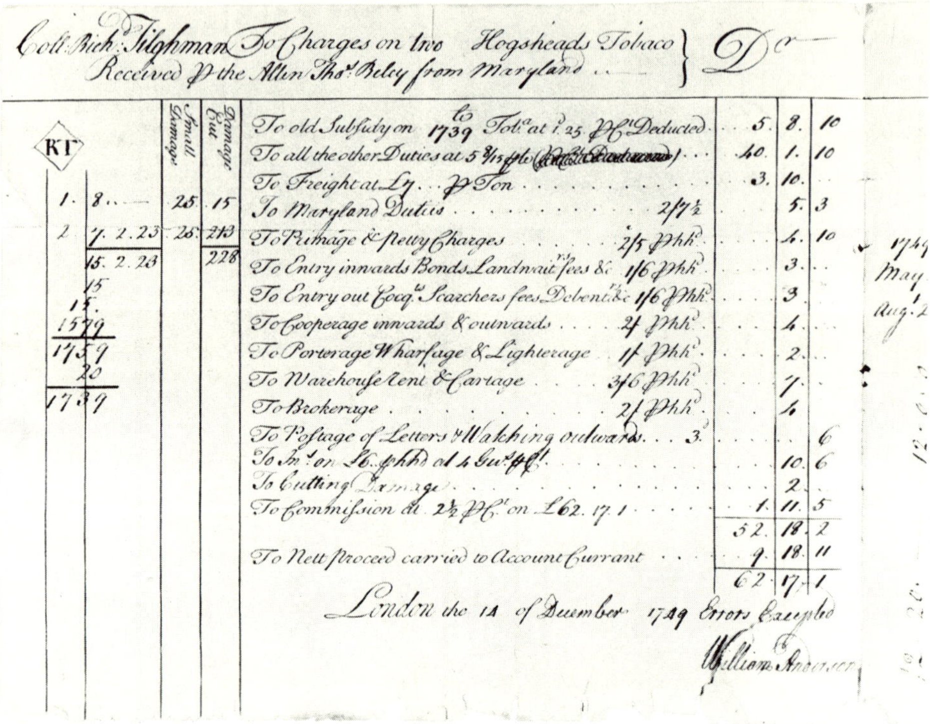

Coll. Rich: Tilghman To Charges on two Hogsheads Tobacco Received ℘ the Allen Thos. Beley from Maryland . . . Dr

To old Subsidy on 1739 Tob.o at 1. 25 ℘Ct Deducted	5.	8.	10
To all the other Duties at 5 1/15 ℘ [illegible]	40.	1.	10
To Freight at £7 . . ℘ Ton	3.	10.	
To Maryland Duties . . . 2/7½		5.	3
To Primage & petty Charges . . . 2/5 ℘ hhd		4.	10
To Entry inwards Bonds Landwait. fees &c 1/6 ℘ hhd		3.	
To Entry out Cocq.t Searchers fees Debent. &c 1/6 ℘ hhd		3	
To Cooperage inwards & outwards . . . 2/ ℘ hhd		4.	
To Porterage Wharfage & Lighterage . . 1/ ℘ hhd		2.	
To Warehouse Rent & Cartage . . . 3/6 ℘ hhd		7.	
To Brokerage . . . 2/ ℘ hhd		4	
To Postage of Letters & Watching outwards . . . 3			6
To In.s on £6. ℘hhd at 4 Gu.s ℘Ct		10.	6
To Cutting Damage		2.	
To Commission at 2½ ℘Ct on £62. 17. 1	1.	11.	5
	52.	18.	2
To Nett Proceeds carried to Account Current	9.	18.	11
	62.	17.	1

London the 14 of December 1749 Errors Excepted

William Anderson

PLATES 19A–B. Account of London Tobacco Merchant William Anderson with Richard Tilghman, Jr., 1749, Tilghman Family Papers. Courtesy, Special Collections, Georgetown University Library, Washington, D.C.

he still had too few workers to improve the bulk of the more than ten thousand acres he owned. Perhaps much of the land was being banked for future inheritances for his children. Modest development through white tenant farmers would augment the value of such reserved land while preserving most of its productive potential for the next generation. In the seventeenth century, many planters in both Virginia and Maryland who owned more land than they farmed themselves turned to developmental leasing as an inexpensive way to get land cleared for planting and some basic structures built on it. This practice became less common in the eighteenth century, when there was little undeveloped land left in older areas and as many large Virginia planters chose instead either to develop all their holdings with slaves or to hold larger undeveloped reserves for their heirs. Tilghman began with two tenants in 1721, and by 1728 he was renting eight farms and the ordinary in Queenstown. Rents brought in between twelve and thirty-four pounds a year across

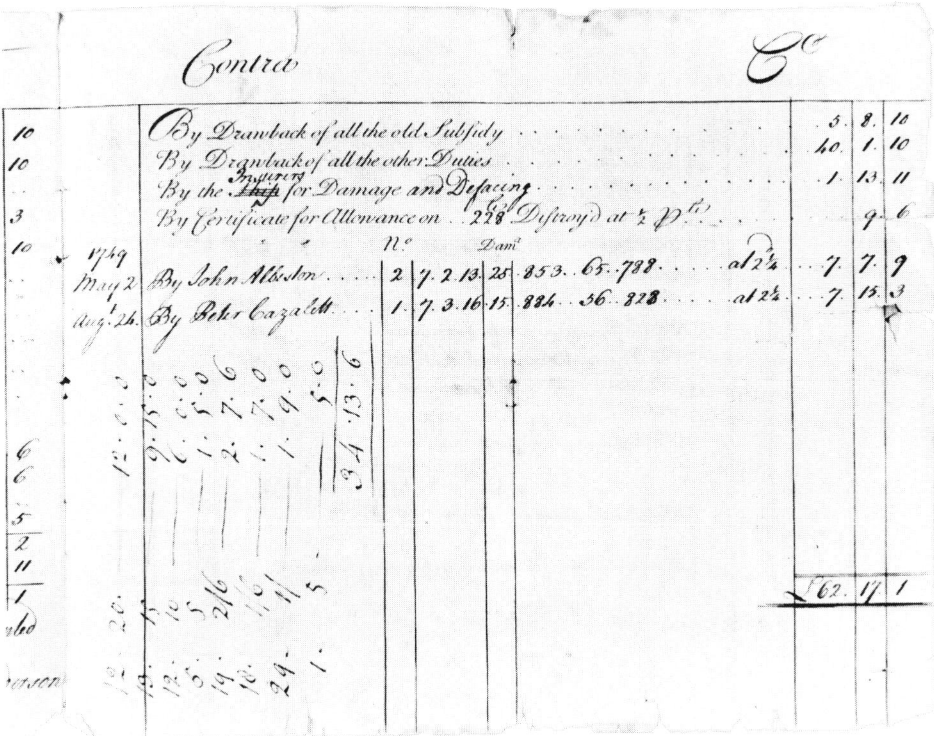
Contra Cr

			No.		Dam.				£	s.	d.
	By Drawback of all the old Subsidy								5.	8.	10
	By Drawback of all the other Duties								40.	1.	10
	By the Insurers for Damage and Defacing								1.	13.	11
	By Certificate for Allowance on 228 lb Destroy'd at ½ P lb									9.	6
1749 May 2	By John Albiston		2	7. 2. 13	25	853. 65. 788	at 2¼		7.	7.	9
Aug. 24	By Peter Cazalett		1	7. 3. 16	15	884. 56. 828	at 2¼		7.	15.	3
									£62.	17.	1

the decade and were fully as important as a source of revenue as was grain or cider.[11]

Tilghman also kept a plantation store and sold goods to tenants, overseers, and neighboring farmers. Because he seldom recorded the cost of the goods imported, it is uncertain how much profit he made from mercantile activities; since the volume of goods sold was relatively small, retailing was certainly not a major source of income. Storekeeping was perhaps as much an expected service as a planned money-making venture; most of the customers lived close by, and many were dependents of one sort or another. Tilghman might have included in his annual invoice to his English factors goods his housekeeper, tenants, and overseers had asked him to order; payments for such special re-

11. Developmental leasing is discussed in Lorena S. Walsh, "Land, Landlord, and Leaseholder: Estate Management and Tenant Fortunes in Southern Maryland, 1642–1820," *Ag. Hist.*, LIX (1985), 373–396; Willard F. Bliss, "The Rise of Tenancy in Virginia," *VMHB*, LVIII (1950), 427–441; Lois Green Carr, "County Government in Maryland, 1689–1709" (Ph.D. diss., Harvard University, 1968), chap. 7.

TABLE 18. Revenues, Expenditures, and Investments on the Richard Tilghman, Sr., Plantations, Queen Anne's County, Maryland, 1719–1729

Category	1719	1720	1721	1722	1723	1724
REVENUES						
Tobacco	—	+	+	+	+	6.45*
Wheat	—	0	0	21.75	0	0
Corn	—	0	0	1.55	0	0
Cider	—	14.84	4.87	21.92	2.51	0
Livestock	—	9.80	14.38	2.44	0	0
Lumber	—	+	0	0	+	0
Rent	—	0	15.99	19.62	27.06	28.90
Hire of servants	—	2.31	1.94	4.75	1.04	2.60
Total	—	26.95	37.18	72.03	30.61	37.95
EXPENDITURES						
Tailor	0	0	0	0	0	0
Spinner/weaver	0	0	0	0	0	0
Leatherworker	0	0	0	0	0	3.91
Cooper	0	0	0	0	0	4.34
Blacksmith	1.05	1.54	0	3.17	0	0
Building repair	0	0	0	9.06	1.13	0
Provisions	0	0	3.53	0	10.04	7.49
Tools	0	0.87	0.43	1.81	0	0
Slave medical	3.12	0	0	8.11	0	0
Hired labor	0	0	0.86	3.03	7.09	0
Supervision	—	—	—	16.26	27.32	20.03
Freedom dues	0	0	0	0	0	0
Slave capture	0	0	0	0	0	0
Total	4.17	2.41	4.82	41.44	45.58	35.77
INVESTMENTS						
Land	0	0	0	0	58.83	0
Buildings	0	0	21.96	0	0	38.26
Cider equipment	0	0	0	0	2.75	0
Boats	0	0	0	0	0	0
Livestock	0	0	0	0	0	10.35
Servants	18.30	0	0	15.39	32.94	0
Total	18.30	0	21.96	15.39	94.52	48.61

Note: Currency values are in constant pounds sterling. An asterisk (*) indicates a partial accounting for that year. A plus sign (+) indicates that the items were present but not valued. Values were recorded in the accounts in both tobacco and Maryland currency. Tobacco was converted to currency using Talbot County, Maryland, tobacco prices in Paul G. E. Clemens, *The Atlantic Economy and Colonial Maryland's Eastern Shore: From Tobacco to Grain* (Ithaca, N.Y., 1980), 226–227, except when a price was available in the accounts. Costs of supervision

1725	1726	1727	1728	1729	Total
138.40	5.36*	9.22*	14.22*	+	173.65
44.58	2.20*	0	0.79*	0	69.32
0	0	0	0	0	1.55
0	34.87	8.99*	6.54*	0	94.54
0	6.13	5.86	0	0	38.61
+	+	0	+	+	+
32.35	33.52	31.48	23.65	11.57	224.14
7.58	4.91	24.79	0	0	49.92
222.91	86.99	80.34	45.20	11.57	651.73
0	0	0	0	3.82	3.82
0	0	0	2.85	0	2.85
5.45	0	0	0	0	9.36
3.95	4.09	0	0	0	12.38
0	0	0	20.54	0	26.30
1.58	0	1.03	0	3.61	16.41
11.38	0.88	5.06	1.21	9.07	48.66
0	0	0	0.79	0	3.90
7.30	0	2.86	0	0	21.39
9.37	2.86	10.98	4.47	0	38.66
26.18	24.19	26.51	20.71	18.30	179.50
7.91	0	0	0	0	7.91
0	2.29	0	0	0	2.29
73.12	34.31	46.44	50.57	34.80	373.43
0	0	0	0	0	58.83
0	12.67	50.48	0	0	123.37
0	0	0	0	0	2.75
0	11.86	0	0	0	11.86
0	6.85	18.52	0	1.46	37.18
0	10.28	0	0	0	76.91
0	41.66	69.00	0	1.46	310.90

include the overseers' shares of tobacco listed in the accounts plus the estimated value of their shares of the corn crops.

Source: Ledger A, 1718–1741, and Loose Papers, Tilghman Family Papers, 1712–1843, Special Collections, Georgetown University Library, Washington, D.C.

quests might, in fact, make up the bulk of the "sales" recorded in the ledger. Otherwise, Tilghman stocked a few staple goods of which neighbors were likely to run short—cloth, thread, buttons, shoes, stockings, hats, blankets, salt, nails, glue, paper, powder and shot, hoes, spades, axes, and an annual supply of a popular local remedy, Bostick's cordial. He also regularly retailed West Indian goods (rum, sugar, molasses, and lime juice), likely the proceeds of grain and lumber shipped to the islands.

Tilghman's accounts permit a more complete reconstruction of expenditures and investments than of revenues. Among recorded outlays, supervision of the workforce topped the list, accounting for nearly half of all expenditures. Next were hired free laborers (10 percent) and provisions (13 percent)—salt, liquor, and the like that could not be produced on the plantation and extra meat and grain that had to be bought to feed teams of carpenters who were putting up houses and barns on the plantation. The major expenses omitted from the accounts were imported cloth and tools for the workforce. The proceeds of Tilghman's tobacco crops (unrecorded except in one year) were returned in the form of clothing, tools, and household goods (also mostly not recorded) designed both for his own family and for the workforce plus additional merchandise that he intended to sell. Consequently, the cost of clothing and tools for the workers cannot be accurately estimated. Table 18 thus shows only a rough approximation of Tilghman's total costs, since some expenses—clothing, quitrents, poll taxes, and tools—were not included in the accounts or were mentioned but not valued, and overseers' wages are missing for some years.

Like Goldsborough, Tilghman invested heavily in plantation improvements. Indeed, over the years of the accounts he spent more on structures (£123) than on indentured servants (£77), additional land (£59), or on livestock (£37). Tilghman had bought all the slaves he thought he needed before 1719 when the account book begins, and he had likely acquired others through his marriage in 1700 to Anna Maria Lloyd, the daughter of a wealthy Eastern Shore planter and merchant. The only bound laborers he purchased in the 1720s were skilled indentured artisans.

Tilghman was almost continually engaged in one building project or another. Most landlords expected tenant farmers to build their own houses and barns in return for a reduction in rent for the first year or so. But, since Tilghman intended his rental farms to be long-term income generating investments, he preferred to build the houses and barns himself and so be assured that the structures would be somewhat more lasting than those tenants were likely to put up on their own. In 1721, he had a tobacco barn and a tenant house built and paid sixteen pounds for other carpenters' and joiners' work.

The next year, he added glass windows and a brick chimney to the kitchen on the home plantation, built an outdoor oven, and made improvements to his rental ordinary in Queenstown. In 1723, he put a stairway in the kitchen and constructed a large cider mill. The next year, he paid an "architect" nearly twenty-seven pounds for 201 days of work on his dwelling house. In 1725, he mended a tenant's tobacco house. The next year, he built a new barn and a granary on the home farm and constructed two new tenant houses. Two years later, he put up another large barn and a corn house, an additional slave quarter or tenant house, and covered a shed. The next year, he spent twenty-four pounds on painting and glazing, surely embellishments to the family's dwelling. Finally, in 1729 he added doors to a barn, blocked up the foundation of an existing house, and put in more windows. The new dwelling Tilghman built in 1724 was sufficiently fashionable to include glass casements. The house is best viewed, though, as part of a larger plan for upgrading all plantation buildings, most of them geared toward generating income rather than display.

A comparison of Goldsborough's and Tilghman's operations between the 1690s and 1729 helps to put both in perspective. The two men's strategies for capital investment differed primarily because they began in very different circumstances. Goldsborough inherited neither land nor labor and thus had to purchase both. He chose to invest relatively more in labor than in land. He bought more — but not a great deal more — land than he could readily develop; his major investment was in bound labor, which he probably expected would produce the quickest returns. Tilghman, in contrast, inherited more land than he could practically develop. In 1694, at age twenty-two, he owned nearly ten times the acreage that Goldsborough would ever acquire, and in subsequent years he nearly doubled his holdings. He, too, had to buy slaves, but he had acquired all he needed to work his established farms before 1719. Nor did he need to buy much additional livestock, aside from horses. Most of his investments were in buildings. Consequently, Tilghman's fortune came more from rising land values and rents than did Goldsborough's.

Had Tilghman followed the practice of similarly endowed planters in Virginia, he would almost certainly have opted to buy more slaves and to put them to work on two or three additional quarters managed by overseers, rather than limiting his enslaved workforce to about twenty adults. We have no way of knowing why Tilghman made the decisions he did. Evidently, he did not think that buying ever-larger numbers of slaves was the best strategy for maintaining and improving his family's fortunes, since he could surely have bought more had he wanted to. Instead, Tilghman limited his bound workforce to no more hands than would staff the home plantation and one quarter, which he could conveniently keep an eye on, if not constantly supervise. He opted to further develop his landholdings by renting improved tenements to free white planting families. Assuming Tilghman could find reliable tenants who could and would pay their rents in full and on time, Tilghman assured himself of a steady annual income without the problems attendant on operating quarters located at an inconvenient distance from the home plantation.

Although their long-term investment strategies differed, in most other ways Goldsborough and Tilghman made strikingly similar management decisions. Both began farming with indentured servants in the early 1690s and then shifted to slave labor. By the mid-1720s, both employed between fifteen and thirty adult slaves and two overseers. Although the upper Eastern Shore was not generally considered a prime tobacco area, Goldsborough and Tilghman seem to have made quality crops. Both normally consigned their tobacco to London factors. Goldsborough dealt primarily with Benjamin Hatley and Thomas Bond. Tilghman shipped primarily to Samuel Hyde and occasionally to John Midford; only in 1725, when the cash price of oronoco reached a record high for the century, did he sell his crop in the country to a Bristol trader. Both lessened their dependence on tobacco in the early 1720s by adding grains to their commercial crop mix. As a result, each of the enslaved workers produced smaller amounts of tobacco, although, since more laborers were added through natural increase, the size of the total crop each man marketed remained relatively constant. The potential tobacco revenue foregone, however, was more than compensated for by profits from selling grains. Goldsborough and Tilghman could have increased their tobacco revenues only by buying more slaves. Goldsborough did not have this option because he did not own enough land to allow him to add more agricultural quarters and still maintain a reserve of fresh land for later generations. Tilghman had enough land so that he could have expanded his workforce, but he chose not to do so. By adding grains to their market crop mix, they could increase plantation revenues without incurring further costs for additional laborers and supervisors or for building new quarters.

Both men also made similar decisions about the management of their slaves and compensation for overseers. They chose to divide their labor forces into small work and residential units of eight or nine adult slaves supervised by an overseer. Such small work units were the norm everywhere along the tobacco coast. Work routines during the growing season and quality control during the crucial postharvest period when the tobacco was processed and packed were sufficiently complicated to require a high ratio of supervisors to workers. Planters who hoped to further increase their total profits per worker

PLATE 20. *An Overseer "Doing His Duty."* By Benjamin Henry Latrobe. Latrobe Sketchbook III. Permission, The Maryland Historical Society, Baltimore. *The overseer is supervising slave women working a field in tidewater Virginia at the end of the eighteenth century.*

could best do so by changing the crop mix to employ all the laborers more intensively throughout the year.[12]

Goldsborough and Tilghman paid their overseers one to one and a half shares of the tobacco and corn crops that they and the enslaved workers assigned to them raised. Additional pay was granted for further responsibilities such as managing more than one plantation. Both men chose their supervisors carefully and well—all of the overseers were able to induce the slaves to make crops of respectable to outstanding size—and most stayed on for relatively long tenures. Overseers' shares and other forms of compensation (corn for their subsistence—usually three barrels but occasionally more if they were

12. Main, *Tobacco Colony,* 128, suggests that small work units might have been a result of decisions to allow cattle and hogs to live off the land and to place the laborers close to the best tobacco lands. To test this interesting hypothesis one would need to identify the location of quarters in counties where both cadastral tract maps for the colonial period and modern farm-by-farm aerial soil surveys are available. Main supplies a good description of conditions on slave quarters in this period (chap. 3).

married — and an allowance of about two hundred pounds of pork or beef a year) were negotiated between planter and prospective supervisor. A modus operandi with the enslaved workers from whom the overseers were hired to extract maximum production was subsequently hammered out day by day in the fields and in the quarters. On plantations where enslaved workers consistently produced average or higher-than-average crops and where overseers enjoyed relatively long tenures, workers and supervisors had likely negotiated agreements about expected work requirements, established entitlements, and discipline. Newly hired overseers who attempted to curtail customary plantation privileges or to change work rules usually met stiff resistance from the slaves and seldom succeeded in making them produce large crops. Exceptionally brutal overseers might increase the intensity or hours of work with the whip, but they usually did not last long in the office either, because masters feared an increase in runaways, injury or loss of valuable human property, and, moreover, did not want to endure a constant barrage of complaints from aggrieved workers.

By the end of the seventeenth century, overseers were customarily compensated with some share of tobacco and corn crops if they were not paid an annual wage. Large planters who diversified their crop mix in the 1710s and 1720s, and who as county magistrates defined and redefined "the custom of the country" from the bench, found therein an opportunity for further profit. Although for some time many middling and large planters had been growing small amounts of wheat, barley, and oats for consumption on the plantation, overseers' contracts had normally not included any share of these minor, secondary crops. Once planters shifted into commercial wheat production, most seem to have managed to cut their supervision costs, for they did not pay their overseers any share of the wheat crop no matter what its size. It was the same with cider. When enterprising planters like Goldsborough and Tilghman turned such — by then — traditional subsistence commodities into commercial products, they were able for a time to force overseers as well as laborers to work harder for no additional reward.[13]

13. For example, Sarah Allaman employed Charles Degge to oversee her Gloucester County, Virginia, plantation in 1707. Degge was to oversee nine slaves and to plant tobacco; cultivate wheat, oats, peas, beans, and whatever else was sown that year; "preserve the stock of cattle, horses, sheep, and hogs; pack and prize the tobacco; house all the fodder that can be gotten"; fence all enclosures; make cider and preserve it; procure firewood for the use of the house; and "use his own Labour and Work . . . if occasion and Necessity shall Require." For these efforts, he was to get one share of the tobacco made, one-half share of the corn produced, plus diet, washing, lodging, and the use of a horse (Francis Lewis Berkeley, Jr., "The Berkeleys of Barn Elms, Planters of Colonial Virginia and a Calendar of the Berkeley Papers, 1653–1767" [master's thesis, University of Virginia, 1940], 77). Cf. York County Deeds, Orders, and Wills, 15: 645, 16: 64, 292–293, microfilm,

Although it is impossible to calculate the revenues or expenditures of either man completely, the surviving documents indicate that both Goldsborough and Tilghman were strikingly successful. Goldsborough, by adding corn and wheat to his mix of staple crops, clearly increased both overall productivity and net revenues per laborer. A partial account of Tilghman's revenues, contrasted with a fuller account of his expenditures, indicates a likely healthy annual surplus. Goldsborough was able to buy additional land and slaves in the 1700s and 1710s, and Tilghman bought more land and indentured servants at a time when the tobacco economy was far from robust, and both men did so without running into debt. In addition, both made considerable investments in buildings and other improvements to their plantations, perhaps in part because they enjoyed relatively long lives and had large families to provide inheritances for but surely also because of their profitable agricultural operations.

There is also evidence of another kind of investment in the future. Tilghman put as much into his children's education—more than one hundred pounds sterling between 1721 and 1724—as he did into houses and barns. Although Goldsborough's expenditures are not so well recorded, he almost certainly spent proportionately. Some years before, Robert Cole's investment in his children's education was probably unusual for a middling planter. After the turn of the century, gentry planters like Goldsborough and Tilghman devoted substantial resources to education as a matter of course.

Both accounts also demonstrate an increasing level of self-sufficiency on large plantations coupled with increasingly dense networks of local exchange. In contrast to the Cole plantation in the 1660s and 1670s, more time was available for things other than clearing land and growing tobacco. Both men had orchards that amply supplied their households, and in Tilghman's case yielded a considerable surplus. Both produced a variety of grains and legumes for market as well as for plantation use. As more grain was grown, they increasingly used plows and carts; more work was thus available for local blacksmiths and wheelwrights. Their wives supervised dairies that supplied their families with butter and cheese, and the Goldsboroughs also kept bees for honey. Both families began to raise sheep for meat and wool. In wartime, when imported cloth was scarce, both employed free white spinners and weavers to make yarn and cloth for plantation use. In addition, Golds-

---

VSL. Similarly, when South Carolina rice planters added indigo as an additional staple, they were able to increase work requirements and erode slaves' customary privileges. Since indigo had no work history in that colony, they forced slaves to cultivate it by the gang-labor method instead of by the task system traditionally employed for rice (S. Max Edelson, *Plantation Enterprise in Colonial South Carolina* [Cambridge, Mass., 2006], 157–162)

borough built tan vats to process slaughter hides for shoe leather, and, since Tilghman paid leather workers for making shoes and saddles in only two years, it is likely that most of the time an indentured servant or a tenant made leather goods for the plantation. Slaves or servants also made most of the casks required for tobacco and cider.

The effect of all these new activities was not increasing isolation; rather, it was increasingly complex links with others in the surrounding community. New enterprises and extensive farm improvements required the services of a wide variety of local artisans. In the 1720s, Tilghman employed one or more of the following: brickmaker, painter-glazier, shoemaker, saddler, spinner, weaver, tailor, plowman, reaper, thresher, boatwright, cooper, carpenter, joiner, architect, wheelwright, blacksmith, tobacco receiver, physician, midwife, schoolmaster, schoolmistress, and dancing instructor. In addition, he helped local artisans by renting small farms to a shipwright and a blacksmith and allowed one skilled tenant to pay his rent by making shoes and building tobacco and dwelling houses. Goldsborough similarly employed a local brickmaker, tanner, shoemaker, spinner, weaver, tailor, thresher, cooper, wheelwright, blacksmith, doctor, and schoolmaster. And both households were large and complex enough to require the services of a full-time free white housekeeper.[14]

The sorts of products that the two families purchased also suggest intensifying local markets. Small planters and tenants found an outlet on the larger Tilghman and Goldsborough plantations for such varied products as corn, wheat, beans, oats, hops, malt, live hogs, pork, horses, sheep, cattle, turkeys, chickens, butter, cider, barrels, timber, and squirrel scalps (which could be delivered to the county court for a bounty on vermin). In turn, the two planters periodically supplied their neighbors with basic English manufactures, butter, pork, bacon, beef, cattle, wheat, corn, and oats, and they occasionally hired out skilled servants and slaves. Such interdependence increased economic opportunities for all free Eastern Shore residents. There was enough of a market for a number of Old World trades that free artisans could make a living from them, and large planters no longer had either to purchase a trained indentured servant or to do without skilled help. Small planters and tenant farmers had incentives for working harder and producing surpluses of products other than tobacco, knowing that neighboring larger planters

14. For the occupational structure of the area, see Jean B. Russo, *Free Workers in a Plantation Economy: Talbot County, Maryland, 1690–1759* (New York, 1989); Christine Daniels, "'Wanted': A Blacksmith Who Understands Plantation Work': Artisans in Maryland, 1700–1810," *WMQ*, 3d Ser., L (1993), 743–767; and Daniels, "Gresham's Laws," *Jour. So. Hist.*, LXII (1996), 205–238.

would buy the overplus. Consequently, they could both increase their family incomes and reduce their dependence on the fluctuating fortunes of a single crop.

The third innovative planter for whom a farm account survives is James Carroll (circa 1680–1729), who lived on the upper Western Shore in Anne Arundel County. Residents of this county farmed soils similar to those on the more tobacco-centered lower Western Shore. They also had increasing contact with alternative trade networks through Annapolis, however, which became Maryland's major port of entry after the capital was moved to this more central location in 1695. Carroll, a member of an exceptionally able, energetic, and close-knit, though impecunious, Irish Roman Catholic family, arrived in Maryland in the 1690s. In Ireland, the Carrolls had once been substantial landowners in the midlands, but they lost most of their holdings in the wars of the 1600s against the English. Carroll was well educated, probably on the Continent, and read French and Latin as well as English. James's uncle, Charles, had immigrated to Maryland in 1688 with almost no property, but by 1720 he had become indisputably the richest man in the colony. James intended to follow his uncle's example.[15]

Carroll's "daybook" is a combination of waste-book and ledger in which overseers and clerks as well as Carroll made entries. Most of the accounts are with individuals and are arranged in standard double-entry form like the accounts of Robert Goldsborough and Richard Tilghman Senior. Carroll also followed, rather unsystematically, the advice of bookkeeping treatises that recommended beginning a waste-book or journal by taking an inventory of all movable and immovable property and financial assets. At the start and scattered throughout the daybook are lists of Carroll's landholdings, slaves, household furnishings, wardrobe, and library. The entries cover only the years 1714 to 1721 (two preceding and at least one subsequent daybook have not survived), but they are especially valuable in documenting a diversified approach to farming, much like that of Goldsborough and Tilghman, adopted in more typical tobacco country. Carroll's accounts, supplemented by an in-

15. The relevant documents are James Carroll Daybook, 1714–1721, Maryland Province, Society of Jesus Archives, Special Collections, Georgetown University Library, Washington, D.C.; Wills, 19: 791, 797, Testamentary Papers, box 35, folder 39, Accounts, 15: 218, MSA; and [Thomas] Colmore Account Book, MS 254, MHS. For Carroll, see Ronald Hoffman, *Princes of Ireland, Planters of Maryland: A Carroll Saga, 1500–1782* (Chapel Hill, N.C., 2000), prologue, chap. 1 (esp. 53, 56), 101–102; and Charles M. Flanagan, "The Sweets of Independence: A Reading of the 'James Carroll Daybook, 1714–21'" (Ph.D. diss., University of Maryland, College Park, 2005). For Anne Arundel County soils, see Carr, "Diversification in the Colonial Chesapeake," in Carr, Morgan, and Russo, eds., *Colonial Chesapeake Society,* 348–349.

ventory taken in 1729, also illuminate the career of a man who relied much less on agriculture for income than did the Eastern Shore planters, a pattern similar to that adopted by great Virginia planters like William Fitzhugh, William Byrds I and II, and Robert "King" Carter.[16]

Since he was a Roman Catholic, Carroll's religion precluded an active political career, although he did advocate for Catholic interests between 1706 and 1720. He also made good use of the sorts of proprietary offices that he was eligible to hold. As deputy surveyor of Baltimore and Anne Arundel counties at the turn of the century, Carroll was able to locate choice tracts of undeveloped land and, indeed, took out warrants for ten thousand acres within a few years, most of which he sold. In 1707, he took charge of keeping the colony's rent roll, a service for which the proprietor rewarded him with a salary and additional land warrants. Carroll also practiced law. Although legislation prohibited Catholic lawyers from practicing in any but the proprietary courts, Carroll might have prepared legal documents and advised Anglican lawyers.

In 1704, Carroll patented four hundred acres in southern Anne Arundel County where he established his home plantation, "Fingaul," a name presumably taken from the seat of a leading Jacobite family in Ireland. There he built a two-story dwelling, commodious for the times, with four rooms on the first floor and two above, an attached study, and a separate kitchen. In some ways, Carroll used consumer goods to assert his social status in defiance of a political establishment bent on eliminating Catholic rights. He dressed in fashionable clothing and furnished his house with imported chairs, gaming tables, bed curtains, pictures, and silver plate. Once he had acquired what he thought necessary, however, he spent little on new furniture in the last fourteen years of his life, aside from the acquisition of more specialized dining equipment and tea and drinking wares.[17]

By 1715, Carroll had purchased eleven adult slaves from five different Maryland planter-merchants and also owned at least five indentured servants, including a joiner and a plowman. He had also accumulated 5,421 acres of land in Anne Arundel and Prince George's counties. Clearly, agriculture figured prominently in Carroll's strategies for building an estate, but commerce contributed even more. He owned town lots in Annapolis (where he began building a town house in the late 1720s), Londontown, and Queen Anne, the last two rented. Like many other large planters, Carroll owned much more land than he had laborers to work. He rented at least three farms and prob-

16. Flanagan, "Sweets of Independence," chap. 1.

17. Ibid., chaps. 3, 6, 7.

ably would have leased more could he have found reliable tenants. By 1717, Carroll was supervising the operation of a store (he called it a "factory") on the Patapsco River for London tobacco merchant Thomas Colmore and also managed the sale of a shipload of 105 slaves brought from Sierra Leone for another London merchant, Samuel Bonham. Carroll subsequently fell out with Colmore (who was likely in financial difficulty), but his probate inventory demonstrates that he continued to sell imported dry goods shipped to him by other merchants. By 1729, Carroll had acquired other land in Baltimore, Somerset, and Calvert counties and was lending money to a number of individuals. Indeed, in 1729 sperate debts accounted for nearly half of his personal estate worth £2,877 sterling constant value.

Carroll for a time operated three plantations, the home farm, "Fingaul" in All Hallows parish, a 1,000-acre second plantation, and another property, the "Out Quarter," twenty miles to the west on a 990-acre tract at Elk Ridge. The home farm, which six to ten adult slaves worked under the direction of an overseer, was a diversified unit producing tobacco, corn, and wheat for market plus oats, barley, buckwheat, beans, malt, cider, brandy, meat, wool, and dairy products for household use. Carroll stocked the farm with horses, cattle, sheep, and hogs and had a tobacco house, granary, and barn built there. Carroll also seems to have followed English styles of husbandry more closely than many Chesapeake planters. He might have acted as an estate manager in Ireland, or might have at least spent time with men who did. Carroll imported English plows, shod cart wheels, sickles, and scythes. He hired an English plowman as his overseer and a German thresher to teach the slaves how to thresh barley and oats. He also hired some Irish laborers to do ditching, scouring, and "hassocking" (cutting and removing dense root systems from swamps), probably in an attempt to create water meadows that would afford livestock better pasture than abandoned old fields or woodlands. These efforts at more intense agriculture had some success. Only in 1715 did Carroll record the size of his wheat crop, which yielded a surplus over seed of eighty bushels. This yield amounts to a little more than eleven bushels per hand, an impressive output for so early in the century. The laborers also harvested fifty bushels of barley, forty-six bushels of oats, and forty bushels of buckwheat. Supposing 1715 was a typical year (his crop mix appears to have been similar in 1729), Carroll derived around 60 percent of his gross farm revenues from tobacco and around 40 percent from grains.[18]

18. James Carroll Daybook, 1714–1721, Maryland Province, Society of Jesus Archives, Special Collections, Georgetown University Library; see also Flanagan, "Sweets of Independence," chaps. 4, 5, 7.

Beginning in 1716, four adult slaves working under an overseer farmed the more traditionally managed Elk Ridge quarter, where only tobacco and corn were grown. Four years later, Carroll calculated that the quarter (like the recently established Loyd plantation discussed previously) was still losing money—capital costs included. Carroll was ahead of his time in the way he assessed the profitability of his investment in the out quarter. For him, plantation expenses included the capital costs of slaves and servants; the expenses of "clothing, tooling, and victualling" the slaves; the costs of putting up a tobacco barn, shed, and granary; and the expenses of boarding the overseer. Few other planters included capital costs in their calculations or considered food consumed on the plantation as a form of revenue. Difficulties in finding reliable overseers probably persuaded Carroll to sell off the out quarter after four years. Crops per hand on all the plantations were only "ordinary"—just over one thousand pounds of tobacco and seven to eight barrels of corn. The accounts indicate, however, that Carroll was grossing between eight and thirteen pounds sterling per hand at the quarter and between thirteen and eighteen pounds per hand on the home plantation, returns equal to or greater than many more established Virginia planters realized in the 1730s and 1740s. Like Goldsborough and Tilghman, Carroll consigned his tobacco to leading London merchants Micajah Perry, John Hyde, and Robert Tyler, in addition to Colmore, expecting to achieve higher returns in England than he could get by selling in the country.[19]

Carroll's labor force followed a transition typical for Maryland plantations from a mix of servants and slaves in the 1710s to all enslaved by 1729. At first, indentured servants worked both as agricultural laborers and as craftsmen. Carroll normally employed one or more overseers (at least one was a relative whom Carroll presumably encouraged to emigrate from Ireland) and secured an experienced indentured servant plowman to handle his wheat and likely to instruct the slaves in raising this English grain. At his death, Carroll owned no indentured servants, and Tom, an enslaved carpenter and cooper, was his only craftsman. Interestingly, although Carroll sold cargoes of new African slaves, only once did he buy new Africans for himself, borrowing the seventy-eight pounds he paid for two men and a woman from a fellow lawyer and planter. The rest of the slaves he purchased appear to have been already well-seasoned hands; several were thirty years of age when he bought them, and some were married couples with several children. Although a number of

19. James Carroll Daybook, 1714–1721, Maryland Province, Society of Jesus Archives, Special Collections, Georgetown University Library; see also Flanagan, "Sweets of Independence," chaps. 4, 5, 7.

infants died shortly after birth during the seven years for which there are accounts, there seem to have been fewer losses among adults than many planters experienced. Carroll might have maintained his slaves more generously than was the norm. He estimated he spent nine pounds Maryland currency a year on clothing, food, and tools for each adult and five pounds for the "little ones," about twice as much as most other planters' spent for adults and more than double what most allowed for unproductive children.[20]

Carroll died unmarried, and as a devout Roman Catholic he left the bulk of his property, including most of his thirty-five slaves, to the church and made bequests to various relatives and provisions for educating three of his nephews living in Ireland. Unlike most slaveowners of the period, Carroll did pay attention to the slaves' spiritual as well as physical welfare, either instructing them in Catholic dogma himself or arranging for priests to do so when they came to say Mass in his house. In his will, Carroll rewarded such enslaved men "as are honest and have a sense of Christian dutyes" with gifts of whatever of his clothing might be suitable for their station, and he wished all the slaves to be further instructed "in the Christian Doctrine." Carroll's career was strongly influenced by his Catholic upbringing, by prior experience with farming practices in Ireland, and by his extraordinarily strong dedication to the welfare of the Carroll family. According to his great nephew Charles Carroll of Carrollton, James Carroll's strategy emphasized "the sweets of independence, and, as a necessary means of attaining it, a well regulated oeconomy."[21]

Carroll's 1729 inventory and the ensuing account of his estate reveal he had practiced as well as advocated a "well regulated oeconomy." Ninety percent of his £2,877 estate (sterling constant value) was invested in capital goods. More than £1,200 (48 percent of the estate) was in sperate debts (an additional £1,200 of questionable debts were not included in the value of the inventory), and the rest was allocated primarily among slaves, merchandise, livestock, and tobacco crops. Carroll's own debts were "few and small." In Carroll's case, honor, wealth, good management, family welfare, and independence were inextricably linked. He seems never to have questioned a sys-

20. James Carroll Daybook, 1714–1721, Maryland Province, Society of Jesus Archives, Special Collections, Georgetown University Library; Testamentary Papers, box 35, folder 39, MSA; Flanagan, "Sweets of Independence," chaps. 4, 5.

21. Charles Carroll of Carrolton to the Countess d'Auzouer, Sept. 20, 1771, in Ronald Hoffman, Sally D. Mason, and Eleanor S. Darcy, eds., *Dear Papa, Dear Charley: The Peregrinations of a Revolutionary Aristocrat, as Told by Charles Carroll of Carrollton and His Father, Charles Carroll of Annapolis . . .*, 3 vols. (Chapel Hill, N.C., 2001), II, 598; Wills, 19: 791, 797, MSA.

tem in which he gained independence for himself and his family by denying it to those he held in perpetual bondage.[22]

Scrambling for a Living on the Periphery

Planters in places where soils were marginally or not at all suited to tobacco had no choice but to fundamentally restructure the way they made a living during the era of hard times. Estimated average income per laborer from tobacco fell from just two pounds in the late 1670s to less than one pound after 1709. Most planters on the periphery abandoned the weed entirely, as it was quite impossible to clothe their families and meet other unavoidable expenses with such paltry sums. In the late seventeenth and early eighteenth centuries, residents of the lower Eastern Shore and of the area around Norfolk increasingly shifted assets from agriculture to craft, mercantile, and maritime activities. Decedents in Somerset County, Maryland, for example, invested a higher proportion of movable assets in shipping and merchandising than did residents of any other Maryland county and invested the least in bound labor. Landowners who raised no tobacco neither needed nor could afford large numbers of slaves.[23]

Building up great landed estates was also less common in peripheral areas. By the turn of the century, no land was left unclaimed on the lower Delmarva Peninsula that anyone thought worth taking up. Residents of the lower James basin were not quite so constrained, since they had access to undeveloped land farther to the south and, moreover, shared a yet-to-be surveyed boundary with the Carolina proprietary, whose barely functional government welcomed settlers on easy terms. A goodly proportion of this land, however, consisted of swamps or sandy wastes, and most undeveloped tracts that were arable lay far from any navigable rivers. Many peripheral-area settlers chose instead to exploit natural resources on unclaimed commons rather than to patent and to pay taxes on tracts of little worth for farming. A combination of self-sufficient agriculture, local exchange of goods and services, and tapping into intercolonial trades were the way most peripheral residents survived, since they produced little that they could sell in Britain. The strategies mainlanders devised to remedy the woes of the Chesapeake tobacco economy became increasingly irrelevant for all but a few great lower Eastern Shore and lower James basin planters whose store of uncharacteristically well-drained and fertile soils encouraged them to continue growing tobacco.[24]

22. Wills, 19: 791, MSA.

23. Main, *Tobacco Colony,* 80–81.

24. Carr, "Diversification in the Colonial Chesapeake," in Carr, Morgan, and Russo, eds.,

Francis Makemie, an Irish Presbyterian minister who had lived on the lower Eastern Shore since 1684, described the economic situation there in 1705 in a tract directed to newly arrived Virginia governor Edward Nott and to Virginia and Maryland legislators. Local tobaccos were deficient both in quality and weight. Planters with poor land and inconvenient storehouses were ruining the trade for larger, well-situated growers. Some farmers with land unsuited to tobacco had already abandoned the staple, becoming instead "half Tradesmen half Planters," and "many poor People would produce more, by selling sundry things which now turn to little account, than they now make of whole Tobacco Crops." England provided no market for local products:

> *Carolina, Barbadoes, Pensilvania, New-York,* and *New-England* carry from us the little scattered Corn we have among us, they buy up our old Iron, Brass, Copper, Pewter, Hides and Tallow, which we often want and might use our-selves. They carry away our wheat and return it again to us in Bread and Flower, and make us pay for trans porting, grinding, boulting, and baking; but which is worst of all, they prey upon that little money we have in *England* by purchasing Bills of Exchange.

Makemie intended his analysis to be a condemnation of current local practices; it was part of his brief, "For Promoting Towns and Cohabitation," a tract that advocated reducing dependence on tobacco by developing an economy based on mixed farming and crafts. Nonetheless, his observations do reveal that Eastern Shore residents had already gone a long way toward establishing a wide-ranging, if somewhat irregular, trade in provisions outside the dominant staple economy.[25]

Makemie emphasized the absence of assured markets as the main reason

---

Colonial Chesapeake Society, 348–349; Jean Elliott Russo, "'The Interest of the County': Population, Economy, and Society in Eighteenth-Century Somerset County, Maryland" (Ph.D. diss., University of Minnesota, 1999), chaps. 1–2.

25. [Francis Makemie], "A Plain and Friendly Perswasive to the Inhabitants of Virginia and Maryland for Promoting Towns and Cohabitation" (1705), *VMHB,* IV (1896–1897), 252–271 (quotations on 261-262, 264, 266); Hugh Jones, *The Present State of Virginia, from Whence Is Inferred a Short View of Maryland and North Carolina* (1724), ed. Richard L. Morton (Chapel Hill, N.C., 1956), 89. See also Lewis Cecil Gray, *A History of Agriculture in the Southern United States to 1860,* 2 vols. (1932; rpt. Gloucester, Mass., 1958), I, chaps. 6, 9; W. Neil Franklin, "Agriculture in Colonial North Carolina," *NCHR,* III (1926), 539–574; Margaret Shove Morriss, *Colonial Trade of Maryland, 1689–1715,* Johns Hopkins University Studies in Historical and Political Science, XXXII, no. 3 (Baltimore, 1914), 16–18; J. Elliott Russo and John C. Coombs, "The Chesapeake without Tobacco? Slavery and Peripheral Economies in Early Maryland and Virginia" (paper presented at the Social Science History Association, 2006).

discouraging lower Eastern Shore residents from exploiting products other than tobacco and from following trades. He did touch on the problem of labor shortages in the region brought on by the decline in European immigration but mentioned "Negroes" only in passing. Planter elites on the lower Eastern Shore and lower James River had been well supplied with slaves in the 1650s, primarily through connections with Dutch merchants trading to the West Indies and New Amsterdam. When the British government prohibited that trade in 1660, however, direct supplies of slaves to the Eastern Shore dried up, and thereafter planters in need of extra labor had to rely almost entirely on indentured European servants or, among the elite, on slaves whom relatives or other agents purchased for them from dealers on the York or lower James rivers. Before 1688, nearly half of Somerset County decedents who left inventories farmed with one or more indentured servants and no slaves. But, as lower Eastern Shore planters shifted from tobacco, the supply of servants also diminished, since the British merchants who transported them wanted few or none of the agricultural products peripheral-area planters could offer in exchange. Consequently, the transition to enslaved labor was long delayed in Somerset, Accomack, and Northampton counties. In Maryland's census of 1712, Somerset had the lowest proportion of slaves of any Maryland county, and almost no planters held enough enslaved laborers to justify setting up a separate quarter. Lower James planters were more successful after 1660 in establishing consistent ties with West Indian trading partners, and through them and the proceeds of the attendant provisioning trade they continued to acquire small lots of slaves from various of the islands. Indeed, the lower James was the only area in Virginia where middling planters who were not members of the governing elite were able to acquire slaves in any numbers in the early eighteenth century. Most of these workers were employed in lumbering and mixed farming, however, rather than in growing tobacco.[26]

The situation Makemie described had not altered much by 1722 when Thomas Cable (?–1743), an aspiring merchant and planter, came from Lon-

26. Joseph Douglas Deal III, "Race and Class in Colonial Virginia: Indians, Englishmen, and Africans on the Eastern Shore during the Seventeenth Century" (Ph.D. diss., University of Rochester, 1981), 192–210; J. Elliott Russo, "'Fifty-Four Days Work of Two Negroes': Enslaved Labor in Colonial Somerset County, Maryland," *Ag. Hist.*, LXXVIII (2004), 466–492; John C. Coombs, "Building 'The Machine': The Development of Slavery and Slave Society in Early Colonial Virginia" (Ph.D. diss., College of William and Mary, 2004), 92–99; April Lee Hatfield, "Dutch and New Netherland Merchants in the Seventeenth-Century English Chesapeake," in Peter A. Coclanis, ed., *The Atlantic Economy during the Seventeenth and Eighteenth Centuries: Organization, Operation, Practice, and Personnel* (Columbia, S.C., 2005), 205–228; Russo and Coombs, "The Chesapeake without Tobacco?"

don to Virginia intending to settle. He first worked as an agent for London merchant Jonathan Forward, selling white convict servants to small planters living at the heads of the York and Rappahannock rivers in exchange for tobacco. A year later, Cable bettered his fortunes by marrying into the powerful Custis family and settled down to planting on his wife's lands at Cherrystones in Northampton County, Virginia.[27]

Cable's letterbook for 1722–1732 shows what it was like to be a large farmer in an area where low-grade oronoco leaf grown on soils poorly suited to tobacco could no longer provide a living for many small planters. Extremely low prices and a shortage of interested buyers forced many families into a highly diversified subsistence agriculture. Having little with which to buy European manufactures, they clothed themselves with homespun and fed their families on corn, beans, and peas. They bought salt and tools and paid their rents and taxes with the proceeds of a small patch of tobacco or from the sale of whatever other products the land afforded. Orchards supplied large quantities of fruit and cider. Sheep could graze in old fields, while cattle and hogs lived and multiplied in the marshes without any need for supplemental winter feed. From forest and swamp planters gathered myrtle berries for candle wax and cut timber for plank, staves, and shingles. Indeed, by the early eighteenth century lumber products and livestock were at least as important to the local economy as tobacco.[28]

Cable, however, and doubtless most other large neighboring planters did few of these things. Possessed of a tract of good land and sufficient credit to buy some slaves, he continued to produce large quantities of tobacco. He cultivated a sweet-scented strain for the London market but also shipped oronoco grown on his own or other farms to various British outports. In order to save on freight, he prized his tobacco into heavy hogsheads of about one thousand pounds gross weight and experimented with stemming (sometimes the top leaves and sometimes only the less desirable under leaves). When he could get freight, Cable preferred to ship his tobacco on consignment to

27. Thomas Cable Letterbook, 1723–1733, MS 171, MHS, microfilm, M-1681, CWL; Ralph T. Whitelaw, *Virginia's Eastern Shore: A History of Northampton and Accomack Counties*, 2 vols. (1951; rpt. Gloucester, Mass., 1968), 149, 255.

28. Morriss, *Colonial Trade of Maryland*, 12–19, 70–71; Russo, "Enslaved Labor in Somerset County," *Ag. Hist.*, LXXVIII (2002), 466–492. Robert Beverley, *The History and Present State of Virginia* (1705), ed. Louis B. Wright (Chapel Hill, N.C., 1947), 137–138, noted that the use of bayberries for candles was a recent discovery in Virginia. Michael G. Kammen, ed., "Maryland in 1699: A Letter from the Reverend Hugh Jones," *Jour. So. Hist.*, XXIX (1963), 369, mentions "a berry with which they make in the Eastern Shore green wax very proper to make candles if mixed with tallow."

the prominent London firms of Micajah Perry and Robert Cary, assuming greater risk in the hope of obtaining higher prices in England. Few of his smaller neighbors were willing to run this kind of hazard, and most sold their tobacco in the country to outport or Scottish traders. As Cable explained to Perry in 1727, only with a consistent run of good prices would Eastern Shore planters "be spirited to ship [consign] tobacco." The low reputation of the area's staple, however, was a major handicap. Cable took "all the care possible to . . . make these Tobacco's exactly fine as to their being clear of trash or any Scrubby short Tobo and as to their Cullour Scent and smoak . . . to answer exactly the London Market." Had he been able to conceal the tobacco's place of origin from buyers, he was certain he would consistently have gotten better prices.[29]

Since tobacco did not generate much revenue and Cable's professional income—from a post as naval officer in a district in which few ships entered or cleared and later from the county clerkship—was not large, he cast about for other ways to make money. Unfortunately, by settling on the lower Eastern Shore, he had sacrificed any hopes he might have had of becoming a prosperous native tobacco merchant. Planters in his area grew too little tobacco and almost none of good quality. Although Cable continued to deal in the weed on a small scale, he usually had trouble securing freight for his purchases. When crops were plentiful throughout the Chesapeake, few merchants regularly sent ships south of the Choptank River. The lower Eastern Shore had become a place of last resort, a final chance to fill out a short cargo. Only by slooping the tobacco to the Western Shore could he be sure of freight, and the added transport cost of five shillings per hogshead often made it not worth the effort.

Cable tried, too, to tap into other trades including slaves, West Indies goods, and produce for the mainland. He supplied his brother-in-law's Williamsburg town house and those of some other officials with sheep, oats, and feathers for beds, and he sent wheat, in part his own crop, to other Western Shore buyers. In 1723, he sold two hundred barrels of corn to a ship captain bound for Jamaica. Three years later, he offered to deliver a load of corn to a West Indies trader anchored in the James; at that point, he had five hundred barrels stored at his landing and no buyer for it. The same year, he accepted a consignment of refuse slaves who could not be sold on the York River, but he was able to dispose of them only "after a great deal of trouble and Expense."[30]

29. Thomas Cable to Micajah Perry, June 1, 1727, Cable to Capt. John Jones, Feb. 27, 1727, Cable Letterbook, CWL.

30. Cable to Capt. John Jones, Dec. 16, 1726, ibid.

Evidence from the Cable letterbook and other account books demonstrates that the lower Eastern Shore was capable of producing surpluses of a variety of products besides tobacco, but many planters were discouraged by a lack of organized and certain markets. Cable, for example, might easily have bought more grain for resale. He had few reliable Western Shore contacts, however, aside from his brother-in-law. And he had no regular correspondents in the West Indies, relying instead on itinerant ship captains to conduct his business. The only way to pursue the West Indian trade effectively, he thought, was to buy his own sloop; he intended to do so once a stepson, then apprenticed to an English merchant, was ready to take over as captain. In August 1726, Cable wrote a James River slaver that local planters were eager to buy teenaged (and, hence, relatively inexpensive) blacks—all they lacked was specie. Some could pay in tobacco, but slave dealers did not care to be paid in Eastern Shore leaf. Many more could pay part specie, part corn, and part wheat. Most Eastern Shore planters, though, were not willing to make a trip across Chesapeake Bay to buy slaves unless they could be assured in advance that their grain would find a "certain market." Cable faulted these cautious planters as "very lazy." Cable himself had to scramble to find buyers for Eastern Shore produce. Still, the situation seems gradually to have improved. Lower Eastern Shore naval office returns for 1727 show regular exports of corn, wheat, and staves. By 1730, Cable advised mainlanders that there was a local scarcity of sheep, as demand from the Western Shore had depleted stocks and driven up the price; in 1733, it was the same with oats.[31]

The accounts of Stephen Loyd, John Tayloe, and Edmund Bagge, discussed previously, also reveal an expanding market on the Western Shore for Eastern Shore and lower James River products, but one which was uncertain and highly dependent on personal contacts. Stephen Loyd, for example, dealt once with a merchant of Curatuck, North Carolina, for tar. In 1719, John Tayloe made two trips to the Eastern Shore, collecting corn, wheat, pork, bacon, lard, and myrtle wax for resale along the Rappahannock, and he made "sundry trips to Elizabeth River" for shingles, plank, tar, and tallow. This seems, however, to have been a short-term venture that did not develop into a regular trade. Similarly, Tayloe once turned over the remnants of a cargo of English merchandise to a James River dealer in exchange for a shipment of tobacco and tar. Tayloe refused to accept the cargo on the grounds that the shipment was late and the price of tar had since fallen. He returned the tar, directing that it be sold to "the negro ships below," and demanded

31. Cable to Capt. John Jones, Aug. 22, 1726, ibid.; Peter V. Bergstrom, *Markets and Merchants: Economic Diversification in Colonial Virginia, 1700–1775* (New York, 1985), table 5.4.

a remittance in more vendable pork or West Indian goods. About the same time, Edmund Bagge was getting wheat, oats, corn, pork, tallow, lard, butter, sheep, plank, and myrtle wax from Eastern Shore relatives. Although he sometimes bought West Indian goods directly from ship captains arriving from the islands, Bagge also bought rum, sugar, and locally made cart wheels from a Nansemond County trader.[32]

Across the Chesapeake Bay south of the James, an economy similar to that of the lower Eastern Shore was developing. In Norfolk and Princess Anne counties, Edmund Jenings informed the Commissioners for Trade and Plantations in 1704, most of the land was "not Agreable for tobacco." Little was grown, and even that "is of the worst Esteem, and soe little Value, that discourages the Inhabitants to plant." Unable to buy enough manufactures, the residents were clothing themselves with locally produced wool and leather.[33]

Pitch and tar were the most readily salable James River products. At the turn of the century, the market was haphazard, and planters were unwilling to commit large amounts of resources or labor. Tar, Jenings reported in 1704, was made from knots and pieces of fallen trees that incurred no costs beyond the labor invested in gathering them. If the market were more certain, planters would begin to cut live trees, which would make a better-quality product in treble the quantity afforded by fallen timber. Naval office records confirm Jenings's pessimistic assessment. In 1701, the Lower James district exported only small amounts of corn and staves and no naval stores. Such low levels of exports may not be representative of the whole period, however, as long years of conflict had curtailed a more robust provisioning trade with Barbados and other West Indian islands carried on before the outbreak of King William's War.[34]

The area was well situated both for ship provisioning and for Caribbean trade, and by the 1710s planters and local merchants had begun to forge

32. Stephen Lyde Account Book, 1708–1711, and Accounts and Letters of John Tayloe I and John Tayloe II, 1714–1778, Tayloe Family Papers, MSS1T2118, b1, VHS; Loyde Account Book, 1708–1710, UVa; Edmund Bagge Account Book, 1726–1733, CWL. See also Maryland Mercantile Ledger, 1710–1713, Force Collection, 8D, no. 99, reel 51, LC.

33. "Pitch and Tar in Virginia, 1704," *WMQ,* 2d Ser., III (1923), 209–210. John Bagge reported that near the Carolina border planters made little or no tobacco and that the little produced was of mean quality so that ministers' salaries were paid in tar, pitch, and pork. Consequently, few parishes had pastors (William Stevens Perry, ed., *Historical Collections Relating to the American Colonial Church,* 3 vols. [Hartford, Conn., 1870], I, 316–317).

34. "Pitch and Tar," *WMQ,* 2d Ser., III (1923), 209–210; Bergstrom, *Markets and Merchants,* table 5.4.

better mercantile connections. Both contemporary observers and probate inventories indicate that lower James River planters were by then producing tar, pitch, pork, hides, tallow, butter, myrtle wax, beeswax, plank, clapboards, shingles, wheat, corn, peas, beans, and cider and harvesting animal skins and snake root for export. The bounties that Parliament offered between 1705 and 1729 for tar, pitch, turpentine, resin, and masts might have stimulated production of naval stores. Naval office returns for 1727 show a substantial trade in corn, wheat, pork, pitch, and tar. Norfolk was by then becoming—by Chesapeake standards—a respectable town and was beginning to function as the center of the export and local provisioning trades. At first, Norfolk merchants tapped only the produce of the Elizabeth River. Trade with North Carolina settlers, however, became increasingly important, especially when Carolina began expanding rapidly after the end of the Tuscarora War in 1712. In the 1710s and 1720s, a number of Bermudian merchant mariners set up shop in Norfolk, bringing with them valuable knowledge and connections developed in the intercolonial carrying trade to which these islanders had turned after they stopped growing tobacco in the 1690s. Bermudians, who either imported small, fast-sailing vessels from their former home or brought in island shipwrights to build them locally, displaced New Englanders as the primary carriers of lower Chesapeake Bay produce to markets in the Caribbean and to ports all along the North American coast where by then other emigrant Bermuda merchants often lived. More reliable marketing networks encouraged residents of the periphery to expand production of a variety of alternative products.[35]

35. "Journal of John Fontaine," in Ann Maury, ed., *Memoirs of a Huguenot Family* (New York, 1853), 292–293; Edmund Berkeley and Dorothy S. Berkeley, eds., "Another 'Account of Virginia' by the Reverend John Clayton," *VMHB,* LXXVI (1968), 424; [Makemie], "A Plain and Friendly Perswasive," *VMHB,* IV (1896–1897), 267; John Oldmixon, *The British Empire in America* (1741), 2 vols. (New York, 1969), 450; Thomas J. Wertenbaker, *Norfolk: Historic Southern Port* (Durham, N.C., 1931), chaps. 1, 2; "Miscellaneous Colonial Documents: From the Originals in the Virginia State Archives (Continued)," *VMHB,* XIX (1911), 12–13; Bergstrom, *Markets and Merchants,* table 5.4 (see also tables 5.1, 5.2); Charles George Gordon Moss, "The Virginia Plantation System: A Study of Economic Conditions in That Colony for the Years 1700 to 1750" (Ph.D. diss., Yale University, 1932), chaps. 11, 12. For bounties on naval stores, see John J. McCusker and Russell R. Menard, *The Economy of British America, 1607–1789* (Chapel Hill, N.C., 1985), 179–180. Inventories from Lower Norfolk County recorded in Lower Norfolk County Deeds, 4–9, microfilm, VSL, document increasing production of naval stores. By the 1720s, these products were appearing frequently in inventories. Pitch boiling required expensive "kettles and appurtenances" that cost between sixteen and thirty pounds. Consequently, only large planters produced pitch, whereas anyone could make tar without purchasing special equipment. For a description of agriculture and society in Surry County, see Kevin Peter Kelly, "Economic and Social Development

A Time to Sow and a Time to Reap

Although seventeenth-century planters had concentrated primarily on clearing land and growing tobacco, by the early eighteenth century throughout the tidewater region many began to put more effort into other activities. As more time and attention came to be devoted to ancillary crops and to the care of livestock, and as planters began to make somewhat greater use of the winter months, the Chesapeake's seasonal work calendar became fuller and more complex. In all but the peripheral areas, the rhythms of tobacco growing continued as the base around which other economic and public and private social activities had to be fitted. Once the hard tasks of farm building were completed, farm families had more free time to devote to a variety of income-generating pursuits. Others chose to put more effort into improving their standards of housing, and Eastern Shore residents who cut back on the size of their tobacco crops or left off cultivating it altogether were among the first to build relatively permanent, fully framed dwellings in any number.[36]

The most pronounced changes in the agricultural calendar during this period were in livestock raising. One difference was dramatic. By the late 1680s, horses—scarce and dear only a decade or so before—had multiplied sufficiently so that they were widely affordable and could be found on almost all farms. Though little used for draft, horses made travel increasingly easy for all. The second change was more gradual. In the most densely settled areas, cattle had less and less room to roam and by the 1690s began to require supplemental feeding, most commonly either corn or corn fodder (tops and blades) that could be gathered in September once the tobacco was housed and before the corn harvest proper began. Some large planters like Robert Carter and William Byrd II also began to grow clover and to mow marsh hay for additional food, and others pastured livestock on sprouting wheat in winter. This change was a matter of necessity, not choice. Faunal archaeologists have discovered that during the first half of the eighteenth century cattle decreased significantly in size in the Virginia tidewater. Planters had either to provide extra feed or to reduce the size of their herds, and most favored the first alternative. Similarly, although hogs continued to forage for themselves with great efficiency, in older areas a shortage of forest mast meant that those

of Seventeenth-Century Surry County, Virginia" (Ph.D. diss., University of Washington), 1972. For Bermuda immigrants, see Michael J. Jarvis, *In the Eye of All Trade: Bermuda, Bermudians, and the Maritime Atlantic World, 1680–1783* (Chapel Hill, N.C., forthcoming), chap. 6.

36. Cary Carson et al., "Impermanent Architecture in the Southern American Colonies," *Winterthur Portfolio,* XVI (1981), 135–196. Cf. Beverley, *History and Present State of Virginia,* ed. Wright, 289.

destined for winter slaughter had to be caught and penned in the fall and fattened on corn.[37]

Cattle and hogs continued to provide the bulk of the colonists' protein, especially as large game—deer and wild turkeys—grew scarcer in older areas. As the century wore on, there was more consistent management of domestic livestock. Planters more regularly gelded most male animals before they reached sexual maturity, intentionally preventing inferior stock from reproducing. Castration made male cattle, hogs, and sheep more docile and manageable and enabled males and females of each species to graze together peacefully. It also hastened fattening of animals intended for meat. Slaughter schedules became more systematic. Hogs were butchered and salted in late December and early January, and beeves (cattle intended to be used for meat) were killed, depending on need, either in late fall (when grass-fed animals were fattest) before hog butchering or in late February (when stocks of fresh pork had been used up). The least-promising calves (called "veals") could then be killed in May or September to supply additional fresh meat, but in the form of small carcasses that could be used quickly before the flesh spoiled. In addition to regular seasonal slaughter schedules, cattle were more consistently killed according to a standard age schedule. Archaeological evidence demonstrates that by the 1680s cattle were usually maintained to older ages than in earlier periods, usually for up to five or six years. And with more regular penning and feeding, cows were easier to catch and milk, and more

37. Philip Alexander Bruce, *Economic History of Virginia in the Seventeenth Century* . . . (New York, 1896), chaps. 6–7; [Durand, of Dauphiné], *A Huguenot Exile in Virginia; or, Voyages of a Frenchman Exiled for His Religion; with a Description of Virginia and Maryland, from the Hague Edition of 1687,* ed. Gilbert Chinard (New York, 1934), 117, 119, 122, 152; "A Letter from Mr. John Clayton, Rector of Crofton at Wakefield in Yorkshire, to the Royal Society, May 12, 1688 . . . ," *Force Tracts,* III, no. 12, 18, 25–26, 35; William J. Hinke, ed. and trans., "Report of the Journey of Francis Louis Michel, from Berne, Switzerland, to Virginia, October 2, 1701–December 1, 1702," *VMHB,* XXIV (1916), 21–22, 36–37; Kammen, ed., "Maryland in 1699," *Jour. So. Hist.,* XXIX (1963), 362–372; "Journal of John Fontaine," in Maury, ed., *Memoirs of a Huguenot Family,* 263; Berkeley and Berkeley, eds., "Another 'Account of Virginia,'" *VMHB,* LXXVI (1968), 419; Jasper Danckaerts, *Journal of Jasper Danckaerts, 1679–1680,* ed. Bartlett Burleigh James and J. Franklin Jameson, Original Narratives of Early American History (New York, 1913), 134; Robert Carter Diary, Sept. 14, 1724, June 6, 1727, Robert ("King") Carter Papers, MS 3807, UVa; Louis B. Wright and Marion Tinling, eds., *The Secret Diary of William Byrd of Westover, 1709–1712* (Richmond, Va., 1941), 28, 362, 365, 580; Beverley, *History and Present State of Virginia,* ed. Wright, 291; Jones, *Present State of Virginia,* ed. Morton, 78, 137; "Narrative of a Voyage to Maryland, 1705-06," *AHR,* XII (1906-1907), 329. Tops and blades began to appear in probate inventories in a number of counties during the 1690s, but especially in York County, Virginia. For the decrease in the size of cattle, see Cary Carson et al., "New World, Real World: Improvising English Culture in Seventeenth-Century Virginia," *Jour. So. Hist.,* LXXIV (2008), 42–49.

women took up dairying, at least to the extent of making butter. Poultry raising probably also became more common in this period, supplying another reliable source of protein.[38]

The debris from plantation trash pits throughout older areas demonstrates that one result of more consistent livestock management was that most families increasingly relied on domestic animals for food and spent less time hunting and fishing. Cattle and hogs were controllable, dependable resources that offered relatively large quantities of meat per animal. From the 1680s, most independent landowners and tenant farmers, whatever their wealth level, had enough livestock to maintain a meat diet comparable to that of their richer neighbors. They no longer had to rely on much less predictable supplies of wild game, fish, or shellfish. Slaves, in contrast, always relied more heavily on wild foods to supplement scanty meat rations.[39]

Perhaps only in marginal tobacco areas did farmers now supplement their diets by more than casual seasonal harvesting of oysters, fish, crabs, turtles, water fowl, and small game. Indeed, most marginal areas lay near the mouth of the Chesapeake Bay where the water is most saline. Here families were not only spending less time cultivating tobacco; they also had access to great concentrations of oysters and to large-sized species of marine fish that were less common or entirely absent from the fresher waters that predominate in most prime tobacco country.[40]

38. [Durand], *A Huguenot Exile,* ed. Chinard, 122–124; John Lawson, *A New Voyage to Carolina* (1709), ed. Hugh Talmage Lefler (Chapel Hill, N.C., 1967), 87; Beverley, *History and Present State of Virginia,* 317; Hinke, ed. and trans., "Journey of Michel," *VMHB,* XXIV (1916), 32, 36–37; Carter Diary, Dec. 17–18, 1722, Jan. 9, 1723, Dec. 15, 18, 1724, Feb. 2, 13, 1725, Jan. 24, 1726, Feb. 14, May 29, Sept. 17, Dec. 5, 20, 1727, Carter Papers, UVa; Wright and Tinling, eds., *Secret Diary of Byrd,* 130, 464, 470; Maude H. Woodfin, ed., *Another Secret Diary of William Byrd of Westover, 1739–1741* (Richmond, Va., 1942), 25, 31–32, 38, 72, 93, 102, 111–112, 122, 130, 145, 148, 157, 168, 184; Alan L. Olmstead and Paul W. Rhode, *Creating Abundance: Biological Innovation and American Agricultural Development* (New York, 2008), 286; "Interview with James Freeman, 1712," in H. Roy Merrens, ed., *The Colonial South Carolina Scene: Contemporary Views, 1697–1774* (Columbia, S.C., 1977), 43–44; Henry Michael Miller, "Colonization and Subsistence: Change on the Seventeenth Century Chesapeake Frontier" (Ph.D. diss., Michigan State University, 1984), chaps. 7–8. Slaughter weights of hogs remained constant at about one hundred pounds across the period. If all hogs were then normally killed at an early age, however, there would have been some savings in feed. James Freeman observed that, by penning only the calves, planters could allow milk cows to range freely to gather wild food during the day but could be certain that they would return to the calf pens where they could then be milked (49–50).

39. Miller, "Colonization and Subsistence," chaps. 7–9.

40. Ibid. For the use of game, see [Durand], *A Huguenot Exile,* ed. Chinard, 123–124; Danckaerts, *Journal,* ed. James and Jameson, 123; Beverley, *History and Present State of Virginia,* ed. Wright, 129–133, 292, 309–310; "Letter from Clayton to the Royal Society," *Force Tracts,* III, no. 12, 36; "Narrative of a Voyage to Maryland," *AHR,* XII (1906–1907), 328–335; Hinke, ed. and trans.,

By the 1690s, most big planters also began to raise sheep, an activity that became more feasible once wolves were eradicated and more cleared grazing land was available. At first, sheep supplied more meat than fiber, since the scrawny creatures typically produced only one or two pounds of coarse wool per year. These small animals, however, could be killed for fresh meat in late spring when it was too warm to slaughter larger species. Few women had yet found the time to make thread or cloth; Durand hardly exaggerated when he said in 1687 that "not one woman in the whole country knows how to spin." In most counties in Maryland, for example, before the 1690s fewer than 10 percent of households owned spinning wheels. As international warfare cut off regular supplies of imported cloth, planters began to pay more attention to wool, and housewives began acquiring yarn if not cloth-making equipment. By the 1720s, sheep were washed in April or May and sheared in May or June. Since most enslaved women were then occupied in the tobacco fields and cornfields, slaveowners distributed the wool to the wives and daughters of overseers and tenants to spin and weave into cloth for winter clothing.[41]

Probate inventories demonstrate that everywhere large planters took the lead both in import replacement activities and in introducing new export crops. "These hard times," Robert Carter wrote in 1723, "will Drive us into All the ways of husbandry we are capable of." Small landowners and tenant farmers were much less likely to risk new commercial crops, but they did shift some of their energies from tobacco into crafts, household industry, and a more varied agriculture. Such changes were most common in places where tobacco growing proved no longer profitable, but, even in areas where to-

---

"Journey of Michel," *VMHB,* XXIV (1916), 33–38, 42; Kammen, ed., "Maryland in 1699," *Jour. So. Hist.,* XXIX (1963), 368.

41. [Durand], *A Huguenot Exile,* ed. Chinard, 113; Berkeley and Berkeley, eds., "Another 'Account of Virginia,'" *VMHB,* LXXVI (1968), 419; "Letter from Clayton to the Royal Society," *Force Tracts,* III, no. 12, 35; Beverley, *History and Present State of Virginia,* ed. Wright, 295; Carter Diary, Feb. 26, 1723, Apr. 18, 1724, Feb. 5–6, 1725, Feb. 19, May 10, 1726, May 29, 1727, Carter Papers, UVa; Wright and Tinling, eds., *Secret Diary of Byrd,* 41, 42, 326, 349–350, 355, 472, 527; Woodfin, ed., *Another Secret Diary of Byrd,* 72, 164; George Melvin Herndon, "The Story of Hemp in Colonial Virginia" (Ph.D. diss., University of Virginia, 1959), chap. 1; Hinke, ed. and trans., "Journey of Michel," *VMHB,* XXIV (1916), 37; "Narrative of a Voyage to Maryland," *AHR,* XII (1906–1907), 336; Jones, *Present State of Virginia,* ed. Morton, 78; "Interview with James Freeman," in Merrens, ed., *The Colonial South Carolina Scene,* 43; David Scott Hardin, "'Alterations They Have Made at This Day': Environment, Agriculture, and Landscape Change in Essex County, Virginia, 1600–1782" (Ph.D. diss., University of Maryland, 1995), 394; Main, *Tobacco Colony,* 62–63, 72–74; Henry M. Miller, "An Archaeological Perspective on the Evolution of Diet in the Colonial Chesapeake, 1620–1745," and Carr, "Diversification in the Colonial Chesapeake," both in Carr, Morgan, and Russo, eds., *Colonial Chesapeake Society,* 183–190, 348–349; Olmstead and Rhode, *Creating Abundance,* 294.

bacco remained the dominant staple, if there was still arable land available for patenting or renting increasing numbers of young, native-born householders just getting started offered a market for basic livestock and foodstuffs. In Charles County, Maryland, for example, in the 1680s and 1690s only about seven new households were formed each year. By the 1710s, the number rose to around seventeen, and by the mid-1720s it increased to more than twenty a year. In this tobacco-centered county, few small planters ventured into new kinds of crops or were able to make a living solely as artisans. Still, population growth afforded small planters a market for short-term labor and surplus cattle, hogs, corn, and cider, which both supplemented income earned from tobacco and provided some hedge against downturns in staple prices.[42]

Large planters everywhere began as well to grow small grains in the 1680s and 1690s. Because there was little demand for wheat or oats in overseas markets, most produced only enough to supply their own tables and to feed their own horses. Normally, wheat was sown broadcast on old fields in October and November and harvested at the end of June. All family laborers and any available neighbors were recruited when the grain became ripe, and harvest took at most only one to three days. The grain was then cleaned when there was spare time between June and September. Oats were sown in March and harvested in late June or early July, just after the wheat was in. A few farmers also tried barley, but low yields discouraged most. Some planters, especially those living on the upper Eastern Shore where markets were more certain and levels of production higher, had to spend more time preparing land for grain and, subsequently, harvesting and cleaning. Robert Goldsborough and Richard Tilghman, for example, managed to harvest their grain crops with-

42. Carter to [Thomas] Evans, Aug. 3, 1723, Robert Carter Letterbook, 1723–1734, Carter Papers, UVa; Carr, "Diversification in the Colonial Chesapeake," in Carr, Morgan, and Russo, eds., *Colonial Chesapeake Society,* 342–388; Main, *Tobacco Colony,* chaps. 2, 5, 6; Lorena S. Walsh, "Estimated Charles County, Maryland, Population, 1642–1755," MS in possession of the author, and SPSS file of Charles County probate inventories, 1706–1720, in Historic Saint Mary's City Probate Inventory Database, MSA. The Historic Saint Mary's City Probate Inventory Database, which consists of numerically coded summaries of more than ten thousand Maryland and Virginia probate inventories, was compiled with support from Historic Saint Mary's City, by grants to Historic Saint Mary's City from the National Science Foundation (GS-32272) and the National Endowment for the Humanities (RO-228-72-468, RO-10585-74-267, RS-23687-76-431, and RS-20199-81-1955), by a postdoctoral fellowship given to Lorena S. Walsh by the Regional Economic History Research Center, Eleutherian Mills–Hagley Foundation, and by an Arthur C. Cole research grant to Walsh from the Economic History Association. The file is described in the notes and appendixes to Lois Green Carr and Lorena S. Walsh, "Changing Lifestyles and Consumer Behavior in the Colonial Chesapeake," in Cary Carson, Ronald Hoffman, and Peter J. Albert, eds., *Of Consuming Interests: The Style of Life in the Eighteenth Century* (Charlottesville, Va., 1994), 59–166.

out hiring additional labor, but Tilghman's slaves were sometimes so busy finishing the tobacco crop that he had to hire white laborers to thresh his small grains.[43]

Orchards also required additional time and effort. Most ordinary planters raised at least apples, pears, and peaches, and all could gather wild fruit from nearby woods. Large planters like William Byrd II, Robert Beverley, and Robert Carter cultivated a much wider variety of fruit trees. Early summer cider was made in June, whereas the longer-keeping winter cider was processed in August and September. Any cider remaining into the winter (a rarity in most thirsty Chesapeake households) had to be bottled in February. Stilling for brandy could begin as early as July. Meanwhile, cherries came in May, apricots in June, pears in July, and peaches, nectarines, figs, and grapes in August. Those who attempted wine began brewing in September.[44]

Finally, timber was cut beginning in November. Wherever new fields were required, land clearing was the first step toward spring planting. Planters (especially those living in long-farmed areas and in places where colonists were processing timber for export) might have begun, when clearing new

43. "Letter from Clayton to the Royal Society," *Force Tracts,* III, no. 12, 20; Jones, *Present State of Virginia,* ed, Morton, 137; Hinke, ed. and trans., "Journey of Michel," *VMHB,* XXIV (1916), 32, 114; Carter Diary, Sept. 25, 1722, Aug. 24, 1723, June 6, July 10, 1727, Carter Papers, UVa; Wright and Tinling, eds., *Secret Diary of Byrd,* 15, 54, 196, 365, 500, 545–546, 548, 580; Woodfin, ed., *Another Secret Diary of Byrd,* 79–80, 83, 169, 171–172, 180; Gray, *History of Agriculture,* I, chaps. 8, 9. Henry Hartwell, James Blair, and Edward Chilton, *The Present State of Virginia, and the College (1697),* ed. Hunter Dickinson Farish (Charlottesville, Va., 1940), 9, noted that planters "commonly provide only as much as they expect they themselves shall have Occasion for, for the Use of their Families, there being no Towns or Markets where they can have a ready Vent for them [grain and pulse]." [Durand], *A Huguenot Exile,* ed. Chinard, 116, stated, "The farmers reap only about a bushel of wheat each on their plantations for making pies, because of the great abundance of game and apples which make very good pies also." Beverley, *History and Present State of Virginia,* ed. Wright, 292, indicates that labor costs as well as lack of markets limited wheat growing to the rich. Many poorer people "don't mind to sow the Ground, because they won't be at the trouble of making a Fence particularly for it." Similarly, he noted, "It is thought too much for the same Man, to make the Wheat, and grind it, bolt it, and bake it himself" (316). For the slaveowners' need to hire harvest labor for wheat when slaves were fully employed in tobacco, see Daniels, "Gresham's Law," *Jour. So. Hist.,* LXII (1996), 205–238.

44. "Journal of John Fontaine," in Maury, ed., *Memoirs of a Huguenot Family,* 264–265; [Durand], *A Huguenot Exile,* ed. Chinard, 116, 127, 152; Danckaerts, *Journal,* ed. James and Jameson, 135; Beverley, *History and Present State of Virginia,* ed. Wright, 314–316; "Miscellaneous Colonial Documents," *VMHB,* XIX (1911), 12–13; Carter Diary, Aug. 15, 30, Sept. 7, 22, Dec. 6, 1722, Aug. 9, 1723, July 10, 24, Aug. 18–19, Sept. 14, 1724, Feb. 17, 1725, Carter Papers, UVa; Wright and Tinling, eds., *Secret Diary of Byrd;* Moss, "The Virginia Plantation System," chap. 10; "Narrative of a Voyage to Maryland," *AHR,* XII (1906–1907), 329; Hinke, ed. and trans., "Journey of Michel," *VMHB,* XXIV (1916), 33; Jones, *Present State of Virginia,* ed. Morton, 78; "Maryland in 1699," *Jour. So. Hist.,* XXIX (1963), 370.

fields, to fell larger trees for lumber, fencing, and firewood rather than simply leaving girdled trees to fall and rot. Cut timber had then to be hauled off the new fields, and enough had to be split before spring planting began to make rail fences to surround the crops. Those who started raising wheat for market might also have favored more thorough clearing techniques, since fallen trunks might still impede plowing when land initially planted in tobacco and corn was used several years later for wheat. Much wood was required as well just to keep the fireplaces stoked, but the best cut timber was often saved to be processed into plank, shingles, and staves when bad weather precluded work in the fields.[45]

The duration if not the pace and intensity of work likely also increased. This aspect is difficult to document, as most accounts of the period are those of elite men who did not have to curtail leisure. The voices of those who were made to sacrifice the most—enslaved workers—are entirely missing from the written record. Absent as well is the perspective of small planters, their wives, and their children, who might have been increasing levels of self-exploitation in order to stay even. Indentured servants' complaints of overwork diminished in proportion as their numbers decreased. By the 1710s and 1720s, most remaining European servants were skilled artisans, who were likely exempted from any ratcheting up of workloads, or adolescent, female, or Irish workers, who were serving according to the custom of the country. Customary servants were less likely to bring formal complaints to county courts than had the somewhat older English youths, more of whom had written indentures, who had immigrated in the 1660s and 1670s. As a result, it was mostly creole orphans and a few immigrant European wage laborers who complained about the elimination of holidays and cutbacks of leisure time to which they had been accustomed at home or to which they thought themselves entitled.

Response to Changing Markets

One of the main differences between the period 1680–1729 and earlier years was that everywhere declining tobacco prices led to falling revenues, and by then few ways remained to further cut the costs of tobacco production. The prices that could be obtained locally for grain and meat stopped falling in the 1690s, an advantage for more established planters producing a surplus, but the cost of manufactured goods relative to tobacco continued to

45. Carter Diary, Nov. 21, 22, 1722, Nov. 11, 1724, Dec. 23, 1726, Dec. 5, 1727, Carter Papers, UVa; "Interview with James Freeman," in Merrens, ed., *The Colonial South Carolina Scene,* 44–45; Woodfin, ed., *Another Secret Diary of Byrd,* 27, 120; Timothy Silver, *A New Face on the Countryside: Indians, Colonists, and Slaves in South Atlantic Forests, 1500–1800* (Cambridge, 1990), 105–106.

rise until about 1710. Meanwhile, labor costs escalated. Indentured servants began to cost more in the late 1670s, and slave prices rose after 1690. The price of prime land was rising, too, and choice acreage was becoming very hard to obtain. Most of the best tidewater tobacco lands were taken up between the 1660s and 1690s. Once improved, such tracts did not often come on the market, as "special good" land was reserved within families as inheritances for subsequent generations. Moreover, in Virginia entailing land on particular heirs prevented a significant proportion of patented acreage from ever coming on the market, thus raising the price of the remaining land that was not so encumbered. One scholar has estimated that by 1730 between one-quarter and one-third of the land in tidewater Virginia was entailed in particular family lines and could not be sold unless heirs obtained a special act from the legislature. In many parts of the tidewater, inexpensive land was still available even in older areas, but either the soils were marginal or the tracts lay far from any navigable streams. Opportunities for anyone not starting out with capital or inherited assets were thus sharply reduced.[46]

One factor contributing to the fall in tobacco revenues between 1689 and 1714 was higher transportation costs. Although planters could do little about high wartime freight rates and insurance charges, they could reduce costs somewhat by increasing the weight of their hogsheads, since charges continued to be levied on volume rather than on weight. The planters' response was uniform throughout the region. Freight rates fell slightly in the 1680s, and planters remained satisfied with the usual four-hundred- to five-hundred-pound container. Once war broke out in 1689, however, freight rates doubled, reflecting both the risk of capture by the enemy and fewer ships coming for the crop. Planters responded both by making bigger hogsheads (those in Virginia disregarded laws stipulating legal dimensions, and those in Maryland acted with lawmakers' blessings) and by packing tobacco more tightly. With the onset of peace (1697–1701 and after 1714), unit freight charges dropped sharply. But freight costs crept gradually upward again as merchants and shippers insisted on hogsheads of uniform size and shape and planters stopped packing their casks so tightly.[47]

46. Russell R. Menard, "The Tobacco Industry in the Chesapeake Colonies, 1617–1730: An Interpretation," *Res. Econ. Hist.,* V (1980), 109–177; Holly Brewer, "Entailing Aristocracy in Colonial Virginia: 'Ancient Feudal Restraints' and Revolutionary Reform," *WMQ,* 3d Ser., LIV (1997), 307–346.

47. Menard, "The Tobacco Industry," *Res. Econ. Hist.,* V (1980), 109–177; Russell Robert Menard, "Economy and Society in Early Colonial Maryland" (Ph.D. diss., University of Iowa, 1975), appendix IV, table A-10; John M. Hemphill II, "Freight Rates in the Maryland Tobacco Trade, 1705–1762," *MHM,* LIV (1959), 36–58, 153–187; Stephen Gregg Hardy, "Trade and Eco-

Large Virginia planters led the way in increasing the weight of their hogsheads, perhaps in part because freight rates tended to run higher in that colony than in Maryland. During King William's War, they upped the weight of the average Virginia hogshead by one hundred pounds (Table 19).[48] In the

---

nomic Growth in the Eighteenth-Century Chesapeake" (Ph.D. diss., University of Maryland, 1999), 439–456; James F. Shepherd and Gary M. Walton, *Shipping, Maritime Trade, and the Economic Development of Colonial North America* (Cambridge, 1972), 49–90, 191–192. The Virginia legislature clamped down on planters who were using casks that were not built according to statutory dimensions in 1686, 1691, 1704, and 1705 (William Waller Hening, ed., *The Statutes at Large: Being a Collection of All the Laws of Virginia . . .*, 13 vols. [1809–1823; rpt. Richmond, Va., 1969], III, 51–53, 435–440; Waverly K. Winfree, comp., *The Laws of Virginia, Being a Supplement to Hening's "The Statutes At Large," 1700–1750* [Richmond, Va., 1971], 19–22, 37–38). Maryland's legislature passed laws allowing larger hogsheads than those of Virginia in 1694, 1699, 1700, 1704, 1707, and again in 1715, arguing that Maryland oronoco tobacco could not be packed as tightly as that of Virginia without damage. The Board of Trade, however, backed by the London tobacco merchants, forced the Maryland lawmakers to acquiesce in requiring casks of equal size to those adopted in Virginia (Morriss, *Colonial Trade of Maryland,* 97–99).

48. Before the implementation of the 1730 Virginia tobacco inspection act, there are problems in determining the average number of pounds of tobacco that a typical tobacco hogshead contained. Some sources list three different figures for each hogshead—*gross* (the weight of both the tobacco and the cask), *tare* (the weight of the empty cask), and *net* (the weight of the tobacco only). Others record only one figure, indicating whether it is a gross or net weight, but not all sources do. The number of pounds of tobacco growers could pack into a hogshead varied owing to differences in the size and density of leaves of the sweet-scented and oronoco strains, to whether all leaves were left whole or some or all were stripped from the stalk and damaged parts discarded, and especially to whether the hogshead was packed by hand or with a mechanical prize. In the seventeenth century, the weights of individual hogsheads might vary by as much as 400 pounds. By the first third of the eighteenth century, when large planters and merchants began using prizes, the range was as much as 900 pounds.

When tare began to be frequently recorded in the early eighteenth century, empty hogsheads usually weighed between 80 and 110 pounds. However, casks could weigh as little as 65 pounds and as much as 135 pounds. Some could have been larger or smaller than the legally specified size, but variations in weight could also have been due to the kind of wood used, to how well the timber was seasoned, to the thickness of the staves and heading, and to the number of hoops used to hold the cask together (Wayne Randolph, Rural Trades, Colonial Williamsburg Foundation, personal communication, Jan. 9, 2009). Incorrectly classifying unspecified hogshead weights as net or gross could result in overestimating or underestimating the weight of the tobacco by as much as one-quarter. Because of the wide variation in net weights created by the use of different packing methods after 1700, coupled with smaller variations in the weights of empty casks, it is not always possible to determine whether recorded weights are gross or net when this is not specified in the document. When observations are numerous, ambiguous entries can simply be omitted from calculations. But data are few for much of the seventeenth century. Separate comparisons (available from the author) of the average weights of hogsheads specified as gross and net with the average for those not specified show that most unspecified observations were probably net weights or a mixture of net and gross but not so biased upward to seriously distort the results presented in Table 19.

TABLE 19. Weights of Tobacco Hogsheads, Virginia and Maryland, 1680–1729

Years	Hogsheads (*N*)	Average pounds per hogshead (% increase)	Low pounds	Median pounds	High pounds
VIRGINIA					
1680–1688	116	486 (−4*)	351	510	700
1689–1696	203	613 (26)	284	620	707
1697–1701	147	621 (1)	308	559	677
1702–1713	660	863 (39)	420	964	1,219
1714–1720	554	604 (−30)	277	522	937
1721–1729	92	760 (26)	429	774	1,350
MARYLAND					
1680–1689	95	401 (1*)	—	—	—
1690–1699	19	413 (3)	—	—	—
1702–1713	228	604 (46)	493	563	800
1714–1720	463	618 (2)	450	616	824
1721–1729	243	780 (26)	400	748	990

Note: 1689–1696 and 1702–1713 were years of war. The other groupings are years of peace. An asterisk (*) indicates an increase over 1670–1679 (see Table 12). The average pounds reported in the second column of the table combines observations for gross (*N*=719), net (*N*=1,784), and unspecified (*N*=203) weights. The last three columns (N=1,987) report the range of weights for each time period, excluding observations specified as gross.

Sources: Edmund Bagge Account Book, 1726–1733, CWL; William Bassett Account Book, 1728–1755, MS1 B2944a, VHS; John Baylor Ledgers, 1719–1755, Baylor Family Papers, 1653–1915, MS 2257, UVa, microfilm, M-62.5, M-62.6, CWL; Robert Bristow Account and Letterbook, 1688–1750, VSL, microfilm, M-1164, CWL; Custis Papers, 1711–1792, VHS, microfilm, M-1073.3, CWL; Stephen Lyde Account Book, 1708–1711, and Accounts and Letters of John Tayloe I and John Tayloe II, 1714–1778, Tayloe Family Papers, MSS1T2118, b1, VHS; Loyde Account Book, 1708–1710, UVa; John Mercer Ledger, 1725–1732, Bucks County Historical Society, Doylestown, Pa., microfilm, M-164, CWL; William Woodford Letterbook, 1723–1737, University of Kentucky, microfilm, M-1060-1, CWL; Ledger A, 1718–1741, Ledger B, 1725–1765, Tilghman Family Papers, 1712–1843, Special Collections, Georgetown University Library, Washington, D.C.; Thomas Colmore Account Book, MS 254, MHS; Robert Gouldesborough Memorandum Book, 1696–1771, Robert Goldsborough Ledgers, 1687–1719, 1718–1742, privately owned, microfilm, MSA. For Maryland, 1680–1699, see Russell Robert Menard, "Economy and Society in Early Colonial Maryland" (Ph.D. diss., University of Iowa, 1975), appendix IV, table A-11.

brief interlude of peace at the turn of the century, everyone left off very tight packing. Merchants' complaints about tobacco damaged in prizing doubtless contributed. In addition, English authorities prohibited shipments of uncasked bulk tobacco in 1699. Thus, inexperienced planters who had not bothered to pack their crops at all now had to do so, and others returned to shipping light hogsheads that they were unable to fill completely.[49]

When war broke out again in 1702, large exporters redoubled their efforts. Any planter could get between four hundred and five hundred pounds of the weed into a cask without using special equipment. Workers apparently compressed the leaf by climbing into the container and pressing it down with their feet, or so the term "foot packt" hogsheads would imply. Tighter packing required a large prize and took more time and labor (Plate 21). With prizing, the tobacco was put into the cask a few layers at a time, weighed down for a few days, then a few more layers were added, and so on. Only large planters had the equipment and labor to accomplish this type of packing, and workers had to learn by trial and error how to tightly prize tobacco without bruising it. By 1710, hogsheads of stemmed sweet-scented, the heaviest sort of tobacco, generally weighed more than one thousand pounds net, but a large operator like William Fitzhugh could also get nearly nine hundred pounds of unstripped oronoco leaves into a standard cask. Most small planters were unable to follow suit. As Richard Tilghman explained to merchant Samuel Hyde in 1712: "Tis Impossible for those who have not good Strength and Convenenceys to prize Tobo (and they are but few) to make it of saving weights, unless they pack it to[o] high in case, which is a Mortall Evill, and now too frequently pract[ic]s[e]d among us."[50]

Large Virginia planters, and especially planter-merchants who like

49. Vertrees J. Wyckoff, *Tobacco Regulation in Colonial Maryland* (Baltimore, 1936), chap. 5. Hemphill, "Freight Rates," *MHM*, LIV (1959), 36–58, 153–187, suggests that a Maryland act of 1704 requiring ship captains to publish their rates might have been responsible for lower peacetime rates in that colony. Table 19 suggests that more uniform weights of Maryland casks also might have been a factor. Robert "King" Carter vented his frustration with merchants' complaints about badly prized tobacco in a 1706 letter to English trader Thomas Corbin: "I find you are got into the Strain of Complaints ag[ains]t the prizing of Tobo. as well as the rest of the Merchts and let the Tobo. be light or heavy or what it will you are all in the Same tune, the design I almost think is to discourage the practice of prizing Tobo. . . . if we must follow yor directions and make such Tobo. as you will think fitt to give a Character to twill do little more then pay the Queen her Customs and the Merchts. their Commission" (Carter to Corbin, Aug. 20, 1706, Robert Carter Letterbook, 1701–1710, Wormely Estate Papers, 1701–1716, Christ Church Parish Records, VSL).

50. Richard Tilghman to [Samuel] Hyde, May 29, 1712, folder 3, Loose Papers, Tilghman Family Papers, Georgetown University Library. Similar observations can be found in letters of 1717 from John Custis Letterbook, 1717–1741, LC, microfilm, M-8, CWL. Hugh Jones, *Present State of Virginia*, ed. Morton, 77, describes prizing.

PLATE 21. Slaves Prizing Tobacco Hogsheads. Detail from William Tatham, *An Historical and Practical Essay on the Culture and Commerce of Tobacco* (London, 1800). Courtesy, Special Collections, John D. Rockefeller, Jr., Library, The Colonial Williamsburg Foundation

Stephen Loyd repacked the lightly prized tobacco of smaller planters, saved the most on freight charges, perhaps as much as 50 percent. A class difference shows clearly in merchant records. When Loyd exchanged manufactures for tobacco from ordinary Rappahannock and Potomac river planters in 1708, the hogsheads he received averaged 590 pounds net before repacking. Two years later, when Loyd sold slaves in the same area (and hence was dealing only with large planters), the hogsheads he purchased averaged 1,148 pounds net. (Both cargos included primarily sweet-scented tobacco.) When peace returned in 1714, planters again ceased to heavily prize their tobacco. In 1717, for example, John Custis was willing to be "cautious in prizing" tobacco, which had been cut before it was fully ripe and hence more liable to damage from overprizing. But, during the 1720s, average weights were again rising, and the weights of Virginia and Maryland casks were converging. By then, many planters had learned how to make heavy hogsheads without damaging the product.[51]

51. Lyde Account Book, 1708–1711, and Accounts and Letters of John Tayloe I and John Tayloe II, 1714–1778, Tayloe Family Papers, VHS; Loyde Account Book, 1708–1710, UVa. Stephen Loyd wrote to [Lawrence Hollister and Company] that the slaves had been sold for 150 hogsheads

Other avenues for cutting the costs of production were exhausted sooner. Through the 1680s in both colonies, the size of the crop a single laborer could produce continued to edge upward, suggesting that planters were still finding ways to increase the number of plants a worker could tend. But, by the turn of the century, most tobacco growers reached a ceiling beyond which production could not be expanded. The results are the same whether one uses the mean of crop sizes recorded in planter account books or calculates average outputs by dividing total tobacco exports by the number of taxable laborers.[52]

Sometime in the 1690s, when wartime trade disruptions left Chesapeake planters facing bleak prospects indeed, tobacco growers in the two colonies began to adopt two very different approaches to staple cultivation and marketing. The first was most common in Maryland's prime tobacco areas, where planters continued to produce as much volume as they could (Table 20). Only in the 1720s did the mean size of oronoco crops decline on the upper Eastern Shore, resulting from a reallocation of labor among planters who added corn and wheat as second and third market crops. Outputs likely remained stable across these years on less diversified lower Western Shore plantations.[53]

The figures from Virginia show a different pattern (Table 21). Outputs per hand began falling in the 1690s in older sweet-scented counties like York and by the 1710s had dropped below one thousand pounds in all areas for which

---

of "choice weighty" tobacco, some of the season's best for substance, length, and neat handling; the receiver was paid as much for rejecting a hogshead as for accepting one in the hope of getting good tobacco (Lyde Account Book, Oct. 10, 1710, VHS). When Thomas Colmore's factors bought Maryland oronoco from ordinary planters on the Patapsco River in 1719 and 1720, their hogsheads weighed on average 592 pounds net (Colmore Account Book, MHS). See also Louis B. Wright, ed., *Letters of Robert Carter, 1720–1727: The Commercial Interests of a Virginia Gentleman* (San Marino, Calif., 1940), 102–103, on reprizing tenant rent tobacco. See John Mercer Ledgers, [1741]–1750, Bucks County Historical Society, Doylestown, Pa., microfilm, M-164, fol. 30, CWL, for a merchant's picking, sorting, and reprizing tobacco purchases; John Baylor Ledgers, 1719–1755, Baylor Family Papers, 1653–1915, MS 2257, UVa, microfilm, M-62.6, CWL, for entry for payment of wages in 1720 to Henry Falwell "for 10 mos 18 da at the steming house at £12 per year"; Custis to ?, 1717, and Custis to Sir ?, 1717, Custis Letterbook, CWL, for fears about prizing too heavily.

52. Walsh, "Summing the Parts," *WMQ*, 3d Ser., LVI (1999), 53–94. Another indication of the exhaustion of technological efficiencies is the finding of the U.S. Department of Agriculture that between 1910 and 1950 every year growers still had to devote between 350 and 450 hours of labor per acre to tobacco crops. In contrast, by 1910 cumulative technological improvements had reduced labor requirements for grains to 35 hours per acre for corn and 15 hours for wheat (Olmstead and Rhode, *Creating Abundance*, 203).

53. Dr. Charles Carroll wrote John Hyde and Company, Apr. 19, 1729, that Maryland's stint act of 1728 would primarily affect only Maryland's Western Shore planters (Dr. Charles Carroll of Annapolis, Letterbook, 1722–1731, MS 208, MHS).

TABLE 20. Tobacco Crops per Laborer in Maryland, 1680–1729

	Western Shore		Eastern Shore	
Years	*N*	Mean	*N*	Mean
1680–1689	6	2,039	1	2,584
1690–1699	4	1,834		
1700–1709	11	1,628	7	1,610
1710–1719	6	1,629	18	1,687
1720–1723	1	1,613	12	1,090
1724–1729			23	1,302

	Six counties		Anne Arundel County		Talbot/Kent counties
Years	*N*	Mean	*N*	Mean	Mean
1660–1699			34	1,914	
1684–1696	171	1,585			
1697–1704	202	1,522			
1700–1709	248	1,609	16	2,239	1,570
1710–1719	276	1,059	23	1,688	
1720–1729			18	1,613	1,370

Note: Comparative observations for the six counties, Anne Arundel County, and Talbot/Kent counties are presented by decade from 1700; the actual year groups from Main, Earle, and Clemens vary slightly from one another.

Sources: Appendix 1. For the six counties, see Gloria L. Main, *Tobacco Colony: Life in Early Maryland, 1650–1720* (Princeton, N.J., 1982), 39. For Anne Arundel County, see Carville V. Earle, *The Evolution of a Tidewater Settlement System: All Hallow's Parish, Maryland, 1650–1783* (Chicago, 1975), 27. For Talbot and Kent counties, see Paul G. E. Clemens, *The Atlantic Economy and Colonial Maryland's Eastern Shore: From Tobacco to Grain* (Ithaca, N.Y., 1980), 173.

figures are available. Sweet-scented growers did respond to rising prices in 1718 and 1719 by pitching larger crops. John Tayloe reported in 1719: "Our Country grows every year more slovenly in their Tobacco sinse there has bin such Demand for it. Every one strives to make the most he can knowing that none of it will be refused at the last. What one won't take of[f] another will." But, by the early 1720s, when prices again sagged, output per laborer dropped even further. Some Virginia planters thus seem to have abandoned the goal of large crops before stint laws forced the issue, and they tried in-

TABLE 21. Tobacco Crops per Laborer in Virginia, 1680–1729

	York Basin		Rappahannock Basin		Potomac Basin	
Years	*N*	Mean	*N*	Mean	*N*	Mean
1680–1689	6	1,980	1	1,582		
1690–1699	8	1,101	1	1,207		
1700–1709	6	1,220	3	1,408	1	661
1710–1719	13	815	7	830	6	812
1720–1723	7	731	10	870	4	843
1724–1729	9	705	7	501		

	Middlesex County	
Years	*N*	Mean
1650–1699	4	1,207
1700–1709	10	969
1710–1719	11	881
1720–1729	36	988

Note: Comparative observations for Middlesex County reflect the time groupings used by Darrett B. Rutman and Anita H. Rutman, *A Place in Time: Explicatus* (New York, 1984), 16.
Sources: Appendix 1; Rutman and Rutman, *A Place in Time: Explicatus,* 16.

stead to make smaller crops of high-quality tobacco specially tailored for the English domestic market. When Robert Carter wrote of the 1724 stint act, "We shall then plant the best of all our Lands, tend it with All Imaginable nicety and probably make more and to be sure much better, than now We do off ten or twelve thousand plants," he was probably describing a change that many planters had already embraced. "Neat and well handled" high-quality crops grown only on the planter's best land became the goal, instead of bigger yields of lesser quality. With such carefully tended crops, tidewater Virginia planters still hoped for, as Carter put it, "some time or another, meeting with a hit for tobacco."[54]

54. Tayloe to [unidentified English merchant], 1719, Lyde Account Book, 1708–1711, and Accounts and Letters of John Tayloe I and John Tayloe II, 1714–1778, Tayloe Family Papers, VHS; Carter to Anonymous, July 2, 1723, Carter Letterbook, Carter Papers, UVa; Carter to [William Cage], in Wright, ed., *Letters of Robert Carter,* 113. In 1726, Carter recorded the number of tobacco plants "told by the countey" on two of his plantations, and the next year he listed the

Scattered evidence indicates that by the later seventeenth century many Virginia planters normally allowed only four to eight leaves to remain on each tobacco plant in order to produce a thick, rounded leaf, whereas in oronoco areas plants were allowed to carry up to twenty or twenty-two thinner, narrower leaves. Different cultivation techniques might then have contributed to the large differences in crop shares between the two colonies. Different methods of processing, discussed below, probably affected the size of individual output even more. The figures for Maryland continue to include all the merchantable, unstripped leaf a laborer could produce. The meaning of Virginia crop shares becomes more difficult to interpret, because many observations are for a crop weighed after additional processing that reduced the weight of the final product by as much as one-quarter. Consequently, for Virginia, declining output per worker cannot be cited as clear evidence for widespread soil exhaustion, for example, nor as an indication of low productivity among enslaved workers.[55]

Further processing of the harvested leaves was perhaps the major way Virginia planters attempted at the turn of the century to make more money without expanding production. This processing was termed "stripping" or "stemming" ("separating the largest stems or fibres from the web of the leaf with adroitness and facility"), a task assigned to both men and women, sometimes at night after they had already put in a full day's work in the fields (Plate 22). Normally, only the more valuable sweet-scented crops intended for the English domestic market were stemmed, since, although a weight reduction of about one-quarter meant considerable savings on freight and duties, extra labor and some skill were involved. Processing practices in sweet-scented and oronoco regions had sufficiently diverged by the 1720s that Robert Carter could claim: "I have hardly a Serv[an]t that knows how neatly to pack a hhd

numbers of laborers on these same quarters. If the workforce was the same size the previous year, Carter tended exactly six thousand plants a hand (Carter Diary, July 11, 1726, Mar. 30, 1727, Carter Papers, UVa). See also Carter to William Dawkins, June 19, July 3, 1723, to Micajah Perry, June 19, 1723, and to Thomas Colmore, June 27, 1723, Carter Letterbook, Carter Papers, UVa.

55. *CSP,* no. 1253, 582–583, cited in Arthur Pierce Middleton, *Tobacco Coast: A Maritime History of Chesapeake Bay in the Colonial Era* (Newport News, Va., 1953), 378; "Letter from Clayton to the Royal Society," *Force Tracts,* III, no. 12, 15–17, 23. In 1705, Robert Beverley estimated that the maximum crop of sweet-scented a prime slave could produce was about thirteen hundred pounds (*History and Present State of Virginia,* ed. Wright, 263). Hugh Jones thought the sweet-scented strain could be "attributed in some measure to the seed and management, as well as to the land and latitude." The most celebrated crops did not necessarily come from the best land (*Present State of Virginia,* ed. Morton, 72–73). John Tayloe associated large crops with low quality (Tayloe to [torn], Feb. 2, 1718/9, Lyde Account Book, 1708–1711, and Accounts and Letters of John Tayloe I and John Tayloe II, 1714–1778, Tayloe Family Papers, VHS).

PLATE 22. Slave Woman Stripping Tobacco. Detail from William Tatham, *An Historical and Practical Essay on the Culture and Commerce of Tobacco* (London, 1800). Courtesy, Special Collections, John D. Rockefeller, Jr., Library, The Colonial Williamsburg Foundation

of Leafe Tobo. My Crops have bin wholly stemd for these thirty Years and More." Conversely, references to stemming in oronoco areas are virtually nonexistent.[56]

How much of the crop was stemmed or simply packed as leaf depended on a variety of considerations, including the prices anticipated for the two sorts, the size and condition of the crop, the labor available, and prevailing freight rates. Some planters habitually had their slaves stem all the tobacco,

56. Carter to Dawkins, Aug. 1, 1723, Carter Letterbook, Carter Papers, UVa (quotation). As William Tatham explained many years later, "The operation is performed by taking the leaf in one hand, and the end of the stem in the other, in such a way as to cleave it *with the grain;* and there is an expertness to be acquired by practice, which renders it as easy as to separate the bark of a willow, although those unaccustomed to it find it difficult to stem a single plant" (*An Historical and Practical Essay on the Culture and Commerce of Tobacco* [1800], in G. Melvin Herndon, ed., *William Tatham and the Culture of Tobacco* [Coral Gables, Fla., 1969], 40–41). For the definition of stemming, see ibid., 40. A Virginia act of 1686 first mentions the practice of stripping (Hening, ed., *Statutes at Large,* III, 34). Hugh Jones in 1724 differentiated stemming (to pull the leaves from the stalk) from stripping (to take out the great fibers) (*Present State of Virginia,* ed. Morton, 77). In correspondence, however, planters often used the two terms synonymously. See, for example, Oldmixon, *The British Empire in America,* 410–415; York County Deeds, Orders, and Wills, 17: 57–58, microfilm, VSL; Middleton, *Tobacco Coast,* 112–113; Gray, *History of Agriculture,* I, 225. John Baylor wrote to John Backhouse, July 5, 1762: "Near 30 years past I was advis'd by one of the best Planters that ever this country produc'd an old uncle and Guardian of mine Colo Bernard Moores Father that in years of Plenty to stem all and [in] those of scarcity to make as much leaf as possible" (Charles Campbell Papers, box 1, folder 3, Earl Gregg Swem Library, College of William and Mary, Williamsburg, Va.).

a few stemmed only prime top leaves, and many more stripped the less valuable under leaves. In 1716 and 1717, for example, William Byrd II wanted his overseers to strip all but the finest leaves and to pack them in bundles, since "all the very large and very small and the discoulourd leafe will answer much better if it be strippt." But the overseers were "for stripping all or none." Byrd acquiesced in their stripping all, concluding that, "if every body wou'd agree to strip all the sweet scented tobacco in the country, twoud be a great advantage to the trade, because twill sink the quantity a fourth part, which proportion they compute the stalks to amount to."[57]

Prime stemmed tobacco could sell for as much as double the price of leaf, and merchants commonly also bought lower-quality stemmed under-tobacco ("lugs") "at half price and none is bad but the Planter can get Something for it." In unfavorable years, stemming might enhance the value of a poor-quality crop; it discouraged rot if the tobacco had been cut while immature, if it had to be packed when partially out of case (that is, when too dry and brittle), or if it had been damaged by storms. When Britain prohibited the import of stripped or stemmed tobacco in 1722 (as had the disallowed inspection law of 1713), Virginians protested mightily. Stemming was, Robert Carter averred, the "only Refuge we had to get a penny by." John Custis complained in 1724 that the season's crop had been heavily damaged by summer storms: "Nothing can bee made of the little tattr'd Tob: left in the field." Not surprisingly, it was a Virginia agent who secured the repeal of the act prohibiting stemmed or stripped tobacco in 1728.[58]

This response among sweet-scented planters parallels that of sugar planters on Barbados who, when they experienced profit squeezes at about the same time, responded in a similar way. By further processing the basic brown muscovado sugar that planters on newer islands produced in ever larger quantities, Barbados planters obtained clayed white sugar. Like stemmed tobacco, clayed sugar required more labor and additional expertise; however, it also sold on the island and at home at a higher price and saved on shipping costs and duties. By the early eighteenth century, planters in more recently settled islands surpassed Barbados in the volume of unrefined sugar they exported. "Looking only at trends in the production of muscovado sugar," David Eltis

57. William Byrd II to Custis, November 1716, Oct. 19, 1717, *Byrd Correspondence,* I, 294, 306–307. Since the proportion of the crop stemmed each year varied from quarter to quarter as well as from one plantation to another, it is very difficult to know by what factor to weight final sweet-scented crop shares to reflect the amount of tobacco actually tended.

58. Carter to Perry, Aug. 1, 1723, to John Stark, May 30, 1728, Carter Letterbook, Carter Papers, UVa; Custis to ?, Aug. 12, 1724, Custis Letterbook, CWL; Warren M. Billings, John E. Selby, and Thad W. Tate, *Colonial Virginia: A History* (White Plains, N.Y., 1986), 235–236.

has argued, "is bound to yield an interpretation of Barbadian decline based on the presence of inefficient production techniques and an entrepreneurial inability to adjust, when the reality was quite the opposite." The same argument can be made for sweet-scented tobacco growers.[59]

Large Virginia planters experimented with attractive packaging as well. In 1721, John Custis sent merchants Perry and Company of London a shipment of fifty-three hogsheads, some tied in bundles and laid straight and some "laid in promiscuously" to see which method best suited the market. The two types were differently marked, and Custis reiterated, "I should bee glad if you would please to tell me by the very first opportunity how I shall order my crops; to please the nice pallates of your Chaps." The next year, he sent Perry eighty-four hogsheads, some bundled in hands, some stemmed and laid straight, and some stemmed and laid in promiscuously. During the same years, Robert Carter experimented both with large bundles tied in the middle and with leaves laid straight without tying. "If neat handling will keep up our reputation," Carter wrote London merchant William Dawkins in 1720, "I resolve to be very nice." Virginia merchant William Woodford added a further nicety: it was advantageous to tie stripped leaf into bundles with wrapper leaves of the same quality. Woodford, however, also shipped some of his hogsheads straight and untied, which "will answer better for those that Buy for Cutting."[60]

Color and scent were also part of the equation, although planters had little control over them. The soil on which the tobacco was grown, the weather, and even the way the barns in which the tobacco was cured were sited and ventilated all affected these qualities. Sometimes it seemed impossible to satisfy fickle London buyers. John Custis was incensed when London merchant Robert Cary complained about his 1728 shipment:

> It was my home Crops under my own eye, and may say I saw allmost every plant from the planting to the prizing and shiping of; and was all choice pickt Tobacco and nothing mean or any like Trash. You say it was to[o] bright. I never had that complaint before; it grows on quick ground and cure'd in open houses, wh[ic]h is the reason of the color; you advise me to make it black; I know no way but by wetting it, or prizing in to[o] high Case.

59. McCusker and Menard, *Economy of British America*, 161–165.

60. Custis to Messers Perry, [1721], to Mr. Perry, 172[2], Custis Letterbook, CWL; Wright, ed., *Letters of Robert Carter*, 23–25, 34–35, 47–48, 56, 62, 71–72, 76–77, 92, 93, 96–97; William Woodford to Friend Taliafero, Dec. 2, 1723, to Friend Welch, May 12, 1723, to Robert Lidderdale, [1733], William Woodford Letterbook, 1723–1737, University of Kentucky, Lexington, microfilm, M-1060-1, CWL.

Mr. Perry, another of Custis's London factors, on the other hand, had complained that his tobacco was *too* dark. How could he please both?[61]

English retail tobacco dealers in London, Bristol, and other major towns in turn commissioned artists to produce advertisements intended to entice consumers with images underscoring the exotic origins of the addictive product and text touting its superior qualities (Plates 23 a–d). Hybrid native American–African figures were common (Plates 23a and 23d), as were idealized plantation scenes. High-quality sweet-scented from the York River, which was marketed largely through consignment merchants who then sold it to tobacconists by the individual hogshead, was the sort most commonly advertised (Plates 23a and 23b). Otherwise, smokers were urged to buy "Best Virginia" (Plate 23c), a designation that could have included less desirable Rappahannock basin tobacco, which (along with Potomac River and Maryland tobacco) was not promoted by identifying the place of origin. Advertisements for less favored oronoco (Plate 23d), of which planters took less care in separating and processing and which was less frequently consigned, were much less common, although this cheaper strain began to find favor with some bargain-hunting smokers toward the end of the century.[62]

In addition to the demands of the different markets for which sweet-scented and oronoco growers were producing, different buying policies among merchants might also have contributed to differing cropping practices. When, for example, Virginia tobacco merchant Stephen Loyd bought a cargo in the Rappahannock for Bristol buyers in 1708, he offered just three prices—2d a pound for stemmed sweet-scented, 12s 6d per hundredweight for sweet-scented leaf, and 10s per hundredweight for oronoco. Similarly, when John Baylor was buying Rappahannock tobacco for English dealers between 1719 and 1721, he usually also purchased at only three prices—one for stemmed sweet-scented, one for leaf, and a lower price for lugs. In contrast, when London merchant Thomas Colmore sent factors to the Patapsco River to buy Maryland tobacco in 1719, they purchased oronoco there at twenty-four different prices, ranging from 8s 4d per hundredweight for the lowest grade up to 30s for the best. The next year, Colmore's factors paid thirteen

61. Custis to Robert Cary, 1725, to Cary, [1729], Custis Letterbook, CWL; Carter to Perry, June 16, 1723, to William Dawkins, June 16, 1723, Carter Letterbook, Carter Papers, UVa; Woodford to Friend Daingerfield, Dec. 2, 1723, William Woodford Letterbook, CWL; Wright, ed., *Letters of Robert Carter,* 1–4.

62. Jerome E. Brooks, *Tobacco: Its History Illustrated by the Books, Manuscripts, and Engravings in the Library of George Arents, Jr.,* 5 vols. (New York, 1937–1952), IV, 194–201. For hybrid images on tobacco advertisements, see Catherine Molineux, "Pleasures of the Smoke: 'Black Virginians' in Georgian London's Tobacco Shops," *WMQ,* LXIV (2007), 327–376.

A

B

C

D

PLATES 23A–D. Four Seventeenth-Century Tobacco Advertisements. By Francis Bedford. Courtesy, The Colonial Williamsburg Foundation

different prices, from 10s up to 26s per hundredweight. Occasionally, the factors paid two different prices for some proportion of individual hogsheads, likely reflecting payment in a combination of cash and of merchandise rated at different values in currency. One cannot generalize very far from only three examples. This evidence does indicate, however, a need to examine further the practices of direct buyers who were playing an increasingly important role in the trade. If merchants trading in Virginia wanted only tobacco of a particular grade, planters would have sought to standardize their product and might have had difficulty finding buyers for inferior leaf. If, on the other hand, traders to Maryland were willing to buy a wider range of grades with an appropriate differential for quality, planters would have been encouraged to raise and to market the largest crops their workers could tend, since they would be paid something for everything they made.[63]

Other studies have asserted that Virginia planters did not grow larger crops because they could not. Soil exhaustion and inexperienced or rebellious slaves are the culprits usually advanced. Neither seems a tenable explanation given strong evidence that yields per hand did not diminish across these years in most oronoco areas. Enslaved Africans, white indentured servants, or some combination of the two were either induced or forced to produce larger outputs of tobacco in the oronoco region in the 1680s and 1690s than had indentured servants, working almost exclusively with fellow Englishmen, in the 1650s, 1660s, and 1670s. Most scholars now agree that by the 1680s the productivity of servants and slaves in unskilled field labor was about equal. Since there is no reason to suppose that Virginia sweet-scented growers were on the whole less competent managers than oronoco growers, explanations for diminished outputs must be sought elsewhere.[64]

63. Loyde Account Book, 1708–1710, UVa; Baylor Ledgers, 1719–1755, Baylor Family Papers, CWL; Colmore Account Book, MHS. In 1725, Stafford County, Virginia, merchant John Mercer also appeared to be buying at only one price (Mercer Ledgers, [1741]–1750, CWL). [Makemie], "A Plain and Friendly Perswasive," *VMHB,* IV (1896–1897), 263–264, suggested that, if all tobacco was brought to a public market, "there would be a general Emulation to out vie one another, especially in Quality, *and the different Prices that would follow, would presently encourage thereunto*" (emphasis added).

64. The argument that slaves were unproductive workers is advanced in Gerald W. Mullin, *Flight and Rebellion: Slave Resistance in Eighteenth-Century Virginia* (New York, 1972), chap. 2. See also Allan Kulikoff, *Tobacco and Slaves: The Development of Southern Cultures in the Chesapeake, 1680–1800* (Chapel Hill, N.C., 1986), chaps. 8, 10. For slaves as close substitutes for servants or interchangeable with them, see Grubb and Stitt, "Liverpool Emigrant Servant Trade," *Exp. Econ. Hist.,* XXXI (1994), 376–405; David W. Galenson, *White Servitude in Colonial America: An Economic Analysis* (Cambridge, 1981), 157–158; and Main, *Tobacco Colony,* 101–106.

By the 1690s, some planters in both colonies had likely already cleared, planted, and temporarily depleted most of their prime tobacco lands and were getting diminished yields either because workers were cropping less desirable land overall or were tending several fertile, but small and widely scattered patches. On the other hand, almost all planters still had a sufficiently high ratio of land to labor that they could continue to rotate fields in long fallow and still produce respectable, if not spectacular, yields. Complaints about "exhausted" lands probably came, not from any extensive long-term damage to the soil (a feat difficult to accomplish so long as laborers planted only a few acres each year, worked primarily if not exclusively with hand tools, and limited cultivation to relatively level fields), but rather from the exploitation of prime virgin soils by planters in newer areas. They therefore had a competitive advantage over those in older areas who had to begin farming new ground of less than optimal fertility or to apply manure to long-cropped lands in order to maintain outputs. As early as the late 1660s, sugar planters in Barbados faced similar problems with depleted soils that needed regular fertilizing. Hence, their production costs, like those of Virginia planters in the 1690s in older tidewater counties, were higher than those incurred by planters on more recently settled islands. A comparison of Table 11 with Tables 20 and 21 suggests that outputs per laborer in both colonies peaked in the 1680s. Thereafter, diminishing returns to land apparently set in, accompanied by changes in the proportions of prime adult and less productive dependent bound laborers.[65]

Geographer Carville Earle has argued that the shift from indentured servants to perpetually enslaved laborers in fact facilitated more systematic rotation of landed resources. The declining supply of indentured servants in the

65. Carville V. Earle, *The Evolution of a Tidewater Settlement System: All Hallow's Parish, Maryland, 1650–1783* (Chicago, 1975), 24–30. Bergstrom, *Markets and Merchants,* chap. 2, calculated the available acreage per tithable for Virginia Naval Districts and found sufficient land available for long-fallow rotation throughout the colony in this period; see, esp., table 2.4. See also Jacob M. Price, "The Economic Growth of the Chesapeake and the European Market, 1697–1775," *JEH,* XXIV (1964), 496–511. John Clayton described a Virginia plantation in 1688 where the crop was scattered in a number of small plots because it was "an old planted Plantation, and being but a small Plot of Ground, was almost worn out, so that they had not Ground all together that would bring forth Tobacco" ("Letter from Clayton to the Royal Society," *Force Tracts,* III, no. 12, 22). For Barbados, see David Eltis, *The Rise of African Slavery in the Americas* (Cambridge, 2000), 200–202; and Susan Dwyer Amussen, *Caribbean Exchanges: Slavery and the Transformation of English Society, 1640–1700* (Chapel Hill, N.C., 2007), 80. For comparative rates of return on Barbados and on newer islands, see J. R. Ward, "The Profitability of Sugar Planting in the British West Indies, 1650–1834," *Economic History Review,* n.s., XXXI (1978), 197–213.

1680s and 1690s forced labor-short planters to cut back on clearing fresh land. Some continually cultivated the same ground and attempted to cope with declining soil fertility by penning cattle on plots intended for tobacco; others interrupted the cropping of tobacco and then corn with only short periods of fallow. The shift to slave labor allowed large and eventually middling planters, as slaveowning trickled farther down the economic spectrum, to maintain a schedule of regular clearing, cropping, and then recycling land back into production after a long fallow of about twenty years. "Perpetually bound labor," Earle contends, "was the perfect complement for an agrarian system which demanded perpetual clearance of land." Slaves provided not only essential labor, environmental historian Timothy Silver notes, but also expertise. Slash and burn clearing was a familiar technique that enslaved Africans, unlike European servants, did not need to be taught. Men, women, and children might have divided the tasks of felling trees, splitting firewood, cutting brush, and tending fires, according to strength, gender, and age, much as they had previously done in West Africa. Economist Gavin Wright has subsequently observed that slaveowners had the additional advantage of being able to move workers "to distant and ofttimes undesirable locations and promptly to set [them] about clearing the land." They could thus realize returns from outlying, previously undeveloped tracts more quickly than could nonslaveholders.[66]

The argument that newly enslaved Africans proved unproductive workers is more compelling, but here, too, once evidence is broken down by subregion, the coincidence evaporates. The Maryland data in fact contradicts the hypothesis. On too many slave-staffed Maryland plantations of the 1690s and early 1700s, recently imported Africans produced crops of tobacco and corn as large as (and sometimes larger than) white indentured servants in the 1660s and 1670s. In Virginia, in the aggregate, a drop in tobacco outputs per laborer does closely parallel massive slave imports, but that is not the case in individual counties. In older places such as York County, planters had turned to slave labor earlier than in most other parts of the Chesapeake. Here blacks became a majority of the bound labor force sometime in the 1660s. Output per worker did not fall, however, for at least another twenty years. Later, throughout Virginia the size of crop shares jumped dramatically in the 1740s,

66. Carville V. Earle, "The Myth of the Southern Soil Miner: Macrohistory, Agricultural Innovation, and Environmental Change," in Earle, *Geographical Inquiry and American Historical Problems* (Stanford, Calif., 1992), 258–299 (quotation on 283); Silver, *New Face on the Countryside*, 106–107; Gavin Wright, *Slavery and American Economic Development* (Baton Rouge, La., 2006), 50, 122; Douglas Helms, "Soil and Southern History," *Ag. Hist.*, LXXIV (2000), 723–730.

just as tobacco prices began a long-term rise. The greatest improvement in yields per field hand appeared in piedmont areas where planters were still packing in large numbers of unseasoned, unacculturated, newly arrived Africans. In older sweet-scented counties like York, where most of the workforce then consisted of either long-resident Africans or a growing proportion of young creoles, yields per laborer showed less improvement.[67]

For the year or so after Africans' initial capture, enslavement, and arrival in the Chesapeake, crop records from individual plantations demonstrate that they did often prove unproductive workers. Sickness, mental trauma, language difficulties, and refusal to accept their new status were doubtless all factors. What they eventually contributed to elite planters' coffers, however, depended upon the managerial and extractive skills of the owners and of their subordinates as well as upon the rigor or laxness of customary work requirements on a given plantation. If a relatively slow work pace became the custom of the plantation early on, it proved difficult later to increase the pace or duration of labor or to increase the size of what was considered an appropriate daily "task," as was the case, for example, on the absentee-owned Bristow plantations. Some overseers or planters found ways for ratcheting up work requirements without triggering overt resistance among their workers, whereas others could extract additional effort only by instituting draconian measures that often led to increased numbers of runaways and to concerted, occasionally violent, opposition.

Owing to the scale and speed with which Virginia planters bought slaves at the turn of the century after the Royal African Company's monopoly collapsed, they might well have purchased more new slaves than they could find competent supervisors to manage. Men inclined to serve as overseers were likely in short supply, as evidence in plantation accounts of extraordinarily high monetary salaries or atypically generous crop shares in the 1700s and 1710s indicates. Many of the young poor whites recruited as managers likely had little or no experience in communicating with or controlling Africans. Overseers assigned to quarters, especially quarters lying far from the owner's immediate direction, might have found it much easier not to push their strange, sometimes rebellious, and frequently sick charges very hard and to

67. For sizes of tobacco crops, see Tables 20 and 21 and Appendix 1. Gavin Wright concludes there were no sustained differences in productivity between servants and slaves. He considers the critical factor to be the property rights that slavery afforded, which permitted planters to move enslaved workers from one location to another at will, to reallocate the labor of women and children from household or nonmarket activities into cash crops, and to reduce risk of shortages of labor at harvest because they possessed a captive labor force (Wright, *Slavery and American Economic Development,* 20–27, 122).

look instead for other ways to feather their nests. The slaves were often adept at quickly transforming minimal work requirements imposed by unskilled or inattentive overseers into the custom of the plantation. Still, one suspects that enslaved creoles might have more adroitly manipulated plantation custom to their advantage than less acculturated Africans.[68]

Additional evidence pointing to supervision as a critical problem is found in the argument that small Virginia planters, who personally supervised their workers and often worked alongside them, were more successful in making good tobacco than were the great planters who never dirtied their hands in the fields. Large planters often complained that smaller sweet-scented growers got better prices than did they. The "common people," Virginia's Governor William Gooch affirmed, "make the best"; careful supervision might have made the difference.[69]

68. See, for example, Carter Diary, Apr. 4, 1727, Carter Papers, UVa, for accusations against an overseer that included keeping his own horses, cattle, pigs, and poultry at the quarter and feeding them with plantation corn while neglecting Carter's livestock; employing quarter slave women to cook, wash, fetch water, and make fires for his family; using the slaves to tend his private patches of corn, cotton, and peas; and making and selling bedsteads and barrel staves fashioned from timber he made the slaves cut from Carter's land. William Byrd II similarly complained of various of his overseers' leaving their farms, stealing cider, being drunk, abusing his cattle, and using Byrd's slaves for their own business (Wright and Tinling, eds., *Secret Diary of Byrd,* 6, 14, 133, 178, 202, 221, 232, 274, 469, 485–486, 489, 497, 580). For overseers' salaries, see ibid., 275; Wright, ed., *Letters of Robert Carter,* 21–22; Lyde Account Book, 1708–1711, and Accounts and Letters of John Tayloe I and John Tayloe II, 1714–1778, Tayloe Family Papers, VHS; Bagge Account Book, CWL; and Robert Bristow to Thomas Booth, Dec. 30, 1719, Robert Bristow Account and Letterbook, 1688–1750, VSL, microfilm, M-1164, CWL. As late as 1724, Hugh Jones said that tenancy was not particularly attractive to landless Virginians, for "they that have not [land] think to make more profit by turning overseers" (*Present State of Virginia,* ed. Morton, 92).

69. Governor Gooch to the Lords of Trade, 1731, PRO CO/5/1322. Cf. PRO CO/5/1321, quoted in Thomas J. Wertenbaker, *The Shaping of Colonial Virginia* (1910; rpt. New York, 1958), 146–147. John Custis complained to Micajah Perry in 172[4] that one James Shelton, "a Carpenter and a man just free," had consigned Perry some tobacco "w[hi]ch was such trash; and wagers laid that it would bring him in debt." Perry sold Shelton's tobacco for 9 ½ and 10d a pound, however, whereas Custis's sold for only 8 ½d a pound. In the same vein, Custis wrote Robert Cary in 1729 that "severall of my poor neighbors, whose ground I am sure cannot make good Tobacco" and who had "not one farthing in y[ou]r Merch[an]ts hands" got better prices than he did (Custis to Perry, [1724 or 1725], to Cary, [1729], Custis Letterbook). Similarly, Robert Carter protested to Richard Perry in 1720, "It troubles me not a little that Mr. Pratt should gett more for his miscellaneous stuff that comes from all parts of their county, low lands as well as high, than I do for mine that is made with my own people and upon as good and proper [lands] as anybody can boast of" (Wright, ed., *Letters of Robert Carter,* 1–4 [quotation on 2]). Almost all the crop shares of more than 1,500 pounds dating from 1710 and after in Appendix 1 come from estates of small planters. One of the few exceptions is a crop of 1,680 pounds per hand produced by slaves on a King William County plantation that William Byrd II inherited from his father-in-law. Byrd

Differences in outputs per laborer between sweet-scented and oronoco regions may be further accounted for by the ways planters often unconsciously made use of the skills and knowledge that enslaved Africans brought with them as well as by the quality of labor inputs that could be extracted through force. All captive Africans had particular expertise in long-fallow hoe culture, the agricultural system then commonly employed throughout West Africa and adopted, following the example of local native Americans, by almost all seventeenth-century Chesapeake planters. Most Africans also likely arrived with an intimate knowledge of tobacco culture, for West African farmers had begun raising the weed (first brought there from Brazil by the Portuguese) in the early 1500s. By the mid-seventeenth century, cultivation of tobacco for local use (and occasionally also as a product raised on a wider scale for interregional trade) had become widely established throughout the areas of West Africa involved in the transatlantic slave trade. Consequently, most Africans transported to the Chesapeake between 1670 and 1730 could draw on accumulated knowledge of small-scale tobacco growing that often antedated that of their masters by one or more generations. Women, especially, could do so, because they were primarily responsible for raising tobacco in West Africa. As a result, slaves, unlike European servants, most of whom had not tended the weed in the Old World, required little or no instruction in raising tobacco, whatever extraordinary coercion might be required to force them to work. No slaves brought directly from Africa had prior experience in curing tobacco for long keeping, however, nor in processing and packing leaf for long-distance transport; slaves had to be taught as well as coerced into performing these tasks. Sweet-scented growers who required high levels of care in cultivation and especially in processing and packing—work efforts not readily extracted solely through fear of immediate physical punishment—might have had less success in forcing unfree workers to take requisite care and might, moreover, have been more vulnerable to acts of intentional carelessness. Indeed, in the mid-1720s one plantation manager reported that, although he could make slaves grow tobacco, he could not get the recently arrived Africans he was overseeing to "doe any thing in Tob[acc]o neatly."[70]

---

noted that the land was exceptionally level and fertile and that the plantation was well managed (Wright and Tinling, ed., *Secret Diary of Byrd,* 484–485, 532).

70. Walsh, "Summing the Parts," *WMQ,* 3d Ser., LVI (1999), 69; Lorena S. Walsh, *From Calabar to Carter's Grove: The History of a Virginia Slave Community* (Charlottesville, Va., 1997), chap. 2; Anthony S. Parent, Jr., *Foul Means: The Formation of a Slave Society in Virginia, 1660–1740* (Chapel Hill, N.C., 2003), 61–65; Linda M. Heywood and John K. Thornton, *Central Africans, Atlantic Creoles, and the Foundation of the Americas, 1585–1660* (Cambridge, 2007), 216–217; Stefano Fenoaltea, "Slavery and Supervision in Comparative Perspective: A Model," *JEH,* XLIV

Irrational Overproducers and Greedy Engrossers?

In the first quarter of the eighteenth century, royal officials in Virginia, British merchants, and sometimes elite Virginians themselves thought that large planters were pursuing a self-defeating course, laying claim to more land than they could develop, buying more new African slaves than they could pay for, and producing so much tobacco as to hopelessly glut European markets. Chesapeake scholars, many relying primarily on official documents generated in the metropolis, have posed similar questions: Why between the 1680s and the 1730s did large Virginia planters continue to amass land and slaves with the goal of continually increasing the amount of tobacco they raised, thus contributing to falling prices to the detriment of everyone? Why did they continue to subject themselves almost exclusively to the vagaries of staple markets and not recognize the wisdom of diversifying into grains, as did Maryland planters around the head of Chesapeake Bay? Or why, if tobacco was so unprofitable, were they unable or unwilling to abandon it? Maryland oronoco growers have been blamed for similar shortsighted overproduction and for sabotaging the allegedly more enlightened attempts of Virginia leaders to limit the amount of tobacco grown in the region. But, since few Marylanders amassed land and slaves on the scale of great Virginia planters like the Carters and Byrds, the actions of Virginians invite the most scrutiny.[71]

The evidence on tobacco production per laborer and on planters' responses to changing markets has demonstrated that large sweet-scented tobacco growers did respond vigorously to the problem of expanding production and falling prices by concentrating on quality rather than quantity and by further processing their leaf to enhance its value. Some of their responses were voluntary, undertaken before any restrictive legislation, and others were mandated by laws of the 1680s, 1700s, and 1720s, specifying the kinds of leaves that could be tended and marketed, and by stint acts in the 1720s, limiting the quantity of tobacco that could be planted. Sweet-scented growers lived closer to the capitals of Jamestown and, from 1699, Williamsburg, than did most oronoco growers, who were concentrated on the more

(1984), 635–668; Roger Jones Family Papers, 1649–1889, MS 18,063, Library of Congress, no. 263, microfilm, M-1397.1, CWL.

71. See, for example, Leonidas Dodson, *Alexander Spotswood, Governor of Colonial Virginia, 1710–1722* (Philadelphia, 1932), chap. 4; Parent, *Foul Means,* chaps. 3, 5; John Mickle Hemphill II, "Virginia and the English Commercial System, 1689–1733: Studies in the Development and Fluctuations of a Colonial Economy under Imperial Control" (Ph.D. diss., Princeton University, 1964), chap. 2.

distant Northern Neck, lower Eastern Shore, and in the counties south of the James. Owing to their generally greater wealth and more central location, sweet-scented magnates dominated the colony's council. And, in the House of Burgesses, representatives from sweet-scented counties outnumbered legislators from oronoco counties up to 1734. Consequently, it was they who determined the final content of laws regulating tobacco growing in Virginia. Oronoco growers objected to some of the most restrictive measures as being contrary to their interests, but they lacked influence with the council and were outvoted in the legislature. How strictly such stint laws were implemented, however, depended on the diligence of county officers rather than on that of provincial officials. In sweet-scented areas, most planters seem to have complied or been compelled to comply with stint acts (Table 22). On the other hand, in the oronoco-growing areas of the Northern Neck and on the lower Eastern Shore, stint laws were poorly enforced.

Direct observations of crop yields found in plantation accounts (summarized in Table 21) show that between 1710 and 1730 in sweet-scented areas the amount of the weed each worker produced fell by as much as half. Estimates of subregional tobacco exports, income, and production, discussed previously, posit a reduction of similar magnitude in output per taxable. Planter correspondence reveals that as much as one-quarter of this drop can be accounted for by the widespread adoption of stemming, rather than by a cut in the number of plants tended. The other quarter or so reduction must have been achieved either intentionally, by having workers tend fewer plants or by allowing fewer leaves to grow to maturity on each plant, or unintentionally, owing to bad weather, to illness or resistance among enslaved workers, or to inattentive or incompetent overseers. The timing of the drop lends support to an intended reduction, although there was also a spate of unfavorable growing seasons, and the disruptions created by an unprecedented influx of unacculturated slaves in the 1700s and by a shortage of experienced overseers cannot be dismissed entirely. Estimated tobacco exports, however, did not fall in the 1690s and 1700s concurrent with the arrival of large numbers of new Africans but plummeted by as much as half only in the 1710s, a decade during which fewer slaves were imported. Sweet-scented producers thus do appear to have taken action to remedy deteriorating markets, albeit the solution was ineffective, since reduced outputs served to diminish rather than to raise revenues per worker.

An ensuing "overproduction" crisis in the late 1720s, moreover, stemmed in part from events that Chesapeake planters could hardly have anticipated. Beginning in 1726–1727, the single French tobacco monopoly buyer forced down prices in the entire European market. This agent, Peter Cavalier, bought

TABLE 22. Regional Tobacco Production in Virginia, 1724, and Population Growth, 1699–1724

County	(A) % of maximum planted	(B) Hogsheads per tithable	(C) Plants per worker (hundreds)	(D) Type	(E) Price (s/d)	(F) % of growth in tithables
HIGH LEVEL						
York	127	1.03	89	S	8/4	63
Stafford	97	1.18	62	O	—	147
Gloucester	96	0.87	57	S	2/0	37
King William	90	0.96	59	S	10/0	—
Essex	90	1.06	60	O/S	4/6	137
Lancaster	89	0.95	57	O/S	—	42
Hanover	88	0.98	58	S	—	—
King & Queen	86	0.93	56	S	8/4	60
Middlesex	86	0.85	56	S	10/6	81
James City	83	0.76	54	S	8/4	25
Westmoreland	83	1.06	56	O	—	114
Richmond	81	0.95	52	O/S	4/6	23
Warwick	80	0.83	52	S	8/6	46
New Kent	80	0.83	54	S	8/6	16
MIDDLE LEVEL						
Henrico	78	0.84	47	O	3/0	208
King George	78	0.78	53	S	—	—
Accomack	75	0.88	47	O	6/0	51
Spotsylvania	65	0.60	39	S	—	—
Charles City	64	0.72	41	O/S	3/0	−14
Northumberland	64	0.79	42	O	—	62
LOW LEVEL						
Elizabeth City	60	0.41	38	S	8/6	82
Surry	59	0.69	35	O	3/1	190
Prince George	53	0.85	47	O	3/0	—
Princess Anne	40	0.48	26	O	3/0	91
Isle of Wight	35	0.43	21	O	2/10	137
Nansemond	33	0.44	21	O	2/6	101
Northampton	32	0.39	20	O	6/0	45
Norfolk	18	0.21	11	O	3/0	179

TABLE 22. Continued

Note:

(A) Number of plants tended/(tithables employed in tobacco × 6,000) + (boys allowed to tend 3,000 plants × 3,000). The assumption is imperfect, as householders without labor were allowed to tend 10,000 plants. In the absence of information on the proportion of laborless households, however, this proved the best measure (in conjunction with other variables) for ranking counties by commitment to tobacco. The percentage of tithables employed in tobacco proved more misleading.

(B) The numbers of hogsheads were recomputed from the original table according to the formulas stated in the document (*N* Oronoco plants/8 plants to a pound/600 pounds per hogshead, and *N* sweet-scented plants/8 plants to a pound/700 pounds per hogshead/*N* of tithables). Differences from the contemporary table were assumed to be the result of arithmetic errors by the original compiler.

(C) The number of plants tended/tithables employed in tobacco.

(D) S = sweet-scented; O = oronoco.

(E) Prices of county quitrent tobacco for 1720 (or the nearest year for which observations were available). No prices are available for counties in the Northern Neck Proprietary. The price of quitrent tobacco for Gloucester County was uncharacteristically low between 1708 and 1720. Data supplied by Peter Bergstrom.

(F) Covers the years 1699–1724. See Edmund S. Morgan, *American Slavery, American Freedom: The Ordeal of Colonial Virginia* (New York, 1975), 412–413.

Source: PRO C.O. 5/1319, fol. 220, microfilm, M-240, CWL.

tobacco in large quantities and deliberately dumped whole shiploads unsuited to the French market in Holland in order to lower the "world" price so that he could buy more cheaply for the monopoly. A London tobacco merchant explained, "What that one Gentleman does, Affects all the Tobacco imported into Great Britain, for all other Buyers govern themselves by what that one man does." The price slump was owing more to the French agent's manipulation of the market than to oversupply. Sweet-scented producers thus stand partly absolved from the charge of "irrational" overproduction.[72]

Most planters in peripheral areas stopped raising tobacco altogether, demonstrating that, when the staple proved truly unprofitable, Chesapeake residents were indeed capable of finding other ways to make a living. After 1710, only in the oronoco subregion did tobacco exports rise well above the benchmarks achieved in the late 1690s. Between 1717 and 1721, more than eight thousand new Africans were brought into Virginia, and an additional sixty-seven hundred were brought in 1726–1727. As in earlier years, most of these captives were purchased by large planters whose home plantations lay

72. Price, *France and the Chesapeake,* 649–655; Memoir from a London tobacco merchant, Cholmondeley (Houghton) MSS 29/7, Cambridge University Library, quoted ibid., 650.

in sweet-scented areas. Many of the new laborers, however, were sent to grow oronoco on new upcountry quarters rather than put to work in the lower tidewater. Exports of the more common variety surged upward in 1718–1719 and again in 1726–1727 to a greater degree than did sweet-scented exports. Oronoco growers thus appear to have been the major "overproducers" of the early eighteenth century, but theirs was an entirely rational response, since the larger outputs yielded higher, not diminished, revenues per laborer.

The drastic fall in income per worker that prevailed in sweet-scented areas beginning in 1710, coupled with fears of a slave insurrection, prompted Virginia legislators to raise duties on imported blacks to five pounds per head in the hope of slowing the accumulation of new Africans. (A more moderate duty of one pound had been imposed beginning in 1699 to generate public revenues without raising unpopular poll taxes.) Another decline in tobacco prices led to the enactment, again accompanied by another insurrection scare, of a similar measure in 1723, which the British government disallowed as harmful to the interests of English slave traders. Since elite planters living in the York and Rappahannock basins were the major purchasers of new Africans, the higher duties on imported slaves, the limitations on the number of plants that a worker could tend enacted in 1723 and again in 1728, and the dispatch of an agent to England to solicit repeal of the prohibition on stripped tobacco were all steps taken to remedy the problems of sweet-scented growers. Just as any rise in tobacco prices prompted calls for repeal of stint acts, the upturns that occurred in 1715 and again in 1724 reversed whatever dampening effect the higher duties might have had. Moreover, the price of a prime African dropped by about ten pounds between the 1710s and the 1720s, a decline that more than offset the four-pound duty assessed in 1723. By then, as Reverend Hugh Jones observed, "This addition in the price of a Negroe made no sensible decrease in the number imported."[73]

The reductions in output per laborer and in total exports that the sweet-scented growers effected in the 1710s and 1720s proved singularly unsuccessful

73. Hemphill, "Virginia and the English Commercial System," 34–35, 65–66, 88–92; Edmund S. Morgan, *American Slavery, American Freedom: The Ordeal of Colonial Virginia* (New York, 1975), 307–308; Darold D. Wax, "Negro Import Duties in Colonial Virginia," *VMHB,* LXXIX (1971), 29–44; Parent, *Foul Means,* 86–95; Jones, *Present State of Virginia,* ed. Morton, 132 (quotation). For the regional distribution of imported slaves, see Lorena S. Walsh, "The Chesapeake Slave Trade: Regional Patterns, African Origins, and Some Implications," *WMQ,* 3d Ser., LVIII (2001), 139–170. Sweet-scented tobacco prices are from Walsh, "Summing the Parts," *WMQ,* 3d Ser., LVI (1999), 92–94. For a comparison of trends in Chesapeake slave imports with tobacco prices, see Lorena S. Walsh, "Mercantile Strategies, Credit Networks, and Labor Supply in the Colonial Chesapeake in Trans-Atlantic Perspective," in David Eltis, Frank D. Lewis, and Kenneth L. Sokoloff, eds., *Slavery in the Development of the Americas* (Cambridge, 2004), 92.

in raising prices. Oronoco growers who made no such reductions fared better. This outcome helps to explain why large tidewater planters were ready to abandon forever mandated quantity and quality controls as well as attempts to impose prohibitively high duties on imported Africans. Reducing outputs had never worked, except for about five years at the turn of the century, so elite planters were ready to try a fresh approach in the form of an untested centralized inspection system by the early 1730s. Once that inspection system produced the hoped-for remediation in quality and Continental markets expanded to absorb increased quantity, sweet-scented planters again reverted to tending as much tobacco per worker as they had in the third quarter of the seventeenth century. But, since profit levels were then considered satisfactory, neither contemporary merchants and politicians nor present-day scholars have castigated mid-eighteenth-century Virginia planters for "overproducing."

On the other hand, the charge that elite Virginia planters engrossed more land than they could develop to the detriment of small planters and landless tenants is entirely merited. Headrights were the main way by which planters obtained land throughout the seventeenth century in Virginia and for most of the century in Maryland. In both colonies, gentry planters and native-born or overseas merchants claimed most of the rights, either from importing large numbers of laborers or by buying unexercised rights from others for a small sum. In 1684, Charles Calvert, third Lord Baltimore, citing flagrant abuses in the trade in headrights, substituted a payment of "caution money" as the only method (aside from a direct grant from the proprietor) for taking up undeveloped land in Maryland. Baltimore intended to put a stop to merchants, ship captains, revenue collectors, and surveyors taking up most of the land with cheaply bought or fraudulent headrights and then reselling it to poor planters at "Excessive Rates." The cost for one hundred acres was initially one hundred pounds of tobacco (around seven shillings), a price Baltimore thought would satisfy "not only the small ones but also the great ones."[74]

Virginians retained the headright system until 1699, although similar abuses prevailed there as in Maryland. As servant immigration declined toward the end of the century, however, the pool of available headrights

74. William Hand Browne et al., eds., *Archives of Maryland*, 72 vols. (Baltimore, 1883–1972), V, 390–391. For the land granting procedures, see Harry Wright Newman, *The Flowering of the Maryland Palatinate* (Washington, D.C., 1961), 61–67; and Aubrey C. Land, "The Colonial Period," in Morris L. Radoff, ed., *The Old Line State: A History of Maryland* (Annapolis, Md., 1971), 13–26. Proprietary revenue agents are discussed in Donnell MacClure Owings, *His Lordship's Patronage: Offices of Profit in Colonial Maryland* (Baltimore, 1953), chap. 8.

began to dry up, except for rights awarded for importing slaves. A change in the land distribution system, uncoupling it from the pace of migration into the colony, was thus becoming inevitable. The British government looked unfavorably both on engrossment and on obtaining tracts with "Negro Rights," which by and large only the great planters obtained, and instructed Governor Francis Nicholson to award land only for importing British subjects. In 1699, the council agreed to end the practice of issuing headrights for bringing in Africans. Apparently, the councillors capitulated to royal pressure over headrights to forestall other reforms to the Virginia land system that British officials were contemplating. Headrights continued to be granted to free "Christian Subjects" who immigrated from Britain or to those who paid for their passage, but so few Europeans migrated in the early eighteenth century that this method of obtaining land became insignificant. Instead the governor and council substituted the "treasury right," whereby any person could survey and patent fifty acres upon paying five shillings to the auditor. Thereafter, in practice, the taking up of land became divorced from trends in immigration, and the amount of land that an individual could acquire was limited only by the depth of his pockets, since legal limits on the size of grants were widely ignored.[75]

In the aftermath of Bacon's Rebellion, large amounts of acreage seized from or sold by the Indians were opened to settlement. In addition to obtaining tracts using fraudulent or duplicate headrights, later-seventeenth-century elite planters devised a number of other ways for engrossing land, including cheating Indians and seizing their land, underestimating the acreage included in a patent, underestimating and often not paying quitrents, bribing county and provincial office clerks and surveyors to certify fake headright claims and false boundaries, and failing to meet the minimal requirements for "seating" lands within three years of obtaining a patent by building a twelve-foot-square wood house, running some livestock on the tract, and tending one acre in corn. Councillors, burgesses, and other provincial officeholders, who controlled all facets of land distribution and quitrent collection, were the

75. Nell Marion Nugent, *Cavaliers and Pioneers: Abstracts of Virginia Land Patents and Grants, 1623–1800,* 7 vols. (Richmond, Va., 1934–1999), II, vii–ix, III, vii–viii; *EJCCV,* I, 420, 457; "An Account of the Manner of Taking Up and Patenting of Land in Her Majesty's Colony and Dominion of Virginia with Reasons Humbly Offered for the Continuance Thereof," *WMQ,* 2d Ser., III (1923), 137–142; Hening, ed., *Statutes at Large,* III, 101–102, 204–209, 304–329, IV, 31–32, 37–42, 81–83; Edmund S. Morgan, "Headrights and Head Counts: A Review Article," *VMHB,* LXXX (1972), 361–371; Parent, *Foul Means,* 41–47. Parent's statement (47) that payment of an immigrant's transportation was required to obtain a treasury right is incorrect.

major perpetrators and beneficiaries of such abuses. By 1700, 60 percent of all estates of more than two thousand acres belonged to families of past or present Virginia officeholders.[76]

Beginning in 1698, the newly established Board of Trade routinely instructed royal governors to restrict the size of individual land grants, to enact laws requiring settlement and development of lands newly taken up within a reasonable time, and to collect quitrents more assiduously, all with the intent of giving small planters more access to land and of increasing royal revenues from quitrents and from customs duties on tobacco. For the next twenty years, successive governors tried to effect reforms, but councillors and burgesses always managed to keep one step ahead, creating new loopholes in the system when old ones were closed, legitimating previous patents however questionably obtained, so diluting requirements for seating and planting that they became virtually meaningless, and continuing to find ways to delay or avoid paying quitrents on their vast reserves. In addition to selling treasury rights, in Virginia the governor and council could also make grants of undeveloped land to worthy recipients — and who more worthy than these selfsame magnates and their relatives? After 1720, the governors allowed the council free rein to make land grants of more than four hundred acres. Large Virginia planter-officeholders awarded themselves vast reserves in the backcountry that cost them little or nothing.[77]

It has been argued up to this point that the best way to understand why large planters chose differing agricultural strategies is to consider the two Chesapeake colonies as either one region dominated by a single staple or as three distinct geographic and economic subregions differentiated by the type of tobacco grown, two of which straddled political boundaries. Also noted was that in both colonies the date of first European settlement in any given county proves an important predictor of the amount of economic opportunity available to new immigrants and of the demographic makeup of the local population. There was a regular progression in all areas from a highly productive, demographically imbalanced immigrant population to a more demographically balanced native-born population with higher proportions of nonproductive dependents. There appears to be little or no relationship, however, between age of settlement and decisions planters made about which crops to grow. Sweet-scented growers farming in the area around Jamestown,

76. Morgan, *American Slavery, American Freedom,* chaps. 11, 12; Morgan, "Headrights and Head Counts," *VMHB,* LXXX (1972), 361–371; Billings, Selby, and Tate, *Colonial Virginia,* 122.

77. Martin H. Quitt, *Virginia House of Burgesses, 1660–1706: The Social, Educational, and Economic Bases of Political Power* (New York, 1989), 144–159; Morgan, *American Slavery, American Freedom,* 358–362; Parent, *Foul Means,* 33–54.

for example, remained firmly wedded to tobacco and engaged in minimal diversification until the middle of the eighteenth century, whereas residents of Virginia's lower Eastern Shore and of the Norfolk area, places seated only a few years after the colony's founding, began shifting from tobacco by the 1670s. Settlement and county formation on Maryland's lower Western and upper Eastern shores shared a nearly identical chronology, but planters on the former were still concentrating on tobacco in the 1820s, whereas the latter adopted corn and wheat as second and third cash crops in the 1720s.[78]

When it comes to estate building and engrossment, however, political boundaries did prove determinant. Few large Maryland planters accumulated landed estates on anything approaching the scale of elite Virginians. Only one-third of the 65 men at the pinnacle of power who served on the Maryland Council between 1660 and 1715, for example, acquired more than five thousand acres of land. Major county and provincial officeholders on Maryland's lower Western Shore in the seventeenth and early eighteenth centuries accumulated landed estates of between just two thousand and three thousand acres. A study of 184 wealthy decedents with movable estates of more than £650 constant value living in Maryland between 1691 and 1742 found that the mean for elite landholding was just over twenty-seven hundred acres in the first quarter of the eighteenth century, decreasing to just over two thousand acres in the second quarter. The very wealthiest men in the colony managed to accumulate estates of only forty to fifty thousand acres, as contrasted to the one hundred or two hundred thousand acres held by the richest mid-eighteenth-century Virginians.[79]

Three factors account for the difference. First, mention has already been made of the geographic constraints faced by upper Eastern Shore residents.

78. P. M. G. Harris, "Integrating Interpretations of Local and Regionwide Change in the Study of Economic Development and Demographic Growth in the Colonial Chesapeake, 1630–1775," *WPREHRC,* I, no. 3 (1978), 35–72, demonstrates the importance of date of settlement to the stage of economic development in particular localities. The argument is more fully developed in Harris, *The History of Human Populations,* I, *Forms of Growth and Decline* (Westport, Conn., 2001).

79. David W. Jordan, "Political Stability and the Emergence of a Native Elite in Maryland," in Thad W. Tate and David L. Ammerman, eds., *The Chesapeake in the Seventeenth Century: Essays on Anglo-American Society* (Chapel Hill, N.C., 1979), 264–265; Lorena S. Walsh, "The Development of Local Power Structures: Maryland's Lower Western Shore in the Early Colonial Period," in Bruce C. Daniels, ed., *Power and Status: Officeholding in Colonial America* (Middletown, Conn., 1986), 58; Trevor Burnard, *Creole Gentlemen: The Maryland Elite, 1691–1776* (New York, 2002), 33–36; Aubrey C. Land, *The Dulanys of Maryland: A Biographical Study of Daniel Dulany, the Elder (1685–1753) and Daniel Dulany, the Younger (1722–1797)* (Baltimore, 1955), 199–200; Hoffman, *Princes of Ireland,* 68–73; Beatriz Betancourt Hardy, "A Papist in a Protestant Age: The Case of Richard Bennett, 1667–1749," *Jour. So. Hist.,* LX (1994), 203.

Although Western Shore planters had the advantage of access to unclaimed land to the west of existing settlements, compared to Virginia possibilities for expansion were exceedingly limited. After 1680, the newly established Pennsylvania colony precluded expansion northward, and a dispute between the two proprietors over the location of the boundary left Maryland residents leery of taking up land near the border. To the south and west, the northward curving Potomac River boundary left Marylanders with fewer undeveloped acres in which to expand than were included in the counties already established by 1700. Members of the Maryland elite seeking to build family empires had perforce to turn fairly quickly to trade, shipbuilding, manufacturing, or moneylending rather than to accumulating acreage on the frontier. In contrast to Virginians, who knew only that the sea constituting the colony's western boundary lay farther west than any European had yet traveled, early-eighteenth-century elite Marylanders faced decided limits to their territorial expansion.

Second, as successive proprietors raised the amount charged for caution money, buying unclaimed land became increasingly expensive in Maryland. By 1738, one hundred acres cost five pounds sterling, as contrasted to ten shillings in Virginia, assuming a grantee there paid anything at all. Third, in Maryland both land and quitrents belonged to the proprietor, and it was he, not the planter-dominated council or lower house, who chose the quitrent collectors. These officials were generally able men (James Carroll was one) whom the Calverts rewarded well for loyal service, and they could retain their positions only by keeping careful, accurate records and collecting rents assiduously. Wealthy Maryland planters were thus much less likely to patent large tracts of land that they had no immediate prospects for developing, given the higher costs both of buying the land and of retaining it. In Virginia, in contrast, the foxes were in charge of the henhouse. Marylanders could have purchased additional land in Virginia, as a number in fact did in the 1670s, but by the early eighteenth century most were deterred by the greater distances involved and by the higher costs they would incur in comparison to Virginians who often obtained land for nothing.[80]

As a direct consequence of Maryland planters' limited chances for accumulating large amounts of land by the early eighteenth century, most also had no need to purchase large numbers of slaves. Those whose holdings did not exceed two thousand acres owned enough land to profitably employ a labor force of up to about thirty working slaves. Beyond that there were better ways

80. Land, "The Colonial Period," in Radoff, ed., *The Old Line State,* 13–26.

to invest their capital. Such land-labor ratios help to explain why planters like Robert Goldsborough and James Carroll stopped buying additional slaves early in their planting careers. These constraints on the size of labor forces in Maryland meant that, in turn, there were very few large tobacco producers in the colony. Elite tidewater Virginians, in contrast, usually held enough acreage to employ all the slaves they could manage to buy and to produce tobacco on a much grander scale.[81]

Exploiting Human Capital

The different financial arrangements that prevailed in the servant and slave trades and the ways that human laborers might or might not be used as collateral had important consequences for the acquisition and maintenance of bound labor forces in the Chesapeake. Little is known about the terms on which indentured servants were sold in the seventeenth and early eighteenth centuries. Ship captains employed in the tobacco trade sailed to the Chesapeake with cargoes of British manufactures and European servants, both of which they traded for tobacco. Apparently, most servants were sold for ready pay. One can imagine that a shipper who arrived late in the season might have had to sell some laborers for next year's pay or that a small planter who did not have quite enough tobacco to buy a servant might have borrowed the difference from a richer neighbor. Still, planters did not usually go into debt to buy servants, presumably because credit was seldom extended, and, if they did, they borrowed only small amounts and only for a short time. Servants were indeed treated as commodities until their terms were up. They could be freely sold to other masters, seized for payment of debts, or devised to others by will. However, owing to their limited time of service—usually between four and seven years—they made poor collateral as security for a loan. By the time a lender decided a borrower had defaulted on a loan, brought suit, obtained a judgment from a court, and got a sheriff to seize the debtor's assets, a mortgaged servant might well have completed his or her term or have only a short time of little value left to serve. Consequently, planters rarely financed the purchase of additional workers using servants as collateral, although provincial lenders regularly accepted cows (with an expected useful

81. For land constraints on the size of labor forces, see Burnard, *Creole Gentlemen,* 36–41. Jacob M. Price similarly remarked on the virtual absence of planters in eighteenth-century Maryland who made thirty thousand pounds of tobacco a year, whereas some Virginia grandees could market more than one hundred thousand pounds ("Merchants and Planters: The Market Structure of the Colonial Chesapeake Reconsidered," in Price, *Tobacco in Atlantic Trade: The Chesapeake, London, and Glasgow, 1675–1775* [Aldershot, Eng., 1995], 21–25).

life of ten to twelve years), featherbeds, and iron pots as security for other kinds of loans.[82]

That owners of servants could buy more laborers only with the profits that servants generated rather than borrowing against their bodies might be one reason why great planters failed to accumulate large numbers of indentured servants. The thirty-nine servants apiece held by sometime Virginia governor George Yeardley and Cape merchant Abraham Peirsey in 1625 are the largest known European labor forces in the Chesapeake. A survey of more than ten thousand Virginia and Maryland probate inventories reveals only a score of seventeenth- or early-eighteenth-century estates with twenty or more servants, and only one with more than thirty. In the 1640s and 1650s, wealthier planters on Barbados encountered similar logistical or financial problems, as estates there with twenty or more servants also appear to have been rare. Another reason might have been the high turnover costs associated with finding and training new workers every few years.[83]

The international slave trade employed different financial arrangements in which payment in bills of exchange or coin rather than in produce predominated and in which credit was often extended. Practices developed by the Royal African Company for slave sales in the West Indies between 1680 and 1700 were adopted by all colonial slave dealers by the end of Queen Anne's War. Merchants in Britain were becoming increasingly specialized in the later seventeenth and early eighteenth centuries, and the slave trade was no

82. For terms of service, see Morgan, *American Slavery, American Freedom,* 216–218; and Lorena S. Walsh, "Charles County, Maryland, 1658–1705, a Study of Chesapeake Social and Political Structure" (Ph.D. diss., Michigan State University, 1977), 157–158. Hilary McD. Beckles, "Plantation Production and White 'Proto-Slavery': White Indentured Servants and the Colonisation of the English West Indies, 1624–1645," *The Americas,* XLI (1985), 42, found that in the 1640s Barbadian planters commonly used servants as property security on two-to-five-year mortgage agreements. John C. Coombs has compiled a database of all entries pertaining to bound labor in extant Virginia county court records from 1634 to 1700. He found only six instances (all but one dating to the 1650s and 1660s) of servants being offered as security for debt. (Personal communication, Mar. 13, 2008; see Westmoreland County Deeds, Wills, and Patents, 1653–1659: 105–106; Westmoreland County Deeds and Wills, 1, 1653–1671: 244–249; Westmoreland County Deeds, Patents, etc., 1665–1677: 265–266; Northumberland County Record Book, 1658–1666: 106; [Old] Rappahannock County Records, 1656–1664: 215, 287–288, all on microfilm in VSL). By the 1750s, merchants did sell servants on credits of four to twelve months (Walsh, "Mercantile Strategies, Credit Networks, and Labor Supply," in Eltis, Lewis, and Sokoloff, eds., *Slavery in the Development of the Americas,* 114–115).

83. Morgan, *American Slavery, American Freedom,* 118–119; Historic Saint Mary's City Probate Inventory Database, MSA; Russell R. Menard, *Sweet Negotiations: Sugar, Slavery, and Plantation Agriculture in Early Barbados* (Charlottesville, Va., 2006), 32–33; Beckles, "Plantation Production and White 'Proto-Slavery,'" *The Americas,* XLI (1985), 34; Christopher Hanes, "Turnover Cost and the Distribution of Slave Labor in Anglo-America," *JEH,* LVI (1996), 307–329.

exception. Slave traders concentrated on buying slaves in Africa efficiently, delivering them to colonial purchasers expeditiously, and, it was hoped, collecting the proceeds within fifteen to eighteen months. Although some merchants were involved in both the slave and tobacco trades, many English slave dealers had little or no experience with either the Chesapeake region or with tobacco and had no desire to dabble in this commodity. A return cargo of tobacco—whether tendered for slaves or shipped on consignment—would add to earnings from freight. But, if collecting the weed delayed the ship's departure for England too long, the slave traders might not be able to maintain the tight seasonal schedule required for completing one round-trip slaving voyage per year, and their overall profits depended on making regular annual voyages.[84]

The preferred remittances were thus always bills of exchange, preferably readily negotiable ones drawn on London merchants rather than on the outports. The only other form of ready pay was hard coin, which was in short supply in the colonies. Moreover, in the first half of the eighteenth century slave traders accepted specie only grudgingly, in part because the investors were loath to risk losing coin to shipwreck or capture on the return voyage to England. (Bills of exchange were normally purchased in sets of two or three duplicate copies and sent separately on different vessels.) In addition, Virginia's and Maryland's currencies differed in value from sterling (and from each other). Establishing the precise relative values of coins in both colonial currency and sterling added one more difficulty to any transaction. "The loss arising on Cash Remittances" remained a problem for investors. Consequently, agents were instructed to pay out locally as much specie received for slaves as they could for duties, for repairing and provisioning the ship, for the ship's captain and for sailors' wages (part of which were paid in local currency

84. For organization of the slave trade, see K. G. Davies, *The Royal African Company* (London, 1957); David W. Galenson, *Traders, Planters, and Slaves: Market Behavior in Early English America* (Cambridge, 1986); Charles L. Killinger III, "The Royal African Company Slave Trade to Virginia, 1689–1713" (master's thesis, College of William and Mary, 1969); and David Richardson, "The British Empire and the Atlantic Slave Trade, 1660–1807," in P. J. Marshall, ed., *The Eighteenth Century*, vol. II of Wm. Roger Louis, ed., *The Oxford History of the British Empire* (Oxford, 1998), 447–450. Transatlantic trade cycles and strategies for return freights are discussed in Stephen D. Behrendt, "Markets, Transaction Cycles, and Profits: Merchant Decision Making in the British Slave Trade," *WMQ*, LVIII (2001), 171–204; and Herbert S. Klein, *The Middle Passage: Comparative Studies in the Atlantic Slave Trade* (Princeton, N.J., 1978), chap. 6. For financial arrangements, see Jacob M. Price, "Credit in the Slave Trade and Plantation Economies," in Barbara L. Solow, ed., *Slavery and the Rise of the Atlantic System* (Cambridge, 1991), 293–339; and Walsh, "Mercantile Strategies, Credit Networks, and Labor Supply," in Eltis, Lewis, and Sokoloff, eds., *Slavery in the Development of the Americas*, 89–119.

in Chesapeake ports), and for the agent's commission. Agents then tried to buy bills of exchange with any remaining coin. Sometimes the investors had no option but to accept some tobacco (and in the most desperate cases even lumber and wheat), but usually only as a last resort when the most unpromising captives could be disposed of by no other means.[85]

English investors hoped to avoid selling slaves on credit, but here their interests diverged from the desires of both agents and planters. To avoid having to wait longer than intended for the proceeds of human cargos, slave traders relied on their agents to sell only to purchasers who had sufficient reserves with British correspondents to back their bills of exchange, and they obligated the agents to make good any bills that were protested. The Royal African Company ostensibly set the most inflexible terms. Slaves were to be sold for bills of exchange, and factors were obliged in theory to make good any protested bills within twelve months from the time of the sale. In practice, factors seem usually to have taken two to three years to collect outstanding debts from a given voyage, and some were never collected. Although some of the bills of exchange that planters drew on British tobacco merchants were fully backed by proceeds of tobacco sales, others were deliberate overdrafts executed in the expectation (or hope) that the tobacco factor on whom the bill was drawn would advance the planter the required amount. The tobacco merchants were not always willing or able to make such advances, so slave sellers could expect that some proportion of remitted bills would be returned protested. Independent traders, particularly those from the outports, more often explicitly authorized their agents to sell slaves on credit, accepting bills of exchange payable, not at the usual sixty days after sight, but six to twelve months later. And often agents sold less desirable slaves on credit despite orders to the contrary, needing to shift the risks of death or escape of unwilling captives and the costs of their maintenance onto buyers as quickly as possible. In addition, the agents had a further incentive to make credit sales because planters were willing to pay up to 10 percent more for slaves sold on credit than for those bought for ready pay, thus increasing the amount the agent received in commission.[86]

The ready availability of credit made it possible for large planters to expand their labor forces to an extent that had not been possible with indentured servants. And, although the problems British merchants faced in col-

85. Walsh, "Mercantile Strategies, Credit Networks, and Labor Supply," in Eltis, Lewis, and Sokoloff, eds., *Slavery in the Development of the Americas,* 96–98; Price, "Credit in the Slave Trade," in Solow, ed., *Slavery and the Rise of the Atlantic System,* 293–339.

86. Walsh, "Mercantile Strategies, Credit Networks, and Labor Supply," in Eltis, Lewis, and Sokoloff, eds., *Slavery in the Development of the Americas,* 96–98; Davies, *The Royal African Company,* 316–325.

lecting these debts in all the colonies to which they traded was occasioned by planters' inability to pay during cyclical economic downturns, it was owing even more to "colonial legal systems, judicial as well as legislative," that "were controlled by those favorable to debtors, and indeed often by the debtors themselves." When between 1705 and 1712 in oronoco regions and in the late 1720s in the sweet-scented region tobacco prices plummeted and planters experienced drastic drops in income that prevented them from repaying large sums borrowed to buy slaves when prices were higher, they turned to their legislatures for relief. The planter-legislators then enacted measures that made it harder for British merchants to collect, such as exempting future earnings from the claims of current creditors and setting strict time limits for legal actions for recovery of debts. The merchants, in turn, requested that the Board of Trade recommend to the Privy Council that such acts be disallowed. During a severe contraction, the activities of slave buyers precipitated a chain reaction affecting all colonists; bills of exchange, which in better times might have circulated as a local currency, were all remitted to England, some of the coin paid for slaves might be shipped home, further contracting the money supply, and large planters cut back on normal levels of lending to smaller neighbors. Large planters had already taken on more risk than they had been accustomed to by consigning their tobacco rather than selling in the country, and borrowing to buy slaves further increased their vulnerability to market busts. As we have seen from the individual planters discussed above, however, by and large those who enjoyed relatively long lives succeeded in making the new Africans they purchased pay for themselves. And among those who died young, heirs and executors gave first priority to paying off mortgages in order to secure recently purchased slaves.[87]

Moreover, not all planters required credit. Factors aimed to remit half the proceeds of a shipload of Africans in the vessel that had brought them to a colonial market. To achieve this goal, normally about a third of the more promising captives were sold for ready pay—either for specie or bills of exchange deemed sound—and another third or so were sold for half ready pay, with a year's credit extended for the balance. Only those who were least promising were sold entirely on credit. When a planter paid for slaves with bills of exchange that were not backed by sufficient remittances of tobacco or by sterling balances in England, he in effect gained a year in which to find another way to pay what he owed, during which time he had the benefit of the new African's labor, and had then only to pay a penalty of 10 to 15 per-

87. Galenson, *Traders, Planters, and Slaves,* 149 (quotation); Main, *Tobacco Colony,* 81–82; Parent, *Foul Means,* 86–95.

cent on the protested bills. Before passage of the Colonial Debts Act of 1732, British creditors could not seize a planter's land or slaves unless he had given a mortgage. If he did not pay from the next year's crop, the merchant or his representative would then go to court to obtain a judgment, or perhaps force the debtor to give bond with securities if he had not already done so, or make him execute a mortgage. Meanwhile, the merchant in Britain could increase pressure by also refusing to send the planter any additional goods or to honor any further bills of exchange. On balance, however, the arrangements prevailing in the colonies before 1732 generally privileged the rights of slave buyers over those of slave sellers.[88]

Throughout the British Empire, slaves, unlike indentured servants, were readily accepted as collateral for loans. British merchants usually took mortgages on land and slaves only as a last resort, generally preferring bonded debts, because managing plantations an ocean away or selling them to others might entail a great deal of trouble and often considerable loss. Local lenders, however, preferred or at least readily accepted mortgages on chattels as well as on land. By the 1720s, as the example of John Bagge's mortgage of his home plantation and twenty slaves to Robert "King" Carter attests, indebted slave buyers had begun financing—or more likely refinancing—their human purchases by borrowing from the planter elite. They could then pay the English merchant what they had borrowed, resume their usual orders of metropolitan goods, preserve their most valuable assets from the clutches of overseas creditors, and perhaps negotiate a longer time for repayment. The extent of such local lending is difficult to determine, since mortgages did not have to be recorded to be legally binding, and mortgage holders went to the trouble and expense of recording them only if the debtor defaulted. But, from the numbers of mortgages that were recorded, it is clear that both great planters and resident merchants with money to lend were becoming deeply involved in not just buying slaves for themselves and in selling cargoes of new Africans but also in financing or refinancing the purchase of slaves by others.[89]

88. Price, "Credit in the Slave Trade," in Solow, ed., *Slavery and the Rise of the Atlantic System,* 308–311. Examples of early-eighteenth-century Chesapeake slave sales can be found in Loyde Account Book, 1708–1710, UVa (*Leopard Galley,* 1710); Lyde Account Book, 1708–1711, and Accounts and Letters of John Tayloe I and John Tayloe II, 1714–1778, Tayloe Family Papers, VHS (*Charfield,* 1717); James Carroll Daybook, 1714–1721, Maryland Province, Society of Jesus Archives, Special Collections, Georgetown University Library (*Margaret,* 1718); and Baylor Ledgers, 1719–1755, Baylor Family Papers, CWL (*Prince Eugene,* 1719).

89. Russell R. Menard, "Financing the Lowcountry Export Boom: Capital and Growth in Early Carolina," *WMQ,* 3d Ser., LI (1994), 659–676; Land, *The Dulanys of Maryland,* 103–106; Hoffman, *Princes of Ireland,* 73–75. For the validity of mortgages between contracting parties, see Hening, ed., *Statutes at Large,* IV, 408–431.

Slaves bought on credit could thus not only pay for themselves but also, through their labor, produce revenues with which to purchase more slaves. As Jacob Price has concluded, credit served to "accelerate the processes of initiation and expansion of slave systems." Money borrowed locally secured by mortgages on land and slaves or on slaves alone similarly financed much of the expansion of rice agriculture in South Carolina between 1710 and 1740. In both mainland colonies, the primary reason for mortgaging plantations and slaves was not a final desperate attempt to delay creditors from seizing productive assets; rather it was a choice that enabled them to further expand their estates by purchasing additional land and, especially, slaves. However reprehensible using human beings as collateral appears to modern eyes, planters who borrowed to buy slaves were behaving as agricultural entrepreneurs, not undisciplined spendthrifts.[90]

Planters' obligation to repay the debts they incurred to buy slaves had other consequences, constraining their choices of crop mix, and was likely one factor retarding diversification into grains in both prime sweet-scented- and oronoco-growing regions. Since there was little market in Britain for imported corn or wheat flour, to retire their debts to merchants in London and elsewhere as quickly as possible recent slave buyers had to concentrate on maximizing tobacco revenues, since, if they shipped provisions to the West Indies, payment was often remitted in sugar, molasses, and rum, rather than in bills of exchange. Upper Eastern Shore oronoco growers who bought fewer slaves and had already paid for them by the 1720s could better afford to risk a shift in crop mix that involved a reduction in metropolitan earnings and several years of experimentation, of training the workforce in new skills, and of further investment in land clearing and draft animals than could those who bought greater numbers of Africans.

As we have seen in earlier chapters, the wealthiest and more powerful Chesapeake planters recognized as early as the 1640s and 1650s that they could realize higher agricultural profits by switching from servants to enslaved Africans. By the mid-1670s, provincial elites had made considerable progress toward that goal before servant emigration declined or tobacco prices reached unprecedented lows. Not until the late 1690s, however, did the numbers of Africans carried into the Chesapeake begin to meet labor-hungry planters' demands and increasingly available credit enable them to

90. Price, "Credit in the Slave Trade," in Solow, ed., *Slavery and the Rise of the Atlantic System*, 294; Menard, "Financing the Lowcountry Export Boom," *WMQ*, 3d Ser., LI (1994), 674–675. For slaves as a form of credit superior even to land, see Wright, *Slavery and American Economic Development*, 69–70.

speed up the pace of acquisition. Until then, the short supply and high cost of Africans forced large Chesapeake planters to shift into slaves gradually while continuing to buy indentured servants to fill up their quarters. Into the 1680s, for example, elite Virginia planters like William Fitzhugh, William Byrd I, and John Carter II made tobacco with a mix of indentured servants and enslaved Africans, a mode of operation that less wealthy and less centrally located Maryland planters like Robert Goldsborough and Richard Tilghman continued to pursue into the 1690s and 1700s.

An argument for a conscious decision among Chesapeake elites to adopt slave labor in the 1660s and 1670s must, however, be reconciled with evidence that the prices of indentured servants also rose beginning in the 1680s. The most plausible explanation for the bidding up of servant prices is not that *all* planters preferred servants; rather, there was a dual market for servants and slaves. Middling and small planters, who made up the majority of labor buyers, likely bid up the price of the diminishing supply of white servants at the same time that more limited numbers of great planters opted primarily or exclusively to buy enslaved Africans. Men who still worked the fields alongside their bound laborers might have been willing to pay a premium for fellow countrymen, rather than taking the risk of a crippling loss should a slave die, and they might have eschewed the trouble of trying to extract work of comparable quality from culturally alien and decidedly unwilling workers.[91]

Still, all aspiring slaveowners in the seventeenth and early eighteenth centuries faced a transitional period after they first acquired Africans during which they could not fully realize the savings in maintenance costs and the intensified workloads that slaveholding promised. So long as a mixture of indentured Europeans, enslaved Africans, and, less frequently, enslaved native Americans shared housing and meals as well as work assignments, it would have proved difficult if not impossible to systematically force some members of the group to work longer and harder than others, to deny some the privilege of going off the plantation while allowing others to leave, to permit some to have sexual relations but expect others to be celibate, to distribute unequal rations to workers sharing a common mess, or to consistently force black women to do the same work as men while sparing white female coworkers from heavy field labor.

As a result, beginning in the 1680s many of the largest planters began assigning black and white field-workers to separate residential and work units,

91. Russell Menard, "From Servants to Slaves: The Transformation of the Chesapeake Labor System," *So. Stud.*, XVI (1977), 373–374, 385–387, argued for a preference for servants but also proposed a dual labor market as an alternative hypothesis.

a change that improved their chances for cutting costs and for minimizing disruptive contacts between Africans and European servants. Rations for a mixed group of servants and slaves, for example, were likely provided communally every day. In the last decades of the century, however, slaveowners shifted to giving a standard allotment of corn and scantier rations of meat (consisting primarily of salted pork) to individual slaves weekly. Archaeological assemblages on slave quarters have higher frequencies of pig remains relative to cow remains than are found on sites occupied by Europeans, suggesting the emergence of a new provisioning strategy based on preserved rather than fresh meat, which slaves might have kept secure in the multiple subfloor pits that they began digging in their quarters near the end of the century. Separate work units also improved slaveowners' chances for imposing identical work requirements for black women as for men, for increasing both the duration and intensity of work, for forcing children into the fields at young ages, and for administering harsher discipline. By separating slaves from servants, slaveowners likely also hoped that the enslaved would have fewer opportunities for learning what European servants considered a fair daily task or a customary ration. Moreover, given the widespread belief among the enslaved even well into the eighteenth century that conversion to Christianity might entitle them to freedom, some slaveowners might also have hoped to limit contact between Christian servants and unconverted Africans.[92]

Another reason for separating slaves from servants was the differing expectations labor owners had about sexual relations among temporarily bound workers and those held for life. Slaveowners soon came to condone, if not to encourage, sexual activity among those who were enslaved. Perpetual celibacy could not realistically be enforced, since transgression could not be punished by extending the term of service. Moreover, slaveowners began to recognize that in the short run the creation of domestic units was an effective strategy for deterring new Africans from running away and for reconciling them to bondage, and in the long run owners stood to benefit financially from any children whom sexually active enslaved women might bear. Indentured servants, by contrast, were expected to abstain from sex during their period of indenture, and magistrates imposed harsh penalties on women who became pregnant and who thus deprived their owners of some weeks of promised service. Owners of servants also resented being burdened with the trouble and expense of accommodating, feeding, and caring for an infant or toddler

92. Main, *Tobacco Colony,* 105–106, 130–136; Coombs, "Building 'The Machine,'" 204–210. For changes in the provisioning system, see Graham et al., "Adaptation and Innovation," *WMQ,* 3d Ser., LXIV (2007), 508–510.

whose presence interfered with the mother's work, so they often bound out these unwanted children, once weaned, to other families. Worse, when black and white bondmen and bondwomen had sex together, the biracial children born of these unions challenged an emerging labor system grounded in racial difference. Landowners doubtless hoped that accommodating workers among whom sexual relations were permitted in living quarters separate from those for whom sex was proscribed might reduce unwanted pregnancies among servant women and discourage interracial liaisons that challenged racial hierarchies in multiple ways. On separate outlying quarters where only black workers lived, barracks gradually gave way to individual cabins or two-family duplexes where men and women who entered into regular unions, or at least mothers and their children, were allowed to live together. Once racially segregated quarters became places where reproduction could be expected, slaveowners had to acknowledge the necessity of making or of allowing residents to make some sort of arrangements for the care and supervision of children who were weaned but too young for field labor.

Spatial isolation was another of the strategies owners adopted for controlling slaves and maximizing profits but was not without advantages for the enslaved. Removed from the direct surveillance of the master and the prying eyes of whites other than the overseer, slaves had more freedom to choose how they would cook their meals, tend their gardens, and arrange their houses and yards. Separate residences afforded a measure of domestic privacy and opportunities for modifying the interior of their dwellings by digging underground storage pits and for pursuing communal activities in quarter courtyards that might resemble compounds in Africa.

Elite planters might have turned away from indentured servants in the late 1670s and 1680s (as did their contemporaries in Barbados), not so much because they feared the social unrest that discontented former servants might stir up after they gained their freedom, but because their presence in mixed work groups limited planters' ability to fully exploit enslaved laborers. Blacks who had been brought into the region before the mid-1670s, a group Ira Berlin has termed the "charter generation," were hardly model slaves. As a result of living and working with indentured servants who shared, at least temporarily, a common bondage, enslaved blacks often gained fluency in English, a mastery of English customs and laws, familiarity with economic exchange and the Christian religion, and an intense desire for freedom. As large planters acquired more slaves, physical separation of black workers enabled them not only to intensify labor requirements and cut maintenance costs but also to deny the new arrivals any opportunity to become part of mainstream

Chesapeake society. As the numbers of blacks in the population rose and their physical isolation increased, interactions with whites were limited to relations of deference and authority.[93]

Africans who belonged to the later-arriving "plantation generations" were more likely to speak pidgin or creole rather than standard English, to have no knowledge of or interest in Christianity, and to learn little about European customs and institutions. Increased numbers made it easier, at least for those living on large plantations, to retain more of their original languages and customs at the same time that increased physical and social isolation limited opportunities for the newly arrived to develop fluency in English or to gain much knowledge about the wider, off-plantation world to which they had been forcibly brought. The same cannot be said for the native born, who spoke better English and had greater understanding of European ways. At the turn of the eighteenth century in Maryland, only about one in ten adult blacks were native born, but the proportion was surely higher in Virginia, where as much as half the black population might have been born in the colony. Whatever their level of proficiency in English or familiarity with the dominant culture, however, in the last quarter of the seventeenth century all slaves in the Chesapeake were restricted in their movements, denied independent economic activities, and usually relegated solely to unskilled field labor. Opportunities to escape slavery were almost entirely closed off, and all blacks were consigned to a debased status defined solely by color. A social system was emerging based upon chattel bondage maintained by force and by white supremacy.[94]

93. Morgan, *American Slavery, American Freedom,* 308–309, 225–234, argues that the dangers posed by freedmen in the aftermath of Bacon's Rebellion were a leading motivation for the labor switch. Amussen, *Caribbean Exchanges,* 69–70, emphasizes the disruptiveness and lower productivity of servants in Barbados. Ira Berlin, *Many Thousands Gone: The First Two Centuries of Slavery in North America* (Cambridge, Mass., 1998), 17–28, 95–108, enumerates the distinctions between societies with slaves and slave societies and, in chaps. 1 and 5, the differences between the charter and plantation generations in the Chesapeake. Coombs, "Building 'The Machine,'" 98–99, 112–125, emphasizes servants as the main agents shaping cultural choices among the enslaved in the seventeenth century.

94. Ira Berlin attributes many of the differences between the charter and plantation generations to changes in the geographic origins and prior cultural experiences of blacks brought into the region. His contention that most blacks arriving in the first three-quarters of the seventeenth century had previously undergone a period of acclimatization and acculturation in the West Indies and hence were less linguistically and culturally isolated (*Many Thousands Gone,* 29, 100–112) is especially problematic, since scholarship on the transatlantic slave trade demonstrates that nine of ten imported laborers either came directly from Africa or had only a brief stopover in the islands. Data from David Eltis et al., *Voyages: The Trans-Atlantic Slave Trade Database* (Atlanta,

The combination of an increased supply of Africans and a pronounced decline in servant emigration during King William's and Queen Anne's wars produced what economists term a path-dependent trend in the use of servants and slaves. Once a planter acquired a workforce in which more were enslaved than not, there was no turning back. Once in the minority, white servants would not work at the same tasks alongside slaves nor accept the more spartan living conditions and harsher discipline planters were learning to impose on permanently bound workers. By the 1690s, many large planters stopped buying servants for ordinary field labor and switched entirely to slaves. Slaves outnumbered servants by nine to one, for example, in labor forces of the Virginia elite in the final decade of the century, and by then the whites were working almost exclusively at trades or in supervisory positions. Had they attempted to revert to a mix of servants and slaves, productivity per worker would have almost inevitably declined, and maintenance costs would have risen. A decade later, in estates with just six or more laborers, planters had ceased buying servants after acquiring slaves, and those who could afford even fewer workers increasingly opted to buy only slaves.[95]

Ga., 2008), http://www.slavevoyages.org, and Lorena S. Walsh Database of Slave Trading Voyages to Virginia and Maryland (November 2008) also raise questions about Berlin's argument (102–104) that the charter generation's background in the African littoral better prepared them to survive and to make their way in a new environment than did the provincial origins of the farmers and herders from the African interior who came later. There is no evidence for any pronounced shift from coast to interior in the geographic origins of Africans captured between 1680 and 1700, and London merchants remained the predominant shippers up to 1718. The shift in the center of the trade from London to Bristol and then Liverpool did not occur until after the Chesapeake plantation regime had been fully developed. On the other hand, starting in the first decade of the 1700s, there was a shift in geographic origins of new Africans. More than half of those arriving in the 1700s, 1710s, and 1720s came from the Bight of Biafra, a region from which few slaves had previously been imported. The effect of the arrival of large numbers of Africans brought from the same region on evolving African American culture in the Chesapeake continues to be debated. How much a posited "Africanization" of the black population, starting in the 1670s and accelerating after 1700, might have led to increasing debasement of all who were enslaved depends in part on the proportion who were African born. Berlin asserts, based on evidence for Maryland that 90 percent of adults were African born, that around 1700 "newly arrived Africans composed nearly 90 percent" of the Chesapeake slave population (*Many Thousands Gone,* 110). Philip D. Morgan, on the other hand, estimates the proportion of Africans in Virginia at only 50 percent in 1700, rising to 52 percent in 1710 and falling thereafter (*Slave Counterpoint: Black Culture in the Eighteenth-Century Chesapeake and Lowcountry* [Chapel Hill, N.C., 1998], 61). For trade patterns, see Richardson, "The British Empire and the Atlantic Slave Trade," in Marshall, ed., *The Eighteenth Century,* vol. II of Louis, ed., *Oxford History of the British Empire,* 440–464. For proportions of slaves brought from the West Indies and for regions of departure in Africa, see Walsh, "The Chesapeake Slave Trade," *WMQ,* 3d Ser., LVIII (2001), 139–170.

95. Main, *Tobacco Colony,* 105–106; Grubb and Stitt, "Liverpool Emigrant Servant Trade,"

As day-to-day living and working conditions worsened and legal repression of the enslaved increased, servants and nonslaveowning whites often became willing participants in a system of racial separation and degradation of blacks. Servants and hired wage laborers increasingly demanded, and obtained, separate living quarters, different and better rations, and separate work assignments, or they insisted that their status be preserved by giving them some measure of supervisory authority over the slaves. By the turn of the century, slaveowners had little choice but to employ indentured servants only as skilled workers or as overseers, the course Robert "King" Carter, for example, followed in the 1720s and early 1730s, or to dispense with servants altogether. And, as all whites came to have an interest in separating and distinguishing themselves from blacks, it became increasingly easy to enlist the help of poor whites in capturing runaway blacks and in policing slave quarters.[96]

In the other major British colonies that became slave societies, Barbados and South Carolina, the introduction of new labor-demanding staple crops—sugar and rice—accelerated a shift from servants to slaves. Intensification of commodity production necessitated the imposition of harsher work regimens and required greater violence to enforce them. In these colonies, worsening living and working conditions led to large organized conspiracies, in 1675 and 1692 in Barbados and in 1739 in South Carolina. In the Chesapeake, where adoption of a staple came first, no alteration in crop mix or pronounced changes in plantation operations followed from the more gradual shift from servants to slaves. Consequently, no conspiracies were reported on the scale of the planned uprisings in Barbados, and no insurrections broke out as bloody as the Stono Rebellion in South Carolina. Period travelers frequently noted the increasing disparity in living standards between servants and slaves, but they commented less about the intensification of work requirements or growing debasement of blacks. These developments are more inferred from the progressively repressive content of Maryland and Virginia laws and from the refusal of white servants to be treated the same as black slaves. Although Bacon's Rebellion momentarily raised the specter of slaves

Exp. Econ. Hist., XXXI (1994), 376–405; Coombs, "Building 'The Machine,'" 99. Grubb and Stitt argue from an analysis of the prices planters in Maryland were willing to pay for servants and slaves that by the 1680s slaves were seen as close substitutes for servants and that by the 1710s slaves were considered more productive.

96. Lois Green Carr and Lorena S. Walsh, "Economic Diversification and Labor Organization in the Chesapeake, 1650–1820," in Stephen Innes, ed., *Work and Labor in Early America* (Chapel Hill, N.C., 1988), 144–188; Main, *Tobacco Colony,* chap. 3; Berlin, *Many Thousands Gone,* chap. 5.

and servants making common cause against large planters in the late 1670s, that moment quickly passed as the composition of the labor force shifted irrevocably over the course of the next few years. By 1680, slaveowners were more worried about the sounds of mysterious drums in the night and clandestine gatherings of Africans holding feasts and funerals for their dead in the forests. In that year, they uncovered the first of many purported black insurrection plots that would surface in the region over the next half century. The economic difficulties that large planters faced during these years heightened their resolve to maximize plantation revenues and to cut costs with a single-mindedness that precipitated both individual and more widely orchestrated resistance in the quarters. It is hardly accidental that this economically troubled era concluded in 1730 with the largest slave uprising, involving some two to three hundred conspirators, that occurred in the Chesapeake during the whole colonial period.[97]

Regional Divergence, Inequality, and the Rise of a Slave Society

The social consequences of the agricultural and labor-force adaptations that Chesapeake planters made during the era of hard times can be summarized by combining the results of county population censuses taken in Maryland between 1704 and 1712, information on the number of tithables in Virginia counties in 1699, and a report Virginia lieutenant governor Hugh Drysdale prepared for the Board of Trade in 1724 listing the number of tithables in each county and the number of tobacco plants tended that year (Map 9, Tables 22, 23). (In Virginia, the number of tithables is not broken down by race, so one must rely on local studies for some idea of the proportions of whites and blacks living in various counties.)[98]

Growth in the white population, one indicator of how much economic opportunity a jurisdiction afforded to ordinary planters, was most pronounced in two sorts of Chesapeake counties. Not surprisingly, the numbers of whites rose in more newly settled areas like Stafford and Surry counties in Virginia, where families could still improve their fortunes through farm building. Rather less expected is the white population's rapid expansion in

97. Richard S. Dunn, *Sugar and Slaves: The Rise of the Planter Class in the English West Indies, 1624–1713* (Chapel Hill, N.C., 1972), 256–258; Edelson, *Plantation Enterprise,* 81–89; Lorena S. Walsh, "The Differential Impact of Free and Coerced Migration to Colonial America," in David Eltis, ed., *Coerced and Free Migration: Global Perspectives* (Stanford, Calif., 2002), 144–146; Parent, *Foul Means,* chap. 5.

98. Drysdale's report appears in PRO C.O. 5/1319, fols. 212–213, 220, microfilm, M-240, CWL.

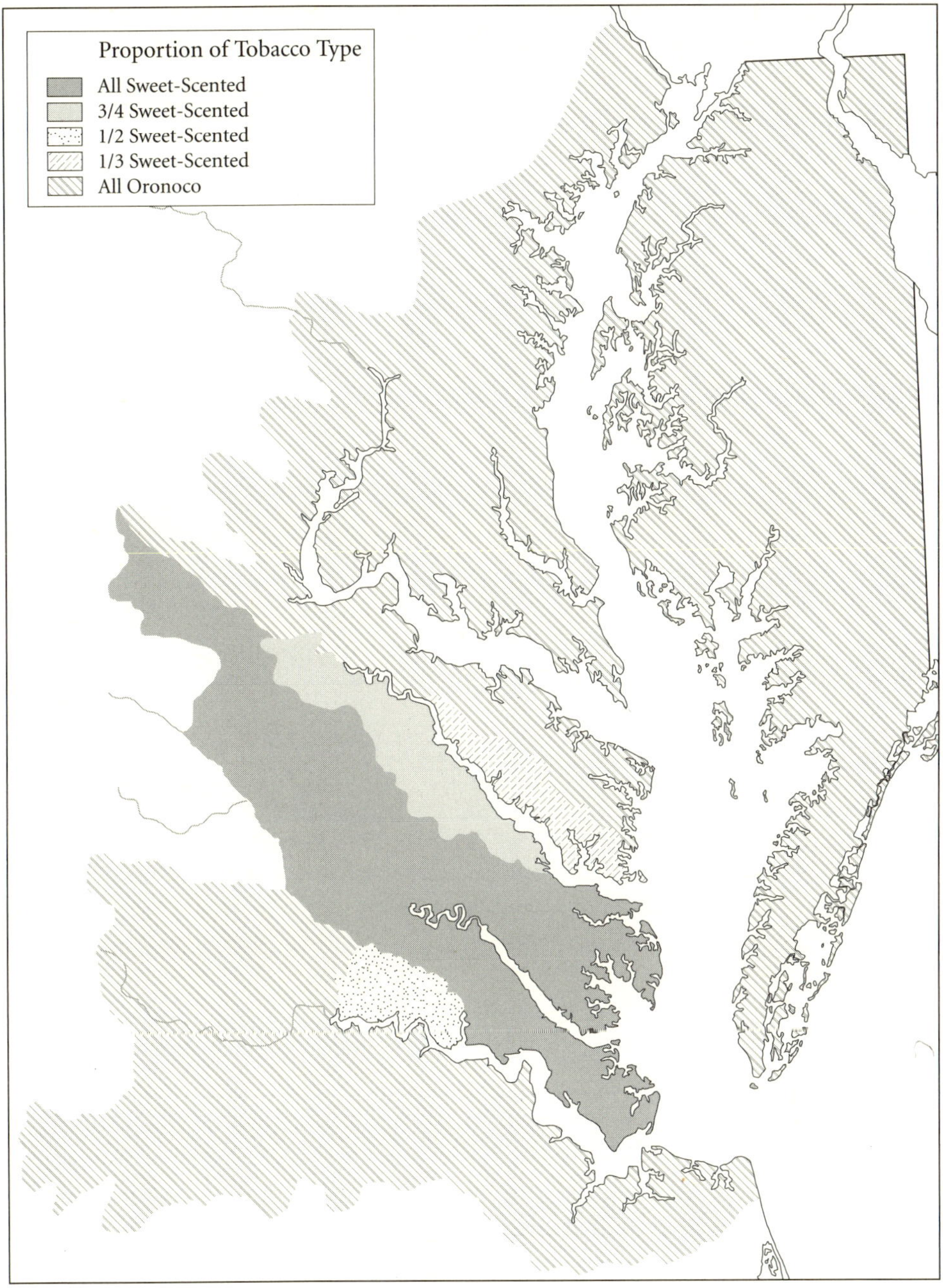

MAP 9. Sweet-Scented, Oronoco, and Peripheral Tobacco Areas in Tidewater Chesapeake, 1724. After David S. Hardin, "'The Same Sort of Seed in Different Earths': Tobacco Types and Their Regional Variation in Colonial Virginia," *Historical Geography*, XXXIV (2006), 137–158.

TABLE 23. Levels of Tobacco Production and Population Growth in Maryland Counties, 1704–1712

County	% growth of white population, 1704–1712	Black population as % of total population in 1710
HIGH LEVEL		
Calvert	−15	34
Prince George's	5	33
Anne Arundel	−9	31
Charles	27	18
Saint Mary's	11	13
Baltimore	21	16
MIDDLE LEVEL		
Kent	28	17
Cecil	−18	14
Talbot	2	12
Queen Anne's		12
LOW LEVEL		
Dorchester	32	11
Somerset	28	9

Sources: For rankings of counties, see Historic Saint Mary's City Probate Inventory Database, MSA; Paul G. E. Clemens, *The Atlantic Economy and Colonial Maryland's Eastern Shore: From Tobacco to Grain* (Ithaca, N.Y., 1980); Gloria L. Main, *Tobacco Colony: Life in Early Maryland, 1650–1720* (Princeton, N.J., 1982). For population figures, see William Hand Browne et al., eds., *Archives of Maryland,* 72 vols. (Baltimore, 1883–1972), XXV, 256–257; these figures were adjusted following Russell R. Menard, "Five Maryland Censuses, 1700 to 1712: A Note on the Quality of the Quantities," *WMQ,* 3d Ser., XXXVII (1980), 616–626.

older counties like Norfolk, Isle of Wight, and Nansemond in Virginia and Somerset and Dorchester counties in Maryland, where tobacco had become a marginal crop. Although opportunities for amassing great wealth had declined significantly in peripheral areas, continued availability of cheap land and growing markets for locally produced goods and services afforded the native born opportunities for marrying and making a living from subsistence farming or craft work. In jurisdictions with a mix of good and poor tobacco land, the numbers of both whites and blacks increased—more enslaved blacks were brought into the best-endowed parts, and enclaves of small planters expanded in less fertile areas.[99]

In contrast, in older counties where tobacco remained king and where there was an abundance of good tobacco land, the black population grew most rapidly, whereas the white population grew slowly, if at all. Examples of this trend are James City and Gloucester counties in Virginia and Calvert, Anne Arundel, and Saint Mary's counties in Maryland. In fact, poorer folk who lived in prime tobacco-growing areas had fewer opportunities for getting ahead than did families of comparable wealth residing in areas less well endowed with natural resources. In counties with a high proportion of land well suited for growing tobacco, wealthy slaveowning planters had taken up most of the prime acreage. These planters put their slaves to work on the most-fertile lands, where the value of the products the bound workers produced was high enough to quickly repay their substantial initial cost. And, where slaves were being packed in, poor whites were often forced out. Large planters held on to any land they had inherited or purchased, reserving it for future use, ceased renting small tracts of one to two hundred acres to tenants and instead farmed them with slaves, and often snapped up any additional acreage that came on the market when someone died or moved away. With land no longer readily available for purchase, new white immigrants did not move in, and sons of men who owned little or no land often moved out. With few new households being established, an important outlet for surplus produce was removed. Small planters who managed to hold on to freeholds or long-term leaseholds that they had inherited could do little to reduce their near-complete dependence on the staple, since markets for alternative goods and services were not expanding. Slave-staffed quarters devoted only to raising tobacco for export and to producing just enough food for subsistence did little to encourage growth in the local economy. Most of the slaves' clothing and tools were imported from Britain, and not much was required in the way of services apart from an overseer. The enslaved consumed little aside from what was either imported or made or raised on the quarter, and they had scant free time for producing goods for sale on their

99. For population trends, see Harris, *The History of Human Populations,* II, *Migration, Urbanization, and Structural Change* (Westport, Conn., 2003), 429–433; and Lorena S. Walsh, "Land Use, Settlement Patterns, and the Impact of European Agriculture, 1620–1820," in Philip D. Curtin, Grace S. Brush, and George W. Fisher, eds., *Discovering the Chesapeake: The History of an Ecosystem* (Baltimore, 2001), 220–248. For the continued concentration of slaves in areas with the most valuable farmland, see Gavin Wright, "Slavery and American Agricultural History," *Ag. Hist.,* LXXVII (2003), 527–552.

own. Moreover, their owners tried to prevent such trading in order to curtail clandestine exchanges of stolen goods. Low demand for goods and services meant, in turn, that small planters and tenant farmers had few opportunities for marketing agricultural products locally or for following crafts or trades as by-employments.[100]

Spatial patterns of growth in the black population—one indicator of rising opportunities for the rich, of diminishing opportunities for the poor, and of increasingly coercive labor regimes—was nearly the reverse of the pattern for whites. The numbers of blacks in the peripheral area remained small, their proportion in the total population was low (as seen in Table 23), and their labor was less essential to the continued viability of the local economy. Many of the enslaved continued to work with whites and to reside near them rather than being housed on isolated quarters. Owing to higher levels of diversification in counties shifting from tobacco, the tasks assigned to slaves in the periphery were more varied, and the degree of autonomy they had in performing them was greater. Slavery in peripheral counties retained some of the characteristics of a society with slaves rather than those of a fully developed slave society.[101]

In prime tobacco-growing counties, in contrast, wealthy tobacco planters imported large numbers of new Africans, increasing the proportion of blacks in the total population to as much as one-third. Examples are Calvert, Prince George's, and Anne Arundel counties in Maryland. In these areas, more of the enslaved were housed separately from white laborers, worked under the supervision of white overseers, and lived in units large enough to permit slaveowners to institute gang labor at least occasionally. Denser and larger black populations in good tobacco areas afforded the enslaved who lived there more opportunities for meeting in numbers, for recreating elements of African culture, and also for discussing ways for resisting their condition. Slaveholders thus had more need for closer supervision of their labor forces and for repressive measures to deter runaways and potential insurrectionists. As a result, the more favorable a jurisdiction was for tobacco, the more fully developed was its local slave society.

If one were to rely solely on planter correspondence and official pronouncements to characterize these years, the period from 1680 to 1729 must be interpreted in gloomy terms indeed. The planting business was just not

100. Walsh, "Land Use," in Curtin, Brush, and Fisher, eds., *Discovering the Chesapeake,* 220–248; Wright, *Slavery and American Economic Development,* 28–30.

101. Russo, "'Fifty-Four Days Work of Two Negroes,'" *Ag. Hist.,* LXXVIII (2002), 466–492; Coombs, "Building 'The Machine,'" 92–99.

what it had been earlier, and everywhere stark necessity forced planters into a great variety of compensatory activities just to keep their heads above water. Chesapeake planters could rightly feel that, with wars, falling prices, and new imperial taxes and mercantile impositions, they had been dealt a particularly bad hand. In England, interest rates and with them the normal rate of profit of most businesses were also falling, and many smaller, less efficient mercantile firms were forced to close. Intimately tied as the Chesapeake was to the fortunes of the English economy, in the normal course of events profit levels could be expected to fall there as well.[102]

Although hard times did for a time check the growth of the free population and, in the 1680s and 1690s, curb expansion into new areas, conditions among whites were clearly not as bad as commonly portrayed in metropolitan accounts or in more recent histories based either on these accounts or on statistics that disregard subregional differences. Age at first marriage for free white males, a generally accepted measure of economic opportunity, remained low across the period. Forming new households, especially for the growing majority of creoles, remained easy, even when it was clearly becoming increasingly difficult to get rich by so doing. Ordinary folk enjoyed no major improvements in their housing or domestic furnishings and in scarce years were unable to obtain imported cloth and tools. Still, there might have been modest gains in comfort; more families could sleep on feather mattresses instead of less comfortable flock or rush bedding, and even tenant farmers might replace their wattle-and-daub chimneys with brick. Getting around became easier as well, since by the 1690s almost everyone could afford a horse, and local governments began building more roads and bridges to accommodate overland travel. And, for all but slaves, the level of diet probably rose; certainly, it did not fall. Most ordinary farm families did not then experience any real decline in material living standards. The major deprivations, ones much less easily measured but nonetheless painful, were the dimming of hopes for

102. Hemphill, "Virginia and the English Commercial System"; Richard Grassby, "English Merchant Capitalism in the Late Seventeenth Century: The Composition of Business Fortunes," *Past and Present,* no. 46 (February 1970), 87–107; Grassby, "The Rate of Profit in Seventeenth-Century England," *English Historical Review,* LXXXIV (1969), 721–751; H. J. Habakkuk, "The Long-Term Rate of Interest and the Price of Land in the Seventeenth Century," *Economic History Review,* 2d Ser., V (1952–1953), 26–45; P. G. M. Dickson, *The Financial Revolution in England: A Study in the Development of Public Credit, 1688–1756* (London, 1967), chaps. 2, 19; Jacob M. Price, *Capital and Credit in British Overseas Trade: The View from the Chesapeake, 1700–76* (Cambridge, Mass., 1980); Ralph Davis, *The Rise of the English Shipping Industry in the Seventeenth and Eighteenth Centuries* (London, 1962), 370–377; and Jacob M. Price and Paul G. E. Clemens, "A Revolution of Scale in Overseas Trade: British Firms in the Chesapeake Trade, 1675–1775," *JEH,* XLVII (1987), 1–43.

ever improving their fortunes and a diminution of the amount of leisure they were either allowed or that they allowed themselves.[103]

Whereas the 1640s through the 1670s can be characterized as the age of the small planter, the 1680s through the 1720s marked the rise of the privileged slaveowning elite. In the peripheral tobacco areas, many small planters lapsed into subsistence farming, always a viable if not especially attractive option. Elsewhere, ordinary planters drove themselves and their families to work harder, mostly in the old ways, just to stay even. Hard times were hard on all, but they enabled the rich to siphon off a larger share of the profits still to be made than could the less fortunate white majority and also allowed the elites to devote more of the proceeds to buying enslaved workers. As a consequence, everywhere across the Chesapeake, albeit with differences in timing tied to the date of initial settlement, the mean worth of the richest planters inexorably increased during these years while the estates of ordinary residents failed to grow.[104]

The rich could buy slaves, produce more, and thus make more, whatever the state of the market. Great Virginia planters built up vast landed estates because the extremely favorable policies their leaders crafted allowed them to do so, whereas wealthier Marylanders bought up a disproportionate share of the much smaller amount of land available in that colony. Well financed both by English merchants and increasingly also by local moneylenders, wealthy Chesapeake planters purchased ever-greater numbers of enslaved blacks to clear and till their expanding estates. Despite short-term disappointments and periodic predictions that utter ruin was imminent, the assets of wealthy Tobacco Coast families continued to grow. Large planters realized anything

103. For ages at marriage, see Russell R. Menard and Lorena S. Walsh, *The Demography of Somerset County, Maryland: A Progress Report,* Newberry Papers in Family and Community History, Paper 81-2 (Chicago, 1981); Darrett B. Rutman and Anita H. Rutman, *A Place in Time: Explicatus* (New York, 1984), chap. 5; and Kulikoff, *Tobacco and Slaves,* chap. 2. On the early age of male household formation, see Kelly, "Economic and Social Development," chap. 4. For a somewhat different perspective on marriage opportunities, see James M. Gallman, "Determinants of Age at Marriage in Colonial Perquimans County, North Carolina," *WMQ,* 3d Ser., XXXIX (1982), 176–191. On diet, see Lois Green Carr, "Emigration and the Standard of Living: The Seventeenth Century Chesapeake," *JEH,* LII (1992), 271–291.

104. Aubrey C. Land, "Economic Behavior in a Planting Society: The Eighteenth-Century Chesapeake," *Jour. So. Hist.,* XXXIII (1967), 469–485; Hemphill, "Virginia and the English Commercial System," chaps. 1, 2; Bergstrom, *Markets and Merchants,* chap. 2; Gray, *History of Agriculture,* I, chap. 5; Kulikoff, *Tobacco and Slaves,* chap. 3; Lois Green Carr and Russell R. Menard, "Wealth and Welfare in Early Maryland: Evidence from St. Mary's County," *WMQ,* 3d Ser., LVI (1999), 95–120; Main, *Tobacco Colony,* chap. 2; Harris, "Integrating Interpretations of Local and Regionwide Change," *WPREHRC,* I, no. 3 (1978), 35–72; Kelly, "Economic and Social Development," chap. 7; Clemens, *Atlantic Economy,* chap. 5.

from average to spectacular returns from their agricultural operations and, by engaging in other revenue-producing activities, could at worst manage to keep their operations afloat during a spate of hard years. Professional fees, official salaries, tobacco dealing, slave trading, moneylending, and the like added substantially to the fortunes of the already well off. Perhaps they could have done better had they lessened their concentration on tobacco, but returns remained sufficiently high so that planters in prime tobacco areas in both colonies were not forced to change. Moreover, they were spurred on by examples such as William Byrd II and Robert "King" Carter who proved that it was possible to build plantation empires on the scale of West Indian sugar magnates provided one could only lay hands on enough land and enough slaves.

CHAPTER 5

The Tidewater Economy Comes of Age

SOUTHERN VIRGINIA

1730-1763

Throughout the British Empire, the 1730s through the late 1750s were generally a time of peace and prosperity, an era that would later be looked back to nostalgically as something of a Golden Age. The British economy experienced vigorous growth, as did the intercontinental trades that linked disparate elements of an emerging global empire. Industrialization and urbanization of the workforce had proceeded further than elsewhere in Europe, and a high proportion of the English population produced goods and tradable services for domestic and Continental markets but also increasingly for consumers in the Americas, Africa, and Asia. This expansion of trade and empire was made possible by a combination of early industrialization, enterprising merchants, and the construction of financial institutions such as the national debt and the Bank of England that enabled British rulers to tax their subjects or to borrow the money required to support a great transoceanic power.[1]

This period also saw greater social, cultural, and economic integration between Britain and its Atlantic colonies. Governance in Britain became less focused on the person of the monarch than in earlier years, since the first Hanoverian kings, limited by their foreign language and continued focus on Continental issues, played diminished roles as national leaders. Parliament became the ultimate arbiter of imperial affairs, with day-to-day operations of government in the hands of a cabinet dominated by patronage politics. Support for the state's foreign, commercial, and imperial strategies was widely diffused throughout the nation. A "cult of commerce," viewed by political philosophers as the highest stage of social development, became an increasingly important part of British identity. Ever more widespread participation

1. Patrick K. O'Brien, "Inseparable Connections: Trade, Economy, Fiscal State, and the Expansion of Empire, 1688–1815," and Jacob M. Price, "The Imperial Economy," in P. J. Marshall, ed., *The Eighteenth Century,* vol. II of Wm. Roger Louis, ed., *The Oxford History of the British Empire* (Oxford, 1998), 53–77, 78–104.

in buying and using British consumer goods knit residents of colonies and metropole more closely together in a shared "empire of goods." Imported cloth, furniture, ceramics, metalwares, and tropical and Asian groceries afforded not just higher levels of comfort but also new ways of life. Elite colonists reinforced their social status and demonstrated their membership in a panatlantic genteel culture through the acquisition and use of amenities and fashionable luxury goods and by building fine, large houses in which to display and use status-laden possessions. A greater sense of imperial community developed, sustained by commerce as well as by a burgeoning array of newspapers, books, and transatlantic travelers.[2]

The colonies themselves enjoyed a period of "salutary neglect" until a renewal of wars for empire with Spain and France led British rulers to take a new view of the proper relationship between the mother country and its North American possessions, which were now growing rapidly in population and wealth. The main laws enacted in this period that directly affected colonists' economic lives were the Colonial Debts Act of 1732 and the Molasses Act of 1733. The first made real estate and slaves in all the colonies liable for book debts and for obligations secured by bond. Although this law made it easier for British merchants and their colonial factors to collect debts, it also made longer-term credit backed by mortgages and bonds more feasible. The Molasses Act departed from the earlier policy of largely ignoring trade with colonies of a rival nation and instead moved to tax British colonial imports of French West Indian sugar and molasses, which was cheaper than what was made in the British colonies, thereby establishing Parliament's right to levy duties on products imported into the colonies. The law, enacted to placate British West Indian planters, was not enforced. Widespread evasion resulted, marking the beginning of colonial resistance to imperial regulations. Indeed, as the colonies reached new levels of economic maturity, the thrust of imperial policy shifted from controlling colonial trade in the interest of increasing imperial customs revenues to proscribing what colonists might do when new provincial enterprises conflicted with the interests of British merchants and manufacturers, a policy that gained considerable traction after the beginning of the Seven Years' War.[3]

2. Ian K. Steele, "The Anointed, the Appointed, and the Elected: Governance of the British Empire, 1689–1784," Jack P. Greene, "Empire and Identity from the Glorious Revolution to the American Revolution," and Richard Drayton, "Knowledge and Empire," all ibid., 105–127, 208–230, 231–252; Linda Colley, *Britons: Forging the Nation, 1707–1837* (New Haven, Conn., 1992), 55–71; T. H. Breen, *The Marketplace of Revolution: How Consumer Politics Shaped American Independence* (Oxford, 2004), chaps. 2–4.

3. Price, "The Imperial Economy," and Steele, "The Anointed, the Appointed, and the Elected,"

Just as Parliament gained ever more control of governance in Britain, in the colonies more and more power devolved to the colonial assemblies. As the size of the Chesapeake gentry increased and well-established families took greater interest in politics, more of the elite pursued political ambitions in the lower houses of assembly rather than in the more exclusive councils. To be effective, royal governors in Virginia had to enlist the cooperation of powerful colonial leaders and to act circumspectly when attempting to enforce crown policies. By 1750 in Virginia, the House of Burgesses eclipsed the council in political power. "The most striking characteristic of Virginia politics" in this era, however, was, according to Jack P. Greene, "its tranquility." There was no serious rivalry between the council and the house, bound as the members were by family and social ties. Disagreements were limited to specific issues, and no sustained political factions emerged. Sectional rivalries rooted in geography had not yet developed, and there was "no evidence of any major dissatisfaction with gentry leadership or of any group intent upon challenging it" up to 1763.[4]

In Maryland, the lower house, composed of large planters, merchants, and lawyers, also gained power. Politics there was more fractious than in Virginia, however, characterized by continued contests between the assembly, dominated by a "country" party, and the proprietor or his governors, who were supported by a smaller "court" faction that enjoyed a monopoly on proprietary favor and offices. There were also tensions over power sharing and representation between the Eastern and Western shores and newer piedmont counties as well as severe repression of Roman Catholics—who were suspected of sympathizing with the French—during wartime. Disputes over how to finance government and defense and over the extent to which English law applied in the colony led to repeated political deadlocks that made it difficult to enact legislation addressing the colony's economic problems.[5]

The era saw two further contests for empire from which Britain emerged greatly strengthened in power and territory. During the 1730s, Anglo-Spanish hostility intensified in the Caribbean and along the Florida-Carolina frontier, where the new Georgia colony encroached on lands claimed by the Spanish and challenged Spain's dominance of the southeastern fur trade. Between 1739 and 1741, the undeclared War of Jenkins's Ear was marked by expensive

both in Marshall, ed., *The Eighteenth Century,* vol. II of Louis, ed., *Oxford History of the British Empire,* 96, 116–117.

4. Jack P. Greene, *The Quest for Power: The Lower Houses of Assembly in the Southern Royal Colonies, 1689–1776* (Chapel Hill, N.C., 1963), 22–31 (quotations on 29–30).

5. Robert J. Brugger, *Maryland: A Middle Temperament, 1634–1980* (Baltimore, 1988), 88–102.

failed attempts by the British to conquer Cuba, Panama, and Saint Augustine and by Spanish raids on Georgia and North Carolina. France and Spain were pledged to defend one another's possessions in Europe and America, so Britain and France again became direct opponents. Britain engaged on the Continent in the War of Austrian Succession between 1744 and 1748 so that the Low Counties would not fall into the hands of either France or Prussia. Fighting in North America, known as King George's War, consisted largely of sporadic raids between frontiersmen and Indians. Colonial troops from New England captured Louisbourg, the French stronghold on the Saint Lawrence, but the peace treaty of Aix-la-Chapelle returned all captured territories to their prewar status. As in earlier contests over empire, the main effect for Chesapeake colonists was disrupted trade, although some colonial soldiers did participate in the failed Caribbean expeditions.[6]

Issues of the balance of power in Europe, of rival Anglo-French claims to territory in North America ranging from Nova Scotia through the Ohio Valley, and for supremacy in India thus remained unresolved. British ministers and the Board of Trade, determined to resist French encroachments in North America, gave charters to companies speculating in Ohio Valley lands and encouraged the activities of Anglo-American fur traders. The inevitable conflict that ensued, known in the colonies as the French and Indian War, started in the Ohio Valley in 1754 and spread to Europe two years later, there called the Seven Years' War. It was, according to Bruce P. Lenman, "truly an Imperial war for Britain in a way that no previous war had been," fought primarily in North America and in India. Early defeats stimulated Anglo-Americans into unprecedented levels of unity and military effort. The British Treasury spent £5.5 million on the army in America, nearly £1 million on the navy, and more than £1 million to reimburse colonies for their troops. Colonial contributions were roughly £2.5 million. British and colonial armies prevailed, conquering New France and removing the French from North America, while victories in India gave the East India Company more territory and greater control in Asia. In this conflict, Virginia troops were heavily involved in the fighting in the Ohio Valley, billeting and provisioning of British troops and naval vessels afforded many residents an alternative outlet for foodstuffs, and political leaders and land speculators were intensely concerned with retaining control over western territories wrested from the French. Marylanders were

6. Bruce P. Lenman, "Colonial Wars and Imperial Instability, 1688–1793," and John Shy, "The American Colonies in War and Revolution, 1748–1783," both in Marshall, ed., *The Eighteenth Century,* vol. II of Louis, ed., *Oxford History of the British Empire,* 151–168, 300–324; Jeremy Black, *Trade, Empire, and British Foreign Policy, 1689–1815: The Politics of a Commercial State* (London, 2007), chap. 7.

mostly concerned with defending western settlements from attacks by Indians, and its assembly, engaged in a struggle with the proprietor over control of revenues, refused to fully support the war effort.[7]

With the return of peace in 1763, Chesapeake planters could anticipate benefiting from a postwar boom in tobacco prices and the arrival of increased supplies of new African slaves. Those who had speculated in western lands also hoped to benefit from the rise in land values that the removal of the French promised. In the first flush of the British triumph, they could hardly have anticipated that ensuing crises over problems of imperial defense and finance would shortly destroy the foundations of their prosperous colonial world.

Between 1730 and the end of the Seven Years' War, a vigorous economic recovery, accompanied by growing political and cultural maturity, similarly sustained a Golden Age in the tidewater tobacco colonies. All free residents, whatever their level of wealth, shared at least to some extent in the general imperial prosperity. Although commerce was disrupted during King George's and the Seven Years' wars, trade quickly rebounded at the close of each of these conflicts. By the mid-1740s, the value of the products Chesapeake residents exported began to rise relative to the costs of the goods they imported. Large planters responded to increasingly favorable terms of trade for export products by diversifying their sources of income and expanding the productive capacities of their estates. Middling planters were unable to take full advantage of economies of scale, but they, too, benefited from this spate of favorable trading conditions. At the same time, the society reached a new level of demographic and social maturity, shedding most of the rough edges characteristic of poor, sparsely populated, socially and politically unstable frontier settlements. Almost all of the European settlers who had migrated to the Chesapeake colonies before 1700 had died, and few newcomers replaced them. The local white population in the tidewater became almost entirely native born and had a more normal age structure and a more even balance between the sexes.[8]

7. Lenman, "Colonial Wars and Imperial Instability," 159; Brugger, *Maryland,* 94–96.

8. For the economic history of the period, see John J. McCusker and Russell R. Menard, *The Economy of British America, 1607–1789* (Chapel Hill, N.C., 1985), chap. 6; Jacob M. Price, *Tobacco in Atlantic Trade: The Chesapeake, London, and Glasgow, 1675–1775* (Aldershot, Eng., 1995); Lois Green Carr, "Diversification in the Colonial Chesapeake: Somerset County, Maryland, in Comparative Perspective," in Carr, Philip D. Morgan, and Jean B. Russo, eds., *Colonial Chesapeake Society* (Chapel Hill, N.C., 1988), 342–388; Marc Egnal, *New World Economies: The Growth of the Thirteen Colonies and Early Canada* (New York, 1998), chap. 5; Peter V. Bergstrom, *Markets and Merchants: Economic Diversification in Colonial Virginia, 1700–1775* (New York, 1985); Warren M.

The Chesapeake's total population rose from an estimated 212,000 in 1730 to just over 500,000 by 1760. In 1730, more than 90 percent of the people still lived in the tidewater. By 1749, however, as new settlements were established above the fall line, the proportion of tithables residing in the tidewater dropped to just over two-thirds, and by 1755 it fell to 60 percent. Twenty-five new piedmont counties were created between 1730 and 1763 (Map 10). This rapid expansion into fresh agricultural land resulted more from continued imports of Africans than from population growth among whites. The number of white settlers doubled over these thirty years from approximately 158,000 in 1730 to 312,000 by 1760, an increment owing primarily to natural increase rather than to immigration (unlike most of the seventeenth century, when opportunities on the Tobacco Coast were often sufficient to overcome the substantial risks immigrants faced in moving from their birthplaces to a raw, unfamiliar colonial environment). Over the same thirty years, the number of black inhabitants increased more than threefold from 53,000 to 190,000, an upsurge occasioned in part by natural increase but also resulting from the forced migration of at least 50,000 additional Africans. Aside from the dominant trade in manufactured goods with Great Britain, Africa became, by a substantial margin, the second most important source of "goods" Chesapeake colonists imported in the eighteenth century. This statistic underscores the extent to which the Chesapeake's golden years were underwritten by the exploitation of thousands of unwilling and uprooted forced migrant laborers at a time when free English, Irish, and Scottish men and women had decided that the potential rewards of moving to the colonies under a labor contract no longer offset the considerable disadvantages of uprooting themselves from their homelands.[9]

In peacetime (1730–1738, 1749–1753, 1763), on average about two thousand new African slaves were brought into the region each year. In wartime (1739–1748, 1754–1762), the number declined to an average of just over one thousand. Through the 1730s, most of the Africans, as in earlier years, were

---

Billings, John E. Selby, and Thad W. Tate, *Colonial Virginia: A History* (White Plains, N.Y., 1986), chap. 9; Stephen Gregg Hardy, "Trade and Economic Growth in the Eighteenth-Century Chesapeake" (Ph.D. diss., University of Maryland, College Park, 1999); and Trevor Burnard, *Creole Gentlemen: The Maryland Elite, 1691–1776* (New York, 2002).

9. McCusker and Menard, *The Economy of British America,* chap. 5; John J. McCusker, "Colonial Statistics," in Susan B. Carter et al., eds., *Historical Statistics of the United States: Earliest Times to the Present,* millennial ed., 5 vols. (New York, 2006), V, 561–563 (which gives a lower number for whites in 1730 than does *Economy of British America*); Hardy, "Trade and Economic Growth," chap. 4; Philip D. Morgan and Michael L. Nicholls, "Slaves in Piedmont Virginia, 1720–1790," *WMQ,* 3d Ser., XLVI (1989), 215.

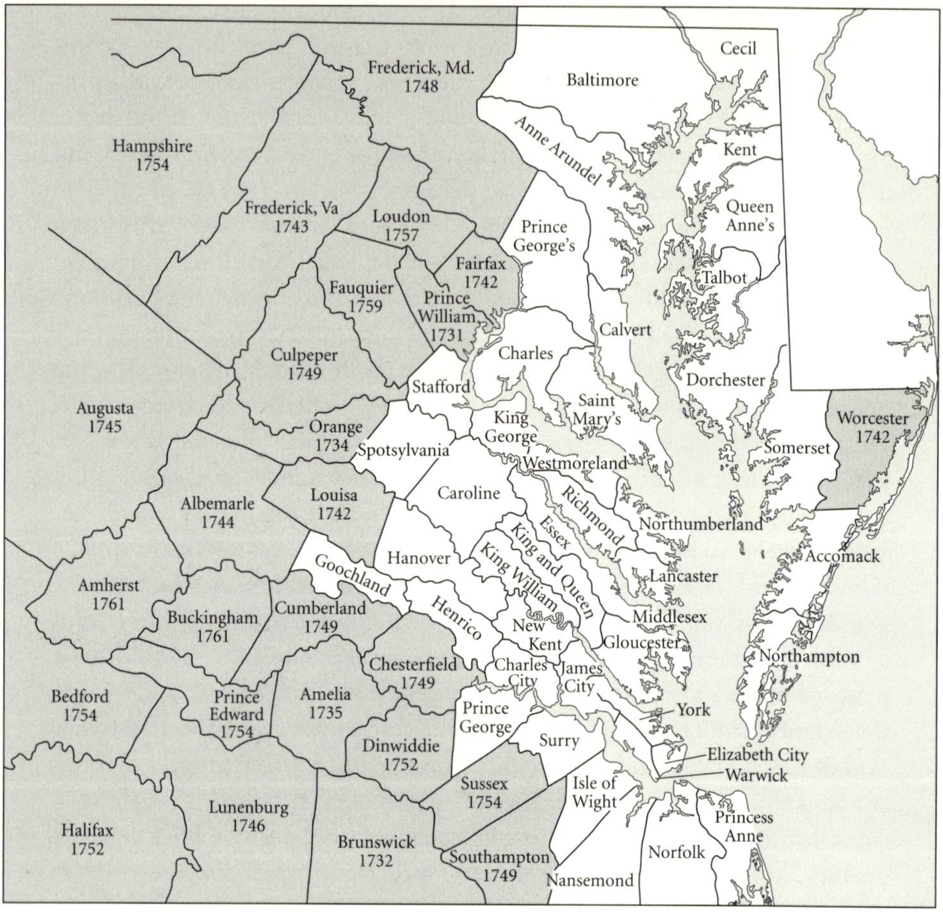

MAP 10. Chesapeake Counties Established through 1763

bought by planters in the lower tidewater. But, by 1750, the majority of new African captives were purchased by piedmont planters who set them to work raising tobacco in the west. Prime African slaves cost on average between sixteen and twenty pounds sterling in the 1730s, twenty-five to thirty pounds in the 1740s, and from twenty-six to thirty-three pounds between 1750 and 1763. The most conservative estimate of the net profits that slaves could be expected to earn from tobacco was four pounds sterling per year. At this rate, planters could expect to recover the cost of their investment in four to five years in the 1730s, in five or six years in the 1740s, and in six to eight years in the 1750s. Should their workers be able to clear an additional two pounds per year from surplus grains, however, the purchasers could recoup their initial costs in only three to five years. This massive importation of forced labor enabled (and, indeed, required) larger planters to greatly expand outputs of both tobacco and grain.[10]

Prosperity at Last

During the 1730s and 1740s, elite creole planters of the second and third generations born in the Chesapeake assumed control of scores of enslaved laborers and of well-developed plantations as well as of additional undeveloped western tracts. These privileged men found themselves in an optimal time and place to capitalize on inherited assets, especially if they were fortunate enough to take possession of their inheritances at or soon after they reached twenty-one, the age of legal majority. Most would, in turn, manage to build up sufficiently augmented estates to permit their children to start out in a similarly privileged economic position.

Elite families used a variety of inheritance strategies to achieve these goals. In Virginia, the practice of entailing land, and after 1705 often enslaved laborers as well, in the male line created great estates. Entail was also, however, a policy that limited the ability of subsequent generations of parents to apportion lands and slaves among all their heirs. Often elite parents retained control of some land and of some enslaved workers held in fee simple of which they could freely dispose, but they also held entailed land and enslaved workers whose disposition they could not direct. As these assets became increasingly valuable and tidewater land became increasingly scarce, more parents transferred such crucial productive assets primarily or exclusively to their sons. Those who had several children to provide for needed either to dock entails on inherited land (an expensive and time-consuming process requiring a special act of the legislature) or to acquire additional undeveloped western tracts on which to set up younger sons for whom there was insufficient land in the tidewater. They also needed to find ways to raise cash dowries for

10. Lorena S. Walsh, "The Chesapeake Slave Trade: Regional Patterns, African Origins, and Some Implications," *WMQ*, 3d Ser., LVIII (2001), 139–170. The numbers for imported Africans are calculated from Lorena S. Walsh Database of Slave Trading Voyages to Virginia and Maryland (November 2008). Slave prices are from the John Mercer Ledgers, [1741]–1750, Bucks County Historical Society, Doylestown, Pa., microfilm, M-164, CWL; Robert Carter Letterbooks, 1723–1724, 1727–1728, 1728–1731, 1731–1732, Robert ("King") Carter Papers, MS 3807, UVa, microfilm, M-113, CWL; Carter Family Letterbook, 1732–1782, Tracy W. McGregor Library, Accession 4996, UVa, microfilm, M-62.8, CWL; Robert Carter Letterbooks, 1723–1724, 1727–1728, 1728–1730, Carter Family Papers, VHS; Charles Stewart Letterbooks, 1751–1753, 1754–1763, HSP, microfilm, M-32, CWL; and David Eltis and David Richardson, "Prices of African Slaves Newly Arrived in the Americas, 1673–1865: New Evidence on Long-Run Trends and Regional Differences," in Eltis, Frank D. Lewis, and Kenneth L. Sokoloff, eds., *Slavery in the Development of the Americas* (Cambridge, 2004), 181–218.

marriage portions for their daughters. Across this period, entails remained a popular estate-preserving strategy among great Virginia planters, but the practice finally began to fall out of favor around 1760, as the problems arising from tying up such a large proportion of a family's assets became apparent.

Elite Marylanders were somewhat less inclined to try to preserve family land over the generations through the use of entails, and Maryland law had no provisions for entailing slaves. Because fewer elite Marylanders had reserves of undeveloped western land with which to endow younger sons than did most Virginia planters, many parents had few options aside from dividing up the family's tidewater landholdings, a strategy that entails would have prevented. By the 1730s, though, middling to elite Maryland parents did begin adding encumbering provisions to final bequests. These provisions were designed to keep family land in the family and to make sons who were fortunate enough to inherit land help pay portions in cash for siblings who did not. When they owned sufficient acreage to provide farms for more than one member of the next generation, Maryland parents most often divided their land among the sons and gave their daughters slaves as marriage portions or as final shares of the parents' estates.[11]

Some third-generation creole planters added to inherited workforces by the continued purchase of new Africans, but increasingly tidewater planters acquired additional labor simply from the natural increase of the enslaved women they already owned. Just as creole white women bore more offspring than did their immigrant mothers, servant or free, enslaved creole women also had more children than did their mothers whom slave traders had carried forcibly out of Africa. Some of the advantages of a self-reproducing labor force were offset by the need to feed and clothe a growing proportion of nonproductive children and of superannuated bondpeople. Still, the costs of maintaining and expanding a labor force from natural increase were almost certainly less than those faced by planters at the turn of the century who had to purchase and break in unacculturated Africans and who lost a higher proportion of unacclimatized bondpeople to both imported and endemic diseases. In the 1760s, planter Henry Fitzhugh made explicit the growing connection between natural increase among the enslaved and not just stable but

11. Holly Brewer, "Entailing Aristocracy in Colonial Virginia: 'Ancient Feudal Restraints' and Revolutionary Reform," *WMQ*, 3d Ser., LIV (1997), 307–346; Jean Butenhoff Lee, "Land and Labor: Parental Bequest Practices in Charles County, Maryland, 1732–1783," in Carr, Morgan, and Russo, eds., *Colonial Chesapeake Society*, 306–341; Lois Green Carr, "Inheritance in the Colonial Chesapeake," in Ronald Hoffman and Peter J. Albert, eds., *Women in the Age of the American Revolution* (Charlottesville, Va., 1989), 155–208; Burnard, *Creole Gentlemen*, chap. 5.

also expanded production. In the early 1750s, Fitzhugh had to buy Africans to clear and plant an undeveloped backcountry quarter, but ten years later he no longer needed to purchase additional slaves because "my negroes increase yearly by which I am enabled to settle new quarters and consequently my exports must increase."[12]

The emergence by the 1750s of a creole slave population in which the majority of adults were native born also had far-reaching implications for labor management. Slaves born and raised in the Chesapeake were strongly attached to their families and to more far-flung kin networks that developed from cross-plantation marriages and from the division of enslaved property when decedents' estates were settled. Slaveowners exploited the concerns of their bondpeople for spouses, children, and extended kin. Threats of separating or selling family members were even more effective for controlling slaves than were whippings and other measures of physical brutality. Acts of resistance among the enslaved continued but were usually undertaken only by individuals rather than by organized groups. Slaves continued to run away to escape punishment or to visit separated family members, but seldom did they go far or stay away for long. Private plantation discipline, county slave patrols, and the apparatus of the courts were sufficient to deal with incidents of theft, running away, or occasional acts of violence against owners or overseers. Now and then slaves committed or were suspected of committing arson and poisoning, and such cases did for a time raise levels of apprehension among slaveholders. Fears among whites of mass slave insurrections, however, all but disappeared in the 1740s and 1750s. During the French and Indian War, some political leaders feared that slaves might revolt if local militias were sent west to fight the enemy, but in fact no major unrest occurred. By and large, slaveholders were sanguine about their ability to control their bondpeople. Likely most slaveowners, as a few explicitly stated, considered themselves "good masters," and they further salved their consciences by comparing favorably

12. Demographic developments and labor-force shifts are summarized in Allan Kulikoff, *Tobacco and Slaves: The Development of Southern Cultures in the Chesapeake, 1680–1800* (Chapel Hill, N.C., 1986), chap. 2; McCusker and Menard, *Economy of British America,* chap. 6. The quotation is from Henry Fitzhugh to Messrs. John Stewart and Campbell, July 29, 1766, Henry Fitzhugh Letterbook, 1746–1774, fol. 66, Henry Fitzhugh Papers, 1746–1786, Duke University Library, Durham, N.C., microfilm, M-1886.29, CWL. For his purchases of new Africans, see Fitzhugh to Thomas and John Backhouse, May 8, 1747, fol. 3, to William Bogle, Sept. 24, 1753, fol. 17, to ?, July 27, [1754], fol. 19, ibid. Many West Indian planters thought it more cost-effective to continue importing new Africans rather than to build workforces through natural increase. Food was scarcer and more expensive in the islands, however, and levels of reproduction among enslaved women were lower than in the Chesapeake.

the condition of enslaved blacks in the Chesapeake to that of the white poor in Europe.[13]

Slaveholding also spread into the ranks of middling planter families through both purchase and inheritance, since most slaveowners divided their human chattels, unlike their land, among all their children. By midcentury, between half and two-thirds of the householders in many tidewater counties owned slaves. Some middling planters bought new Africans — one or at most two at a time from the slavers that arrived in the colony every summer — but such purchases required access to credit from British merchants gained either by consigning tobacco or through the offices of a well-connected relative or other provincial patron. Most likely acquired slaves by buying them locally, especially when they were sold to settle an estate or to pay the debts of a bankrupt planter. A major advantage was that the terms of local sales were quite different from those prevailing in the African trade. Purchasers could buy not only with bills of exchange, specie, or paper currency but also sometimes with tobacco. Those who paid immediately received a discount of 5 percent on the selling price. But most buyers instead purchased slaves on credit, which by the third quarter of the century was routinely extended upon the buyer's giving a bond backed by good security. These terms made the slaves that came on the local market affordable to anyone who had land or other assets he or she could mortgage or who had a more affluent relative or patron willing to stand as a surety. Consequently, more and more ordinary planters were able to increase income earned from exports by acquiring a slave whose work effort, over and above that of family members, allowed them to expand agricultural production. Access to bound laborers, however, continued to be channeled through the filters of wealth, class, and patronage.[14]

13. Philip J. Schwarz, *Twice Condemned: Slaves and the Criminal Laws of Virginia, 1705–1865* (Baton Rouge, La., 1988), chaps. 1–6; Schwarz, *Slave Laws in Virginia* (Athens, Ga., 1996), chap. 1; Gerald W. Mullin, *Flight and Rebellion: Slave Resistance in Eighteenth-Century Virginia* (Oxford, 1972), chap. 3; Ira Berlin, *Many Thousands Gone: The First Two Centuries of Slavery in North America* (Cambridge, Mass., 1998), chap. 5; Sally E. Hadden, *Slave Patrols: Law and Violence in Virginia and the Carolinas* (Cambridge, Mass., 2001); Douglas B. Chambers, *Murder at Montpelier: Igbo Africans in Virginia* (Jackson, Miss., 2005), part 1; James Titus, *The Old Dominion at War: Society, Politics, and Warfare in Late Colonial Virginia* (Columbia, S.C., 1991), 75–76; Mark J. Stegmaier, "Maryland's Fear of Insurrection at the Time of Braddock's Defeat," *MHM*, LXXI (1976), 467–483; Brugger, *Maryland*, 94; Philip D. Morgan, *Slave Counterpoint: Black Culture in the Eighteenth-Century Chesapeake and Lowcountry* (Chapel Hill, N.C., 1998), chap. 5; Lorena S. Walsh, "The Differential Cultural Impact of Free and Coerced Migration to Colonial America," in David Eltis, ed., *Coerced and Free Migration: Global Perspectives* (Stanford, Calif., 2002), 117–151; Burnard, *Creole Gentlemen*, 37–43; Joseph Ball to Joseph Chinn, Jan. 18, 1755, Joseph Ball Letterbook, 1743–1780, LC, microfilm, M-21, CWL.

14. Lorena S. Walsh, "Mercantile Strategies, Credit Networks, and Labor Supply in the Colo-

Most small planters and tenant farmers still had to manage with only the labor family members could supply, occasionally augmented by the work of an indentured or convict servant, forms of bound labor offered primarily in poorer areas such as the Potomac River basin, where few African slaves were marketed, or the port of Annapolis, which served both urban and backcountry buyers. Since the supply of unskilled voluntary servant immigrants remained small to negligible throughout this period, labor-hungry poorer planters on both banks of the Potomac eagerly bought British convicts, a form of labor supplied in much larger numbers beginning in 1718 by a handful of London and Bristol contractors subsidized by the British Treasury. These contractors specialized in human commerce conducted similarly to the trade in Africans. Convicts, like slaves, were often shipped in irons, and, although they were not as tightly packed as were Africans in the Middle Passage, they were not infrequently carried in vessels also used to transport slaves. The shippers readily sold convicts for tobacco and sometimes wheat, corn, and other produce as well. Most small purchasers bought on credit, however, which was almost always allowed to purchasers who could give bond. The proportion of European servants in the total workforce, though, remained small. In 1755, the only year for which there is a census for Maryland, servants and convicts made up only 14 percent of the total bound workforce, with the majority concentrated in towns and in areas where mixed farming prevailed.[15]

Tobacco prices began to rise in the early 1730s and advanced strongly from the late 1740s, affording planters a greater income from their traditional staple (Figure 4). Most planters maintained or even expanded the quantity of tobacco that they grew, and some colonists living in marginal tobacco areas (many of whom had dropped out of the market between the 1690s and 1720s) again took up tobacco when they had some land suitable for growing it. Tobacco imports to Great Britain rose from forty-one million pounds in 1730 to eighty-five million pounds in 1753, declined during the Seven Years' War when the French captured a portion of the crops planters attempted to ship, and then climbed to ninety-eight million pounds in 1763. Expansion of Continental markets, especially in France, where snuff manufactured from cheaper oronoco increasingly was in demand, accounted for most of the growth. Much of the augmented output came from large plantations. As John Wayles observed, "In 1740. no man on this River [the James] made 100 hhds of tobacco; now [1766] not less then six exceed that Number." The

---

nial Chesapeake in Trans-Atlantic Perspective," in Eltis, Lewis, and Sokoloff, eds., *Slavery in the Development of the Americas,* 89–149.

15. Ibid.

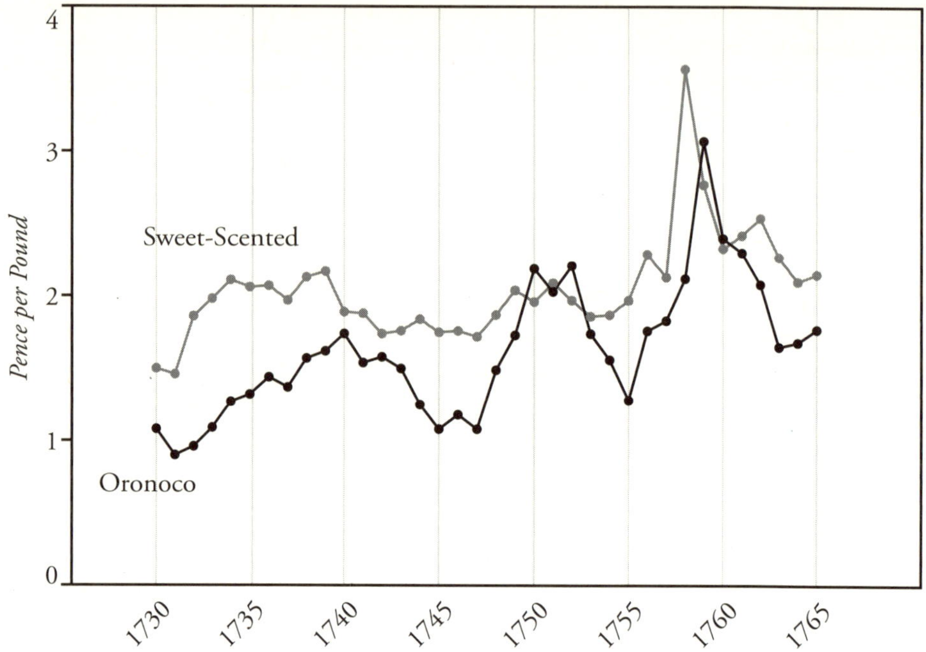

FIGURE 4. Oronoco and Sweet-Scented Tobacco Prices, 1730–1765. Drawn by Kimberly Foley. John J. McCusker, "Colonial Statistics," in Susan B. Carter et al., eds., *Historical Statistics of the United States: Earliest Times to the Present,* millennial ed., 5 vols. (New York, 2006), V, 681–684; price series compiled by Historic Saint Mary's City, MSA.

number of planters exporting one hundred or more hogsheads per year similarly rose on the York and Rappahannock.[16]

16. Jacob M. Price, *France and the Chesapeake: A History of the French Tobacco Monopoly, 1674–1791, and of Its Relationship to the British and American Tobacco Trades,* 2 vols. (Ann Arbor, Mich., 1973), 843–845; Price, *Perry of London: A Family and a Firm on the Seaborne Frontier, 1615–1753* (Cambridge, Mass., 1992), chaps. 5–7; T. M. Devine, *The Tobacco Lords: A Study of Tobacco Merchants of Glasgow and Their Trading Activities, 1740–90* (Edinburgh, 1975), chap. 4; Basel H. Brune, "The Changing Spatial Organization of Tobacco Marketing in the Patuxent River Basin," in Robert D. Mitchell and Edward K. Muller, eds., *Geographical Perspectives on Maryland's Past,* University of Maryland Occasional Papers in Geography, no. 4 (College Park, Md., 1979), 71–89. Quotation from John M. Hemphill II, ed., "John Wayles Rates His Neighbours," *VMHB,* LXVI (1958), 305. The shift from smoking tobacco in pipes to inhaling it in the form of snuff, however, might have reduced the amount of tobacco individual users consumed and so contributed to decelerating growth of demand in the eighteenth century (Jan de Vries, *The Industrious Revolution: Consumer Behavior and the Household Economy, 1650 to the Present* [Cambridge, 2008], 157–158).

The structure of marketing networks, as well as the volume of exports, changed dramatically. Scottish and outport merchants bought an ever-larger share of the region's tobacco, especially of the more ordinary grades, at least two-thirds of which was reexported to the Continent by midcentury. They set up chains of retail stores that operated year-round and that provided small and middling planters with a much more certain supply of basic manufactures and of West Indian goods. These merchants bought tobacco outright, offering cash or merchandise (the "money" price being lower that the "goods" price) or some combination of the two, and, where competition between merchants was intense, they extended credit liberally, and sometimes extravagantly, to secure full cargoes of tobacco for their British employers. By the late 1730s, growing competition from Scottish, Bristol, and Liverpool merchants, who were willing to pay higher prices or to offer merchandise on more favorable terms, forced Londoners who wanted to maintain retail stores in the Chesapeake to adopt similar innovations. Chesapeake residents indeed might have benefited more from reduced prices for imported goods that accompanied increased competition between retailers than they did from higher tobacco prices. With the return of peace in 1748, a number of London and Bristol merchants shifted from the consignment form of trade into a "cargo" trade since, with the proliferation of Chesapeake stores, they had difficulty obtaining enough consignments to fill their ships. These English merchants bought cargoes of manufactured goods on about one year's credit for local Chesapeake merchants with the understanding that the recipients would remit bills of exchange or tobacco to pay for the goods within the year.[17]

In all the major British ports, in the 1730s, 1740s, and 1750s the tobacco trade became concentrated in the hands of fewer, larger importing firms, in part because of the ever-increasing importance of the French market. Large-scale traders in Britain had an advantage over smaller merchants in dealing with the French monopoly buyers as well as greater expertise in negotiating the increasingly complex apparatus regulating customs duties in Britain. The

17. Jacob M. Price, "The Rise of Glasgow in the Chesapeake Tobacco Trade, 1707–1775," *WMQ,* 3d Ser., XI (1954), 179–199; Price, "The Economic Growth of the Chesapeake and the European Market, 1697–1775," *JEH,* XXIV (1964), 496–511; Price, *Perry of London,* 95–101; Paul G. E. Clemens, "The Rise of Liverpool, 1665–1750," *Economic History Review,* 2d Ser., XXIX (1976), 211–225; J. H. Soltow, "Scottish Traders in Virginia, 1750–1775," *Economic History Review,* 2d Ser., XII (1959), 83–98; John W. Tyler, "Foster Cunliffe and Sons: Liverpool Merchants in the Maryland Tobacco Trade, 1738–1765," *MHM,* LXXIII (1978), 246–279; Calvin Brewster Coulter, Jr., "The Virginia Merchant" (Ph.D. diss., Princeton University, 1944); Robert Polk Thompson, "The Merchant in Virginia, 1700–1775" (Ph.D. diss., University of Wisconsin, 1955), chaps. 1–5; Hardy, "Trade and Economic Growth," 29–32; *Byrd Correspondence,* I, 376–377; and Callister Papers, 1741–1776, Maryland Diocesan Library, Baltimore, microfilm, M-142, CWL.

number of London tobacco importers dropped from just over one hundred in 1719 to between forty and sixty in the 1740s, 1750s, and early 1760s, and Bristol dealers declined from about fifty in the 1740s to fewer than thirty in the early 1760s. Since Liverpool and Glasgow became significant players in the trade only after higher duties were imposed, these ports never had large numbers of small importers. However, they, too, experienced a comparable shift toward ever-larger operations.[18]

A similar concentration was occurring in the sugar and slave trades as well, suggesting that increasing specialization and economies of scale in the use of shipping were affecting all overseas trades. For tobacco carriers, average days in port per ton of shipping decreased from 1.61 in 1726 to only .88 in 1756. Scottish shippers working in concert with resident storekeepers could effect somewhat shorter turnaround times than could Londoners involved in the more labor-intensive consignment trade. The Londoners managed to pare down the average time spent crossing the Atlantic, however, offsetting the extra time needed to assemble return cargoes in the Chesapeake, and they were able to reduce the number of sailors required to man their vessels. With these efficiencies, London traders remained competitive against those from the outports.[19]

Growing concentration proved a net benefit to elite colonial planters, for the larger firms were in a better position to offer their correspondents credit, and sufficient numbers of importers survived so that competition rather than collusion remained the norm. Although all small and most middling planters now sold their tobacco to Scottish, English, or indigenous storekeepers or merchants, most large tidewater planters continued to consign their tobacco to English factors, except in a few years when prices offered in the country were exceptionally high. By the 1740s, most had also adopted the habit of dividing their consignments between a combination of London, Bristol, and Liverpool factors. They considered one or more London correspondents essential, for above all elite planters wanted fashionable London goods for their own households. On the other hand, they found that coarse cloth, stockings, hats, blankets, hoes, axes, nails, carpenters' and coopers' tools, millstones, seines, and other utilitarian goods needed to clothe their enslaved workers and to carry on plantation operations could be bought more cheaply

18. Jacob M. Price and Paul G. E. Clemens, "A Revolution of Scale in Overseas Trade: British Firms in the Chesapeake Trade, 1675–1775," *JEH,* XLVII (1987), 1–43.

19. Ibid., 31–37; Egnal, *New World Economies,* 87–88; Hardy, "Trade and Economic Growth," 468–470. Bristol shippers were also able to reduce crew sizes during peacetime but did not manage to reduce turnaround time (Kenneth Morgan, *Bristol and the Atlantic Trade in the Eighteenth Century* [Cambridge, 1993], chap. 3).

in Bristol or Liverpool. These outports could supply shipping and port facilities more cheaply than could London, and Bristol and Liverpool merchants also received extensive credit from nearby Lancashire and Midland manufacturers. In addition, outport ships often arrived earlier than did London vessels, obviating the dilemma of allowing ill-clothed slaves to "either perish, or set up all the winter by the fire." By shipping tobacco to two or three major ports, great planters obtained the range of imported goods they needed or wanted at optimal prices, and they could also test their tobacco in a combination of home, Irish, and Continental markets. In addition, Bristol and especially Liverpool afforded safer and surer markets and sources of supply during wartime, a circumstance that permitted consigning planters to minimize insurance costs and losses of tobacco or essential supplies on the high seas. Sending consignments to two or more ports must have been advantageous, since every large planter for whom records of farm management survive adopted this strategy.[20]

After 1730, Britain experienced a large inflow of capital from northwestern Europe, mainly from the Dutch, which allowed merchants to redirect capital formerly tied up in European trades into Atlantic commerce. A sharp drop in rates of return to capital in Britain also encouraged reallocation of surplus funds from domestic to colonial borrowers. Yields on British government-funded debts from the mid-1730s onward were only about 3 to 3.5 percent, whereas merchants could earn 5 percent on colonial debts contracted in Britain and 6 to 8 percent on loans advanced in the colonies. The larger Bristol and Liverpool firms began extending more and more credit to corresponding planters, ending their dependence for financing on a limited number of sometimes imperious London factors. The more prudent planters maintained positive balances with their principal correspondents that permitted them to take advantage of discounts on customs duties on tobacco if they were paid in cash soon after entry. Others were able to finance the purchase of additional land and slaves or ventures into ancillary activities such as commercial milling, shipping, and iron mining. Some enterprising but less disciplined elites, by playing one merchant off another and by opening new, initially credit-granting correspondences without paying old debts to previous factors, were able to continue high living and favored pursuits such as horse breeding and racing with minimal risk to their estates or to their children's future fortunes.

20. John Custis to Micajah Perry, 1735, John Custis Letterbook, 1717–1741, LC, microfilm, M-8, CWL; Price and Clemens, "A Revolution of Scale," *JEH,* XLVII (1987), 1–43; Morgan, *Bristol and the Atlantic Trade,* chap. 6. For the specialties of the various ports, see Coulter, "The Virginia Merchant," chap. 3.

Occasionally, the most imprudent and profligate did fail to stave off their creditors. But, before 1764, high wartime tobacco prices enabled most big planters and most British mercantile houses to realize sufficiently high levels of profit to avoid bankruptcy, if not to achieve outright prosperity. Moreover, by midcentury local elites rather than British merchants were likely providing the majority of domestic loans. Once large planters could rely on natural increase to expand their enslaved workforces and no longer needed to buy new Africans, they had more capital, previously remitted to slave dealers in Britain, to lend to neighbors.[21]

At the same time, a rising market for corn and wheat, first in the West Indies, the wine islands, and urban areas of the northern mainland colonies and by midcentury also in southern Europe, meant that tidewater planters no longer had to pin all their hopes on the fortunes of a single staple. Most large planters turned to corn and to a much lesser extent to wheat as increasingly important second and sometimes third export crops, and by the 1750s the bulk of middling planters also began growing grains for the export market. Most of the expansion in grain exports occurred between 1744 and 1758, as Britain, which had previously shipped grain to some of these areas, needed most of what it raised for home consumption. A complex multilateral trading system was emerging whereby the North American colonists could import more goods from Britain than they paid for directly since favorable balances of trade with the West Indies, southern Europe, and Ireland compensated for excess purchases from Britain. Tobacco exports increased by only about 16 percent during these years, owing in part to droughts in 1755 and 1758 and also to wartime losses to French privateers, but corn and wheat exports more than doubled.[22]

This dramatic increase in grain production was not the result of a spectacular increase in the prices planters received for grain as compared to to-

21. R. C. Nash, "The Organization of Trade and Finance in the British Atlantic Economy, 1600–1830," in Peter A. Coclanis, ed., *The Atlantic Economy during the Seventeenth and Eighteenth Centuries: Organization, Operation, Practice, and Personnel* (Columbia, S.C., 2005), 125–127; Egnal, *New World Economies,* 93–95; Hardy, "Trade and Economic Growth," chap. 6. T. H. Breen, *Tobacco Culture: The Mentality of the Great Tidewater Planters on the Eve of Revolution* (Princeton, N.J., 1985), provides an extensive discussion of the debt issue. Breen's analysis has sometimes been interpreted to mean that planter indebtedness was a chronic problem across the eighteenth century. Almost all the examples Breen discusses occurred in 1764 and after, however, not in the 1740s and 1750s. Trevor Burnard also found that elite Maryland planters were not heavily indebted before the Revolution and that the majority of their debts were owed to other elite Marylanders, not to British merchants (*Creole Gentlemen,* chap. 3).

22. Charles W. Sargent, "Virginia and the West Indies Trade, 1740–1765" (Ph.D. diss., University of New Mexico, 1964), chaps. 1, 2, 4; Price, "The Imperial Economy," in Marshall, ed., *The Eighteenth Century,* vol. II of Louis, ed., *Oxford History of the British Empire,* 88–91.

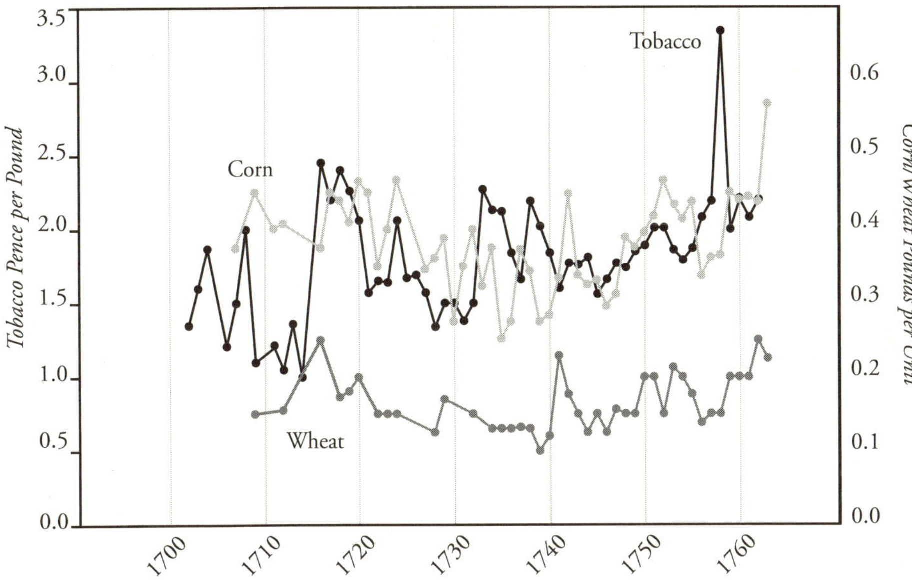

FIGURE 5. Tobacco, Corn, and Wheat Prices, York River Basin, Virginia, 1700–1763. Drawn by Kimberly Foley. Prices for corn are in pounds per barrel, and prices for wheat are in pounds per bushel. John J. McCusker, "Colonial Statistics," in Susan B. Carter et al., eds., *Historical Statistics of the United States: Earliest Times to the Present,* millennial ed., 5 vols. (New York, 2006), V, 681–688.

bacco. The prices for both rose (Figures 5 and 6). The ratio of revenues received from a pound of tobacco or from a bushel of corn or wheat, however, did not tip markedly in favor of one crop over another. Perhaps the more important precipitating change was that reliable markets for surplus grain became increasingly assured. (Lack of a certain market had been one of the factors that had discouraged many planters from going heavily into corn and wheat in the 1720s.) And, during times of war, not only did international demand rise but also planters could send their grains (unlike tobacco, which the Navigation Acts allowed to be shipped only to Britain) to less risky but still lucrative intercontinental destinations.[23]

23. David Klingaman, "The Significance of Grain in the Development of the Tobacco Colonies," *JEH,* XXIX (1969), 268–278; Marc Egnal, "The Economic Development of the Thirteen Continental Colonies, 1720 to 1775," *WMQ,* 3d Ser., XXXII (1975), 191–222; Paul G. E. Clemens, *The Atlantic Economy and Colonial Maryland's Eastern Shore: From Tobacco to Grain* (Ithaca, N.Y., 1980), chaps. 4–6; Harold B. Gill, Jr., "Wheat Culture in Colonial Virginia," *Ag. Hist.,* LII (1978), 380–393; Bergstrom, *Markets and Merchants,* chap. 5; Hardy, "Trade and Economic Growth," chap. 4.

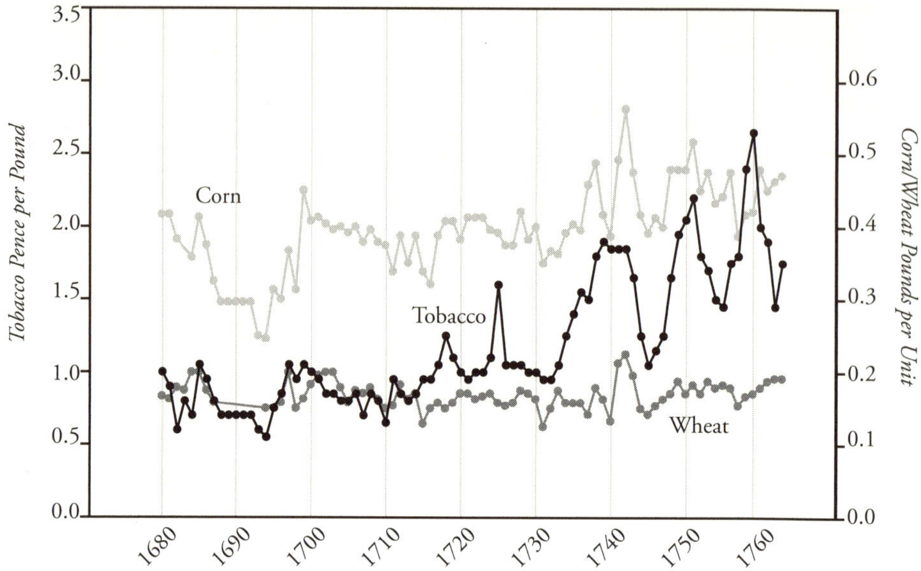

FIGURE 6. Tobacco, Corn, and Wheat Prices, Maryland Upper Eastern Shore, 1680–1763. Drawn by Kimberly Foley. Prices for corn are in pounds per barrel, and prices for wheat are in pounds per bushel. John J. McCusker, "Colonial Statistics," in Susan B. Carter et al., eds., *Historical Statistics of the United States: Earliest Times to the Present,* millennial ed., 5 vols. (New York, 2006), V, 681–688.

Copying the strategies adopted by innovative planters on Maryland's upper Eastern Shore and at the head of Chesapeake Bay in the 1720s, large planters throughout the tidewater Chesapeake, by the 1740s, 1750s, and early 1760s, increasingly emphasized raising farm revenues by expanding grain production. Although less affected by land constraints than upper Eastern Shore farmers, improved market opportunities induced more and more planters to use both land and labor more fully. They discovered by trial and error that by using plows they could increase grain outputs, especially outputs of corn, on land that had been previously cleared and planted in tobacco and that they could do so without acquiring more laborers. Consequently, the expenses involved in expanding tobacco outputs—which required buying more workers, hiring more supervisors, and setting up additional quarters—were greater than those required for raising more corn or wheat. The latter could be accomplished at limited cost to the planters, since most found ways to make overseers and slaves work harder, both in- and off-season for little or no additional reward. The goal, as astute planters or those who employed discerning general managers realized, was to raise enough grains to cover the costs of supervision, taxes, and cloth and tools that had to be purchased locally. The planter could then spend the entire proceeds of the tobacco crop, excluding the cost of any imported cloth, bedding, and tools the workforce still required, on whatever goods or services he and his family desired.[24]

And, by the 1730s, the range of consumer goods elite and middling planters desired rose appreciably. Fashionable textiles, social props for genteel dining, tea services, and specialized, elaborately shaped and carved chairs and tables of mahogany or walnut were among the most popular acquisitions. The proliferation of stores made amenities more available to all, and middling planter families began acquiring ordinary comforts such as feather beds, bed and table linens, and lighting devices, and also newer goods like teawares. Merchants also began to send cargoes more frequently—twice a year rather than once, so the waiting time for orders was cut in half. As more and more sons of the local gentry either returned to Britain to get an education or attended the College of William and Mary in genteel Williamsburg, their knowledge of current metropolitan fashions and prices improved. Along with concerns about price and practicality, planters increasingly insisted that what they bought be new and fashionable as well; secondhand items would no longer do. New methods of English manufacture and improved transportation networks that helped create marketing efficiencies were making a variety of goods more plentiful and often cheaper than before, and competition between Chesapeake merchants kept profit margins on goods low. Thus, the pursuit of comfort, fashion, and genteel display did not lead planters to sacrifice investment in productive goods. Among the rich, the total value of consumer durables in estate inventories had begun to level out or to decline in the 1720s, whereas the number of amenities in those estates continued to rise, and beginning in the 1730s a similar pattern had started among middling and poorer consumers as well.[25]

24. Lois Green Carr and Russell R. Menard, "Land, Labor, and Economies of Scale in Early Maryland: Some Limits to Growth in the Chesapeake System of Husbandry," *JEH,* XLIX (1989), 407–418. To maintain revenues in the face of falling rice prices, many large South Carolina planters similarly added indigo as a second staple in the 1740s and 1750s, requiring enslaved laborers to produce that crop during slack seasons in the rice cultivation cycle. See Joyce E. Chaplin, *An Anxious Pursuit: Agricultural Innovation and Modernity in the Lower South, 1730–1815* (Chapel Hill, N.C., 1993), 187–208; S. Max Edelson, *Plantation Enterprise in Colonial South Carolina* (Cambridge, Mass., 2006), 111–113, 157–162; Peter Mancall, Joshua L. Rosenbloom, and Thomas Weiss, "Agricultural Labor Productivity in the Lower South, 1720–1800," *Explorations in Economic History,* XXXIX (2002), 390–424; and Morgan, *Slave Counterpoint,* 159–164.

25. Lois Green Carr and Lorena S. Walsh, "Inventories and the Analysis of Wealth and Consumption Patterns in St. Mary's County, Maryland, 1658–1777," *Historical Methods,* XIII (1980), 81–104; Walsh, "Urban Amenities and Rural Sufficiency: Living Standards and Consumer Be-

The prosperity and enhanced economic stability that marketing two or more staples afforded was soon also reflected in the built environment. No longer having to invest in land, slaves, or livestock and finding most essential farm structures already in place, elite planters in the mid-eighteenth century turned to building elegant Georgian houses to advertise their status and to facilitate genteel modes of living within (Plate 24). An elite colonial planter who sought "to act his part well," one midcentury essayist averred, ought, among other things, to be "something of an architect." Chesapeake scholars have long celebrated the planters' aesthetic accomplishments, noting, for example, "It was the country house, a symbol of regional power and cultural achievement, on which the Virginia elite lavished their energies and brought to a degree of perfection unsurpassed anywhere in the empire outside the mother nation." Others, placing surviving mansions in the context of the rudimentary dwellings that sheltered the great majority of colonists, have emphasized instead that "the most pretentious of them were outrageous statements of wealth and status and durability that only a few could claim." Chesapeake "great houses" were first and foremost manifestations of their owners' wealth, political power, and social hegemony. "They were also intellectual statements, conscious testimony that the colony's leaders were abreast of the latest current of ideas in the Western world." Built symmetrically of brick with multiple, evenly spaced sash windows and a pedimented central door, many great houses followed a double-pile plan with four rooms above and below and incorporated elements of the classical style then so in vogue in Britain. Interiors featured high plastered ceilings, ornate carved woodwork in public rooms, and formally ordered floor plans. An increased number of rooms permitted greater specialization of the activities that took place within, with partitions serving as social buffers that separated genteel white residents and guests from ordinary whites and enslaved blacks.[26]

---

havior in the Colonial Chesapeake, 1643–1777," *JEH,* XLIII (1983), 109–117; Carr and Walsh, "The Standard of Living in the Colonial Chesapeake," *WMQ,* 3d Ser., XLV (1988), 135–159; Carr and Walsh, "Changing Lifestyles and Consumer Behavior in the Colonial Chesapeake," and Cary Carson, "The Consumer Revolution in Colonial British America: Why Demand?" both in Carson, Ronald Hoffman, and Peter J. Albert, eds., *Of Consuming Interests: The Style of Life in the Eighteenth Century* (Charlottesville, Va., 1994), 59–156, 483–697; Breen, *The Marketplace of Revolution,* chaps. 2–4.

26. Billings, Selby, and Tate, *Colonial Virginia,* 215–229 (quotations on 216, 224); Camille Wells, "The Eighteenth-Century Landscape of Virginia's Northern Neck," *Northern Neck Hist. Mag.,* XXXVII (1987), 4217–4255 (quotation on 4246); Rhys Isaac, *The Transformation of Virginia, 1740–1790* (Chapel Hill, N.C., 1982), 34–42, 65–79; Kevin M. Sweeney, "High-Style Vernacular: Lifestyles of the Colonial Elite," and Edward A. Chappell, "Housing a Nation: The Transformation of Living Standards in Early America," in Carson, Hoffman, and Albert, eds.,

PLATE 24. Carter's Grove, James City County, Virginia. Drawn by Richard Anderson. *Built by Carter Burwell between 1751 and 1755. A midcentury mansion.*

Perhaps in part because conceptions of the proper layout of a country house were borrowed from the mother country, a nonslave society, Chesapeake planters had yet to incorporate structures for housing slaves into their new polite landscapes. Enslaved laborers were typically housed in small earthfast wooden cabins that afforded a slave family or group of housemates little more than two hundred square feet of living space. These cabins (or "quarters," as they were commonly called) were built of logs standing or lying directly on the ground or erected in postholes. The walls and roof were covered with riven clapboards, and the inside was left unfinished. Floors were usually of dirt, chimneys of wattle and daub, doors were small, and windows were equipped with only sliding wooden shutters. Some great planters chose to site slave dwellings and other vernacular agricultural buildings such as to-

Of Consuming Interests, 1–58, 167–232; Brugger, *Maryland,* 79–83; Burnard, *Creole Gentlemen,* chap. 7; Samuel Martin, *An Essay upon Plantership, Humbly Inscrib'd to All the Planters of the British Sugar-Colonies in America,* 2d ed. (Antigua, 1750), vii–viii.

bacco houses at a distance from their new mansions, in effect creating two distinct landscape zones, "one in which to live and the other from which to profit." But others found nothing anomalous in putting up home house quarters, no more finely finished or formally laid out than worker housing on outlying quarters, in plain sight of the great house. This juxtaposition of luxury and deprivation underscored the slaves' degraded position at the bottom of the social hierarchy and the raw power that the slaveowners exercised over their bound workers. The communal work and living spaces that the slaves fashioned for themselves in and around the quarter, however, revealed elements of choice and continuity salvaged from a very different cultural tradition. Great planters' control over the landscape was far from complete.[27]

In addition to building mansions emulating aristocratic country houses in Britain, great planters in this period also began more often according their home plantations pretentious, status-elevating names. The names of outlying quarters continued the conventions that had prevailed in earlier years when planters had devoted most of their energies simply to wresting profit from holdings to which they had little emotional attachment. Many quarters took their names from local geography (Black Swamp, Fleet's Bay, River Quarter, Hickory Thicket), were known by the surname of the current or a previous owner or perhaps even that of a long-resident overseer, reflected the sequence of acquisition (Old House, New Quarter), or sometimes expressed whimsy (Forlone). But, by the mid-eighteenth century, the Chesapeake planter elite (as did their contemporaries in South Carolina) began calling their home plantations by names celebrating family prominence (Carter's Grove, Corbin Hall), ancient Rome (Sabine Hall), or titles that invoked an exotic native American past (Morriatico, Corotoman) as well as continued an older tradition of adopting English place-names (Berkeley, Stratford, Barns Elms).[28]

Most ordinary planters, in contrast, continued to live in impermanent ver-

27. For slave housing, see Wells, "The Eighteenth-Century Landscape," *Northern Neck Hist. Mag.,* XXXVII (1987), 4238–4239; Chappell, "Housing a Nation," in Carson, Hoffman, and Albert, eds., *Of Consuming Interests,* 191–193; Edward Chappell, "Looking at Buildings," *Fresh Advices: A Research Supplement* (November 1984), i–vi; and Lorena S. Walsh, *From Calabar to Carter's Grove: The History of a Virginia Slave Community* (Charlottesville, Va., 1997), 19–20, 102–103, 181–183, 199–203. For plantation layouts (of which few are available for the mid-eighteenth century), see Isaac, *Transformation of Virginia,* 41 (Shirley); Wells, "The Eighteenth-Century Landscape," *Northern Neck Hist. Mag.,* XXXVIII (1987), 4223 (Belle Isle); William M. Kelso, *Kingsmill Plantations, 1619–1800: Archaeology of Country Life in Colonial Virginia* (San Diego, Calif., 1984), 153 (Kingsmill); and "William Byrd Title Book," *VMHB,* XLVII (1939), opposite 285 (Westover). For South Carolina planters' comparable creation of genteel houses and plantation landscapes in this period, see Edelson, *Plantation Enterprise,* 136–151.

28. Isaac, *Transformation of Virginia,* 38–39; Edelson, *Plantation Enterprise,* 141–142.

nacular houses little changed from the labor-saving "Virginia houses" immigrant settlers like Robert Cole had begun building in the mid-seventeenth century. Small and middling planters who worked farms of modest size inherited from their parents or who, less often, cleared new farms on any marginal lands left unclaimed east of the fall line, remained content—or had to be contented—with earthfast wooden dwellings of one or two rooms on the first floor and a loft above that afforded no more than four hundred square feet of living space. At midcentury, the majority of Chesapeake houses were still covered with riven clapboards rather than with sawn weatherboards and rounded shingles, and they lacked glazed windows, plastered interiors, or paint, although brick chimneys did increasingly replace wattle and daub. Outbuildings such as kitchens and barns changed only in that they were more frequently built entirely of logs rather than having posts set in the ground, a new building technology that further reduced the amount of joinery required. On the whole, ordinary planters confined their material aspirations to the acquisition of modest personal and household amenities such as more varied clothing, bed and table linens, a few more chairs and tables, some ceramic dining wares, and teaware. Unlike the rural elite, according to Edward A. Chappell, the procurement and use of new consumer items had yet to enmesh common planters in "a reciprocal process" in which acquisition of portable goods encouraged "investment in conspicuously superior domestic architecture."[29]

Some lesser gentry, Anglican clergymen, and more prosperous middling planters, however, were beginning to build more durable, better-finished houses, with finer exterior and interior finish, glazed windows, and brick walls, chimneys, cellars, or underpinnings. Pear Valley in Northampton County on the lower Eastern Shore is a surviving example of a well-built one-room house constructed about 1740 by a planter who owned 150 acres and served in the mid-level county post of tobacco inspector. The dwelling afforded only 346 square feet of living space on the first floor (which served the family as the primary space for cooking, eating, visiting, and sleeping), but one end wall of the otherwise frame fabric is of fashionable brick (Plate 25). The owner also took the trouble to finish the interior room on the ground floor room with plaster but left the more private attic bedchamber unfinished. This status-conscious middling planter advertised his material achievements, not in the size of the family's new dwelling, but rather in a more finished appearance

29. Chappell, "Housing a Nation," in Carson, Hoffman, and Albert, eds., *Of Consuming Interests,* 216 (quotation); Wells, "The Eighteenth-Century Landscape," *Northern Neck Hist. Mag.,* XXXVIII (1987), 4217–4255.

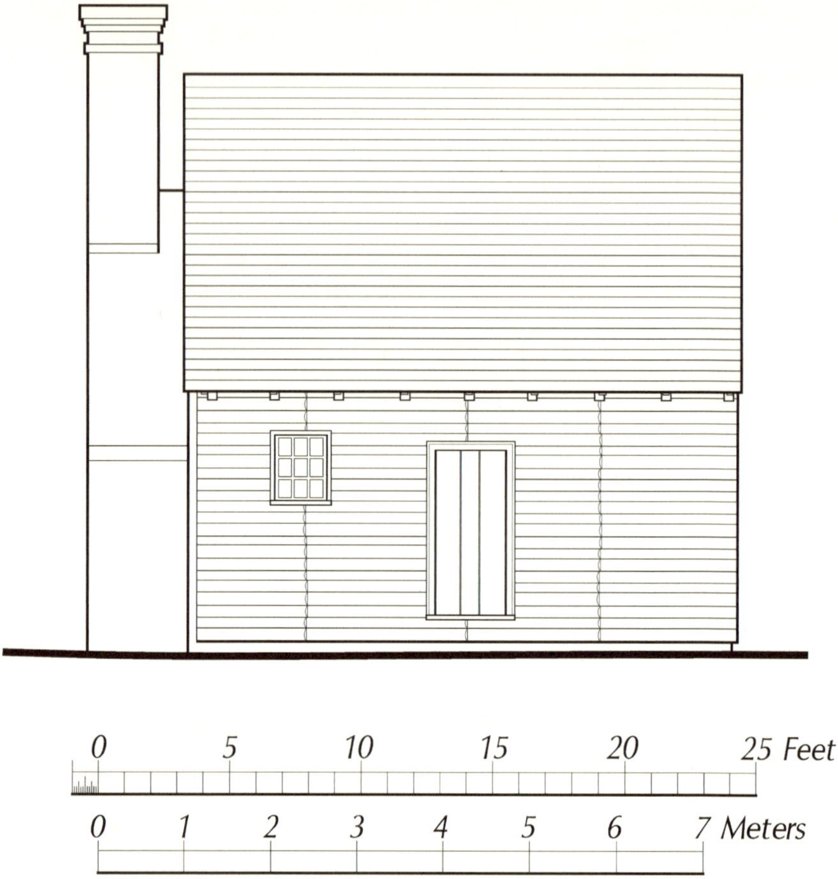

PLATE 25. Pear Valley, Northumberland County, Virginia. Drawn by Richard Anderson.
Built by Robert Nottingham about 1740. A midcentury middling planter's dwelling.

for the parts of the house exposed to public view. Comparison of Pear Valley and Carter's Grove (Plates 24 and 25) suggests that middling planters were coming to share a common understanding of the elements of genteel housing with their economic and social superiors rather than continuing the marked differences in perceptions of adequacy revealed in a comparison between earlier mansions like Fairfield and the vernacular Atkinson dwelling (Plates 11a–b, 13).[30]

30. Edward A. Chappell and Julie Richter, "Wealth and Houses in Post-Revolutionary Virginia," in Annmarie Adams and Sally McMurry, eds., *Exploring Everyday Landscapes,* Perspec-

As the Chesapeake gentry plunged into the increasingly complex transatlantic world of goods and burgeoning print culture from the 1730s to the 1760s, they looked ever more to Britain for cultural models to emulate, not just for the houses they built, the goods with which they furnished them, and the genteel rituals that structured their lives but also for models for managing their estates. During the era of hard times, large planters had learned how to exploit enslaved workforces and to make pragmatic adjustments in the mix of crops they raised and in the ways they processed them. Current British agricultural practices offered no guidance for increasing outputs of New World crops of tobacco and corn tended with labor-saving native American and African hoe technology, nor for improving their quality. The opportunities afforded by expanding international markets for grain beginning in the 1740s, with the accompanying need to incorporate Old World plow culture into plantation operations, however, prompted large planters to learn more about recent changes in British agriculture.[31]

The emphasis on raising new fodder crops in order to carry more livestock that dominated the late seventeenth and early eighteenth centuries was beginning to yield unexpected results. Crop rotations alternating grains with root crops (such as turnips) and with fodder crops (such as clover, sainfoin, and ryegrass) that added nitrogen to previously infertile soils, improved tillage techniques, the increased use of the manure the greater numbers of livestock produced, and better integration of livestock raising with arable farming enabled British farmers working lighter soils to expand their cereal outputs. (The new techniques effected fewer improvements on heavy, clay soils.) Rotating crops permitted farmers both to increase grain yields and to stock more animals per acre, and it also minimized losses to weeds and insect pests. And, although convertible husbandry did require more labor, much of

---

tives in Vernacular Architecture, VII (Knoxville, Tenn., 1997), 4–6; Chappell, "Pear Valley Interpreted," research report, Department of Architectural Research, Colonial Williamsburg Foundation, 1992.

31. The adoption of plowing on large plantations in the York and Rappahannock basins coincided in time with the change among enslaved adult field-workers from predominantly African to predominantly native born. Large planters' response to expanding international grain markets, however, rather than changes in the labor force, offer the best explanation for the midcentury shift in technology. As noted in Chapter 4, upper Eastern Shore planters were the first to make widespread use of plows in the 1720s at a time when their workers were almost entirely African born. Kulikoff, *Tobacco and Slaves,* 406–408, posits a relationship between more widespread use of plows and carts on midcentury tobacco plantations on Maryland's lower Western Shore and the emergence of a creole workforce but does not take into account expanded grain production as the main reason for adopting plows.

that work could be done in the off-season, thus evening the demand for labor over the year.[32]

As large planters throughout the Chesapeake took advantage of rising overseas markets for grain and expanding domestic and intercolonial markets for meat, the innovations in cereal culture and livestock raising taking place in Britain became increasingly relevant to the challenges they faced in expanding colonial outputs. In addition, slaveowners had an overriding interest in making optimal year-round use of their permanently bound workforces. Growing interest in Enlightenment science more generally also fostered a new willingness to experiment and a greater faith in the possibility of effecting improvements in social and material conditions. Beginning in the 1730s, Chesapeake planters imported not only titles on gardening, orchards, and the care of horses but also books on British field husbandry. Some of the books that graced the shelves of great planters' libraries in this period were first published at the turn of the century, such as John Houghton's *Collection of Letters for the Improvement of Husbandry and Trade* (1681–1683) and John Worlidge's *Dictionarium rusticum* (1704), but others were more recent publications, such as Georg Andreas Agricola's *Experimental Husbandman and Gardener* (trans. Richard Bradley, 1721), Stephen Switzer's *Practical Husbandman and Planter* (1733), and one of the most widely read farming books in Britain, William Ellis's *Modern Husbandman* (1731) along with Ellis's *Practical Farmer* (1732). By the 1740s and 1750s, new publications on agriculture appeared in the libraries of elite planters soon after first publication in Britain. Great planters' ability to implement the recommendations contained in such farming manuals was an entirely different matter. Rather than attempting an impractical, wide-ranging transformation of Chesapeake agriculture, they experimented selectively with only a few of the new methods. Some raised new forage crops, others adopted more intensive stock-raising techniques, some concentrated on better accounting and record keeping, and yet others began experimenting with manuring and expanded crop rotations. Like contemporary South Carolina rice planters, elite Chesapeake planters sought to improve on "planting strategies that had developed over decades of experience on the land," rather than rigidly following recommendations better suited to places where land was scarce and labor cheap.[33]

32. E. L. Jones, *Agriculture and the Industrial Revolution* (Oxford, 1974), chaps. 3–4; Mark Overton, *Agricultural Revolution in England: The Transformation of the Agrarian Economy, 1500–1850* (Cambridge, 1996), chap. 3.

33. For British husbandry literature, see G. E. Fussell, *The Old English Farming Books from Fitzherbert to Tull, 1523 to 1730* (London, 1947); Fussell, *More Old English Farming Books: From Tull to the Board of Agriculture, 1731 to 1793* (London, 1950); and G. E. Mingay, ed., *The Agrarian*

In contrast to the planter elite who benefited mightily from inherited assets, the poor during this period of economic recovery had fewer opportunities to improve their lot than during favorable years in the seventeenth century. By 1750, no more cheap land was available for new plantations in older areas, and growing local populations began to press upon existing resources. Despite increased out-migration of some poor, landless whites, everywhere in the tidewater land values escalated, rents rose, and the ranks of tenants increased. In older areas, acquiring a farm of one's own became an ever more distant dream for those not fortunate enough to inherit land. More children had to wait for their parents to provide portions with which to set up households; by the mid-1740s, the mean age at first marriage for both men and women began to rise toward European norms. Once creoles managed to marry and set up housekeeping, however, even landless but well-situated tenants shared to some degree in the growing regional prosperity.[34]

Trade disruptions during the two French wars occasioned short-term distress, especially for those living in or near areas where fighting occurred, but were in balance a blessing in disguise. Shortages of manufactures stimulated local industry, especially the production of shoes and country cloth. Large planters began to order that more coarse cloth for the slaves' clothing be woven on their own plantations, and for the first time most small and middling planter families started to raise some fibers and to spin yarn, if not to weave cloth. After the wars, colonists resumed the import of manufactured goods and for a time placed less emphasis on import replacement. But in the years of falling tobacco prices, rising prices for imported manufactures, and

---

History of England and Wales, VI, *1750–1850,* in H. P. R. Finberg, gen. ed., *The Agrarian History of England and Wales* (Cambridge, 1967–2000), 361–370. For planters' embrace of Enlightenment science, see Billings, Selby, and Tate, *Colonial Virginia,* 215–224; Drayton, "Knowledge and Empire," in Marshall, ed., *The Eighteenth Century,* vol. II of Louis, ed., *Oxford History of the British Empire,* 237–247; and Rhys Isaac, *Landon Carter's Uneasy Kingdom: Revolution and Rebellion on a Virginia Plantation* (Oxford, 2004), 76–84. For South Carolina planter strategies, see Edelson, *Plantation Enterprise,* 166–176 (quotation on 171).

34. For population growth and out-migration, see Kulikoff, *Tobacco and Slaves,* chap. 3; Carville V. Earle, *The Evolution of a Tidewater Settlement System: All Hallow's Parish, Maryland, 1650–1783* (Chicago, 1975), chap. 3; Clemens, *The Atlantic Economy,* 161–166; Jean Elliott Russo, "'The Interest of the County': Population, Economy, and Society in Eighteenth-Century Somerset County, Maryland" (Ph.D. diss., University of Minnesota, 1999), chaps. 1, 3. Economic prospects of ordinary planters are discussed in Kulikoff, *Tobacco and Slaves,* chaps. 2–4; Clemens, *The Atlantic Economy,* 136–167; Robert E. and B. Katherine Brown, *Virginia, 1705–1786: Democracy or Aristocracy?* (East Lansing, Mich., 1964), chap. 1; D. Alan Williams, "The Small Farmer in Eighteenth-Century Virginia Politics," *Ag. Hist.,* XLIII (1969), 92–94; and Lorena S. Walsh, "Land, Landlord, and Leaseholder: Estate Management and Tenant Fortunes in Southern Maryland, 1642–1820," *Ag. Hist.,* LIX (1985), 373–396.

reduced credit that inevitably followed brief postwar booms, they retained a capacity to produce many more basic goods than earlier in the century. Home production of some essentials, especially yarn, cloth, and shoes, increasingly permitted the purchase of more imported amenities.[35]

Other evidence of local economic development could be found along many tidewater streams where large planters were constructing gristmills that served both their own plantations and those of smaller neighbors. By the 1740s, the laborious daily task of pounding corn by hand in mortars had virtually ceased in the older tidewater, freeing time for more productive pursuits. As the grain trade boomed, some planters constructed larger merchant mills that processed flour and ship's bread for export as well as ground grain for local consumption. Along the Potomac and Patapsco rivers, iron mines were developed in earnest, and the more successful of these enterprises provided handsome returns to those tidewater planters who chose to invest in them. Iron developer Richard Snowden of Baltimore County, Maryland, for example, reputedly started out several thousand pounds sterling in debt in the 1730s, but by 1750, according to Annapolis lawyer Stephen Bordley, was "making a little Estate every year, notwithstanding, as I have lately learn't, his Iron is of the Worst sort amongst us." More and more tobacco ships took on bar iron as part of their cargoes, and the ironworks afforded a growing, if localized, market for grain, livestock, and fodder. Finally, the region's towns experienced modest growth during these years, and additional hamlets and villages sprang up around newly established tobacco warehouses. Although the impact of these urban markets was limited, some planters living near towns like Williamsburg and Annapolis began regularly supplying urban residents with grain, livestock products, and firewood. Improvements to overland transport were less marked, since construction and maintenance of roads and bridges remained under the jurisdiction of individual county courts. Although the gentry increasingly invested in private enterprises like mines, they had not yet developed mechanisms for financing public transportation improvements such as turnpikes, canals, or port facilities.[36]

35. Lois Green Carr and Lorena S. Walsh, "Economic Diversification and Labor Organization in the Chesapeake, 1650–1820," in Stephen Innes, ed., *Work and Labor in Early America* (Chapel Hill, N.C., 1988), 144–188; Bergstrom, *Markets and Merchants,* chap. 5; Earle, *The Evolution of a Tidewater Settlement System,* chap. 5; Carr, "Diversification in the Colonial Chesapeake," in Carr, Morgan, and Russo, eds., *Colonial Chesapeake Society,* 342–388.

36. William Fitzhugh Account Book, 1761–1764, MS 1831, MHS, documents a commercial mill and bakery. For ironworks, see Bordley to ?, Nov. 20, 1750, Stephen Bordley Letterbooks, 1727–1759, MS 81, MHS (quotation); John Tayloe Account Book, 1749–1768, MSS1T2118b5, VHS; Robert Morris Ledger B, MS C714-1, MSA; *Byrd Correspondence,* I, 377–378; William Byrd, *A Progress to the Mines in the Year 1732,* in Louis B. Wright, ed., *The Prose Works of William*

The expansion of full-time retail stores at first pushed most creole planters from the trade in European imports, as almost none had the capital to compete with large Scottish firms or enough cash on hand to match the resources of aggressive English outport factors. On the other hand, more resident planters began to speculate in West Indian goods, exchanging plantation produce for rum, sugar, molasses, and salt that they then retailed to small-planter neighbors or divvied up with their relatives. The expanding trade in grain also offered new niches that local merchants began to exploit. Great planters sometimes invested directly in cargoes of grain bound for the West Indies or wine islands. The intercolonial produce trades had become so complicated and time consuming, however, that most planters chose instead to rely on full-time professional merchants rather than personally "dabbling" in these trades. The increased specialization so apparent among British merchants by midcentury spread to colonial traders as well.[37]

Byrd of Westover: Narratives of a Colonial Virginian (Cambridge, Mass., 1966), 339–378; Donald Jackson, ed., *The Diaries of George Washington,* 6 vols. (Charlottesville, Va., 1976), I, 253; Edward Dorsey Account Book, 1761–1765, Peter Force Collection 8D, no. 35, reel 37, LC; Keach Johnson, "The Baltimore Company Seeks English Subsidies for the Colonial Iron Industry," *MHM,* XLVI (1951), 27–43; Johnson, "The Genesis of the Baltimore Ironworks," *Jour. So. Hist.,* XIX (1953), 157–179; Johnson, "The Baltimore Company Seeks English Markets: A Study of the Anglo-American Iron Trade, 1731–1755," *WMQ,* 3d Ser., XVI (1959), 37–60; Michael Kammen, *Empire and Interest: The American Colonies and the Politics of Mercantilism* (Philadelphia, 1970), chaps. 3, 4; Michael Warren Robbins, "The Principio Company: Iron-Making in Colonial Maryland, 1720–1781" (Ph.D. diss., George Washington University, 1972), chaps. 1, 2, 4, 5; Jerrilynn Eby, *Laying the Hoe: A Century of Iron Manufacturing in Stafford County, Virginia* (Westminster, Md., 2003); John Bezís-Selfa, *Forging America: Ironworkers, Adventurers, and the Industrious Revolution* (Ithaca, N.Y., 2004), chap. 3. Hardy, "Trade and Economic Growth," chap. 4, provides a recalculation of the value of nontobacco exports. For town growth, see Jacob M. Price, "Economic Function and the Growth of American Port Towns in the Eighteenth Century," *Perspectives in American History,* VIII (1974), 164–173; Carville Earle and Ronald Hoffman, "Staple Crops and Urban Development in the Eighteenth-Century South," ibid., X (1976), 57–78, and James O'Mara, "Urbanization in Tidewater Virginia during the Eighteenth Century: A Study in Historical Geography" (Ph.D. diss., York University, 1979). The Letterbook of Dr. Charles Carroll of Annapolis, 1752–1755, and of his son Charles Carroll Barrister, 1755–1769, MS 208, MHS, documents investments in mines, mills, and urban development. For trends in urban provisioning, see Lorena S. Walsh, "Provisioning Tidewater Towns," *Explorations in Early American Culture,* IV (2000), 46–80.

37. Clemens, *The Atlantic Economy,* chap. 5; Robert Morris Ledger B, MS C714-1, MSA; Clement Hill Papers, MS 446, box 1, MHS; Samuel Galloway Ledger, 1748–1758, Cheston-Galloway Papers, MS 1994, MHS. Trevor Burnard interprets wealthy Maryland planters' tendency not to engage in trade from the mid-eighteenth century as a decision to specialize in agriculture and to avoid mercantile adventures that "would place their properties at risk and thus abuse their duties to descendants." He found that the majority of wealthy immigrants combined trading with planting but that later generations of native-born residents specialized in agriculture. "They adopted risk averse, conservationist strategies, seeing profit maximization as secondary to the maintenance

Tobacco Inspection to the Rescue

In addition to adopting new crop mixes, investing in new enterprises like milling and mining, and trading in West Indian products, Chesapeake planter elites also exerted greater control over their own colonial economies by embracing inspection systems for their premier staple designed to enhance quality rather than continuing to rely on ineffective measures that attempted to reduce the quantity produced. When markets deteriorated in the later seventeenth century, Chesapeake planters were too few, too disorganized, and too divided to adopt effective countermeasures. By 1730, the planter elite was organized and powerful enough to mobilize support for tobacco inspection. From the perspective of the later seventeenth century, the system adopted represented a radical degree of public interference in private pursuits. It entailed trade-offs between public and private responsibilities. Most of the costs of inspection and construction of centralized warehouses were borne by the entire taxpaying public, whether or not they grew tobacco. On the other hand, individual planters paid additional inspection fees in proportion to the amount of tobacco they marketed, and they had also to assume the costs of transporting their crops to the warehouses. Implementation of inspection acts in Virginia in 1730 and in Maryland in 1747 changed the way planters managed the dominant cash crop, brought higher returns, and also increased the local money supply in the form of receipts for inspected tobacco that circulated as a legal currency.[38]

The Virginia inspection act of 1730 was orchestrated by Virginia Governor William Gooch, who worked assiduously to gain the support of royal officials and leading merchants in Britain as well as the acquiescence of all classes of planters in Virginia. Gooch's plan required that all tobacco intended for sale be brought to public warehouses where it would be examined by sworn officers and any trash found in it removed and immediately destroyed, that inspected tobacco be securely stored in the warehouses until it was exported, and that exact records be kept to deter smuggling. Probably because of deteriorating conditions in the tobacco trade in the late 1720s, the Virginia as-

---

of a comfortable sufficiency." He also argues, however, that profits from planting rose in the third quarter of the century, and trading was often less profitable and required more capital to make a great fortune than did planting (*Creole Gentlemen,* chap. 2 [quotations on 22, 30]).

38. Lewis Cecil Gray, *History of Agriculture in the Southern United States to 1860,* 2 vols. (1932; rpt. Gloucester, Mass., 1958), I, chaps. 6, 7, 10; Thompson, "The Merchant in Virginia," chap. 4. For the role of tobacco notes as currency, see Mary McKinney Schweitzer, "Economic Regulation and the Colonial Economy: The Maryland Tobacco Inspection Act of 1747," *JEH,* XL (1980), 551–569.

sembly proved remarkably cooperative in drafting a law acceptable to a majority in both houses. After a year or so of turmoil—ranging from petitions to the legislature urging modification or abolition of the law to scattered riots and warehouse burnings, primarily in the less politically powerful oronoco areas—the inspection system indeed soon produced the desired improvement in prices.

Critical differences between Gooch's proposals and those of Governor Alexander Spotswood's failed scheme of 1713 were that inspectors were chosen jointly by the governor and the council and that inspectors could not serve as burgesses. These stipulations precluded a governor from using these critical posts to pack the lower house with placemen beholden to him. Inspectors (there were between two and eight per county) were also paid a fixed annual salary rather than a fee for each hogshead of tobacco they passed. The salary eliminated any temptation for inspectors to increase their earnings by passing bad tobacco. The duties of the office were sufficiently onerous, and the annual salary (initially sixty pounds, subsequently reduced to thirty to forty pounds, depending on the amount of tobacco a warehouse handled) was sufficiently low to deter great planters from seeking inspectorships. Most of the men named to this critical post would thus be respectable middling planters who were "reputed skilful in tobacco." Such officers were much more likely to share common interests with small planters and so were more acceptable to the majority of ordinary tobacco growers, who feared prejudicial treatment from great men. Common planters were further reassured by the transfer during the 1730s of much of the responsibility for oversight of the inspection system to local county magistrates. A revised act of 1734 empowered justices of the peace to remove negligent inspectors, and an act of 1738 gave them the duty of nominating candidates for inspectorships from among whom the governor and council would choose. While the legislature worked to iron out deficiencies in the original law, nature conspired to favor the act. Bad growing seasons plagued the colony across the decade, and repeated short crops soon led to rising prices for whatever planters could make. In the 1740s, largely because of international wars, prices again fell. Virginia sweet-scented tobacco, however, continued to fare much better than did Maryland oronoco.[39]

39. Virginia inspection laws are in William Waller Hening, ed., *The Statutes at Large: Being a Collection of All the Laws of Virginia . . .* (1809–1823; rpt. Richmond, Va., 1969), IV, 87–88, 247–271 (1730), 329–340 (1732), 380–393 (1734), 478–482 (1736), V, 9–16 (1738), 124–160 (1742), VI, 154–193 (1748), 222–227 (1752), 472–485 (1755), VII, 387–392 (1761), 530–534 (1762). On implementation of the law, see the Carter Letterbook, 1728–1730, Carter Family Papers, VHS; Carter Letterbooks, 1728–1731, 1731–1732, Carter Papers, UVa; and Custis Letterbook, 1717–1741, LC,

As discussed previously, Maryland planters had seen less reason in the 1720s than Virginia sweet-scented growers to drastically change their ways. The oronoco market recovered more quickly and more vigorously from the 1680–1715 slump. With prices on the rise and production per hand nearly double that of Virginia sweet-scented growers, income per laborer in Maryland rose. But after 1730 the tables turned. The sterling price of sweet-scented tobacco remained high, and with a spate of better weather Virginia planters were able to begin making larger crops. Maryland oronoco prices also began a steady rise—in terms of Maryland currency—in 1733. Much of the increase, however, was illusory, caused by that colony's first emission of paper money, a currency that rapidly depreciated. At first, Maryland planters saw only higher prices, but shortly they realized that there was a disheartening gap between what they and what Virginians could buy with the proceeds of their crops. In 1725, Anne Arundel County planter Dr. Charles Carroll of Annapolis was optimistic: "However tob. may answer att home I'm sure it bears Extravigant rates here." In 1730, however, Eastern Shore planter James Hollyday was complaining to his compatriot Edward Lloyd of "the miserable low market for tobacco, wch hardly maintains the people that make it." By 1733, Carroll warned merchant Philip Smith that "the Lowness of Tobacco has ruined me and unless you sell what's in yr Hands and what goes home this year better I shall not recover in haste."[40]

M-8, CWL; L. C. Gray, "The Market Surplus Problem of Colonial Tobacco," *Ag. Hist.,* II (1928), 1–34; Charles George Gordon Moss, "The Virginia Plantation System: A Study of Economic Conditions in That Colony for the Years 1700 to 1750" (Ph.D. diss., Yale University, 1932), chaps. 6–7; Vertrees J. Wyckoff, *Tobacco Regulation in Colonial Maryland* (Baltimore, 1936), chaps. 6–8; David Alan Williams, "Political Alignments in Colonial Virginia Politics, 1698–1750" (Ph.D. diss., Northwestern University, 1959), chap. 6; John Mickle Hemphill II, "Virginia and the English Commercial System, 1689–1733: Studies in the Development and Fluctuations of a Colonial Economy under Imperial Control" (Ph.D. diss., Princeton University, 1964), chap. 4; Williams, "The Small Farmer in Eighteenth-Century Virginia Politics," *Ag. Hist.,* XLIII (1969), 98–99; Janis M. Horne, "The Opposition to the Virginia Tobacco Inspection Act of 1730" (senior honors thesis, College of William and Mary, 1977); and Stacy L. Lorenz, "'To Do Justice to His Majesty, the Merchant, and the Planter': Governor William Gooch and the Virginia Tobacco Inspection Act of 1730," *VMHB,* CVIII (2000), 348–392.

40. For prices, see Figures 1 and 4; Carr, "Diversification in the Colonial Chesapeake," in Carr, Morgan, and Russo, eds., *Colonial Chesapeake Society,* 342–388; and Schweitzer, "Economic Regulation and the Colonial Economy," *JEH,* XL (1980), 551–569. For paper money, see John J. McCusker, *Money and Exchange in Europe and America, 1600–1775: A Handbook* (Chapel Hill, N.C., 1978), 189–204. For Maryland planters, see Charles Carroll to Richard Burbydge, 1725 (quotation), to John Hyde, June 28, 1725, to William Hunt, Aug. 8, 1725, and to Philip Smith, Sept. 11, 1733 (quotation), Letterbook of Charles Carroll of Annapolis, 1722–1734, MS 208, MHS; and James Hollyday to Edward Lloyd, 1730, James Hollyday Papers, MS 1317, box 1, MHS.

In the 1730s and 1740s, small planters on Maryland's lower Western Shore continued to make as much tobacco as they could, but some large planters, like those in Virginia two decades before, might have cut back, as did tobacco growers on the Eastern Shore. Upper Eastern Shore merchant Henry Callister reported, "These times [1746] are really discouraging to them." The initial legislative responses were traditional—attempts at quality control in 1721, then stinting in 1728, followed by a requirement attached to Maryland's paper money act of 1733 that in the next two years every planter must burn 150 pounds of his worst tobacco for every taxable he employed in the crop. Owing to a short crop in 1734, however, many planters opted to pay a fine of twenty shillings and to keep their bad tobacco, and others refused to comply, hoping, as was often the case, that they would not be prosecuted. When these measures failed to cut production, and with Virginia's inspection law working well, Maryland planters finally also bit the bullet. Drafting an inspection act was especially difficult for Maryland legislators since, given the anticipated rise in the price of tobacco, this action involved altering one of the colony's legal currencies and hence was tied to the politically hot issues of debt scaling and changes in the fees and salaries of public officers and clergymen. In 1747, under pressure both from their constituents and from the colony's proprietor, the Maryland legislature adopted an inspection act closely modeled on that of Virginia.[41]

As had happened earlier in Virginia, legislators from Maryland's lower Eastern Shore counties, which produced the worst tobacco, initially opposed the law, as did some large planters from prime tobacco areas. Representatives from counties on the Potomac bordering Virginia supported inspection from the outset. Similarly, there was also a year or so of faltering implementation.

41. Henry Callister to Foster Cunliffe, May 4, 1745, Callister Papers (quotation), CWL. The production of Maryland lower Western Shore planters is further discussed in Chapter 7, below. For the 1733 paper money act, see Clarence P. Gould, *Money and Transportation in Maryland, 1720–1765* (Baltimore, 1915), 63. While there was inspection in one of the colonies but not in the other, there were many opportunities to circumvent the law. Virginia and Maryland legislators worried about the smuggling of Virginia trash into Maryland and North Carolina. At the same time, at least one enterprising Maryland planter "run our Tobo over Potowmack and got it Inspected there" (in Virginia) to get a higher price (Stephen Bordley to Joseph Chase, Feb. 13, 1750/1, Bordley Letterbooks, MHS). Wyckoff, *Tobacco Regulation,* chap. 7, reports the opinions of legislators that the trash reduction and stint acts had not been enforced. Under Maryland's stint laws between 1727 and 1730, the crops of households without bound labor should have amounted to not more than about 1,250 pounds, and crops produced by slaves should not have been more than 875 pounds. Tables 30, 31, and 35, below, show higher amounts being produced. For the laws, see William Hand Browne et al., eds., *Archives of Maryland,* 72 vols. (Baltimore, 1883–1972), XLIV, 595–638, XLVI, 489–502, L, 303–367, LVI, 128–129, LVII, 433–497.

But, again, nature helped out; short crops in 1748 and 1749 contributed to an immediate rise in prices, helping to convince most Maryland planters that they had made the right decision. As early as December 1747, Henry Callister informed Liverpool merchant Foster Cunliffe, "Tobacco is now on the advance upon the Strength of the Inspection Law, and many Planters that are able will keep their Tobo on hand or some of the late Crop in order to pass inspection." Eastern Shore resident James Hollyday noted, however, that in 1748, the first year the law was in effect, it was late in the season before inspectors began acting, and many who were initially appointed refused to serve, "some from an apprehension of the great difficulty they shoud be under in the Exec[utio]n of the office, being quite a new thing here, and others from an opinion that the Salarys were not a sufficient reward for the trouble and expense they must necessarily be at." The next year, Callister reported that planters "have got this year almost 4 times the price in Goods than they had got some of the six last years." Stephen Bordley observed that influential people on the Eastern Shore attributed the price rise to the inspection law, but he thought two short crops and the peace of 1748 might have contributed more. By 1750, Hollyday could confidently write London merchant William Anderson, "The inspected Tobacco hath sold pretty well and in All Appearances will be Scarce here before the next Crop comes to Market." In 1753, when the inspection law came up for renewal, it easily passed the lower house.[42]

Contributing to planters' initial fears about how inspections laws would affect them was that in both colonies the legal specifications for acceptable tobacco were vague indeed, requiring only that the product be "sound, well-conditioned, and merchantable, and free from Trash." As one writer put it, "There is no Standard for Tobacco, it depends altogether on Humour and fancy." Although some feared that all inferior tobacco, including "seconds" (whole leaves of inferior quality) and ground leaves (the bottom leaves of the plant most likely to suffer damage from heavy rains), would be rejected,

42. Callister to Cunliffe, Dec. 25, 1747, Nov. 12, 1749 (quotations), Callister Papers, CWL (see also Callister to Cunliffe, Aug. 6, 1747, to Evan Callister, Nov. 12, 1749, and to Anthony Bacon, Oct. 13, 1750); Hollyday to William Anderson, Nov. 12, 1749, James Hollyday Letterbook, 1748–1754, Hollyday Family Papers, Special Collections microfilms 436 and 437, MSA; Hollyday to Anderson, Feb. 17, 1749/50, box 1, Hollyday Papers, MHS; Bordley to William Hunt, Nov. 13, 1749, Bordley Letterbooks, MHS. On implementation, see Wyckoff, *Tobacco Regulation,* chaps. 6–8; Schweitzer, "Economic Regulation," *JEH,* XL (1980), 551–569; Gould, *Money and Transportation in Maryland,* chaps. 3, 5; Arthur Pierce Middleton, *Tobacco Coast: A Maritime History of Chesapeake Bay in the Colonial Era* (Newport News, Va., 1953), chap. 4; and Charles Albro Barker, *The Background of the Revolution in Maryland* (New Haven, Conn., 1940; rpt. n.p., 1967), chap. 3. Barker interprets the period much more pessimistically.

merchant and planter account books and estate accounts demonstrate that at least in short years these inferior grades did pass inspection. With time, a community consensus of what constituted acceptable tobacco did develop. As tobacco inspectors gained experience, complaints about arbitrary decisions diminished. And, once planters had a better idea of what tobacco would pass, they could adjust processing methods accordingly and need not, in "alarm," throw away an "abundance of what we us'd to get good money for."[43]

One of the major advantages of the inspection acts was the collection of tobacco in centralized warehouses (Plate 26). This innovation dramatically reduced turnaround time in the Chesapeake. By 1762, ships were spending only half the time loading tobacco that they had in 1700, which especially benefited small planters. Before inspection, as Eastern Shore factor Henry Callister explained, merchants had paid large planters a premium for "the advantage of shippable heavy Tobo and the Convenience of their situation." But, in order to retain the custom of less privileged tobacco growers, "We always do it as privately as we can, for the Common Planters will not allow any preference due."

> It is impossible to persuade them that it sh'd be to their disadvantage to live remote or on barren lands, since as they say they make as good Tobo as another, at least as good as their soil will produce. They will not allow for the charges of prising, rolling, the Quantity or Quality of their Crops, while they are conscious they take as much pains in cultivating it as another. But as they are of the weaker sort, they are manageable one way of another and much dissimulation if not downright lying is sometimes the way to hook 'em and since reason will not do, they must be deceived into a fair distinction.

After the adoption of inspection, all planters had to bear the costs of transporting their tobacco to the warehouse. Once it was delivered, however, the price received was determined solely by the quality of the crop. Large planters could no longer command an extra premium for "shippable" crops assembled at private wharves. Smaller producers could have underweight hogsheads reprized at the warehouse using public equipment, and parcels too small to fill a hogshead were accepted as transfer tobacco that the inspectors were responsible for packing. Inspection thus served to level the playing field between big

43. Hening, ed., *Statutes at Large,* IV, 247–271; Brown et al., eds., *Archives of Maryland,* XLIV, 595–638; John C. Rainbolt, ed., "The Case of the Poor Planters in Virginia under the Law for Inspecting and Burning Tobacco," *VMHB,* LXXIX (1971), 314–321; Robert Carter to William Dawkins, Nov. 15, 1731, Carter Letterbook, UVa (quotation). The types of tobacco accepted after passage of inspection laws are discussed further in Chapter 7, below.

PLATE 26. Public Tobacco Warehouse. Detail from William Tatham, *An Historical and Practical Essay on the Culture and Commerce of Tobacco* (London, 1800). Courtesy, Special Collections John D. Rockefeller, Jr., Library, The Colonial Williamsburg Foundation.
One of many such warehouses mandated by Virginia's tobacco inspection laws.

producers and ordinary planters who took the trouble to "make as good Tobo as another." Those who farmed inferior soils, however, were adversely affected by inspection laws, since, no matter how carefully they managed their crops, their poorly nourished, thin tobacco was at higher risk of being rejected than tobacco grown on richer ground.[44]

Inspection also helped to average out annual tobacco exports and thus

44. Callister to Cunliffe, Dec. 25 (quotation), Aug. 6, 1747, Callister Papers, CWL. For turn-around times, see James F. Shepherd and Gary M. Walton, *Shipping, Maritime Trade, and the Economic Development of Colonial North America* (Cambridge, 1972), chap. 5, appendix 3.

to maintain more consistent prices. When crops were plentiful, inspectors were more severe, rejecting leaf of questionable quality. In short years, they relaxed standards and passed lower-grade hogsheads. Eastern Shore planter James Hollyday reported that in 1749, an unfavorable year, "the Inspectors were generally less careful . . . than they ought to have been." Merchant Henry Callister observed that, when the short crop made in 1750 was brought to Maryland warehouses for inspection in 1751, "the Planter and Inspector will have a fellow feeling and endeavour to make the most of it." In contrast, a large crop in 1753 led to the stricter enforcement of Virginia's inspection law. Norfolk trader Charles Stewart estimated that Virginia inspectors burned approximately one-third of that year's crop. But, when the short crop of 1759 came to the warehouses, "trash that would have been burnt many other years [was instead passed, and] sells for 20/ to 25/ [per hundredweight]."[45]

By the 1750s, inspection systems, coupled with the disproportionate impact of the French reexport market, had together brought about a marked homogenization of the Chesapeake's premier product. Within geographic subregions, prices for ostensibly different sorts of tobacco varied little. After Maryland adopted its inspection act in 1747, the gap between sweet-scented and oronoco prices began to close, and by 1760 it diminished to almost nothing. In addition, changes in cultivation practices, the introduction of new strains of oronoco, and a decline in the quality of sweet-scented led to the further homogenization of Chesapeake tobacco. The inspection acts produced a more standard product of the minimal quality most in demand and so proved markedly more effective than earlier restrictions on output. Similar inspection systems were subsequently adopted for flour, barreled beef and pork, and naval stores, bringing greater regularity to these ancillary trades.[46]

45. Hollyday to Anderson, Mar. 9, 1750, Hollyday Letterbook, Hollyday Family Papers, MSA; Callister to Cunliffe, July 2, 1751, Callister Papers, CWL. In a similar vein, John Custis wrote in 1732, a year that began with a drought and ended with a deluge: "How it [spotted tobacco] will set on the inspectors stomacks, I can not say but if they will not pass spotted Tob: they must burn half the Colony" (Custis to John Randolph, Dec. 4, 1732, Custis Letterbook, CWL). When the Virginia law was first implemented, Governor William Gooch ordered inspectors to pass indifferent tobacco so long as it was well handled and honestly packed (Gooch to Board of Trade, Dec. 22, 1731, CO5/1322/174, quoted in Williams, "Political Alignments," 242). Still, in 1751 Norfolk merchant Charles Stewart would advise a London correspondent that "the Tobacco Law has no effect here yet but probably will the next year as the crop is very large" (Stewart to James Buchanan, Sept. 16, 1751; see also to Walter Lutwidge, Esq. and Son, Aug. 10, 1754, to William Bowden, June 29, 1758, Charles Stewart Letterbook, CWL). Francis Jerdone similarly reported that, owing to a short crop the previous year, in 1758 inspectors on the Potomac River were letting poor tobacco pass (Jerdone to Spiers and Brown, June 10, 1758, Jerdone Letterbook, Special Collections, Earl Gregg Swem Library, College of William and Mary, Williamsburg, Va.).

46. For tobacco prices, see Schweitzer, "Economic Regulation," *JEH,* XL (1980), figs. 1, 2; and

Developments in the Heartland: The Lower Peninsula

Before turning to the management strategies of individual planters of the Chesapeake's Golden Age, it should first be noted that after 1730 large tidewater plantations, wherever they were located, came to look more and more alike once owners of the third and fourth generation born in the region assumed control of inherited farms. Native-born elites learned farming techniques suitable to the region from fathers and other kinsmen as well as from consulting with and observing the farming practices of well-reputed tobacco growers in their neighborhoods. For day-to-day supervision, gentleman planters could now turn to an adequate pool of native-born overseers experienced in growing tobacco and more comfortable managing slaves than had been the case in the first quarter of the century. Since most gentry planters soon also assumed county and often provincial offices, they continually exchanged information at county courthouses, at militia musters, and at meetings of the Virginia and Maryland assemblies about markets, optimal crop mixes, strategies of labor management, new investment opportunities, and recent trends in British agriculture. By 1750, there was general agreement about the efficacy and profitability of expanding production of either corn or wheat or both to increase overall plantation revenues by more fully using existing labor forces. Shared conceptions also emerged that home plantations should be sites for genteel living as well as for agricultural and domestic production. Elite planters increasingly displayed their inherited wealth, status, and power in the form of classically inspired mansions. They also began competing with one another in providing established plantations, enslaved workforces, and new fashionable houses for their children. The few immigrants who arrived with enough capital to start planting on a large scale were by this time at a decided disadvantage, since they had to buy both land and labor rather than starting with inherited assets. Those who attempted to build big estates and at the same time to live in the lavish style of the established creole elite frequently ended up badly in debt.[47]

Our survey begins in the heart of Virginia's oldest sweet-scented region on the Peninsula between the lower James and York rivers, the area where planter wealth was most concentrated. Lower Peninsula planters, along with those farming on the north side of the York in Gloucester County, grew the

---

Carr "Diversification in the Colonial Chesapeake," in Carr, Morgan, and Russo, eds., *Colonial Chesapeake Society,* 342–388. For inspection systems for other products, see Sargent, "Virginia and the West Indies Trade," chap. 2; and Gray, *History of Agriculture,* I, 154, 166, 212.

47. Clemens, *The Atlantic Economy,* chap. 5; Aubrey C. Land, "Economic Behavior in a Planting Society: The Eighteenth-Century Chesapeake," *Jour. So. Hist.,* XXXIII (1967), 469–485.

most celebrated and profitable strains of sweet-scented tobacco, the sorts widely advertised by upscale retail tobacco dealers in Britain. English merchants considered "E Dees," raised by the Digges family in York County, the best of all. The tobacco produced by successive generations of the Burwell family across the river on Fairfield plantation in Gloucester ranked not far behind. Elite planters lucky enough to inherit land in this area thus found it relatively easy to amass sizable fortunes from tobacco growing alone.[48]

As a result, lower Peninsula planters were amply supplied with enslaved workers. As early as 1687, when a Bristol interloper arrived on the James River with 130 slaves obtained in Angola in violation of the Royal African Company's trading monopoly, the Virginia officials who seized the ship and its cargo of sickly captives for the British crown marketed the Africans along the York rather than the James, the former "being the ablest men to purchase for money" (that is, with coin or bills of exchange). York basin planters also bought the majority of slaves imported by the Royal African Company in the 1680s and 1690s. In the 1720s, the York River continued to offer the best market for new Africans. As slave dealer John Tayloe informed Bristol merchant Isaac Hobhouse in 1723, there was "most money Stir[r]ing in that river." In the 1710s and 1720s, about two-thirds of all the slaves carried into Virginia were first disembarked along the York River, more than 9,000 in the 1720s alone. In the 1730s and 1740s, half of all imported slaves were first offered for sale along the York, about 8,400 in the 1730s, dropping to 5,500 in the 1740s. By the mid-1730s, most elite lower Peninsula planters had acquired all the new African hands they needed. Then middling planters and buyers from more distant areas were likely the major customers of the slave traders. By 1752, even this middling market had become saturated, and thereafter in many years not one new slave was sold in the York.[49]

An account book for one of the area's larger plantation complexes, Carter's Grove, estate records detailing Daniel Parke Custis's even more expansive lower tidewater operations, and a series of guardian's accounts for several

48. Hugh Jones, *The Present State of Virginia, from Whence Is Inferred a Short View of Maryland and North Carolina* (1724), ed. Richard L. Morton (Chapel Hill, N.C., 1956), 72–73; Robert Carter to Micajah Perry, June 16, 1723, Carter Letterbook, UVa; William Beverley to Micajah and Richard Perry, July 12, 1737, William Beverley Letterbook, 1737–1744/5, Beverley Family Papers, 1654–1901, MS, NYPL, microfilm, M-1334, CWL.

49. Quotation from deposition of Nathaniell Bacon, Oct. 31, 1691, in papers concerning the *Society* of Bristol, CO5/1308, ff. 14–45; Charles L. Killinger III, "The Royal African Company Slave Trade to Virginia, 1689–1713" (master's thesis, College of William and Mary, 1969), appendix F; Walter E. Minchinton, ed., "The Virginia Letters of Isaac Hobhouse, Merchant of Bristol," *VMHB,* LXVI (1958), 297; Walsh Database of Slave Trading Voyages to Virginia and Maryland (November 2008).

middling estates in York County allow us to explore the management of plantations of various sizes in this most favored region. Carter's Grove was a fourteen-hundred-acre plantation in James City County that Robert "King" Carter bought sometime before 1720 as a gift for his daughter Elizabeth (1692–1734), who had married Nathaniel Burwell of Gloucester County, an elite planter and second-generation Virginian, in 1709. She was to have the income of the plantation and of the enslaved laborers and livestock that her father also provided for her life. After she died, all was to go to her second son, Carter Burwell (1716–1756), and, subsequently, to his male heirs. Carter's father, Nathaniel Burwell, managed the plantation during his lifetime from his Gloucester County estate just across the York River, marketing "fine Stemd Tobo" that had a high reputation among London merchants. After Nathaniel died in 1721, Robert "King" Carter, as acting executor and guardian of the Burwell estate, resumed management of the plantation for the benefit of his daughter and her children. (Elizabeth made a second, imprudent marriage in 1724 to a condemned felon transported for forging a banknote; clearly, Carter did not intend that his grandson's stepfather tamper with the property.) After Robert "King" Carter died in 1732, a Burwell uncle managed the estate until the heir, who was then studying at the College of William and Mary, came of age.[50]

Carter Burwell turned twenty-one in the fall of 1737 and married two months later, soon setting up housekeeping at Carter's Grove. He had a more than favorable start in his career as a planter. In addition to Carter's Grove, Burwell inherited more than five thousand acres of undeveloped land in Prince William County and in what would later become Frederick County, Virginia, from his grandfather and two other quarters of more than two thousand acres each in York County from his father. Burwell's holdings straddled the entire peninsula; his home plantation was on the James, one quarter was on the York, and other units were on the ridge between the rivers. An uncle and a cousin owned home plantations nearby of a thousand acres or more as well as additional scattered quarters, so much of the lower peninsula to the east of Williamsburg was a virtual Burwell family preserve. Carter Burwell's estate was unencumbered, for his mother had died three years before he came of age. Besides the land with its complement of live-

50. Carter Burwell Ledger, 1738–1755, Burwell Family Papers, MS 64.4, CWL; Robert Carter to Micajah Perry, June 16, 1723, Carter Letterbook, 1723–1724, Carter to John Falconer, June 25, 1729, Carter Letterbook, 1728–1731, Carter to Alderman Perry and Company, July 9, 1731, Carter Letterbook, 1731–1732, Carter Papers, UVa; Mary A. Stephenson, *Carter's Grove Plantation: A History* (Williamsburg, Va., 1964), chaps. 1–3, appendices 2–4, 10; Walsh, *From Calabar to Carter's Grove,* chaps. 1, 4.

stock and twenty to twenty-five primarily African slaves that his grandfather Carter had supplied, Burwell would add to his labor force about thirty first- or second-generation Virginia-born slaves willed to him by his father. Apparently, Burwell inherited all the workers he required. His accounts mention some eighty-two adult slaves present between 1740 and 1745, but only two are identified as "new" negroes; he had little need to purchase additional hands. His wife Lucy Grymes's dowry of £1,250 sterling and an additional bequest of £100 sterling from his grandfather Carter provided other capital. With these assets, Burwell would acquire another plantation in James City County and later seat three tracts in the west in Frederick and Prince William counties given to him by Robert "King" Carter. Political offices, too, soon followed from rank and wealth. Burwell shortly became a justice of the peace and a burgess and aspired to (though he failed to get) a seat on the Virginia Council, because his brother and two of his brothers-in-law were already members.[51]

Burwell kept an account book from 1738, when he settled in at Carter's Grove, until shortly before his death in 1756. From it one can gain some insight into the operation of his farms. Unfortunately, Burwell's accounts are too incomplete to provide any systematic information on revenues and expenditures. Like many other large planters, he failed to record the sales of his tobacco crops in his account book (probably instead saving the written accounts his English factors sent him or keeping a separate crop record book), and only occasionally did he note sales of all his surplus corn, wheat, and livestock. Income from cider, wool, and fodder are not mentioned, but some must have been marketed. Similarly, major outlays, including taxes, clothing for the slaves, and tools, are either incompletely recorded or impossible to disentangle from unrelated family expenses. Accounts with his overseers are also incomplete, and even from those that survive it is difficult to estimate either total crop production or production per worker. Neither do we know the size of Burwell's labor force, although it is probable that by the 1750s he held at least 150 slaves, young and old.[52]

51. Stephenson, *Carter's Grove Plantation,* chaps. 1–3; Walsh, *From Calabar to Carter's Grove,* chap. 4; Curtis Chappelear, "The Carter Burwell Tract and Carter Hall," *Proceedings of the Clarke County Historical Association,* VI (1946), 4; York County Project Biographical files, CWR; "Carter Burwell," in Sara B. Bearss et al., eds., *Dictionary of Virginia Biography,* 3 vols. (Richmond, Va., 1998–2006), II, 429–431. One can get a sense of Burwell's prominent role in the House of Burgesses from excerpts from Landon Carter's Diary discussed in Isaac, *Landon Carter's Uneasy Kingdom,* 131–132, 146.

52. The accounts of Carter Burwell's estate kept by William Nelson Senior and William Nelson Junior (Burwell Ledger 2, 1764–1776, Burwell Family Papers, CWL; Yorktown Ledger, 1766–1772, MS 2005.10, CWL) document such an established trade in cider, wool, and fodder

Yet there is still much that we can learn. Like most other large slaveowners, Burwell divided his hands into small work groups. Field laborers were quartered at the home house, on three outlying quarters in York County, and on the newly purchased plantation in James City County. There were also two smaller quarters near the home plantation, one managed by an enslaved man, Abraham, and the other at the plantation mill. A group of five carpenters seem to have lived and worked separately from the other slaves. Three overseers supervised the larger quarters for shares of the crops produced or sometimes for a yearly salary of fifteen to eighteen pounds Virginia currency.

Although tobacco remained Burwell's most important single crop, he ran diversified farms that produced goods for several markets. Livestock was an important part of the operation. Workers on each of Carter Burwell's quarters tended hogs, cattle, and horses, and by 1753 there were at least 250 sheep on two of the farms. Hogs were fattened with surplus corn, as were cattle and sheep destined for market. In winter, the stock was fed corn tops and blades. Burwell also had at least one meadow from which he had hay cut in summer. He raised many more animals than were needed for consumption on the plantation. Each year, excess cattle were sold or older animals slaughtered for butcher's meat. Burwell also marketed as much as ten thousand pounds of pork a year. Given an average slaughter weight of one hundred pounds apiece, he had a surplus of as many as one hundred hogs every year. In the later 1740s and early 1750s, when the accounts are fullest, sales of livestock and meat

---

that it seems highly probable that they were also marketed in Burwell's lifetime. Surviving accounts give the number of pounds and the price per pound of the portion of the total crop Carter Burwell's overseers received in lieu of an annual salary, but they do not indicate how many shares of the crop this payment represented. If an overseer supervising ten enslaved field hands, for example, was paid one share of the crop, the planter divided the total number of pounds accepted by tobacco inspectors by eleven. The resulting "share" counts the labor of the overseer as equivalent to one additional hand, the weight I used in the crop shares reported in Appendix 1. If the same overseer was paid two shares, the planter divided the weight of the crop by twelve, and the result must be standardized by dividing the total pounds by eleven instead of twelve in order not to understate the productivity of the field-workers. In the 1760s, Burwell's executor was paying the overseers either one and a half or two shares of the crop, depending on the overseers' experience and the number of workers they supervised. I have assumed that Carter Burwell paid no more. Shares per laborer presented in Table 24 are the mean of a high estimate of output per worker that assumes all overseers got a share and a half of the total crop and of a low estimate that assumes the overseers each got two shares. In 1758, Burwell, the biggest slaveholder in Yorkhampton Parish (which included the part of both York and James City counties where most of Burwell's holdings were located) had sixty-one tithable slaves on three plantations. Only Edward Digges, with fifty-six tithables on two quarters, was operating on a similar scale (Rev. John Camm, *A Single and Distinct View of the Act Vulgarly Entitled the Two Penny Act* . . . [Annapolis, Md., 1763], pamphlet in Maryland Diocesan Library, microfilm, M-142, CWL).

TABLE 24. Tobacco and Corn Crops and Revenues per Laborer, Carter Burwell Plantations, James City and York Counties, Virginia, 1738–1755

Year	Tobacco (lbs.)	Tobacco revenue (£)	Corn (barrels)	Corn revenue (£)	Total revenue (£)
1738	769	7.75	10.5	4.12	11.87
1739	566	5.74	11.7	3.26	9.00
1740	986	10.14	10.7	3.67	13.81
1741	865	8.81	10.4	3.54	12.35
1742	874	7.67	4.9	1.73	9.40
1743	943	8.28	9.5	3.00	11.28
1744	856	7.40	5.1	1.59	8.99
1745	1,154	10.74	12.4	3.95	14.69
1747	721	6.52	7.8	3.14	9.66
1748	1,096	9.39	12.9	3.67	13.06
1749	901	8.24	19.9	8.08	16.32
1754	678	6.36	14.5	6.64	13.00
1755	862	8.09	12.0	6.22	14.31
Mean	867	8.09	10.95	4.05	12.13

Note: Currency values are in constant pounds sterling.
Source: Carter Burwell Ledger, 1738–1755, Burwell Family Papers, MS 64.4, CWL.

brought in between fifty and one hundred pounds sterling constant value annually. Some of the meat might have been destined for the West Indies or as ship's provisions. The bulk, however, went to residents of the nearby town of Williamsburg.

Burwell's enslaved workers produced on average between 800 and 900 pounds of tobacco per hand per year (Table 24). In bad seasons, the output on individual quarters might drop to 550 pounds, whereas in a few favorable years some of the workers made more than 1,000 pounds each. In addition, on the home farm and adjacent quarters the slaves raised about eight hundred barrels of corn annually. Yields per hand between 1738 and 1755 ranged from a low of five barrels to a high of twenty barrels; the average crop was eleven barrels per hand. Once food for the slaves and corn for fattening livestock were set aside, between a quarter and a half of the crop was marketable surplus. By the late 1740s, Burwell was also growing 500 to 800 bushels of wheat a year, almost all of which was sold. Either Burwell had exceptionally good land or he was using better cultivation techniques than did most tidewater planters. In 1755, he was getting 36.6 bushels of wheat to the acre from old tobacco fields, as much as six times the usual yield for the region.

In the 1740s, tobacco accounted for between two-thirds and three-quarters of Burwell's revenues from field crops. By the mid-1750s, grains were bringing in more, and tobacco's share dropped to about half of gross revenues. This shift was more a result of a rise in the price of grains relative to tobacco than of shifts in the crop mix. The enslaved laborers seem to have tended similar amounts of tobacco across most of Carter Burwell's lifetime, and only toward the end did they put slightly more effort into corn. Possibly Burwell put more acres into wheat in the late 1740s and 1750s, but we cannot be certain. Only on one plantation in one year, 1750, did Burwell not make any tobacco. And, since he pitched large crops there in 1749 and 1751, it is likely that Burwell also had tobacco planted in 1750 but that the crop failed owing to drought.

Burwell was producing for several markets, a strategy that spread risk and minimized the impact of price declines for any one of the three staple crops. Almost invariably he consigned his high-quality tobacco to John Hanbury, the largest tobacco importer in London in the 1740s and 1750s, in return for English goods; in only two years — 1749 and 1752 — did he choose to sell tobacco in the country. For corn and wheat there were several outlets. At least once Burwell sold most of his surplus to a ship captain, presumably bound for the West Indies. In years for which the accounts record few grain sales, Burwell was likely selling most of his grains in the form of meal and flour processed at the plantation mill. In other years, he turned to the nearby town of Williamsburg, which had a population of 885 living in about one hundred households in 1747–1748, and new households were steadily added in subsequent years. The College of William and Mary was his single best customer, sometimes buying more than five hundred bushels of wheat annually and paying promptly in cash. Tavern keepers, the royal governor, and assorted town tradespeople and professionals also purchased grain and meat. Burwell, himself a graduate of William and Mary, doubtless made use of local connections first established during school days.[53]

The Burwell plantation was unusual in its access to a nearby urban market.

53. Alan Simpson, Helen Byrd, and Mary Simpson, "The Discovery of Burwell's Mill, Carter's Grove Plantation, Williamsburg, Virginia," typescript, CWL; Lorena S. Walsh, Ann Smart Martin, and Joanne Bowen, "Provisioning Early American Towns, the Chesapeake: A Multidisciplinary Case Study" (Final Performance Report, National Endowment for the Humanities, Grant RO-22643-93, 1997), typescript, CWL. For similar responses among large South Carolina planters living near Charleston to that city's expanding market for meat, grains, produce, and firewood, see Edelson, *Plantation Enterprise,* chap. 4.

The crop mix, however, was typical of most large tidewater plantations of the period. Perhaps more surprising is an investment strategy devoted almost wholly to agriculture. There is no evidence that Burwell put any capital into mines, ships, merchandise, or rental properties — forms of investment that appealed to other large planters of this period. Instead, he began to develop frontier lands in Prince William County, as early as 1745 sending some of the slaves to grow tobacco there.[54]

Burwell's decision to concentrate on agriculture, however, yielded good returns. Plantation proceeds were sufficient to pay for the construction of an elegant two-story, 72-by-43-foot brick house that "surpassed all but a handful of its contemporaries in size, elaboration, and classical rigor," built between 1751 and 1755 at a cost of about fifteen hundred pounds (Plate 24). The power house Burwell built reflected recent trends in metropolitan architecture, some design elements being adapted from William Salmon's *Palladio Londinensis* (1734) (which Burwell bought the year he began building) and Batty Langley's *City and Country Builder's and Workman's Treasury of Designs* (1740). A floor plan designed to facilitate formal entertainments was apparently borrowed from Burwell's uncle Charles Carter's newly built mansion of Cleve. Williamsburg brick masons and master carpenters who had worked on public buildings in the capital completed the exterior and structural carpentry, but Burwell imported a skilled English artisan to finish the interior. Matching brick dependencies and a formal, terraced garden extending 540 feet toward the James River further enhanced the impression of a carefully ordered and orchestrated arena of gentility. Creating a refined material expression of his family's prominence and prosperity mattered deeply to Burwell. Three communal earthfast dwellings housing the thirty to forty slaves who lived on the home plantation, located just outside the bounds of the polite landscape and in full view of the mansion, however, permitted no illusions about the basis of Burwell's fortune.[55]

In addition to covering the costs of building a mansion, Burwell could apparently also absorb the loss of £348 lent in 1750 to Speaker John Robin-

54. Burwell owned two lots in Yorktown in 1739, but there is no indication that he developed them (York County Project files, CWR). For Burwell's western quarters, see Walsh, *From Calabar to Carter's Grove,* chap. 7.

55. Mark R. Wenger, *Carter's Grove: The Story of a Virginia Plantation* (Williamsburg, Va., 1994), 4 (quotation); Wenger, "The Great House at Carter's Grove: An Interpretive Essay," Colonial Williamsburg Foundation Research Report Series, no. 1655, CWL; Thomas Tileston Waterman, *The Mansions of Colonial Virginia, 1706–1776* (Chapel Hill, N.C., 1945), 180–192; Waterman and John A. Barrows, *Domestic Colonial Architecture of Tidewater Virginia* (1932; rpt. New York, 1968), 99–107; Chappell, "Looking at Buildings," *Fresh Advices,* iii–iv; Stephenson, *Carter's Grove Plantation,* 33–48; Walsh, *From Calabar to Carter's Grove,* 15–20.

son (whose abuse of public funds caused a major political scandal), a sum that was likely never repaid. Administrator's accounts survive for Burwell's estate starting in 1763, and by that year the estate was free of debt. Burwell's strategy for preserving the family estate and for endowing his two sons and five daughters was to devise land and slaves only to the sons and to pay the daughters' portions in cash to be raised either from the profits of the estate or from the sale of lands. Having docked the entail on Carter's Grove at a time when he anticipated having only daughters to provide for, Burwell subsequently left his eldest son that estate in fee simple, but he did entail western lands willed to his second son. Burwell survived only to the age of forty, but he had already transformed the outlying, lightly developed, tobacco-centered quarter he inherited at twenty-one into a seat of power surrounded by ancillary farms producing a variety of products for transatlantic, intercolonial, and local markets.[56]

Daniel Park Custis (1711–1757), a large planter who owned multiple plantations all along the York River, adopted a crop mix and income-generating activities similar to Carter Burwell's, although he operated on an even grander scale (Plate 27). A fourth-generation Virginian, probably like Burwell educated at the College of William and Mary, Custis was the family's sole legitimate male heir. When he reached age twenty-five, his father deeded over family lands in New Kent County to which Daniel moved and commenced farming on his own. In 1749, at the death of his domineering father, John Custis IV, Daniel inherited an additional seventeen thousand acres in New Kent, King William, King and Queen, Hanover, Northampton, and York counties (where eventually he had six quarters on 3,074 acres), scores of enslaved laborers, and stock in the Bank of England as well as entailed land descending from his grandfather Parke. Given the wide geographic spread of the lands he had inherited, Custis needed to adopt flexible management strategies that took into account the differing opportunities afforded by holdings in older marginal areas, in long-settled but still prime lands, and of more recently established quarters.[57]

56. Brewer, "Entailing Aristocracy," *WMQ*, 3d Ser., LIV (1997), 331.

57. Sources for Daniel Parke Custis and for the operation of his estate include Daniel Parke Custis Letters, 1755–1756, MS, HSP, microfilm, M-148, CWL; Custis Papers, 1711–1792, VHS, microfilm, M-58.1, M-1073.3, CWL; Jackson, ed., *Diaries of Washington,* I, 309–310; George Washington, Diary, Jan. 1–Apr. 11, 1760, George Washington Papers, 1741–1799, fols. 2–12, LC; Joseph E. Fields, comp., *"Worthy Partner": The Papers of Martha Washington* (Westport, Conn., 1994); Jo. Zuppan, ed., "'Father to Son': Letters from John Custis IV to Daniel Parke Custis," *VMHB,* XCVIII (1990), 81–100; and "Daniel Parke Custis," in Bearss et al., eds., *Dictionary of Virginia Biography,* III, 265–266. See also James B. Lynch, Jr., *The Custis Chronicles: The Virginia Generations* (Camden, Me., 1997), chap. 7.

PLATE 27. *Daniel Parke Custis* (1711–1757). By John Wollaston. Courtesy, Washington-Custis-Lee Collection, Washington and Lee University, Lexington, Virginia

Custis's main crop was tobacco, which he consigned to the London firms of Robert Cary and John Hanbury and to James Gildart in Liverpool. (After an unsuccessful trial of the Bristol market, he ceased trading with that port.) By the late 1750s, each enslaved laborer made about eight hundred pounds of tobacco a year, and Custis was shipping out more than one hundred hogsheads annually. Custis concentrated on raising high-quality tobacco, and by the 1750s his "DPC" mark was well regarded in England. He achieved this reputation by paying close, personal attention to seed selection, to careful sorting and packing of harvested crops, and to rectifying deficiencies in his tobacco that were pointed out by his British factors. As Custis informed Robert Cary, "If every one would take the same pains with their Tobacco and Fling away as much as I do there woud not be such Complaints of the Inspectors as ther are." Unlike many lower tidewater planters, Custis marketed two-thirds to three-quarters of his sweet-scented crops in the form of carefully chosen but minimally processed leaf tobacco; only inferior plants were further processed into stemmed tobacco. Indeed, Custis paid attention not only to the size and number of leaves packed to a bundle but also to such niceties as whether the leaves should instead be laid loose in the hogsheads, since tying them in bundles sometimes bruised the heads of the leaves. In the late 1750s, Custis ranged from pleased to at least satisfied with the prices his tobacco fetched in England. In 1758 and 1759, tobacco accounted for 53 percent of the Custis estate plantation revenues, a proportion similar to what Carter Burwell received (Table 25).[58]

Like Burwell, Custis raised corn and wheat (and also peas) for local and West Indian markets as well as substantial quantities of oats made on his Eastern Shore plantations, most of which were transshipped to lower Western Shore buyers. Annual output per laborer was about 13 barrels of corn and nine bushels of wheat net of seed. Total production on the far-flung Custis plantations amounted to more than 2,000 barrels of corn, five hundred bushels of wheat, and at least five hundred bushels of oats per year. (Workers on the quarters in York County, for example, produced twenty-five hogsheads of tobacco and 528 barrels of corn in 1759, crops worth just over £340 sterling constant value.) Overall, surplus corn accounted for 17 percent of agricultural revenues, and wheat and oats 12 percent. Surprisingly, Custis made little or no use of plows, which would have enabled him to increase grain production at a lower cost in time and labor. There were a few oxen and work horses on most of his quarters, but these draft animals seem to have been used primarily for carting produce. Only on the Eastern Shore grain-growing quarter were four old plows inventoried, and on the mainland farms there were no plows and only one harrow. Enslaved workers were apparently expected to raise ever increasing quantities of corn and wheat primarily if not exclusively by working longer hours using traditional hoe culture. Custis

58. Custis to [Robert Cary], May 5, 1755, Daniel Parke Custis Letters, CWL. The inventory of Custis's estate is reproduced in Fields, comp., *"Worthy Partner,"* 61–76, as are George Washington's subsequent accounts of the estate (96–146). Outputs per hand were calculated from Washington, Diary, Jan. 1–Apr. 11, 1760, fols. 6–8, LC.

TABLE 25. Revenues, Expenditures, and Investments on the Daniel Parke Custis Estate, 1757–1759

Category	Amount	Percent
REVENUES		
Tobacco	2,441.49	53
Corn	792.90	17
Wheat	356.11	8
Oats	165.78	4
Peas	8.75	<0.5
Brandy	12.34	<0.5
Livestock	473.00	10
Butter	43.81	1
Boats hired out	4.25	<0.5
Rents	120.41	3
Sundries	162.31	4
Total	4,581.15	100
EXPENDITURES		
Imported cloth and clothing	367.09	19
Imported bedding	21.14	1
Shoes	75.28	4
Tools	72.33	4
Provisions for slaves	205.20	11
Weaving	10.38	1
Blacksmith	24.42	1
Medical care for slaves	41.81	2
Building repair	11.38	1
Seed	8.00	<0.5
Tobacco inspection fees	53.02	3
Freight	13.82	1
Insurance	124.87	6
Local travel	4.03	<0.5
Taxes	336.17	17
Supervision	566.00	29
Total	1,934.94	100
INVESTMENTS		
Livestock	13.66	33
Boats	28.25	67
Total	41.91	100

Note: Currency values are in constant pounds sterling.
Source: Joseph E. Fields, comp., *"Worthy Partner": The Papers of Martha Washington* (Westport, Conn., 1994), 96–103, 108–109, 112–117, 120–123.

PLATE 28. *View of the Fishing Shore on York River at Airy Plains, Looking to the East.* By Benjamin Henry Latrobe. Latrobe Sketchbook III. Permission, The Maryland Historical Society, Baltimore.
Black fishermen sketched in 1797 near the town of West Point in King William County.

did, however, take the innovative step of draining a marsh in King William County to produce corn and other grains on the highly fertile soil.

Custis also paid special attention to raising cattle, hogs, and sheep. At his death, there were 941 cattle, 117 or more sheep, and at least 625 hogs on the various quarters. In addition to fattening animals on corn, Custis had as much hay made as weather and time permitted and apparently experimented with raising clover on his home plantation in New Kent. Sales of surplus meat, butter (made under the direction of overseers' wives), wool, and breeding stock accounted for 11 percent of plantation revenues. Three toll gristmills were also integral parts of Custis's operations as well as a commercial fishery on the York River. This operation would have been similar to a later York River fishery sketched by Benjamin Latrobe (Plate 28). Custis also rented several small farms and city lots in Jamestown and Williamsburg to tenants. As at Carter's Grove, Williamsburg provided a major outlet for produce raised on both the York County and the Eastern Shore quarters.[59]

59. Fields, comp., *"Worthy Partner,"* 100; Jackson, ed., *Diaries of Washington,* I, 309–310.

The extent to which Custis attempted to employ elements of the new British husbandry is uncertain. The novel emphasis British writers placed on agricultural profits should have appealed to a planter entrepreneur such as Custis. His library, however, is known to have included only the standard texts on gardening and horse care found on many large planters' bookshelves. The importance Custis attached to livestock and to experiments with fodder crops suggests that he had perhaps adopted some of the methods that English farmers had devised earlier in the century to increase herd size. Conversely, the absence of evidence for any extensive plowing shows that he had not yet attempted to adopt European methods for increasing grain yields.[60]

As a fourth-generation colonist, Custis had a great advantage over planters of only the second or third generation in that he inherited all the enslaved workers he needed. At his death, he held title to 283 enslaved men, women, and children. Many of them were Virginia born, resulting in balanced sex ratios on the older quarters and relatively high rates of reproduction. Custis was thus able to expand his operations without incurring the considerable expense of purchasing new hands, and his young heirs could anticipate eventually inheriting even larger workforces. Although many Chesapeake slaveholders were indifferent or opposed to Anglican missionary efforts, both Carter Burwell and Daniel Parke Custis (as well as New Kent County planter William Massie, discussed below) chose to employ religion as a means for creating a more acculturated and servile workforce. By the 1740s, these three planters at least saw that slave children born on their plantations were baptized in the Anglican Church, if not catechized in Anglican tenets. Perhaps because many of the slaves were creoles, Custis relied heavily on enslaved male and a few female artisans, including shoemakers, carpenters, coopers, a

60. Between 1763 and 1764, George Washington compiled a list of books at Mount Vernon that included both titles that he owned and books that his stepson John Parke Custis had inherited from his father. Washington inserted the initials "JC" or "GW" after many of the titles to indicate the owner. The author of "Catalogue of the Library of Daniel Parke Custis," *VMHB*, XVII (1909), 404–412, assumed the 1763/4 list represented a division of Daniel Parke Custis's library between Washington and his stepson. The editor of the George Washington Papers has since compared this list with an earlier list of Daniel Parke Custis's books that Washington made before 1761, with an inventory of books in John Parke Custis's estate in 1782, and with orders Washington placed for books in 1763–1765. He concludes that Daniel Parke Custis's library included books on gardening and horse care but that Washington, not Custis, purchased all the books on English husbandry, with the exception of *A Dictionary of Husbandry, Gardening, etc.* that also touched on "trade, commerce, and all sorts of country affairs," published in 1717. See Inventory of Books in the Estate, Document C0106d175, and List of Books at Mount Vernon, C0107d192, in Theodore J. Crackel, ed., *The Papers of George Washington Digital Edition* (Charlottesville, Va., 2007), http://rotunda.upress.virginia.edu:8080/pgwde/dflt.xqy?mode=menu&keys=menu-info-home-scholar (accessed January 2009).

miller, a spinner, and a seamstress. Like most other period masters, however, he paid white women to spin and weave and to make clothes for the workers rather than withdrawing able-bodied enslaved women from the fields.[61]

Custis managed his plantations through nine overseers, to whom he paid shares of the tobacco, corn, and wheat crops. He personally supervised the home plantation, but he employed an agent, Joseph Valentine, to oversee the more distant quarters, rewarding Valentine with one pistole (16s 6d) for every one thousand pounds of tobacco and every twenty barrels of corn produced, plus an allowance of corn, wheat, beef, pork, and butter for the Valentine family's consumption. Both overseers and the general manager sold surplus produce, except for the tobacco, whose sale Custis alone oversaw. After Custis's death, Valentine managed the whole estate for an annual salary of £80 Virginia currency. Supervision constituted the major plantation expense, accounting for 29 percent of the outlays recorded in the first two years after Custis's death (see Table 25). Imported clothing and tools, including stocks on hand and new supplies ordered by Custis's widow, Martha, amounted to 23 percent of expenditures. Taxes constituted the third-greatest reported outlay, 17 percent. High wartime insurance on consigned tobacco accounted for a further 6 percent of expenses. Between mid-1757 and mid-1759, total recorded gross plantation revenues were £4,581 sterling constant value, and expenditures were £1,935, leaving a net profit of about £1,300 per year.

Although Custis had diversified his farming operations to take advantage of growing regional and international markets for grain and meat, in the late 1750s he still hoped to expand tobacco cropping, a strategy he likely chose because of the high quality of the tobacco his workers produced. Following his father's lead, Custis had abandoned tobacco for grain on his Eastern Shore plantations, since the elder Custis had found that whatever its actual quality, all Eastern Shore tobacco was discredited in the British market. Daniel resumed tobacco culture on some uncharacteristically rich Northampton County land in 1755, sending fifteen trial hogsheads to Liverpool in 1756, intending to continue cropping on the Eastern Shore if the product "answered" the outport market. How the trial fared is not recorded, but it was likely unsuccessful, since after Custis died no more than nine working slaves remained on the Eastern Shore quarters, and they seem to have reverted to raising only grain and livestock.[62]

61. For baptisms of slaves, see Anthony S. Parent, Jr., *Foul Means: The Formation of a Slave Society in Virginia, 1660–1740* (Chapel Hill, N.C., 2003), chap. 8, and appendix 2, table 5; Walsh, *From Calabar to Carter's Grove,* 151–159; and Bruton Parish Church Register, 1662–1797, Special Collections, Swem Library, microfilm, M-120, CWL.

62. Custis to James Gildart, July 13, 1756, Daniel Parke Custis Letters, CWL.

Perhaps because of the financial difficulties his kinsman William Byrd II faced from assuming the obligations of Custis's maternal grandfather Daniel Parke, Custis abhorred debt and strenuously avoided borrowing. Owing in part to a long-running suit brought by Parke's illegitimate heirs, Custis decided not to "meddle with one farthing I have in England until my Law suit there was over, which if should go against me, all that I have in the whole World wou'd scarce do." However, Custis did much more than simply balance plantation revenues and expenditures, maintain a positive balance with his London factors that allowed him to take advantage of discounts on customs dues, and preserve the principal of his inherited bank stocks. Sales of grain and livestock generated considerable cash income in the colony, which Custis regularly lent to other Virginia planters. When his widow Martha discovered substantially more cash on hand than she required for immediate needs, the family strategy of not letting money lie idle was so well established that on her own initiative she lent more than two thousand pounds Virginia currency at interest soon after Custis's death. Since Daniel died suddenly and unexpectedly, he had not composed a will laying out a strategy for the preservation of family assets. But with only one surviving son, Daniel's last wishes, had he had a chance to express them, might not have differed significantly from the way Virginia law divided intestate property, with all land going to the son (reserving to the widow one-third as dower during her lifetime) and the movable property divided between the widow and all the children.[63]

The careful accounting of Custis's estate compiled by Martha's new husband, George Washington, and family lawyer John Mercer provides a full picture of how Custis distributed his productive assets. Capital goods made up 95 percent of the value of his estate inventory, approximately £27,440 sterling constant value. On most large period estates, enslaved labor accounted for at least three-quarters of capital assets and livestock for most of the remainder. Custis's 283 human chattels represented only a third of his capital goods, and more than sixteen hundred head of livestock constituted only 8 percent. Money loaned at interest (good debts secured by legally enforceable bonds), cash, and bank stocks constituted an astounding 52 percent. Few other Chesapeake planters had the capital, fiscal discipline, and financial expertise to benefit from a similarly distributed asset portfolio.[64]

63. Fields, comp., *"Worthy Partner,"* 31, 98–101.

64. I divided the items in Custis's inventory into capital and personal (or consumer durable) goods following the procedures originally developed by Alice Hanson Jones, *Wealth of a Nation to Be: The American Colonies on the Eve of the Revolution* (New York, 1980), 14–27, which were also employed in the Historic Saint Mary's City Probate Inventory Database, MSA, and by Gloria L. Main, *Tobacco Colony: Life in Early Maryland, 1650–1720* (Princeton, N.J., 1982), 49–

After the marriage of George Washington and Martha Custis, the quality of the Custis tobacco crops suffered. From 1758 to 1764, manager Joseph Valentine succeeded in maintaining the same levels of production of tobacco, corn, and wheat as had prevailed during Daniel Custis's lifetime, but he did not manage to replicate the quality. In 1762, tobacco production per hand on the estate was 818 pounds, in the midrange of the output Carter Burwell extracted from his workers. The firms of Cary, Hanbury, and Gildart to whom the crops were consigned, however, found the tobacco packed loosely, carelessly, or in too high a case, ill scented, sometimes containing mixtures of whole and stemmed leaves, and generally of indifferent quality; as one factor put it, "The Buyers could not believe that it was the right Crops of thy marke." The absence of Custis's careful oversight in sorting and packing was likely one reason for the inferior quality of crops. So, too, was almost certain high levels of anxiety among the enslaved, who surely knew that Daniel's death and Martha's remarriage would entail separation of families and, for those allotted to Martha Washington as her dower, a long, forced move to distant Fairfax County. Lacking both the substantial inherited assets and the unusual fiscal discipline of the Custises, George Washington would subsequently, like most other area great planters, plow back all available assets into his farms and then have to rely for income on plantation revenues alone.[65]

Guardian accounts for estates of five middling York County planters (Lewis Davis, Jane Moore, Anthony Robinson, James McCartey, and John Hay) administered between 1735 and 1765 allow us to compare the operation of smaller farms on the lower Peninsula with those of a large planter like Carter Burwell (Tables 26a–e). All of the decedents were landholders who owned about three hundred acres each, who held between four and eight enslaved adult field hands, and who left personal estates of between one hundred and four hundred pounds sterling constant value.[66] These accounts reflect the

---

51. For proportions of capital and consumer durable goods, see Lois Green Carr and Lorena S. Walsh, "Changing Lifestyles and Consumer Behavior," in Carson, Hoffman, and Albert, eds., *Of Consuming Interests,* 114–115; and Burnard, *Creole Gentlemen,* 48.

65. Fields, comp., *"Worthy Partner,"* 79, 93; Crops of Corn, Wheat, and Tobacco Made by Joseph Valentine, 1760–1770, Capel and Osgood Hanbury to John Parke Custis, Aug. 23, 1763, both in Custis Papers, CWL.

66. All surviving guardian accounts for York County that had a starting date between 1730 and 1763 were examined from transcripts in the York County Project files, CWR (originals, microfilm, VSL). The five selected were those that supplied the most complete information on costs and output. Additional information on these decedents came from the York County Project Biographical Files, CWR, and from transcripts of York County probate inventories, CWR (most transcribed York County inventories are available at http://research.history.org/DigitalLibrary./Browse Probates.cfm).

TABLE 26A. Revenues, Expenditures, and Investments on the Lewis Davis Estate, York County, Virginia, 1739–1743

Category	1739	1740	1741	1742	1743	Total	Percent
REVENUES							
Tobacco	7.40	0	9.93	4.79	8.63	30.75	28
Corn	5.68	4.46	4.35	0	2.11	16.60	15
Wheat	0	1.14	0	0	0	1.14	1
Fodder	1.40	0	0.76	3.78	0	5.94	5
Butter	1.37	0	1.20	2.55	0	5.12	5
Livestock	3.83	2.86	9.84	4.47	1.84	22.84	21
Slaves hired out	0	6.85	6.78	6.31	6.32	26.26	24
Wool	0	0	0	0	0.52	0.52	<0.5
Miscellaneous	0.19	0	0	0	0	0.19	<0.5
Total	19.87	15.31	32.86	21.90	19.42	109.36	99
EXPENDITURES							
Clothes and shoes	5.36	9.15	10.56	7.87	2.35	35.29	63
Supervision	—	0	0	0	0	0	
Taxes	—	0.22	0.22	—	0.44	0.88	2
Local services	0	3.51	0	0.25	0.90	4.66	8
Building repair	1.52	0.06	0.13	3.29	0.85	5.85	10
Tobacco casks	0	0.52	0.35	0.20	0.21	1.28	2
Household goods	1.83	0	0	0.11	0.47	2.41	4
Provisions	0.28	0.50	0.04	0.02	0.38	1.22	2
Tools	1.84	0.70	0.04	1.13	0.78	4.49	8
Total	10.83	14.66	11.34	12.87	6.38	56.08	99
INVESTMENTS							
Livestock	0	1.14	0	0	0	1.14	100
Est. no. laborers	4	4	4	4	4		
Est. gross revenue per laborer	4.97	3.83	8.22	5.78	4.86		
Est. net revenue per laborer	2.26	−1.55	3.69	0.68	1.68		

Note: Currency values are in constant pounds sterling. Percentages for individual categories do not add up to one hundred owing to rounding. Estimated gross and net revenues per laborer exclude the hiring out of slaves. An overseer was employed in 1739, but the accounts do not specify what was paid; no overseer was employed thereafter. The mean for the estimated gross revenue per laborer is 5.53; for the estimated net revenue, 1.35.

Source: Transcripts in the York County Project files, CWR, from York County Guardian Accounts, microfilm, VSL.

TABLE 26B. Revenues and Expenditures on the Jane Moore Estate, York County, Virginia, 1735–1742

Category	1735	1736	1737	1738	1739	1740
REVENUES						
Tobacco	8.30	16.47	19.98	25.45	13.17	10.88
Corn	7.32	0	9.02	5.90	1.64	3.94
Wheat	1.52	0	0	0	0	0
Fodder	0	0	0	0	0	1.17
Livestock	0	0	0.86	0	0	0
Slaves hired out	0	1.42	0	1.40	1.67	2.86
Miscellaneous	0.63	0	0	0	0	0
Total	17.77	17.89	29.86	32.75	16.48	18.85
EXPENDITURES						
Clothes and shoes	4.10	5.07	2.97	6.86	5.17	1.55
Supervision	0	3.64	5.00	6.36	3.30	2.66
Taxes	6.94	6.99	6.48	8.10	6.97	6.98
Building repair	0	0.08	7.22	0.07	0	0.33
Household goods	0.66	2.94	0	0	0	0
Provisions	3.35	3.49	0.35	2.83	2.91	3.47
Tools	0.74	0.68	0.51	0.51	0.50	0.52
Miscellaneous	0	2.95	0	0	0	0
Total	15.79	25.84	22.53	24.73	18.85	15.51
Est. no. laborers	3	4	4	4	4	4
Est. gross revenue per laborer	5.92	4.12	7.47	3.24	3.70	4.00
Est. net revenue per laborer	0.66	−2.34	1.83	1.66	−1.01	0.12

Note: Currency values are in constant pounds sterling. Percentages for individual categories do not add up to one hundred owing to rounding. No investments were made by this estate. Estimated gross and net revenues per laborer exclude the hiring of slaves. The mean for the estimated gross revenue per laborer is 4.22; for the estimated net revenue, −0.42.

Source: Transcripts in the York County Project files, CWR, from York County Guardian Accounts, microfilm, VSL.

1741	1742	Total	Percent
13.78	15.13	123.16	68
4.34	0	32.16	18
0	0	1.52	1
0	0.36	1.53	1
0	0	0.86	<0.5
6.21	7.38	20.94	12
0	0	0.63	<0.5
24.33	22.87	180.80	100
8.70	1.31	35.73	21
3.45	3.76	28.17	17
7.19	9.33	58.98	35
0	0	7.70	5
0.38	0.06	4.04	2
2.84	2.85	22.09	13
0.67	0.64	4.77	3
0.05	5.27	8.27	5
23.28	23.22	169.75	101
4	4		
4.37	0.90		
1.45	−5.72		

TABLE 26C. Revenues, Expenditures, and Investments on the Anthony Robinson Estate, York County, Virginia, 1737–1752

Category	1737	1738	1739	1740	1741	1742	1743
REVENUES							
Tobacco	—	37.62	20.16	41.78	16.80	37.24	—
Corn	—	0	0	6.63	10.62	11.85	—
Wheat	—	10.01	0	3.56	4.91	0	—
Livestock	—	10.24	0	4.17	2.67	0	—
Slaves hired out	—	0	0	12.92	0.76	0.78	—
Cider	—	0	0	0	0	0	—
Total	—	57.87	20.16	69.06	35.76	49.87	—
EXPENDITURES							
Clothes and shoes	1.79	5.21	13.15	12.76	10.99	11.62	1.78
Taxes	0	3.19	3.17	2.46	3.22	1.49	0.16
Local services	0	0.73	0.13	0	0	0	0
Building repair	0	0.88	0.83	0.12	2.68	1.05	0.08
Tobacco casks	0	0	2.50	0	0.50	0	0
Household goods	0	0	0	0.74	2.42	0.28	0
Provisions	14.95	0	4.13	0.16	0.06	0	0.06
Tools	0	1.48	5.07	4.82	1.79	0	1.42
Miscellaneous	0	0.39	0	1.09	0.28	0	0.21
Total	16.74	11.88	28.98	22.15	21.94	14.44	3.71
INVESTMENTS							
Livestock	2.13	1.95	0	0	3.28	0	0
Est. no. laborers	—	6	7	6	6	8	—
Est. gross revenue per laborer	—	9.65	2.88	9.36	5.83	6.14	—
Est. net revenue per laborer	—	7.67	−1.26	5.67	2.18	4.33	—

Note: Currency values are in constant pounds sterling. Percentages of individual categories may not add up to one hundred owing to rounding. No account was recorded for 1744. Estimated gross and net revenues per laborer exclude the hiring of slaves. The mean for the estimated gross revenue per laborer is 6.30; for the estimated net revenue, 3.26.

Source: Transcripts in the York County Project files, CWR, from York County Guardian Accounts, microfilm, VSL.

1745	1746	1747	1748	1749	1750	1751	1752	Total	Percent
—	31.55	46.96	27.18	34.29	29.00	0	33.46	356.04	72
—	2.83	1.20	0	9.87	0	0	0	43.00	9
—	1.54	1.51	0	8.90	0	0	3.27	33.70	7
—	9.58	0	5.51	0	0	0	10.75	42.92	9
—	0	0	0	0	0	0	0	14.46	3
—	0	1.51	0	0	0	0	0	1.51	<0.5
—	45.50	51.18	32.69	53.06	29.00	0	47.48	491.63	100
6.31	15.88	19.27	17.84	14.65	11.88	11.61	2.35	157.09	61
0	0.72	0	0	0	0	0	0	14.41	6
0	2.09	0	0	0	0	0	1.03	3.98	2
0	0.38	0.45	5.04	0.86	2.80	0.19	0.05	15.41	6
1.14	0.77	0.96	0	0	0	0	0	5.87	2
0	0	0	0.07	0	0	0	0	3.51	1
0	0.25	0.02	0	0	0	0	0	19.63	8
0	3.35	1.84	8.13	2.34	1.51	0	1.37	33.12	13
0	1.49	0	0	0	0	0	1.07	4.53	2
7.45	24.93	22.54	31.08	17.85	16.19	11.80	5.87	257.55	101
0	0	0	0	0	0	0	0	7.36	100
—	8	8	8	8	—	—	—		
—	5.69	6.40	4.09	6.63	—	—	—		
—	2.57	3.58	0.20	4.40	—	—	—		

TABLE 26D. Revenues, Expenditures, and Investments on the James McCartey Estate, York County, Virginia, 1757–1765

Category	1757	1758	1759	1760	1761	1762
REVENUES						
Tobacco	—	0	16.50	0	9.32	9.22
Corn	—	0	3.06	0	5.33	16.76
Wheat	—	0	0	0	1.84	0
Fodder	—	0	8.55	3.49	13.66	0
Livestock	—	9.63	12.82	0	0	0
Slaves hired out	—	0	44.01	29.35	38.81	39.52
Cider	—	0	2.54	0	0	0
Interest	—	1.27	5.43	0	4.06	5.11
Miscellaneous	—	0	22.32	0	2.12	1.20
Total	—	10.90	115.23	32.84	75.14	71.81
EXPENDITURES						
Clothes and shoes	6.70	5.83	6.72	6.10	3.60	6.69
Supervision	0	0	2.03	1.88	1.78	0
Taxes	0	0	4.18	0	0	0
Local services	2.69	1.54	0.25	5.62	1.07	0.77
Building repair	0	0	0	0.50	0	0
Household goods	1.23	0	0.55	0.85	0.38	0
Provisions	2.21	0.19	0	0	0	0
Tools	0	2.27	2.03	1.91	0	0
Miscellaneous	0	0.04	24.31	18.30	14.43	35.00
Total	12.83	9.87	40.07	35.16	21.26	42.46
INVESTMENTS						
Livestock	0	3.10	0	4.04	0	0
Est. no. laborers	—	8	5	4	3	3
Est. gross revenue per laborer	—	1.36	14.24	8.21	12.11	10.76
Est. net revenue per laborer	—	0.13	6.23	−0.58	5.02	−3.39

Note: Currency values are in constant pounds sterling. Percentages for individual categories do not add up to one hundred owing to rounding. Estimated gross and net revenues per laborer exclude the hiring of slaves. The mean for the estimated gross revenue per laborer is 10.20; for the estimated net revenue, 4.28.

Source: Transcripts in the York County Project files, CWR, from York County Guardian Accounts, microfilm, VSL.

1763	1764	1765	Total	Percent
21.89	0	5.48	62.41	12
19.15	0	5.48	49.78	10
0	0	0	1.84	<0.5
6.40	0	5.48	37.58	7
0	0	0	22.45	4
51.15	45.43	16.45	264.72	52
0	0	0	2.54	<0.5
0	1.35	0	17.22	3
0	25.40	0	51.04	10
98.59	72.18	32.89	509.58	98
1.95	1.24	0.44	39.27	23
0	0	0	5.69	3
0	0	0.79	4.97	3
1.24	0	1.45	14.63	8
0	0	0	0.50	<0.5
0	0	0	3.01	2
0	0	0	2.40	1
0	0	0	6.21	4
1.84	0.38	1.15	95.45	55
5.03	1.62	3.83	172.13	99
0	0	0	7.14	100
3	3	—		
15.81	8.92	—		
14.14	8.38	—		

TABLE 26E. Revenues, Expenditures, and Investments on the John Hay Estate, York County, Virginia, 1759–1764

Category	1759	1760	1761	1762	1763	1764	Total	Percent
REVENUES								
Tobacco	0	—	21.84	9.43	12.79	0	44.06	17
Corn	26.46	—	44.17	73.27	36.53	0	180.43	68
Wheat	0	—	1.83	0.77	0	0	2.60	1
Slaves hired out	0	—	0	0	0	21.72	21.72	8
Land rents	0	0	7.74	2.88	0	2.72	13.34	5
Miscellaneous	1.94	—	0	0	0	0	1.94	1
Total	28.40	—	75.58	86.35	49.32	24.44	264.09	100
EXPENDITURES								
Clothes & shoes	3.26	8.87	6.31	6.50	1.53	0	26.47	20
Supervision	8.96	14.48	9.32	1.84	0	0	34.60	26
Taxes	2.73	5.07	5.62	7.82	1.64	0	22.88	17
Local services	2.03	1.72	2.86	0.96	0	0	7.57	6
Building repair	0.92	1.01	3.51	0	0	0	5.44	4
Tobacco casks	0	0	0.61	0	0	0	0.61	<0.5
Tools	0	0.65	1.13	0	0.82	0	2.60	2
Miscellaneous	9.30	3.03	10.31	3.45	4.86	0.91	31.86	24
Total	27.20	34.83	39.67	20.57	8.85	0.91	132.03	99
INVESTMENTS								
Buildings	0	0	3.62	0	0	0	3.62	100
Est. no. laborers	6	—	6	6	5	—		
Est. gross revenue per laborer	4.73	—	12.60	14.39	9.86	—		
Est. net revenue per laborer	0.20	—	5.99	10.96	8.09	—		

Note: Currency values are in constant pounds sterling. Percentages for individual categories may not add up to one hundred owing to rounding. Estimated gross and net revenues per laborer exclude the hiring of slaves. The mean for the estimated gross revenue per laborer is 10.40; for the estimated net revenue, 6.31.
Source: Transcripts in the York County Project files, CWR, from York County Guardian Accounts, microfilm, VSL.

most conservative management practices of the period, for guardians were required only to maintain the value of the estate, to produce enough income to maintain the orphans, and to avoid any risks that might jeopardize the orphans' future inheritance. Since the 1660s when Luke Gardiner administered Robert Cole's estate (discussed previously), plantations were better established, risks had diminished, and opportunities to improve an estate through the purchase of additional assets like livestock had declined. The York County guardians thus pursued conservative courses; none invested in additional slaves, and almost nothing was put into new buildings or livestock. By the 1730s, even small York County plantations had become somewhat diversified operations. Dairy products, cider, fodder, and wool were probably produced on all the farms, and some sold modest surpluses of these products. Most, too, raised some excess livestock. All of the estates grew a little wheat, and most made a surplus of corn over and above the amounts required to provision the slaves and feed the stock. Revenues from these sources, however, were small, for the surpluses were never very large. Tobacco was by far the most important source of income for four of the five estates, accounting for 55 to 82 percent of recorded gross revenues from field crops and up to 72 percent of total gross revenues. Only on one estate in the early 1760s, John Hay's, did income from surplus grains exceed that from tobacco.

Clothing for the slaves was a major cost on these middling estates, accounting for as much as 63 percent of estate expenditures, as was the case with Lewis Davis's estate. Once this basic necessity had been met, the slaves got little else. Almost nothing was spent on bedding and other household equipment, and, except on one plantation, only a few shillings went for provisions like salt that could not be produced on the farm. Expenses had to be kept at a minimum, for quitrents and head taxes, where fully accounted for, bore much more heavily on these smaller estates than on larger operations.

Supervision was another headache. Three of the five guardians chose to let the enslaved laborers toil on their own most or all of the time. This seems to have been a wise choice if the estate's four to eight adult field hands could be kept working with only occasional visits from the guardians and if they were not prone to running away. On the two plantations where overseers were regularly retained, their shares ate up much of the plantation's annual profit. Slaves allowed to manage their own time seem to have produced smaller crops than those working under a resident overseer, but the difference in output between closely and loosely supervised enslaved laborers was only about two hundred pounds of tobacco per hand, an amount insufficient to cover the cost of an overseer on farms where there were only a few workers. In some cases, the guardians hired out any skilled or adolescent slaves the heirs possessed, thus passing on much of the burden of supervision to others and providing an assured income for maintaining the orphans.[67]

Comparison of these middling farms with the much larger operation at Carter's Grove suggests that there were economies of scale. Supervision is a prime example. Only when there were enough field hands to make up work groups of six or more, as in Anthony Robinson's estate, was it likely to pay to employ an overseer. In addition, the variety of tasks that had to be performed on these somewhat diversified farms clearly cut into the time that could be devoted to tobacco. Where there were only a few hands to plow the land, look after the livestock, run the dairy, shear the sheep, spin yarn, gather fodder, catch fish, make cider, plant a vegetable garden, sow and harvest wheat, beans, and peas, tend corn, make casks, and keep fences and buildings in repair, few of the enslaved could tend a full crop of tobacco. In the seventeenth century, many such secondary activities had been foregone on orphans' plantations, and free whites had been hired to do all the carpentry and coopering. On the York County estates in the 1740s, 1750s, and 1760s, however, diversified activities like dairying, wheat culture, and sheep raising continued even in the absence of the white family. (The orphans were typically boarded with neighboring families if their mother was also dead.) Normally, too, enslaved Africans and African-Americans, not white craftsmen, took care of all the rough carpentry and coopering needed on the farms. On plantations of similar size where the owner and his family were present, the whites likely performed many of these ancillary tasks, keeping the enslaved laborers in the fields whenever possible. On large plantations like Carter's Grove, slave artisans who worked exclusively at their trades did most of the ordinary carpentry and coopering, whereas old or disabled adults and children were assigned to a variety of other, less arduous chores. The energies of prime hands then could be reserved exclusively for the major field crops.

A comparison of yields and income per hand is also instructive. Carter Burwell's field laborers produced roughly 850 pounds of tobacco, eleven barrels of corn, and perhaps ten bushels of wheat a year. Annual gross revenues per field hand from the three major crops were approximately £12 sterling constant value. During the same period on the orphans' estates, adult slaves made only about 385 pounds of tobacco, a bushel or so of wheat, and probably about five barrels of corn per hand. Total gross revenues from all planta-

67. Many other estate guardians did not try to keep orphans' plantations in operation. Instead, they hired out the slaves and rented the farms to tenants and used these rents to support the orphans. Probate procedures are discussed in Harold B. Gill, Jr., and George M. Curtis III, "Virginia's Colonial Probate Policies and the Preconditions for Economic History," *VMHB*, LXXXVII (1979), 68–73.

tion products averaged only about £7 per hand per year. (This figure would increase to about £8 if the estimated value of the corn consumed on the plantation, an item not mentioned in these accounts but included in the estimates for larger plantations, was added in.) Total net revenues per laborer fluctuated between £14 and –£6 sterling, averaging just under £3 sterling constant value over the whole period. Both gross and net revenues were lowest on the orphans' estates managed in the 1730s and 1740s, but they doubled in value on the two estates administered between 1757 and 1765. In the two later estates that had the highest returns, on one many of the slaves were hired out, and on the other the workers concentrated on growing more corn and less tobacco.

Given the usual impossibility of calculating the full amounts large planters spent on imported clothing and tools for their enslaved workers and the atypicality of the orphans' estates, especially the frequent absence of oversight and the lack of labor inputs from the white family, the preceding analysis is at best only suggestive of differences in net revenues between large and middling York basin plantations. Middling planting families who worked alongside or at least kept close watch over their workers likely realized somewhat higher returns. Since it proved impossible to estimate all relevant expenditures for Carter's Grove plantation, it is uncertain how much large planters like Carter Burwell might have been netting per laborer. Since Burwell's gross revenues per hand were double those produced by the orphans' enslaved workers, however, he clearly was regularly extracting a profit from the labor of his human property. His operating costs per capita were probably similar to those on the smaller estates, given that there is no evidence that he provisioned his slaves more generously and that Burwell could purchase imported clothing and tools at lower prices than could lesser planters who had to pay a country storekeeper's markup on goods imported from England. The increased costs that large planters incurred from hiring several overseers were more than offset by the increased outputs per hand that economies of scale afforded.

Capitalizing on Corn: The Upper York Basin

Our survey now proceeds further up the York River to later-settled New Kent County, where planter-merchant William Bassett (1709–1743) operated four separate plantations. Bassett was of the second generation of his family born in Virginia and had inherited land and slaves from his father, a member of the Virginia Council who had married into the powerful Burwell clan. The younger Bassett was elected a burgess and married into another propertied Middlesex County Virginia family. Accounts survive for Bassett's farms from 1732—shortly after he came of age—until his death in 1743 and then continue for his widow, Elizabeth Churchill Bassett (1710–1779), through 1754. The size of Bassett's workforce is uncertain, but it likely included at least forty adult laborers.[68]

Like the lower York basin planters, Bassett grew tobacco, with grains as a strong second staple. His herds of cattle and hogs yielded an annual surplus, and the family also raised sheep. Bassett and later his widow responded much more vigorously to rising corn prices, however, than did Carter Burwell and the estate guardians in York County. Like the lower York basin planters, they did not cut back on tobacco; Bassett's enslaved laborers continued to produce about eight hundred pounds a hand across the period of the accounts. But Basset did dramatically increase the amount of corn each worker was expected to tend. In 1732–1734, the slaves each were raising about eight barrels of corn a year. Virginia corn prices began to rise in 1735, and production on all the Bassett plantations increased. Between 1738 and 1750, annual output per hand averaged seventeen and a half barrels. The slaves also grew about 6 bushels of peas each and tended some wheat. Because overseers did not get shares of wheat (as they did of tobacco, corn, and peas), the sizes of the wheat crops are uncertain. An account of the profits of the estate in 1754 indicates that in that year 132 bushels of wheat were produced, bringing in only thirty-two pounds sterling constant value, whereas a surplus of two hundred barrels of corn sold for seventy-eight pounds. Corn then, not wheat, was the second staple. This strategy enabled the Bassetts to maintain a reasonably steady annual gross income of about twelve and a half pounds sterling constant value per hand from corn and tobacco. When the price of Virginia tobacco fell in the 1740s, increased corn production made up for declining tobacco revenues. In the early 1730s, tobacco accounted for about two-thirds of gross revenues from the two major crops (Table 27). After 1740, it contributed just over half.

The account book of another New Kent County planter-merchant, William Massie (1718–1749), for 1746–1748 shows a crop mix almost identical to Bassett's. Massie, like Bassett, was of the third generation of his family in the colonies, had inherited several thousand acres of land and twenty-one slaves, had married into a distinguished Virginia family, and served his county as a burgess. Massie held three plantations totaling more than three thousand acres in New Kent, Hanover, and Goochland counties, and by 1746 he owned a total of sixty-one slaves, twenty-eight of them laboring adults. Tobacco yields averaged 766 pounds a hand and corn seventeen barrels, with

68. William Bassett Account Book, 1728–1755, MS1 B2944a, and Elizabeth Bassett Accounts, 1744–1754, MS1 B2944 a8–16, VHS; "William Bassett," in John T. Kneebone et al., eds., *Dictionary of Virginia Biography* (Richmond, Va., 1998), I, 387–388.

TABLE 27. Tobacco and Corn Crops and Revenues per Laborer, William Bassett Plantations, New Kent County, Virginia, 1732–1749

Year	Tobacco (lbs.)	Tobacco revenue (£)	Corn (barrels)	Corn revenue (£)	Total revenue (£)
1732	621	6.07	8.0	2.34	8.41
1733	590	7.20	9.0	4.12	11.32
1734	895	9.10	6.0	2.75	11.85
1735	793	8.71	9.0	4.94	13.65
1736	779	6.47	—	—	[6.47]
1737	722	7.40	11.0	4.70	12.10
1738	997	10.04	17.5	6.47	16.51
1739	845	10.62	17.0	4.73	15.35
1740	719	6.86	17.5	6.61	13.47
1741	813	7.12	17.5	6.93	14.05
1742	882	6.50	17.8	6.56	13.06
1743	748	6.30	17.2	5.44	11.74
1744	841	6.55	14.1	4.39	10.94
1745	704	5.45	18.7	5.97	11.42
1746	813	6.07	15.3	4.73	10.80
1747	990	7.48	19.5	5.89	13.37
1748	704	5.03	23.0	6.57	11.60
1749	788	6.23	17.4	5.31	11.54
Mean	791	7.18	15.0	5.20	12.42

Note: Currency values are in constant pounds sterling. The total revenue for 1736 (in brackets) is incomplete.
Sources: William Bassett Account Book, 1728–1755, MSS1 B2944a3, and Bassett Family Papers, MSS1 B2944, VHS.

the value of the two crops about equal. Massie grossed about ten and a half pounds sterling constant value per worker. Both Massie and Bassett kept their enslaved laborers working full-time in the fields raising staple crops. They hired local white women and men to spin yarn, weave cloth, curry leather, make shoes, and repair boats.[69]

69. William Massie Account Book, 1747–1748, MS 21, VHS; "Massie Family," *WMQ*, 1st Ser., XIII (1905), 196–203; Oliver Morris Refsell, "The Massies of Virginia: A Documentary History of a Planter Family" (Ph.D. diss., University of Texas, 1959).

Like most other large planters, the fortunes of Bassett and Massie did not depend on agriculture alone. Bassett had a servant blacksmith who performed both plantation and custom work. (In 1744, the smith shop did business worth £117 sterling constant value and made a net profit of £74.) He also rented out a tavern and ran a gristmill. Even more important were his mercantile operations. In his New Kent County store, Bassett retailed British dry goods; West Indian molasses, sugar, and rum; lower James River specialties like myrtle wax and tar; and pickled fish and cranberries procured from New England ship captains. He purchased tobacco for, or consigned his own crops to, London merchants John Hanbury, Jonathan Forward, and Haswell and Hunt, to Bristol traders John King and Lidderdale and Harmer, to the Liverpool firms of Thomas and John Backhouse and of Charles Goore and William Bulkeley, and also to the Buchanan family in Glasgow. In partnership with another local merchant, he imported West Indian goods and exported cargoes of local plank, staves, pork, beef, corn, and peas as well as hired out a large decked sloop and several flats to carry produce to Yorktown and Norfolk. Massie owned a house and lot in Newcastle Town and, like Bassett, operated a country store. He shipped lumber and corn to the West Indies and tobacco to England, and by his death in 1749 he had a sizable sterling balance in the hands of his London correspondents. Massie had but two surviving children, both sons, so all the family land eventually went to them and their heirs. His wife, Martha, got the use of the home plantation and eighteen slaves during her lifetime, but Massie instructed that she had to make a will dividing the slaves between the sons.

These two accounts suggest that large planters at the head of the York were responding in the 1740s and 1750s to the combination of a robust local grain market and a good supply of prime corn lands. (Similarly, about the same time on the lower peninsula, Carter Burwell saw opportunity in a combination of good wheat land and a strong local urban market for wheat and flour.) For the New Kent planters, the major corn market was almost certainly the West Indies. Bassett traded there directly and also, like Massie, exchanged goods with Yorktown and Norfolk merchants as well as itinerant New England ship captains. Both also traded in wheat, but on a much smaller scale, and most of this grain was sold locally. By expanding corn production, Bassett and Massie were able to maintain the same level of gross revenues per laborer despite falling tobacco prices.[70]

70. Another example of this trade is Neil Buchanan's factor, Francis Jerdone, who was buying New Kent County and Hanover County corn and wheat either for resale in the colony or for shipment to the wine islands in 1739–1741. Francis Jerdone Account and Letterbook, 1736–1737,

The Joneses Try to Keep Up

During the Tobacco Coast's golden years, many large native-born planters on the lower Peninsula and the upper York basin like Carter Burwell, Daniel Custis, William Bassett, and William Massie employed inherited assets wisely and by diversifying their plantation activities and keeping out of debt were able not only to maintain plantation revenues but also to improve their estates. Not all, however, were equally successful, as the example of the Jones family demonstrates. Thomas Jones (by 1677–1758), the son of a Virginia settler who had returned to England, arrived in the colony in 1698 as a factor for London merchants. Settling in the town of Williamsburg by 1708, Jones began acquiring land by patent and purchase and also bought twelve city lots. He served a term as a burgess in 1720–1722 and became a town alderman in the latter year, at which time he ceased active trading, instead seeking to become a planter on the grand scale then practiced by elite native-born Virginians.[71]

Thomas and his brother Frederick, also a merchant, invested jointly in a tract in King William County in 1713 and started buying African slaves to work their lands. After Frederick's death in 1723, Thomas began courting a widow at least twenty-five years younger than he and in 1726 succeeded in marrying Elizabeth Cocke (1701–1762), daughter of a politically prominent (albeit bankrupt) Virginia physician and the widow of Walter Pratt of Gloucester County, a prosperous merchant. Elizabeth brought a fortune of more than two thousand pounds sterling and an additional eight lots in Williamsburg to the marriage. Thomas immediately used these funds to purchase his brother's share of the King William land, additional tracts in Caroline and Hanover counties, and many more slaves, at the same time making a marriage settlement granting Elizabeth twelve hundred pounds sterling, a sum that he had apparently promised to deliver in the form of cash or bills of exchange to her uncle in London. Instead, Thomas settled seventy-eight slaves on Elizabeth, human assets of which he had full use during his lifetime, and by then he had bought an additional fifty-six enslaved workers not included in the marriage settlement.[72]

Always short of capital, Jones financed most of his extensive purchases on credit. His strategy was to try first to get credit from English merchants

---

1738–1744, Special Collections, Swem Library. Daniel Park Custis was also shipping New Kent County corn to the West Indies in the late 1740s (Custis Papers, CWL).

71. York County Project Biographical Files, CWR.

72. Roger Jones Family Papers, 1649–1889, MS 18,063, LC, microfilm, M-1397-1-15, nos. 18, 123, 143, 164, 168, 177, 256, 282, 304, 379, 800, CWL.

by drawing bills of exchange on them for much more than any tobacco he shipped was worth. Should these merchants fail to pay his overdrafts (as most did), he then borrowed from large planters and merchants in Virginia, offering mortgages on the recently purchased land and slaves as collateral. In 1721, for example, he bought slaves costing £657 from Royal African Company agents Charles Chiswell and Francis Willis. He paid with bills of exchange drawn on two London merchants, John Maynard and Thomas Colmore, who subsequently refused to honor Jones's overdrafts. To pay his debt to Chiswell and Willis, in the following year Jones mortgaged two thousand acres and twenty-seven slaves (probably those whom he had just purchased) to Virginia planter Peter Beverley in return for a loan of £638, which he promised to repay a year later along with 4 percent interest. Through this strategy, Jones got a year's labor from the twenty-seven slaves without paying anything. And should he fail to repay Beverley in full at the end of the following year, making annual payments of the interest alone would likely deter Beverley from foreclosing on the mortgage for at least another five years.[73]

Throughout the 1720s, Jones continued to buy more land, slaves, and imported consumer goods, although he still owed a balance on a cargo of goods he had sold for London merchant Robert Cary some years before, had outstanding debts to other London merchants, and had also borrowed money from Norfolk trader Samuel Smith (to whom he mortgaged other land and slaves in 1725), Yorktown merchant Thomas Nelson, Gloucester County planter Nathaniel Burwell, and several other James and York river traders and planters. By the late 1720s, the sixty to seventy hogsheads of tobacco that Thomas shipped each year did little more than pay for the imported English goods that he continued to order and the interest on the principal sums he owed. He had yet to find a means for paying for most of the minimum of eighty-two African slaves he had recently purchased.

In 1729, Jones's primary London factor (and creditor) was Edward Randolph, who was himself financially pressed by too many correspondents drawing bills of exchange on tobacco yet unsold. Randolph turned the collection of his Virginia debts over to his brother John, the attorney general of Virginia and a prominent member of the House of Burgesses. When John Randolph learned that Jones had that year drawn bills of exchange on Edward for a further £851 pounds, about twice the value of the tobacco he usually shipped, John pulled the plug on Jones's credit line, demanding that he mortgage more land and slaves. As another Virginia merchant explained, "No man of any Account makes a Mortgage here but 'tis soon known, his Credit is at an end

73. Jones Family Papers, nos. 148, 179, 217, CWL.

and it Affects him like an Act of bankruptcy." The aggrieved Jones, who suspected Edward Randolph of selling his tobacco too hastily for only half the price that his other tobacco was fetching in Bristol, protested that he "was no intruding Correspondent," since Randolph had accepted the consignments on the condition that he advance money to Jones.[74]

Between 1729 and 1732, mortgaging of more land and slaves (some for which Jones did not have a clear title) and the outright sale of other slaves raised insufficient funds to forestall his creditors. They won several suits against him in the General Court, forcing Jones to sell more land and to turn over seventy some slaves for their use. Neither his English nor his Virginia creditors were willing any longer to accept Jones's excuses that it was not "Suitable to his circumstances" to pay "at that time." The creditors valued the labor of the seized slaves at four pounds sterling per adult worker per year, a sum that rather accurately reflected the net proceeds of their tobacco crops after deducting expenses for imported clothing and tools, but which did not take into account the value of the surplus grain and livestock that they also continued to produce. Whether Jones's English creditors were ever fully repaid is unclear. On the other hand, his Virginia creditors almost certainly recouped all and perhaps more than what Jones owed them.[75]

Jones had set up at least seven quarters in King William and Hanover counties by the mid-1720s, which he attempted to manage at a distance through overseers and a general manager, John Cosby. Perhaps owing in part to his financial difficulties, Jones had developed an unenviable reputation for stinting his slaves. Cosby complained in the 1720s that the laborers lacked adequate winter clothing. William Byrd II, on a trip to Fredericksburg and Germanna in late September 1732, visited some of Jones's plantations. Byrd reported, "The poor Negroes upon them are a kind of Adamites, very scantily supplied with clothes and other necessaries." At the very time that the slaves' condition was bad enough to scandalize Byrd, the Joneses had stockpiled in their Williamsburg house 140 "suits of Negroe's clothes," 230 "Negro shirts and shifts," and more than 700 yards of coarse cloth. These supplies were, however, seized by some of Jones's creditors rather than delivered to the slaves, who continued to suffer for their owner's improvidence. Another of the creditors, William Nelson, noted that the slaves whom he seized in

74. Ibid., nos. 389, 398, 407, 410, 430–431, 437, 445, 453, 476, 545–546, 642 (quotation), 1672; Hemphill, "John Wayles Rates His Neighbours," *VMHB,* LXVI (1958), 304 (quotation).

75. Jones Family Papers, nos. 476 (quotation), 506, 568–569, 574, 576, 591, 606, 611, 646, 654, 677, 768, 843, 850, 971–972, 1174, 1536, CWL. John Wayles observed in 1767 that mortgagees postponed payment of other creditors until their debt was satisfied (Hemphill, "John Wayles Rates His Neighbours," *VMHB,* LXVI [1958], 304).

1737 "can't be deliver'd [back to their owner] more destitute of Cloaths and Beding than they were [when taken]."[76]

In other ways, Jones was fortunate in that his human property proved exceptionally healthy and prolific, perhaps in part because the enslaved field-workers might have escaped the hard-driving characteristic of resident masters. (The domestics, however, over whom Jones retained control, were often subjected to the lash.) William Byrd further noted on his 1732 visit to the Jones quarters: "Nevertheless (which is a little incomprehensible), they continue in perfect health and none of them die except it be of age. However, they are even with their master and make him but indifferent crops, so that he gets nothing by his injustice but the scandal of it."[77]

Byrd was again at least partly correct. Some of the slaves seized by Jones's creditors in 1737 continued to work for them for fifteen years. In one group of eighteen adults, many of whom were married with several children and so likely already in their thirties, six men and five women survived these fifteen years, and a few lived much longer. Only five of the men and two of the women died between 1745 and 1752, constituting a record for relative longevity that other Chesapeake slaves surely did not enjoy. Also, already in 1726, the eighty-two adult Africans Jones then owned were parents to fifty-four surviving Virginia-born children. By 1732, many of the enslaved couples had two, three, or four children, reflecting much higher levels of reproduction among transported Africans than those prevailing on many other big estates like that of Robert "King" Carter, inventoried in 1733. Although a number of the children either died in infancy or before reaching age five, their parents somehow managed to keep the children relatively well fed. Jones recorded the heights of thirty-one enslaved children who ranged in age from three to thirteen in 1726. The stature children attain by a given age is a measure of net nutrition, that is, caloric intake minus the physical toll wrought by various infectious and dietary deficiency diseases and parasitic infestations. Children on Jones's quarters proved taller at a given age than enslaved Virginia youngsters at the end of the eighteenth century, suggesting that they were better fed, suffered from fewer diseases than average, or a combination of the two. The creditors were clearly surprised at the large number of children, an encumbrance that, from their perspective, could not "be expected to raise much money." Through natural increase, the size of the enslaved workforce that Jones managed to keep from the clutches of his creditors rose from thirty-

76. Byrd, *A Progress to the Mines,* in Wright, ed., *Prose Works,* 349; Jones Family Papers, nos. 262–263, 440, 969, CWL.

77. Byrd, *A Progress to the Mines,* in Wright, ed., *Prose Works,* 349–350.

five tithables in 1744 to forty-two in 1754. When Jones died in 1758, the total number of slaves he retained had climbed to more than eighty.[78]

Whatever the relative trade-offs between relaxed oversight and scant provisioning the slaves on Jones's quarters might have experienced, some of them strenuously resisted their bondage. One man stole meat to supplement his supposedly "adequate" rations, which consisted solely of "bread and mush." More ominously, in 1726 or 1727 a group of enslaved men initially considered poisoning the general manager, but then thought better of such a risky stratagem and instead plotted to run into the mountains where they hoped to find refuge with the "Cannyde Indians."[79]

Jones was well aware of the slaves' family ties, keeping unusually detailed records of husbands, wives, and their respective children. However, there is little indication that he and, as we will see later, Elizabeth, regarded these people as anything other than commodities. In settlements he made for his wife and children, Thomas was always careful to specify that all "future increase" of any enslaved females belonged to the family estate. He seems to have had few scruples about selling some children away from their parents or about moving all the children born to the slaves seized for debt onto other plantations as soon as they were weaned.[80]

William Byrd's third observation that the slaves made "but indifferent crops" is also corroborated by the manager's letters and by Jones's later plantation accounts. John Cosby laid out the reasons for the poor quality of some of the tobacco produced in the mid-1720s: "There may be a bad leaf or bundle put up some times which is not to be admired at Considering what Cursed deceatful villons I have to deal with such as Ralph and Mabe and a parsel of new negroes that know as little of Croping or to doe any thing in Tob:o neatly as they did when they Came first out of their own Contrey."[81]

In addition, Cosby had to train inexperienced overseers who could learn only by first making serious mistakes in sorting and packing. The only remedy Cosby could propose was to have all the tobacco made on the various quarters brought to his own house, where he would overhaul and repack each hogshead. Jones's overseers continued to perform poorly, for when William Byrd revisited Jones's King William quarters on his return journey from Ger-

78. Jones Family Papers, nos. 227, 431, 547, 550 (quotation), 574, 913, 1122, CWL; Walsh, *From Calabar to Carter's Grove,* 83–89; Richard H. Steckel, "A Peculiar Population: The Nutrition, Health, and Mortality of American Slaves from Childhood to Maturity," *JEH,* XLVI (1986), 721–741.

79. Jones Family Papers, no. 263, CWL.

80. Ibid., nos. 217, 569, 800, 801, 913, 1779, 1798, 1800, 1803, 1887.

81. Ibid., no. 263.

manna in October 1732, he found "the Colonel's overseers were all abroad, which made me fearful I should find mine [on adjoining plantations Byrd owned] as idle as them." In 1735, the mean tobacco crop each of Jones's enslaved workers raised was a scant six hundred pounds. Once the seized laborers were returned to the family in the 1750s, the tobacco crops they produced were still often less than those other great planters managed to extract from their workers. Between 1755 and 1763, the enslaved field-workers were making just over seven hundred pounds apiece, yielding a gross revenue of about eight pounds sterling constant value per hand.[82]

By 1725, Jones was already following the strategy adopted by other successful great planters in the region, growing tobacco as the primary crop and at the same time producing some surplus corn and peas and small but marketable amounts of wheat on all of his larger quarters. He also had a plantation mill and a large shoemaking operation on Tanyard quarter in King William County, supervised by a white man but otherwise staffed by enslaved men and boys who produced several hundred pairs of coarse shoes a year. Corn tops and blades and wheat straw were saved to carry cattle through the winter, permitting the sale of surplus fattened beef in addition to surplus pork and the making of considerable amounts of butter, provided that the overseers' wives were willing to undertake the task of supervising the dairies. But with the bulk of his assets in the hands of his creditors in the 1740s and early 1750s, Thomas benefited little from these innovations by which other planters were augmenting their tobacco revenues.[83]

The Joneses lived in Williamsburg during the years when their creditors held their most productive human assets, at first begging the creditors for surplus hogs and a few cows to make butter. Meanwhile, some of Thomas's land purchases and a hoped-for inheritance from the Pratt family were tied up for years in litigation. In 1743, however, Jones somehow managed to buy at least two new African slaves. In 1751, expecting shortly to recover possession of the seized workers, the family sold their Williamsburg house and moved to a rented plantation in Hanover County. Five years later, Jones bought a 300-acre plantation in the same county. Thomas still hoped to recoup the family fortunes by marketing high-quality tobacco raised on the 3,675 acres he had managed to hold onto, by selling corn and wheat to grain export merchants operating from Williamsburg, and by retailing surplus beef, pork, mutton,

82. Byrd, *A Progress to the Mines,* in Wright, ed., *Prose Works,* 375. Outputs and revenues are derived from accounts with overseers in Jones Family Papers, nos. 524, 586, 1192, 1193, 1225, 1227, 1230, 1231, 1233, 1620, 1621, 1810, 1878, 1885, 1952, 1986, 2020, 2032, 2037, 2128, 2267, 2278, 2295, 2297, 2299, 2507, 3854, 4044, CWL.

83. Jones Family Papers, nos. 440, 444, 577, CWL.

butter, cider, and brandy locally. Jones also rebuilt the plantation mill, continued the tanning and shoemaking operation in King William County, added small amounts of oats to the crop mix, and began renting a few small tracts to tenants.

Like Daniel Parke Custis, the Joneses concentrated on making high-quality sweet-scented leaf, seldom marketing any further-processed stemmed tobacco. In the late 1750s, their crops sometimes commanded good prices, but relatively low output per worker limited their returns, as did occasional less well managed consignments of "ordinary" tobacco not suited for the British domestic market. Thomas also tried to resume his mercantile career, importing cargoes of British manufactures on credit and shipping tobacco from his own plantations and other tobacco that he purchased to London merchants John Norton, John Bland, James Buchanan, and Robert Bogle, Bristol merchants Thomas Knox and Farrell and Jones, Liverpool merchant Charles Goore, and Glasgow traders Andrew Cochran and James Donald as well as at least once selling some tobacco in the country. Seeming to have learned little from his previous misfortunes, by 1757 Thomas was already in debt both to Knox and to the Glaswegians. When negotiating a marriage settlement for his second son, Frederick, Jones complained of "the charge of my son's Education in England, high freights and Insurances, the loss of about £100. Ster in [the West] Indies the last season taken by the French [and] a very bad crop this year." Thomas still protested that "from my station in life" he was "obliged to live at a considerable expense which can't easily or at once be retrenched."[84]

Thomas died in 1758, leaving his wife and older sons Thomas (1726–circa 1785) and Frederick to manage the family assets. Elizabeth immediately tackled Thomas's creditors, who still retained the services of some of the family slaves. She protested that Thomas had power to assign their use only during his lifetime, whereas the creditors maintained that there was still a balance outstanding on Thomas's debts. While trying aggressively to regain control of all the adult workers, Elizabeth attempted to cut costs by giving some of the oldest slaves to two of the overseers on the condition that they post bond to indemnify her of any maintenance charges. And to discharge debts owed to Glaswegian Robert Donald, the Joneses sold "as many of the worst of our negros as amounts to 400£." Struggling to manage the estate herself with the periodic help of her sons, Elizabeth had doubts about the probity of the overseers, about one of whom she wrote, "I have reason to think he always means to impose if he can." In early 1759, she reported that

84. Ibid., nos. 570, 577–578, 807, 830, 1076, 1143, 1152, 1267, 1387, 1436 (quotation).

the previous years' corn crop had still not been apportioned and that "the overseers doe what they think fitt without controul." The quality of their tobacco suffered as well, for Farrell and Jones reported their last consignment was at best "but ordinary," including hogsheads in which several sorts of tobacco had been mixed indiscriminately. Then most of the 1761 crop was captured by the French (Elizabeth decided to risk shipping it uninsured).[85]

Meanwhile, Thomas Junior moved to Northumberland County, having secured the position of county clerk. He took charge of some lands the family owned in that jurisdiction as well as of a tobacco-producing plantation in distant Cumberland County that he had recently bought, and he attempted to make money by selling West Indian goods to neighbors and tobacco to local merchants, ventures that were, of course, financed on credit. Elizabeth, who disliked living on the isolated Hanover farm, then followed Thomas Junior to Northumberland and contracted to buy a plantation of her own in that county. She died soon thereafter, dividing in her will of 1762 the two thousand pounds that Thomas Senior had bequeathed to her between the eight surviving Jones children and some of her grandchildren.

Neither Thomas Senior nor Elizabeth adopted any of the dynastic strategies that established families such as the Burwells and Carters employed to keep productive assets in the family. Rather, both expected that some or all of the family land and slaves, none of which were entailed, would be sold and that many of the children would use the cash they received to set up new establishments elsewhere. The family appears to have had little emotional investment in particular tracts of land, instead viewing acreage and enslaved workers as mere commodities. Thomas Senior simply divided the residue of his estate equally between the children, leaving it to Elizabeth to reward some of the children more than others from the two thousand pounds he left to her. She did favor their eldest son, specifying that Thomas should have first choice of any of the slaves he might wish to buy, being "Intitled as his Birthright," and any siblings who refused him were to be disinherited.[86]

Elizabeth's monetary bequests, along with Thomas's provision for the equal division of any assets remaining in the estate after payment of his debts between all the children, forced the sale of both her own and Thomas's estate, which had been appraised at £10,961 sterling constant value including land. Frederick Jones was opposed to selling the slaves, partly for humanitarian reasons, arguing that "for my part it will give me concern to see them separated and squandered about the Country." His reluctance also stemmed from

85. Ibid., nos. 569, 1453–1454, 1460, 1563, 1681, 1709, 1841, 2170.

86. Ibid., nos. 801, 1418, 1575, 1807, 1817, 1912–1913, 2021, 2169, 2279.

financial motives, however, "as all of us, must lay out the Money; I suppose, in Negroes, [and I] don't think we shall be able to supply ourselves with so valuable ones." Frederick's preference unfortunately did not prevail. Thomas Junior bought some of the slaves belonging to the estate, but most were exposed for sale in several different counties at auctions, where "Rum and Bomboe" were provided to encourage potential bidders.[87]

The younger Joneses found themselves in the unenviable position of having to catch up with their contemporaries rather than setting a standard that other families might wish to emulate. Thomas and Elizabeth bequeathed to their offspring not only an estate encumbered by debt but also aspirations for living in a style beyond their means, along with the assumption that they had an inherent right to the credit required for pursuing this goal. Thomas Jones's problems stemmed, not from his initial borrowing to buy land and slaves to expand the family's productive capacities, but rather from a consistent failure to devote the proceeds of the crops his mortgaged workers produced to repaying outstanding debts. In the event borrowers encountered hard times, both British and domestic lenders were willing to accept payments of interest alone for only a limited period of time, usually no more than three or four years. They could not afford to indulge indefinitely any who made no good-faith effort to make payments on the principal owed, nor would they allow irresponsible borrowers to avoid coming to some composition with them. Beginning in the 1730s, drawn-out court procedures and a tendency among the justices to sympathize with borrowers afforded the chronically indebted only a limited respite. By then, many of the lenders were themselves members of (or in the case of indigenous merchants, allied by marriage to) the ruling gentry. The ready availability of credit from multiple sources did allow some overreaching entrepreneurs like Jones opportunities for purchasing more land and more slaves than they could afford and thus for amassing significant levels of debt. The temptations to engage in such behavior were widely understood but not widely condoned by the majority of large planters, who valued a reputation for prudent financial management and who were averse to taking irresponsible risks with family assets.

87. Ibid., nos. 2184, 4044, 5051.

CHAPTER 6

Managing for Posterity

RAPPAHANNOCK AND POTOMAC VIRGINIA

1730–1763

Farther up Chesapeake Bay on Virginia's middle Peninsula and in the counties of the Northern Neck adjoining the Rappahannock River, other great planters, like those living on the lower Peninsula, benefited from a near monopoly of prime river bottomlands and large enslaved workforces. This combination of fertile land and numerous laborers enabled them to take maximum advantage of rising tobacco prices after Virginia adopted a system of tobacco inspection. Middling planters in the area prospered as well, although to a lesser extent than their big neighbors, while many poorer planters were either forced to move out or were relegated to farming less productive uplands or to working rented tenements.

Like those in the upper York basin, large and middling Rappahannock River planters also made corn an increasingly important second staple while still raising significant amounts of tobacco. Elite planters continued to consign some of their tobacco crops to London in order to obtain the fashionable metropolitan goods they coveted, but they turned as well to Bristol and Liverpool for markets for their staple, for cheaper sources of more ordinary goods, and for alternative sources of credit. By the 1740s, however, Bristol merchants produced such poor returns for high-quality Rappahannock tobacco that most planters confined their consignments to London and Liverpool, although a few made occasional trials of Whitehaven and Dublin markets. Those who also grew lower-grade oronoco on holdings where the sweet-scented strain did not thrive, or who received rents and fees in this less valuable strain, increasingly found that selling such tobacco in the country produced better results than consigning it to England.

To increase overall agricultural production, large Rappahannock basin planters also began paying closer attention to incorporating livestock into their farming strategies. Since the amount of good bottomland in this area was quite limited, Rappahannock planters were especially conscious of the need to prolong the fertility of their most productive soils. They sought both to improve the quality of their animals and the quantity of their herds and to make use of animal manures on their fields, an innovation lower Peninsula planters had yet to adopt.

Since Rappahannock basin planters could not increase tobacco outputs by farming more tidewater acres, they, like the York basin planters, began raising more corn, about half of which was sold to the West Indies and the other half to New England. In general, the proportion that tobacco contributed to the great planters' incomes decreased to about half, and revenues from grains rose to equal importance. By the 1750s, most large Rappahannock planters were as a standard practice raising twice the amount of corn that they needed for consumption on the plantation. Much of the surplus can be accounted for by changes in cultivation techniques, for planters who regularly used plows produced twice as much corn as those who employed only hoes. Wheat played little part in expanded grain exports from the Rappahannock before the end of the Seven Years' War. Only a minority of elite planters raised this more risky crop and then only in years when prices were high. Although most resident planters adjusted their crop mix to take advantage of new markets for grain, however, some absentee plantation owners continued to pursue a more stubbornly traditional, tobacco-centered agriculture.[1]

Resident elite planters devised this more intensive and more profitable agricultural regimen, but it was primarily enslaved workers who implemented the program, since it was a system that required resources—land, labor, forage, tools, and draft animals—that planters who did not own slaves seldom possessed. Between 1730 and 1763, nearly seven thousand new African slaves were brought into the Rappahannock basin. Close to thirty-five hundred arrived between 1730 and 1745, a time when elite Rappahannock planters were expanding their workforces in order to increase tobacco production and when, moreover, magnates like Robert "King" Carter secured other new hands from York River slavers. By the late 1740s, however, most elite planters had fully staffed their own plantations and usually bought new Africans only to provide appropriate portions of laborers for their children. Almost no slaves arrived in the Rappahannock directly from Africa during

1. David Scott Hardin, "'Alterations They Have Made at This Day': Environment, Agriculture, and Landscape Change in Essex County, Virginia, 1600–1782" (Ph.D. diss., University of Maryland, 1995), chaps. 6, 7; Robert Anthony Wheeler, "Lancaster County, Virginia, 1650–1750: The Evolution of a Southern Tidewater Community" (Ph.D. diss., Brown University, 1972), chap. 4; James Blaine Gouger III, "Agricultural Change in the Northern Neck of Virginia, 1700–1860: An Historical Geography" (Ph.D. diss., University of Florida, 1976), chaps. 3, 4. See also "Journal of Col. James Gordon, of Lancaster County, Virginia," *WMQ*, 1st Ser., XI (1902), 98–112, 195–205, 217–236, XII (1903), 1–12.

MAP 11. *A Survey of the Northern Neck of Virginia . . . as Surveyed According to Order in the Years 1736 and 1737.* By John Warner, London, 1745. Black-and-white line engraving. Courtesy, The Colonial Williamsburg Foundation.
Land-short Rappahannock River and Potomac River planters took up tracts even farther to the west in the 1730s, prompting surveys such as this one to settle the bounds of the Northern Neck proprietary.

either King George's War or the Seven Years' War, forcing any large planters who still wanted to expand tobacco production to buy some combination of African and more acclimated native-born slaves at estate sales.

By the 1750s, as the surviving children of Africans purchased in the 1730s and early 1740s grew old enough to work in the fields, most elite Rappahannock basin planters, like those on the lower Peninsula, were growing their labor forces through natural increase rather than by purchase. None of the gentry planters who left letterbooks mentioned that they purchased new Africans after the 1740s. In 1761, 1762, and 1763, as the Seven Years' War wound down, a further two thousand captive Africans were brought into the Rappahannock, but by then most were purchased by middling planters rather than by the elite. By the mid-1750s, the numbers of young enslaved children had so increased on some gentry estates that their owners could endow not only their children but also some or all of their grandchildren with young workers. Tellingly, after 1763, only a trickle of new Africans were offered for sale in the Rappahannock, most of them brought in small consignments transshipped from the West Indies rather than in large shipments sent directly from West Africa. By war's end, the market for new Africans in the tidewater Rappahannock basin had dried up, just as it had along the York River ten years earlier.

The region's elite planting families who had acquired vast tracts of undeveloped land in the 1720s and early 1730s were bent on increasing their overall revenues by producing ever more tobacco on these lands in the 1730s and early 1740s. Then, in the later 1740s and 1750s, more astute land-rich planters, as well as those who had failed to accumulate large reserves of frontier land, seized upon grains as an alternative avenue for increasing revenues without further expanding their workforces. Some needed the extra income to sustain their own conspicuous consumption, but many more felt compelled to generate additional revenues to finance increasingly generous portions for their privileged and sometimes numerous offspring.

As at the turn of the century—and as continued to be the case with many lower Peninsula residents—a number of gentry planters succumbed to an early death, leaving as heirs only a widow and one or two very young children. An increasing proportion of Rappahannock area planters, though, at least among those for whom family papers survive, instead enjoyed longer life spans and so found themselves with five, ten, or even more surviving children for whom they were obliged to provide. (A similar situation existed in the higher and healthier piedmont hinterlands.) By the middle of the eighteenth century, competition for social preeminence among the gentry came to entail not just rivalries with members of their own generation in the building of large, fashionable houses and the acquisition of up-to-date interior furnishings, fine clothing, and costly riding carriages but also competition in the provision of marriage portions for their children. This development can best be described as "portion inflation," rather than the more common phenomenon of "dowry inflation," since competition over provision for sons outstripped any competition over provision for daughters.

Gentry parents' obligation to provide all their offspring with the resources to replicate their own privileged economic and social status, at a time when living standards were rising perceptibly and at a historically unprecedented pace, fueled an intense desire to extract ever greater returns from their enslaved laborers. The letters and diaries of Rappahannock basin planters compellingly document the ratcheting up of work requirements, the elimination of much of the bound laborers' already scant leisure time, and an apparently callous indifference to sundering enslaved families by parceling out slave children among widely scattered heirs.

Clinging to Tradition: Absentee Planters

Robert Bristow III (1688–1737), the third-generation absentee landowner-tobacco merchant whom we encountered previously, and his grandson Robert Bristow IV (d. 1776), who took over his grandfather's London firm, ran stubbornly traditional operations on their inherited plantations in Gloucester and Lancaster counties. The goal of both was simply to generate revenues to augment their incomes in England, so they were not inclined to finance risky innovations or to approve any new activities that might lessen the time the enslaved workers spent in the tobacco fields. By the 1730s, the family was operating three agricultural quarters on the more fertile Gloucester County lands (2,250 acres) and only one quarter on the 2,000-acre Lancaster County plantation. By 1753, Bristow IV had a total of fifty-eight adult field hands, a sizable labor force comparable to those held by the Massie and Basset families.[2]

Clinging to a style of management reminiscent of that which prevailed when Bristow III's grandfather had set up his Virginia estate in the 1660s, the later Bristows continued to make tobacco their primary crop and cultivated only enough corn to feed the slaves and livestock. In 1737, Bristow III hired Gloucester County planter Francis Willis, a large planter with seventy working slaves of his own, as general manager. Willis advised Bristow to begin raising surplus corn, wheat, beef, and peas, which might together bring between £150 and £200 Virginia currency per year, enough to pay managers and overseers and to clothe the slaves and pay their taxes. Willis supposed that Bristow was a stranger to such activities, "for the old planters still going on that way but a considerable sum may be got yearly from your Estate for those things and not make less Tobs, nor abuse your Negroes." Larger grain crops would be realized by substituting plowing for hoe culture, and the proceeds from these crops would cover all contingent expenses, leaving the proceeds of the tobacco crop as pure profit.[3]

2. Robert Bristow Account and Letterbook, 1688–1750, VSL, microfilm, M-1164, CWL; Robert Bristow Letterbook, 1747–1770, PRO/A.O. 13/84, microfilm, M-494, CWL.

3. Willis to Bristow, Oct. 15, 1736, Bristow Account and Letterbook, CWL.

Although Willis's advice was sound, his performance was less satisfactory. Under his management, the plantations yielded little profit, and all the experiments in diversified agriculture came to nothing. Willis's own operations likely took up most of his time and attention, a problem the Bristows shared with absentee plantation owners throughout the slave South. Bristow IV replaced Willis in 1752 with James Pasley. Bristow wrote Pasley in that year that he was "very Certain I have been most abominably Cheated and my plantations greatly hurt." Bristow added: "You have some letters of Mr. Willises with you, where he talk'd very largely of what he should doe, I desire you will not imitate that stile, but give me a very exact account of every thing."[4]

Willis's major problem was his inability to extract much work from Bristow's enslaved laborers. They had lived and worked, raised children and died for the better part of a century on quarters where there was no resident master. Over time they had apparently developed and refined techniques for minimizing work that were difficult to combat. In 1747 and 1748, the field hands made only about 700 pounds of tobacco apiece, a modest crop size that by 1753 declined to only 620 pounds per worker. The plantations' total output remained stuck between thirty and forty hogsheads of tobacco per year until the early 1760s. The slaves also made very little corn, not more than three barrels apiece in 1748, at a time when other area planters were getting about nine to twelve barrels from each worker. The least valuable estate revenue, the rent tobacco, thus had to be sold in the country to buy corn for workers and livestock. In the fall of 1747, the laborers on one quarter received no winter clothing from their overseer. During the winter, they also went without salt for three months and ran short of corn. Though Willis lived but one mile from the quarter, no one came to report the deficiency. Bristow's slaves, Willis concluded, "are of such a Nature if they are allowed to live Lazy they will not complain." That same winter, a mulatto "wench" assigned to make clothes for the other slaves stole a silver cup and escaped on a ship sailing for England.[5]

Quarter overseers were paid one-tenth of the net proceeds for which their tobacco sold in England, one-tenth of any surplus grain or cattle sold, and one-tenth of any increase in livestock. This proportion proved insufficient

4. Bristow to Willis, Mar. 27, 1751, to Pasley, Nov. 30, 1752, Mar. 24, 1753, Bristow Letterbook, CWL. For the lower profitability of absentee-owned plantations in South Carolina, see S. Max Edelson, *Plantation Enterprise in Colonial South Carolina* (Cambridge, Mass., 2006), 172–173.

5. Willis to Bristow, Jan. 1, 1747/8, Bristow Letterbook, CWL. Willis had more success with his private affairs. In 1750, he sold more than 1,700 bushels of wheat and 270 bushels of corn to a Norfolk merchant (Alexander McKenzie Account Book, 1748–1750, HSP, microfilm, M-1349, CWL).

to motivate or retain supervisors. Overseers were fired or voluntarily quit in 1747, 1748, 1749, and 1751. In 1748, the home house overseer in Gloucester County refused to stay despite new incentives Willis offered, claiming he could have made one-third more elsewhere, "if they [the slaves] would have worked like other Negros." In frustration, some of the overseers (and perhaps Willis himself) in turn mistreated the enslaved. Bristow received several anonymous letters from Virginia acquaintances reporting abuses.[6]

In contrast, Bristow's "Brenttown" tract of seventy-five hundred acres in Prince William County was farmed by white tenants, as initially the family did not have enough enslaved workers to develop the land. This property apparently provided secure returns (even though the Prince William tobacco was of a lower grade than that of Gloucester and Lancaster) with less trouble and expense than the plantations worked with slaves. Willis advised Bristow in 1746 that he must grant leases for three lives, the accepted practice in the area, "or the people will not Build nor stay." The next year, Willis optimistically estimated he could clear one hundred pounds sterling a year from rents, providing wells were dug for the tenants who did not have sufficient water. The cost to do so was thirty-two pounds, and by 1753 Bristow had thirty tenant families settled on his tract. Rents then yielded, by Bristow's estimate, a gross income of fifty-five pounds per year. By 1758, all the arable land on the Brenttown tract was parceled out among forty-four tenants, and Bristow could contemplate either raising the rents or selling the tract. (The value of the tract seems to have increased substantially; in 1750, he was seriously considering an offer of one thousand pounds sterling for the seventy-five hundred acres, but by 1759 he expected between three thousand and four thousand pounds for the land.) Bristow judged tenancy the best option for making a profit from his least valuable holdings, but, despite the problems he encountered in finding competent managers, he resolved to continue working his better tracts with slaves.[7]

Willis's other schemes failed in part because of short corn crops and poor management of both tobacco and wheat. Not only did Willis have to sell tobacco to buy food; cattle and hogs suffered heavy losses because there was not enough corn fodder to sustain them. Consequently, no surplus beef could be sold for several years after a disastrously short corn crop in 1747/8 (or, as Bristow maintained, because the overseers regularly embezzled livestock). In the early 1750s, the livestock herds remained below the normal size. Much of the tobacco shipped home was ill scented, and some of it was ill sorted. When

6. Willis to Bristow, Sept. 9, 1748, Bristow Letterbook, CWL.

7. Ibid., Apr. 27, 1746.

tobacco obtained from other sources was mixed in the shipment, Bristow expostulated in 1751, "the Reputation of the Mark will be lost, and Price and Plantation destroyed." Some wheat, too, was raised—about two bushels a hand in the late 1740s—but the crops were usually poorly managed, and by marketing time the grain was often eaten by weevils.[8]

Given this record, Bristow IV considered selling his Virginia estate in the early 1750s. Potential purchasers wanted long credits, however, a condition to which he would not agree. Unable to sell and discouraged by poor returns, Bristow was unwilling to invest anything in further improvements. There were to be no new buildings, he wrote Pasley, unless it was something the enslaved could do at "leasure times" and at no additional cost. Although more grain was at last being produced, Bristow objected to building a structure to house it, finally grudgingly agreeing in the mid-1750s to the building of a corn barn, provided it could be erected at a cost of no more than two pounds in nails.[9]

Similarly, Bristow vetoed Pasley's suggestion of devoting one of the Gloucester County plantations to grain. He saw his main problem as one of poor management and embezzlement by overseers, not a result of a less than optimal crop mix. Neither was he willing to switch from hoe culture to plowing in order to produce more grain. In 1753, he advised Pasley to "consider well about plowing the Lands," as Francis Willis had to abandon a similar scheme. He did authorize Pasley to buy a Scot indentured servant plowman, but only if he could be procured "cheap."[10]

Bristow also squelched a scheme to establish a new quarter on the Brenttown tract in Prince William County. Pasley considered this the only plantation "capable of improvement," and he did send some enslaved workers there to start a new farm. They refused to be separated from their families, however, and promptly decamped to their original homes. Bristow could not agree to Pasley's solution of buying additional slaves to staff the new quarter. Instead, he advocated draconian measures for the existing workforce. The slaves were to be reallocated among the existing quarters, "to put them all in a method to doe Something." The most refractory were to be sold, along with any that were superannuated, as were some of the young children and any excess of "Woomen and Girls, not capable of the Servitude of men." "I

8. Bristow to Willis, Mar. 27, 1751, ibid. The Palmer hydrological drought index indicates severe water shortages in Virginia during both the 1747 and 1748 growing seasons. The index is described in David W. Stahle et al., "The Lost Colony and Jamestown Droughts," *Science,* CCLXXX (1998), 564–567 (Dennis Blanton kindly supplied the index numbers).

9. Bristow to Pasley, Mar. 24, 1753, Bristow Letterbook, CWL.

10. Ibid.

conclude they have been neglected so long," Bristow wrote in 1753, "they are become their own Masters and Mistresses, and will work when they please." "When they see the method you take with them, I shou'd hope the remainder will submit, for I am told here one with another ought at least to make 800 weight of Tobacco." Apparently, the disruption (or threats of disruption) of the slave community had the effect Bristow desired, for by 1757 he noted that the quantity (if not the quality) of the tobacco had increased.[11]

By 1755, Bristow was more sanguine; at least he was more amenable to making improvements, so long as tobacco was not neglected as a result. After losses among the slaves to bloody flux in 1750, their numbers were again increasing. The herds of cattle and hogs in Gloucester and Lancaster had risen to the maximum the land could support, and there were some sheep at the home plantation in Gloucester. Bristow directed that on the Lancaster plantation the ablest hands were to grow tobacco on the best lands, while some "near past labor" were to be set to manuring poorer ground for grain. There was some surplus corn for sale, and Bristow had hopes of marketing more wheat, especially since his farms were located on the water where the grain could be readily put on ships. In 1757, he even authorized Pasley to use six or seven hands to try a new crop, indigo, then a fashionable innovation among large Virginia planters. South Carolina rice planters had adopted this semitropical crop as a second staple in the 1740s to combat declining rice prices and to diversify their export base. The industry expanded there when supplies from the French West Indies were cut during the Seven Years' War, encouraging Virginia planters to try their hand at indigo as well. This experiment must have failed (indigo was difficult to process, and the Chesapeake climate was marginal for growing it), for nothing more was said about it. Nonetheless, Bristow was still intent on cutting corners—for example, sending Pasley a box of medicines that he was directed to administer himself in order to cut down on doctor bills for the enslaved.[12]

Pasley died in 1758, and Bristow hired Mordecai Booth, a grandson of an earlier Virginia agent for the family, as the next general manager, paying him one-sixth of the profits of all the plantations. Booth proved a better manager, for Bristow's complaints diminished, and by 1759 the estate was turning a good profit. In 1758, Booth began having Virginia cloth woven to clothe the women and children, thus reducing outlays for imported cloth. By 1759, he

11. Ibid., Mar. 24, June 1, 1753.

12. Ibid., Mar. 26, 1755, November 1757; Joyce E. Chaplin, *An Anxious Pursuit: Agricultural Innovation and Modernity in the Lower South, 1730–1815* (Chapel Hill, N.C., 1993), 187–208; Edelson, *Plantation Enterprise,* 111–113.

had a large surplus of corn for sale and had also marketed some plank. The slaves seem to have responded to a switch from threats to incentives. Bristow wrote in 1760 that he approved of Booth's "encouragement" to the enslaved, agreeing that they should be allowed new bed rugs "when they have sufficiently earned them." Booth died in 1762, however, and, without close supervision, that year's tobacco crop proved badly cured and poorly packed. From 1755, after continued prodding from his managers, Bristow did slowly adapt to changing conditions, but only partially and with grudging reluctance.[13]

The letterbook of another absentee planter, Joseph Ball (1689–1760) of Morattico (a plantation named for a local Indian group) in Lancaster County, covering the years 1744–1760, documents a conservative but much less reactionary management style than that of the Bristows. It also provides one of the more intimate descriptions of master-slave relations for the colonial period. Ball, whose great-grandfather immigrated from England in 1657, had inherited eighteen hundred acres in Lancaster and four hundred acres in King George (later Stafford) counties from his grandfather, who had entailed the family land, as well as received slaves and a newly built manor house from his father. In 1709, he wedded an English woman, Frances Ravenscroft, and thereafter made his home near London, his wife's substantial inheritance being contingent on her remaining in England. Subsequently, Ball practiced law in London but made periodic trips back to Virginia and, until his death fifty years later, revealed in his letters a minutely detailed knowledge of the workings of his four separated quarters that was seldom matched in the surviving correspondence of resident planters. Unlike the Bristows, Ball could envision his own children returning to the colony and taking up planting where Ball had left off. So he had more incentive at least to maintain, if not to increase, the value of his Virginia estate.[14]

In 1744, when the letterbook opens, Ball had just hired his nephew Joseph Chinn as general manager for his Virginia concerns. Chinn would serve his employer long and faithfully, relinquishing the post only after Raleigh Downman, Ball's son-in-law, took over the estate following Ball's death in 1760. Although Ball's letters to Chinn suggest a relationship that was usually trusting and respectful, his attitude toward his lesser overseers was often contemptuous. Ball considered overseers in general "a parcell of Slubbering

13. Bristow to Booth, Feb. 11, 1760, Bristow Letterbook, CWL.

14. Joseph Ball Letterbook, 1743–1780, LC, microfilm, M-21, CWL; Peter Walne, "The English Ancestry of Colonel William Ball of Millenbeck," *VMHB,* LXVII (1959), 399–409; Robert N. McKenney, "Morattico Plantation, Lancaster County," *Northern Neck Hist. Mag.,* XXXVIII (1988), 4318–4336.

Sons of bitches." He objected to any who did not "stay home and work," to those who stayed home but entertained too much company, and especially to married overseers whose wives were forever "taking the hands to wait upon them." He made one overseer agree "to make good all Dammages" before he would let him move into a refurbished dwelling, "for I see no reason why I should keep a fine House with sash windows for an overseer." Solicitude for his slaves made Ball suspect that, when some promising children died, Chinn had "let unmerciful overseers beat, and abuse them Inhumanly, and Break their hearts." Ball expected Chinn to effect a delicate balance; "let not the overseers abuse my People: nor let them abuse their overseers," he wrote in 1744.[15]

Tobacco remained Ball's major crop, and he expected that nothing should interfere with its careful cultivation and processing. In the mid-1740s, he made leaf oronoco at Morattico and at Little Falls quarter in King George County. At Forrest and Pig Hill, the two other Lancaster quarters, the tobacco was to be "Right Sweet Scented True Townsend Clean Stemm'd and laid Strait." Ball asserted he could make money of True Townsend, and Chinn "must take Good Care that the overseers don't Shark nor Sham, in making Some like it, in Stead of the True Thing." Ball authorized Chinn to sell part of the crop in the country, provided he could get a satisfactory price, and to consign one-half of the remainder to the Liverpool firm of Goore and Bulkeley and the other half to Ball himself in London. (Occasionally, ship captains working for London merchants would not take tobacco consigned to anyone other than the ship owners, and then Chinn had to ship the tobacco intended for London to the firms of Hanbury or Athawes.) In 1749, Ball found that stemmed tobacco was selling for less than leaf, and he ordered Chinn to stop stemming. By 1751, he was growing only enough sweet-scented to make one hogshead for himself; the rest was oronoco leaf, more of which was sold in the country. Ball decided that country sales yielded more money more quickly.[16]

Unfortunately, Ball's tobacco seemed no longer to suit the English market. As Ball protested in 1755 to Liverpool merchant Charles Goore, "When I lived in Virginia, my Tobacco was so far from being accounted the worst, that it was always very well liked, and sought after"; why was Goore now

15. Ball, Instructions for My Nephew Joseph Chinn, Feb. 18, 1743/4, Ball to Chinn, Mar. 19, July 17, 1745, June 27, 1749, Ball Letterbook, CWL.

16. Ball to Chinn, Aug. 30, 1746, ibid.; see also Ball to Goore and Bulkeley, Nov. 22, 1746, to Chinn, July 17, 1745, Mar. 16, 1746/7, Aug. 18, 1747, Nov. 9, 1749, Mar. 21, 1749/50, Aug. 9, 1751, to Charles Goore, June 13, 1751.

reexporting it to Holland as unsalable in England? Ball's son-in-law Raleigh Downman had similar problems in the early 1760s when he took control of the estate. With high wartime freight and insurance costs, it did not pay to consign Lancaster County tobacco; Downman cleared little more than four pounds sterling per hogshead in 1760 and only six to seven pounds in 1761 and 1762, when he sold the crop at the mast to tobacconists, thus avoiding merchants' commission charges. In contrast, Rappahannock planters with more favored crops were netting a hefty fifteen pounds a hogshead in these years.[17]

Corn, Ball's second most important crop, was grown primarily for plantation use; even when Ball was in residence he counted on a surplus only in bumper years. On the other hand, he did not want the overseers to tend too little, for extra grain was needed to fatten hogs and to carry the rest of the livestock over the winter. Any surplus might also be used to buy salt and molasses for the slaves. Some wheat was sown, but apparently not much, for Ball gave this crop scant consideration. Like many other planters, Ball offered his overseers no share of this grain: "It is a very pretty fancy, that they shall have a Share of Every thing that is made for me; and besides I must find them Ground to make wheat for themselves; and I don't doubt but my people must do all the work about it from End to End." If overseers wanted to eat wheat, they could swap some of their corn for it. Ball's parsimony probably resulted in part from an inability to grow much small grain, for he continued to cultivate his grounds almost entirely by hand with hoes.[18]

Ball put special emphasis on careful management of the livestock. Stocks were never to be allowed to exceed the natural carrying capacity of a plantation. Old cows, "ill thriven" cattle, and scrawny lambs in danger of dying in severe winter weather were to be killed in the fall to supply additional meat for the workforce. Ball advised Chinn that as a general rule he should give the slaves a share of most animals slaughtered; when calves or sheep were killed, for example, the slaves were to have "the Head, and Pluck, or the value of it out of the meat." They were also to have for bedding the skins of any sheep slaughtered, and cattle hides were to be processed for shoe leather.[19]

As cattle were likely to provide the greatest revenues, Ball was most specific about their management. Overseers were allowed to milk only two or three cows, and only so long as their wives, "out of Covetousness to make

17. Ball to Charles Goore, Mar. 15, 1755, Raleigh Downman to Chinn, Jan. 22, 1761, Aug. 12, 1762, Mar. 3, 1763, Downman to William Downman, Jan. 22, Aug. 12, 1762, ibid.

18. Ball to Chinn, Mar. 19, 1745, ibid.

19. Ibid., Feb. 18, 1743/4, Feb. 21, 1755.

butter," did not starve the calves; the slaves were to have the rest of the milk to manage as they saw fit. In the spring "miring time," when animals sometimes perished in swamps into which they ventured too far in search of early grass, all cattle were to be penned to prevent unnecessary losses. The secret of good stock, Ball told Chinn, was not to have too many breeders. Steers, he thought, were more easily kept and would sell better when fattened for market at ages four through seven. A pair of oxen was to be kept at each quarter to pull a cart. As for horses, Ball wanted a good bay broodmare on every plantation, a stallion at Morattico, and a horse or two kept at each quarter for riding. Colts were to be well cared for until ready for sale. In winter, all horses were to be stabled and well fed. The colts, when fit to break, and any horses past their prime were to be sold, for "I don't want to feed a Gang of useless Horses."[20]

In the 1740s, pigs were kept primarily for plantation meat. Each winter, all "Killable Hogs" were slaughtered for the overseers' pork allowance, and a ration of sixty pounds per year of "fat Backs, Necks, and other coarse pieces" was allotted for each adult slave and thirty pounds for each child along with all the hog fat. The rest of the desirable pork was made into "good Bacon" to feed any white workers who might be hired, and the surplus was sold. Only a small breeding stock was to be retained on each quarter, consisting of just a few hogs including two brood sows, a young one "coming on," a boar "fit for business," and another young boar. When Ball decided in 1756 that pork was selling better than beef, he directed Chinn to grow more corn and to "Endeavor after a stock of Hogs."[21]

Ball was less enthusiastic about sheep. He was willing to maintain a flock on the home plantation for wool and meat but would not allow them to be kept on the quarters—when poorly fenced and inadequately supervised, the damage they might do to field crops and orchards far outweighed their usefulness. Even on the home plantation Ball saw the main benefit of sheep as a cheap source of fresh meat for enslaved workers and overseers. If a scrawny lamb or old ewe that no one else wanted could keep an ailing slave alive through cold weather, so much the better. If necessary to keep the flock size down, some sheep might be sold, but Ball seems to have preferred that the plantation family eat them. Ball began trying to improve the stock by keeping only the best ewe lambs for breeding and castrating all but one or two rams. By 1750, he was adding new blood, ordering ten white ewes and a new ram from the Eastern Shore.

20. Ibid., Feb. 18, 1743/4, Nov. 13, 1746, Apr. 11, 1748, Dec. 3, 1759.
21. Ibid., Feb. 18, 1743/4, Oct. 22, 1756.

Careful livestock management was essential because Ball relied for the animals' food only on a combination of natural pasture and corn and corn fodder. Ball did have fenced pastures at each quarter that included woodland where the cattle and hogs could forage and streams where the cattle could drink and the hogs wallow. In 1744, Ball agreed that one small field might be cut for hay, "if it can be made without prejudice to the [tobacco] crop, else not." Regular haymaking was clearly not part of the seasonal routine.[22]

Orchards were another important part of the operation. Young trees were to be well fenced, with special rail arrangements that would admit shoats and calves who would keep down the weeds while excluding older stock likely to damage the trees. Hogs were fattened on fallen fruit, and sheep grazed in the home orchard. Chinn made as much cider as he could without neglecting the tobacco. After measuring the overseer's share and saving some for the slaves, he had the rest distilled into brandy and sold.

Recognizing that resources were limited on an estate that was by Virginia standards quite small, Ball sought to conserve soil and timber. Depleted tobacco ground at Morattico was, not to be further worn out with corn or wheat, but was to be harrowed smooth, sown in rye, and then left fallow until, with cow penning, it could again be used for tobacco. Because herd size was closely controlled, all animal manure was reserved for the tobacco fields; discarded tobacco stalks were used to dress the orchards. Similarly, Ball wanted no deciduous trees to be cut for anything but board timber required for plantation buildings, except trees cleared for new fields. Fence rails were to be made from the tops of trees cut for timber, from trees that blew down, or from pine. Although not an aggressive improver, Ball was a prudent manager.

Ball was more willing than some of his contemporaries to build the value of his estate by making structural improvements and by keeping any existing buildings worth maintaining in good shape. The great house at Morattico remained boarded up and unoccupied in Ball's absence, but Chinn kept it in minimal repair. On the home plantation in 1744, there was also a kitchen, a covered well, slave cabins, and an overseer's dwelling; houses for cider, corn, hens, carts, prizing, and tobacco; a fenced pasture and several orchards; and Ball was building a shed to shelter small boats. At each quarter, there were slave cabins and an overseer's house; corn, cart, and tobacco houses; fenced pastures and orchards; and at the Forrest there was a stripping house where the slaves stemmed the sweet-scented tobacco.

When Ball built, he built well, hiring trained carpenters to put up new

22. Ibid., Feb. 18, 1743/4.

buildings and directing Chinn to see that they were made of good wood and properly constructed. Fences, gates, and a bridge were to be kept in repair, and Chinn was to supervise the projects himself—if left to the overseer's discretion, a new fence "won't be worth a farthing when it is done." Where new materials were required for repairs, Ball wanted only the best: white oak, chestnut, locust, or cedar for sills and brick or stone to underpin houses when rotting posts in the ground had to be replaced. Conveniences were not slighted either: "shades" were added to buildings to shelter busy entrances from inclement weather, and, when new doors were needed, Chinn was to make certain they were wide enough to serve their intended purpose. Roofs were to be made as tight as possible and coated with tar to stop leaks. On the other hand, Ball conserved materials whenever he could. Used bricks from remodeled structures at the home farm were used to make fireproof chimneys in buildings on outlying quarters or to provide insulation in the walls of the slave cabins. Sound plank was salvaged from decaying outbuildings for cabin lofts. Buildings not worth fixing were burned down to recover reusable nails.[23]

The Ball correspondence is most explicit about details of slave management. A classic patriarch, Ball prided himself on being a "good master," continually reiterating that his people were to be (by the standards of the time) well housed, well clothed, well warmed ("a Good Fire is the Life of a Negro"), well fed, and well doctored. He was astute enough to realize that local whites often took advantage of enslaved blacks and that he must include in his instructions details such as the length of blankets to be distributed (four and a half yards of cloth each), the construction of suits ("not too scanty nor bobtail'd"), the time clothing was to be distributed ("not for the Winter to be half over before they get their Winter Cloths, and the Summer to be half over before they get their Summer Cloths; as the Common Virginia Fashion is"), and the size and quality of slave dwellings. He regularly supplied Chinn with English medicines and sent detailed instructions on how to use both these and local herbs. He expected Chinn to do most of the doctoring, not in order to save money as Robert Bristow did, but because he thought a concerned manager was more likely to cure sick slaves than were local physicians. "Breeding Wenches," on the other hand, were always to have the assistance of a midwife, and they were to be given old linen to tear up to make baby clothes. Not surprisingly, a visit to France convinced Ball that the lives of French peasants "are worse than that of Negroes that have a Good Master."[24]

23. Ibid., Feb. 18, 1743/4, July 17, 1745, Nov. 13, 1746, Oct. 7, 1758.

24. Ibid., Feb. 18, 1743/4, July 17, 1745, June 30, 1749, Jan. 18, 1755. For a similar observation

Ball insisted, too, that the enslaved workers have a certain degree of autonomy. They were to be allowed to manage their rations of meat and milk as they saw fit. When they needed corn, the overseer was to give them the corn house key and let them get grain for themselves. Cabins were equipped with locks and keys, so families could feel more secure in their possessions. Preferences about which of the four plantations they might live on were to be respected whenever possible, and favored slaves were allowed to ride the least valuable plantation horses to go visiting.

On the other hand, Ball realized so much liberty might encourage the slaves to make off with more than he intended if Chinn did not keep careful watch. He ordered Chinn, for example, to put tar in a pot of lard saved to grease cartwheels to prevent the slaves from eating it. As much cider was to be distributed as "will do them good; but none for Feasts; unless it should be for a Burying, or wedding; and then but sparingly." The overseers were to see that the slaves returned the keys of the quarter corn houses once they had taken out their rations, "for else they will sell it, and starve themselves; as they did when I was in England before."[25]

Although long separated from all but a few house servants whom he took with him to England, Ball continued to remember individual slaves left in Virginia. Yearly he shipped back a parcel of his old clothes refitted for various favorite male slaves, often enclosing additional presents of pocket knives, tobacco boxes, needles, and thread. Mrs. Ball occasionally included old bedding and rags that could be used for baby clothes. The family even saved the stubs of their used clay tobacco pipes and shipped them to Morattico, where Chinn was to distribute them in the quarters.

Ball made explicit his concern for an individual slave when in 1754 he sent Aron, a young Virginia-born waiting man (to whom he had given the surname Jameson), from England to Morattico. Aron arrived equipped with a generous assortment of bedding, clothing, household goods, and even a violin with spare strings. Aron was to be used kindly the first year, and, although he was to learn to work at the hoe, he was not to be kept at it constantly, as he was not used to hard labor. He was to have a framed house twelve feet by ten feet, the privilege of raising fowl, and the use of a horse to ride to church, "or where else you shall think it to be Convenient, and not to be too nice about

about the comparative lots of Chesapeake slaves and European peasants, see *Byrd Correspondence*, II, 464. For the rationalization of the treatment of slaves using an ideology of "moderate" government, see Alexander B. Haskell, "'The Affections of the People': Ideology and the Politics of State Building in Colonial Virginia, 1607–1754" (Ph.D. diss., Johns Hopkins University, 2005), 339–350.

25. Ball to Chinn, Feb. 18, 1743/4, Ball Letterbook, CWL.

it," so long as this did not include going "Stroling on nights." "With Prudent Management," Ball concluded, "he will be a very good hand."[26]

In a subsequent letter to Chinn, Ball revealed the other, darker side of "Prudent Management":

> Aron I believe will be too Saucy. If he offer's to be unruly or strike his overseer, he must be tyed up and Slasht Severely, and pickled. I suppose if you go to do it he will Resist you; for he Resisted me twice: If you have occasion to Slash him get two or three Good white men to help; and give it him heartily at once. And if he Should Runaway, he must wear a Pot hook about his Neck. And if that won't bring him under, he must wear Iron Spancels upon his Legs till you are pretty sure he will be orderly; for as he is my Slave he must and Shall be obedient.

Ball wrote Aron a similar letter (although omitting the lines about getting additional whites to subdue him), stressing the necessity of obedience and submission, promising a reward of "my best old Cloths; and other things" for proper progress, and closing with, "If you will be good I shall be yr Loving Master."[27]

Embracing Change: The Rappahannock Elite

Plantation accounts for six other resident Rappahannock basin planters generated between 1737 and 1763 give a better idea of actual levels of production and gross revenues earned than do the letterbooks of the absentee planters. All six were native-born Virginians of the third generation in the colonies and all but one benefited from the substantial amounts of land, enslaved workers, and political influence that they inherited or were given when they came of age. All faced the challenge, not of building an estate from scratch, but of maintaining their family's economic, social, and political status and of expanding agricultural production to generate enough revenues to enable them to endow their children with fortunes equal to those they had received. Some of them might have tried to adopt recent innovations in British agriculture that they had learned about while studying in England or had read about in British farming manuals that they acquired after returning to Virginia. If so, they made no mention of it in existing correspondence, and no listings of their libraries have survived. On the whole, the changes these planters implemented seem to have evolved from local experience rather than been borrowed from imported models.

26. Ibid., Apr. 23, 1754.

27. Ibid., Aug. 31, 1754; Ball to Aron, Aug. 31, 1754.

Members of the Berkeley family of Barn Elms, Middlesex County, were particularly preoccupied with keeping family lands and enslaved workers within the male line and with maintaining the family's fortune and status across generations. Edmund Berkeley III (1704–1767) and his eldest son Edmund (1730–1802) inherited a considerable estate from the elder Berkeley's father, Edmund II (circa 1665–1718), a member of the Virginia Council. Indeed, Edmund III was a member of the third (and perhaps fourth) generation of a family who had invested in the Virginia colony since its inception. His grandfather Edmund (?–by 1674) married well, enabling him to establish several plantations in Middlesex and possibly Gloucester counties. His father, Edmund II, inherited land from both parents and married into the powerful Burwell family, acquiring additional land and enslaved workers through his wife.

In the early 1700s, Edmund Berkeley II had begun purchasing further acreage in Gloucester, King and Queen, Caroline, and King William counties and more slaves to work it. Unlike many similarly situated gentry planters, Edmund II acquired no speculative grants but instead intended that all the plantations he bought remain in the family, ensuring that they did so by entailing the land on his sons and their male heirs. He moved to Middlesex County about 1712 when the heir to his stepfather's plantation came of age; by the time of his death six years later, he was building a fashionable new brick house, which he directed be left unfinished until his oldest son and namesake came of age. Conservative in his finances, despite substantial purchases of land and slaves, Edmund II maintained a positive balance of fifteen hundred pounds with London merchant Micajah Perry, his principal factor. He intended that this money, along with profits made by the enslaved workers during Edmund III's minority, go to endow his daughters, and all the working slaves and the land were to go to his two sons.[28]

Thus, Edmund III, who was probably educated in England, in 1725 at age twenty-one inherited all the family plantations in Middlesex County, nearly four thousand acres in King William County, and an unknown number of slaves, more than enough resources to maintain a privileged economic, social, and political status. Although he never became a member of the council, Edmund III did attain the offices of sheriff, militia lieutenant, and burgess.

28. Francis Lewis Berkeley, Jr., "The Berkeleys of Barn Elms, Planters of Colonial Virginia and a Calendar of the Berkeley Papers, 1653–1767" (master's thesis, University of Virginia, 1940); Berkeley Family Papers, MS 38–113, microfilm, Records of Ante Bellum Southern Plantations, Series E, pt. 2, UVa; "Edmund Berkeley (ca. 1671–by 3 March 1719)," and "Edmund Berkeley (5 December 1730–8 July 1802)," in John T. Kneebone et al., eds., *Dictionary of Virginia Biography* (Richmond, Va., 1998), I, 448–450.

He also secured a grant for six thousand additional acres in Hanover County before his marriage to Mary, daughter of planter-merchant Thomas Nelson of Yorktown, and in 1745 inherited the rest of the family's land upon his brother Lewis's death.

Edmund III, like his father, concentrated on raising high-quality sweet-scented tobacco for the British home market. He continued to ship tobacco to his father's factors, the Perrys, until that firm foundered in the 1740s. Subsequently, he sent consignments to other London firms—Robert Cary, Edward Athawes, Osgood and Capel Hanbury, Flowerdew and Hatley, and Bosworth and Griffith—to Liverpool merchants James Gildart and Charles Goore, and at least once made a trial of the Dublin market. Generally, Berkeley had about a third of the crop stemmed and marketed the rest as leaf. This combination addressed the varying tastes of the different ports and gave Berkeley flexibility in packaging his tobacco in a form most likely to attract buyers.

By 1757, when his oldest son, Edmund IV, became engaged to Judith Randolph, a member of another wealthy and politically prominent Virginia family, the father was in a position to settle on the couple four plantations and forty slaves, more than half of them adult workers, and to build them a new house. In addition, upon Edmund III's death the couple was to receive forty more slaves, again half of them laboring adults. Following the same estate-retaining strategies as his father, Edmund III intended that his three daughters have cash legacies of one thousand pounds sterling and one enslaved girl apiece. He gave his younger son, Nelson, a three-thousand-acre plantation in Hanover County along with slaves to work it and built for him a second new brick house with flanking dependencies. His wife, Mary, was to have the use of the Middlesex plantations and dower slaves for her life. Afterward, all the core family lands and enslaved workers were to go to the two sons. These substantial legacies, along with a spate of poor-quality tobacco, forced Berkeley in 1763 to borrow money from some of his factors, naming his father-in-law, merchant Thomas Nelson, as his security. For the last three years of his life, Edmund III, then sixty, managed only the plantations in Middlesex, turning the direction of the remaining family lands over to his sons.

Berkeley employed five or six overseers to manage the enslaved workers, paying those supervising smaller farms one share of the tobacco and corn crops and those managing larger plantations, where there were as many as twenty to twenty-eight field hands, two to two and a half shares of the major crops. The enslaved workers made on average more than nine hundred pounds of tobacco a year, a higher level of output than most other period planters achieved. Better-than-average land and use of manure may account for some of the higher productivity. Altogether, Berkeley was marketing more than one hundred thousand pounds of tobacco per year, which produced an annual income of roughly eight hundred pounds sterling constant value; tobacco earnings per worker averaged more than eight pounds, a sum that few other great planters were able to realize in this period (Table 28).

Penning cattle on tobacco lands raised outputs of both tobacco and of corn, which was subsequently grown on the manured plots. Corn outputs per hand on the larger farms sometimes reached thirty to forty barrels in favorable years in the early 1740s. The level of corn production apparently dropped in the 1750s and early 1760s (information about the size of crops is sparse), but corn continued to contribute three-quarters of the gross revenues generated by tobacco. Berkeley also raised small amounts of wheat, most of which he sold. Cattle and hogs were kept on all the farms, and sheep on some of them, with surplus livestock providing further revenues, as did a shad fishery in the spring. Edmund III kept the slaves working full time in the fields, employing whites to make shoes for them from local leather and clothes from imported cloth. Indentured servant women continued to make up part of the domestic labor force on the home plantation. Five enslaved men did work as carpenters, but Berkeley still hired white artisans to make the tobacco casks, to build houses, tobacco barns, and small boats, and to do blacksmithing. Berkeley's main aim was clearly to maximize revenues from two staple crops. In the 1740s and early 1750s, this strategy paid off exceedingly well. The provision of exceptionally generous portions to five children as they came of age or married, however, strained even this wealthy family's resources.

John Baylor (1705–1772) of Caroline County, another third-generation Virginian who had been educated in England, left more detailed accounts of his farming operations on the nearly twenty thousand acres he owned in Caroline and King and Queen counties. Some farms (most were entailed) he inherited from his father John (circa 1660–1720), who had been a prominent planter, land speculator, merchant, and slave dealer, and the rest Baylor purchased or patented on speculation. Like his father, Baylor quickly rose to high political office, becoming a vestryman, justice of the peace, a lieutenant in the Virginia militia, and a burgess. The elder Baylor had been in a good position to purchase new African slaves from the cargoes whose sales he managed, and by 1742, two years before he married, John already owned seventy-nine enslaved laborers of tithable age and an additional sixty-two children. Thirty-five slaves lived on Newmarket, Baylor's home plantation, another twenty-six lived on six thousand acres "at the mountains" in Orange

TABLE 28. Tobacco and Corn Crops and Revenues per Laborer, Berkeley, Baylor, Beverley, and Mercer Plantations, Rappahannock and Potomac Basins, Virginia, 1737–1763

Year	Tobacco (lbs.)	Tobacco revenue (£)	Corn (barrels)	Corn revenue (£)	Total revenue (£)
BERKELEY					
1741	969	7.67	10.0	2.92	10.59
1742	1,204	8.85	14.0	5.31	14.16
1743	1,256	8.60	14.0	4.42	13.02
1744	1,062	8.81	12.6	3.23	12.04
1745	996	8.84	39.6	12.65	21.49
1746	1,019	8.75	28.8	8.90	17.65
1747	626	4.38	—	—	[4.38]
1748	1,049	7.86	13.9	6.53	14.39
1749	1,007	9.82	—	—	[9.82]
1752	952	7.19	—	—	[7.19]
1754	769	5.92	—	—	[5.92]
1755	964	7.72	—	—	[7.72]
1756	1,327	10.70	—	—	[10.70]
1758	1,262	22.81	12.0	4.40	27.21
1760	561	5.65	—	—	[5.65]
1761	671	6.82	—	—	[6.82]
1763	1,138	10.27	15.0	6.85	17.12
Mean	990	8.86	17.8	6.13	16.41
BAYLOR					
1737	800	7.59	10.0	4.55	12.14
1738	705	6.58	11.1	3.68	10.26
1739	674	6.40	9.8	4.10	10.50
1740	705	6.87	10.4	2.98	9.85
1741	720	6.10	10.7	3.63	9.73
1742	903	7.18	9.8	3.10	10.28
1743	678	5.39	6.7	2.33	7.72
1744	924	7.68	8.0	2.49	10.17
1745	757	6.23	—	—	[6.23]
1746	763	6.19	—	—	[6.19]
Mean	763	6.62	9.6	3.36	10.08

TABLE 28. Continued

Year	Tobacco (lbs.)	Tobacco revenue (£)	Corn (barrels)	Corn revenue (£)	Total revenue (£)
BEVERLEY					
1742	785	6.07	—	—	[6.07]
1743	739	5.80	15.7	4.14	9.94
1744	628	5.21	12.0	3.11	8.32
1745	673	7.17	10.0	2.66	9.83
1746	754	6.99	—	—	[6.99]
1748	685	5.43	—	—	[5.43]
Mean	711	6.11	12.6	3.30	9.36
MERCER					
1746	881	4.54	5.5	1.70	6.24
1747	895	5.15	6.0	1.39	6.54
1748	781	3.72	8.8	2.51	6.23
1749	1,004	7.66	—	—	[7.66]
1750	850	5.10	—	—	[5.10]
Mean	882	5.23	6.8	1.87	6.34

Note: Currency values are in constant pounds sterling. Values were calculated using prices appearing in the account books or from a regional price series. Mean annual tobacco and corn crops for Berkeley, Baylor, and Beverley include data from plantations in Spotsylvania and Hanover counties not reported in Appendixes 1 and 2. Total revenue in brackets is incomplete.

Sources: John Baylor Account Books, 1717–1749, UVa, microfilm, M-62.5, M-62.6, CWL; Berkeley Family [Plantation and Estate Account Book, 1745–1767] and [Plantation Account Book, 1734–1758], Berkeley Family Papers, MS 38-113, microfilm, Records of Ante Bellum Southern Plantations, Series E, pt. 2, UVa, M-1816.18, CWL; William Beverley, Accounts with Tenants, 1742–1743, Account Book of Estate of Henry Beverley, 1743–1744, Beverley Family Papers, 1654–1901, NYPL, microfilm, M-1334, CWL; John Mercer Ledger, [1741]–1750, Bucks County Historical Society, Doylestown, Pa., microfilm, M-164, CWL.

County, and the others were scattered among six quarters in Caroline and King William counties. Although the elder Baylor had been a major wheeler and dealer in the Rappahannock and upper York basins, he had not been the best of financial managers. His estate amounted to a hefty sixty-five hundred pounds but was heavily "envolv'd" with large outstanding debts from the sale of slaves due to Bristol merchants.[29]

Baylor retained a general manager who lived on the home plantation and supervised the work of four overseers for a salary of forty pounds Virginia currency per year. Enslaved men ran five smaller quarters, where they directed the work of between five and ten other adult slaves, avoiding the expense of hiring more overseers. Like Edmund Berkeley, Baylor concentrated on raising tobacco and corn. Between 1737 and 1746, the field hands made about 750 pounds of tobacco and ten barrels of corn apiece, enough grain to ensure some surplus for sale. Gross revenues from tobacco were about six and a half pounds sterling constant value a hand, and corn contributed just over three pounds. If any wheat was grown, it was not mentioned in the plantation accounts. Baylor generally ordered the finer tobacco packed as whole leaves and had the inferior tobacco stemmed. The proportions shipped as leaf or stemmed depended on the quality and quantity of the crop and on the varying preferences of the British merchants to whom he consigned his crops. (London merchants wanted stemmed, Bristol merchants leaf, and Liverpool merchants a combination of the two.) Further income came from horse breeding, a plantation mill, a blacksmith's shop that served a wide neighborhood clientele, a few small rented tracts, and a store where Baylor sold neighboring planters imported merchandise in exchange for tobacco.

Not satisfied with this level of profit, in 1749 Baylor decided to expand his operations. He had a great quantity of fresh lands in Orange County, which he thought "most Prudent to settle . . . as fast as I can." He bought ten additional Virginia-born slaves with the intention of raising his tobacco output to more than one hundred hogsheads a year and, moreover, hoped "to establish a Character [on] my Tobacco." In 1758, he was temporarily sidetracked by a trial of indigo, but he dropped that experiment after only one year, explaining

29. John Baylor Ledgers, 1719–1755, Baylor Family Papers, 1653–1915, MS 2257, UVa, microfilm, M-62.5, M-62.6, CWL; John Baylor Letterbooks, 1749–1753, 1757–1763, MS 22456, VSL, photocopy, CWL; "The Baylor Family," *VMHB,* VI (1898–1899), 197–199, 307–309, XXV (1917), 315–323; "Will of John Baylor of New Market," *VMHB,* XXIV (1916), 367–373; "John Baylor," in Kneebone et al., eds., *Dictionary of Virginia Biography,* I, 406–407; Walter Minchinton, ed., "The Virginia Letters of Isaac Hobhouse, Merchant of Bristol," *VMHB,* LXVI (1958), 278–301; Louis B. Wright, ed., *Letters of Robert Carter, 1720–1727: The Commercial Interests of a Virginia Gentleman* (San Marino, Calif., 1940), 53–54.

that low prices had dashed his "sanguine expectations" for the alternative crop, while rising tobacco prices prompted him to return to that which he knew best.[30]

No accounts survive for the years after 1749 to indicate the extent to which Baylor might have managed to increase production on his plantations. He was likely more successful in playing British merchants off one another than in producing quality tobacco. His purchase of additional slaves in 1749 along with his building a new house, establishing a stable of racehorses, and sending at least two of his sons to England for their educations soon got him into debt.

Baylor habitually split his consignments between several London, Bristol, and Liverpool factors. He was furious when in 1750 London merchant Edward Athawes, who was himself teetering on the brink of bankruptcy, refused to pay bills of exchange that Baylor had drawn for a greater sum than the tobacco he consigned was likely to fetch. Baylor was unwilling to forgo shipping some tobacco to London because he wanted fashionable London goods. Leaving the outstanding debt to Athawes unpaid, Baylor began instead sending tobacco to two other London firms, Flowerdew and Norton and Robert Cary, to both of whom he soon also became indebted. In addition, Baylor sent consignments to the Bristol firms of Lyde, Cooper, and Knox and to John Backhouse in Liverpool. From them he received primarily coarse goods designed for the slaves. Even more crucial to Baylor, Lyde and Backhouse continued to allow him to draw bills of exchange that exceeded the value of the tobacco he sent, crops that commanded no more than ordinary returns since they were often of uneven quality. These merchants' generosity enabled Baylor to continue buying pedigreed racehorses on credit. Baylor made fulsome promises to all of his factors that he would ship sufficient quantities of tobacco to pay outstanding balances, but in fact he shipped each one just enough to keep the credit flowing. In 1762, his tobacco proved so bad that he fired his steward and several of the overseers. With the return of peace in the following year, Baylor decided to send even more slaves to further expand production on the mountain quarters. He was then sufficiently sanguine about his financial prospects to import additional expensive English horses, confident that by one means or another he could continue to persuade or manipulate his factors into financing his extravagant way of life.

Records of the third generation of the Beverley family document the operation of several family estates in Caroline and Spotsylvania counties

30. Baylor to John Backhouse, [1749], to Flowerdew and Norton, July 18, 1759, Baylor Letterbook, CWL.

and of a piedmont quarter in Orange County between 1737 and 1749; they also record the strategies employed on other family lands in Essex and York counties. For this family, land speculation and rental properties provided important supplements to agricultural revenues. The estate belonged to Harry Beverley (son of Robert, the original settler), who died in 1730, and to Harry's son Robert, who died just three years later, shortly after patenting twenty-four thousand acres in the piedmont. The younger Beverley arranged that his estate be kept intact to pay his and his father's debts and legacies and to provide support for his widow, three unmarried sisters who were living with them, and an infant son. His cousin William (1696–1756), a son of Harry's brother Robert, the historian, was the executor. William's task was further complicated by the need to dock entails on some of Robert's land, which he wanted to sell in order to purchase more slaves for the estate and to pay the debts. William kept very garbled estate records but also left a more informative letterbook covering the same years.[31]

Both sweet-scented and oronoco tobacco was grown on several of the farms. Whereas the Beverleys ordered the best sorts stemmed and shipped them to London or Liverpool factors on consignment, they left the more ordinary hogsheads in leaf to market in the country. William Beverley wished, for example, that higher-grade tobacco made on a quarter in York County be sold separately "that it may obtain a good name and become as famous as Mr. [Lewis] Burwells tobacco." Hoping to obtain the prices in England he thought he deserved, or that some of his neighbors received, Beverley sent trial consignments to a number of merchants—the Perrys, William Black, Charles Smith, and Neil Buchanan in London, James Crosbie and Thomas Backhouse in Liverpool, Charles Scandret in Bristol (the tobacco not suiting

31. Accounts with Tenants, 1737–1748, 1742–1743, Account Book of Estate of Henry Beverley, 1743–1744, Sales of Lands in Beverley Manor, 1737–1740, Account Book, 1742, William Beverley Rents, 1748, and William Beverley Letterbook, 1737–1744/5, all in Beverley Family Papers, 1654–1901, NYPL, microfilm, M-33, M-1334, CWL; Beverley Manor Accounts, 1738–1745 (MSS1B4678a 4783), Account Book, 1750 (MSS1B4678a 4781), Account Book, 1747 (MSS1B4678a 4782), Rents, 1745 (MSS1B4678 b11), Rents, 1750 (MSS1B4678a 4777), Rents, 1769 (MSS1B4678a 4776), all in Beverley Family Papers, VHS; John M. Weisner, ed., "Account Book of William Beverley, 1752," *Magazine of Virginia Genealogy,* XLI (2003), 5–17; W. G. Stannard, "Major Robert Beverley and His Descendants," *VMHB,* II (1894–1895), 405–413, III (1895–1896), 47–52, 169–176, 261–271, 383–392, XX (1912), 213–214, 332–333, 437–438, XXI (1913), 97–102, 212–214, XXII (1914), 297–301, which traces the family and reprints wills; "Harry Beverley," and "William Beverley (ca. 1696–28 February 1756)," in Kneebone et al., eds., *Dictionary of Virginia Biography,* I, 467–468, 477–478. See also "Diary of William Beverley of Blandfield during a Visit to England, 1750," and "Robert Beverley, the Historian of Virginia," *VMHB,* XXXVI (1928), 27–35, 333–344; and Charles E. Kemper, ed., "The Early Westward Movement of Virginia, 1722–1734," *VMHB,* XIII (1905–1906), 113–138.

the Bristol market, that correspondence was discontinued), and to Tubman and Hartley in Whitehaven.[32]

By 1740, selling locally became increasingly attractive for all grades of the weed. English merchants were failing to send timely accounts, whereas, when Beverley sold in the country, "I should then know what I had to depend on." Often, too, English merchants sent shoddy merchandise or failed to make proper allowance for "damnified" goods; one such shipment in 1740, Beverley protested, had filled his house with cockroaches. In addition, labor expended in carefully preparing the crop for home consumption was not always repaid. Stemming, Beverley noted, reduced the weight of the tobacco by 30 percent, but he could not count on getting 30 percent more in England for stemmed as opposed to leaf. And neatly arranged hogsheads sometimes brought no more than "common . . . Tobo. packed at the warehouse in a most slovenly way." Perhaps the final straw came in 1743 when merchant Neil Buchanan remitted less for tobacco consigned to him in London than Buchanan's own factor had earlier offered Beverley in Virginia. William Beverley indeed needed to get good prices for the estate crops, for the family had an abundance of land and too few enslaved laborers to work it. In 1743, William, following the directions of his brother Robert's will, purchased twenty-two new African slaves (twelve men, seven women, and three girls) from slave dealer George Braxton at a cost of £457 sterling. These bondpeople were to go to Robert's son. Consequently, William had to continue consigning tobacco to England, even when he might have done better by selling in the country, to cover the bills of exchange he had drawn to pay for the slaves.[33]

Between 1742 and 1748, enslaved workers on the largest Beverley quarters made about ten thousand pounds of tobacco, 350 barrels of corn, and 150 bushels of wheat a year. Small amounts of peas and beans were also grown. Surplus corn was traded to Barbados and wheat to Madeira in return for salt, sugar, rum, and wine. Yields per hand were about seven hundred pounds of the weed along with 12 barrels of corn and 5 bushels of wheat. Gross revenues per laborer were thus about six pounds sterling constant value from tobacco and more than three pounds from the two grains. Sales of surplus cattle, hogs, and butter contributed additional income.

Once balances in England declined as less tobacco was shipped on consignment, the Beverleys turned more seriously to import-replacement strategies. Much of the cloth needed for their enslaved workers was purchased

32. William Beverley to Micajah and Richard Perry, July 12, 1737, Beverley Letterbook, Beverley Family Papers, CWL.

33. Ibid., July 11, 1738, Oct. 20, 1740.

from country merchants, and local women were hired to make up finished garments. Shoes, too, were locally made by white craftsmen. The slaves, however, were still kept working primarily in the fields. They did not make clothes or shoes, and whites were hired, too, to make all the tobacco casks and to put up new buildings.

The family, characteristically, did not depend solely on farming for their fortune. Harry Beverley rented out houses and lots in the town of Urbana, and William owned lots in Port Royal, Falmouth, and Fredericksburg. William peddled hundreds of acres of western lands in "the Manor of Beverley," a grant originally of more than one hundred thousand acres in present Augusta County, and also rented land in Caroline, Culpepper, Essex, King and Queen, and King George counties to tenants. By 1748, William was collecting rent tobacco worth more than four hundred pounds sterling from 119 tenants who held about one hundred acres each along with an additional annual assessment of four chickens or two turkeys apiece. For the first three years, Beverley charged low rents of 44 to 88 pounds of tobacco per year while new tenants were getting established. Thereafter, rents ranged from 365 to 574 pounds per year for most tracts, an amount that was not increased before the Revolution. To prevent clearing the tenements too rapidly, tenants who employed more than three tithables were charged an additional 100 pounds of tobacco a year for each extra laborer.[34]

In the 1740s, Beverley also built a water mill and a windmill for the custom trade that produced steady annual rents, as did a smith shop and a tavern at the Caroline County courthouse. Attempts to mine iron, copper, and silver and a trial shipment of furs, in contrast, seem to have met with little success. Beverley aspired, too, to become a factor for Barbadian merchants, hoping to join with his son-in-law James Mills, who had a store at Hobbs Hole, in trading African slaves, sugar, molasses, rum, and salt for corn, lumber, and cash. Beverley succeeded in soliciting only one cargo of salt from the Barbadians, however, in a year when corn unfortunately proved difficult to procure on the Rappahannock. Public office also brought in other income. William was clerk of Essex County between 1717 and 1745, a fee-paying office, and from 1751 he served on the council, a position providing other avenues for advancing the family fortunes. A combination of tobacco produced in quantity along with land speculation and rents were the core management strategies for this land-abundant family.

34. Account Book, 1750 (MSS1B4678a 4781), Beverley Family Papers, VHS; John M. Weisner, trans., "Account Book of William Beverley, 1752," *Magazine of Virginia Genealogy,* XLI (2003), 5–17.

Richard Corbin (1714–1790), another elite middle Peninsula planter and second-generation native-born Virginian, through prudent management became one of the largest and most solvent tobacco growers in the region. Rather than getting an education in England, he probably went to the College of William and Mary in the late 1720s. In 1737, he married into the prominent Tayloe family and commenced planting family lands in Middlesex County. His wife's dowry of two thousand pounds sterling gave the couple an ample head start along with whatever marriage settlement Corbin received from his father, Gawen. Upon Gawen's death in 1744, Richard moved to his parents' home plantation, Lanesville, in King and Queen County. Corbin also operated quarters on older family lands and on acreage he purchased in Middlesex County, as well as on other holdings in Caroline, and had inherited a reserve of six thousand acres farther west in Spotsylvania County. He divided a large inherited enslaved workforce, which had grown to at least 213 laboring adults by 1761, among fourteen quarters. The smallest farm had only 5 adult laborers. The biggest, Corbin's home plantation, housed 29 workers. There were seven other major quarters on each of which 14 or more field hands and a total population of about 30 slaves lived and worked. Nine overseers, who were paid in shares of the crop, and a general manager, paid an annual salary of fifty pounds Virginia currency, supplied supervision.[35]

Like other large Rappahannock River planters, Corbin raised tobacco and corn on all the farms and wheat and peas on at least some of them. In 1760 and 1761, tobacco production per laborer varied from a low of about 470 pounds to a high of more than 900 pounds; the mean was 667 pounds. (Since most of the tobacco was stemmed, this amount is equivalent to about 800 pounds of unprocessed tobacco.) Enslaved workers on the smaller quarters grew somewhat more tobacco and somewhat less corn than did those assigned to bigger quarters. Corbin might well have adjusted crop acreage to landed resources, concentrating on tobacco on those farms with the most productive tobacco soils and growing more grain on farms with a higher proportion of soils less suited to tobacco. He might also have been taking advantage of economies

35. For the Corbin family, see Darrett B. Rutman and Anita H. Rutman, *A Place in Time: Middlesex County, Virginia, 1650–1750* (New York, 1984), 266–267 n. 47; and "Richard Corbin," in Sara B. Bearss et al., eds., *Dictionary of Virginia Biography,* 3 vols. (Richmond, Va., 1998–2006), III, 466–468. Primary sources include Richard Corbin Letterbook, 1758–1768, Corbin Tobacco Book, 1746–1795, and Corbin Family Loose Papers, microfilm, M-55.1, M-55.2, CWL; and Edmund Wilcox Ledger, 1760–1776 (vol. 3), Wilcox Farm Journal, 1761–1762 (vol. 4), and Wilcox Account Book, 1761–1764 (vol. 6), in the E. W. Hubard Papers, 1741–1907, University of North Carolina, Chapel Hill, microfilm, M-24.1 and M-24.2, CWL. Wilcox served Corbin as general manager between 1761 and 1764.

of scale on newer quarters where there was more likely to be enough fresh acreage to keep a large number of hands employed in both tobacco and corn and where sufficient food for working livestock could easily be raised.

Corbin was a big producer, indeed, in the late 1740s shipping between 85,000 and 119,000 pounds of tobacco a year, and in 1760 he marketed a record high of 156,900 pounds. On average, he was consigning more than 150 hogsheads per year. Corbin had the enslaved workers stem most of the tobacco, although each year he also marketed a few hogsheads of choice leaf. His crops were probably of varying quality, and each year he split his consignments between several London and Liverpool merchants. An early trial of the Bristol market with the firm of Cooper and Lyde proved unpromising, and thereafter he sent his crops to the London firms of Edward Athawes, Capel and Osgood Hanbury, Robert Cary, Humphrey Bell, James Buchanan, and James Russell. In the late 1740s, Corbin sent nearly half of his tobacco to Liverpool merchants Charles Goore and William Bulkeley, John Backhouse, and Thomas Rumbold and John Walker, but by the 1750s he was again shipping three-quarters or more to London factors. Only in 1749 and again in 1759 did high prices offered by local merchants prompt him to sell some or all of his crop in the country. The returns of one year determined which of the factors would receive the largest shipment in the next. Often Corbin opted to risk shipping tobacco without insurance but began regularly insuring cargoes after losing one to the French. "We must not look back upon our losses but forward to repair them," he remarked to one correspondent. Corbin habitually maintained positive balances with most of his factors and demanded that they either pay him interest or use the funds to pay customs duties, thus securing him the discount for early payment.[36]

So much of Corbin's time was taken up with official duties that he could not oversee day-to-day plantation operations. He began his public career in the 1730s as a justice of the peace and a burgess, served on the council from 1750 until 1776, and was receiver general of the colony between 1754 and 1776. He was also a trustee of the towns of Port Royal and Falmouth and was involved in a project to improve navigation on the Mattaponi River. Indeed, Corbin relied on local tobacco inspectors ("desired by me to examine them with Strictness") and on his English factors to judge the quality of the tobacco his various overseers produced. Corbin did keep a tobacco crop book where he recorded the size, disposition, and proceeds of annual tobacco shipments (Plates 29a–b). And he ordered most of the cloth, bedding, and tools needed for the enslaved workers from London and Liverpool merchants.

36. Richard Corbin to James Buchanan, July 9, 1760, Corbin Letterbook, CWL.

Corbin left it to the general manager, however, to keep records of the disposition of grain crops and to keep lists of the livestock, tools, and slaves on the various quarters.[37]

The manager (between 1760 and 1764, one Edmund Wilcox, a well-educated but impecunious son of a London tobacco buyer) sold grain and surplus livestock and purchased rum, salt, and small amounts of cloth and tools locally, as did some of the more experienced overseers with whom the manager accounted each year. From Wilcox's accounts, Corbin could see at a glance the disposition and profits of corn crops. The manager's method of accounting, sometimes reporting only balances of accounts with overseers rather than individual transactions, did not lend itself to a clear reckoning of actual expenses or of total profit or loss. No record was kept of the costs of the clothing and tools Corbin imported, only of items purchased locally, which were relatively insignificant. As on other period plantations, supervision was the major local expense. Taxes, blacksmithing, medical care, and tobacco-inspection fees entailed regular but much less significant expenditures.

In 1760, a year when there was probably too much rather than too little rain, the corn crop was completely accounted for on nine quarters (Table 29). On four of them, about 50 percent of the crop was salable surplus, and two others made surpluses of between 15 and 38 percent. But on three other quarters, only enough was made to satisfy farm needs. A better year for corn was 1761, when there are crop records for ten quarters. On all of the quarters there was some surplus, and on seven of the quarters more than half of the crop was sold. Approximately 13 percent of the corn went to pay the overseers' shares of between one-ninth to one-seventh of the crops produced by workers under their direction. A further 7 to 16 percent was used to feed working horses and oxen and to fatten beef and hogs, whereas only 1 to 5 percent was needed for seed. The slaves consumed between 19 and 82 percent of what they raised; on most quarters, the slaves' ration of two barrels of corn per working adult, one and a half barrels per younger half sharer, and one barrel for nonworking children accounted for about one-third of the crop.[38]

These records show that once corn production reached ten barrels a hand, half that amount was clear surplus (on the quarters with no surplus, the workers were making only three to four barrels apiece). Corbin explained that extra corn "is always ready Money," since buyers usually paid cash upon or soon after delivery of the grain. Indeed, corn contributed about two-thirds of nontobacco plantation revenues. Some was sent to local merchants in ex-

37. Corbin to Robert Cary, Aug. 8, 1763, Mar. 3, July 14, 1764, ibid.

38. Wilcox Farm Journal, 1761–1762 (vol. 4), E. W. Hubard Papers, CWL.

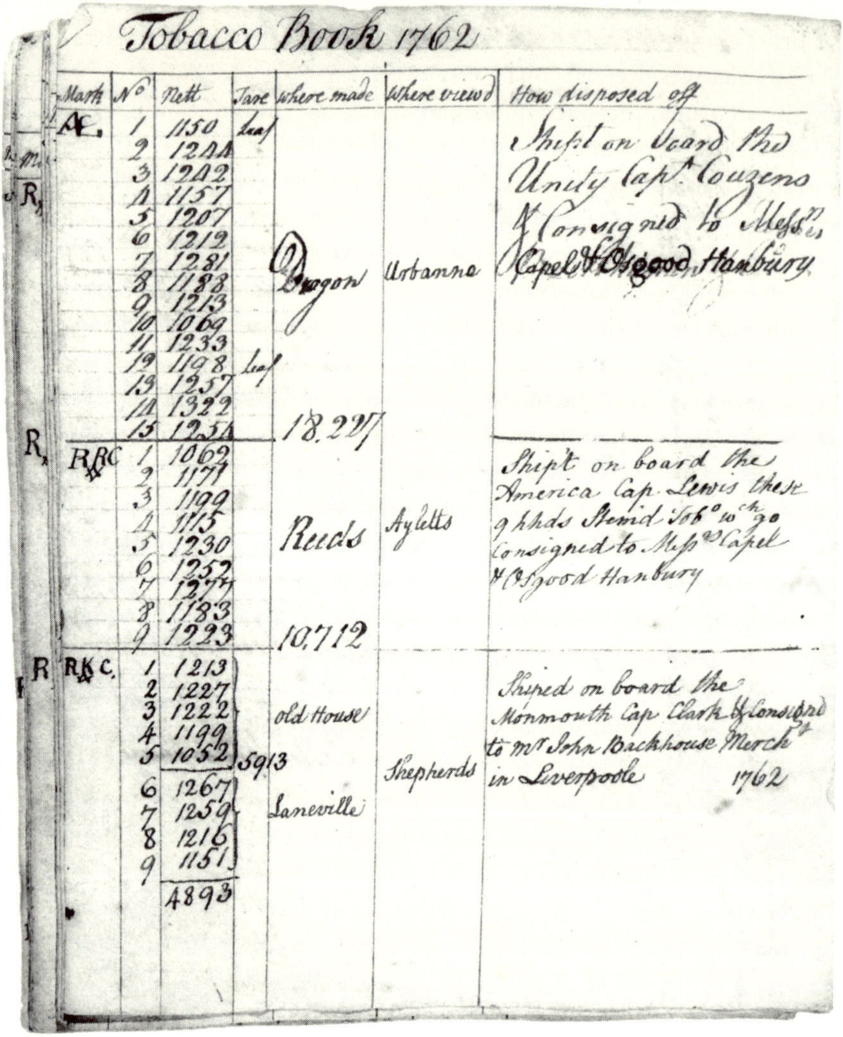

Tobacco Book 1762

Mark	No	Nett	Tare	where made	Where view'd	How disposed off
AE	1	1150	leaf			Shipt on board the Unity Capt. Couzens & Consigned to Messrs. Capel & Osgood Hanbury
	2	1244				
	3	1202				
	4	1157				
	5	1207				
	6	1212				
	7	1281		Dragon	Urbanna	
	8	1188				
	9	1213				
	10	1069				
	11	1233				
	12	1198	leaf			
	13	1257				
	14	1322				
	15	1254		18.227		
RRC	1	1069				Shipt on board the America Cap. Lewis these 9 hhds Stemd Tob.o w.ch 90 Consigned to Messrs Capel & Osgood Hanbury
	2	1171				
	3	1199				
	4	1115		Reeds	Aylets	
	5	1230				
	6	1252				
	7	1277				
	8	1183				
	9	1223		10,712		
RKC.	1	1213				Shiped on board the Monmouth Cap. Clark & Consigned to Mr John Backhouse Merch.t in Liverpoole 1762
	2	1227				
	3	1222		old House		
	4	1199				
	5	1052	5913		Shepherds	
	6	1267				
	7	1259		Laneville		
	8	1216				
	9	1151				
		4893				

PLATES 29A–B. Pages from Richard Corbin's Tobacco Book, Richard Corbin Papers, 1746–1825, Entries for 1762. Courtesy, Special Collections, John D. Rockefeller, Jr., Library, The Colonial Williamsburg Foundation

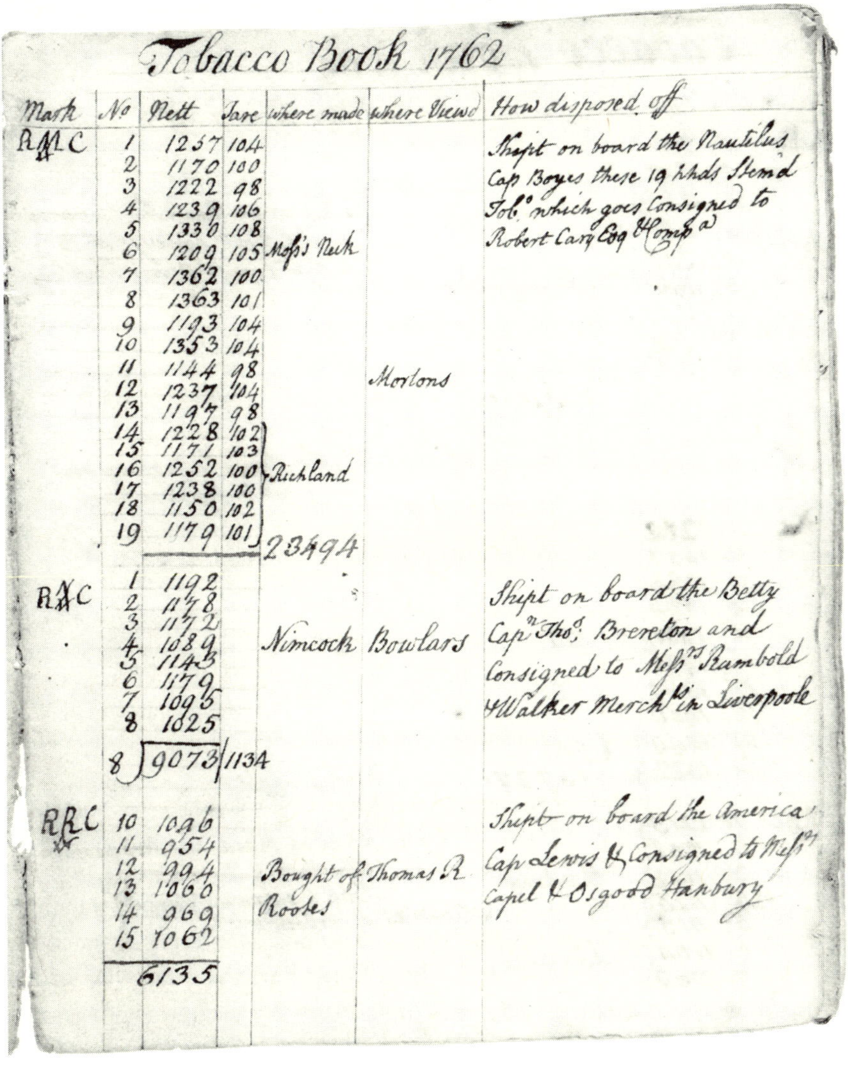

Tobacco Book 1762

Mark	No	Nett	Tare	where made	where View'd	How disposed off
RMC	1	1257	104			Shipt on board the Nautilus Cap Boyes these 19 hhds Stem'd Tob.o which goes Consigned to Robert Cary Esq & Comp.a
	2	1170	100			
	3	1222	98			
	4	1239	106			
	5	1330	108			
	6	1209	105	Moss's Neck		
	7	1362	100			
	8	1363	101			
	9	1193	104			
	10	1353	104			
	11	1144	98		Mortons	
	12	1237	104			
	13	1197	98			
	14	1228	102			
	15	1171	103			
	16	1252	100	Richland		
	17	1238	100			
	18	1150	102			
	19	1179	101			
				23494		
RNC	1	1192				Shipt on board the Betty Capn Thos. Brereton and Consigned to Messrs Rumbold & Walker Merchts in Liverpoole
	2	1178				
	3	1172				
	4	1089		Nimcock	Bowlars	
	5	1143				
	6	1179				
	7	1095				
	8	1025				
	8	9073	1134			
RRC	10	1096				Shipt on board the America Cap Lewis & Consigned to Messrs Capel & Osgood Hanbury
	11	954				
	12	994		Bought of	Thomas R	
	13	1060		Rootes		
	14	969				
	15	1062				
		6135				

TABLE 29. Disposition of Corn Crops, Richard Corbin Plantations, Rappahannock Basin, Virginia, 1760–1761

	Plantation name							
	Moss Neck		Richland		Nimcock		Old House	
Category	1760	1761	1760	1761	1760	1761	1760	1761
Total crop (barrels)	265	344	24	130	208	106	94	164
No. of laborers	21	[21]	7	[7]	[10]	10	[12]	12
Share per laborer (barrels)	12.6	[16.4]	3.4	[18.6]	[20.8]	10.6	[7.8]	13.7
Slave rations (%)	23	19	82	15	21	22	62	33
Overseer's share (%)	13	15	14	0	14	[15]	0	0
Livestock (%)	13	8	0	0	[6]	6	0	4
Seed (%)	1	1	5	1	[1]	[1]	?	?
Misc. (%)	0	0	0	0	0	0	0	0
Surplus (%)	50	56	0	84	57	55	38	[63]

Note: The figures in brackets are estimates. The crops are reported partly in barrels and partly in hogsheads. Hogsheads were converted to barrels at the rate of 1 barrel equals 1.078 hogsheads, a formula found in Wilcox's account book. Percentages do not add up to one hundred owing to rounding.
Source: Edmund Wilcox Account Book, 1760–1761 (vol. 4), E. W. Hubard Papers, University of North Carolina, microfilm, M-24.4, CWL.

change for goods needed on the quarters, but most was sold to Norfolk merchants trading to the West Indies or the wine islands. Marketing connections were well enough established that Corbin did not have to bother with selling corn and wheat; the general manager could dispose of the grain without difficulty.[39]

In 1760 and 1761, Corbin grossed between £1,500 and £1,800 sterling constant value per year from tobacco and corn. Each field hand produced gross revenues of at least £11.50 apiece, with corn contributing about £4.75 and tobacco £6.75. The amount of revenue generated from wheat is uncertain, but apparently it brought in less than 10 percent of nontobacco income. On one quarter, laborers produced twelve bushels of wheat per hand in 1760, but yields of fewer than five bushels were probably more typical. Among the Rappahannock basin planters discussed above, Corbin ranked second (£11.50)

39. Corbin to James Semple, Jan. 1, 1759, Corbin Letterbook, CWL.

Lanesville/ Gate		Dragon		Parrots		Weeks		Hill		Corbin Hall
1760	1761	1760	1761	1760	1761	1760	1761	1760	1761	1761
35	173	98	183	190	184	136	141	116	85	262
?	18	[23]	23	17	19	[13.5]	13.5	[19]	19	24
?	9.6	[4.3]	[8.0]	11.2	9.7	[10.1]	10.4	[6.1]	4.5	10.9
78	23	62	32	43	27	26	25	37	51	17
0	1	13	13	13	12	13	13	0	3	0
18	3	24	16	16	12	[3]	[6]	6	18	7
?	1	1	1	2	1	[1]	[1]	?	3	1
0	0	0	0	11	0	2	0	9	0	4
3	72	0	39	15	49	55	55	48	27	70

after the Berkeleys (£15) in annual gross revenues realized per worker. John Baylor, who was not raising as much corn as Berkeley and Corbin, averaged about £10. The Beverleys and John Mercer (discussed below), both of whom made less valuable tobacco and who also grew less corn, averaged only £7 to £9 per worker per year.

Like these other planters, Corbin kept most of his enslaved workers in the fields, relying on local whites to supply services like tanning leather and weaving cloth. Rather than training slave artisans, he hired white wheelwrights, bricklayers, blacksmiths, and midwives. The only healthy adults who did not do field labor were two to five carpenters who built and mended plantation buildings, cut timber for tobacco hogsheads, and occasionally worked for others. Boys and girls were typically sent to work in the fields at age ten or eleven. For the first year or so, they were rated as half sharers, but by no later than age fourteen they were expected to perform the same work as adults.

Corbin hoped to induce the enslaved workers to raise large crops of tobacco and grain through a combination of what he perceived to be kind and fair treatment and systematic planning. Overseers were to give the slaves adequate allowances of corn or ground corn meal at regular times but were to be careful that no grain was wasted. He further instructed the overseers, "The Breeding Wenches more particularly you must be kind and Indulgent to, and not force them when with child upon any Service or hardship that will be Injurious to them," and he wished the children "to be well looked after." Following his employer's instructions, the general manager hired doctors to tend sick slaves and bought rum for the invalids, arranged for midwives to attend birthing women, supplied blankets for the babies, and distributed molasses and worm medicine to the children. Corbin also allowed the slaves to raise chickens and ducks, which the enslaved women sold to Elizabeth Corbin for cash. He expected the overseers "to keep the negroes in good order and enforce Obedience by the Example of their own Industry," a "more effectual Method in any respect of Succeeding and in making good Crops, than Hurry and Severity." Foresight, he declared, was easier on the people and would prevent the overseers from rushing about in confusion.[40]

Large herds of livestock were on all the quarters, and surplus stock made a significant contribution to plantation revenues. By 1760, each farm had between 70 and 90 adult cattle, 40 to 50 sheep (except Corbin Hall with 137), 30 to 50 hogs, and 30 to 80 geese. The cattle got extra food in the form of corn tops and blades, and the hogs were fattened on shattered corn. There were about ten oxen on most quarters, and at four of the quarters there were also two to five horses, but some had no draft animals at all. The use of plow horses instead of slow-moving oxen to prepare ground and to help with weeding clearly increased the amount of corn one worker could tend, but it was an innovation that Corbin did not yet fully employ, perhaps because there was not enough land available for expanded grain production on the smaller quarters. After setting aside meat for the owner's table and for the general manager, overseers, and slaves, between 20 and 150 surplus hogs were marketed each year, the occasional old cow was sold to provision slave ships calling in the Rappahannock, and small amounts of mutton and lamb were sold to neighbors. Livestock accounted for about 20 percent of nontobacco revenues. Other income came from some small tracts rented to a handful of tenants and from a mill that was rented for thirty-five pounds Virginia currency per year. Corbin once invested fifty pounds in a grain-trading voyage to Tenerife organized by a Norfolk merchant, but proceeds were apparently insufficient to induce him to make further ventures into trade.

40. Ibid.

By the 1750s, Corbin was coming to recognize that his landed resources were not infinite. To preserve some farms for an inheritance for his children, he called a halt to indiscriminate clearing. Instead, he regularly had cattle penned on older fields. These temporary enclosures confined one hundred cattle on plots containing one thousand tobacco hills, and the animals were moved to another part of the field each week. As Corbin remarked to his general manager in 1759, "A proper use of the Cattle will answer every purpose of making Tobo without the distruction too Commonly made of the Timber'd land." Corbin also sought to further diversify the crop mix, raising minor amounts of peas and by 1763 ordering some English clover seed in an attempt to improve pastures.[41]

Like Carter Burwell and Daniel Parke Custis, Corbin's financial conservatism allowed him to live well and to endow his children generously without putting the family estate at risk. He sent at least one son to school in England and had sufficient reserves with factors to promptly pay a daughter's one-thousand-pound dowry. In 1758, he built a new fifty-two-by-twenty-foot brick house when his original dwelling burned and promised shortly to build a similar house for one of his sons. At the same time, Corbin cautioned the young man to emulate his own prudence and never to run into debt nor to buy anything but what he had money in hand to pay for. Corbin's emphasis on careful planning and record keeping and on systematic management was unusual for the period. Not until the 1790s would many other large planters such as George Washington and John Tayloe III adopt similar management practices.[42]

Coping with Scarcity: The Lesser Gentry

William Fauntleroy, Junior (1713–1793), of Lancaster County was a third-generation native Virginian born into a solid gentry family, but one less wealthy than the Berkeleys, Baylors, Beverleys, and Corbins. Since William's long-lived father, Colonel William (1684–1757) of Richmond County, continued to manage plantations that he had inherited or bought until he died

41. Ibid.

42. Corbin's attention to time management as well as to record keeping might have reflected his reading of bookkeeping manuals directed to landed gentlemen such as Roger North, *The Gentleman Accompt; or, An Essay to Unfold the Mystery of Accompts, by Way of Debitor and Creditor* . . . (London, 1715, 1721), that emphasized "managery" of all facets of a landed estate and the attainment of honor through "leaving an Example of Prudence, Oeconomy, and Industry for their [heirs] Imitation and Encouragement" (5–6). See also Rob Bryer, "The Roots of Modern Capitalism: A Marxist Accounting History of the Origins and Consequences of Capitalist Landlords in England," *Accounting Historians Journal*, XXXI, no. 1 (June 2004), 1–56 (esp. 19).

at age seventy-three, William Junior, though he was the eldest son, in order to marry, start a family, and live an independent life initially had to build up an estate on his own.[43]

Between the early 1730s and 1757, William Junior accumulated 1,141 acres in Lancaster County by purchasing ten separate parcels of between 60 to 150 acres each from smaller planters. In the third quarter of the eighteenth century, the majority of Lancaster County landowners had farms of 200 acres or fewer, so Fauntleroy's ostensibly modest holdings still put him in the top 5 percent of landowners. William's father gave him at least three enslaved women to help him get started. In 1735, Fauntleroy began buying slaves on his own, making a first purchase of nine new Africans from slave dealer John Tayloe in partnership with another area planter. These captives might have been refuse slaves, some or all of whom the partners sold to finance the cost of other slaves they bought for themselves. Fauntleroy bought two African boys, Summersett, whose age was judged by the county court to be ten years in December 1735, and Teaboy, age nine (both were probably brought to Virginia on the *Ann* of Bristol) at a total cost of £27 sterling. By buying two young boys instead of one adult man, Fauntleroy sacrificed initial revenue from their labor but minimized costs and spread his risks. The next year, William purchased a girl, Affoe, twelve years old, and a boy, Bristol, age ten, who probably arrived on the *Indian Queen* of Bristol, paying Tayloe £28 sterling. The proceeds of these slaves' labor helped Fauntleroy to buy twelve more Africans costing £186 sterling from slave dealer George Braxton in 1739. These captives, who arrived on either the *Betty* of London or the *Vine* of Liverpool included nine adults and three boys all judged to be ten years old.[44]

Fauntleroy seems to have raised oronoco tobacco almost exclusively, and, like the larger Rappahannock basin planters discussed here, when he consigned tobacco he split his cargoes between London, Liverpool, Bristol, and other outport factors. Only in 1753 and 1754 did he ship a few hogsheads of sweet-scented, but this tobacco might not have been from his own crops. Fauntleroy, following the example of producers of the more valuable strain,

43. Col. William Fauntleroy, Jr., Letterbook and Account Book, 1735–1774, CWL. For the Fauntleroy family, see *VMHB*, XXXII (1924), 128–129, XXXVIII (1930), 383–384; and "Fauntleroy Family," http://homepages.rootsweb.com/~mershall/esmd48.htm (accessed July 5, 2005).

44. Ages of slaves Fauntleroy purchased are recorded on a loose page in the Letterbook, CWL. For the ships on which they arrived, see Walter Minchinton, Celia King, and Peter Waite, eds., *Virginia Slave-Trade Statistics, 1698–1775* (Richmond, Va., 1984), 83, 89, 101. For distribution of landholdings in Lancaster County, see Camille Wells, "The Eighteenth-Century Landscape of Virginia's Northern Neck," *Northern Neck Hist. Mag.*, XXXVII (1987), 4219–4227.

did attempt to tailor his product to the tastes of the different markets. Usually he had the slaves stem about half of the tobacco, and he experimented with packing some in bundles and some laid straight. A drought in the summer of 1755, followed by a hailstorm and an early frost greatly curtailed that year's crop. William reported that the field hands, who usually made thirty to forty hogsheads in normal years, in 1755 made only three. The 1759 and 1762 crops also proved short. In the latter year, William got a crop of only one thousand pounds from a plantation on which the slaves usually made thirty thousand pounds, and he did "as Bad with my Other Quarters and I never took more paines about a Cropp then about the last."[45]

In years when resident factors offered high prices, when his crops were small, or when shipping in time of war proved too risky, he sold most or all of his tobacco in the country. William traded with an even larger number of merchants than did the great Rappahannock River planters. Between 1737 and 1769, he dealt with London factors Joseph Younger, William Bland, John Bland, Jeremiah Smith, Sydenham and Hodgson, James Buchanan, John Taylor, Robert Fauntleroy (a relative), James Concannon, Mathias Gale, Thomas Ribright, Stewart and Campbell, and a Mr. Maxwill. Interestingly, most of the London merchants with whom Fauntleroy dealt were relatively small traders, rather than the large firms favored by great planters who shipped one hundred hogsheads or more annually. In 1739, William made his first trial of the Liverpool market, and he continued thereafter to send part of his crop almost every year to one or more of the following factors: Thomas Backhouse, Robert Seale, James Gildart, Charles Goore, Rumbold and Walker, and Clay and Midgeley. In the 1740s, William also tested Whitehaven, dealing with Edward Tubman and John Younger. Bristol proved the least favorable market, and only between 1753 and 1760 did Fauntleroy send consignments there addressed to Thomas Knox.

One reason for William's having so many correspondents was that, in years when shipping was scarce, he could not always find captains willing to take consignments for any but the ship's owners. Unlike John Baylor, however, William did not play one merchant against another in order to borrow more money and to avoid paying debts. Generally, he ordered no more goods than his tobacco would pay for, and, if he ended up in debt to one of the factors, he promptly paid the balance either with the next year's shipment of tobacco or with bills of exchange. Only in 1739 when he made the extraordinary purchase of twelve slaves did William request sixty to seventy pounds credit from London factor Joseph Younger to help cover the 50 percent of

45. William Fauntleroy to Clay and Midgely, July 10, 1763, Fauntleroy Letterbook, CWL.

the purchase price he had to pay when he took delivery of the slaves, and he also drew twelve pounds on William Black before the tobacco shipped to him was sold. The next year, he drew fifty-five pounds in advance on tobacco shipped to Liverpool merchant Thomas Backhouse to cover the remaining sum William owed for the slaves.

Despite having so many English correspondents, William judged selling tobacco in the country more advantageous in 1738, when his crop was small. And, during most of King George's War, he found prices in England too low, freight and insurance charges too high, and chances for capture by the French too great to warrant risking many shipments. In 1744, he informed Joseph Younger, "We run a verry Great Risk in shipping at Present and for Inshuring [*sic*] one may as Well Loose it almost or better run that Risk." "The times is so Precarious," he wrote Sydenham and Hodgson in 1748, "that the less a man Ships the Better as we gitt neare as much for what Tob. we dispose in the Cuntry and that's without any Risque as if it was shipt and have the Best of the market." With the return of peace in 1749, Fauntleroy resumed consigning his crops and made small, unsuccessful trials of the Bristol and Glasgow markets. Otherwise, he resumed regular shipments to London and Liverpool. William continued to ship tobacco to England throughout the Seven Years' War and fared relatively well until 1761, when some of his tobacco was taken. He concluded: It "will not answer to shipp in vessells of noe force [that is, unarmed]. I Loose in such Every year and am Determin'd to shipp no more in such shipps."[46]

When merchant Edward Tubman of Whitehaven first approached William in 1748 about selling a cargo of goods for him in return for tobacco, William was initially reluctant, explaining that his experience was limited to selling small amounts of goods that he had imported on his own account. Tubman persisted, however, and the next year Fauntleroy built a new storehouse and accepted cargoes of English manufactures and of West Indian rum and sugar that Tubman ordered for him. William intended to sell them at a substantial advance, and he asked Tubman to send him an invoice for the goods that he could show his customers on which the prices would be marked up 15 percent ("or something more on small matters that will bare it") over the prime cost that Fauntleroy actually paid. However, Tubman was "to lay on Noe advance

46. Fauntleroy to Younger, Aug. 6, 1744, to Sydenham and Hodgson, Sept. 15, 1748, and to Thomas Ribright, July 15, 1762, Fauntleroy Letterbook, CWL. For wartime freight rates, see Stephen Gregg Hardy, "Trade and Economic Growth in the Eighteenth-Century Chesapeake" (Ph.D. diss., University of Maryland, College Park, 1999), 450–456. The usual rate in peacetime was £8 sterling per ton. During King George's War, the cost rose to between £9.5 and £16 per ton, and during the Seven Years' War it rose from £10 to £14 per ton.

on Iron work or aney sort of Brown lining for Fear of the Planters finding me out."[47]

The next year, Tubman also shipped to Fauntleroy from Barbados ten refuse African slaves. This slave-dealing venture proved profitable for neither man, as two prime adult males died on board the ship, and five of the remaining slaves were boys and girls, instead of men and women as the invoice stated. Moreover, the surviving captives were "Bad with the Flux and much out of Order other waies," and they infected Fauntleroy's bondpeople with these distempers when he temporarily housed them at his quarter. The twenty pounds a head that Tubman charged him, William considered "a Monstrous Price for such Slaves as these are." So he refused to sell more consignments of slaves for anything less than a 10 percent commission on gross sales (but later suggested that 5 percent on gross sales or 10 percent if he took responsibility for making good any bad debts received in payment for the slaves would be acceptable). In fact, William had problems selling not only the sick slaves but the other merchandise that Tubman sent. In 1750, Virginia was overstocked with West Indian rum, and "the Cuntrey has more [English] Goods in it now by half then was Ever known before." Fauntleroy and Tubman seem to have mutually agreed to terminate their dealings. Subsequently, Tubman sent at least one other shipment of nineteen refuse slaves from Barbados to the Rappahannock, but he found someone other than Fauntleroy to handle the sales. William continued to keep a store, but thereafter stocked it only with goods that he imported on his own account from Liverpool and other Whitehaven merchants. These goods consisted of crates of earthenware, powder, shot, fishing gear, metalware, blankets, and "negro cottons." Searching for additional ways to augment his income, in 1753 William built a custom gristmill. He also requested the Liverpool firm of Clay and Midgeley to send him a tailor and a ship carpenter bound to work for wages for a term of years.[48]

The difficulties that William encountered building an estate while waiting for an inheritance from his father, along with the burden of supporting eleven children for whom, by the late 1750s, he also had to provide inheritances, persuaded him to send his two eldest sons to school in Scotland, where education was less expensive than in England. They stayed in Scotland for nine years, studying law or medicine so that they could make five hundred to six hundred pounds a year from their professions in Virginia while they, in

47. Fauntleroy to Tubman, Mar. 2, 1747/8, n.d. [July 1749], July 16, 25, 1749, Fauntleroy Letterbook, CWL.

48. Ibid., Apr. 15, July 11, Aug. 8, Sept. 6, n.d. [September or October], 1750, June 25, 1751; Minchinton, King, and Waite, *Virginia Slave-Trade Statistics,* 153. The shipment of slaves that Tubman sent to Fauntleroy does not appear in the *Slave-Trade Statistics.*

turn, waited to collect an intended inheritance. (In the third quarter of the eighteenth century, only lesser gentry had to resort to starting sons in professions other than planting. Not until the 1790s did land shortages force some elite planter fathers to adopt similar practices.) William and other members of his family had been schooled only in Virginia, so Fauntleroy continually reminded his sons that they had "advantages never offered to others" in the family. To improve his daughters' eligibility for advantageous marriages, he asked a London merchant friend to send an English schoolmistress to teach the girls reading and sewing.[49]

William inherited his parents' dwelling plantation and additional slaves when his father died in 1757, and then he could allow that he had achieved "a Midling good Living," but the responsibility of providing for eleven children still led him to eschew "extravagance." William did order a chariot emblazoned with the family coat of arms from London in 1753, but he cut costs by buying a secondhand vehicle. The only other extravagances he allowed himself among the goods he ordered each year from England were two dozen new mahogany chairs and a mahogany couch ordered from London in 1761.[50]

Fauntleroy recorded little about how he managed his workforce. A list of tithables that he made in 1758 does reveal that six white men were then in his employ. Some were presumably overseers, but they also might have included some of the skilled indentured servants or workmen hired on wages that he had asked his Liverpool factors to procure for him. William reduced supervision costs by putting enslaved men in charge of his smaller quarters. In the mid-1740s, Fauntleroy made note of tobacco produced on two farms that slaves Generall and Liverpool managed. By 1758, his enslaved workforce had grown to thirty-eight adults (and presumably about an equal number of children). These bondpeople seem to have fared better in terms of longevity than many Virginia slaves. All but one of the ten enslaved children whose names and ages William recorded between 1735 and 1739—seven Africans and three native-born infants—still survived twenty-odd years later. William seemingly needed to buy no new hands after 1739, and repetition of some names on the tithable list, including those of Generall and Liverpool, likely sons of the quarter overseers, suggests substantial natural increase and family continuity.

William's letters, along with his father's 1757 will, document the differ-

49. Fauntleroy to Sydenham and Hodgson, Aug. 10, 1756, to James Buchanan, Sept. 10, 1758, to Dear Boys, Sept. 10, 1758 (quotation), to Captain James Concannon, July 18, 1759, Fauntleroy Letterbook, CWL.

50. Fauntleroy to [Alexander] Gray, Nov. 20, 1761, ibid.

ent family inheritance strategies that lesser-gentry planters employed, at least those families whose heads enjoyed exceptionally long life spans as did William and his father. The limited amount of land the family held forced it to choose other courses than those followed by wealthier neighbors who owned thousands of acres. The elder Fauntleroy, like his son and namesake, inherited insufficient land from his parents to pursue large plantation agriculture and so had to build an estate in the tidewater through the purchase and piecing together of a number of small tracts. This land he left to his three sons, but, since William Senior lived into his seventies, his heirs did not get control of the tracts intended for them until they were between thirty-three and forty-four years of age. In the interim, assuming that they did not want to move west, as neither William nor his younger brothers were apparently inclined to do, the sons had to piece together farms of one thousand or so more or less adjoining acres by buying up small parcels whenever they came on the market and also to purchase enough slaves to work them. Otherwise, as William Junior arranged for his sons, the prospective heirs would have to make an independent living by practicing some remunerative profession until their fathers passed on.[51]

William Senior gave the two younger sons their land in fee simple, but he did not entirely lack dynastic pretensions. He entailed the dwelling plantation that he devised to William Junior and stipulated that it "continue in the name of the family forever and not be disposed of." William reserved one tract for the support of a married daughter and her children, but her three brothers were to manage the farm for her; the other five daughters received slaves and fifty-pound cash legacies. For families like the Fauntleroys who had no surplus land to give to sons when they came of age, slaves proved a much more readily available form of wealth that parents could use to set up their children of both sexes and to endow their grandchildren. The sons initially had to purchase land for themselves, but by then William had given all of them a few slaves with which to commence farming.

Subsequently, a lesser-gentry father could most effectively assist mature sons and daughters possessed of growing families of their own by giving slaves to the grandchildren. Fauntleroy Senior divided some of his enslaved workers among his sons and daughters, but he devised one or more young enslaved children to all of his by then numerous grandchildren. Each of William Junior's children, for example, received one young slave of about their own age, workers whose labor their father would manage for them during their minority. For one favored grandchild, William Senior directed that his execu-

51. Richmond County Wills, 6, 1753–1764: 111, microfilm, VSL.

tors lay out two hundred pounds in young enslaved girls about ten years old, clearly intending that the girls' future increase would relieve the child's parents of the burden of providing him with bound laborers. The eventual sundering of the families of the enslaved boys and girls so devised was apparently not a matter of great concern to those divvying up the family property. Long life spans and land scarcity in the tidewater were forcing members of the lesser gentry who did not choose to move west to develop altered strategies for maintaining family fortunes. Delayed or encumbered bequests of land could, however, be offset to some extent by early gifts of slaves, a change made possible by the greater numbers of children native-born enslaved mothers bore.

Almost on the Periphery: Along the Potomac

John Mercer (1704–1768) managed newly established plantations farther north in the less fertile Potomac basin. Planters in this area could grow only less valuable oronoco tobacco, and, given their smaller incomes, had less access to credit in Britain, fewer chances to buy African slaves locally, and fewer connections to intercontinental and European grain markets than did those living on the James, York, and Rappahannock rivers. Mercer is only the second immigrant among the group of large Virginia planters for whom detailed farming records have survived for this period, a reflection of greatly diminished immigration from Britain to older tidewater counties. Mercer had come to Virginia from Ireland in 1720 at age sixteen possessed of a good education from Trinity College in Dublin but little else. Like Thomas Jones, the other immigrant of this period discussed previously, Mercer started as a minor merchant and had to build a landed estate and an enslaved workforce almost from scratch. His rule, he explained near the end of his life, was "that every thing I attempted ought to be set about in earnest and not carried on by peicemeal *[sic].*" An early marriage in 1725 into the propertied Mason family and an active career as an attorney provided him with the necessary capital to go into planting.[52]

52. John Mercer Ledgers, 1725–1732, [1741]–1750, Bucks County Historical Society, Doylestown, Pa., microfilm, M-164, CWL. C. Malcolm Watkins, *The Cultural History of Marlborough, Virginia* . . . (Washington, D.C., 1968), provides a biography of Mercer and a description of an archaeological investigation of Mercer's home plantation. The quotation is from Lois Mulkearn, ed., *The George Mercer Papers Relating to the Ohio Company of Virginia* (Pittsburgh, Pa., 1954), 194. Mercer's accounts show that in 1748 he had sixty-one tithables in Stafford, Prince William, and Fairfax counties. For period Potomac basin trade, see Daniel Payne Account Book, Dumfries, Va., MS 635, MHS; William Robinson Account Book, Bunker Hill, Westmoreland County, 1736–1823, Robert E. Lee Memorial Foundation, Stratford Hall, Va., microfilm, M-141, CWL; Philip Ludwell Lee Ledger, Westmoreland County, 1743–1783, Duke University, microfilm, M-2236,

Mercer began in 1728 with the purchase of a number of vacant lots in the then defunct port town of Marlborough, and in 1730 he acquired one thousand undeveloped acres farther west. By 1753, Mercer had invested nearly five thousand pounds sterling constant value in lands and had accumulated an impressive estate of sixty thousand acres scattered across several counties. Especially interested as a speculative investor in western lands, Mercer became a leading member of the Ohio Company, organized by Virginians in 1747 to claim land west of the Appalachians. By the late 1740s, on the home farm at Marlborough Mercer had built an elegant 108-foot-long new brick house complete with an arched portico and other outbuildings at a cost of more than six hundred pounds sterling along with a wharf, warehouse, and windmill. He also farmed two other quarters in Stafford and established a new quarter farther up the Potomac in Fairfax County.

Mercer had much poorer luck with his human speculations than with his landed ones. He bought his first slaves in 1731 and thereafter added from 1 to 20 new laborers annually. By 1753, he had purchased a total of 112 slaves, most of them newly imported Africans, including a number from the Bight of Biafra, at a cost of more than thirty-two hundred pounds sterling. His effective labor force never approached 100, however. Between 1731 and 1750, 46 of the 89 slaves he had purchased by that year had died, and an additional 3 had been sold. Although there were offsetting additions from births, the children were not part of Mercer's calculations of profits and losses on his investment in enslaved laborers. Mercer's neatly kept ledger, tallying up his financial losses, in no way addresses the misery and appalling loss of life his involvement in the slave trade entailed (Plate 30). Moreover, death rates among infants and young children were also shockingly high. In 1768, near the end of his life, Mercer lamented to his son George: "All the slaves I now have, accounting those you and your brother had upon the division, do not make up the number I bought, tho I have had 98 born. Your sister Selden attributes it to the unhealthiness of Patomack Neck, which there may be something in." Mercer's mostly imported slaves were not only sickly but also desperately unhappy. One hanged himself, and several were chronic runaways. Another sign of trouble was frequent turnover among Mercer's overseers.[53]

CWL; and Business Book of a Merchant [probably William Campbell] Trading in Westmoreland County, Scottish Record Office, GD1/455/1–3, microfilm, M-1475, CWL. For the Ohio Company, see Kenneth P. Bailey, *The Ohio Company of Virginia and the Westward Movement, 1748–1792: A Chapter in the History of the Colonial Frontier* (Glendale, Calif., 1939), chaps. 1–10.

53. Mulkearn, ed., *Mercer Papers,* 213. For the origins of some of the Africans Mercer purchased, see David Eltis et al., *Voyages: The Trans-Atlantic Slave Trade Database* (Atlanta, Ga., 2008), http://www.slavevoyages.org, voyage nos. 17047, 17150.

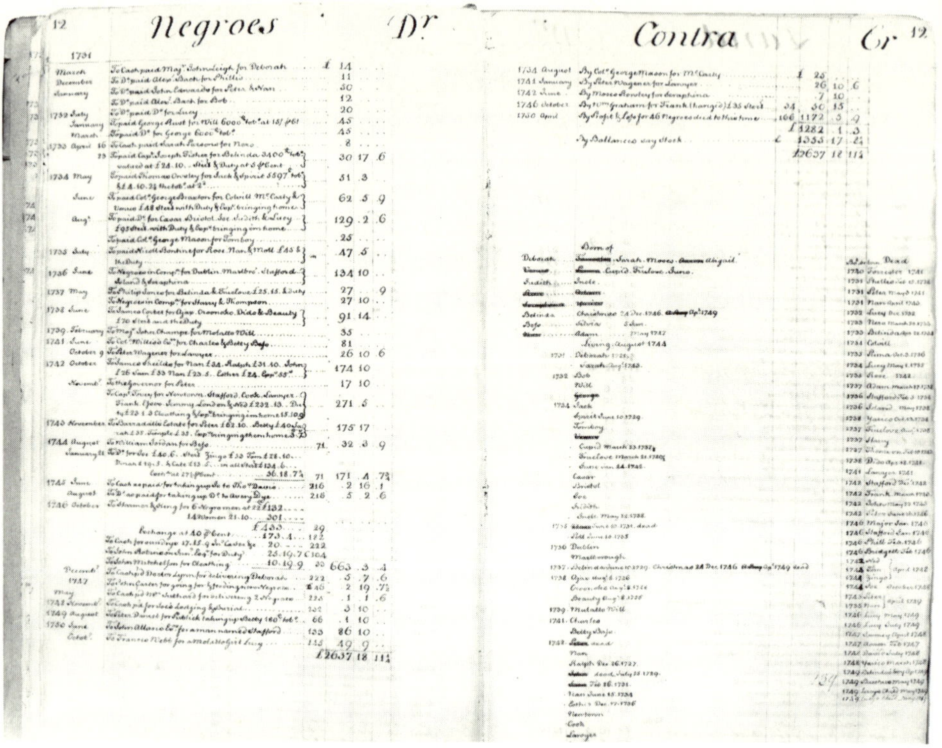

PLATE 30. John Mercer's Ledger, 1741–1750. Courtesy, Bucks County Historical Society, Doylestown, Pa.
The ledger shows Mercer's record of profit and loss from the purchase of more than one hundred enslaved Africans.

The enslaved workers made about eight hundred pounds of tobacco apiece between 1746 and 1750—the norm for the area—but they grew little corn, only about six barrels a hand, an output unlikely to leave a surplus for sale. Instead, any nontobacco farm revenues came from extra cattle, hogs, and sheep. Mercer made gross revenues of less than £2 sterling constant value per worker from corn and about £5.25 from tobacco. His tobacco revenues were lower than those of most large Rappahannock basin planters because Mercer was growing only less valuable oronoco. After trials of London, Bristol, and Whitehaven markets in the late 1740s, Mercer usually sold most of his crop in the country, and, by the mid-1740s, he found, like William Beverley, that any tobacco consigned to England netted less than what could be got from local merchants. Also, like the Beverleys, Mercer began to replace some imports with homemade goods, regularly hiring white weavers to make wool sheared from the plantation flocks into cloth for the slaves. (For comparative outputs and gross revenues, see Table 28.)

Mercer's attempts at estate building would not have been very successful had he depended solely on income from planting. He could not have borrowed enough to buy so much additional undeveloped land—and so many slaves—nor could he have spent £650 sterling between 1744 and 1750 on a mansion along with around £250 a year for books imported from England. Mercer was also an accomplished if irascible lawyer, however, attracting numerous clients from throughout Virginia while working on retainer for a number of prominent provincial merchants. By the 1740s, his legal business was worth £1,000 sterling annually. These revenues, coupled with enough financial discipline to keep annual expenditures within the limits of his income, initially proved the basis for a considerable fortune.[54]

Between 1730 and 1750, Mercer built a large library. He bought some books secondhand in Virginia but ordered many more from England. Titles not just on law but also on history, geography, natural history, literature, politics, mathematics, astronomy, and physics demonstrate wide-ranging intellectual interests. Mercer clearly sought to keep abreast of literary, scientific, and cultural developments in Britain. When building his house, for example, Mercer consulted Andrea Palladio's *Four Books of Architecture* (1570; trans. 1716) and Batty Langley's *City and Country Builder's and Workman's Treasury of Designs* (1750). In the 1760s, he decided to undertake beer brewing on a large scale; in preparation, he bought a number of books on brewing and distilling. Mercer also read widely in English agricultural literature. He owned most of the important manuals by William Ellis, John Houghton, and Richard Bradley, first published in the 1720s and 1730s, and he continued to buy new titles through the 1750s. There is no indication, however, that Mercer applied these authors' recommendations to his own plantation operations. His continued emphasis on tobacco and limited grain production suggests that up to 1763 he relied primarily on colonial farming practices.

Mercer's story does not end so favorably. In the early 1750s, he began to go into debt, frequently spending money he anticipated receiving but had not yet collected. For example, he represented the Custis family in a suit brought by the heirs of Daniel Parke's illegitimate children that dragged on for sixteen years; eventually, he borrowed large sums from the Custises in anticipation of fees for which a final accounting was long delayed. Meanwhile, unlike the Beverleys, Mercer seems to have had scant success in attracting tenants to

54. Watkins, *The Cultural History of Marlborough*, 3–64.

work his undeveloped holdings. During the Seven Years' War, he was unable to collect his legal fees, and, when ill health subsequently forced him to quit his practice, his ruin was sealed.

In 1759, pressed by two prominent Virginia planters who had stood surety for Mercer's substantial debts, John and his second wife, Ann, deeded much of their land and fifty slaves to two of their sons, George and James, on the condition that the sons pay half of their father's debts. Having conveyed so much property to his children, Mercer's chances of getting further credit were greatly diminished. Like the Jones family, the Mercers had become accustomed to a style of life that farm income alone could not support. John used his legal expertise, however, to devise a way to keep the estate he had worked so hard to build in the family. His sons rather than his creditors got control of the land, but their legacy was heavily encumbered, since they had to assume responsibility for their father's financial imprudence.[55]

The Pinnacle of Privilege: The Carter Dynasty

No account of plantation management in mid-eighteenth-century Virginia would be complete without reference to the activities of the Carter family. Robert "King" Carter (1663–1732) and subsequently his sons, John (1690–1742) of Corotoman and Shirley, Charles (1707–1764) of Cleve, and Landon (1710–1778) of Sabine Hall, managed some of the largest estates in the colony. Charles and Landon attempted to adopt more innovative agricultural strategies than many of the Rappahannock basin planters discussed above, but they had trouble expanding production of small grains while at the same time still cultivating large amounts of labor-intensive tobacco. Whatever the practical outcome of their efforts to transform Chesapeake agriculture into something more closely approximating Old World practices, they came to pride themselves on being pioneering "improving farmers" who were following recent British innovations. For them, maintaining or increasing family wealth through agriculture, land speculation, and trade and sustaining family power through office holding was no longer enough. In addition to providing for their own families and serving as traditional political leaders, they now believed that gentleman planters ought also "to serve mankind" as agricultural innovators.[56]

55. Ibid.; Joseph E. Fields, comp., *"Worthy Partner": The Papers of Martha Washington* (Westport, Conn., 1994), 29–32, 39–46, 56; Mulkearn, ed., *Mercer Papers,* 40–45. John Wayles advised in 1767, "If you trust a man after he has made a Conveyance to his children etc. and that is Recorded, you do it with your Eyes Open and can have no Relief in Case of Insolvency" (John M. Hemphill II, ed., "John Wayles Rates His Neighbours," *VMHB,* LXVI [1958], 305).

56. Sources include Robert Carter Letterbooks, 1723–1724, 1727–1728, 1728–1731, 1731–1732,

In Britain in the 1730s, 1740s, and 1750s, a new generation of gentleman improvers embraced the agendas promoted by earlier generations of patriotic projectors—the manufacture of silk and wine, for example, along with the cultivation of naval stores and of tropical products with the aim of advancing Britain's interests by producing at home or in the colonies commodities normally purchased from other nations. There was, however, a new emphasis on the role of mechanical inventions and on the role the state might play in patronizing scientific endeavors and agricultural reforms. Members of newly formed learned societies in England, Ireland, and Scotland hoped to replicate the formal alliances between monarchy and natural science that had been forged in other European countries, especially France, Prussia, and Sweden. As Joyce Chaplin has argued, elite colonial improving planters like the Carters used transatlantic scientific, "learned-culture and patronage networks" "to seize control of an expensive, prestigious, and often exotic form of knowledge" to enhance their own prestige and power. Recognition came not just from cultural achievements but also from the new public role gentleman improvers assumed.[57]

In the 1720s, "King" Carter had used assets gained from agriculture, trade, and office to amass an estate in land and slaves unmatched anywhere in the Chesapeake. By the time he died in 1732 at the advanced age of sixty-nine, he owned outright 829 enslaved men, women, and children scattered among forty-seven quarters on more than three hundred thousand acres throughout the colony, although concentrated in the Northern Neck and adjacent interior. Moreover, Carter retained control of additional slaves he had willed to various family members, so the contemporary estimate that he commanded

---

all in the Robert ("King") Carter Papers, MS 3807, UVa, microfilm, M-113, CWL; Carter Family Letterbook, 1732–1782, Tracy W. McGregor Library, Accession 4996, Special Collections, UVa, microfilm, M-62.8, CWL; Robert Carter Inventory, 1733, Robert Carter Will, 1732, Robert Carter Letterbooks, 1728–1730, Papers of the Third Generation of the Carter Family, Part 3, MSS1 C2468a27-144, Carter Family Papers, VHS; Carter Family Papers, 1659–1797, UVa, microfilm ed., 1967, M-1139.1, M-1139.2, CWL, some published as "Letters and Other Papers, 1735–1829, from the Collections of the Virginia Historical Society," *VMHB,* XXIII (1915), 162–192; Jack P. Greene, ed., *The Diary of Colonel Landon Carter of Sabine Hall, 1752–1778,* 2 vols. (Charlottesville, Va., 1965), II, 880 (quotation); Louis Morton, *Robert Carter of Nomini Hall: A Virginia Tobacco Planter of the Eighteenth Century* (Williamsburg, Va., 1941), chaps. 1, 3, 5; "Robert Carter (ca. 1664–4 August 1732)," in Bearss et al., eds., *Dictionary of Virginia Biography,* III, 84–86; and Edmund Berkeley, Jr., ed., *The Diary, Correspondence, and Papers of Robert 'King' Carter of Virginia, 1701–1732,* http://etext.virginia.edu/users/berkeley.

57. Chaplin, *An Anxious Pursuit,* chap. 5 (quotation on 132); Raymond Phineas Stearns, *Science in the British Colonies of America* (Urbana, Ill., 1970), 258–273; Richard Drayton, *Nature's Government: Science, Imperial Britain, and the "Improvement" of the World* (New Haven, Conn., 2000), chap. 3.

the labor of 1,000 bondpeople is no exaggeration. Carter also held 26 white indentured servants, most of them artisans working on the home plantation. With so much labor at his command, Carter could outproduce any of his contemporaries, whether or not his plantations were especially efficient. Indeed, Carter used labor lavishly and technology hardly at all, since he did not believe in using either plows to prepare ground or carts to carry crops in from the fields.[58]

As we saw previously, in addition to growing tobacco, Carter pursued a diversified regimen, raising and marketing surplus corn, wheat, livestock, and orchard produce. By the late 1720s, he was simultaneously engaged not just in farming on an unprecedented scale but also in grain milling, foreign trade, slave dealing, land speculation, and mining, and he served as president of the Virginia Council from 1726 until a few weeks before he died. Obsessed with firmly establishing a family dynasty, in 1726 Carter began drafting his last testament, which, with four codicils added in the next four years, eventually covered forty densely written pages. Robert divided his immense estate among his ten sons and daughters and several grandsons (the second sons of his daughters who had been christened "Carter"). To keep productive assets in the Carter family in perpetuity, he entailed both the lands and slaves bequeathed to his five sons in the male line. Carter did also transfer some land and slaves to three married daughters, but these were also entailed and could subsequently pass only to the daughters' male heirs. Carter further hedged these bequests with elaborate schemes of descent among children and grandchildren should any of the beneficiaries die without male heirs. Two unmarried daughters received only gifts of cash.[59]

But low tobacco prices between 1727 and 1730, coupled with heavy losses of new Africans, forced Carter to reduce some of the cash legacies he had intended to leave to his children and grandchildren. Carter wrote London merchant William Dawkins in 1730: "You are not unsen[si]able of the vast Expense we live at Small crops poor prizes [prices] vs great Salarys [to the merchants] and great mortalitys [among the slaves] and less for the maintainance of my Familye I cannot subsist with. I can too truely say I grow poorer and poorer Every Year I live." Carter's complaints, however, in no way imply that he died in debt. Rather, his income turned out to be less than he had anticipated when he wrote the first draft of his will in 1726. Consequently,

58. Robert Carter Inventory, Carter Family Papers, VHS; *Gentleman's Magazine,* II (December 1732), 1082.

59. Robert Carter Will, Carter Family Papers, VHS; Holly Brewer, "Entailing Aristocracy in Colonial Virginia: 'Ancient Feudal Restraints' and Revolutionary Reform," *WMQ,* 3d Ser., LIV (1997), 328–332.

his financial reserves of nearly six thousand pounds stock in the Bank of England and more than two thousand pounds lent at interest on mortgages in Virginia might have been somewhat depleted or at least did not grow at the rate Carter expected. Still, rather than being reduced to "beggars," his ten surviving children and a number of favored grandchildren would begin their planting careers with the advantage of being among the largest landowners and slaveowners in the region.[60]

The achievement that all of Carter's obituaries celebrated was that he, in the words of the Philadelphia *American Weekly Mercury,* "died immensely Rich," or (in the more tastefully phrased inscription on the elegant marble and limestone tomb that his heirs imported) was "possessed of ample wealth, honorably acquired." Moreover, the epitaph attested, as a prominent provincial officer Robert had "equally upheld the regal dignity and public freedom." The problematic contemporary understandings of honorable accumulation and of public freedom are, however, underscored by the initial notice of Carter's death published in an earlier edition of the *American Weekly Mercury.* According to the paper, Carter "died of the Flux, which 'tis supposed he caught on board a Vessel from which he bought several Negroes." Carter's unparalleled success in estate building entailed the forced transportation of at least a thousand African men, women, and children whose labor and that of their progeny Carter brutally exploited over his nearly fifty-year career as a planter.[61]

Once his estate was divided up, "King" Carter's children never amassed enslaved labor forces of equal size, but the sons did inherit about one hundred bondpeople apiece, and they increased their holdings by heavy purchases of additional Africans in the 1730s. By the early 1760s, as a result of natural in-

60. Carter to Col. John Tayloe, Jan. 29, 1729, Carter Family Papers, VHS; Carter to Dawkins, Apr. 16, 1730, Carter Letterbook, Carter Papers, CWL.

61. *American Weekly Mercury* (Philadelphia), Aug. 10–17, Aug. 31–Sept. 7, 1732, and Tomb Top Inscription (quotations), in Robert Carter's "Obituaries and Other Records of His Death, 1732 and Later," in Berkeley, ed., *Diary, Correspondence, and Papers of Robert "King" Carter,* http://etext.virginia.edu/users/berkeley. Carter's will charged John, the oldest son, with writing his epitaph. Third-quarter naval office shipping lists for 1732 for the Rappahannock District have not survived, so the arrival of a slaving ship in the first days of August cannot be corroborated (Minchinton, King, and Waite, eds., *Virginia Slave-Trade Statistics,* 200). Carter could have imported well over one thousand captives, since most of the approximately one thousand slaves Carter owned or controlled in 1732 appear to have been Africans. Carter had, of course, inherited some slaves from relatives and had realized further additions to the workforce from births. But, given the high death rates new Africans sometimes suffered on his plantations (Carter lost seventy adults in the winter of 1726–1727 alone), over Carter's nearly fifty-year planting career it is likely that among the Africans deaths exceeded births and that the number imported exceeded the number present at his death.

crease, Charles and Landon each owned more than three hundred enslaved men, women, and children. On the Carter sons' plantations, blacks were still usually employed only as agricultural laborers, and indentured servants continued to do most skilled work. In 1735, for example, John Carter wrote to an English factor for two new servants, a gardener and a smith. His previous craftsmen had left at the end of their terms, "during which I neglected to have some young Negroes instructed." Yet, in the late 1750s, Landon Carter still employed white servant plowmen, seedsmen, and gardeners, although he also trained some enslaved men in these occupations.[62]

"King" Carter could tend personally to the running of only a few of his farms in the immediate vicinity of the home plantation in Lancaster County, and such a vast operation could get out of hand. As he admitted in 1727, "Some of my remote Overseers have played the rogue and have not done their duty in neat handling of their tobo." After his death, his sons would agree that the quality of Carter's crops had declined once their father "labour'd under age and infirmitys." Some of the decline must have been due as well to the sheer scale and wide scattering of a business that required at least forty-six different overseers. Indeed, the Carters had problems with total output as well as with quality. The Carter Family Papers give scant information on the size of the crops each working slave typically produced, but it is probable that, as the number of slaves grew, including as they did so many new Africans, yields per hand slumped; in the 1730s, production apparently dropped to less than five hundred pounds of tobacco per laborer.[63]

Similarly, undeveloped interior lands produced less than they might have, for the Carter sons, unlike the contemporary Beverley family, were so preoccupied with concerns in the tidewater that they had neither the time nor the energy to locate suitable tenants for backcountry holdings. In 1738, John Carter reported to Edward Athawes that his brother George's and nephew Robert's backlands were "almost intirely useless and uncultivated" and that the quitrents were becoming "an intolerable charge." Attempts either to sell the lands or to secure tenants had so far been unsuccessful. In the 1740s and

62. Carter to ?, Aug. 8, 1734, Carter Family Letterbook, CWL. Allan Kulikoff, *Tobacco and Slaves: The Development of Southern Cultures in the Chesapeake, 1680–1800* (Chapel Hill, N.C., 1986), chap. 8, analyzes quarter sizes and laborers' occupations on "King" Carter's farms, as does Lorena S. Walsh, *From Calabar to Carter's Grove: The History of a Virginia Slave Community* (Charlottesville, Va., 1997), 81–93.

63. Carter to Micajah Perry, n.d. [December 1727?], Carter Letterbook, 1727–1728, Carter Papers, CWL; John Carter to Robert Cary, Aug. 1, 1734, John Carter to ?, Jan. 12, 1735, Carter Family Letterbook, CWL; Cf. Carter to William Dawkins, Feb. 13, 1723/4, Carter Letterbook, Carter Papers, CWL.

1750s, Landon Carter attempted to cut costs by failing to pay or by paying only tardily the quitrents due to Lord Fairfax for some of the forty-four thousand acres he owned in the Northern Neck. By 1760, Carter owed Fairfax £433 and claimed, probably incorrectly, that he was being charged for too much acreage.[64]

"King" Carter had concentrated on producing quality sweet-scented tobacco for the English market in the 1720s; he heavily processed the final product, shipping either leaves stemmed and "strait laid" in the hogsheads or "well pickt clean lugs," all tightly prized to make freight-saving heavy hogsheads. Since Carter had all his tobacco brought to one location, ship captains paid him a higher price for convenience in loading. John, Charles, and Landon Carter continued producing for the domestic market (although after the passage of the inspection act they lost the advantage of a premium for convenient loading), "believing," as John put it in 1735, "that a particular Care in the choice of my overseers and my own additional trouble" should produce a grade of tobacco especially suited to English palates. In the 1750s, Landon experimented with sweet-scented seeds obtained from a number of other large Virginia growers "to see what it is that has made so much noise" as well as with other strains including "Orleans" and "Parkers Broad Green." Charles Carter maintained a concern for neat handling, in 1757 thanking Liverpool merchant Charles Goore "for the hint you gave of tying over Stem'd tobo which mistake I shall rectify for the future." Three years later, Charles wrote Capel and Osgood Hanbury of London, "No Body takes more Pain than I do, in the management of their Tobo, and that none is in higher Esteem at the Warehouses than my Crops."[65]

Unfortunately, English merchants did not always share the Carters' own high opinion of their tobacco. Between 1735 and 1738, London factor Edward Athawes regularly upbraided the Carters for shipping foul-scented, flabby, rotten, house-burnt, poorly stemmed, ill-sorted, and sloppily packed tobacco. During the next three decades, the Carters would try a large number of London, Liverpool, and other outport factors, periodically opening new accounts and breaking off old correspondences when individual merchants failed to obtain the high prices that the Carters thought they deserved. By

64. John Carter to Edward Athawes, Aug. 1, 1738, Carter Family Letterbook, CWL; George William Fairfax to Landon Carter, Apr. 10, 1760, Carter Family Papers, UVa, microfilm ed., CWL; Wells, "The Eighteenth-Century Landscape," *Northern Neck Hist. Mag.*, XXXVII (1987), 4227.

65. John Carter to ?, Jan. 12, 1735, Charles Carter to Charles Goore, Mar. 10, 1757, Charles Carter to Capel and Osgood Hanbury, Feb. 6, 1760, Carter Family Letterbook, CWL; Greene, ed., *Diary of Carter*, I, 197, 200.

the 1730s, they were getting better returns for their tobacco in Liverpool than in London. They were compelled to continue sending some consignments to London, however, to pay debts and to obtain metropolitan goods.[66]

John, Charles, and Landon also followed their father's example in pursuing other avenues for making money. From the 1730s, John and Charles were involved in selling new African slaves for English merchants, maintaining mills, promoting towns, trying to mine copper, and holding various lucrative public posts. They also shipped corn and wheat to the West Indies and to the wine islands, often in ships they owned. Assembling a cargo was not a problem, for they could load an entire vessel with the surplus of their plantations alone. Since surviving papers document their grain marketing much less fully than their tobacco selling, the amount of income each of these trades generated is unclear.

What is certain is that, along with prime productive assets, the Carters also inherited and further developed habits of lavish conspicuous consumption. And they did not always follow "King" Carter's advice about balancing expenditures with actual income rather than relying on overly optimistic projections based on the most favorable years. Among grandees like themselves, a growing taste for large brick houses, rich clothing, stylish furniture, silver plate, fine wines, elaborate carriages, fast horses, and expensive English educations for sons sometimes outstripped annual disposable income. "King" Carter's four surviving sons—Charles, John, Robert, and Landon—as well as two sons-in-law, Mann Page and Benjamin Harrison, for example, all built large, fashionable new great houses in the second quarter of the century. The costs of constructing and maintaining the Pages' three-story Rosewell mansion, started in the 1720s and completed around 1737, "probably the grandest of the new large and refined private houses built in the continental Anglo-American colonies," bankrupted his heirs. In 1735, the Carter brothers would complain that, though their nephew Mann Page (son and heir of Rosewell's initiator) had abundant workers and land as fine as any in Virginia, "yet we have scarce been able to make tobo enough to support the Expense of the Estate, tho my father used to say he had raised great profits by it." London merchant Edward Athawes thought poor management the culprit. "Tis strange to me that above 100 working negroes on fine land should not be able to raise a greater Quantity of Tobo in a year not remarkable for bad seasons or short crops." He advised the Carters that the slaves' low productivity "intimates the necessity of a more industrious management." Charles

66. Edward Athawes to John and Charles Carter, June 8, Dec. 20, 31, 1736, Jan. 12, 1736/7, July 5, 1737, Mar. 10, 1737/8, Dec. 22, 1738, Carter Family Papers, UVa, microfilm ed., CWL.

Carter's seven-bay Cleve mansion was equally as pretentious as Rosewell in exterior decoration, if not in size. Although Charles could well afford the cost of its construction, the further expenses of the lavish ways of life such buildings encouraged put strains on the family's fortune.[67]

Year after year, one or another of the Carter clan would overdraw their accounts with English factors, bemoan unexpected expenses or short crops, and enclose an invoice for additional costly clothing, furniture, and wine that they explained were absolute necessities of housekeeping. Extravagant living, especially among Carter's grandchildren, coupled with indifferent management of plantations and workers could bring even well-endowed planters such as these into debt. And, since much of the land and many of the enslaved workers each owned were entailed, "King" Carter's foresight precluded their selling land or slaves to pay outstanding balances. Hence, there was always pressure either to expand tobacco production or to add additional cash crops to augment tobacco revenues.

In addition, the Carter sons succumbed to the widespread tendency among elite families after midcentury to escalate the portions parents gave their children and to advance such portions earlier in the heirs' lives. Passing on family lands and laborers to work them upon the death of one or both parents no longer sufficed. Most of the Carters felt increasingly obligated to set up their sons with plantations and enslaved workforces and even with fashionable new brick dwellings, if not as soon as the sons turned twenty-one, at least by the time they married. The size of daughters' marriage portions also increased, although, when fathers fell upon hard times, it was usually the girls' dowries that were reduced first. Attempting to provide so generously for their offspring might well have placed greater burdens on the Carters' estates than did the parents' own conspicuous consumption.[68]

67. Carter to Dawkins, Apr. 16, 1730, Carter Letterbook, Carter Papers, CWL; Athawes to John and Charles Carter, Jan. 12, 1735/6, Carter Family Papers, UVa, microfilm ed., CWL. For Carter family mansions, see Thomas Tileston Waterman and John A. Barrows, *Domestic Colonial Architecture of Tidewater Virginia* (1932; rpt. New York, 1968), 111–116; Thomas Tileston Waterman, *The Mansions of Virginia, 1706–1776* (Chapel Hill, N.C., 1945), 111–123; and Edward A. Chappell, "Rosewell's Architecture Reevaluated," in Rachel Most, ed., *Discovering Rosewell: An Historical, Architectural, and Archaeological Overview* (Gloucester, Va., 1994), 9–38 (quotation on 9).

68. Rev. Andrew Burnaby, *Travels through the Middle Settlements in North-America, in the Years 1759 and 1760; with Observations upon the State of the Colonies,* 2d ed. (1775; rpt. Ithaca, N.Y., 1960), 22–23, observed: "The display of a character thus constituted ["indolent, easy, and good-natured; extremely fond of society, and much given to convivial pleasures"], will naturally be in acts of extravagance, ostentation, and a disregard of oeconomy; it is not extraordinary, therefore, that the Virginians outrun their incomes."

Although the Carters remained heavily reliant on tobacco, like other large Rappahannock basin planters they increasingly attempted to defray the "contingent charges" of their plantations by selling surplus corn, wheat, and pork. In the 1730s, there was "not always a tolerable price for these," but the situation had changed by the 1750s. The Carters differed from most other planters, however, in the greater emphasis they put on wheat and other small grains instead of on indigenous corn. Rather than opting to greatly expand production of a crop the seasonal cycle of which meshed most readily with tobacco, they devoted more of their attention and energy to Old World grains. These grains were more difficult to grow successfully in the Virginia tidewater, and to be very profitable required unfamiliar cultivation and processing techniques that conflicted with the old ways both overseers and slaves were determined to follow.[69]

The Carters' motives were surely in part to profit, but equally important was the enhanced prestige they associated with practicing up-to-date European-style agriculture. Both Charles and Landon were well read in recent British agricultural literature as well as in works of history, natural history, science, and political economy that argued for human perfectibility and that advocated a cult of improvement. Like Cambridge-educated Antiguan pioneering "improving" sugar planter Samuel Martin, the Carters had come to believe that a good planter "ought at least to know the rudiments of all the sciences, if he attains not the mastery of them." Among other requisite accomplishments that Martin enumerated for a gentleman planter were that he be "something of an architect, and well skilled in mechanics, because there is in every plantation a variety of buildings and machinery, upon the right contrivance and use of which, much of the planter's success depends," and he should also have enough "physical skill" to treat his slaves. In addition to designing their own great houses, Charles Carter tried to mobilize the colonial government to promote mechanical arts, and Landon devised cattle-feeding stalls and tide gates for his plantation. Landon, at least, also spent a great deal of time and energy doctoring both the slaves and his own children. (On the other hand, neither Charles nor Landon ever became, as Martin recommended, "an adept in figures, and in all the arts of oeconomy.") Both Carters, however, like Martin (whose *Essay upon Plantership* appeared in multiple editions between 1750 and 1802), advocated adopting English husbandry methods, including plowing, manuring, crop rotations, and new and better tools to raise outputs of colonial staples. And, like contemporary large

69. John Carter to Alderman Perry, Aug. 1, 1734, Carter Family Letterbook, CWL.

South Carolina planters, the Carters sought to make their own plantations "a showplace for the profitability of commodity diversification."[70]

Landon (1710–1778), "King" Carter's fourth son, returned to Virginia in 1727 to commence planting after seven years of schooling in England. After a brief time at the College of William and Mary, he began managing some of his father's quarters in Northumberland County, and in 1732 he married the daughter of another Virginia grandee. (After his first wife's death in 1740, Landon took in rapid succession second and third wives in the 1740s.) Shortly after "King" Carter died, Landon moved to one of the plantations in Richmond County that he inherited. Almost immediately he was appointed a justice of the peace and soon also became an Anglican vestryman and colonel in the militia. Upon receiving a final share of his father's estate in 1738, Landon used some of the windfall to build an elegant brick Georgian great house, Sabine Hall, that reflected and reinforced the high social rank that was also part of the assets he inherited. Sited atop a ridge in order to give a commanding view of the Rappahannock River and the milewide valley below, the mansion's rusticated stone trim, geometric proportions, and elements of Greek and Roman decoration displayed Landon's knowledge of up-to-date British architecture and his reverence for the classics as well as his ambitions (often thwarted) for controlling everything and everyone within his domain (Plate 31).[71]

Carter became both an aggressive estate builder and an extraordinarily

70. Rhys Isaac, *Landon Carter's Uneasy Kingdom: Revolution and Rebellion on a Virginia Plantation* (Oxford, 2004), 85–103, 359, discusses Landon Carter's library and his reading of it. Among the agricultural manuals Landon owned were Georg Andreas Agricola, *Experimental Husbandman and Gardener . . .* , trans. Richard Bradley (London, 1726), and Stephen Switzer, *Practical Husbandman and Planter . . .* (London, 1734). For Landon's scientific and medical interests, see 76–84, 105–120. For a catalog and discussion of Landon's surviving library, now at the University of Virginia, see Carol Edith Curtis, "The Library of Landon Carter of Sabine Hall, 1710–1778" (master's thesis, College of William and Mary, 1981). The contents of Charles Carter's library is unknown, but the language of his writings clearly demonstrates his familiarity with British agricultural literature. For Samuel Martin, see [Martin], *An Essay upon Plantership, Humbly Inscrib'd to All the Planters of the British Sugar-Colonies in America,* 2d ed. (Antigua, 1750), iii, vi–vii (quotations); and Richard B. Sheriden, "Samuel Martin, Innovating Sugar Planter of Antigua, 1750–1776," *Ag. Hist.,* XXXIV (1960), 126–139. For South Carolina, see Edelson, *Plantation Enterprise,* 204.

71. Greene, "Introduction," in Greene, ed., *Diary of Carter,* 3–61; "Landon Carter," in Bearss et al., eds., *Dictionary of Virginia Biography,* III, 76–78. Isaac, *Landon Carter's Uneasy Kingdom,* 85–103, also discusses Carter's management practices. For Sabine Hall, see ibid., 4–7; and Waterman, *The Mansions of Virginia,* 130–136. Waterman believed Robert "King" Carter built Sabine Hall for his son, but dendrochronology has subsequently proved Landon to be the builder between 1738 and 1742. The layout of slave quarters and other domestic and agricultural outbuildings in this period is unknown.

PLATE 31. Sabine Hall, Richmond County, Virginia. Drawn by Richard Anderson. *Built by Landon Carter between 1738 and 1742.*

attentive plantation manager. To eight operating farms in Richmond, Northumberland, and King George counties inherited from his father, he would add other plantations in York, Charles City, King and Queen, Stafford, and Westmoreland counties and lots in the town of Williamsburg that he obtained though his three marriages, other inheritances, or purchase. By 1750, he owned more than thirty-five thousand acres in the Northern Neck alone. Having also inherited at least one hundred enslaved workers, Landon, like his father, tended to use labor lavishly, at least on his home farm and on nearby quarters. In 1756, when his diary begins, Carter owned no fewer than two hundred enslaved Africans and African Americans.[72]

72. Carter inherited about one hundred slaves from his father and other bondpeople from his younger brother George (who died without issue), all of whom were entailed. His older brothers bought additional Africans for him in the late 1730s, and he might have acquired other slaves

By the mid-1750s, when he had succeeded, after several failed attempts, in getting himself elected to the House of Burgesses (1752–1768), Landon adopted what he called "English husbandry." Abandoning the extensive system of agriculture based on hoe culture, slash-and-burn clearing, and long fallows that Chesapeake colonists had developed over more than a century, Carter hoped instead to adopt up-to-date intensive British husbandry methods relying on plow culture and continuous cultivation of fields whose fertility was to be maintained with a combination of animal and vegetable manures. His main concession to regional agricultural traditions was to retain tobacco as one of his major cash crops. Carter laid out permanent, regularly shaped fields to facilitate plowing in divisions of fifty acres each in which he began alternating crops of tobacco with small grains. The shift to continuous cultivation of large swaths of land, however, meant that Carter raised tobacco on inferior as well as prime soils. Consequently, he expected that his tobacco crops would be less than those of a planter "who culls his land." He averred, however, that "yet by our grain we more than equal him and profit and 'tis evident that our lands run better into grass after such a method than theirs and of Course our creatures will improve and in time add to our profit."[73]

Carter had cattle and sheep penned on his tobacco grounds to improve yields and sometimes ordered that manure be carted to and spread on some portion of his corn and wheat lands. Although Carter then got more tobacco to the acre, annual manuring, he found, required "a greater labour than I thought," because litter from the cow pens had to be plowed under several times if it was to rot properly, something that he could not always manage to get done. Neither did he have enough manure to regularly dress all the tobacco ground the slaves tilled, much less the fields planted in grain. He was thus encouraged to keep more livestock than the land could well support. Clearly, the weakness in Landon's scheme was insufficient grain or pasture to carry enough manure-producing animals. He did try to grow a little clover, timothy, and turnips but, like Joseph Ball, relied largely on natural pasture, marsh hay, and corn fodder to feed his cattle and sheep. In bad years,

---

through his three marriages. Carter's biographers have been decidedly incurious about how many slaves Carter held in the 1750s. There were 401 listed in his 1779 probate inventory, at which time he had already given at least 56 others to his sons (Philip Morgan, "Three Planters and Their Slaves: Perspectives on Slavery in Virginia, South Carolina, and Jamaica, 1750–1790," in Winthrop D. Jordan and Sheila L. Skemp, eds., *Race and Family in the Colonial South* . . . [Jackson, Miss., 1987], 43, 148–149 n. 9). Isaac, *Landon Carter's Uneasy Kingdom,* 59, estimated the number of slaves at Sabine Hall and five nearby quarters at around 100. This number does not include bondpeople living on more distant plantations or those deeded to Carter's eldest son.

73. Greene, ed., *Diary of Carter,* I, 163–164.

he scarcely made enough provender to keep the stock alive over the winter, much less to bring them through in good health.[74]

Poorly nourished livestock, in turn, hampered Landon's efforts to make wheat an important second staple. Since wheat yields per acre remained low, to make a crop large enough to bring in much money a planter needed to prepare many acres of at least minimally smoothed ground. Well-fed draft animals were required to pull the plows, harrows, and rollers and at harvest time to cart the grain to the barns. Landon was not able to make enough corn, fodder, oats, or hay to adequately feed working horses and oxen. Often his exhausted animals either refused to work or, if pressed too hard, simply dropped in their traces. Consequently, planting was often delayed because the ground was not ready in time, or the grains did poorly because of inadequate soil preparation. In addition, Landon seems to have miscalculated the extra effort required to harvest and thresh so much grain. Sometimes more oats came up than he could get mowed. Enslaved threshers, bored with the tedium of their assignment and exhausted by the long work hours Landon demanded, retaliated with work slowdowns. In 1757, he was able to get his oats out only by daily whipping the threshing crew. Even a Carter did not always have enough workers to grow and process two major and several minor cash crops each year while also implementing labor-intensive improvement projects, developed for British farms of no more than 150 acres, on plantations of 500 to 1,500 acres.[75]

Landon also made other, less systematic attempts to further diversify his crop mix. Like Robert Bristow and John Baylor, he made a trial of indigo in 1757 and 1758 in response to high prices occasioned by the Seven Years' War. He was unable to get clear directions for raising the crop, and in the fall of 1757 the seed from the first year's planting failed to mature. In 1758, Carter noted that he had invested eighteen days of labor from three enslaved men and one boy in making hills for indigo and sowing the seed, which he recorded in his diary so that he could later estimate his profit. As no further mention is made of the crop, Carter, like Baylor, must have abandoned this

74. Ibid., 149–150, 172, 204–205. An attempt to plow under pea vines for manure was also unsuccessful (129–130, 147–148). For problems with manuring and livestock food, see 147–150, 163–164, 201, 207–208, 210, 226, 236, 252, 254, and 260–261. Ulrich Bonnell Phillips, *Life and Labor in the Old South* (1929; rpt. Boston, 1963), 221, also observed the problem of insufficient forage.

75. For example, Greene, ed., *Diary of Carter,* I, 129–131, 133, 138–139, 146–147, 149–151, 155–156, 158–159, 160–162, 179, 190, 208, 272. For English farm size, see Robert Allen, "Agriculture during the Industrial Revolution," in Roderick Floud and Donald McCloskey, eds., *The Economic History of Britain since 1700,* I, *1700–1860,* 2d ed. (Cambridge, 1994), 99–100.

risky staple, likely when the second year's planting came to naught. By 1763, Landon, emulating his brother Charles, had also begun experimenting with growing grapes for making wine. This project, too, he undertook both for profit and for prestige. As merchant Thomas Nelson wrote Carter, "I have ever been of opinion, that this business will be attended with great profit for the adventurer as well as to the Country in general." Carter likely had to settle for prestige, since there appears to have been little or no profit.[76]

In any event, Landon would have been hard-pressed to assess the overall profitability of his agricultural operations. Although he did plan to calculate the profit from experiments such as raising indigo and feeding cattle in stalls, wanting "to see how things would turn out before I engaged myself in too much expense about them," keeping systematic records of total plantation revenues and expenditures had no role in his schemes for agricultural improvement. Carter continued to calculate plantation profits from the size of his cash crops. Two inventories of Landon's papers reveal that he practiced "pigeonhole" bookkeeping, saving loose accouts from English factors, local merchants, overseers, and artisans, records of the size and disposition of corn and tobacco crops, land records, and miscellaneous correspondence in at least forty unwieldy bundles stored in his desk. No mention is made in his diary of keeping regular double-entry accounts.[77]

Whether Carter gained, lost, or broke even by adopting "English" husbandry, it is clear that his enslaved workers lost leisure time with the shift to continuous land cultivation. Although they no longer had to clear new fields each year, they were still required to raise full crops of tobacco along with tending many additional acres in wheat and oats. The amount of time devoted to plowing increased dramatically, and enslaved field hands still had to do the hard labor involved in turning over, loading, hauling, and digging in more than 33,000 bushels of animal manure in March and April when the pressure for making enough hills for tobacco and corn was already intense. Moreover, Carter was obsessed with maintaining steady output throughout the thirteen to sixteen hours slaves were required to work during the growing season, at times subjecting those who failed to keep pace to daily whippings. For intensive English-style husbandry to turn a profit, more work had to be

76. Greene, ed., *Diary of Carter,* I, 180–185, 187–188, 203, 220, 228; Thomas Nelson to Landon Carter, May 26, 1763 (quotation), Charles Carter to Landon Carter, August 1763, Carter Family Papers, UVa, microfilm ed., CWL.

77. Greene, ed., *Diary of Carter,* I, 149–150 (quotation on 150), 220; Landon Carter, A List of All My Papers in the House, Mar. 30, 1773, Robert Wormley Carter, A List of Papers of L. C. Decd, May 1784, Carter Family Papers, UVa, microfilm ed., CWL.

extracted from the enslaved than had ever been required with traditional farming methods.[78]

Charles Carter (1707–1764), "King" Carter's third son, started his career as a planter in 1724 at age seventeen upon returning from three years of schooling in England. He married in 1728 as soon as he reached majority, and his father set the couple up on a plantation in Middlesex County. Like his brother Landon, Charles sought to capitalize on the high social status he inherited as well to project a refined, genteel image, as shown in a portrait commissioned around the time of his marriage (Plate 32). Charles was shortly appointed naval officer for the Rappahannock district and justice of the peace. He also became a long-serving and influential member of the House of Burgesses (1736–1764), although his first two elections were contested, and in the second it was alleged that he had bribed some of the voters. Upon "King" Carter's death, Charles inherited land in eight counties and eventually settled on a tract he purchased in King George County near the town of Port Royal. There he built his great house, Cleve, in the late 1740s, employing as construction supervisor the man who subsequently built the new capitol in Williamsburg (Plate 33). Rusticated stone dressings on the corners and around windows and doorways contributed to the impressive appearance of the exterior, and an elaborate Georgian stair and identically sized, richly ornamented ground-floor rooms demonstrated Carter's concern for order and symmetry and his desire for creating spaces designed to facilitate genteel entertainments. Grand buildings and high social aspirations, in turn, required that inherited plantations generate high revenues. Charles's strategies for enhancing the value of his estate included not only the sorts of agriculture on which brother Landon concentrated but also a greater reliance on commerce and manufacturing. Carter invested in shares in oceangoing ships and in milling and mining ventures, was actively involved in marketing grain to the West Indies and the wine islands, and also engaged in selling African slaves on commission for English merchants.[79]

78. Greene, ed., *Diary of Carter,* I, 147, 172, 212, 226. For the increased work entailed by continuous cultivation and plow culture, see Justin Roberts, "Sunup to Sundown: Plantation Management Strategies and Slave Work Routines in Barbados, Jamaica, and Virginia, 1776–1810" (Ph.D. diss., Johns Hopkins University, 2008).

79. "Charles Carter (ca. 1710–26 April 1764)," in Bearss et al., eds., *Dictionary of Virginia Biography,* III, 55–57; Fairfax Harrison, ed., "The Will of Charles Carter of Cleve," *VMHB,* XXXI (1923), 39–45. For the mansion, see Waterman and Barrows, *Domestic Colonial Architecture,* 111–116; Waterman, *Mansions of Virginia,* 178–180; and Mark R. Wenger, "The Great House at Carter's Grove: An Interpretive Essay," Colonial Williamsburg Foundation Research Report

PLATE 32. *Portrait of Charles Carter of Cleve* (1707–1764). By an unidentified American artist, circa 1725–1730. Courtesy, The Colonial Williamsburg Foundation

PLATE 33. Cleve, King George County, Virginia. Drawn by Richard Anderson. *Built by Charles Carter around 1747.*

Charles had even more need to manage his plantations well and to adopt other income-producing activities than did Landon, since he sired thirteen surviving children (ten of them daughters whose dowries would drain family assets) by three successive wives. By the early 1760s, Carter had added saw-milling, merchant grain milling, distilling, and the manufacture of ship's bread to the family's income-generating enterprises. But, in order to enhance a larger public reputation as an improving farmer and "benefactor of mankind," Charles also experimented with making saltpeter, raising silk worms, gathering medicinal plants, curing fish, growing English grasses, raising olives, and making wine. Most of these schemes were labor-intensive, hopelessly un-

---

Series, no. 1655, 64–65, CWL. For Cleve after Charles's death, see Charles Carter to Landon Carter, Mar. 31, 1765, Aug. 4, 1766, Carter Family Papers, UVa, microfilm ed., CWL.

profitable, and, in the case of olives, climatically impossible ventures that only an extremely wealthy man like Carter had the resources to pursue.[80]

Carter was committed to diversifying the colony's economy and to developing western lands, through both private example and public agency. In 1759, he sponsored a bill in the Virginia assembly creating a committee to promote economic diversification and to award monetary bounties to those who perfected "any art or manufacture of service to the public." As committee chair, Carter initiated a correspondence with the Royal Society of Arts, an organization founded in London in 1754 to award premiums for the production of commodities in short supply in the British Empire and for the discovery of new techniques and inventions. Carter's proposals for colonial economic diversification were only marginally more practical than those advocated more than a century before by the Virginia Company and were quite similar to the ones promoted by Governor William Berkeley in the 1660s. They included products such as iron, tar, pitch, turpentine, hemp, flax, and lumber (which some Chesapeake entrepreneurs were beginning to turn into viable export commodities), preserved fish (which was at least a possibility) but also less practical, labor-intensive wines, potash and pearl ash, and that perennial will-o'-the-wisp, silk. The London Society of Arts awarded Carter a gold medal in 1763 "for care and industry to promote in Virginia the purposes of this Society in the British colonies," a recognition that, according to Carter, made other members of the Virginia committee jealous so that in the end "all my endeavours were little regarded."[81]

Like his brother Landon, Charles also attempted to implement "English husbandry" on his tidewater farms. Concerned with the depletion of soil fertility that indigenous methods wrought, Charles turned to continuous cultivation of already cleared acreage, rotation of crops of tobacco with small grains, and penning cattle for manure. In his will he states:

> Whereas I have endeavoured to put my Manor of Cleve into a situation different from the common method of cultivating lands in the Colony, I do, therefore, desire if the plan begun is not completed, that my ex[ecut]ors will carry on the same agreeable to my directions in writing, entitled "A new system of Virginia Husbandry, or the Little farm improved wherein the business of making Tobo., farming, improving lands and making wine, are largely treated of and earnestly recommended," all which, I have wrote for the benefit of my children.

80. Robert Leroy Hilldrup, "A Campaign to Promote the Prosperity of Colonial Virginia," *VMHB*, LXVII (1959), 410–428.

81. Ibid., 428.

Unfortunately, this document has not survived, although some of its content can be inferred from letters Charles Junior wrote in 1765 and 1766. Charles Senior started out by systematically penning cattle on fields intended for tobacco and then plowing in the manure; the next year, cattle were to be penned again on areas insufficiently fertilized the first year to "make all the Tobo ground good." In addition, slaves carted out dung from the stables and farmyard to fertilize each newly made tobacco hill. Other fields were rotated in corn, wheat, oats, barley, and rye. By the second year after Charles's death, his "new system" was proving difficult to implement in the face of repeated summer droughts. The tobacco crop of 1766 was a total failure, the corn crop "very mean," the wheat destroyed by rust, and the oats hardly worth mowing. The overseer at Cleve quit in despair. Sarcastically noting, "A fine prospect there truly, of extricating an Estate over head and Ears in Debt," Charles Junior had serious doubts about the practicality of his father's ambitious improvement schemes.[82]

Charles Senior wanted to educate all his sons in England until they reached age twenty-one and then to set them up with plantations and slaves of their own. He also hoped to give his youngest daughters dowries of two thousand pounds each, as he had done when the two oldest had married, but "this long and ruinous war [and] the unfavorable seasons has rendered [this plan] impossible." Consequently, Carter could give the eight younger daughters portions of only one thousand pounds and two slaves apiece and, moreover, directed that they be maintained during their minorities "with great frugality." (His definition of "frugality," however, included keeping, at the expense of the estate, a coach, six horses, a coachman, and a postilion.) Since no account books or systematic records of crop production have been preserved, it is impossible to determine whether Charles's financial difficulties stemmed from wartime shipping losses, poor farm management, overspending on improvement schemes and on portions for the older children, or simply from overly optimistic projections of future income. When Charles cautioned his executors that "the extravagance of the present days and the flattering hopes of Great fortunes may be a temptation to run into unnecessary expenses in living," he identified a weakness to which many members of the Carter family were prone. Had Charles lived long enough to recoup his fortunes during the brief economic boom following the war, his younger children might have fared better.[83]

As preoccupied as his father with maintaining the family estate and mem-

82. "Will of Charles Carter of Cleve," *VMHB*, XXXI (1923), 45–69 (quotation on 64).

83. Ibid., 58, 63.

ory, Charles enjoined the son who would inherit the dwelling plantation and his subsequent heirs to keep the family burial vault in repair and each year to pay the parish minister to preach a sermon on the anniversary day of Charles's death "so long as my manor of Cleve shall remain in the hands of my Male issue." Since "King" Carter had entailed the land he devised to Charles, this land and some of the slaves were required by law to descend to his eldest son, Charles Junior. Charles left all the land he had acquired since his father's death to his two other sons, along with 130 slaves to the second son and 100 more to the third, but similarly tied their hands by entailing their land in the male line and the slaves to their heirs. In the end, Charles's intentions could be fulfilled only by the eldest son's agreeing to secure an act from the legislature docking the entail of some of his inherited tracts so that he could sell land rather than slaves to pay his father's debts and legacies.[84]

Initially, Charles planned to give almost all of the slaves remaining in his control to his two younger sons, since Charles Junior had already been given his share of bondpeople. Charles later reconsidered this act of parental generosity. Anticipating that "there may be a considerable increase by birth" among the slaves before the two younger sons came of age, Charles decided that these sons would have to buy at their appraised value any "overplus" that exceeded the number of 230 allotted to them in Charles's will. This money was then to go to paying their unmarried sisters' reduced portions of one thousand pounds apiece.

As for the enslaved men, women, and children upon whose labors and biological increase the family's prosperity depended, Charles was far less generous. His sole remembrance to them was to require of his executors "that my slaves may be treated with humanity and supplied with necessary food, cloathing and bedding, and, in case of sickness or other misfortune, I desire they may be properly attended." Neither Charles Carter, nor other members of his privileged family, had as yet found reason to question the system of human bondage that sustained their fortunes.[85]

Among the large Rappahannock basin planters whose management strategies have been discussed here, Richard Corbin emerges as the most successful, both as a farmer and as an estate builder. Corbin embraced a pragmatic style of management that emphasized systematic planning, careful use of time, and stringent financial discipline. Not until the end of the century were such practices widely prescribed or routinely implemented on the largest and most

84. Ibid., 46–57, 60–62, 68; Brewer, "Entailing Aristocracy," *WMQ*, 3d Ser., LIV (1997), 329–330.

85. "Will of Charles Carter of Cleve," *VMHB*, XXXI (1923), 63, 68.

complex plantation operations. Corbin, however, made no public claims for the efficacy of his practices, and his accomplishments remain little noted. Charles and Landon Carter, in contrast, were less successful than Corbin in wresting the highest possible returns from their plantations or in managing their estates wisely. But, since they were among the first great tidewater planters to add a pronounced public dimension to large plantation agriculture, they remain much better known. In addition to obligations to maintain or expand family fortunes and to bequeath to one's heirs estates of the same or greater value than those the parents inherited, the Carters enlarged a gentry planter's duties to include the study and promotion of "improved" (that is, metropolitan) farming methods and the promotion of useful arts and manufactures. Such "endeavor to serve mankind," as Landon Carter later phrased it, served during the Golden Age in the Chesapeake to enhance reputation and to bring imperial honors to those progressive gentlemen who embraced the cult of improvement. Paradoxically, however, the noble aim of promoting improvement among part of the population also served to rationalize and justify the continued oppression and increased exploitation of perpetually bound laborers of African origin who were considered neither capable nor worthy of sharing in any meaningful way in that better future.[86]

86. Greene, ed., *Diary of Carter,* II, 880.

CHAPTER 7

Maryland, the Periphery, and Agricultural Change 1730–1763

In the early 1730s, it was not immediately apparent that Maryland planters were about to enter a period of unparalleled prosperity. As more African slaves were imported into the colony during the 1720s, the increased amounts of tobacco they were put to raising had led to a drop in price. Planters strove to compensate for lower prices by buying more laborers and clearing ever more land to grow even more tobacco, depleting forest reserves. As satirical poet Ebenezer Cook put it in *The Planter's Looking-Glass,* published in 1730, all tobacco growers

> *At Ax and Hoe, like Negroe Asses tug,*
> *To glut the Market with a poisonous Drug:*
> *Destroy sound Timber, and lay waste their Lands,*
> *To head a Troop of* Aethiopian *Hands.*

Cook was less concerned about the deleterious properties of the weed than about the effect its culture had on the colony's economy. He had no doubt about the connection between rising slave imports and overproduction of the staple:

> *From scorching* Africa's *burnt Shore,*
> *Brought* Aethiopian *Slaves great Store.*
> *More* Weeds *turn out, to Heat inur'd,*
> *Than by the* Populace *are cur'd,*
> *Makes it a Drug, as* Merchants *feel,*
> *Whose Chance it is in Trash to deal.*

Increased customs duties on tobacco in Britain, petty charges exacted by metropolitan merchants, and "other Frauds us'd in the Trade" worked especially to the detriment of small planters who, unlike large slaveowners, lacked the labor to expand production. *"Plebians,"* Cook asserted, "by it scarce can live, / To naked Brats Subsistence give." Planters who continued a shortsighted overdependence on tobacco coupled with a wasteful, extensive agriculture were depleting assets for future generations and, he maintained, *"will by their Heirs be curst for these Mistakes, / E'er* Saturn *thrice his Revolution makes."*[1]

Marylanders did not adopt Cook's proposed remedy of limiting the amount of tobacco planters would be allowed to produce, nor did they switch entirely to raising hemp, flax, cotton, and small grains as he advised. Many also failed to adopt plow culture or integrated stock farming, whereby animals fed on improved pastures would supply manure for the fields. As will be demonstrated in the case studies that follow, however, they did not wholly neglect agricultural diversification and import-replacement activities. The Maryland legislature did increase the colony's money supply by issuing a limited amount of paper currency, a measure that Cook favored. And, by the 1750s, the public warehouses and port towns that facilitated trade, local shipbuilding and iron industries, and expansion of intercolonial trades conducted by independent merchants—all also advocated by Cook—had become realities.

During the period in which management strategies, crop mixes, and typical outputs per worker of the major staples were converging on large tidewater Virginia plantations, gentry planters in Maryland chose different strategies for maintaining or expanding agricultural revenues depending on where they lived. Those on the lower Western Shore continued to concentrate on maximizing tobacco production and were slow to take much advantage of expanding grain markets. In contrast, those living at the head of the Chesapeake Bay and on the upper Eastern Shore turned aggressively to both corn and wheat as second and third staples. Although they continued to produce for and retain ties to the British tobacco market, Eastern Shore farmers also increasingly produced for and established ties to Philadelphia and local Maryland merchants trading to other mainland colonies, the West Indies, and Europe. Planters in peripheral areas—the lower Eastern Shore and the south bank of the James River—benefited less from improving markets for tobacco. Instead, they turned to making the most of whatever natural resources their region offered and began selling a broad array of products in regional and intercontinental markets.

Since most Maryland planters lacked the opportunities so readily available to elite Virginians to invest in or move to not too distant western lands,

1. E[benezer] C[ook], *The Planter's Looking-Glass, in Verse* (1730), in Bernard C. Steiner, ed., *Early Maryland Poetry: The Works of Ebenezer Cook . . .*, Maryland Historical Society Fund Publication, XXXVI (Baltimore, 1900), 33–52 (quotations on 36, 43).

they had to farm the acres they had more intensively. Land constraints, in turn, imposed limits on the number of slaves gentry planters could employ profitably, so most maintained comparatively small workforces. In the 1740s and 1750s, elite Maryland planters determined to build large family estates looked to trade, mining, and moneylending to augment income from agriculture. As in Virginia, profits from farming and nonagricultural investments enabled Maryland gentry families to build fine new houses and to acquire the requisite trappings of genteel culture. Middling and small planters benefited as well from higher tobacco prices and from intensified local exchange. The enslaved workers on whose labor the region's prosperity rested, in contrast, realized only longer workdays and more intense year-round labor.

Still Betting on Tobacco: Maryland's Western Shore

The management strategies of large planters living in the prime oronoco tobacco areas of Maryland's Western Shore are poorly documented for the Golden Age, since no plantation account books have apparently survived. Two studies of more ordinary planters based on probate inventories and the accounts of landlords of the Society of Jesus, however, provide some insights into the decisions Western Shore planters were making about crop mixes and allocation of labor and into how the tobacco inspection acts affected their fortunes. Moreover, since small planters in both colonies shared similar land and labor constraints, the experiences of these Maryland farmers are more broadly representative of the strategies common oronoco growers were adopting along both banks of the Potomac River.

In Saint Mary's County, a jurisdiction where commitment to tobacco was perhaps the most complete of any locality in the Chesapeake, mean crop per laborer remained between 1,000 and 1,100 pounds across the first half of the century. Among small planters who had neither bound labor nor, probably, additional hired hands, annual crops were higher—1,200 to 1,300 pounds. The size of the crop almost certainly reflects both the efforts of the planter and the additional, unrecorded part-time contribution of his wife or children to making the crop. Families such as these, at the bottom of the economic ladder, tried to compensate for especially low returns between 1740 and 1747, when Maryland oronoco prices plummeted, by planting more of the weed; in these years, the annual crops of the smallest planters jumped 150 pounds (Table 30).[2]

2. Although estimates of crop shares from inventories and accounts (in Saint Mary's County, crops were almost always reported in the account rather than in the inventory) should reveal very gross shifts in output over time, such estimates are subject to considerable error and hence are less

TABLE 30. Tobacco, Corn, and Wheat Crops per Laborer, Estimated from Saint Mary's County, Maryland, Inventories, 1718–1754

	No bound labor		1–4 laborers		5 or more laborers		All estates
Years	*N*	Mean	*N*	Mean	*N*	Mean	Mean
TOBACCO (Pounds)							
1718–1719	3	1,250	2	935	2	807	1,033
1720–1729	43	1,167	38	1,115	10	1,009	1,120
1730–1739	79	1,219	54	1,065	19	909	1,126
1740–1747	36	1,374	31	1,100	7	414	1,169
1748–1754	45	1,194	43	923	14	767	1,021
CORN (Barrels)							
1718–1719	2	5.5	—	—	—	—	5.5
1720–1729	6	10.6	11	14.8	9	9.6	12.1
1730–1739	8	11.8	6	13.8	5	12.5	12.6
1740–1747	8	11.6	10	12.5	2	7.0	11.6
1748–1754	8	10.6	11	9.1	5	11.5	9.6
WHEAT (Bushels)							
1718–1719	1	0.5	—	—	—	—	0.5
1720–1729	1	10.0	10	8.9	3	4.5	8.3
1730–1739	2	9.3	3	2.6	3	3.9	4.8
1740–1747	2	2.5	5	7.6	1	0.9	5.5
1748–1754	2	19.0	4	3.2	4	6.6	7.7

Note: For methods of estimation, see Note 2 in this chapter.
Sources: Inventories 1–60, Accounts, 1–37, MSA. These records include decedents from all Maryland counties. Inventories and accounts for Saint Mary's County residents were identified and coded by Historic Saint Mary's City in research conducted under grants from the National Endowment for the Humanities (RO-6228-72-468, RO-10585-74-267, RS-23687-76-431). Card files of Saint Mary's County decedents and printouts that include information on the size of the labor force are in MSA.

In slightly larger Saint Mary's County households, where there were between one and four indentured servants or slaves of age to work in the fields, tobacco shares per laborer averaged 925 to 1,100 pounds, perhaps closer to the amount that one regular field hand ordinarily produced. Among larger slaveholders who owned five or more field-workers, however, shares per hand were consistently smaller and much more variable. Between 1718 and 1739, estimated shares per hand on these estates were 800 to 1,000 pounds. During the 1740–1747 price slump, tobacco shares on the larger slave estates sunk to a mere 414 pounds per hand. This drop mirrors the one in tobacco outputs in Virginia sweet-scented counties in the 1710s and 1720s during a price fall in that region, suggesting that reduced emphasis on tobacco was a conscious strategy. Once Maryland's tobacco inspection law came into effect in 1748 and prices began to improve, these slaveholders seem to have begun to

---

reliable than crop shares reported in plantation accounts where the number of laborers working in the crop and the number of shares the overseer received is clearly stated. I used the following procedures: 1) For estates with no bound labor, I counted the decedent as one share. This assumption clearly underestimates labor, for other family members might have worked in the crop as well. I eliminated all inventories in this category with crops of more than twenty-five hundred pounds because additional free adult laborers must have been present. The figures for this category most accurately reflect the output of one adult man, his wife, and minor children. 2) In all estates with bound labor, I added an additional share for a supervisor, either the decedent or an overseer. (The exception was a few widows' estates where the account made clear that no supervisor was present.) In a handful of very large estates where the inventory specified that there were two quarters, I allowed for two supervisors. These assumptions overestimate the number of supervisors, especially on farms with fewer than five or six laborers. The decedent might have died before the crop was finished or might have been too ill to contribute any labor to it. Since overseers usually contracted for a full crop cycle, it was difficult to secure replacement supervisors when a planter died in midseason. 3) Especially in larger estates, the number of field laborers is probably overestimated. I used the definition of a crop share that period planters employed: all adult male and female hands were counted as one share and adolescents ages twelve to fifteen (determined by ages in the inventory when given, otherwise inferred from prices) as half sharers. I eliminated from the calculations all slaves specifically listed as skilled, superannuated, or disabled and all those with exceptionally high or exceptionally low valuations that indicate one of these conditions. Even with these deductions, the number of field laborers is doubtless still overestimated, as some slaves on larger estates almost certainly worked at trades or as domestics. In the absence of special notations or extraordinarily high valuations, there is no way to identify such nonfield laborers.

Another problem is that crop outputs recorded in inventories might have been uncharacteristically low compared to previous years. Where plantation accounts survive over long runs of time and especially where they bridge two generations, outputs were especially low in the year the original owner died (and perhaps in the year before if he suffered from a lingering illness) and for a season or so after until the next owner gained some management experience. Consequently, estimates of crop shares based on inventories might well be biased downward owing to the circumstances that precipitated the record.

push their laborers harder; shares per hand on larger estates then rose to 767 pounds a hand.[3]

Starting in the 1720s, even in an eminently tobacco-centered county like Saint Mary's, planters also began to produce more grains, especially maize. Families without any additional bound labor made only about eleven barrels of corn per year, a crop that probably afforded little marketable surplus. These planters were not necessarily unresponsive to new market opportunities, as confirmed by the exporting of wheat and corn by some Saint Mary's County merchants by the 1750s and by their contracting many of their debts in grain as well as in tobacco. So long as farmers readied their fields for planting using only hoes, however, time constraints in the spring precluded any significant expansion of their cornfields. Nor did small Saint Mary's County farmers bother much about wheat. Those who did sowed no more than an acre or two, from which they would have gotten but five to ten bushels of grain, a product likely intended primarily for family consumption. In contrast, planters who held any bound laborers, be it only one, began to raise some corn for the market. Households with one to four enslaved field hands made more than twelve barrels per laborer, thus producing a total crop that consistently afforded some salable surplus. On plantations with even more workers, corn production per hand (like that of tobacco) was slightly lower but still averaged about ten barrels, again a level that indicates production for market. On farms where there were adult slaves or servants, planters were more likely, too, to raise some wheat. The amount produced was not large, ranging from one to nineteen bushels a hand, again much of it probably intended for home consumption, although on a few farms crops were large enough to provide some grain for sale. Access to bound labor thus seems to have been the most crucial prerequisite for raising grains for market.[4]

3. These figures doubtless suffer from the problems of estimating labor-force size discussed in Note 2. There is no reason to suppose, however, that estimation errors obscure the direction of change from lower to higher outputs after 1747. A later letter of Virginia landlord James Mercer to his estate agent demonstrates that enslaved blacks were assumed to tend more acreage in crops than were white laborers. Mercer stated that, when he rented farms to tenants who owned no bound laborers, "a white man, his wife and child, the whole three" did not work as much land as one adult slave doing full-time fieldwork. Mercer wanted to charge an extra rent of 250 pounds of tobacco apiece for additional white laborers working a tenement but 400 pounds apiece for blacks (Mercer to Battaile Muse, Sept. 7, 1780, Muse Papers, Duke University Library, Durham, N.C., quoted in Willard F. Bliss, "The Rise of Tenancy in Virginia," *VMHB,* LVIII [1950], 432). This observation underscores the effect of bound labor on farm output and the greater intensity with which enslaved field hands worked.

4. According to the seventeenth-century rule of thumb (which seems to have held true in

A study of levels and distribution of inventoried personal wealth in Saint Mary's County across the eighteenth century highlights some of the consequences of the inability of small planters to take much advantage of the new opportunities afforded by rising grain markets. The personal wealth of both large planters and of those in the upper middling ranks, 90 percent of whom owned slaves by the 1750s, rose between 1730 and 1755. Between the 1730s and 1760s, the share of total personal wealth held by the poorest planters with estates valued at less than fifty pounds fell from about 7 percent to less than 4 percent. The authors concluded nonslaveowners were the Golden Age's chief losers. A lack of resources to adopt new methods of husbandry left small planters "trapped in an outmoded method of farming, unable to participate as much as others in the wave of prosperity."[5]

Farther up the Potomac in adjacent Charles County, tenant farmers holding life leases on Cedar Point Neck, a large rental property owned by the Jesuits, achieved similar levels of production and, through intensified local exchange, realized improved standards of living. Two-thirds of the tenants were family farmers who owned no bound laborers, and, of the third who did, most had only one indentured servant. Their farms ranged from 50 to 230 acres, the mean being 130 acres. Attached to the locality by family ties, most tenants taking up leases after 1720 lived their lives on their tenements and passed on the leaseholds to their children. Since life leases gave the tenants an interest in their farms, most planted orchards and were careful to hus-

later years), if each taxable worker tended two acres in corn (yielding about five barrels of shelled grain) the total amount would suffice to feed any dependent women and children living on the plantation. Since enslaved women field hands were expected to grow as much corn as the men, there was likely to be some built-in surplus produced on most slaveholders' farms (Lois Green Carr, Russell R. Menard, and Lorena S. Walsh, *Robert Cole's World: Agriculture and Society in Early Maryland* [Chapel Hill, N.C., 1991], 36–37; for spring time constraints, see 69–70). Paul G. E. Clemens, *The Atlantic Economy and Colonial Maryland's Eastern Shore: From Tobacco to Grain* (Ithaca, N.Y., 1980), table 24 and 191–193, found that, on upper Eastern Shore farms that produced both tobacco and wheat, tobacco crops per hand were higher than on farms that grew only tobacco. He concluded that the difference probably reflected a greater use of hired labor among grain growers. Whether this was also the case in Saint Mary's County is uncertain, as estate accounts seldom specify the nature of estate expenses. For diversification and marketing networks in Saint Mary's County, see Lois Green Carr and Lorena S. Walsh, "Economic Diversification and Labor Organization in the Chesapeake, 1650–1820," in Stephen Innes, ed., *Work and Labor in Early America* (Chapel Hill, N.C., 1988), 144–149, 166–175.

5. Lois Green Carr and Russell R. Menard, "Wealth and Welfare in Early Maryland: Evidence from St. Mary's County," *WMQ*, 3d Ser., LVI (1999), 95–120 (quotation on 106). See also Aubrey C. Land, "The Tobacco Staple and the Planter's Problems: Technology, Labor, and Crops," *Ag. Hist.*, XLIII (1969), 79–81.

TABLE 31. Tobacco, Corn, and Wheat Crops per Adult Male Tenant Laborer, Charles County, Maryland, 1700–1776

Year	*N*	Tobacco (lbs.)	Corn (barrels)	Wheat (bushels)
to 1700	2	1,447	—	0
1700–1709	2	1,546	—	0
1710–1719	5	1,206	6.9	9.6
1720–1729	4	1,309	—	0
1730–1739	3	953	—	0
1740–1747	6	1,365	7.7	3.8
1748–1759	5	1,171	5.1	4.3
1760–1769	4	762	4.9	2.0
1770–1776	7	1,164	7.8	2.0

Source: Lorena S. Walsh, "Land, Landlord, and Leaseholder: Estate Management and Tenant Fortunes in Southern Maryland, 1642–1820," *Ag. Hist.*, LIX (1985), 373–396.

band their resources, keeping a reserve of land in fallow for future planting and maintaining wood lots sufficient to keep their households supplied with firewood and timber for building and fencing. Between 1710 and 1747, the tenants grew about thirteen hundred pounds of tobacco per adult male hand. After the inspection act took effect, production dropped 15 percent to eleven hundred pounds per hand. Apparently, the tenants began either discarding some inferior leaf when they packed their tobacco or the inspectors burned minor amounts. But the law occasioned no drastic reduction in outputs, an indication that these tenant farmers had already been taking reasonable care in processing. That the Cedar Point Neck tenants farmed especially fertile soils might, however, have permitted them to raise crops of higher quality than could tenants working more marginal land (Table 31).[6]

Although none of the tenants became rich, this level of production consistently enabled leaseholding families in which all members worked to some

6. For documentation and further detail, see Lorena S. Walsh, "Land, Landlord, and Leaseholder: Estate Management and Tenant Fortunes in Southern Maryland, 1642–1820," *Ag. Hist.*, LIX (1985), 373–396. This study is based on records of the Jesuit landlords that include Day Books, 1741–1743, 1763–1768, 1770–1785, Rent Book, 1755–1818, and Loose Papers, Maryland Province, Society of Jesus Archives, Special Collections, Georgetown University Library, Washington, D.C. The less favorable conditions tenants on poorer lands faced are described in Gregory A. Stiverson, *Poverty in a Land of Plenty: Tenancy in Eighteenth-Century Maryland* (Baltimore, 1977).

extent in the fields to pay rents of eight hundred to one thousand pounds of tobacco plus one to two hundred pounds of corn-fed pork per year and still to have enough left to live in relative comfort. The mean and median values of tenant's personal estates inventoried between 1700 and 1739 remained stable at about £75 Maryland currency constant value. But, between 1740 and 1759, the mean value of tenant estates rose to £110, and median values to £87, reflecting a standard of living comparable to that of small freeholders. These Charles County tenants depended on tobacco for most income from staple crops. Almost none used plows for cultivation, and they grew only about six barrels of corn per hand, a crop that should have left little extra after the needs of the family and their livestock were satisfied. Some of the tenants did sell small amounts of surplus corn to their landlords, however, who in turn had begun a minor trade in grain by midcentury.

Livestock raising, import-replacement activities, and working for their landlords provided tenant families with additional income that allowed them to purchase a few imported amenities for their households. All the tenants raised cattle and hogs, and some began to keep sheep and to grow a little wheat for family use and local trade. Most produced their own casks for tobacco and cider and did their own carpentry work. The women spun wool yarn for knitting stockings and gloves. Only one tenant family wove cloth, however. Most purchased cloth, clothing, and metal products from the English merchants who bought their tobacco.[7]

The leaseholders used any spare income to buy more livestock, tools, and consumer goods, rather than to invest in markedly better houses. At midcentury, most still lived in small impermanent dwellings similar to one sketched at the turn of the century on a survey of a Cedar Point Neck tenement (Plate 34). Still, around 1750 tenant houses in the area typically measured only sixteen by twenty-four feet, affording space for one or two rooms below and a sleeping loft above, with two end chimneys. Earlier in the century, wattle and daub, as depicted in the plat, would have predominated. By the 1750s, however, leases more often mentioned chimneys made of brick. The leaded glass windows shown in the sketch would have been an unusual amenity in the 1710s, but by the 1750s and 1760s more prosperous

7. Walsh, "Land, Landlord, and Leaseholder," *Ag. Hist.*, LIX (1985), 373–396. When George Washington, who lived just across the Potomac River from Cedar Point Neck, needed to buy large quantities of corn to supply newly staffed plantations in 1760, he could not locate any sellers in his immediate neighborhood but was able to buy enough corn from these tenants' Charles County landlords. See Donald Jackson, ed., *The Diaries of George Washington,* 6 vols. (Charlottesville, Va., 1976), I, 211–215, 231–232.

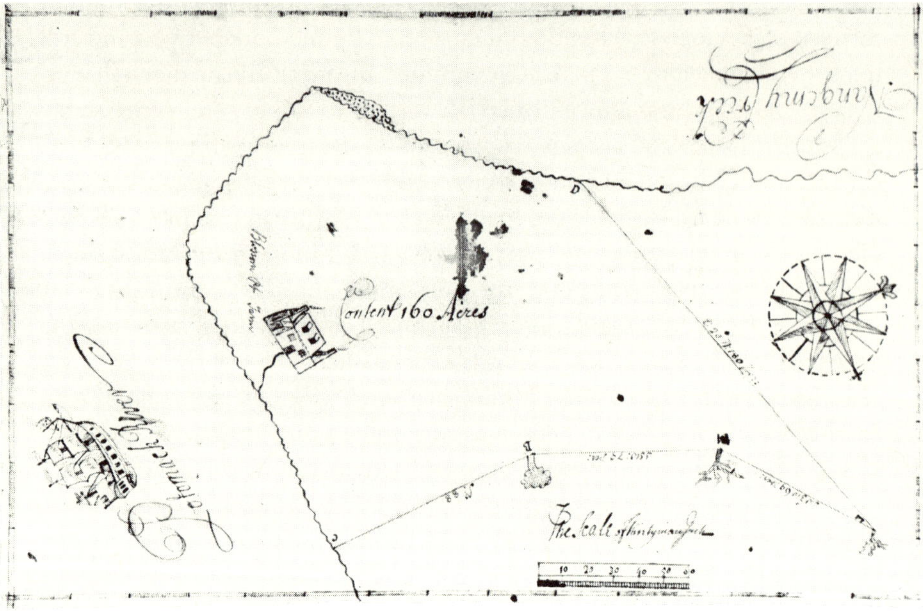

PLATE 34. Plat of Blossom Point Farm, 1712, Cedar Point Neck, Charles County, Maryland. Maryland Provincial Archives. Courtesy, Special Collections, Georgetown University Library, Washington, D.C.
Sketch of a prosperous tenant farmer's house.

tenants could contemplate adding such comforts to their unpretentious dwellings.

The tenants also supplied foodstuffs, building materials, and labor needed by the Jesuits for their manor farm, church, and enslaved workforce, which numbered in the thirties by the 1760s. The presence of this concentrated population afforded the tenants a ready market for produce and a chance for both men and women to earn extra income from part-time work. They sold butter, beans, wheat, livestock, meat, poultry, honey, and wax to the Jesuits; collected stud fees for breeding animals; did blacksmithing and tanning; repaired boats; sawed timber; gathered oyster shells for mortar; and helped with building on the Jesuits' manor farm and with repairing the church and mill. They also made casks, baskets, and hoes; threshed wheat; hired out oxen and carts; and moved goods and livestock by water. Tenant wives knitted stockings, dressed flax, made clothing for the slaves, and did washing, mending, spinning, and weaving; occasionally, they instructed and supervised the labor of the Jesuits' enslaved women and children at these tasks. Whereas large planters increased their incomes by taking advantage of expanding international markets, enterprising small freeholders and favorably situated tenant farmers like those on Cedar Point Neck improved their living standards by taking advantage of expanding local markets.[8]

Farther to the north in Anne Arundel County, still a prime tobacco area, four large planters for whom letterbooks and family papers survive—John Galloway, his son Samuel Galloway, Charles Carroll of Annapolis, and his cousin Dr. Charles Carroll of Annapolis—combined the raising of tobacco and grains with a wide variety of lucrative nonagrarian enterprises. These men were somewhat atypical of other large Maryland planters in this period, most of whom remained much more dependent on agriculture for their revenues. The careers of the Galloways and the Carrolls demonstrate, however, the extent to which a few talented entrepreneurs were able to augment their fortunes by taking advantage of a rapidly diversifying colonial economy. Both the Galloways and Charles Carroll were precluded by their religious faiths—the Galloways were Quakers and Carroll a Roman Catholic—from participating in Maryland public affairs. Their exclusion from offices of power—and, in the case of Charles Carroll, an acute sense of vulnerability—heightened their drive to build and secure family fortunes and might have encouraged these men to develop more innovative strategies for doing so.

Unlike many large Virginia planters, the Galloway family's estate building and maintenance strategies in the mid-eighteenth century did not involve the acquisition of vast acreage or of scores of slaves. The architect of the family's fortune, Samuel (1659–1721), the son of a small planter who had immigrated to Maryland in 1649, aggressively bought hundreds of acres of land at the turn of the century with which he endowed his five surviving sons. Samuel was not quite so keen to buy slaves, however, and, at his death in 1721, the family workforce numbered but twenty-nine slaves, including children, and one servant man. The bulk of Samuel's £3,595 personal estate (constant value) lay in outstanding mercantile credits. Thereafter, Galloway men farmed inherited tracts of one thousand to two thousand acres with workforces of twenty to forty slaves. Fathers who had several sons had to buy additional land, some of it in adjoining counties, that was later sold to raise portions for younger children; one kinsman who had no children kept his land in the family by passing it on to brothers' sons. Since the sizes of their farms were modest by elite standards, the Galloway offspring had no need to purchase large numbers of new hands in addition to the ten or so slaves most

8. Walsh, "Land, Landlord, and Leaseholder," *Ag. Hist.*, LIX (1985), 373–396.

inherited from their parents. The enslaved women's natural increase apparently met any additional labor needs.[9]

Samuel's third son, John Galloway (1693–1747), benefited from having been set up with a tract of land and a few slaves to work it during his father's lifetime and from inheriting more land and ten more slaves upon his parents' deaths. Like his father Samuel, uncle Richard (1663–1736), cousin Richard (1691–1741), and brothers Richard (1701–1731) and Joseph (?–1753), John combined diversified agriculture—tobacco (which the appraisers of his estate commended for "well handling good Quality and great weights"), corn, wheat, barley, oats, beans, orchard products, cattle, hogs, and sheep raised on fifteen hundred acres in southern Anne Arundel County—with mercantile enterprises. This combination of planting and trading was so successful that five of six Galloway family estates probated between 1721 and 1753 were valued between £1,627 and £3,596 (constant value), and the other, of John's widowed mother who died in 1723, at £871. John's second marriage in 1744 to Jane Fishbourn, the widow of a Philadelphia grain merchant, brought further useful trade connections. John and Jane Galloway, like John's parents, uncle, brothers, and cousin, enjoyed a genteel style of living. They furnished their house with fine furniture, clocks, sets of china, glassware, and silver plate; they drank imported wines and traveled the countryside in a chaise. John also sported a large and expensive wardrobe. The Galloways were careful, however, not to spend beyond their means. Final accounts were filed with the probate court for four of the seven family estates, and in none of them were there significant outstanding debts owed by the estate. Founder Samuel had more than £850 sterling lodged with English merchants in 1721, cousin Richard had more than £800 of sound debts owed his estate in 1741, and brother Joseph had more than £2,000 in sperate debts in 1753.

John's eldest son Samuel (1720–1785) inherited a quarter of the residue of his father's £2,319 personal estate and immediately commenced trading on his own account. Like other family members, Samuel raised tobacco, corn,

9. John Galloway Papers, Samuel Galloway Ledgers, 1748–1751, 1753–1779, and Samuel Galloway Papers, all in Galloway/Maxcy/Markoe Papers, 1–6, LC; Samuel Galloway Ledger, 1748–1758, Cheston-Galloway Papers, MS 1994, MHS. Biographical information for members of the Galloway family comes from biographical files compiled by Historic Annapolis, MSA. For Samuel Galloway (d. 1721), see Inventories, 4: 313, Accounts, 5: 2, Wills, 16: 393. For Ann, widow of Samuel (d. 1723), see Inventories, 8: 286, Accounts, 6: 108, Wills, 18: 158. For Richard (d. 1736), see Inventories, 30: 384, 392, Wills, 21: 804. For Richard (d. 1741), see Inventories, 29: 120–129, Accounts, 20: 220, Wills, 22: 410. For John (d. 1747), see Inventories, 37: 335, Wills, 25: 125, 313. For Joseph Galloway (d. 1753), see Inventories, 57: 133, 151, Accounts, 36: 426, Wills, 28: 412. Deflated inventory values are from Historic Saint Mary's City SPSS inventory files, MSA. See also "Journal of a French Traveller in the Colonies, 1765, II," *AHR,* XXVII (1921), 72, 76.

and wheat for market along with oats and beans on his home plantation and on quarters on West River and at the Ridge, all managed by overseers. The major source of Galloway's revenues, though, came from trading in tobacco, grain, servants, and slaves and from shipbuilding, ship provisioning, and milling. Unlike most native-born merchants, Galloway engaged in both the tobacco and West Indian trades. He bought tobacco on both the Western and Eastern Shores that he shipped to London, Bristol, and Ireland and filled out his cargoes with iron, lumber, and flaxseed. Galloway also traded to the city of Norfolk, Virginia, and to Barbados and Antigua, shipping grain, flour, meat, and lumber in return for cargoes of James River naval stores and West Indian rum, sugar, and molasses that he sold in Anne Arundel County along with British dry goods and groceries. Some of the ships Galloway built in his boatyard were sold to other merchants, but he also employed two brigantines and several schooners to carry his own cargoes of tobacco to Britain and shiploads of provisions to the West Indies. Further income came from the sale of European indentured and convict servants and of African slaves. Galloway profited from sales of laborers on consignment and, in partnership with Thomas Ringgold, a Chestertown merchant and a kinsman, also invested directly in slaving voyages to Africa. Although the Seven Years' War offered new marketing opportunities for grain and provisions, Galloway also lost several ships to French privateers, losses that substantially reduced the tidy profits he would otherwise have made during the war.

Charles Carroll of Annapolis (1702–1782), the oldest surviving son of Charles Carroll the settler (1661–1720), seventeenth-century Maryland's wealthiest planter, lawyer, merchant, and land speculator combined, expanded his inheritance into "one of the great fortunes of colonial America." Indeed, the scale of the Carroll enterprises in Maryland approached those of the preeminent Carter family in Virginia. But, unlike the Anglican Carters, almost all of the Carrolls were members of a disadvantaged and often vulnerable religious minority. Denied powerful political connections, public influence, and the right to practice law because of his faith, Carroll had no incentive to gain the approval of his contemporaries and a compelling reason to try to augment and protect the family's fortune, especially during the French wars when Maryland's Roman Catholic gentry feared for the safety of their property. His father's death forced Carroll to end his studies in France at the Jesuit school at Douai, a center for the education of English Catholics, and to return to Maryland in 1721 at age eighteen to begin managing the vast estate he and other family members inherited from Charles Carroll the settler. Young Carroll learned plantation management and commerce from his uncle James Carroll, whose estate-building strategies were discussed earlier. Charles took over his father's mercantile business in Annapolis and advertised his favorable economic status by building an imposing brick house on the Annapolis waterfront. A few years later, he would build a second great brick house on more distant Doorhoragen Manor.[10]

During the 1730s, Carroll developed new strategies for augmenting the family estate rooted in agriculture, mining, rental properties, and banking. He discontinued the family's mercantile business and seldom engaged in land speculation. Instead, in 1731 he invested £700 sterling in the Baltimore Company Ironworks, a successful enterprise that by the mid-1750s yielded Carroll revenues of more than £300 sterling per year. He also began renting out farms he owned in Prince George's and Baltimore counties and patented another ten thousand acres on the Maryland frontier. He recruited tenants from Ireland and Germany to develop his western manor of Carrollton and by 1733 had 52 renters. By 1765, the number of rented tenements had risen to 195 and total acres under development to nearly twenty thousand. Leases were granted for twenty-one years at cash rents that Carroll doubled every seven years. The long-term fertility of the tenements was protected by restrictions on timber cutting and subsequently by restrictions on the number of laborers tenants could use to work the land. In the late 1750s, when most of the original leases expired, Carroll relet most of his tenements on annually renewable leases (leases "at will") and at higher rents, then at the rate of 1,000 pounds of tobacco for each one-hundred-acre tenement. By the early 1760s, Carroll was realizing at least £369 per year from rents. Further revenues came from the sale of lots in the new town of Baltimore.

Nor did Carroll neglect agriculture. Having inherited ten-thousand-acre Doorhoragen Manor in Anne Arundel County (later augmented by another inheritance from a deceased brother), one-thousand-acre Poplar Island in the Chesapeake Bay, twenty-four hundred acres in Prince George's County, and eventually more than 100 slaves, Carroll also managed a vast agricultural operation based on tobacco, grain, and livestock. By 1764, the number of bondpeople had risen to 285, primarily and perhaps entirely through natural increase. With the exception of a few domestics living in Annapolis and a few field hands working on quarters near Annapolis and on Poplar Island, nearly

10. This summary is based primarily on Ronald Hoffman, *Princes of Ireland, Planters of Maryland: A Carroll Saga, 1500–1782* (Chapel Hill, N.C., 2000), chaps. 3, 7 (quotation on 98); Carville Earle and Ronald Hoffman, "Genteel Erosion: The Ecological Consequences of Agrarian Reform in the Chesapeake," in Philip D. Curtin, Grace S. Brush, and George W. Fisher, eds., *Discovering the Chesapeake: The History of an Ecosystem* (Baltimore, 2001), 279–290; and Charles Carroll of Annapolis Account Book, 1754–1784, MS 211, MHS. For Charles Carroll, see also "Journal of a French Traveller," *AHR,* XXVII (1921), 74.

all Carroll's bound laborers were concentrated on Doorhoragen Manor, where he was placing increased emphasis on the raising of wheat. Although Carroll occasionally used plows, like Robert "King" Carter he continued to rely primarily on hand hoeing, hilling, and rotation of fields rather than crops.

Finally, by the early 1760s, Charles Carroll had also become Maryland's premier moneylender. At that time, he had twenty-four thousand pounds sterling on loan at 6 percent interest, loans to which he added any unpaid interest to the principal remaining due at the end of each year. This practice raised the complaint that he illegally charged compound interest; however, Carroll avoided violating Maryland's usury laws. Lacking the social pressures that constrained so many Virginia lenders, Carroll did not hesitate to foreclose on mortgages or to pursue debtors who fell in arrears. By 1756, Carroll could realistically assert, "There is but one Man in the province whose Fortune equals mine." In January 1764, he calculated his net worth at more than eighty-eight thousand pounds sterling that yielded a clear revenue of at least eighteen hundred pounds per year.[11]

Charles Carroll's cousin, Dr. Charles Carroll (1691–1755), who immigrated from Ireland to Annapolis by 1716, like his kinsman also practiced an aggressive, diversified estate-building strategy that combined agriculture with several other enterprises. Unlike Charles, Dr. Carroll attempted to improve his political situation by converting to Anglicanism and during the 1750s even advocated stringent anti-Catholic measures in the Maryland legislature, steps that earned the enmity of his cousin Charles. In the 1720s, Dr. Carroll became a tobacco factor for several London merchants, invested in ocean-going ships, and traded to the West Indies. He also owned a merchant mill, warehouses, a bakehouse, and a shipyard in the vicinity of Baltimore Town. In addition, Carroll speculated in frontier land, purchased slaves to raise tobacco on several Western Shore plantations, and engaged in moneylending. Involved with his cousin Charles in the Baltimore Company Ironworks, by the 1750s Dr. Carroll was investing in other ironworks in Frederick County,

11. Carroll to son Charles Carroll, July 26, 1756, Carroll-McTavish Papers, MHS; Carroll to Carroll, Jan. 9, 1764, Carroll Papers, MS 206, MHS, both quoted in Hoffman, *Princes of Ireland,* 122. Between 1688 and 1747, fellow Roman Catholic Richard Bennett of Talbot County amassed, despite legal disabilities, an estate of similar size through the same means as did Charles Carroll—agriculture, commerce, shipping, slave trading, land rentals, milling, and moneylending (Beatriz Betancourt Hardy, "A Papist in a Protestant Age: The Case of Richard Bennett, 1667–1749," *Jour. So. Hist.,* LX [1994], 203–228). For the role of moneylending in estate-building strategies, see also Aubrey C. Land, "Economic Base and Social Structure: The Northern Chesapeake in the Eighteenth Century," *JEH,* XXV (1965), 639–654; and Land, "Economic Behavior in a Planting Society: The Eighteenth-Century Chesapeake," *Jour. So. Hist.,* XXXIII (1967), 478–480.

Maryland, borrowing money from Eastern Shore magnate Edward Lloyd to get sufficient capital. In 1753, Carroll advised his oldest son, "I cannot see that by making Tobacco, I should better my own Yours or Your Bro[ther]s Fortunes and that induces me to go upon the Iron Businness and make grain to Support it." Thereafter, he turned his plantations into provisioning farms for the ironworks. Although not as spectacularly successful as his kinsman, Dr. Carroll succeeded in building a diversified estate worth an estimated fifteen thousand pounds sterling at his death in 1755.[12]

A New System of Chesapeake Agriculture: Maryland's Upper Eastern Shore

Modes of plantation management on Maryland's upper Eastern Shore are more fully documented for this period than those of typical Western Shore planters. In an area where London, Bristol, Liverpool, and Whitehaven merchants were aggressively competing for the region's tobacco, primarily making direct purchases through stores, and in a place well situated for both the West Indian provision trade and the Philadelphia wheat market, Eastern Shore planters could choose among a greater variety of options than could planters in areas less accessible to markets. By the 1730s, the region was well supplied with slaves delivered either through the port of Oxford or shipped across the bay from Annapolis. Many of the region's tobacco dealers were also involved in the transportation of English and Irish indentured servants and of British convicts. Indentured laborers as well as slaves continued to command a ready market on the upper Eastern Shore. Some wealthy planters like Robert Lloyd (discussed below) employed large enslaved workforces in the manner of great Virginia planters, but others such as the Richard Tilghmans Senior and Junior remained satisfied with a smaller number of enslaved workers augmented by hired free white laborers.[13]

Plantation accounts and probate inventories and wills of the Tilghman family of Queen Anne's County continue to document the somewhat un-

12. Hoffman, *Princes of Ireland,* 229–230, 269–273; Letterbook of Dr. Charles Carroll of Annapolis, 1752–1755, and his son Charles Carroll, Barrister, 1755–1769, MS 208, MHS.

13. For the Eastern Shore economy, see Clemens, *The Atlantic Economy;* John W. Tyler, "Foster Cunliffe and Sons: Liverpool Merchants in the Maryland Tobacco Trade, 1738–1765," *MHM,* LXXIII (1978), 246–279; "State of the Province of Maryland in 1758," *MHM,* XXXI (1938), 228–247; Stephen Bordley Letterbooks, 1727–1759, MHS; Robert Morris Ledger B, MS C714-1, MSA; and Callister Papers, Maryland Diocesan Library, Baltimore, microfilm, M-142, CWL. For the Eastern Shore labor supply, see Lorena S. Walsh, "Mercantile Strategies, Credit Networks, and Labor Supply in the Colonial Chesapeake in Trans-Atlantic Perspective," in David Eltis, Frank D. Lewis, and Kenneth L. Sokoloff, eds., *Slavery in the Development of the Americas* (Cambridge, 2004), 104–117.

conventional and essentially conservative approach to planting discussed earlier. The elder Richard Tilghman (1672–1738), a first-generation Maryland native, made few changes in the operation of his Hermitage farm in his later years. He had added to an already large inheritance of land, had acquired enough slaves to work part of his farms and attracted white tenants who were developing other portions, and had built himself a comfortable house. He found little else left to do to improve the family estate. Tobacco, grains, cider, and lumber all remained important sources of income, and rents now contributed a steady twenty-five to thirty pounds a year (Table 32).[14]

14. Ledger A, 1718–1741, Ledger B, 1725–1765, and Loose Papers, Tilghman Family Papers, Special Collections, Georgetown University Library; Wills, 22: 35, MSA; Clemens, *The Atlantic Economy,* 127–129, 134; Christopher Johnston, "Tilghman Family," *MHM,* I (1906), 181–184, 280–284, 290, 369–376; James Bordley, Jr., *The Hollyday and Related Families of the Eastern Shore of Maryland* (Baltimore, 1962); Edward C. Papenfuse et al., *A Biographical Dictionary of the Maryland Legislature, 1635–1789,* 2 vols. (Baltimore, 1985), I, 332, II, 830–831.

The following conventions were followed in compiling Tables 32 and 33a–e:

1. An em dash indicates that either an item was mentioned in the account but not valued or that it did not appear but is known to have been produced or disbursed (for example, tobacco and taxes).
2. Values in the accounts were recorded in tobacco, Maryland hard currency, Maryland paper currency, and, occasionally, in sterling and Pennsylvania currency. Tobacco was converted to Maryland currency using the Talbot County price series published in Clemens, *The Atlantic Economy,* appendix A3. Maryland and Pennsylvania currencies were reduced to sterling using the commercial exchange rates published in John J. McCusker, *Money and Exchange in Europe and America, 1600–1775: A Handbook* (Chapel Hill, N.C., 1978). All values were then deflated to a constant value by a series supplied to Historic Saint Mary's City by P. M. G. Harris with base years of 1700–1709. The index is described in Harris, "Inflation and Deflation in Early America, 1634–1860," *Social Science History,* XX (1996), 469–505.
3. Only agricultural revenues are included; sales of imported merchandise were not quantified.
4. Only the costs of maintaining the plantation workforce are included in the table; family expenditures for clothing, household goods, schooling, medicine, and so forth, were not quantified.
5. Costs of clothing for the slaves, probably the largest plantation expenditure, was not reported in the accounts. Imports continued to be listed only as unspecified "new goods," which included a variety of items besides clothing and tools and which might have been intended for the white family, for the slaves, or for resale.
6. Most tools were imported; expenditures reported are for only those made or purchased locally.
7. Most investment in buildings was either for outbuildings on the home farm, for structures on tenant farms, or for a rented inn.
8. Tobacco revenues and payments made to overseers were not recorded in the account book. Richard Tilghman, Jr., kept track of tobacco revenues by saving the loose paper accounts sent him by his factors and loose paper accounts he made of transactions with the overseers. Where these have survived, the information was incorporated into the tables. Tobacco accounts are

TABLE 32. Revenues, Expenditures, and Investments on the Richard Tilghman, Sr., Plantations, Talbot County, Maryland, 1730–1737

Category	1730	1731	1732	1733	1734	1735	1736	1737	Total
REVENUES									
Tobacco	19.69	39.98	39.52	—	—	—	—	—	99.19
Corn	0	12.43	0	4.09	1.28	0	0	0	17.80
Wheat	0.55	3.32	0.33	0	18.58	2.80	5.22	0.50	31.30
Oats, etc.	0.28	0	0.30	0	21.02	0	0	0	21.60
Cider	8.09	19.81	14.31	8.84	0	3.15	17.30	13.71	85.21
Livestock	5.32	0	2.22	0	4.23	0.51	0	0	12.28
Lumber	1.16	0.05	1.28	4.65	0	4.01	16.24	31.22	58.61
Rent	39.94	24.98	24.99	25.77	25.56	20.72	27.11	30.36	219.43
Servant hire	0	0	0	0	1.64	2.84	11.06	8.98	24.52
Land sold	0	0	0	25.77	0	0	0	0	25.77
Labor sold	0	11.10	31.22	0	0	0	0	0	42.32
Total	75.03	111.67	114.17	69.12	72.31	34.03	76.93	84.77	638.03
EXPENDITURES									
Tailoring	0	0	0	1.01	0	7.96	0	0	8.97
Spinning	0	0	0.93	0.45	1.01	0	0	0.42	2.81
Knitting	0.19	0	0	0	0	0	0	0	0.19
Weaving	0.82	1.04	0	0	0	0	0	0	1.86
Tanning	0	0	0	0	0	0	0	3.09	3.09
Shoemaking	0	4.78	0	2.02	2.97	2.44	0	0	12.21
Cooper	0	9.62	0	1.03	0	8.56	0	1.83	21.04
Carpenter	0	0	0	0.49	0	0	0.94	0	1.43
Boatwright	0	0	0	0	1.84	0	0	0	1.84
Blacksmith	2.68	2.55	2.55	2.96	2.02	1.59	1.08	0	15.43
Sawyer	0	0	0	0	0	9.32	0	0	9.32
Meat	0	2.98	11.72	15.16	3.79	0	0	0	33.65
Corn	0	0	2.34	0	0	0.28	0	0	2.62
Wheat	0	0	0.11	0	0.24	1.04	0	0	1.39
Misc. food	0	0	14.28	0	0	0	0	0	14.28
Tools	0	0	0	0	0	0.30	0	0	0.30
Marketing	0	0.32	0.58	0	0	0	0	0	0.90
Slave medical	0	1.11	1.11	0	1.54	0	0	1.24	5.00
Hired labor	22.28	2.73	0	0	5.02	2.35	0.24	0.61	33.23
Supervision	9.37	9.54	12.03	11.28	0	0	0	0	42.22
Total	35.34	34.67	45.65	34.40	18.43	33.84	2.26	7.19	211.78

TABLE 32. Continued

Category	1730	1731	1732	1733	1734	1735	1736	1737	Total
INVESTMENTS									
Land	0	0	0	0	2.97	0	0	0	2.97
Buildings	3.48	10.85	0	17.73	0	13.83	0	0	45.89
Livestock	0	0	2.22	0	0	1.68	0	9.92	13.82
Total	3.48	10.85	2.22	17.73	2.97	15.51	0	9.92	62.68

Note: Currency values are in constant pounds sterling.
Source: Ledger A, 1718–1741, and Loose Papers, Tilghman Family Papers, 1712–1843, Special Collections, Georgetown University Library, Washington, D.C.

In his will of 1738, Tilghman divided about fifteen thousand acres of land, three town lots, a warehouse and landing, and unspecified personal property between his wife, five sons, one daughter, and the child of a second daughter who had died in 1736. The eldest surviving son, Richard (1705–1766), received title to Forlorne Hope plantation, already in his possession, and to his parents' dwelling plantation, the Hermitage, which he was to share with his mother during her lifetime. Since the elder Tilghman had already endowed all his children generously, his namesake inherited only an additional forty-five hundred acres of land at his father's demise and the slaves and livestock already lent to him to stock Forlorne quarter. Having carefully provided for his progeny during his lifetime, Richard Senior could be magnanimous at his death. He directed his family to make no inventory of his estate (since it was clearly solvent, an inventory was not required) and to account only for sound debts; insolvent debts were to be passed over, and very poor debtors were to be forgiven what they owed.[15]

The younger Tilghman's accounts, begun in 1724 when he was only nineteen, document the process by which the heir of long-lived and prolific parents built up an estate in his own right. Richard started by putting a generous colonial education to good use; by 1724, he had obtained the county

missing for some years, however, as are accounts with overseers, and some of the latter that survive are incomplete. Thus, both tobacco revenues and supervisory costs are understated.

9. Tobacco revenues may be overstated if the tobacco the Tilghmans consigned included the rent tobacco produced by tenants.

10. Almost no records of major sales of grain survive (these were apparently also kept in the form of loose papers). The grain revenues and expenditures recorded in the tables reflect local exchanges only, not sales for export.

15. Wills, 22: 35, MSA.

clerkship, a position that brought in a steady, if modest, income. By 1728, his father had set him up at Forlorne quarter, and over the next four years Richard proceeded to build himself a dwelling house, shed, tobacco prize, and barn (festively celebrated with rounds of rum at a barn raising). Richard had observed his father's style of management closely indeed, for over his lifetime he, too, would make few changes. Fortunately for him, there was yet little need for change at the Hermitage.

Unlike many Chesapeake planters, young Tilghman did not seek to acquire a large gang of slaves. He began farming at Forlorne with four or five adult workers given him by his father. Between 1736 and 1748, years for which he kept records, he paid taxes on only six to twelve laborers. In 1731, he bought one indentured servant, and in 1739 he purchased five family slaves whom Richard Senior had willed to a younger brother. Thereafter, Richard Junior bought no more servants or slaves, relying instead on the natural increase of the blacks to eventually augment the size of his workforce, meanwhile hiring both full-time and casual free white men and women when he had further labor requirements. Preoccupied from the outset with the clerkship, Tilghman always employed an overseer to supervise the workers at Forlorne. Once he married Susanna (1718–?), daughter of wealthy and well-connected Cecil County planter Peregrine Frisby, and inherited the Hermitage (these events seem to have happened in the same year), Tilghman hired a second overseer for that plantation. In 1746, he followed in his father's footsteps to a seat on the colony's provincial court, a position that would take up time and energy until his death in 1766. By then, Tilghman's workforce had grown to sixty slaves, accounting for half of his net worth. Among these were only nineteen tithable laborers, just fourteen or fifteen more than he had held nearly forty years before. Six of the slaves were age sixty or older, and thirty-two were children age thirteen or younger. This unusually high proportion of dependent children and old people surely elevated Tilghman's maintenance costs. Nonetheless, he did not consider selling any of the slaves; instead, Richard reserved the youngsters as inheritances for his own children.[16]

16. James Hollyday of Readbourne, another large Queen Anne's County planter who had a similar crop mix, also employed a number of free whites, especially to help harvest and thresh wheat (James Hollyday Account Book, 1746–1784, MS 454.1, MHS). John Gresham of Kent County, another planter of similar scale, acquired a small enslaved workforce in the 1730s and thereafter bought no more new slaves. He also relied instead, when he needed extra labor, on hired free white workers and occasionally on indebted men bound over by county courts to serve for a short term to pay their debts. See Christine Daniels, "Gresham's Laws: Labor Management on an Early-Eighteenth-Century Chesapeake Plantation," *Jour. So. Hist.*, LXII (1996), 205–238; John Gresham, Sr., and John Gresham, Jr., Account Book, 1716–1750, SC 1999-1-1, MSA.

There is other evidence of financial conservatism in addition to Richard's decision to buy no additional adult slaves in order to expand production of cash crops. With cheap land no longer available, Tilghman purchased little additional acreage. In 1747, he bought one tract worth only £13 sterling constant value (possibly an adjacent parcel bought for convenience). Only near the end of his life when he had seven children to provide for did he invest heavily (£553 sterling constant value) in two additional improved plantations, one with a new gristmill that could be rented out. Apparently satisfied with the house his father had built on the Hermitage in the 1720s, Tilghman made no major status-enhancing improvements to his main dwelling.

Tilghman could almost certainly have spent more lavishly had he chosen to—his was not an enterprise in financial trouble. Accounts with British factors indicate that he was careful not to order more goods or to draw more bills of exchange than his tobacco consignments and remitted bills would cover. Richard's records fail to document his grain dealings in any detail, but there is no evidence that he borrowed money locally (interest payments, if any, would almost certainly have appeared in his account book). Still, the Tilghmans unquestionably lived well. They had plenty of mahogany furniture, china, crystal, and silver plate; they could boast of a spinet and two riding carriages; and Richard had not stinted on his wardrobe. His estate was never formally accounted, but his inventory valued his movable goods (excluding crops and debts receivable) at £2,242 sterling constant value. Just before he died, Tilghman had also given his eldest son five slaves, livestock, and tools worth about another £200.[17]

Like many other Eastern Shore planters, Tilghman raised the trio of tobacco, corn, and wheat. Annual totals of his crops are almost never listed; perhaps like his neighbor James Hollyday, he kept a separate tobacco book that has not survived. From Tilghman's accounts with overseers, however, it appears that in the 1740s and 1750s gross revenues from grains and tobacco were about equal (Tables 33a–f). In the late 1730s and early 1740s, Tilghman consigned his tobacco to London firms—James Buchanan, Lawrence Williams, Samuel Hyde, and William Anderson. By the mid-1750s he was dealing

17. Annual accounts with merchants, Loose Papers, Tilghman Family Papers, Georgetown University Library. See also Inventories, 91: 63, Wills, 34: 407, Queen Anne's County Deeds, RT#G: 266–267, and Queen Anne's County Administration Papers, no. 2418, all in MSA. Talbot County planter Robert Goldsborough (son of the planter of the same name whose operations were examined earlier) demonstrated similar care in maintaining a favorable balance with Liverpool, Bristol, and London factors between 1745 and 1775. See Robert Goldsborough Ledger, 1739–1789, privately owned, microfilm, MSA.

TABLE 33A. Revenues, Expenditures, and Investments on the Richard Tilghman, Jr., Plantations, Talbot County, Maryland, 1728–1733

Category	1728	1729	1730	1731	1732	1733
REVENUES						
Tobacco	14.22	0	19.07	40.00	38.90	—
Corn	5.55	0	0	12.80	0.09	10.35
Wheat	5.81	2.30	0	2.11	0.62	0
Oats, etc.	0	0	0	0.41	0	0
Cider	0	0	0	0	0	0
Livestock	0	0	0	11.66	15.32	12.40
Lumber	0	0	0	0	0	0
Rent	0	0	0	0	1.48	1.72
Servant hire	0	0	0	0	0	0
Miscellaneous	0	0	0	0	0	0
Land sold	0	0	0	0	0	0
Total	25.58	2.30	19.07	66.98	56.41	24.47
EXPENDITURES						
Tailoring	0	0	0	3.81	1.13	3.44
Spinning	0.61	0.31	0	0	0	0
Knitting	0.82	0	0	0	0	0
Weaving	0	0	0	0.37	0.88	0
Tanning	0	0	0	0	0	0
Shoemaking	0.99	0	0	0.74	0	0
Cooper	0.28	0	3.03	0	0	0
Carpenter	0	0	0	0	0	0
Blacksmith	0	—	0	0	0	11.55
Sawyer	0	0	0	0	0	0
Ditcher	0	0	0	0	0	0
Meat	0	0	0	0	0	0
Fish	0	0	0	0	0	0
Corn	0.63	0	0	0	0	0
Wheat	0	0.56	0	0.58	0	0
Small grains	0	0	0	0	0	0
Misc. food	0.14	0	0	0	0.19	0
Seed	0	0	0	0	0	0
Livestock care	0	0	0	0	0	0
Tools	0.79	0	0	0	0	0
Marketing	0.86	0	0.06	0	0	0.43
Slave medical	0.78	0	0	0	0	0

TABLE 33A. Continued

Category	1728	1729	1730	1731	1732	1733
Hired labor	4.46	0	0.42	1.66	0	0
Supervision	4.72	—	11.28	10.32	11.85	—
Rent	0	0	0	0	0	0
Taxes	2.81	3.38	2.55	—	—	—
Miscellaneous	0	0	0	0	0	0
Total	17.89	4.25	17.34	17.48	14.05	15.42
INVESTMENTS						
Slaves/servants	0	0	0	11.10	0	0
Livestock	2.38	17.36	1.56	0.37	0	0
Buildings	15.86	0	0	4.32	0	0
Land	0	0	0	0	0	0
Boats	0	0	0	0	0	0
Total	8.24	17.36	1.56	15.79	0	0

Note: Currency values are in constant pounds sterling.
Source: Ledger B, 1725–1765, and Loose Papers, Tilghman Family Papers, Special Collections, Georgetown University Library, Washington, D.C.

almost exclusively with Anderson, a relative by marriage. In the years between, he tried some shipments to a Glasgow firm and to Sedgely and Cheston in Bristol but established no long-term correspondence with these ports; nor did he switch to selling his crops in the country. Tilghman's unwavering consignments to London merchants suggest that he continued to make relatively high-quality tobacco suitable for the English domestic market. During the 1730s and 1740s, the Tilghmans culled their hogsheads carefully, discarding "trash" and packing less valuable ground leaves separately.[18]

Tobacco production per hand on the Tilghman plantations remained high across the period—about thirteen hundred pounds per worker. Neither father nor son curtailed the amounts tended, and apparently they had no insoluble problems in maintaining yields over time. By the 1750s, the slaves were also making twenty to thirty barrels of corn apiece. At this level of production, half to three-quarters of the crop would have been salable surplus. The size of Tilghman's wheat crops are less certain, although in the 1750s output amounted to between three hundred and four hundred bushels per

18. For the typical regional crop mix, see Clemens, *The Atlantic Economy,* 150–151.

TABLE 33B. Revenues, Expenditures, and Investments on the Richard Tilghman, Jr., Plantations, Talbot County, Maryland, 1734–1739

Category	1734	1735	1736	1737	1738	1739
REVENUES						
Tobacco	—	—	—	—	—	241.97
Corn	1.01	0	1.14	0	0	1.22
Wheat	0	0	0.17	5.16	0	2.72
Oats, etc.	0	0	0.12	0	8.96	0.26
Cider	0	0	0	0	0	0
Livestock	5.89	2.11	11.12	2.04	14.51	24.40
Lumber	0	0	0	0	0	0
Rent	1.70	3.27	1.47	4.24	5.66	16.82
Servant hire	0	0	0	0	0	0
Miscellaneous	0	0	0	0	0	0
Land sold	2.98	0	0	0	0	0
Total	11.58	5.38	14.02	11.44	29.13	287.39
EXPENDITURES						
Tailoring	0	0.37	0	0	0	0
Spinning	0	0	0	0	0	0
Knitting	0	0	0.29	0.43	0	0
Weaving	0	0	0	0	0	0
Tanning	0	0	0	0	0	0
Shoemaking	0	0	0	0	0.51	7.12
Cooper	0	0	0	0	1.09	0
Carpenter	0	0.63	0	0	0	0
Blacksmith	0	0	0	0	0.15	0
Sawyer	0	0	0	0	0	0
Ditcher	0	0	4.71	0	0	0
Meat	0	0	2.33	0	0	17.92
Fish	0	0	0	0	0	0
Corn	0	0	0	0	0	0.67
Wheat	0	0	0	0	0	0
Small grains	0	0	2.13	0	0	0
Misc. food	0	0	0.60	0.78	0	0
Seed	0	0	0	0	0	0
Livestock care	0	1.09	0	0	0	0.67
Tools	0	0	0	0	0	4.02
Marketing	0	0	0	0	0	0
Slave medical	5.97	0	0	0.69	0	0

TABLE 33B. Continued

Category	1734	1735	1736	1737	1738	1739
Hired labor	0	13.09	12.40	13.73	16.22	30.55
Supervision	—	—	—	—	—	—
Rent	0	0	0	0	0	0
Taxes	3.14	3.80	3.03	2.80	3.54	—
Miscellaneous	0	0	0	0	0	0
Total	9.11	18.98	25.49	18.43	21.51	60.95
INVESTMENTS						
Slaves/servants	0	0	0	0	0	31.90
Livestock	0	9.35	11.35	4.41	0	66.43
Buildings	0	0	0	0	0	11.38
Land	0	0	0	0	0	0
Boats	0	0	0	0	0	0
Total	0	9.35	11.35	4.41	0	109.71

year. Tilghman's enslaved workers were thus producing gross revenues from the three major crops of about £450 a year at midcentury.[19]

Livestock, especially surplus cattle and after 1750 sheep, which were sold for breeding stock and for meat, provided a steady annual supplement to income from crops. Pork was mostly consumed on the plantation, and the size of the herd was strictly limited; aside from a small breeding stock, all other hogs were castrated, fattened, and killed within eighteen months. The flock of sheep (82 in 1766) was probably also regularly cropped. Cattle (196

19. For the proportion of probable corn surplus, see the discussion of the Corbin plantation in Chapter 6. Tilghman and his overseers kept a very close watch, indeed, over all available corn and the uses to which it might be put. Tilghman was careful to subtract from his overseers' shares of corn any grain they might have used over and above their annual free allowance of three barrels of shelled corn. On the other hand, overseers seem to have retained certain gleaning privileges. In 1745, Tilghman promised one overseer some extra meat on the condition that his wife not raise any geese on the quarter, and in 1752 Tilghman had to pay another overseer 7s. 6d in compensation for letting Tilghman's own cattle fatten on the quarter cornfield. In the same period, James Hollyday was selling sheep, hogs, poultry, and beef locally and was driving cattle to New Castle County, Delaware. Between 1760 and 1763, Hollyday made annual crops of 297 (1760), 801 (1761), 730 (1762), and 527 (1763) bushels of wheat with twenty to twenty-three tithable slaves, with shares per hand ranging from 15 to 59 bushels. These unusually large wheat crops were made at the expense of the tobacco crop; Hollyday was making only between seven hundred and eight hundred pounds of tobacco per hand at this time (James Hollyday Account Book, MHS).

TABLE 33C. Revenues, Expenditures, and Investments on the Richard Tilghman, Jr., Plantations, Talbot County, Maryland, 1740–1745

Category	1740	1741	1742	1743	1744	1745
REVENUES						
Tobacco	195.25	47.34	260.38	215.54	177.09	184.68
Corn	8.81	0	0.87	0	0	0
Wheat	5.41	0.27	0	0.14	18.96	0
Oats, etc.	0.81	0	0.59	0.06	12.57	0
Cider	0	7.81	4.71	4.23	13.12	4.87
Livestock	25.92	3.17	32.38	20.65	19.24	28.79
Lumber	0	1.18	0	0	0.12	0
Rent	20.79	40.24	38.71	31.00	72.18	50.25
Servant hire	0	0	0	0	0	1.30
Miscellaneous	0	0	0	0	0	0
Land sold	0	0	0	0	0	0
Total	256.99	100.01	337.64	271.62	313.28	269.89
EXPENDITURES						
Tailoring	3.99	0.36	0.10	1.37	0.22	0.53
Spinning	0	0.72	0	1.81	0	0
Knitting	0	0	0	0.65	0	0
Weaving	0	0	0	0	0	0
Tanning	0.16	0	0	0.88	0	2.37
Shoemaking	2.31	2.10	2.76	5.14	33.19	11.00
Cooper	0	0	0	0	0	0
Carpenter	0	0	0	0	0	0
Blacksmith	0	0.58	0	0.89	0	0
Sawyer	0	0	0	0	0	0
Ditcher	0	0	0	0	0	0
Meat	7.00	0	1.15	0	17.03	0
Fish	0	0	0	0	0	0
Corn	0	0	4.27	8.42	11.29	0
Wheat	0.56	0	0	0	0	0
Small grains	0	0	0.45	0	0.37	0
Misc. food	0	0	0	0	0	0
Seed	0	0	0	0	0	0
Livestock care	0.27	0	0	0	0	0
Tools	1.89	0	0	0	0	0
Marketing	0	0	0.69	0	1.71	0
Slave medical	0.32	0	0	0	0	0

TABLE 33C. Continued

Category	1740	1741	1742	1743	1744	1745
Hired labor	24.07	30.25	45.70	23.98	63.49	27.36
Supervision	4.57	4.90	5.78	15.33	15.82	10.71
Rent	0	0	0	0	8.55	3.07
Taxes	—	5.99	—	—	—	—
Miscellaneous	0	0	0	0	0	0.28
Total	45.14	44.90	60.90	58.47	151.67	55.32
INVESTMENTS						
Slaves/servants	0	0	0	0	0	0
Livestock	49.37	0.26	7.67	19.85	0.21	0
Buildings	0	13.95	11.85	16.95	9.30	0
Land	0	0	0	0	0	0
Boats	0	10.99	0	0	0	0
Total	49.37	25.20	19.52	36.80	9.51	0

in 1766) too were carefully managed to produce both meat and milk, and the size of the herd was tailored to the amount of forage available. After a bull or two was selected for breeding, other male cattle were castrated and either sold immediately or butchered at about six years of age. Most females, on the other hand, were retained for breeding and for the dairy, although a few of the least promising were neutered and then fattened for meat. In 1766, livestock was Tilghman's second most important capital asset, worth £437 sterling constant value, 19 percent of his movable property.[20]

The inventory and the plantation accounts show that Richard and his wife Susanna engaged in other kinds of diversification. Once Tilghman moved to the Hermitage with its well-established orchard, there was plenty of cider made each year and often a surplus for sale. Susanna Tilghman supervised dairying, beer brewing, and candlemaking. By 1766, the family was raising a little flax and cotton. Mrs. Tilghman or their daughters used one linen wheel located in the family's quarters, and older slave women must have spun on three other linen and woolen wheels listed in the outbuildings. There was no

20. Factor Henry Callister explained that Talbot County and Queen Anne's County pork was preferable to that of Dorchester County to the south, as hogs fattened with corn tasted better than those fattened on mast alone; "This being a Corn Country it may be relied on" (Callister to Robert Morris, Aug. 6, 1747, Callister Papers, CWL). In 1747, Tilghman was selling corn-fattened pork to a ship captain at a premium price.

TABLE 33D. Revenues, Expenditures, and Investments on the Richard Tilghman, Jr., Plantations, Talbot County, Maryland, 1746–1751

Category	1746	1747	1748	1749	1750	1751
REVENUES						
Tobacco	36.25	19.06	146.55	119.03	4.56	305.01
Corn	0.43	0.12	0.32	0.22	0	2.42
Wheat	0	0.11	0	0	54.53	45.33
Oats, etc.	0	0	0	1.31	14.14	0.42
Cider	18.82	13.57	7.25	2.57	0	0
Livestock	9.12	24.45	20.14	53.62	54.14	21.98
Lumber	0	0	0	1.75	0	0
Rent	41.42	53.64	71.86	102.21	109.77	110.18
Servant hire	0.88	0.87	1.13	6.83	5.02	9.33
Miscellaneous	0	0	3.87	5.06	7.27	7.75
Land sold	0	0	0	0	0	15.51
Total	106.92	111.82	251.12	292.60	249.43	517.93
EXPENDITURES						
Tailoring	0	0	0	0	0	0
Spinning	2.39	0.87	0	0	0	0
Knitting	0	0	0	0	0	0
Weaving	4.82	0	1.54	0.80	0.36	0
Tanning	0	2.49	26.15	6.24	7.27	7.75
Shoemaking	4.01	0.87	8.94	5.01	4.18	3.49
Cooper	0.16	0	0	0.82	0	0
Carpenter	0	0	0	0	0	0
Blacksmith	0	0	0	7.98	0	0.05
Sawyer	0.67	0	0	0	0	0
Ditcher	0	0	0	0	0	0
Meat	0	0	0	0	0	0
Fish	0	0	3.93	0	0	1.12
Corn	4.28	0	0	0	24.26	4.71
Wheat	1.50	0	5.11	0	5.96	0
Small grains	0	0	0	0	0	0
Misc. food	0	0	0	1.67	0	0
Seed	0	0	0	0	0	0.19
Livestock care	0	0.31	0	0	0	0
Tools	0.43	0	0.23	0.08	0.37	0
Marketing	0	0	1.27	0	0	0
Slave medical	0	0	0	0	0	0

TABLE 33D. Continued

Category	1746	1747	1748	1749	1750	1751
Hired labor	27.66	0.35	21.26	21.20	65.66	51.64
Supervision	19.72	12.00	17.45	14.40	10.00	14.93
Rent	3.20	3.25	4.43	5.69	12.42	7.11
Taxes	—	—	5.28	—	—	—
Miscellaneous	0	0	0	0	0	0
Total	68.84	20.14	95.59	63.89	130.48	90.99
INVESTMENTS						
Slaves/servants	0	0	0	0	0	0
Livestock	0	0	4.69	0.22	0	0
Buildings	29.49	8.61	9.88	3.81	30.22	9.28
Land	0	12.99	0	0	0	0
Boats	0	0	0	0	0	0
Total	29.49	21.60	14.57	4.03	30.22	9.28

attempt to make much yarn, however. Most of the bedding, clothing, and table linen were imports. Much of the wool was sold or put out to poor white women for processing shortly after the sheep were sheared each year. There was no weaving equipment on the plantation.

The family had a kitchen garden and also grew oats, rye, barley, beans, onions, and potatoes for plantation use. Tilghman attempted to improve his pastures by sowing clover and timothy. He needed to make all the forage he could in order to feed his draft horses and oxen, for he relied heavily on plows and harrows for tilling. Tilghman owned ten plows—five furrow, one double-winged, and four new bar shares appeared in his inventory—and he had just supplied his oldest son with another "good plow," a tobacco plow, and a corn harrow. He also used a "large, well-fitted ox cart" and several smaller carts to haul his grain. Altogether there were agricultural tools worth twenty-six pounds sterling constant value on the plantation. Investment in craft tools was only slightly lower, twenty pounds. One of the enslaved men was a carpenter, and others ran a blacksmith shop. In winter, the slaves also cut and processed timber; when Richard died, there was an abundance of plank and shingles on hand and a large timber cart for hauling the wood.[21]

21. In neighboring Talbot County where the crop mix was similar to that of Queen Anne's, by 1740 agreements between planters and overseers were specifying the number of plows and plow

TABLE 33E. Revenues, Expenditures, and Investments on the Richard Tilghman, Jr., Plantations, Talbot County, Maryland, 1752–1758

Category	1752	1753	1754	1755	1756	1757	1758
REVENUES							
Tobacco	205.58	14.15	170.26	121.57	173.54	194.98	—
Corn	0	0	4.73	0.07	0.14	0.21	0.79
Wheat	0	0	1.76	0.30	0	0.37	0.21
Oats, etc.	0.09	0.46	0	0	0.18	0	0.62
Cider	0	0	0	0	0	0	0
Livestock	28.95	16.35	14.49	4.42	16.66	14.67	16.14
Lumber	0	0	0	0.09	0	0.12	0
Rent	91.14	103.64	100.29	86.63	105.76	121.31	131.48
Servant hire	8.86	12.43	4.97	28.83	27.90	0.96	16.61
Miscellaneous	82.47	8.65	7.63	7.44	7.09	3.32	0
Land sold	0	0	0	0	0	0	0
Total	417.09	155.68	304.13	249.35	331.27	335.94	165.85
EXPENDITURES							
Tailoring	0	0	0	0	0	0	0
Spinning	0	0	2.74	0.93	0	0	0
Knitting	0	0	0	1.80	0	0	0
Weaving	0	0	3.58	3.10	0	0	0
Tanning	7.54	7.73	7.63	10.06	0	0	0
Shoemaking	0	6.65	3.58	6.29	0	0	0
Cooper	0	0	1.56	0	0	0	0
Carpenter	0	0	0	0	0	0	0
Blacksmith	0	1.18	0	0	0	0	0
Sawyer	0	0	0	0	0	2.04	1.40
Ditcher	0	0	0	0	0	0	0
Meat	1.05	0	0	0	0	0	0
Fish	0	0	0	0	0	0	0
Corn	0	0.77	0	0	3.54	0	0
Wheat	0	0	0	0.74	0	0	0
Small grains	0	0.81	0.38	2.34	0	0	0
Misc. food	0	0	0	0	0	0	0
Seed	0	0	1.60	0	0	0	0
Livestock care	0	0	0	0	0	0	0
Tools	0	0	0.95	0	0	0	0
Marketing	0	0	0	0	0	0	0
Slave medical	0	0	0	0	0	0	0

TABLE 33E. Continued

Category	1752	1753	1754	1755	1756	1757	1758
Hired labor	32.66	32.06	37.25	33.70	20.43	37.09	20.91
Supervision	3.16	—	0	0	0	—	9.81
Rent	5.66	5.48	4.77	0	0	0	0
Taxes	—	—	—	—	—	0	0
Miscellaneous	0	0	0	0	0	0	0
Total	50.07	54.68	64.04	58.96	23.97	39.13	32.12
INVESTMENTS							
Slaves/servants	0	0	0	0	0	0	0
Livestock	0	0	11.45	1.38	2.30	0	6.76
Buildings	0	0	0	6.23	20.20	0	0
Land	0	0	0	0	0	207.22	345.35
Boats	0	0	0	0	0	0	0
Total	0	0	11.45	7.61	22.50	207.22	352.11

After his father's death in 1738, the younger Tilghman initially took over his retail trade, selling cloth, nails and tools, groceries, liquor, and an annual shipment of "bostick's cordial" to tenants and neighbors. Once retail stores proliferated in the area at the end of King George's War, however, Tilghman ceased trading in any volume. He kept on hand only a few tools and some salt, rum, and sugar, apparently more of a convenience for hired laborers than anything else. And, although the family continued to buy many English imports with their tobacco, the wheat trade brought increasing connections with Philadelphia merchants, and the Tilghmans began to acquire tools and a few other coarse goods from that port.[22]

The Tilghman family accounts document dense local economic interconnections and, moreover, reveal the substantial amounts one large planter expended locally for goods and services. Both father and son regularly hired neighbor women to spin and to knit stockings, tanners to process hides, weavers and shoemakers to make coarse cloth and shoes for the slaves, doc-

horses as well as the number of slaves the planter was to supply. Talbot County Judgments, JL3: 147, 234, JL6: 253, *Charles Goldsborough v John Love,* March 1750, Civil Judgments, 1750–1751, MSA. These references were kindly supplied by Jean Russo.

22. For Philadelphia connections, see Henry Callister to Foster Cunliffe, Dec. 25, 1747, July 2, Sept. 26, 1751, Apr. 5, 1752, Callister Papers, CWL. By the early 1750s, James Hollyday also began buying goods in Philadelphia (James Hollyday Account Book, MHS).

TABLE 33F. Revenues, Expenditures, and Investments on the Richard Tilghman, Jr., Plantations, Talbot County, Maryland, 1759–1763

Category	1759	1760	1761	1762	1763	Totals (1728–1763)
REVENUES						
Tobacco	—	—	—	13.86	3.59	2,962.43
Corn	1.14	1.37	4.03	2.11	0	59.94
Wheat	0	0	0	0	0	146.28
Oats, etc.	0	0	0	0	0.23	41.23
Cider	2.86	0	0	0	0	79.81
Livestock	0.24	5.54	11.03	4.78	0	546.32
Lumber	0	3.46	0	0	0	6.72
Rent	158.20	109.56	106.85	80.54	64.25	2,038.26
Servant hire	6.95	0	0	0	0	132.87
Miscellaneous	0	0	0	0	0	140.55
Land sold	0	0	0	0	0	18.49
Total	169.39	119.93	121.91	101.29	68.07	6,172.90
EXPENDITURES						
Tailoring	0	1.54	0	0	0	16.86
Spinning	0	0	0	0	0	10.38
Knitting	0	0	0	0	0	3.99
Weaving	0	0	0	0	0	15.45
Tanning	0	0	0	0	0	86.27
Shoemaking	0	0.96	0	0	0	109.84
Cooper	0	0	0	0	0	6.94
Carpenter	0	0	0	0	0	0.63
Blacksmith	0	0	0	0	0	22.38
Sawyer	0	0	0	0	0	4.11
Ditcher	0	0	0	0	0	4.71
Meat	0	0	6.44	0	0	52.92
Fish	1.18	0.79	0	0.37	0	7.39
Corn	0	1.49	0	0	0	64.33
Wheat	0	0	0	0	0	15.01
Small grains	0	0	0	0	0	6.48
Misc. food	0	0	0	0	0	3.38
Seed	0	0	0	0	0	1.79
Livestock care	0	0	0	0	0	2.34
Tools	0	0	0	0	0	8.76
Marketing	0	0	0	0	0	5.02
Slave medical	4.13	0	4.87	0	0	16.76

TABLE 33F. Continued

Category	1759	1760	1761	1762	1763	Totals (1728–1763)
Hired labor	26.03	19.65	13.90	32.77	32.15	833.75
Supervision	8.39	41.69	—	—	21.05	267.88
Rent	0	0	0	0	0	63.63
Taxes	0	0	0	0	0	36.32
Miscellaneous	0	0	0	0	0	0.28
Total	39.73	66.12	25.21	33.14	53.20	1,667.60
INVESTMENTS						
Slaves/servants	0	0	0	0	0	43.00
Livestock	0	10.32	0.34	0	0.42	228.45
Buildings	0	0	0	36.87	48.35	286.55
Land	0	0	0	0	0	565.56
Boats	0	0	0	0	0	10.99
Total	0	10.32	0.34	36.87	48.77	1,134.55

tors and midwives for both white and black families, and smiths to repair plows and tools. Increasingly, the family turned to local fishermen and poultry keepers to supply their table. Occasionally, they also hired carpenters, coopers, tailors, joiners, sawyers, wheelwrights, boatwrights, ditchers, woodcutters, and a "quilting woman." One suspects that the Tilghmans had too few adult field hands to put much time or effort into import-replacement activities. Instead, they relied on local whites with skills and time on their hands. The Tilghmans spent £273 over thirty-six years for hired skilled labor (see Tables 32 and 33a–f). By far the most, nearly £200, went for tanning leather and making shoes. Richard Junior spent four times more, however, more than £800, for agricultural labor to supplement the efforts of the family slaves. Beginning in 1739, after he had moved to the Hermitage and begun raising more wheat, he paid between £20 and £65 per year in annual wages to two free white men (who might have undertaken or at least supervised plowing and the seeding and threshing of wheat) and to a combination of free whites and neighbors' slaves hired to help with the harvest. The addition of wheat to the crop mix on large plantations thus clearly created significant new opportunities for small planters, tenants, and free laborers to earn extra income that were never afforded by large plantations on which slaves tended only tobacco and corn.[23]

As we saw in an earlier chapter, the elder Tilghman had made rental properties an integral part of his operation; his son would continue this strategy. During the 1740s, fourteen tenant farms and an ordinary in Ogletown (a port town on the Chester River in Queen Anne's County that has since disappeared) brought in at least forty pounds sterling constant value a year. After Richard's mother died in 1748 and her share of the home plantation came to him, rental revenues jumped to more than one hundred pounds per year with the addition of several more tenements and a tanyard. Rather than requiring tenants to build their own houses and barns, Tilghman did the work himself, perhaps wanting to be certain of the quality of the improvements. Between 1739 and 1759, he built sixteen new tenant houses, a millhouse, and a stable for the ordinary. Although he seldom raised rents on resident tenants (in 1740 these rents ranged from four hundred to six hundred pounds of tobacco per year), as farms fell vacant in the 1740s and 1750s he relet them at higher rents of one thousand to twelve hundred pounds.

The tenant farmers and their wives as well as the wives of Tilghman's overseers often earned extra income by doing work for the Tilghmans – spinning, making clothing and shoes, cutting wood, making casks, building houses, and doing midwifery and housework. They also sold the family fish, poultry, and eggs. Relations between landlord and tenants did not, however, revolve around only economic exchanges. Tilghman took an active interest in the personal lives of his renters. He, for example, lent money to one to get a marriage license and to pay the minister presiding at the wedding, and he supplied coffins when there were deaths in tenant families. These renters, like the tenants with life leases in Charles County, were a surprisingly stable group. Tenants who took up farms between 1721 and 1749 rented farms for a mean of ten years. Although a few failed to prosper and left after one or two seasons, others remained for as long as thirty-six years. Renting a farm was clearly no way to get rich; as Tilghman informed an English relative in 1737, tenants Simon Knox and his wife "by their Industry . . . Injoy a comfortable tho mean life." On the other hand, Tilghman's tenants had better-improved

23. Robert Goldsborough's Ledgers, 1718–1742, 1751–1755, privately owned, microfilm, MSA, show almost identical dependence on local free workers in Talbot. See also Jean B. Russo, "Self-Sufficiency and Local Exchange: Free Craftsmen in the Rural Chesapeake Economy," in Lois Green Carr, Philip D. Morgan, and Jean B. Russo, eds., *Colonial Chesapeake Society* (Chapel Hill, N.C., 1988), 389–432; Christine Daniels, "'Wanted: A Blacksmith Who Understands Plantation Work': Artisans in Maryland, 1700–1810," *WMQ*, 3d Ser., L (1993), 743–767; and Daniels, "Gresham's Laws," *Jour. So. Hist.*, LXII (1996), 205–238.

farms than did most renters, and they were not impoverished; few failed to pay their rents, and most were able to do so on time.[24]

The Tilghman family papers, like those of the Carter dynasty in Virginia, demonstrate considerable continuity between generations both in styles of life and in styles of farming. In the 1720s, the elder Tilghman set up his son with a plantation, livestock, and laborers to work it shortly after he came of age. In the 1760s, Richard Tilghman would do the same for his oldest son. Both generations of Tilghmans lived well, but not so well as to outstrip their incomes; Richard Junior's injunction to "avoid vain pomp and unnecessary charge" about his funeral embodied the attitudes that had governed his and his father's planting careers. Richard also duplicated most of his father's farming practices and even his sometimes frustrating accounting methods. So little change was called for in part because father and son used their land more lightly than did many other planters. By keeping a relatively low ratio of labor to land and by steadily improving rental farms, the family had a better chance of maintaining their level of prosperity over the long run than did planters who both employed more slaves and paid less attention to the long-term productivity of the land.[25]

Robert Lloyd (1712–1770), a third-generation Maryland native and a member of one of Maryland's wealthiest creole families, kept a record between 1756 and 1762 of the management of one such more labor-intensive operation. Lloyd was closely connected to the Tilghmans, having married Anna Maria, one of the elder Tilghman's daughters. However, he did not follow the Tilghmans' conservative approach to land use. By the 1730s, Lloyd had settled on Hope plantation in Queen Anne's County. He had inherited more than four thousand acres in that county and in adjacent Talbot and a lot in Wye Town from his father, and he controlled additional farms in Queen Anne's belonging to the estate of his wife's first husband. A justice of the peace and a very active burgess, Lloyd entrusted the supervision of a large inherited slave workforce to five white overseers and to one slave, Dick, who directed twelve workers on a quarter on the home plantation. By 1756,

24. Richard Tilghman to unnamed kinsman, Aug. 18, 1736, Loose Papers, Tilghman Family Papers, Georgetown University Library. For the condition of tenants in the area, see Clemens, *The Atlantic Economy,* 146–147, 160–163, and for returns from rent, 157–158. For tenant stability, see Walsh, "Land, Landlord, and Leaseholder," *Ag. Hist.,* LIX (1985), 373–396. Like Tilghman, Charles Carroll also built tenant dwellings and tobacco barns himself (Earle and Hoffman, "Genteel Erosion," in Curtin, Brush, and Fisher, eds., *Discovering the Chesapeake,* 284).

25. Similarly, in 1769 Robert Goldsborough set up his son Robert with a complement of slaves and livestock almost identical to what Richard Tilghman Senior deeded Richard Junior in 1766 (Goldsborough Ledger, 1739–1789, MSA; Queen Anne's County Deeds, RT#G: 266–267, MSA); Wills, 34: 407, MSA.

Lloyd had five separate quarters of his own plus other farms that he managed for his stepchildren. These farms were worked by more than fifty adult field hands and were home to about an equal number of dependent slaves, young and old.[26]

Like the Tilghmans, Lloyd combined tobacco culture with extensive fields of corn and wheat. Additional land was put into oats to help feed draft animals. Tobacco yields varied widely from a low on one plantation of about 450 pounds per worker in 1763 (a bad year) to a high of more than 1,800 pounds. The mean was 1,000 pounds, a respectable crop, considering the other work the slaves had to do. Lloyd grossed between three and a half and nine and a half pounds sterling constant value per hand per year from tobacco (Table 34).[27]

Corn proved to be Lloyd's biggest cash crop. Yields per hand varied from seventeen to more than forty-one barrels; the mean was twenty-six barrels. This output produced gross revenues of between £6.5 and £11 sterling constant value annually. Shares of seventeen to nineteen barrels a hand (large crops by anything but upper Eastern Shore standards) Lloyd dismissed as "poor doings." By his lights, it took forty barrels per hand to make a "good" crop. Wheat yields were also quite variable; output per laborer ranged from thirteen to an extraordinary eighty-seven bushels, with the mean crop per hand hovering at about forty bushels. Gross revenues from wheat added an additional £1.75 to £9 sterling a year. All told, Lloyd was grossing about £20 a hand from the three crops, an exceptionally high return.

Although Lloyd was extracting high outputs, as were the two Edmund Berkeleys and Richard Corbin in Virginia's Rappahannock basin, there were

26. Robert Lloyd's Plantation Crop Book and Accounts, 1743–1762, Lloyd Papers, 1658–1910, MS, 2001, microfilm, reel 6, MHS; Papenfuse et al., *Biographical Dictionary of the Maryland Legislature,* II, 543–545.

27. See also Paul G. E. Clemens, "The Operation of an Eighteenth-Century Chesapeake Tobacco Plantation," *Ag. Hist.,* XLIX (1975), 517–531; and Clemens, *The Atlantic Economy,* 188–191. My interpretation of the profitability of Lloyd's farms differs from Clemens's. He estimated revenues per hand by dividing total reported crops by total reported hands, concluding that some farms grew only tobacco and no corn or wheat, since none was recorded. In contrast, I decided the accounts were probably incomplete. Full crops of tobacco, corn, and wheat are recorded for the home farm where Lloyd resided. Only tobacco was accounted for on the stepchildren's property (possibly because tobacco was the only crop that Lloyd was responsible for marketing), whereas, on outlying farms that Lloyd owned outright, corn and wheat crops appear sporadically; in two years the account for one quarter is noted as being "incompleted." It seems unlikely that corn was not grown on all farms and also unlikely that perhaps as much as fifty acres would be prepared for wheat one year and then abandoned the next. I assumed that both tobacco and grains were being raised on all the plantations, whether or not all were recorded, and calculated gross revenues per hand using the mean annual output of all three crops.

TABLE 34. Estimated Tobacco, Corn, and Wheat Crops and Revenues per Laborer, Robert Lloyd Plantations, Talbot and Queen Anne's Counties, Maryland, 1756–1762

Year	Tobacco (lbs.)	Corn (barrels)	Wheat (bushels)	Total revenue
1756				
Mean	1,056	28.7	62.8	
£	5.46	9.66	7.79	22.91
1757				
Mean	951	—	31.0	
£	5.93	—	4.51	[10.44]
1758				
Mean	979	20.5	14.5	
£	7.35	6.41	1.81	15.57
1759				
Mean	1,172	26.5	70.5	
£	9.72	8.37	9.04	27.13
1760				
Mean	1,102	30.3	32.2	
£	7.07	11.18	4.34	22.59
1761				
Mean	1,139	21.8	25.0	
£	6.51	7.08	3.38	16.97
1762				
Mean	793	28.0	33.0	
£	3.55	9.61	4.69	17.85

Note: Currency values are in constant pounds sterling. The total revenue for 1757 (in brackets) is incomplete.

Sources: Robert Lloyd's Plantation Crop Book, Lloyd Papers, MS, 2001, microfilm, reel 6, MHS. Prices from Paul G. E. Clemens, *The Atlantic Economy and Colonial Maryland's Eastern Shore: From Tobacco to Grain* (Ithaca, N.Y., 1980), appendix, table A3

warning signals nonetheless. With the assets he inherited—a large enslaved labor force and, by comparison with either land-rich Maryland families like the Tilghmans or Virginians with more extensive holdings, very little land—Lloyd was at risk of wearing out his soils through overcropping. Land could be planted in tobacco for only three or, at most, four seasons, after which leaching so depleted soil nutrients that both quantity and quality declined precipitously. Old tobacco fields could then be planted in corn for a year or two more but subsequently had to be rested for about twenty years before burning the trees restored nutrients and the calcium content of the ash helped neutralize acidity. A high proportion of arable land thus had to be kept in long fallow, which, given the size of his labor force, Lloyd might have been unable to do. Old tobacco and cornfields might be kept longer in production by planting for another year or so in wheat, although yields were usually only about six bushels per acre. Given the large amounts of corn and wheat Lloyd's slaves were producing, he must have been clearing additional, previously untilled acres for grains, but he was likely also replanting old fields without sufficient time in fallow. In 1761, for example, he calculated that, from sixty acres, of which "not more than twelve acres in the whole [were] but what might be called much worn old corn ground," he averaged just over four barrels of corn per acre, a normal but hardly spectacular yield. Wheat growing allowed Lloyd to use a higher proportion of his land, since much shorter rotations were customarily employed on fields devoted exclusively to that crop. With two harvests followed by only one season in fallow, wheat growers could keep two-thirds of acreage devoted to small grain in continuous production. This intensive style of agriculture clearly could not be maintained indefinitely, however, unless Lloyd undertook extensive manuring, a practice not evident from his plantation records.[28]

Since he lived on the densely settled, geographically isolated upper Eastern Shore, Lloyd could not count on buying more productive land for his slaves to work. By the 1780s and 1790s, some large Eastern Shore planters would begin to think that they owned more slaves than they could profitably employ on their plantations. Still, in the early 1760s, neither Lloyd nor other slaveholders living on the Eastern Shore (or anywhere else in the tidewater) complained that they were burdened with "too many" bound laborers. Although there were limits to how much planters could expand tobacco culture in older areas, they had not yet reached diminishing returns to labor invested in expanding grain production.

Pork and Plank Pay: Prosperity in the Periphery

Agricultural practices in the peripheral tobacco-growing area stretching along Maryland's and Virginia's lower Eastern Shore and crossing the bay to

28. Clemens, *The Atlantic Economy,* 189, 195–198; David Scott Hardin, "'Alterations They Have Made at This Day': Environment, Agriculture, and Landscape Change in Essex County, Virginia, 1600–1782" (Ph.D. diss., University of Maryland, 1995), 121–122. For characteristics of southern soil that made continuous cultivation more difficult than in the northern colonies, see Douglas Helms, "Soil and Southern History," *Ag. Hist.,* LXXIV (2000), 723–758; and John Majewski and Viken Tchakerian, "The Environmental Origins of Shifting Cultivation: Climate, Soils, and Disease in the Nineteenth-Century South," ibid., LXXXI (2007), 522–549.

embrace the south bank of the James River and the Albemarle Sound and Roanoke River settlements of North Carolina are much more sparsely documented than those of the Virginia tidewater and Maryland's upper Eastern Shore. Such records as there are suggest that a few planters who had better-than-average land increased tobacco production in the 1730s but that most farmers living in the periphery turned to exploiting a variety of local natural resources in order to develop other exportable products. A highly diversified export base is a reflection both of more varied natural resources and of the scarcity of large plantations. Small planters, who lacked the resources to undertake large-scale enterprises, dominated in the periphery to a greater extent than they did in other older tidewater areas. More of the region's small farmers owned land than did those living in prime tobacco areas, and that land was somewhat more equitably distributed than in other parts of the tidewater. Similarly, slaveholding was spreading among established landowners, but even big planters did not own large gangs of laborers. At the turn of the century, white servants had also been an important component of the workforce, but after the 1720s few indentured or convict servants were marketed in an area that no longer grew much tobacco. Families without bound laborers were not in a position to produce a large volume of any particular product; instead, they made small amounts of a wide variety of products. By the 1740s, more lower Eastern Shore planters found means for acquiring slaves. In Somerset County, for example, the proportion of inventoried decedents owning slaves had by then become comparable to the proportions found in other Maryland counties, and tax lists for the 1740s reveal that more than a third of all households in Somerset included at least one taxable slave.[29]

As the French tobacco monopoly began buying up ever larger amounts of low-grade oronoco, the price of the weed in this region—which had fallen well below that of other areas by the 1680s—rose to parity with other strains

29. Jean Elliott Russo, "'The Interest of the County': Population, Economy, and Society in Eighteenth-Century Somerset County, Maryland" (Ph.D. diss., University of Minnesota, 1999), chaps. 1, 2; Russo, "'Fifty-Four Days Work of Two Negroes': Enslaved Labor in Colonial Somerset County, Maryland," *Ag. Hist.*, LXXVIII (2004), 466–492; Harry Roy Merrens, *Colonial North Carolina in the Eighteenth Century: A Study in Historical Geography* (Chapel Hill, N.C., 1964), 74–81; Robert E. Gallman, "Influences on the Distribution of Landholdings in Early North Carolina," *JEH,* XLII (1982), 549–575; Marvin L. Michael Kay and Lorin Lee Cary, "A Demographic Analysis of Colonial North Carolina with Special Emphasis upon the Slave and Black Populations," in Jeffrey J. Crow and Flora J. Hatley, eds., *Black Americans in North Carolina and the South* (Chapel Hill, N.C., 1984), 71–121; Kay and Cary, *Slavery in North Carolina, 1748–1775* (Chapel Hill, N.C., 1995), chap. 1; Walsh, "Mercantile Strategies, Credit Networks, and Labor Supply," in Eltis, Lewis, and Sokoloff, eds., *Slavery in the Development of the Americas,* 110–117.

of oronoco. Consequently, planters who had good bottomland well suited to tobacco now had an incentive to reenter the market. Much of the land, however, was marsh, swamp, and sandy seashore, unsuitable either for tobacco or grains. Thus, residents had to make use of what resources they had as well as to learn how to deal with overabundant mosquitoes, snakes, ticks, and horseflies. Somerset County, Maryland, planters grew tobacco on their best lands and ever larger amounts of corn on the remainder. In the area of present-day Wicomico County, where there was much arable land but little of it suited to tobacco, corn became the major cash crop. In Worcester and Dorchester counties, where swamps and sandy seashore predominated, most families had to be content with subsistence farming. Farther south, in Northampton and Accomack counties, Virginia, planters combined limited tobacco production on their best lands with large crops of corn, wheat, oats, peas, and beans and extensive sheep raising. Across the bay, James River and North Carolina farmers produced tobacco, corn, wheat, legumes, rice, cotton, barley, oats, buckwheat, and potatoes; raised numerous hogs; and mined forests and swamps for lumber, staves, shingles, and naval stores.[30]

The extent to which large lower Eastern Shore planters had diversified by midcentury is illustrated in the will and inventory of merchant-planter Thomas Cable (?–1743), whose comments on the region's economy in the 1720s were discussed previously, and in the will of his widow, Sorrowful Margaret Custis Kendall Cable (?–1750). The Cables retained their dwelling plantation on Sorrowful Margaret's land in Northampton County, but by his death Cable had also acquired five hundred acres on Chincoteague Island, around five hundred acres on King's Creek, where they had set up an

30. For tobacco, see Lois Green Carr, "Diversification in the Colonial Chesapeake: Somerset County, Maryland, in Comparative Perspective," in Carr, Morgan, and Russo, eds., *Colonial Chesapeake Society,* 342–388; Alan D. Watson, "Society and Economy in Colonial Edgecombe County," *NCHR,* L (1973), 231–255. Henry Callister reported that Dorchester County farmers made little tobacco after 1750 (Callister to Foster Cunliffe, Oct. 2, 1750, Callister Papers, CWL). For other crops, see "Observations in Several Voyages and Travels in America in the Year 1736 (from the London Magazine, July, 1746)," *WMQ,* 1st Ser., XVI [XV] (1907), 1–17 [143–159]; *Itinerant Observations in America: Reprinted from the London Magazine, 1745–6,* Georgia Historical Society, *Collections,* IV, pt. 2 (1878); Gregory A. Stiverson and Patrick H. Butler III, eds., "Virginia in 1732: The Travel Journal of William Hugh Grove," *VMHB,* LXXXV (1977), 37–38; John Brickell, *The Natural History of North Carolina* (1737; rpt. Murfreesboro, N.C., 1968); "Journal of a French Traveller," *AHR,* XXVI (1921), 736–741; and Custis Papers, 1711–1792, VHS, microfilm, M-58.1, M-1073.3, CWL. Increasing grain production in northern North Carolina is suggested by the appearance of gristmills. In the 1730s, Brickell found only two or three in the region (*Natural History,* 263), but Watson ("Society and Economy in Edgecombe County," *NCHR,* L [1973], 231–255) found that court records indicated that the pace of mill building had picked up by the 1740s.

outlying quarter, and another small tract on Old Plantation Neck. Thomas owned eleven slaves outright, but the family's total workforce likely included others whom Margaret had inherited from her father or her first husband. The names of two of the men, Dominac and Spannish Jack, suggest Cable acquired at least some of the slaves through West Indian trading connections. By the time he died, however, Cable seems to have given over most mercantile pursuits.

Since some of the family's land was better than average for the area, Cable and later Sorrowful Margaret continued to raise tobacco into the 1750s. Their crops, along with tobacco due Thomas for his fees as county clerk, were sold in the country for cash. The enslaved workers also raised large amounts of corn (250 barrels in 1744) as well as wheat and oats, with the help of two or more pairs of workhorses, six oxen, four plows, two harrows, a cart, and a wagon. They also made cider, some of which they distilled into brandy. In winter, the men cut timber for plantation use and possibly for sale and made casks for tobacco and cider. Two canoes and a pettyauger (a long canoe hollowed from a single tree trunk) were available for fishing and for transporting goods. The family pastured more than sixty cattle on the home plantation, and another forty-seven—likely intended for market—were left to forage on their own on the barrier islands. Enslaved women on the home plantation were heavily involved both in dairying and in textile production. Wool came from a flock of fifty-five sheep, and the women carded and spun the wool into yarn. They also raised flax, apparently in fairly large amounts, which they broke, spun on six linen wheels, and then wove into cloth.

The proceeds of their farms allowed the Cables to enjoy a genteel lifestyle in a house graced with fashionable furniture, but in styles influenced by provincial as well as by metropolitan tastes. There were a variety of elaborately carved tables, chests of drawers, bureaus, linen presses, desks, large looking glasses, sconces, and canopy bedsteads (some of which the Cables proudly identified as "new") along with family portraits hanging in the hall. Some of the furniture fashioned from pine, maple, and wild cherry was surely locally made, and some desks and chests of drawers of New England design had been bought from northern traders. The Cables probably purchased other pieces made of walnut from craftsmen working "over the Bay," or perhaps they were imported from England, as were a wide assortment of china punch bowls, sets of plates, decorative delft teawares, monogrammed silver, books, and a map of parts of America. Entirely missing, however, were any examples of the high-style mahogany furniture then becoming popular among elites on the mainland. And, although upscale fabrics like calico and "china" were clearly also imported, many of the linen tablecloths and sheets had almost certainly been made on the plantation. On the periphery, families acquired genteel furnishings from a wide variety of sources, but they saved scarce resources by making do with a higher proportion of locally produced textiles and furniture than did families elsewhere in the region. Since Thomas's inventory was not valued and his estate was not accounted, the state of his finances is uncertain. His estate seems not to have been heavily encumbered, though, since Thomas expected that the proceeds of one year's tobacco crop, together with his clerk's fees, would be sufficient to discharge any outstanding debts, and county magistrates, who decided that the value of his movable goods did not have to be appraised, concurred.[31]

Settlers on the periphery made use of every available resource. Few testators in prime Maryland and Virginia tobacco areas made the complex arrangements for use of marshland, pasture, orchards, and distilling equipment that often appear in Somerset County, Maryland, wills. Slaveholding families in Somerset also devised complicated schemes for distributing scarce labor among as many family members as possible. Few planters in the periphery had much need for large gangs of slaves to work their fields, but, just as on the tobacco-centered lower Western Shore, the addition of one or two adult workers to a household could significantly increase outputs of a variety of minor products. Dying parents divided their slaves among most or all of their children. If there were not enough to endow all family members with at least one laborer, then testators devised the future increase of enslaved women to those heirs who received no share or an unequal share of the family's human property.[32]

An economy based on diversified agriculture and resource extraction also required that labor arrangements be more flexible than those in prime tobacco areas where most slaves were field laborers working year-round in tobacco. Long- and short-term slave hiring and slave sharing between family members were much more common on the lower Eastern Shore. In addition to farming, enslaved men and boys were frequently set to cutting, sawing, and hauling timber and to tending livestock. Tellingly, in the 1740s more than half of estate inventories of slaveowners in Somerset County also included

31. For Thomas Cable, see Northampton County Wills and Inventories, 19, 1740–1750: 101–103 (will) and 205–206 (inventory), and for Sorrowful Margaret Cabel, see 525–527 (will), microfilm, VSL. The 1746 inventories of elite planter Col. John Custis, who owned sixty-five slaves, shows much higher levels of agricultural diversification and import replacement activities (ibid., 273–276, 312–313).

32. Russo, "'Fifty-Four Days Work of Two Negroes,'" *Ag. Hist.,* LXXVIII (2004), 466–492; Carr, "Diversification in the Colonial Chesapeake," in Carr, Morgan, and Russo, eds., *Colonial Chesapeake Society,* 370.

oxen, whereas these draft animals appeared in less than one-fifth of the estates of nonslaveowners. In the periphery, acquisition of one or more slaves encouraged farmers to employ more draft animals in a variety of extractive industries, just as it encouraged planters in prime tobacco areas to employ plows and carts in expanded grain production.[33]

Every account of the products of the lower Eastern Shore, the lower James River, and Albemarle Sound catalogs a cornucopia of carefully harvested natural resources. Both cultivated orchards and wild fruit provided the ingredients for making beer, cider, and brandy. Wild animals supplied skins, furs, and feathers; stranded whales were stripped of marketable bone and oil; abundant runs of herring were caught with seines and weirs and pickled and packed for export to the West Indies, so long as the supply of salt held out. Bees, wild and increasingly domesticated, supplied honey and wax, and myrtle berries were gathered and boiled down to make high-quality candles. Snakeroot and ginseng were harvested in the woods. Settlers in the longleaf pine areas made tar, turpentine, pitch, and resin in the winter months and cut timber for plank and masts, and those with access to especially durable hardwoods—red oak, white cedar, and cypress—made barrel staves and shingles. Some local craftsmen began to produce canoes, furniture, and treenware for export to other areas as well as for local use. And a few enterprising boatwrights began building oceangoing ships that they either sold to local merchants employed in intercolonial trade or exported to Britain. A painting of a Maryland Eastern Shore boatyard of the period gives an idea of the wide variety of vessels required by burgeoning intercolonial and transtlantic trade (Plate 35).[34]

Livestock probably accounted for a greater proportion of planter income in the periphery than elsewhere, not only for subsistence but for export as well. Hogs foraged in marshes and swamps in summer and fattened on upland mast in the fall. The diet on the Virginia-North Carolina border, William Byrd II frequently complained, was "pork upon pork, and pork again upon that," and still there was a surplus for export. Cattle could be wintered in swamps and marshes where they needed no extra fodder, and enough were penned to enable industrious housewives to make some surplus butter and cheese. Extra horses were shipped to the West Indies. Enough animals were

33. Russo, "'Fifty-Four Days Work of Two Negroes,'" *Ag. Hist.*, LXXVIII (2004), 466–492.

34. See Note 30, above; Merrens, *Colonial North Carolina,* chaps. 5–6; and James O'Mara, "Urbanization in Tidewater Virginia during the Eighteenth Century: A Study in Historical Geography" (Ph.D. diss., York University, 1979), 163–170. Brickell, *Natural History,* 272–276, noted slaves were the main suppliers of snakeroot, which they gathered on Sundays. The price would be too high if whites gathered it.

PLATE 35. *Grey's Inn Creek Shipyard.* By an unknown artist, 1750. Overmantel panel from Spencer Hall, Kent County, Maryland. Permission, Maryland Historical Society, Baltimore.
Quickening intercolonial and transatlantic commerce required vessels of many sizes and designs.

killed each year that slaughter by-products—hides and tallow—became important exports as well. Families in this region could sell more livestock because abundant venison, oysters, crabs, fish, and waterfowl supplied them with other kinds of protein over a good part of the year. Some slaughtered their animals and made pickled pork and beef for ships' provision and export; many others drove their surplus cattle and hogs overland to Norfolk for sale, thus saving on both the cost of salt and of transporting bulky containers by water. Some Norfolk women did an extensive business producing large quantities of tallow candles from a by-product of these slaughtered animals, as did rural housewives in Somerset County.[35]

Among small planters, such activities were part of an integrated family enterprise. Many visitors to the region commented unfavorably on the supposed indolence of the white men ("slothful in everything but getting of chil-

35. William Byrd II, *A Journey to the Land of Eden Anno 1733,* in Louis B. Wright, ed., *The Prose Works of William Byrd of Westover: Narratives of a Colonial Virginian* (Cambridge, Mass., 1966), 384, 408 (quotation). Cf. Byrd, "The Secret History of the Line," ibid., 59–60, 63, 100, 203–204, 311–312; John Oldmixon, *The British Empire in America* (1741), 2 vols. (New York, 1969), 460–523; Brickell, *Natural History,* 35, 39–40, 43–44, 51–56, 81–82, 227–241, 255, 262; "Observations in Several Voyages," *WMQ,* 1st Ser., XVI [XV] (1907), 1–17 [143–159]; "Journal of a French Traveller," *AHR,* XXVI (1921), 736; and Russo, "'The Interest of the County,'" chap. 2.

dren," in William Byrd II's words) in much the same way earlier observers had criticized the presumed indolence of Indian men. But all agreed that the women and children worked hard indeed. "The ordinary Women," John Oldmixon noted, "take Care of Cows, Hogs and other small Cattle, make Butter and Cheese, spin Cotton and Flax, help to sow and reap Corn, wind Silk from the Worms, gather Fruit, and look after the House." Children went to work at an early age both in the fields and around the house. They helped their mothers harvest myrtle berries in the fall, caught fish and crabs, and did not neglect to collect a few pence by trapping cardinals and mockingbirds to sell to ship captains bound for Europe.[36]

Although this mode of living attracted the scorn of gentry observers, once markets improved such strategies did produce a higher standard of living for all but the poorest residents. Import-replacement activities—everything from home cloth production to perfecting methods for making buildings and casks without nails—left families with enough spare cash to buy a few nonessentials from country storekeepers or itinerant traders. Most necessi-

36. Oldmixon, *The British Empire,* 518; Byrd, "Secret History," in Wright, ed., *Prose Works,* 192, 311–312; Byrd, *Journey to the Land of Eden,* ibid., 409; Brickell, *Natural History,* 32–33, 81–82, 103, 185–186, 190–191, 246; *Byrd Correspondence,* II, 445–446.

ties were so abundant that, as William Byrd II put it, only "law, physic, and strong drink" were dear. Increased production for multiple markets provided established householders on the lower Eastern Shore, Albemarle Sound, and the Roanoke River with enough income to buy a slave or two. Growing prosperity was reflected as well in improved housing. In the mid-1730s, John Brickell noted that planters along the Chowan River and the Roanoke Inlet had begun to build brick houses. The Wicomico area in Maryland also saw a growing number of substantial dwellings.[37]

All this increased prosperity was fueled by expanding markets. In the late 1720s and early 1730s, for example, Northampton County merchant Thomas Cable had trouble finding buyers for all the grain lower Eastern Shore farmers sold him, and William Byrd II observed that North Carolina settlers, hampered by a lack of good deep-water harbors, were producing less meat and naval stores than they might because they were unsure of a market. (On the other hand, continued lax enforcement of the Navigation Acts in that colony furthered a lively clandestine trade.) Shortly, increasing numbers of local merchants, either independent or allied to large Scottish firms, were forging diverse market connections outside the traditional channels of the dominant tobacco trade—with the British outports, New England, the British and French West Indies, the wine islands and Portugal, and with other independent Western Shore merchants. These smaller-scale trades required less capital than did dealing in tobacco and hence afforded the best opportunities for indigenous merchants. By the 1750s, intercolonial and transatlantic demand was so great that some merchants were unable to fill all the orders they received for local products.[38]

37. Carr, "Diversification in the Colonial Chesapeake," in Carr, Morgan, and Russo, eds., *Colonial Chesapeake Society,* 370; Byrd, "Secret History," in Wright, ed., *Prose Works,* 206–207; Brickell, *Natural History,* 2–3, 37; Watson, "Society and Economy in Edgecombe County," *NCHR,* L (1973), 231–255; Kay and Cary, "A Demographic Analysis," in Crow and Hatley, eds., *Black Americans in North Carolina,* 71–121; Merrens, *Colonial North Carolina,* chap. 4; L. A. Fenn and Peter H. Wood, *Natives and Newcomers: The Way We Lived in North Carolina before 1770* (Chapel Hill, N.C., 1983), chaps. 3–6; Cary Carson et al., "Impermanent Architecture in the Southern American Colonies," *Winterthur Portfolio,* XVI (1981), 171–173.

38. Byrd, "Secret History," in Wright, ed., *Prose Works,* 208, 210; Brickell, *Natural History,* chap. 1; Charles Stewart Letterbooks, 1751–1753, 1754–1763, HSP, microfilm, M-32, CWL. See also Robert Polk Thompson, "The Merchant in Virginia, 1700–1775" (Ph.D. diss., University of Wisconsin, 1955), chaps. 1, 5; *Byrd Correspondence,* II, 539–540; David C. Klingaman, *Colonial Virginia's Coastwise and Grain Trade* (New York, 1975), chaps. 2–3; Charles George Gordon Moss, "The Virginia Plantation System: A Study of Economic Conditions in That Colony for the Years 1700 to 1750" (Ph.D. diss., Yale University, 1932), chaps. 11–12; Charles W. Sargent, "Virginia and the West Indies Trade, 1740–1765" (Ph.D. diss., University of New Mexico, 1964), chap. 2.

Some examples give an idea of the scope and pace of trade. Independent merchants and English and Scotch factors stationed in the tobacco-growing areas of the upper Eastern Shore traded in lower Eastern Shore lumber, meat, and grain as a sideline. In Dorchester County, Maryland, in the 1740s and 1750s, native merchant Joseph Ennalls bought tobacco, plank, shingles, staves, wheat, corn, cattle, hogs, and pork from local planters. These he sold to other local merchants, to traders in Talbot, Anne Arundel, and Saint Mary's counties, to a cattle dealer in Pennsylvania, and to Whitehaven (capitalizing on local family connections) and London merchants. George Yuille, an agent for a Glasgow firm, traded on the Appomattox River in the 1750s. He shipped a mixture of Virginia and North Carolina produce, including tobacco, corn, naval stores, and ginseng to Glasgow, Dumfries, Whitehaven, Bristol, Liverpool, London, and Ireland in return for British manufactures, West Indian rum and sugar, and Madeira. He also sold slaves consigned to him by Antiguan planters (some of the sellers were related to Yuille) to James River planters.[39]

The bulk of the regional trade, however, was increasingly concentrated in Norfolk. This James River port captured both the early intercolonial coastal trade and, with the construction of better roads from the Perquimans and Chowan districts in the 1730s, overland trade with North Carolina as well. To this trade in regional products, Norfolk merchants added shipbuilding and wholesaling of dry goods and West Indian products to interior storekeepers. In the 1750s, for example, Norfolk merchant Charles Stewart dealt with merchants in Barbados, Antigua, Montserrat, and Saint Kitts, the Canary Islands, Madeira, Portugal, New England, North Carolina, New York, England, and Scotland, and with Maryland wheat growers at the head of Chesapeake Bay. He exchanged a diverse mix of tobacco (mostly dark leaf destined for the Caribbean), corn, wheat, peas, pork, beef, tallow, flour, bread, barrel staves and headings, shingles, plank, myrtle and beeswax candles, pitch, tar, oysters and snakeroot for slaves, sugar, rum, molasses, salt, Madeira, coffee, indigo, pimento, and ginger. Similarly, Alexander McKenzie, a merchant operating in Hampton and Norfolk between 1748 and 1750, sold beef, pork, hogs, tallow, butter, myrtle and beeswax candles, flour, bread, rice, peas, cider, apples,

39. Joseph Ennalls Account Book, 1729–1762, MS 937, MHS; George Yuille Account Book, 1754–1757, Special Collections, Duke University Library, Durham, N.C., microfilm, M-36.4, CWL. See also Mercer Family Papers, MSS1 M5345, a113–122, VHS; and John Caile Account Book, 1759–1767, Dorchester County, Maryland, microfilm, MSA. Caile operated a subsidiary store in the town of Cambridge for the Liverpool firm of Foster Cunliffe and Sons under the direction of their principal factor, Robert Morris.

tar, pitch, turpentine, pig iron, plank, and staves to London, Glasgow, Rhode Island, Philadelphia, the Canary Islands, and the West Indies for return cargoes similar to Stewart's.[40]

Although less important than these intercolonial and international markets, the counties on the mainland also offered outlets for the products of the periphery. Western Shore planters continued to trade directly with Eastern Shore and lower James River producers and traders as they had done in the 1710s and 1720s. For example, in the 1740s John Mercer of Stafford County purchased pork, myrtle wax, beeswax, shingles, and plank from Eastern Shore contacts. And, in 1760, George Washington got shingles, tar, turpentine, tallow, and myrtle wax from dealers in Norfolk and Nansemond and bought oats on the lower Eastern Shore.[41]

Planters on the periphery thus were taking advantage of expanding markets within the region, within the British Empire, and, in the case of the wine islands and southern Europe, outside the Empire as well. A combination of subsistence farming, home manufacturing, and trade in a variety of products afforded residents of the periphery some protection from cyclical downturns in the tobacco economy. But, like all other Chesapeake colonists, they were not immune to the periodic disruptions of wars in the Atlantic. Slaves were less essential to planter prosperity in the periphery than elsewhere. Slave labor there, however, as everywhere in the mainland colonies, was increasingly critical to participation in and expansion of the export economy.

Changes in Tobacco Culture and Marketing during the Golden Age

As we saw earlier, the adoption of tobacco-inspection acts in Virginia and Maryland produced significant changes in the way Chesapeake tobacco was handled and marketed. Elite planters on the whole favored the new system, but, in order to gain the assent of ordinary growers, they had to make compromises that eroded some of the advantages they had previously enjoyed as large producers. Great planters could no longer command premiums for convenient loading or for making heavier hogsheads, and they could no longer market unsorted, substandard crops that they had grown on marginal land or that inexperienced or poorly supervised overseers on distant quarters not

40. Thomas J. Wertenbaker, *Norfolk: Historic Southern Port* (Durham, N.C., 1931), chaps. 1–2; Byrd, "Secret History," in Wright, ed., *Prose Works,* 173–174; Brickell, *Natural History,* 262; Charles Stewart Letterbooks, 1751–1753, 1754–1763, CWL; Alexander McKenzie Account Book, 1748–1750, HSP, microfilm, M-1349, CWL. See also Francis Carroll Huntley, "The Seaborne Trade of Virginia in Mid-Eighteenth Century: Port Hampton," *VMHB,* LIX (1951), 297–308.

41. John Mercer Ledger, [1741]–1750, Bucks County Historical Society, Doylestown, Pa., microfilm, M-164, CWL; Jackson, ed., *Diaries of Washington,* I, 235, 266, 274.

infrequently produced. Moreover, the quality of elite crops became subject to the scrutiny of local inspectors, men who were almost invariably their social inferiors but whom it appears they could not effectively intimidate. Large planters who considered themselves "crop masters" believed they lost face and reputation when inspectors publicly condemned some of their output. Initially, it was small planters who most feared that their crops would be disproportionately subject to rejection. Evidence on subsequent levels of tobacco production, on the economic fortunes of tenant farmers, and on the proportion of rejected tobacco deducted from overseers' shares of a crop presented below demonstrates, however, that this was not necessarily the case. There were winners and losers among all classes of tobacco growers, but overall more planters benefited rather than lost from inspection.[42]

A few large planters believed inspection was detrimental to their interests, a point of view seconded by Virginia's governor William Gooch, who stated that "rich Man's trash" had always been the greatest cause of low prices. John Custis, a large planter and member of the Virginia Council living in Northampton County, a marginal tobacco area, anticipated the worst. In 1732, he wrote a Bristol merchant: "This fine Tob: Law wch you are so very fond of has prevented me [from making more tobacco], and I must beleive that it will bee impossible to make Tob: there being so many difficulties in the Law and so much subjection to the arbitrary inspectors." Among Custis's numerous objections to the law was that he could no longer disguise the origins of his presumably lower-grade Eastern Shore tobacco by transporting it to the Western Shore in order to include it in his prime York County consignments. Four years later, Custis threw in the towel, telling a London correspondent, "I find it very difficult to make Tob: under our Law: and seeing no prospect of an end to it," he was turning over direction of his plantations to his son: "I believe I should still have gone on with planting if I could have had the liberty to make the most of my slaves labor; but to have it destroy'd by burning after so much trouble in making is a disheartening thing."[43]

Other great planters, some of whom like Robert "King" Carter originally opposed inspection, had less difficulty adjusting to the law. In 1731, Carter wrote a correspondent, "All people and especially our Overseers are wonderfully alarum'd in the neat handling of tobo[;] it occasions the throwing away

42. T. H. Breen, *Tobacco Culture: The Mentality of the Great Tidewater Planters on the Eve of Revolution* (Princeton, N.J., 1985), 59–75.

43. Gooch's opinion is discussed in Thomas J. Wertenbaker, *The Planters of Colonial Virginia* (1922), rpt. in Wertenbaker, *The Shaping of Colonial Virginia* (New York, 1958), 146–147; John Custis to Mr. Loyd, [1732], Custis to Perry, 1736, John Custis Letterbook, 1717–1741, LC, microfilm, M-8, CWL.

abundance of what we us'd to get good money for." On his own plantations, Carter complained that the new law had cost him "a full month to get them managed . . . when otherwise I could have very Well done it in a fortnight." But by the next year Carter had changed his mind about the desirability of inspection and reported with satisfaction that his crop "hath pass'd the fiery tryall of our new Laws" and that "the Country are wonderfully fond of the tobo Law."[44]

The assertion of a Virginia writer that "our planters are now got into such a method, that they make as much good Tobacco, as they used to make good and bad before the Law was made" was largely true for that colony, as crop outputs per laborer demonstrate (Table 35). Possibly, as Francis Makemie had predicted at the turn of the century, with public inspection there developed "a general Emulation to out vie one another, especially in Quality." On the other hand, as a mid-nineteenth-century agricultural writer observed, since all tobacco that passed inspection "was generally sold at the same price, or with little difference," planters might have found "but little inducement to improve the quality of the staple, either by method of cure or after-assortment." Consequently, even in sweet-scented areas quantity rather than quality might have assumed ever greater importance after passage of the inspection laws, as had always been the case in oronoco areas, especially among those who normally sold their crops in the country rather than consigning them to Britain. Production per hand did fall in tidewater Virginia in the 1730s, the decade the law was implemented. By the 1740s and early 1750s, though, it rose well above preinspection levels, not just in new areas where soils were fresh but in older counties as well. Since the reported outputs consist only of the portion of the crop that passed inspection, the Virginia laws clearly did not result in any permanent reduction in the amount of tobacco a full-time hand could produce.[45]

Small planters in both colonies who lacked the resources to sustain their families through the next year should their cash crop be rejected and destroyed had the most to lose. Some scholars have argued that in Maryland small planters lost more from the rejection of a substantial portion of their

44. Carter to William Dawkins, Nov. 15, 1731, to Edward Tucker, Aug. 16, 1731, May 11, 1732, to Alderman Perry, May 12, 1732, Robert Carter Letterbook, 1731–1732, Robert ("King") Carter Papers, MS 3807, UVa, microfilm, M-113, CWL.

45. Letter to the editor, *Maryland Gazette* (Annapolis, Md.), May 19, 1747; [Francis Makemie], "A Plain and Friendly Perswasive to the Inhabitants of Virginia and Maryland for Promoting Towns and Cohabitation" (1705), *VMHB,* IV (1896–1897), 263; N. F. Cabell, "Some Fragments of an Intended Report on the Post-Revolutionary History of Agriculture in Virginia," *WMQ,* 1st Ser., XXVI (1917), 153–154.

TABLE 35. Tobacco Crops per Laborer, Virginia and Maryland, 1720–1763

	Virginia								Maryland			
	York River Basin		Rappahannock River Basin		Potomac River Basin		James River Basin		Western Shore		Eastern Shore	
Years	*N*	Mean pounds	*N*	Mean pounds	*N*	Mean pounds	*N*	Mean pounds	*N*	Mean pounds	*N*	Mean pounds
1720–1729	16	717	17	718	4	843	—	—	1	1,613	35	1,229
1730–1739	46	677	9	556	—	—	6	668	4	1,009	29	1,360
1740–1747	69	820	21	770	14	787	15	901	6	1,275	16	1,060
1748–1753	15	891	7	851	11	872	3	966	8	941	6	1,170
1754–1763	32	732	37	731	37	704	4	672	16	792	42	926

Note: Time periods are divided as closely as possible into decades, but one is split at 1747 to look for effects of the Maryland tobacco inspection law. Tobacco crops are classified by the river basin from which they were most likely exported. Crops from the Burwell plantations in James City and York counties were assigned to James River. Shares for plantations in Caroline and King and Queen (which might be sent to either the York or the Rappahannock) were assigned to the York or the Rappahannock rivers according to the location of the owner's dwelling plantation.

Source: Appendix 1 and Tables 34 and 35.

crops than they gained from higher prices offered for what remained. Most of these assertions stem from Carville V. Earle's estimate that about half of a laborer's crop was likely discarded before inspection or refused by the inspectors. Earle drew much of his evidence for the destruction of inferior grades (seconds) as well as of trash from the observations of one opponent of inspection, "Mr. Q in the Corner," who stated that both ground leaves and seconds would be destroyed. In fact, however, the law required the destruction only of trash. Other contemporaries estimated that just one-quarter to one-third of most crops would be rejected (Table 35 corroborates these more optimistic contemporary estimates). Outputs per laborer did decrease on Maryland's Western Shore just after the passage of the inspection law, but only by about one-quarter, not the 50 percent Earle posited. Later reductions in tobacco outputs appearing in both colonies during the Seven Years' War were more likely owing to shifts of some labor time from tobacco into grains rather than to more rigorous enforcement of the inspection law.[46]

The evidence from rental properties suggests that even tenant farmers more likely benefited than lost. Everywhere the amount of rent tenants paid rose with the implementation of the inspection acts. Even those whose nominal annual rents remained unchanged paid more, since payments had subsequently to be made in more valuable inspected tobacco. Moreover, beginning about 1750, many landlords did raise the amount of rent tobacco charged for tidewater tenements. But in most cases tenant families were able to pay more without undue hardship. Accounts for large rental estates in both Virginia and Maryland show no increase in tenant defaults after passage of the inspection acts. Since most tenants owned no bound labor, they could not expand the size of their crops by very much. Had they been forced to discard up to a third or a half of their output, one would expect to find more evidence of tenant distress than in fact appears.[47]

46. For the fears of small planters, see John C. Rainbolt, ed., "The Case of the Poor Planters in Virginia under the Law for Inspecting and Burning Tobacco," *VMHB*, LXXIX (1971), 314–321. For the argument that 50 percent of tobacco would be rejected, see "Q. in the Corner," Letter to the Editor on Mercator's "Reflections on an Inspection Act," *Md. Gaz.*, Apr. 28, 1747; and Carville V. Earle, *The Evolution of a Tidewater Settlement System: All Hallow's Parish, Maryland, 1650–1783* (Chicago, 1975), 25–27, 96–100. In contrast, Allan Kulikoff, *Tobacco and Slaves: The Development of Southern Cultures in the Chesapeake, 1680–1800* (Chapel Hill, N.C., 1986), 116, concurred with estimates of other contemporary observers that between one-quarter and one-third would likely fail inspection.

47. See discussions of crop shares in Saint Mary's and Charles counties and of the Tilghman rental properties, above; Walsh, "Land, Landlord, and Leaseholder," *Ag. Hist.*, LIX (1985), 373–396; Bliss, "The Rise of Tenancy in Virginia," *VMHB*, LVIII (1950), 427–441; and Mary McKinney Schweitzer, "Economic Regulation and the Colonial Economy: The Maryland Tobacco

In addition, under the terms of the inspection laws, overseers were required to make good from their shares of the crop any tobacco burned by the inspectors. A few examples from Virginia survive in planter accounts, where less than 5 percent of the weed presented for inspection was put to the torch. The real question is what proportion of the crops the planters discarded before inspection. Here they were not entirely helpless, for, as newspaper discussions over the inspection law made clear, most unmerchantable tobacco resulted from careless curing and prizing, not from deliberate passing off of trash. More careful processing could preserve more of the crop.[48]

Much of the controversy over the likely impact of inspection laws stems from confusion over what was meant by "seconds" as well as ambiguity about whether seconds were automatically rejected under the inspection acts. As noted previously, only trash was explicitly prohibited. Especially in years when crops were short, inspectors did accept "well-conditioned" inferior grades of tobacco and occasionally passed even ground leaves. One kind of "seconds" resulted from the practice of allowing plants that were already cut to produce additional shoots late in the season, making for a low-quality product often damaged by frost before the secondary leaves matured. Cultivating second-growth leaves was illegal in Virginia during most of this period and periodically prohibited in Maryland. Some planters, though, probably continued to tend this prohibited type whenever they thought it unlikely that local constables (who were charged with viewing fields after the first cutting to prevent the practice) would catch them. The other meaning of "seconds" was legally cultivated first-growth leaf of inferior quality. Changes in estate appraisal practices make it difficult to determine the fate of the latter. In Maryland before the inspection law, appraisers in the predominantly bright-leaf growing counties of Calvert, Prince George's, Anne Arundel, Baltimore, and parts of Talbot frequently recorded crops separated into "top" (high-priced) and lower grades (variously styled "second," "bad," "under," "ground leaf," or "trash"), valued at one-third to one-half the price of "top" grades. In the dull leaf growing counties of Saint Mary's, Charles, Somerset, and the Poto-

Inspection Act of 1747," *JEH,* XL (1980), 551–569. Rental accounts of Maryland landlords show neither a rise in tenant failures nor any increase in overdue rent payments before 1763. In addition, the value of inventoried tenant estates on the lower Western Shore either held constant or rose slightly in this period, further evidence that tenants were not suffering undue distress (Historic Saint Mary's City, Analysis of Lower Western Shore Inventories, compiled by the author).

48. In the John Mercer Ledger, [1741]–1750, see Mercer's account with overseer William Wheland, 1748, in which 204 of 7,140 pounds were burned, and his account with James Savage, 1748, in which 342 of 10,000 pounds were burned (CWL). See also Robert Carter to William Dawkins, Nov. 15, 1731, and to Hasswell and Brooks, May 29, 1732, Carter Letterbook, Carter Papers, CWL; and Mercator, *Md. Gaz.,* Apr. 7, 1747.

mac River side of Calvert, planters seldom separated their crops, and estate appraisers used a lower, court-rated "current" price that took into account a mix of "good and bad." At that time, both grades were clearly considered merchantable. After the passage of the inspection law, appraisers throughout Maryland began valuing tobacco crops in all counties at a single court-rated "current" price. That oronoco buyers continued to purchase some tobacco at rates below the established "current" price does demonstrate that some inferior grades did pass inspection, at least in years of short crops.[49]

In addition to trash, the other sort of tobacco that the inspection laws explicitly prohibited was "lug" tobacco—hogsheads packed promiscuously with the usable portions of "Ground leaves and all the shatter'd and broken peices." This product was judged too easily adulterated with trash, so under the inspection regime planters could market only whole leaves. These might be stemmed, but they had to be sufficiently intact to be packed in "straight laid" bundles. Virginia sweet-scented growers, who were in the habit of stemming most of their crops, made most of the lug tobacco exported in the early eighteenth century, although, before 1747, a few planters on Maryland's upper Eastern Shore also made it. Although Virginia planters were most affected by the prohibition on this product, there was little contemporary comment. Dull-leaf growers on the lower Eastern and Western shores, who seldom did any stemming or separating by grade, probably rarely, if ever, made this more labor intensive sort.[50]

49. Accounts, 22: 5, 10, 45, 188, 190, 197, 222, 276, 278, 343, 23: 41, 164, 224, 250, 263, 277, 288, 289, MSA; Prince George's County Accounts, D6: 171, MSA. Robert Goldsborough noted that in 1749 and 1750 he had got hogsheads of ground leaves past inspection (Memorandum Book, 1696–1771, privately owned, microfilm, MSA). For marketing areas, see Jacob M. Price, *France and the Chesapeake: A History of the French Tobacco Monopoly, 1674–1791, and of Its Relationship to the British and American Tobacco Trades,* 2 vols. (Ann Arbor, Mich., 1973), I, 667–668. When Maryland's inspection law lapsed during the Revolution, division of crops into tops and seconds quickly reappeared in probate inventories. Whether valuing crops at a single price (a practice adopted after implementation of the inspection act) indicates appraisers thought seconds were no longer merchantable or whether appraisal procedures simply changed merits further investigation.

50. For descriptions of lug tobacco, see Robert Carter to Micajah Perry, Aug. 1, 1723, Carter Letterbook, Carter Papers (quotation), and John Custis to ?, Aug. 12, 1724, Custis Letterbook, CWL. Custis reported that a storm had damaged the crop, and without stemming "nothing can be made of the little tatter'd Tob." When an Act of Parliament in 1723 prohibited shipment of stemmed tobacco, there was considerable protest. Robert Carter, for example, advised Micajah Perry that planters would have to "heave away all the Ground leaves and all the shatter'd and broken peices," which would lessen output by the same amount "as it is now by heaving away the stems." Stems were calculated at 25 to 30 percent of gross weight (Carter to Perry, Aug. 1, 1723, Carter Letterbook, Carter Papers, CWL). In 1731, Carter wrote to William Dawkins, "I am sure it [stemming] very much lessens my Quantity in the Main" (July 9, 1731, Carter Letterbook,

The trash tobacco that appears in Maryland probate inventories in the 1740s was more likely whole, unprocessed, but damaged or inferior leaves. Elimination of such low-grade tobacco should have reduced crops in Maryland, and, indeed, as happened in Virginia just after passage of that colony's inspection act, after 1747 output per hand did decline on the Western Shore. By midcentury, few laborers anywhere in the tidewater could now produce marketable crops of more than one thousand pounds. But whether this was due to the forced discarding of a portion of the crop that had been sold before the inspection acts or to a drop in yields on long-cropped land is uncertain. In fact, outputs per worker on the upper Eastern Shore did not decline in the years immediately following passage of the inspection law. In the early 1740s, some area merchants might have imposed more rigorous private systems of inspection than the public law later required. Consequently, after 1747 Eastern Shore planters might have been able to market a higher proportion of their crops than they had been able to sell before passage of the public act.[51]

As the price gap between sweet-scented and oronoco tobacco diminished after Maryland's tobacco inspection act came into effect in 1748 (see Figure 4), fewer planters continued to be concerned about maintaining the distinctive quality of particular crops, and more abandoned the cultivation of particular strains, careful sorting by grade or color, and discrete labeling that they had used earlier in an effort to obtain better prices. Fine product distinctions were most characteristic of the consignment trade and were effective only for the English domestic market, where tobacconists and their customers might become attached to particular "brands" of the weed (Plates 36a–b). Only commission merchants continued to sell hogsheads individually in the traditional manner at prices differentiated to a sixteenth of a penny per pound to reflect

---

Carter Papers, CWL). See also William Woodford Letterbook, 1723–1737, University of Kentucky, microfilm, M-1060-1, CWL, and *Byrd Correspondence,* I, 306–307, in which stems were also estimated at 25 percent of gross weight. Since the inspection law allowed stemming (which seems to have been continued in the same proportions after inspection as before) but prohibited lug tobacco, Virginians' crops should also have been reduced by about a quarter. [William Gooch], *Dialogue between Thomas Sweet-Scented, William Oronoco, Planters* . . . (Williamsburg, Va., 1732), John Carter Brown Library, microfilm, M-30, CWL, also suggested a probable 30 percent reduction. Virginia planters nevertheless soon found ways to more than compensate for the loss of lugs.

51. Edward C. Papenfuse, Jr., "Planter Behavior and Economic Opportunity in a Staple Economy," *Ag. Hist.,* XLVI (1972), 305, like Earle, *Evolution of a Tidewater Settlement System,* 25–27, 97–100, also found, in a comparison of a small number of probate inventories for 1746 and 1750, some drop in production in Charles County, Maryland. For rigorous private inspection systems, see Tyler, "Foster Cunliffe and Sons," *MHM,* LXXXIII (1978), 260–264.

PLATES 36A–B. Later-Seventeenth-Century and Mid-Eighteenth-Century Tobacco Advertisements. By Francis Bedford. Courtesy, The Colonial Williamsburg Foundation.
In the later seventeenth century, tobacco dealers like Robert Smith enticed consumers with distinctive, high-quality tobacco. By the mid-eighteenth century, "Virginia Best" was as likely to be better-grade oronoco grown in the James River piedmont as sweet-scented grown in the lower tidewater.

differences in quality. Once more planters began to sell their tobacco in the country, there was little reason to bother with these niceties. In the Chesapeake, factors representing reexport houses bargained largely over price and over the amounts of actual cash as opposed to store goods they would pay, and they were little concerned about any special characteristics of the product. They were interested only in getting large quantities of tobacco of a minimally standard quality that they could then sell to the French at a flat price. Weed destined for the Continental market, much of which was ground into snuff, passed through so many hands before it got to the final consumer that brand loyalty was not a consideration.[52]

52. Preferences of English tobacconists for particular crops or particular types of tobacco are discussed in William Woodford to George Mason, Nov. 23, 1723, to Mrs. Tarrant, Dec. 2, 1723, to James Gillison, Dec. 2, 1723, to Charles Chiswell, July 31, 1725, to William Whitaker, Aug. 7,

Even those merchants who remained interested in buying particular grades or types of tobacco for English, Irish, or Dutch markets began to encounter problems, since tobacco inspectors, rather than buyers, examined, classified, and weighed planters' crops. Often the only thing the buyer saw was the inspectors' note certifying the weight of a hogshead, whether the tobacco was in leaf or stemmed, and whether it was "crop" tobacco delivered by a planter as a full hogshead or "transfer" delivered in smaller parcels and packed after inspection by warehouse employees. Virginia merchant William Woodford explained that, after passage of that colony's inspection act, buyers depended entirely "on the Inspectors Judgmts and seldom look on any [of] it being of no Use for they must take it." In 1751, Maryland Eastern Shore factor Henry Callister informed Liverpool trader Foster Cunliffe, "What you recommend as to chusing the Quality of our Tobacco is impracticable here, we cannot know it without residing at the inspecting houses and even then can only judge of the colours of and other qualities of Crop h[ogs]h[ea]ds." Between one-third and one-half of inspected tobacco, Callister estimated, consisted of transfer, which "will be made up of 3, 4 or more sorts in a hhd—for this Reason the Inspectors will not trouble themselves to distinguish the quality in their Notes, and as often p[er]haps are not capable to distinguish or if they undertook it, as little to be relied on."[53]

Furthermore, many planters had decided that separating different grades and colors was no longer worth the effort. Callister explained to Cunliffe, "The People of this part of the Country had long let drop the Practice you say first invited you to send a Store to the head of this River [Chester]; that is to distinguish the bright from the dull." Initially it had been different:

> In Capt. Pritchard's time I remember some little notice taken, as marking p[ar]ticular hhds "B" or "D" to signify bright or dull, but never more of that since the fr[ench] War began, either in this concern or any other on our side the Bay—and as there had been no Consideration offer'd the Planter for sorting his Cropp and there's no Law to bind him or the Inspector to distinguish or separate the diff[eren]t sorts his Grounds produce, we cannot expect [he'd] take more pains than will fetch him more mony, for the bright Tobacco being lighter he will have a third part more trouble with [a] hhd of this sort than a hhd of the dull.

1735, to Edward Randolph, [August 1733?], William Woodford Letterbook, CWL. For buying practices of the French agents, see Price, *France and the Chesapeake,* chaps. 20–23.

53. Woodford to William Whitaker, June 27, 1736, Woodford Letterbook, CWL; Callister to Cunliffe, July 2, 1751, Callister Papers, CWL.

Callister had offered area planters thirty shillings extra per hogshead for certified bright tobacco and had gotten for his pains only twelve hogsheads, all of questionable quality.[54]

Passage of inspection laws had no apparent effect on the different pricing policies buyers of sweet-scented and of oronoco tobacco employed. In the sweet-scented areas of tidewater Virginia and in the bright-leaf areas of Maryland, merchants continued to buy tobacco at only two or three prices — the lowest price was offered for transfer tobacco, and one to two shillings more per hundredweight was given for crop tobacco. Prices for the latter were often differentiated by a shilling or two for "ordinary" and "good" grades. Further negotiations could then occur, however, over whether a planter would be allowed "the benefit of rise of the market" after selling his tobacco and over the proportion of the proceeds that would be paid in ready cash, in merchandise rated at the cash price, or in merchandise rated at a "goods" price based on the prime cost of the goods in sterling plus the merchants' markup of 75 to 100 percent. There appears to have been no regular market in the sweet-scented region for inferior grades. Some planters might have used poor-quality leaf for green manure or, in transactions of dubious legality, sold it to sailors. Others likely used lower-grade tobacco that could nonetheless pass inspection to pay taxes or to purchase goods locally.[55]

In contrast, in dull-leaf oronoco areas in Virginia and Maryland, merchants continued to purchase tobacco at a wider range of prices both before and after passage of inspection laws, a possible indication that the quality of this strain varied more widely than did that of sweet-scented. During the period of "ordinary" trading from early November, when inspection began, into the following spring, merchants made most purchases at or near a current market price established for each county or for tobacco carried to particular warehouses in early trading. Then, near the end of the trading season from April into June, however, if the merchant became desperate to complete his cargo he might pay extraordinarily high prices for the additional hogsheads needed to fill the ship. Prices also varied widely depending on whether the planter was paid entirely in cash, in goods, or in some combination of the two. So perhaps pricing policies for the two strains were in fact more similar than they initially appear, with the greater range of prices oronoco dealers paid reflecting the outcomes of individual bargains over the mode and com-

54. Callister to Cunliffe, Aug. 7, 1752, Callister Papers, CWL.

55. Ledgers of an unidentified merchant in Louisa County, Virginia, Frederick Hall Plantation Books, Southern Historical Collection, University of North Carolina, microfilm, M-245, CWL; Everard Taylor Account Books, 1731–1742, MS 804, MHS; Calvin Brewster Coulter, Jr., "The Virginia Merchant" (Ph.D. diss., Princeton University, 1944), chap. 4.

position of payment that sweet-scented dealers did not always record in their formal accounts.[56]

Moreover, the differences between sweet-scented and oronoco leaf, which stemmed from the soils in which they grew, were also diminishing. As the fertility of their best tidewater bottomlands declined, Virginia growers of sweet-scented were forced both to use animal manure on their older fields (which might have affected the taste of the tobacco) and to crop less fertile upland tracts. Yields obtained from a single sweet-scented plant dropped considerably. Between 1700 and 1730, it took six to eight plants to make a pound of cured sweet-scented tobacco. In 1763, on Martha Washington's dower plantations in York County, it took twelve plants to make one pound of cured leaf. The following year on Landon Carter's Rappahannock River plantations, it took fifteen plants (grown on ground of indifferent quality) to make one pound. Many sweet-scented growers found it no longer worthwhile to cultivate a strain that required that they tend fewer leaves per plant. Growers were carefully evaluating the advantages and disadvantages of various strains and were quick to note, for example, that "Long Green," a new variety of sweet-scented in vogue in England in the 1730s, lost more weight in stemming than the more commonly raised strains. On Maryland's upper Eastern Shore, bright-leaf growers were likely encountering similar problems with declining yields, and some switched entirely to dull leaf because the less valued strain weighed more. As the prices of good-quality oronoco rose to parity with sweet-scented, almost everywhere planters were opting for varieties that yielded greater quantities over those of ostensibly higher quality.[57]

In addition, a deterioration in the quality of sweet-scented was also quite apparent by the early 1760s, perhaps a result of genetic modifications that occurred in the strain as seed was ever more widely exchanged among planters living in different river basins. Robert Beverley protested to London merchant Edward Athawes in 1762: "The Tobo; w[hic]h I tend upon all my Plantations is what is generally call'd, and I believe, is the genuine best sweet Scented, and w[hic]h my Father ordered strictly to be cultivated during his

56. Business Books of a Merchant Trading in Westmoreland County, Virginia, 1731–1733, 1734–1735, 1748–1749, Scottish Record Office, GD1/L155/1–3, microfilm, M-1475, CWL; Francis Jerdone Journal, Sept. 15, 1750–June 2, 1760, VSL, microfilm, M-1169-15, CWL; Daniel Payne Account Book, Dumfries, Virginia, MS 635, MHS; Samuel Galloway Ledger, 1748–1758, MS 1994, MHS.

57. PRO C.O. 5/1319, fol. 220, microfilm, M-240, CWL; Jackson, ed., *Diaries of Washington*, I, 309–310; Account with Joseph Valentine, 1762, Custis Papers, CWL; Jack P. Greene, ed., *The Diary of Colonel Landon Carter of Sabine Hall, 1752–1778* (Charlottesville, Va., 1965), 316; William Byrd II, *A Progress to the Mines in the Year 1732*, in Wright, ed., *Prose Works*, 372; Callister to Cunliffe, Aug. 7, 1752, Callister Papers, CWL.

Life; and w[hic]h has been continued ever since his Death." Nonetheless, since the sweet-scented crops on some of his Rappahannock River farms were consistently judged "indifferent," Beverley was opting "for another of an inferior sort, because a double Quantity may be tended, and it will always bring the Market Price, and of Course be more advantageous to me." A few years later, Lancaster County planter Raleigh Downman was trying another strain: "I can grow large Tobo but not thick Tobo[.] My new sort came from near York River tis not the true sweet scented but rather near it." Even in premier sweet-scented counties like York, planters were having problems with the seed stock. In 1763, some new seed that Joseph Valentine, who managed the Custis plantations in the York River basin for George Washington, procured from a famous local tobacco grower miscarried, producing only "Rank" oronoco "no more like graveses tobaco then Chalk is like Chees[e]." Valentine reacted fervently to Washington's subsequent accusation that he was intentionally growing oronoco. "Reather the ground should ly unplanted then to plant arranoco," he averred. Joseph then defensively explained to Washington that "Rail tru sweet sented" seed had to be exchanged among different plantations, "for it will alter and ware out if it is planted two long upon one plantation with out shifting."[58]

Besides experimenting with different strains, after passage of inspection acts, as in earlier years, many tobacco growers did make short-term adjustments in crop size in response to changing market conditions. Fluctuations in the amount of tobacco produced and in the prices planters received in the later 1730s and 1740s were more often related to international warfare and to short crops caused by bad weather than to deliberate changes in the amount of tobacco planters tended. War with Spain beginning in 1739 and with France as well starting in 1744 increased the risk of cargoes being captured and led to a rise in freight and insurance rates. During the war, some planters in both colonies hedged their bets by allocating more labor time to preparing additional acreage for grain. Once a general peace was concluded in 1748 and prices began to rise, both Maryland and Virginia growers returned to maximum tobacco production. In 1751, for example, Henry Callister reported from the Chester River that "the Planters set heartily to making Tobacco, a trade they had neglected for some time past our way." The importation of more than eleven thousand new African slaves between 1750 and

58. Robert Beverley to Edward Athawes, Mar. 3, 1762, Robert Beverley Letterbook, 1761–1793, LC, microfilm, M-3, CWL; Rawleigh Downman to Athawes, Aug. 10, 1768, Joseph Ball Letterbook, 1743–1780, LC, microfilm, M-21, CWL; Capel and Osgood Hanbury to John Park Custis, Aug. 23, 1763, and Joseph Valentine to George Washington, July 29, 1764, Custis Papers, CWL.

1753 also contributed significantly to a jump in production. In the summer of 1752, some planters set out more tobacco plants than they could tend in anticipation of buying more slaves.[59]

Even planters in older areas like York County, Virginia, most of whom were no longer in the labor market, seem also to have set their slaves to tending more tobacco plants. The number of hogsheads exported from that river, for example, averaged about eleven thousand per year from 1745 to 1749, but rose to between twelve thousand and fifteen thousand between 1750 and 1755. Higher prices and cutthroat competition between London, outport, and Scottish tobacco buyers doubtless contributed to decisions to expand tobacco production. The ensuing bumper crops of the early 1750s quickly glutted the market. In 1751, for example, London merchant Silvanus Grove informed a Prince George's County, Maryland, planter, "I have used my utmost endeavour to get rid of them [last year's shipment] but the Market is over loaded with low Tobacco of w[hi]ch Sort the last Years Crop chiefly consisted." Similarly, the next year Eastern Shore factor Henry Callister advised that "the Trade is overdone and undone." With the resumption of international war in 1756, planters again stepped up production of other crops, and tobacco outputs per worker dropped by 20 percent throughout the tidewater. In addition, severe droughts in 1755 and 1758 and wet summers in 1759, 1760, 1761, and 1762 that made proper curing difficult further reduced outputs.[60]

Planters also made adjustments in how tightly they packed their tobacco, balancing a greater chance of damaging the finished product by prizing it too hard against the savings in freight charges they could obtain from heavier hogsheads. The inspection laws prescribed minimum weights (800 pounds in Virginia in 1730, 850 pounds in 1738, and 950 pounds in Maryland and Virginia in 1747–1748) and so produced much more uniform packaging.

59. Callister to Cunliffe, July 2, 1751, Callister Papers, CWL; Charles Stewart to Messrs Manville and Co., Mar. 26, 1752, Charles Stewart Letterbooks, CWL. Cf. Stewart to Manville, Moore, Depriza, and Massah, and Stewart to Blackman and Adams, July 5, 1751, Stewart Letterbooks, CWL.

60. Silvanus Grove to Roger Brooke, 1751, Brooke Papers, MS 1162, MHS; Stewart to Charles Craven, Sept. 18, 1752, Callister Papers, CWL. For volume of exports from individual Virginia naval office districts, see John J. McCusker, "Colonial Statistics," in Susan B. Carter et al., eds., *Historical Statistics of the United States: Earliest Times to the Present,* millennial ed., 5 vols. (New York, 2006), V, 760. For the state of trade, see also correspondence dating to the 1750s and early 1760s between Edmund Jennings and John Brice, Brice-Jennings Papers, MS 1997, MHS, and of William Anderson to James Hollyday, box 2, Hollyday Family Papers, microfilm, MSA. For weather conditions, see Hollyday Papers, box 2, MSA; Coulter, "The Virginia Merchant," chaps. 5, 8; and William Waller Hening, ed., *The Statutes at Large: Being a Collection of All the Laws of Virginia . . .* (1809–1823; rpt. Richmond, Va., 1969), VI, 567–569, VII, 44–45, 240–241.

TABLE 36. Weights of Tobacco Hogsheads, Virginia and Maryland, 1730–1763

Years	Hogsheads *N*	Average net pounds	Low pounds	High pounds	% increase
VIRGINIA					
1730–1738	162	822	358	1,020	8
1739–1743	122	941	834	1,123	14
1744–1748	463	983	769	1,191	4
1749–1753	148	1,006	836	1,112	2
1754–1763	1,045	1,145	827	1,409	14
MARYLAND					
1730–1738	101	856	455	1,178	10
1739–1743	127	990	645	1,242	16
1744–1748	84	1,048	946	1,226	6
1749–1753	191	1,050	853	1,293	0
1754–1763	358	993	814	1,464	−5

Note: The periods 1730–1738 and 1749–1753 were years of peace. In 1739–1743, there was war between Britain and Spain; in 1744–1748, war between Britain and France; and, in 1754–1763, there was the French and Indian or Seven Years' War.

Sources: Hogshead weights were collected from plantation and merchant accounts from a variety of areas in each colony. For pre-1730 weights, see Table 19.

More planters seem to have built prizes to get their hogsheads up to required weights, and small planters, who might have lacked the proper equipment, could use public prizes at the warehouses. (Crop tobacco, however, "made heavy by Planters themselves," was considered better than "that carried to the warehouses light and prised there, as it is less liable to Damage and generally more carefully prised.") Merchants and ship captains also benefited, since the inspectors packed loose parcels of transfer tobacco for them. Hogshead weights increased all across the period (Table 36). As in previous years, there were bigger gains in wartime—when freight rates were higher—than in peacetime. Earlier in the century, when Virginia tobacco prices slumped, planters and merchants there had turned to tight packing to cut costs. Now there were additional cost savings since peacetime freight rates remained what they had been in 1713. Maryland oronoco growers, who had previously made lighter hogsheads, followed suit after 1730. Like some Virginia merchants such as Stephen Loyd and John Tayloe, encountered earlier, outport traders in Maryland took to shipping exceptionally heavy hogsheads that they had reprized at their own warehouses. Once more of these merchants set up permanent stores and warehouses, ordinary Maryland planters might have observed and duplicated their prizing equipment.[61]

Changes in the Use of Land, Labor, and Time

This period also saw a large increase in grain production throughout the region. A few planters, like William Bassett in New Kent County, seem to have followed the market closely indeed, stepping up grain production as prices rose. Others on Maryland's upper Eastern Shore made even more pronounced long-term adjustments in crop mix in response to market opportunities not so readily available elsewhere. Few however, even on the upper Eastern Shore, abandoned tobacco entirely. As Paul G. E. Clemens explains, although there were occasional shifts in the relative prices of tobacco and grain, on the whole returns from the two crops did not change much between 1720 and 1760. "While the market pushed farmers toward grains," he states, "it did not pull them away from tobacco."[62]

61. Earlier prizing practices are discussed in Chapter 4, above. For the advantages of crop tobacco, see James Hollyday to William Anderson, Apr. 19, 1750, James Hollyday Letterbook, 1748–1754, Hollyday Family Papers, MSA (quotation). For freight rates, see John M. Hemphill II, "Freight Rates in the Maryland Tobacco Trade, 1705–1762," *MHM,* LIV (1959), 36–58, 153–187; and Stephen Gregg Hardy, "Trade and Economic Growth in the Eighteenth Century Chesapeake" (Ph.D. diss., University of Maryland, College Park, 1999), 439–436. David O. Percy, "The Production of Tobacco along the Colonial Potomac," The National Colonial Farm Research Report No. 1, the Accokeek Foundation, Md., 1979, traces legislation on Maryland casks. The Maryland tobacco hogshead remained two inches larger in diameter than the Virginia hogshead (which had been the case earlier in the century, except for 1708–1715 when the Board of Trade forced the colony to use the smaller Virginia gauge). The dramatic increase in hogshead weights that planters achieved can be seen in one of the objections Virginia planters had about the inspection act of 1730 that the minimum weight of 800 pounds stipulated in the law was too high (Moss, "The Virginia Plantation System," chap. 6). In contrast, by 1747 merchant Henry Callister would complain that the higher legal minimum of 950 pounds stipulated in Maryland's inspection law was by then too *low.* If the inspectors made up hogsheads of transfer tobacco weighing no more than the required minimum, then merchants buying transfer tobacco would be at a disadvantage because "Croppers"—planters who packed their own casks and who were by then making much heavier hogsheads—"will be enabled while Freights are high to undersell us at the Mast abt 2 or 3 Guineas p[er] hhd" (Callister to Cunliffe, Aug. 6, 1747, Callister Papers, CWL).

62. David Klingaman, "The Significance of Grain in the Development of the Tobacco Colonies," *JEH,* XXIX (1969), 268–278; Klingaman, *Colonial Virginia's Coastwise and Grain Trade;* Marc Egnal, "The Economic Development of the Thirteen Continental Colonies, 1720 to 1775," *WMQ,* 3d Ser., XXXII (1975), 191–222; Peter V. Bergstrom, *Markets and Merchants: Economic Diversification in Colonial Virginia, 1700–1775* (New York, 1985), chap. 5; Hardy, "Trade and Economic Growth," chap. 4; Lewis Cecil Gray, *History of Agriculture in the Southern United States to*

In the central tidewater, the tilt toward grains was less pronounced and more gradual than on the upper Eastern Shore. Between the mid-1720s and mid-1740s, many planters shifted from growing corn only for self-sufficiency (production of three to five barrels per hand) to planting some surplus for market. In part, this might have been a response to the changing composition of the labor force. In the mid-seventeenth century, young apprenticed orphans and chronically sick indentured servants who were of limited use in cultivating tobacco were put to raising corn and processing it for food. By the mid-eighteenth century, there were enough gristmills and windmills in the tidewater so that most planters could either grind their own grain or pay to have it done at a neighbor's mill rather than laboriously pounding it by hand each day. Small planters might have found that the younger children could help grow extra corn even if they were still too little to tend much tobacco. This was likely the case among big slaveowners whose workforces included growing numbers of enslaved adolescents. Whoever tended the extra corn, as the region's population continued to grow even small shifts in crop mix began to have a large cumulative effect.

In addition, by this time enough land had been cleared of forest for tobacco—and then abandoned—that most planters had untended land that could be sown in grain without having to pull out the roots and stumps of large trees. Planters found that they could increase grain output without reducing or by only slightly reducing the amount of tobacco they produced. This might have been one reason, too, why many planters began regularly growing small amounts of wheat. Large-scale wheat production, however, except in the Maryland counties closest to Philadelphia, remained a rich man's project. Yields per acre were low, ratios of harvested grain to seed were many times higher for wheat than for corn, and too much land and animal power was needed to make wheat a viable alternative staple for ordinary planters.[63]

1860, 2 vols. (1932; rpt. Gloucester, Mass., 1958), chap. 7; Gaspare J. Saladino, "The Maryland and Virginia Wheat Trade from Its Beginnings to the American Revolution" (master's thesis, University of Wisconsin, 1960); Carville Earle and Ronald Hoffman, "Staple Crops and Urban Development in the Eighteenth-Century South," *Perspectives in American History,* X (1976), 7–80; James Blaine Gouger III, "Agricultural Change in the Northern Neck of Virginia, 1700–1860: An Historical Geography" (Ph.D. diss., University of Florida, 1976), chap. 4; O'Mara, "Urbanization in Tidewater Virginia," chap. 2; Clemens, *The Atlantic Economy,* chap. 6.

63. Earle, *Evolution of a Tidewater Settlement System,* 128–129; Lois Green Carr and Russell R. Menard, "Land, Labor, and Economies of Scale in Early Maryland: Some Limits to Growth in the Chesapeake System of Husbandry," *JEH,* XLIX (1989), 407–418; Clemens, *The Atlantic Economy,* chap. 6. Clemens also argues that returns from wheat were likely to be higher than returns from tobacco on marginal or worn lands. Earle and Hoffman, "Staple Crops and Urban Development," *Perspectives in American History,* X (1976), 7–80, present some cost comparisons between

The uneven progress of the shift toward grains shows that access to markets was a critical factor. In addition to traditional markets in the northern mainland colonies, increased demand was fueled by expansion of the sugar industry in the West Indies, by grain shortages in the Old World, and by a growing market for military provisions during times of war. Philadelphia merchants and Delaware River merchant-millers trading to the Caribbean and southern Europe began buying ever larger quantities of northern Maryland wheat, offering special opportunities to planters on the Eastern Shore and at the head of the Chesapeake Bay. Annapolis merchants engaged in both the West Indian and Pennsylvania flour reexport trades afforded other outlets, as did local merchants operating from the port of Oxford in Talbot County. In addition, a class of specialized wheat brokers emerged who acted as middlemen between Maryland growers and Pennsylvania flour manufacturers. They aggressively sought supplies in the northern Chesapeake, either sending ships into Maryland waters to collect the grain or arranging for transport overland to the Delaware River. Pennsylvania buyers especially prized Eastern Shore soft white wheat; this wheat had a lower proportion of bran, which speeded milling and produced a whiter flour. By midcentury, upper Eastern Shore planters were marketing as much as one hundred thousand bushels of corn and wheat a year through Annapolis and Philadelphia. Farmers living closest to marketing networks made the most pronounced changes in crop mix. Between the 1740s and 1760s, for example, planters in Kent County, Maryland, near the Pennsylvania border shifted from tobacco to wheat as their principal staple, whereas those in adjoining counties to the south marketed a more balanced mix of tobacco, corn, and wheat. Meanwhile, itinerant New England merchants and West Indian traders operating first from Norfolk and then on most of the major Chesapeake rivers encouraged more grain production along the James, York, Rappahannock, and Patuxent. Planters living along the Potomac—the river farthest from either market—were the last to shift from self-sufficiency to market production (Table 37).[64]

If grains came to offer planters a sound supplement to tobacco cultivation, especially for covering more of the local "contingent" expenses of their plantations, some still also bit on the sorts of risky schemes for getting rich quick that had been so attractive in the seventeenth century. In 1728, it was

tobacco and wheat. Their estimates understate the prices of Eastern Shore tobacco, however, and overstate the amount of wheat that Chesapeake planters could produce.

64. David E. Dauer, "The Expansion of Philadelphia's Business System into the Chesapeake" (paper presented at the annual meeting of the American Historical Association, 1981); Bergstrom, *Markets and Merchants,* table 5.4; Klingaman, *Colonial Virginia's Coastwise and Grain Trade;* Clemens, *The Atlantic Economy,* 174–183, 198–204.

TABLE 37. Grain Crops per Laborer, Virginia and Maryland, 1640–1763

Crop	Virginia								Maryland			
	York River Basin		Rappahannock River Basin		Potomac River Basin		Peripheral areas		Western Shore		Upper Eastern Shore	
	N	Mean	*N*	Mean	*N*	Mean	*N*	Mean	*N*	Mean	*N*	Mean
Corn												
1640–1649	—	—	—	—	—	—	—	—	4	7.3	1	4.0
1650–1659	2	10.4	—	—	—	—	3	6.0	1	7.0	—	—
1660–1669	3	7.4	—	—	—	—	3	6.8	—	—	—	—
1670–1679	1	4.5	—	—	1	18.0	1	7.0	—	—	—	—
1680–1689	1	13.3	—	—	—	—	1	5.0	3	6.2	—	—
1690–1699	1	5.6	1	4.5	—	—	—	—	—	—	—	—
1700–1709	6	8.6	—	—	1	4.0	—	—	5	7.5	—	—
1710–1719	7	5.2	3	8.3	—	—	—	—	—	—	2	7.0
1720–1729	14	7.8	6	11.3	—	—	—	—	2	6.9	15	12.3
1730–1739	17	9.4	8	9.7	—	—	1	8.0	3	12.2	6	17.7
1740–1747	38	13.3	19	11.9	4	8.6	—	—	—	—	5	19.9
1748–1753	10	16.6	1	2.5	2	8.8	1	10.0	5	14.5	3	24.4
1754–1763	29	15.3	47	12.4	3	15.5	1	15.4	17	18.4	17	22.3
Wheat												
1730–1739	3	6.6	—	—	—	—	—	—	—	—	6	11.1
1740–1747	5	3.0	2	6.5	—	—	—	—	—	—	2	20.5
1748–1753	1	5.6	—	—	—	—	—	—	5	4.7	—	—
1754–1763	12	8.1	10	8.2	5	10.7	1	4.8	11	8.3	27	39.0

Note: Corn outputs are measured in barrels, and wheat outputs are measured in bushels. Corn and wheat crops are classified by the river basin in which they were primarily consumed or most likely marketed. Crops from the Burwell plantations in James City and York counties were assigned to York River. Crops for

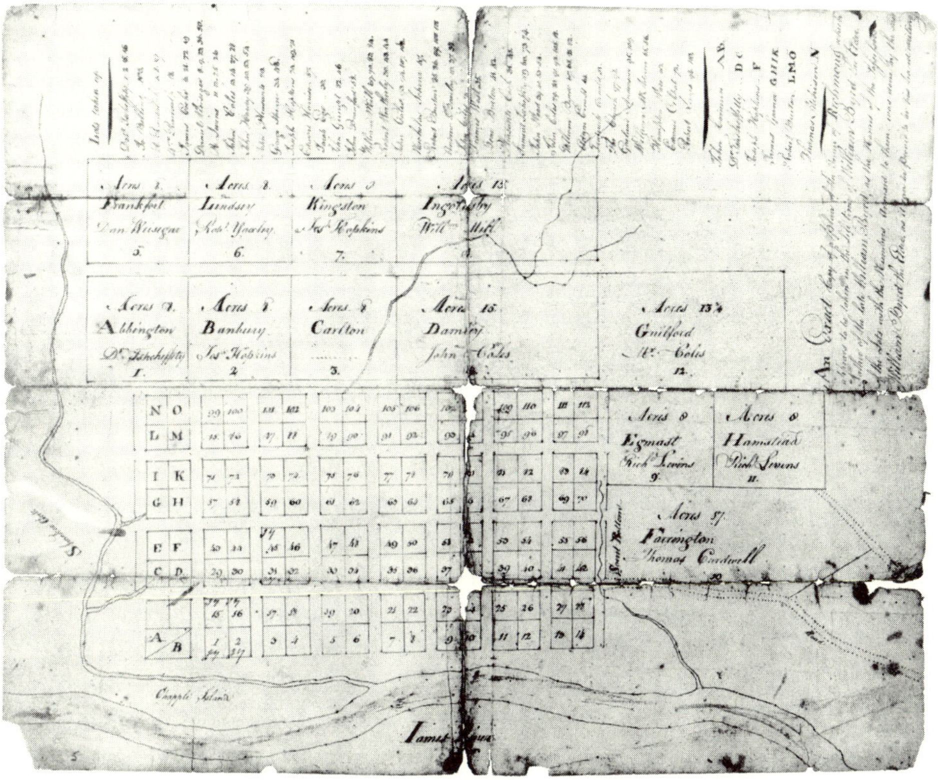

PLATE 37. A Plat of the Town of Richmond. Peter Force Papers, series IX, Library of Congress, Washington, D.C. Photograph courtesy, Virginia Historical Society, Richmond.

This town plan, drawn by William Byrd II in 1737, shows that more than forty people had already taken up lots in this fall line town.

hemp. A few years later, it was copper, as in the 1720s it had been iron. In 1733, William Byrd II made a trip along the Nottoway, Roanoke, and Dan rivers, finding "so many appearances of copper in these parts that the inhabitants seem to be all mine-mad and neglect making of corn for their present necessities in hopes of growing very rich hereafter." Having just been burned with hemp, Byrd resisted the copper. As a result of his trip, however, he and an associate became engrossed in laying out plans for cities on the present sites of Richmond and Petersburg, where falls in the James and Appomattox rivers required that produce from the upcountry be unloaded and stored before transshipment downriver (Plate 37). Here Byrd's financial instincts were sounder, but, as he admitted, at the time "we did not build castles only, but also cities in the air."[65]

In 1757, indigo was all the rage. The Seven Years' War cut off England's supplies of high-grade indigo from the French West Indian colonies, driving up prices and encouraging colonists in Virginia, as well as in South Carolina and Georgia, to try to profit from the resulting shortages of the coveted dye. As seen in previous chapters, William Byrd II, John Baylor, Landon Carter, Philip Ludwell, and even a hidebound traditionalist like Robert Bristow IV were among the great planters who were tempted to try this semitropical plant. A meager output in the first year of a crop both difficult to process and wholly unsuited to the Virginia climate quickly dampened their enthusiasm. Most Chesapeake planters had dropped indigo by 1759, when tobacco prices began to rise as a result of a short crop in the previous year. John Baylor would remark with some relief in 1760, "I have now dispos'd Indigoe and shall vigorously pursue my favourite imployment Tobo. which you are sensible I can make a great deal of very choice."[66]

There was increasing diversification as well into import replacement activities. Probate inventories demonstrate that big planters led the way with ventures into a wide variety of activities including tanning, shoemaking, smithing, spinning, weaving, soapmaking, and candlemaking. Individual plantation accounts demonstrate, though, that before 1764 most big planters had made only limited moves toward self-sufficiency. Wealthy planters could easily afford to spend a few pounds on equipment that would keep young, old, or disabled slaves busy processing raw plantation materials. What they were usually unwilling to do was to take productive slaves from the fields, except in the rare cases where a slave had somehow acquired enough skill in a craft that he (almost always a male) could make more by carpentry, smithing, or shoemaking, for example, than by growing crops. Small-planter families did not plunge into so many disparate activities, but they, too, were making

65. *Byrd Correspondence,* I, 379, 397–398, 410–411, II, 513; Byrd, *Journey to the Land of Eden,* in Wright, ed., *Prose Works,* 387–388.

66. John Norton to Flowerdew and Norton, Sept. 16, 1760, in Frances Norton Mason, ed., *John Norton and Sons, Merchants of London and Virginia, Being the Papers from Their Counting House for the Years 1750 to 1795* (Richmond, Va., 1937), 9–10. See also *Byrd Correspondence,* II, 629–631; William Byrd III, Memorandum Book, 1757–1758, LC, microfilm, M-97, CWL; Greene, ed., *Diary of Carter,* I, 180–185, 187, 220, 228; Robert Bristow to James Pasley, November 1757, Robert Bristow Letterbook, 1747–1770, PRO/A.O. 13/84, microfilm, M-494, CWL; Gray, *History of Agriculture,* chap. 8; Coulter, "The Virginia Merchant," chap. 5. For wartime dye shortages, see John J. McCusker and Russell R. Menard, *The Economy of British America, 1607–1789* (Chapel Hill, N.C., 1985), 187; and Joyce E. Chaplin, *An Anxious Pursuit: Agricultural Innovation and Modernity in the Lower South, 1730–1815* (Chapel Hill, N.C., 1993), 199–201.

more goods for themselves and also selling surplus products to large planters. Import replacement was very much a community affair in which the income-generating activities of planters large and small, tenant farmers, free white male artisans, and poor white women became ever more interconnected.[67]

Textile production, for example, remained largely the province of many white female spinners and knitters and of a few white male and female weavers. As enslaved black women became increasingly associated, in the minds of whites, with hard manual labor, white females of all stations sought or were pressured to avoid regular fieldwork. In addition, as the black population grew, more ordinary planters were inheriting, buying, or hiring a slave or two—especially relatively inexpensive adolescents—who were too young to do much fieldwork but who could be put to manual chores previously performed by white youngsters. Even overseers' families, by appropriating the services of one of the master's adolescent slaves, escaped some forms of household drudgery. More white housewives could assign garden digging, firewood chopping, water hauling, laundering, and the like to blacks. Small and middling planters' wives and daughters would then turn increasingly to spinning, knitting, and making clothing, both for their own families and often as a source of outside income.[68]

By the 1740s and 1750s, enslaved families in the tidewater also began to find ways to take advantage of the quickening and thickening of local trade. Raising poultry and other perishables on their own time and harvesting wild resources became means for earning small amounts of cash that they could use to acquire small comforts. Most planters continued to try to restrict slaves' trading activities to their own or at most to neighboring plantations, so the scope of their internal economy was decidedly constrained. Poultry and eggs

67. Lois Green Carr and Lorena S. Walsh, "Changing Life Styles in Colonial St. Mary's County," *WPREHRC,* I, no. 3 (1978), 73–118; Carr and Walsh, "Inventories and the Analysis of Wealth and Consumption Patterns in St. Mary's County, Maryland, 1658–1777," *Historical Methods,* XIII (1980), 81–104; Walsh, "Urban Amenities and Rural Sufficiency: Living Standards and Consumer Behavior in the Colonial Chesapeake, 1643–1777," *JEH,* XLIII (1983), 109–117; Carr and Walsh, "The Standard of Living in the Colonial Chesapeake," *WMQ,* 3d Ser., XLV (1988), 135–159; Carr and Walsh, "Changing Lifestyles and Consumer Behavior in the Colonial Chesapeake," in Cary Carson, Ronald Hoffman, and Peter J. Albert, eds., *Of Consuming Interests: The Style of Life in the Eighteenth Century* (Charlottesville, Va., 1994), 59–156; Carr, "Diversification in the Colonial Chesapeake," in Carr, Morgan, and Russo, eds., *Colonial Chesapeake Society,* 342–388; Carr and Walsh, "Economic Diversification and Labor Organization in the Chesapeake, 1650–1820," in Stephen Innes, ed., *Work and Labor in Early America* (Chapel Hill, N.C., 1988), 144–188; Earle, *Evolution of a Tidewater Settlement System,* chap. 5.

68. See note 67, above; Cary Carson and Lorena S. Walsh, "The Material Life of the Early American Housewife" (paper presented at the conference, Women in Early America, Williamsburg, Va., Nov. 5–7, 1981), CWL; Kulikoff, *Tobacco and Slaves,* chap. 5.

were the main items of trade slaveowners sanctioned, followed by fish, fruits, and vegetables. Although such household purchases are seldom enumerated in planters' account books of the 1740s and 1750s, archaeological evidence from slave quarters dating to this period demonstrates that by midcentury slaves were acquiring some consumer goods such as utilitarian kitchenwares, sewing items, shoe buckles, combs, mirrors, and jewelry as well as accumulating hard currency by one means or another. Slaveowners found that permitting trading within the plantation "family" was in their own interest, as it allowed them to display their benevolence while also providing bondpeople with incentives to pursue productive activities in their own time. Through this small beginning, Chesapeake slaves began to develop an understanding of the cash economy and to develop marketing skills that they would turn to greater advantage in the next quarter century.[69]

In addition to changes in crop mix and the intensification of import-replacement activities, planters also began paying more attention to livestock husbandry, seeking both to integrate stock raising with field agriculture and also to manage stock to take better advantage of growing markets for livestock products. Hogs—prolific breeders and notorious destroyers of crops—were more closely confined and much more carefully managed and harvested. In the seventeenth century, when hogs normally roamed the woods at will and semiwild ones were sometimes difficult to catch, one can find animals of varying ages up to four years listed in inventories. By the 1730s, planters kept better track of their ranging hogs. They were more closely cropped, and fewer animals were kept after they reached maturity. After the best breeders had been selected in the spring, other hogs approaching breeding age were castrated, fattened in the fall, and put to the knife at about eighteen months of age. As corn production rose, more was fed to livestock, including hogs. Hog slaughter weights, however, show no increase before 1750, despite evidence of more regular feeding, owing to more limited natural forage and the reduced slaughter age (Table 38). In most areas, a growing demand for pork for local markets, ships' provisions, and the West Indian trade encouraged early slaughter; if more hogs were killed than were needed for plantation use, the surplus could always be sold. Moreover, a small number of brood

69. Lorena S. Walsh, *From Calabar to Carter's Grove: The History of a Virginia Slave Community* (Charlottesville, Va., 1997), 90–92, 181–184, 191–199; Walsh, "Fettered Consumers: Slaves and the Anglo-American 'Consumer Revolution'" (paper presented at the annual meeting of the Economic History Association, 1992), CWL; Philip D. Morgan, *Slave Counterpoint: Black Culture in the Eighteenth-Century Chesapeake and Lowcountry* (Chapel Hill, N.C., 1998), 193–194, 358–361; Ira Berlin, *Many Thousands Gone: The First Two Centuries of Slavery in North America* (Cambridge, Mass., 1998), 119.

TABLE 38. Hog Slaughter Weights in Virginia and Maryland, 1678–1763

	Net pounds		Gross pounds		Unspecified	
Years	*N*	Mean	*N*	Mean	*N*	Mean
1678–1699	24	156	1	90	23	106
1700–1719	2	147	—	—	21	95
1720–1729	8	112	11	142	20	108
1730–1739	190	104	13	168	22	122
1740–1749	45	105	1	170	—	—
1750–1759	7	143	11	149	—	—
1760–1763	482	123	54	129	—	—

Source: Database of hog slaughter weights compiled from planter accounts and correspondence available from the author.

sows, along with their offspring who tended to stay with their mothers, were doubtless easier to control than larger herds of mature hogs, and weight gains among older animals were judged insufficient to justify the amount of food they would consume if they were kept alive an additional year.[70]

Faunal remains from tidewater Chesapeake trash pits demonstrate that, across the eighteenth century, the proportion of hogs killed at the target market age of between eighteen and twenty-four months rose. In the first half of the eighteenth century, this age group was but 11 to 17 percent of all hogs slaughtered in rural areas; in the third quarter of the century, the proportion of hogs of prime marketable age jumped to 38 percent. Moreover, proportionately more hogs of between eighteen and twenty-four months began to be sold in town markets such as Williamsburg and Annapolis (and likely in the export market as well), whereas more younger animals less than one year of age were consumed on the plantation. These numbers reflect a more careful culling of herds and a rationalization of feeding practices.[71]

As local wolf populations were eradicated and more cleared pastures became available, many more planters were also raising sheep, and these, too,

70. In 1747, fresh corn-fed pork was selling for 16s. 8d to 18s. per hundredweight in the upper Eastern Shore county of Talbot, whereas pork fattened only with acorns, raised primarily on the lower Eastern Shore, brought only 12s. 6d to 14s. (Callister to Cunliffe, Dec. 25, 1747, Callister Papers, CWL).

71. Lorena S. Walsh et al., "Provisioning Early American Towns: The Chesapeake: A Multidisciplinary Case Study," Final Performance Report to the National Endowment for the Humanities, Grant RO-22643, Sept. 30, 1997, 43–47, CWL.

required careful management. Once breeders had been selected, excess rams were regularly castrated and fattened either for the planter's table or for the market. Old ewes and weak lambs afforded convenient small parcels of meat, and they could be killed in all seasons without fear of the meat spoiling before all could be used. In addition to meat, large planters found a ready market for breeding stock among smaller planters just beginning to keep sheep. Archaeological evidence demonstrates that 44 percent of sheep were killed at ages greater than three and a half years, a pattern of slaughter that suggests the sheep were being raised primarily for meat. (Where sheep are kept principally for their wool, slaughter is typically deferred until the animals are six years old, which is when wool quality tends to drop.) Wool was an important by-product, however, that was not wasted, as it had sometimes been in earlier years. The fiber was either processed on the farm to help cut down on clothing imports or sold to local women who had more time to spin yarn than the planter was willing to allow for his slaves.[72]

Horse herds were generally less carefully regulated, although some planters became increasingly interested in improving the quality of their stock. But, for every planter who sought to avoid keeping "a gang of useless horses," there were scores who kept more animals than were really required for riding or for pulling carts, wagons, and carriages. Horses were more often used than in earlier years for carting crops and for plowing, the latter a task for which the horses' heavier front ends provided an advantage over oxen. But the continued lack of adequate grain and fodder during summer months still curtailed intensive use of horses for draft. By the 1740s and 1750s, some large planters, especially those living along the Rappahannock, did begin to import Arabian stallions from England. Their aim, however, was to breed better racehorses, not to improve their stock of work animals.[73]

Management of cattle, the most valuable of plantation livestock, altered least during the Golden Age. Woodland pastures remained an integral ele-

72. Ibid., 37–39, 57–59; Hardin, "'Alterations They Have Made at This Day,'" chap. 5; Hardin, "Laws of Nature: Wildlife Management Legislation in Colonial Virginia," in Lary M. Dilsaver and Craig E. Colten, eds., *The American Environment: Interpretations of Past Geographies* (Lanham, Md., 1992), 137–162.

73. David O. Percy, "Of Fast Horses, Black Cattle, Woods Hogs, and Rat-Tailed Sheep: Animal Husbandry along the Colonial Potomac," the National Colonial Farm Research Report No. 4, the Accokeek Foundation, Md., 1979; "The Equine FFVs," *VMHB*, XXXV (1927), 329–370; Alan L. Olmstead and Paul W. Rhode, *Creating Abundance: Biological Innovation and American Agricultural Development* (Cambridge, 2008), 266–268. For the 1802 report of the cost of feeding plow horses, see Leo Rogin, *The Introduction of Farm Machinery in Its Relation to the Productivity of Labor in the Agriculture of the United States during the Nineteenth Century* (1931; rpt. New York, 1966), 52–53.

ment in even the wealthy planters' herding systems. Light grazing probably increased the productivity of wild pastures, but too many animals eventually reduced the quantity and quality of grazing resources in forests and old fields. Moreover, as grain production increased, the supply of second-growth fallow fields diminished. Overgrazing leads to slower rates of growth and of weight gain and to decreases in fertility and overall health. Cow bones excavated on lower tidewater Virginia sites dating to the first half of the eighteenth century indicate that there adult animals attained significantly smaller sizes than they had during the second half of the seventeenth century. By the 1740s, large planters recognized the effect too many animals had on their pastures and consequently limited the size of the herds they kept. The contradictory goal of raising more animals in order to get more manure, however, encouraged "improving" farmers to maintain larger herds than they could adequately feed. Grain growers also began to make more use of steers trained for the cart or the plow. Although oxen were slower than horses, they were sturdier, could work for eight hours a day if given a break for grazing, and had a decided advantage in that they required less supplemental food.[74]

As local urban and intercolonial demand for meat rose in the 1740s and 1750s, wealthier planters did begin to introduce husbandry techniques aimed specifically at producing cattle for market. Analysis of tidewater faunal remains demonstrates that the slaughter population became younger as the eighteenth century progressed, and it included larger proportions of cattle between ages twenty-four and forty-eight months, well below the minimum age of four years, when purely range-fed cattle mature to optimal slaughter weight. By the second quarter of the century, some planters began fattening cattle intended for sale in stalls. Much of this meat was marketed in Chesapeake towns. Urban faunal assemblages show that town dwellers were consuming more very young animals less than twelve months old along with younger beef in the twenty-four- to forty-eight-month age range. In contrast, plantation owners slaughtered higher proportions of older animals more than three years of age for on-site consumption, and they seldom ate veal, instead preferring to sell it to town dwellers. Larger planters, then, were beginning to consciously alter their herding systems in response to the emergence of commercial markets.[75]

74. Walsh et al., "Provisioning Early American Towns," 34–36, 39–42, 47–57; Olmstead and Rhode, *Creating Abundance,* 266–268; Joanne Bowen, "Historical Ecology and the British Landscape" (paper presented at the annual meeting of the Society for Historical Archaeology, Mobile, Ala., Jan. 10, 2002, CWL); Cary Carson et al., "New World, Real World: Improvising English Culture in Seventeenth-Century Virginia," *Jour. So. Hist.,* LXXIV (2008), 45–49.

75. Walsh et al., "Provisioning Early American Towns," 34–36, 39–43, 47–57; David B. Landon

Cattle were increasingly valued as well for their manure because more planters were using manure to extend the productivity of worn lands. Lower Eastern Shore planter John Custis, for example, by 1737 was growing tobacco in Northampton County "on the richest ground I have and nothing is there tended but strong cowpens." Most period sources indicate that cowpenning was usually reserved for the tobacco fields, and manure was spread on cornfields only if supplies and available labor permitted. Although the use of manure had been discouraged at the turn of the century because it was believed to produce ill-tasting tobacco, by the 1740s and 1750s manuring had become an essential ingredient in the regimens of "improving" farmers. Extensive manuring involved a delicate balancing act—large herds could be sustained only with the fodder and straw that were by-products of increased grain production. The clearing of more land for grain, however, steadily ate into available natural forage. Some planters tried to improve pasture by planting artificial grasses such as clover and timothy. Few succeeded in these years owing to lack of experience, unsuitable plant species, inhospitable soils for imported grasses, or because they were forced to overgraze available pastures in order to keep stock alive in years when the seasons proved unfavorable. Unfortunately, the low nutritional quality of forage in the southern colonies meant that cattle there produced much less manure than did northern livestock.[76]

These changes in cropping, clearing, and livestock husbandry underscore

---

and Joanne Bowen, "Microstructural Analysis of Cattle Teeth (*Bos taurus* L.) from Chesapeake Archaeological Sites: Interpreting Histological Preservation and Slaughter Patterns" (paper presented at the Society for Historical Archaeology Conference on Historical and Underwater Archaeology, Quebec City, Quebec, 2000), CWL.

76. Custis to Perry, 1737, Custis Letterbook, CWL. According to Earle, cattle manure tends to keep tobacco green and growing too late in the season and affects color and quality when cured (*Evolution of a Tidewater Settlement System,* 28 n. 3). Ester Boserup, *Population and Technological Change: A Study of Long-Term Trends* (Chicago, 1981), chap. 3, points out that, where there is fallowing, most fodder production can be avoided, as animals can feed both in fallow fields and in natural pastures. For corn fodder, see Callister to Mr. Greenway, Oct. 1, 1759, explaining that Eastern Shore planters were unable "to set heartily about treading out" their wheat in September; "as they are not hay makers, they must take care of their fodder." Both timothy and clover perform better in heavy, alkaline soils than in the sandy, acid soils of the tidewater (Alfred L. Wingo, *Virginia's Soils and Land Use,* Commonwealth of Virginia State Board of Education, *Bulletin,* XXXI, no. 8 [June 1949], 209–213). For differential dung output, see Majewski and Tchakerian, "Environmental Origins of Shifting Cultivation," *Ag. Hist.,* LXXX (2007), 530. Planters on older Barbados were similarly perplexed about how to maintain sufficient cattle to produce essential dung and about how to use it most efficiently. See, for example, [Samuel Martin], *An Essay upon Plantership, Humbly Inscrib'd to All the Planters of the British Sugar-Colonies in America,* 2d ed. (Antigua, 1750), 20–29.

the point that everywhere in the tidewater the landscape was being filled up. In Saint Mary's County, Maryland, for example, still in 1705 about a third of the county's land was unclaimed. By 1750, all acreage suitable for cultivation had been taken up, as planters patented most remaining marginal lands and regularized boundaries by taking up odd bits and pieces between established farms that were a legacy of initial sloppy surveys. As early as the 1720s, timber was becoming scarce in some older areas, especially on tenements too small to allow planters to maintain sufficient woodlots for building, fencing, and firewood and still to rotate arable fields in long fallow. In 1730, traveling through adjoining Calvert County, Maryland, Ebenezer Cook bemoaned "mangled Oaks, laid blended on the Plains; Cut down for Fuel by unthinking Swains." By midcentury, timber shortages were a widespread problem in areas that had been farmed for up to two centuries. In a description of the Chesapeake prepared for the Board of Trade in 1753, Dr. Charles Carroll noted that "many parts are becoming waste wanting both Fencing Fireing and Timber for Building or Repairing."[77]

Not surprisingly, signs of potential land shortages also began to appear during these years. In 1728, for example, in tidewater Virginia there had been more than eighty acres of land for each tithable, but by 1749 the proportion had dropped to sixty-five acres per worker, a ratio of land to labor that left little margin for error, given that a high proportion of arable acres needed to be rested in long fallow. For Maryland's upper Eastern Shore, Clemens observed that a land to labor ratio of forty to one still allowed a planter to grow both grains and tobacco but that careful planning was required. So long as small landholders in either colony who owned farms of a hundred acres or so employed only family labor, they could maintain reasonable levels of soil fertility through rotations and fallows. Smallholders were apparently well aware of optimal land-labor ratios, and, when they parceled out their estates among children, most were careful not to subdivide holdings into units too small to be viable farms. Corroborating this generalization, between 1749 and 1755 in the Virginia tidewater the land-labor ratio of sixty-five to one did not decline any further. Absent marked changes in cultivation techniques, in the aggregate planters judged it imprudent either to further subdivide plantations or to significantly increase the number of workers on existing farms. Small landowners or better-off tenants who managed to acquire one or two

77. Saint Mary's County tract maps for 1705 and for circa 1750, Historic Saint Mary's City, MSA; Earle, *Evolution of a Tidewater Settlement System,* 30–34; C[ook], *The Planter's Looking-Glass,* in Steiner, ed., *Early Maryland Poetry,* 36; Carroll to Charles Carroll, February 1753, Letter-book of Dr. Charles Carroll of Annapolis, MHS.

slaves to increase crop outputs, however, ran a significant risk of overcropping their land.[78]

Large planters, especially those living in the lower Rappahannock basin where shortages of arable land and problems of overcropping and soil exhaustion seem to have been most acute, responded by cutting back on clearing and by cowpenning their worn, but better, grounds. As their labor forces continued to grow through natural increase, some had either to open up marginal land or to continue cultivating better soils without adequate fallowing. Even rich alluvial soils on lower terraces along the rivers could bear good crops of tobacco for no more than six to eight years, and such naturally fertile soils were in limited supply. And, even with proper rotations, some decline in productivity was inevitable. Seventeenth-century settlers had a keen eye for the best land, and they exploited it first. By the mid-eighteenth century, many tidewater planters were finding that any lands left that had not previously been tilled were undesirable indeed. It was becoming increasingly attractive — for those who had the option — to send some slaves to fresh western holdings where good crops could be raised with less effort.[79]

By mid-eighteenth century, tidewater planters like the Bristows and Beverleys in Virginia and Charles Carroll and the Tilghmans in Maryland — who still had more land than they had slaves to farm it — turned very seriously to making the most of rental properties. Growing land scarcity, rising land prices, continued population increase, and restrictive inheritance practices

78. Bergstrom, *Markets and Merchants,* chap. 1; Clemens, *The Atlantic Economy,* 147, 197; Earle, *Evolution of a Tidewater Settlement System,* 24–30; Walsh, "Land, Landlord, and Leaseholder," *Ag. Hist.,* LIX (1985), 373–396; O'Mara, "Urbanization in Tidewater Virginia," chap. 3; Lois Green Carr, "Inheritance in the Colonial Chesapeake," in Ronald Hoffman and Peter J. Albert, eds., *Women in the Age of the American Revolution* (Charlottesville, Va., 1989), 155–208.

79. Among the planters who mentioned cowpenning as a vital part of their agricultural regimen in this period, one, John Custis, was farming on the lower Eastern Shore where land had long been scarce, and another, George Washington, lived on the Potomac. All the rest — Bristow, Ball, Landon and Charles Carter, and Corbin — managed Rappahannock basin lands. Planters in this area, a big growth region in the mid-seventeenth century and one that received a large number of slaves, might have been among the first to wear out their lands. As early as the 1720s, Robert Carter would complain: "I us'd to have as much for my Tob. [grown in the Rappahannock Basin] as the finest crops of York river. How it comes to be other ways now I can't Imagine the reason," as "neither their Grounds nor their care exceed ours" (Carter to Thomas Colmore, June 27, 1723, Carter to William Dawkins, Aug. 10, 1728, Carter Letterbook, Carter Papers, CWL). Price, *France and the Chesapeake,* 669–671, observes that tobacco shipments from the Rappahannock region stagnated after 1750. One reason was the limited market, as most Rappahannock tobacco continued to be sold in England and not exported to the Continent, but land constraints also might have begun to affect production levels. For discussions of the effects of the sweet-scented strain on land usage, see Hardin, "'Alterations They Have Made at This Day,'" chap. 4.

that excluded some children from a share in the land all contributed to an increase in the ranks of landless whites. Everywhere in the tidewater tenancy increased, and by the mid-1750s rental properties seem to have been divided into all the tenements the land would support. Earlier, landlords had to grant long-term leases for several lives at low rents in order to attract tenants. Once such leases began to expire at midcentury, landowners often granted new ones only for shorter terms and at higher rents. Moreover, planters with uncultivated lands farther west could now count on a supply of tenants to begin developing additional farms there.[80]

Just as the land was filling up, so, too, was the agricultural calendar. In large part, planters produced additional output while continuing to put the same effort into tobacco by increasingly using off-season labor (Table 39). If the fall and winter had been seasons of relative leisure for many seventeenth-century planters and indentured servants, by the mid-eighteenth century that was less often the case, especially for slaves who had little choice in the matter.[81]

The addition of wheat as a third staple intensified late summer, fall, and winter work, since the slaves had at minimum to thresh enough grain between mid-July and early September to make seed for early fall planting. During this time, tobacco continued to require repeated worming and suckering, almost up to the critical and hectic harvest in early to mid-September. Moreover, workers were pressured to clean as much grain as possible in late summer and early fall, since that which remained unthreshed was at greater risk of being devoured by weevils or rotting should it get too wet. The ideal scenario was to get the next year's wheat sown by the end of September, but, in practice, plowing and sowing often dragged on into late October or early November. When harvests of wheat (and less often barley, another grain that matured in late summer) were large, threshing could continue intermittently from October through the following February. Oats, a spring-sown grain increasingly cultivated to feed draft horses, added extra work for ground preparation and sowing whenever weather permitted in January, February, and March. In addition, increasing the acreage devoted to any crop required that more

80. Bliss, "The Rise of Tenancy in Virginia," *VMHB,* LVIII (1950), 427–441; Walsh, "Land, Landlord, and Leaseholder," *Ag. Hist.,* LIX (1985), 373–396; Earle and Hoffman, "Genteel Erosion," in Curtin, Brush, and Fisher, eds., *Discovering the Chesapeake,* 280–290.

81. Carr and Walsh, "Economic Diversification and Labor Organization," in Innes, ed., *Work and Labor in Early America,* 144–188; Earle, *Evolution of a Tidewater Settlement System,* chap. 5. For seasonal routines, see also Percy, "The Production of Tobacco"; Percy, "Of Fast Horses"; and Percy, "'English' Grains along the Colonial Potomac," the National Colonial Farm Research Report No. 3, the Accokeek Foundation, Md., 1977.

TABLE 39. Agricultural Calendar for the Rappahannock River Basin, 1730–1763

Month	Field Crops: Tobacco	Corn	Wheat	Barley	Oats	Rye	Pasture
Jan.	sow beds; make hills	plow ground; husk	thresh		thresh; plow ground		
Feb.	sow beds; plow & grub ground; make hills		thresh		plow ground; harrow		plow for clover
Mar.	plow ground; make hills	plow ground			sow		plow for clover
Apr.	manure ground; make hills; plant	plow ground; make hills; plant	roll	sow	sow; roll		sow seed; harrow clover
May	make hills; plant; replant; weed; sell	plant; replant					make hay; sow & roll clover
June	plant; replant; weed; sell	plant; replant; plow ground	reap		mow	reap	make hay
July	plant; replant; weed; sucker; sell	weed; lay by	reap; plow ground; thresh		mow	reap	make hay; grub
Aug.	replant; worm; sucker; sell		thresh; grind				make hay
Sept.	cut; house; strike; strip; stem	gather fodder	sow; thresh	sow		sow	sow timothy; make hay
Oct.		gather fodder	plow ground; sow; tread; thresh	sow		sow; thresh	sow timothy
Nov.	sow seed; strike; hoe & plow ground; prize	gather; measure; house; plow ground	sow; thresh				
Dec.	strip; stem; prize	plow ground; gather; husk; measure; house	thresh		thresh		

Sources: "Journal of Col. James Gorden, of Lancaster County, Virginia," *WMQ,* 1st Ser., XI (1902), 98–112, 195–205, 217–236, XII (1903), 1–12; Ralph Emmett Fall, ed., *The Diary of Robert Rose: A View of Virginia by a Scottish Colonial Parson, 1746–1751* (Verona, Va., 1977); Donald Jackson, ed., *The Diaries of George Washington,*

	Livestock			Other Activities				
Other	Cattle	Hogs	Sheep	Orchard	Timber	Building	Fishing	Craft
		kill	lambs born	plant	cut firewood	fence		
clean vegetable seed			feed	bottle cider	cut timber & firewood	fence; make carts	catch	
plant vegetable garden	turn dung; cart dung			plant graft		fence	catch	
sow indigo; plant potatoes; plant vegetable garden	cart & spread dung					fence	catch	
plant peas; plant vegetable garden	cart & spread dung		shear	weed				make clothes
tend garden								
plow for & sow turnips				make cider			catch	
plow for & sow turnips; sow buckwheat				make cider, brandy				
	plow cow pens	kill					catch	
sow spelts; pick cotton	pen	fatten; kill						make clothes, shoes
harvest potatoes	pen; kill	fatten; kill						
	pen in tobacco ground; heap dung; hoe in dung	fatten; kill			cut firewood; get staves	repair		

6 vols. (Charlottesville, Va., 1976), I, 211–315; Gregory A. Stiverson and Patrick H. Butler III, eds., "Virginia in 1732: The Travel Journal of William Hugh Grove," *VMHB*, LXXXV (1977), 18–44; Richard Corbin Letterbook, 1758–1768, CWL.

fences be built anew or existing ones repaired between January and the completion of spring planting in April.

Why improving planters like Landon Carter had difficulty raising enough meadow hay and imported grasses to feed their livestock, despite a large enslaved workforce and the availability of plows and draft animals, becomes clear when one learns that preparing ground for clover had to be accomplished in the spring, at the same time as thousands of hills had to be made and planted in tobacco and corn. The first cutting of meadow hay came during the time for planting and replanting corn and tobacco and often conflicted as well with the wheat harvest in late June or early July. A second cutting had to be fitted in later in the summer around the weeding of corn and weeding and cutting of tobacco, both of which took greater priority. One of the main reasons for feeding and raising more animals was to obtain their manure. But carting out, spreading, and then hoeing in manure likewise interfered with planting in the spring and with tobacco processing and corn harvesting in the fall. Attempts to integrate Old World livestock and small-grain farming with the traditional New World crops of tobacco and corn was an immense undertaking frought with so many difficulties that even the greatest planters were ill equipped to overcome them. Another indication of the intensity of spring and early summer work requirements is that Virginia slaves most frequently ran away during the high-pressure months of April and June.[82]

The impact of endemic disease on seasonal work demands should also be considered. Both the *vivax* strain of malaria, probably imported from Europe early in the seventeenth century, and the more lethal *falciparum* strain, likely introduced into the region by enslaved West Africans in the 1680s, sickened workers from mid- to late summer until fall frosts killed the mosquito vectors. Henry Callister, for example, wrote on August 19, 1747, "Now it is a busy time again with the planters, cutting and housing Tobo pressing Cyder and nursing the fever and ague, one half of the poorer sort are down with it." Thus, all Euro-American Chesapeake settlers, whether or not they owned slaves, paid a heavy price for their economic betters' participation in the trade in Africans. Malaria also diminished the productivity of enslaved workers. In the 1750s and 1760s, for example, a goodly proportion of Landon Carter's enslaved workforce was incapacitated by "fevers and agues" from August to early October nearly every year, frequently interfering with the tobacco harvest. Planters, especially those farming in unhealthy areas, could count on part of their workforce being disabled at one of the most critical points in

82. For runaways, see Morgan, *Slave Counterpoint,* 152, 167.

the agricultural year. Local disease environments might then have had a substantial impact on both the sizes and quality of crops.[83]

Not surprisingly, given the growing complexity of the agricultural calendar, large planters were beginning to appreciate the value of time, not just at peak seasons but also throughout the year. At Sabine Hall, for example, Landon Carter, at the end of March 1758, required a thirteen-hour workday "clear of eating time." Since on March 31 the sun rose just before 6:00 AM and set at 6:30 PM, the slaves were required to get up at first light and to work not just from sunrise to sunset but also from dawn (5:30 AM) to dusk (7:00 PM) with only a half hour break for midday meal. Near the end of December, the workday on Carter's plantations would have decreased to ten hours (7:00 AM to 5:30 PM), but it increased to fifteen hours (4:15 AM to 8:00 PM) in June and July. Indeed, careful planning was increasingly necessary as more crops had to be meshed into increasingly crowded seasonal calendars. When he prepared instructions for a new manager in 1759, for example, Richard Corbin explained how fully he expected labor to be utilized. The old practice of using "Hurry and Severity" at the last moment when crops suddenly had to be planted or harvested was both hard on the slaves and left the overseers rushing about in a state of ineffectual confusion. Instead, "foresight"—steady work throughout the year—was Corbin's watchword: "To imploy the Fall and the Winter Well is the foundation of a Successful Crop in the Summer."[84]

Slaveowners also attempted to extract maximum work by conducting time-motion studies in order to set daily output requirements. Landon Carter, for

83. Darrett B. Rutman and Anita H. Rutman, "Of Agues and Fevers: Malaria in the Early Chesapeake," *WMQ*, 3d Ser., XXXIII (1976), 31–60; Callister to Robert Morris, Aug. 19, 1747, Callister Papers, CWL; Greene, ed., *Diary of Carter*, 128, 165, 179, 328, 431, 459, 494, 504–505, 507, 513, 524, 765, 771, 778, 939, 943. For the effects of malaria, see also Philip R. P. Coelho and Robert A. McGuire, "Diets versus Diseases: The Anthropometrics of Slave Children," and Richard H. Steckel, "Diets versus Diseases in the Anthropometrics of Slave Children: A Reply," *JEH*, LX (2000), 232–246, 247–259.

84. Corbin to James Semple, Jan. 1, 1759, Richard Corbin Letterbook, CWL; Greene, ed., *Diary of Carter*, I, 212. Times of sunrise and sunset and of civil twilight for Tappahannock, Virginia, the town nearest Carter's home plantation for 1758, were retrieved from http://www.usno.navy.mil/USNO/astronomical-applications (accessed June 2009). Civil twilight begins in the morning and ends in the evening when the center of the sun is geometrically six degrees below the horizon. This is the limit within which ordinary outdoor activities can be carried on in good weather, with natural light, absent moonlight or other artificial illumination. On March 31, the period of civil twilight, rather than sunrise or sunset, corresponds with Carter's thirteen-hour workday, leaving just a half hour for "eating time." Complete darkness ends in the morning before civil twilight begins and comes in the evening after the end of civil twilight.

example, on March 31, 1758, "set by" his "weaving boy" (who was producing fewer yards of cloth per day than Carter expected) for a half hour, during which the slave "mended 15 threads in the time and moved his roller 6 times and yet he wove 11 inches and had many hindrances with his shuttle which is 22 [inches] the hour." From this observation, Carter concluded that the boy ought to weave eight yards in a thirteen-hour workday, "but I ask no more than 7 which is better than an hour and a half leisure time."[85]

Fuller use of all available resources was the key to tidewater planters' success in maintaining fairly constant gross revenues per laborer throughout the Golden Age. Although mid-eighteenth-century plantation accounts provide a reasonable idea of revenues, at least for the main crops of tobacco and corn, they provide less information about final profitability, for most costs remain poorly documented. The plantation accounts discussed in this and the preceding chapters suggest that great planters' gross revenues (in constant value) from major field crops in the central tidewater were in the range of ten to twelve pounds per laborer in the 1730s and 1740s and that they had risen to about thirteen to fourteen pounds per worker by the 1750s. During the Seven Years' War, the revenues of optimally situated planters with enslaved workforces of fifty or more who managed to raise substantial quantities of both corn and wheat rose even higher. The most productive area was Maryland's upper Eastern Shore, where Robert Goldsborough was realizing fifteen pounds per hand by the late 1720s, and by the 1750s at least one planter, Robert Lloyd, was averaging more than twenty pounds per field hand. Oronoco growers along the Potomac River did not fare nearly so well, however, some realizing only half the revenues of sweet-scented planters. Everywhere planters' willingness and ability to translate rising revenues into sound family fortunes varied. Some, like the Berkeleys, Richard Corbin, Daniel Parke Custis, and the Carrolls succeeded admirably. Others, including some members of the Carter and Baylor families, had both the economic clout and the acumen to manage to live beyond their means without seriously risking their children's future inheritances. Recent immigrants like Thomas Jones and John Mercer, who tried both to build large estates and to live in the same style as the native-born elite, foundered in the race to keep up.

Did increases in agricultural productivity occur during the Chesapeake's Golden Age that might have led to higher living standards, as economic historians now agree was the case in contemporary South Carolina and Georgia, where the introduction of tidal rice irrigation dramatically increased yields per worker? For the region's premier staple, the answer is clearly no. Tobacco

85. Greene, ed., *Diary of Carter*, I, 212.

outputs per laborer were lower in the 1750s than at the turn of the century and dropped below the midcentury level during the Seven Years' War. Growers had clearly already exhausted any possibilities for raising productivity per worker with the technologies available to them and, indeed, were unable to maintain production at the levels reached in the 1690s. The only positive change was a modest improvement in packing and prizing.[86]

The main productivity advances instead occurred in grains, especially maize. By midcentury, outputs per worker rose in all areas, although increased yields were most pronounced in the York River basin (where outputs per worker doubled over those prevailing in the early 1700s) and on Maryland's upper Eastern Shore (where yields per hand rose threefold). These advances were limited primarily to large plantations where use of plows allowed field hands to tend more acres in grain. Small operators with few or no bound laborers, almost all of whom continued traditional hoe culture, were unable to expand outputs. The doubling of revenues per laborer on large upper Eastern Shore plantations over those ordinarily obtained in other areas is the clearest evidence for significant increases in agricultural productivity in one part of the Tobacco Coast.

Less clear is how much of the increase was due to technological changes and more efficient use of labor and how much was simply due to greater labor inputs. The addition of new crops — wheat and briefly indigo — along with the expansion of corn production markedly increased the amount of work slaves had to perform, since planters were not adding new workers to mixed tobacco and grain farms. Slaves were increasingly required to strip and bundle tobacco and to husk and shell corn at night after a full day's labor of ten or more hours in the fields. Late fall and winter were no longer seasons of relative leisure, since workers were sent to plow fields, cut timber, and make fences whenever weather permitted or set to threshing grain indoors. Among wheat growers, higher yields per full-time field hand could be obtained only by also increasing inputs of hired seasonal labor. Grain farmers hired extra

86. R. C. Nash, "South Carolina and the Atlantic Economy in the Late Seventeenth and Eighteenth Centuries," *Economic History Review,* XLV (1992), 677–702; Peter Mancall, Joshua L. Rosenbloom, and Thomas Weiss, "Slave Prices and the South Carolina Economy, 1722–1809," *JEH,* XLI (2001), 616–639; Mancall, Rosenbloom, and Weiss, "Agriculture Labor Productivity in the Lower South, 1720–1800," *Explorations in Economic History,* XXXIX (2002), 390–424; David Eltis, Frank D. Lewis, and David Richardson, "Slave Prices, the African Slave Trade, and Productivity in Eighteenth-Century South Carolina: A Reassessment," and Mancall, Rosenbloom, and Weiss, "Slave Prices, the African Slave Trade, and Productivity in Eighteenth-Century South Carolina: A Reply," *JEH,* XLVI (2006), 1054–1065, 1066–1071; Russell R. Menard, "Plantation Empire: How Sugar and Tobacco Planters Built Their Industries and Raised an Empire," *Ag. Hist.,* LXXXI (2007), 309–332.

workers, at a minimum, for the summer harvest but often also employed one or more free plowmen or threshers annually or semiannually. Since the new grain-centered husbandry required both more locally produced goods and services and more hired unskilled labor, productivity increases on large diversified plantations also expanded opportunities for free workers, tenants, and small planters to increase their incomes. That these groups experienced modest improvements in living standards during the Golden Age has been amply demonstrated in probate inventory studies.[87]

For enslaved laborers, in contrast, agricultural change and advances in productivity led, not to improved living standards, but instead to increased levels of exploitation. Although gross revenues per worker rose as outputs of grains increased, slaveowners did not augment the food rations or clothing allowances allotted to enslaved field hands. They also attempted to limit the scope of the slaves' internal economy and to prohibit trading off the plantation. So long as tobacco and corn were the chief crops—cultivated largely with hoes, harvested and stored without the aid of carts and draft animals, and prepared for consumption or packed for market with the most primitive equipment—work routines varied little for men and women. But the new crops and routines often involved some degree of skill—sowing and mowing grains, plowing, harrowing, carting, ditching, lumbering, fishing, and milling, for example. These new tasks fell primarily to men. Agricultural change entailed considerable losses for slave women, and, unlike enslaved men, almost no gains. A few became more involved in textile production and house service than in earlier years. But free white women filled most of the slots created by expanded home industry. The great majority of black women continued to perform unskilled manual field labor, such as hand hoeing and weeding, planting and thinning corn, worming tobacco, cutting corn tops, and carrying corn to the barn, now more often without the help of men. The new jobs assigned to slave women (or old jobs formerly shared with men) included many of the least-desirable chores: building fences, grubbing swamps in the dead of winter, cleaning winnowed grain of weed seed, breaking up new ground too rough or weedy to plow, cleaning stables, and loading and spreading manure. New agricultural routines thus entailed a greater sexual division of labor, not between field and domestic production as in the white population, but between semiskilled and unskilled agricultural labor, with women relegated to the latter. And in the Chesapeake, just as in Barbados and South Carolina, planters used new crops and changes in agricultural

87. Paul G. E. Clemens, "The Consumer Culture of the Middle Atlantic, 1760–1820," *WMQ*, 3d Ser., LXII (2005), 577–624.

methods to justify heavier workloads and to erode customary definitions of "reasonable" requirements.[88]

The rewards flowing from increased agricultural productivity during the Golden Age were thus unevenly distributed by economic status, by geographic region, and, especially, by race. Large planters with enough bound labor to undertake cultivation of three staple crops and to take advantage of economies of scale benefited the most, followed by slaveholders of middling rank. Small planters who lacked the resources to expand agricultural outputs benefited only slightly or not at all. Rewards were also unevenly distributed among geographic regions, as productivity increases were much easier to achieve in areas with better soils and with good access to commercial and transportation networks. In areas where agricultural diversification advanced the most, increased demand among grain growers for local goods, services, and hired labor led to more widespread sharing of the benefits. The enslaved workers who produced most of the advances, however, lost rather than gained.

88. Carr and Walsh, "Economic Diversification and Labor Organization," in Innes, ed., *Work and Labor in Early America,* 176–177; S. Max Edelson, *Plantation Enterprise in Colonial South Carolina* (Cambridge, Mass., 2006), 156–161; J. R. Ward, "The Amelioration of British West Indian Slavery, 1750–1834: Technical Change and the Plow," *Nieuwe West-Indishche gids / New West Indian Guide,* LXIII (1989), 41–57.

CHAPTER 8 *Reassessing the Golden Age*

Rising prices for the region's staple beginning in the 1730s, a result of expanding European markets and quality improvements effected by Virginia's tobacco inspection act, marked the opening of an era of unprecedented prosperity for Chesapeake planters. Throughout the region, plantation revenues rose, and tobacco revenues were often by then augmented with returns from grains. Unlike tobacco, there were decided economies of scale in grain production that benefited large operators. Most tidewater planters, however, were unable to take maximum advantage, since few were able to overcome constraints imposed by an inadequate supply of forage for draft animals. Nonetheless, through careful seasonal meshing of crops and some implementation of plowing, they were able to increase outputs of grains, especially of maize. They began instituting more systematic crop rotations and taking steps to alleviate problems of declining fertility of long-farmed soils. By adding corn and wheat as second and sometimes third export staples, great planters succeeded in extracting higher revenues per worker than they had managed in previous decades. Middling planters, who by midcentury had also acquired slaves, benefited along with large producers from the opening of new markets and from the quickening pace of trade. Even small planters managed to modestly improve their standard of living, primarily through various family members' working for wages at harvest and in slack seasons of the agricultural calendar and by producing goods for local markets. Like contemporary peasants and proletarians in Europe, small-planter and tenant-farmer families in the Chesapeake increased household labor-force participation and market-oriented activities, thereby increasing total household incomes.[1]

Large-scale importation of enslaved Africans, augmented by the onset of natural increase among the slaves, alleviated the labor shortages that had curtailed expansion of farming operations in earlier years. Moreover, slaveowners learned how to exploit enslaved workers more efficiently and thoroughly by increasing the length of the workday and by keeping the labor force fully employed throughout the year. Plantation agriculture expanded rapidly into

1. Jan de Vries, "Between Purchasing Power and the World of Goods: Understanding the Household Economy in Early Modern Europe," in John Brewer and Roy Porter, eds., *Consumption and the World of Goods* (London, 1993), 85–132.

the Chesapeake piedmont, as younger sons for whom there was insufficient land in the tidewater moved west to develop new upland plantations. Even the disruptions occasioned by the seven-year global conflict between Britain and France failed to stop the region's economic momentum. Although their growing prosperity was rooted in a system of extraction of forced labor that depended on violent coercion and the maintenance of strict racial boundaries, few planters had reason to question the morality or efficacy of holding Africans in perpetual bondage before the close of the Seven Years' War.[2]

Elite planters who lived through the 1730s, 1740s, and 1750s thought they had their fair share of troubles with all the problems that crafting and implementing tobacco inspection laws entailed as well as with having to cope with droughts, floods, and periodic international warfare. There were no severe sustained economic downturns, however, and most large planters realized greater revenues than they had in the first quarter of the century. Threats of organized collective slave revolts faded as creoles began to outnumber Africans in enslaved tidewater workforces. Instead, resistance was largely confined to individual acts of pilfering, shirking work, and running away, acts of disobedience that planters thought they could readily control through a combination of plantation discipline, slave patrols, and occasional punishments meted out by county courts, which the slaveowners controlled. Gentry planters also enjoyed virtually unchallenged political and cultural dominance. Deference from social inferiors remained the norm, and challenges from evangelicals to the established Anglican religion were not yet widespread. Chesapeake elites also increasingly enjoyed the pleasures afforded by their now culturally sophisticated and architecturally distinguished capitals of Williamsburg and Annapolis as well as by rounds of visits to the country mansions of neighboring gentry. From the perspective of the chaotic last third of the eighteenth century, they would look back nostalgically to this time as a sort of untroubled Golden Age, a characterization that most historians of the Chesapeake region have subsequently also embraced.[3]

2. The morality of slavery in British colonies was seldom (and never effectively) questioned on either side of the Atlantic before 1760. Contests with the French during the Seven Years' War for control of sugar-producing islands in the West Indies raised awareness in both Britain and in the colonies of the dual vulnerability of slave societies to invasion from without and to rebellion from within. The political contest between mainland colonists and the British government that followed led to the first sustained questioning of the morality of slavery by both imperial administrators and by slaveowners themselves. See Christopher Leslie Brown, *Moral Capital: Foundations of British Abolitionism* (Chapel Hill, N.C., 2006), chaps. 1–2; Bruce A. Ragsdale, *A Planter's Republic: The Search for Economic Independence in Revolutionary Virginia* (Madison, Wis., 1996), chap. 4.

3. Aubrey C. Land, "The Tobacco Staple and the Planter's Problem: Technology, Labor, and

An unusual degree of stability within the British Empire was a necessary precondition for midcentury colonial prosperity. In contrast to the uncertainties and disruptions of the earlier reigns of the Stuarts, according to Ian K. Steele, "governance of the disparate kingdoms, companies, and colonies of the British Empire was not seriously challenged" between the Glorious Revolution and the onset of resistance to tighter imperial rule in the mainland colonies in the wake of Britain's decisive defeat of France in the Seven Years' War. During this period of salutary neglect, "effective administrative power remained diffuse, flexible, and limited." Aristocrats and merchants joined forces to promote policies favoring commercial expansion, freer trade, and enhanced naval power. These were initiatives that also furthered the interests of colonial planters and merchants, including providing them with a larger supply of enslaved Africans. Since Britain could seldom enforce laws regulating colonial trade or interfering in matters of internal government with fleets or armies (both of which concentrated on defending the home country against rival powers in Europe) or through courts, "most decisions were negotiated, moderated, appropriated, evaded or even resisted through layers of governance." Expanding international trade and enhanced naval power in turn led to unparalleled levels of affluence for free citizens of the Empire, both in the mother country and in the North American colonies.[4]

Political stability at both provincial and local levels of government similarly prevailed in the Chesapeake during the golden years. Appointment or election to positions of power by virtue of inheritance of place, rather than, as had often been the case in the seventeenth century, by default owing to lack of leaders with traditional qualifications, "enhanced the legitimacy of political authority." The gentry faced no serious internal challenges to their rule. And British indifference to colonial affairs before the Seven Years' War left colonial leaders freer "from the customary political tug-of-war with imperial officials than at any time since the end of Bacon's Rebellion." The spread of slaveholding among the majority of planters strengthened a sense of common racial and economic interest among whites. Expansion of western

---

Crops," *Ag. Hist.*, XLIII (1969), 69–70, 76–81; Lorena S. Walsh, "The Differential Cultural Impact of Free and Coerced Migration to Colonial America," in David Eltis, ed., *Coerced and Free Migration: Global Perspectives* (Stanford, Calif., 2002), 147–149; Rhys Isaac, *The Transformation of Virginia, 1740–1790* (Chapel Hill, N.C., 1982), part 1; Warren M. Billings, John E. Selby, and Thad W. Tate, *Colonial Virginia: A History* (White Plains, N.Y., 1986), chap. 9.

4. Ian K. Steele, "The Anointed, the Appointed, and the Elected: Governance of the British Empire, 1689–1784," in P. J. Marshall, ed., *The Eighteenth Century*, vol. II of Wm. Roger Louis, ed., *The Oxford History of The British Empire* (Oxford, 1998), 105.

settlement in Virginia afforded new opportunities for landownership. And the regular creation of new counties as western populations grew opened new political positions for aspiring frontier leaders and assured western settlers of adequate representation in the lower house, thus lessening the possibility of social disruption or of political confrontations between backcountry and tidewater.[5]

Elite Chesapeake colonists began to consider themselves equal participants in a Western civilization whose intellectual achievements were believed to be matched only by those of the ancient world; material progress was coming to be considered not just possible but also a positive good. "Within a generation," a recent history of Virginia contends, "the upper class in the older settled areas of the colony turned from provincial frontier barons into close replicas of the eighteenth-century British country gentry." Many of the third-generation planter elite had obtained a university education either in Europe or at the College of William and Mary. They shared a common background in classical literature and in Enlightenment philosophy, including a belief in the efficacy of scientific methods and in the possibility of improving the human condition. They participated fully in the new world of metropolitan consumer goods, adopting the genteel manners and refined ways of life that the acquisition of fashionable clothing, dining wares, and household furnishings facilitated. Cultural symbols—mansions in a classical style and hierarchically ordered brick courthouses and churches—along with elaborate social ceremonies legitimated and enforced distinctions of rank and encouraged deference from social inferiors. Imitation of the British gentry thus served to rationalize and reinforce the hierarchical social and political structure of the region.[6]

One manifestation of cultural borrowing was the emergence by the 1750s of a select group of agricultural improvers like Charles and Landon Carter, who were bent not just on increasing family wealth but also on transforming Chesapeake agriculture into something more closely resembling European practices. Great planters who embraced these goals added a new cultural and political dimension to large-scale plantation agriculture. Significant agricultural innovations flowing from the application of scientific methods in fact

5. Martin H. Quitt, "Immigrant Origins of the Virginia Gentry: A Study of Cultural Transmission and Innovation," *WMQ,* 3d Ser., XLV (1988), 653 (quotation); Billings, Selby, and Tate, *Colonial Virginia,* 251 (quotation); Jack P. Greene, *The Quest for Power: The Lower Houses of Assembly in the Southern Royal Colonies, 1689–1776* (New York, 1963), 28–31.

6. Billings, Selby, and Tate, *Colonial Virginia,* 215–229 (quotation on 215); Isaac, *Transformation of Virginia,* chaps. 2–6; Richard Drayton, "Knowledge and Empire," in Marshall, ed., *The Eighteenth Century,* vol. II of Louis, ed., *Oxford History of the British Empire,* 237–243.

lay years in the future. Initially, this new cult of improvement served more to reinforce the prestige of the great planters through their privileged access to transatlantic intellectual and cultural networks and interaction with European learned societies than it did to advance husbandry in the region. To the extent that improving planters held all things British superior to colonial practices better suited to the region's climate and resources, they often ended up advocating impractical and at times harmful changes. And, in dismissing entirely the traditional wisdom of unlettered ordinary planters and of enslaved Africans, they ignored altogether alternative sources of agricultural knowledge gained from direct experience with growing crops. During the Golden Age, less self-conscious, more practically oriented agricultural innovators on Maryland's upper Eastern Shore achieved greater productivity advances than did most of the more publically active, self-proclaimed tidewater improvers. Nonetheless, among those convinced that agricultural experiments and the application of scientific methods would also lead to higher profits, the new role appealed both to their vanity and to their pocketbooks.[7]

At the beginning of the eighteenth century, enterprising tidewater planters had amassed land and slaves in order to become rich. By midcentury, the goals of their children and grandchildren, who started out rich by virtue of inheritance, shifted from estate building to estate maintenance and revenue enhancement. Those with reserves of undeveloped land continued to expand tobacco production by setting up additional quarters. But in older, more crowded areas, emphasis shifted to simply maintaining existing levels of tobacco outputs while markedly expanding grain production, an alteration in crop mix that increased revenues per worker without requiring any expansion of the workforce. Shifts in strategy in turn led to changes in the mix of assets in which large planters typically chose to put their resources. As in the earlier estate-building period, investments in western land for speculation, renting, or future development as ancillary farms remained widespread. After the 1730s, however, most established planters had to invest little in new slave labor or in additional livestock—inheritance and natural increase among both human and animal property supplied most of their needs. These changes allowed more fiscally disciplined elite planters to increase moneylending, thus making more capital available to planters of middling rank. Most of the planter elite had also inherited sufficient land to employ their growing labor forces, if not enough to endow all their children. Rising land prices that accompanied population growth increased the value of large planters'

7. Joyce E. Chaplin, *An Anxious Pursuit: Agricultural Innovation and Modernity in the Lower South, 1730–1815* (Chapel Hill, N.C., 1993), chap. 5.

estates without their having to expend effort to improve them. A number did, however, make improvements in a period when there was more money available to put into farm buildings and fine homes as well as into other productive enterprises such as mills, mines, ships, warehouses, and town properties that further augmented their incomes while also contributing to regional economic development. In contrast, overall, investments in mercantile assets declined. For most, agriculture became a full-time occupation; no longer did many large planters combine farming with extensive trading activities. Intercolonial commerce was becoming too complex to pursue profitably part-time. Withdrawal from active commercial life, coupled with a rising gentry ethos, fostered the self-conscious agrarianism that would play an increasingly dominant role in the region's culture.[8]

One area in which Chesapeake elites did not follow British precedents, however, was in family inheritance strategies. Restrictive inheritance practices like entail were employed to keep wealth in the family, but not to force younger sons to make their own way in the world in occupations other than planting or to curtail the dowries of younger daughters. Relatively equal portions were the rule for gentry sons, regardless of birth order, and so all were attached to a common culture and shared the expectation that they would follow in their fathers' footsteps as substantial landholding and slaveowning planters and would be virtually assured of appointment to important political offices. During the Golden Age, increased plantation profits, augmented with nonagricultural revenues, indeed enabled gentry planters to provide their children with marriage portions sufficient to allow them to replicate their parents' standards and styles of living.[9]

Although the considerable successes that great Chesapeake planters achieved in building and maintaining family fortunes have long been recognized, the strategies that they chose for managing their estates have been less well understood. At the same time, some scholars who have celebrated the political and cultural accomplishments of the elite planters of the mid-eighteenth century, many of whom later played a critical role in creating the American nation, have also tended to offer excuses for the means they employed to achieve their economic, social, and political preeminence and to romanticize the nature of the racially divided society that resulted. This study has sought, through analysis of the strategies that individual plantation managers employed, to offer more concrete measures of their achievements over time and place and to provide a better understanding of the reasons for their

8. For agrarianism, see Quitt, "Immigrant Origins," *WMQ*, 3d Ser., XLV (1988), 654–655.

9. For cultural integration and expectations, see ibid.

successes and failures. It has also sought to document the exploitation and violence that underlay their economic achievements.

The world of the creole Berkeleys, Beverleys, Burwells, Byrds, Carters, Carrolls, Corbins, Galloways, Goldsboroughs, and Tilghmans had changed markedly from the strange, challenging, but brightly promising New World that early English immigrants like Robert Wintour and Robert Cole had crossed an ocean to embrace more than a century before. Although few of the early settlers reaped the windfall rewards many had hoped for, a privileged, plucky, and lucky minority managed to wrest profits from farming and from farm building that enabled them or, more likely, their children to acquire acreage and bound workforces on a level Wintour could scarcely have envisioned. For about fifty years, more ordinary migrants who were flexible enough to abandon labor-intensive English-style agriculture for New World food and cash crops and to adopt labor-saving cultivation methods (and who managed to survive long enough) also realized some of the advantages that prompted gentlemen like Wintour to immigrate. Landowning and the independence it afforded (the first of the "pleasures" Wintour enumerated), along with the ability to produce grain and meat in plenty, was much more widespread in the Chesapeake than in the mother country. With time, more colonists would also enjoy the pleasure of "friendly neighbourhood" in the small rural communities that emerged in longer-settled areas throughout the region. And, as Wintour had projected, profits realized from raising export crops allowed most farm families to purchase necessities, if not always amenities, although wars and economic downturns periodically occasioned shortages.[10]

In the following half century from about 1680 to 1730, low tobacco prices and shortages of land and labor brought an end to the opportunities that had induced English men and women to emigrate and left small native-born planters struggling just to stay even. Ordinary planters could still subsist, but seldom handsomely. Only the already rich grew richer, as the profits they accumulated from the labors of enslaved Africans, who by 1730 had almost entirely replaced temporarily bound indentured servants, along with the privileges flowing from political position, allowed them to continue to accumulate more land and more slaves.

Once Chesapeake settlements were firmly established, high-minded pro-

10. The motives for emigrating are laid out by Robert Wintour in John D. Krugler, ed., *To Live Like Princes: A Short Treatise Sett Downe in a Letter Written by R. W. to His Worthy Friend C. J. R. concerning the Plantation Now Erecting under the Right Honorable Lord Baltemore in Maryland* (Baltimore, 1976). For eighteenth-century living standards, see Lois Green Carr, "Emigration and the Standard of Living: The Seventeenth Century Chesapeake," *JEH,* LII (1992), 271–291.

moters no longer urged potential investors and immigrants in Britain to join in "honourable and pious" colonial adventures. Wealthy planters turned instead to the more mundane "prospect both of Pleasure and Profit" afforded by keeping somewhat better records of their business affairs in order to maintain or further inherited fortunes. According to mid-eighteenth-century colonial conceptions, the "ample" wealth accumulated through the exploitation of enslaved Africans by prominent gentry families such as the Byrds, Carters, Carrolls, and Corbins was "honorably acquired," and patriarchs sought additional honor by providing "an Example of Prudence, Oeconomy, and Industry" for their "Posterity."[11]

Moreover, the quickening and expansion of international, intercolonial, regional, and local trade that accompanied the Golden Age brought some resumption of opportunities for profit for ordinary native-born colonists. By the mid-eighteenth century, even poor planters could participate to some extent in the burgeoning world of consumer goods, even though most still lacked the resources to expand farming operations. Meanwhile, as slaveholding became ever more widespread, the ranks of middling Anglo-European families possessed of comfortable estates increased. On the eve of the Seven Years' War, Chesapeake residents could rightly consider that they had also gained honor by materially advancing the interests and prosperity of the British Empire. The greater honor of becoming "Founders of a nation," however, would come only when the colonists chose to declare independence from that Empire.[12]

But offsetting these considerable accomplishments were pronounced social failures. Instead of, as Wintour envisioned, English settlers' acquiring honor by converting the Indians to Christianity and "civilized" European ways, when he disembarked in Maryland in 1637 Virginia colonists had already exchanged a policy of accommodation and conversion for one of destruction and dispossession. Expelling Indians from areas Europeans coveted and expropriating the land of any native American groups not decimated by European diseases or warfare became the standard practice. Few European settlers dealt honorably with the peoples whose country they had seized, and most no longer thought native Americans worthy or capable of becoming members of a "Christian" Anglo-American polity.

11. Krugler, ed., *To Live Like Princes,* 36; Alexander Malcolm, *A New Treatise of Arithmetick and Book-Keeping* . . . (Edinburgh, 1718), appendix, 34; Tomb Top Inscription, in Robert Carter's "Obituaries and Other Records of His Death, 1732 and Later," in Berkeley, ed., *Diary, Correspondence, and Papers of Robert "King" Carter,* http://etext.virginia.edu/users/berkeley; Roger North, *The Gentleman Accomptant* . . . (London, 1721), 5–6.

12. Krugler, ed., *To Live Like Princes,* 28.

The early decision of Chesapeake elites to adopt African slavery and to embrace it wholeheartedly whenever trade patterns in the Atlantic World permitted produced the second social failure. Convinced that perpetually bound "non-Christian" workers could be made to produce higher profits in New World agriculture than could temporarily bound European indentured servants, Chesapeake elites had eventually to craft a legal apparatus that protected their property rights, supported the private tyranny they exercised on their own plantations, and legitimated the violent measures to which they often had to resort. Slavery, as it would be later argued, was not only incompatible with free labor; it was incompatible with indentured servant labor as well. Once voluntary European migration slowed to a trickle, buying slaves became the only feasible way of augmenting family labor year-round. As more and more planters acquired human property, and many of those who had none hoped to obtain it in the future, it became increasingly easy for elites to enlist the aid of all free members of the society in upholding both a system of racial bondage and a society based on racial differences. The labor system from which elite Chesapeake planters so handsomely profited, and from which most ordinary white Virginians and Marylanders benefited or hoped to benefit, compromised the honor of everyone involved.

EPILOGUE

Through the mid-1760s, Chesapeake residents were exceedingly optimistic about their future economic prospects. Tobacco prices had surged upward at the onset of the Seven Years' War, and, although the market sagged between 1762 and 1764, by 1766 prices were again on the rise, a trend planters hoped would continue. So they borrowed heavily from both British merchants and local lenders to buy more land and, among those few still needing to expand their workforces, more slaves. Factors working for competing British tobacco firms drove up prices offered in the country, extended credit lavishly, and made more direct cash purchases, all putting more money into planters' pockets. Rather than saving, in the late 1760s most families spent the additional earnings on consumer goods and on improving their houses or building new ones. Fortuitously, harvest failures in Europe and growing urban populations in the more northerly colonies created a boom market for corn and wheat in the late 1760s and early 1770s. Most big planters tried to maintain tobacco production at prewar levels while growing ever larger amounts of one or both staple grains, but a few dropped tobacco altogether and shifted into wheat as their major cash crop. In addition, the decisive defeat of the French in the Ohio Country fanned the hopes of land speculators, Indian traders, and prospective western settlers. Nothing, it seemed, would at this time prevent larger planters from becoming immensely rich very quickly.

The Chesapeake's golden years had passed, however, and in the later 1760s and early 1770s political and economic crises escalated. The British government's unprecedented efforts to force the colonists to help pay for the costs of the Seven Years' War initiated the political crisis. At first Britain's quick capitulation when residents of all the mainland colonies vigorously resisted the Stamp Act encouraged colonists to expect further accommodations. But instead the king and Parliament responded with a new round of more extreme coercive measures that threatened the colonists' political independence and economic base. Colonial resistance stiffened, precipitating an imperial crisis that would ultimately be settled only after seven long years of war. Political problems were exacerbated by a second major crisis in the tobacco market in 1772–1773 that drove a number of British firms into bankruptcy, leading to a credit crunch in the colonies that some of the most overextended planters could not weather, and under which almost all suffered.

The war that followed would test the coping skills of even the best plan-

MAP 12. *A Map of the British and French Dominions in North America. . . .* By John Mitchell, London, 1755. Black-and-white line engraving with period color. Courtesy, The Colonial Williamsburg Foundation.
Britain's defeat of France in North America potentially opened vast new territories to English settlement. Virginia claimed a major share of the conquered territory.

tation managers. The British blockade of Chesapeake Bay stopped almost all external trade, leaving planters with few outlets for their crops of tobacco and grain and no means for obtaining the European manufactured goods on which the region's prosperity depended. Planters then had to devote enormous effort to making cloth, shoes, tools, and salt and to scale back their planting of export crops. Farming families earned but scant revenues because they produced few surplus crops, and they planted few surpluses of anything, because they could not be sure of finding a market. All classes experienced material hardships, ranging from inconvenience to severe depravation. British raiders destroyed crops, livestock, and buildings and seized enslaved workers; resident planters no longer maintained farm infrastructure; and plantations deteriorated further when their owners left to fight in the army or to serve in government in distant places, or when loyalist planters abandoned their colonial estates and went back to Britain. With the outcome of the struggle always in doubt until the American victory at Yorktown, optimism alternated with pessimism, individual sacrifice in pubic service with unrestrained private greed. Strategies for long-term family survival predominated, but sometimes a chance to make fast money proved irresistible, even by questionable means.

When the fighting finally ended, tidewater Chesapeake planters, large and small, attempted to return to business as usual. They resumed high-volume tobacco production, hoping in the ensuing postwar boom to make up for years of lost revenues. Renewing prewar connections with English merchants, the bigger operators cut back on self-sufficient activities and returned to importing manufactured goods, reactivated gristmills and merchant mills, invested in ships, and, now freed from imperial trade restrictions, also tested new venues for international trade. By the late 1780s, revenues per laborer on the best-managed tidewater plantations were again equaling those of the early 1760s. Other big planters aggressively expanded their operations on piedmont plantations and speculated in land farther to the west, now, with the removal of the British, open for settlement.

But a series of misfortunes soon dampened their initial postwar optimism. The British Parliament closed crucial West Indian markets to American shipping, the postwar boom in tobacco soon turned into another cyclical bust, trade with new continental European partners proved disappointing, taxes remained high, and specie was scarce. Hard times spurred out-migration, depressing land values in older tidewater counties. By 1789, with the inauguration of a new government under the Constitution, some great Chesapeake planters again turned more sanguine, convinced that a strong federal government would create another era of prosperity. These optimists began to plan new agricultural and commercial ventures at home in the tidewater, in the west, or in nearby cities that were by then gaining rapidly in population. Others had more trouble accommodating themselves to new conditions. Suffering a real or perceived decline in status as a result of the war, they continued to look back longingly to the brighter days of the early 1760s when economic and social conditions were more promising and decidedly less complicated.

Some tidewater areas did indeed become economic backwaters, with the residents who chose to remain rather than migrate never recovering anything akin to their pre-Revolutionary economic status. In contrast, other elite planters responded to the opportunities created by free international trade and the formation of an integrated domestic market with unparalleled dynamism. As a result, between 1790 and 1819 some few great Chesapeake planters who were able to respond to changing market conditions by quickly altering their crop mix, and who were ruthless enough to pare down the maintenance costs of prime field hands and to sell off surplus enslaved women and children, realized higher agricultural returns than were ever achieved during the colonial era.

Overall, however, planters in the tidewater area experienced a marked decline in economic and political stature relative both to more westerly counties within the region and to New England, the mid-Atlantic, and the tramontane west. The region's total population ceased growing as whites moved out in droves, and even newer western counties lost nearly as many residents to out-migration as they gained in-migrants from the tidewater. Slow or nonexistent population growth contributed to economic stagnation in older areas and everywhere reinforced perceptions of declining fortunes. And, although increasing numbers of slaves were forced to move west or were sold south, even more whites were leaving voluntarily so that everywhere the proportion of blacks in the total population rose. The crops these primarily agricultural workers now tended, however, changed dramatically.

Following the decision of the revolutionary French Constituent Assembly to abolish their national tobacco monopoly in 1791, many tidewater planters, fearing the collapse of crucial Continental markets, permanently abandoned the region's traditional staple. Starting in 1792, tobacco reigned king only in the James River piedmont, the Southside, and in southern Maryland. Everywhere else the majority of planters turned to grain farming with generally mediocre to disappointing results. Moving agricultural operations to Kentucky, Tennessee, or the lower South or abandoning farming to pursue the good life in urban centers became increasingly attractive, especially after 1819 when prospects for so much as maintaining previous levels of farm revenues evaporated in the worldwide depression that followed the end of the Napoleonic Wars.

The consequences of the Revolution entailed more than acute economic turmoil and political changes that sometimes threatened the power of the gentry class. When in 1774 Virginia's last royal governor, Lord Dunmore, offered freedom to slaves who would join the British cause, the threat of an internal slave revolt became a frightening reality. During the war, some thousands of slaves fled to the British, and most large tidewater planters lost some of their human property. The egalitarian philosophy that whites used to justify their revolution forced slaveholders to examine, many for the first time, the legitimacy and morality of the institution of forced labor on which their fortunes depended. Most continued to defend slavery as an economic and social necessity, but some began thinking about possible ways to end it or at least to curtail slavery's spread. Others took the more radical step of manumitting some or all of their slaves. The rapid growth of a free black population raised whites' fears of social disruption, however, and manumissions were soon severely curtailed. Through inheritance and purchase, slavery continued to spread ever wider throughout the white population, as many middling and small planters eagerly embraced chattel bondage as a means for enhancing their agricultural revenues and hence their families' welfare.

Meanwhile, the enslaved embraced the white Revolutionaries' egalitarian political philosophy for their own ends. Some escaped slavery entirely during the war, and many others died trying. Those who remained on farms and plantations would no longer passively or hopelessly accept their unfree status. Slaves could now appeal to universal natural rights to protest their condition. Possessed of potent ideological justification for demanding an end to bondage, individual and collective resistance became endemic in the quarters. Blacks began running away in numbers unprecedented before the war, going farther, staying out longer, and more often eluding capture altogether. Those who did not take such drastic measures stepped up passive resistance to work demands and continually confronted owners and overseers with their failure to carry out the logical implications of Revolutionary political philosophy by granting freedom to all. Plantation managers found themselves dealing with a workforce that most discovered grew more troublesome every day.

Managing plantations in the years following the American Revolution thus posed new challenges to the Chesapeake planter elite, a number of whom had finally won the title of founder of a nation (a goal that had motivated migration two centuries before) in the struggle for independence. Some responded to changing market opportunities by altering their crop mix and by adopting new technologies and ostensibly more modern methods in places where colonial crops and practices no longer produced satisfactory returns. The outcome was decidedly mixed, however, as the changes yielded substantial short-run profits at the cost of serious long-run environmental problems. Others chose the more cost-effective option of moving to new areas where farm building and old-style staple agriculture continued to produce high rewards. Slaveholders responded to the challenges of managing more articulate and more restive workers by adopting paternalistic rhetoric. But the drive to maintain profit levels by wringing ever more work from the enslaved usually trumped ameliorative impulses. The violence underpinning the labor system abated only a little, as threats of sale to the Southwest became an equally or more effective instrument of control than physical punishment. Slaveholders' failure to resolve the contradictions between claiming natural rights for themselves and denying all rights to those they enslaved tarnished the legacy of their political and economic achievements. Instead, an ever-growing need to defend a system that the wider world was coming to view as indefensible increasingly warped and corroded Chesapeake society.

APPENDIX 1 *Tobacco Crop Shares per Laborer*

YEAR	POUNDS	COUNTY	SOURCE, COMMENTS
1635	800–1,000	All Md.	*Relation of Maryland,* 82. crop share
1637	1,000+	Saint Mary's	*Calvert Papers,* I, 206. est. share per hand
1639	570	Saint Mary's	*Archives of Md.,* IV, 89. crop share
1640	878	Saint Mary's	*Archives of Md.,* IV, 90. crop share
	500	Saint Mary's	*Archives of Md.,* IV, 92. inmate's share
	400	Kent	*Archives of Md.,* IV, 92. crop share
1641	150	Northampton/Accomack	Ames, *County Court Records,* I, 71. freedman's share
1644	275	Lower Norfolk	Deeds, B: 20. crop in inv., no labor
	1,000	All Va.	Clarendon MS, 24.52. crop share
1645	425	York	DOW, 2: 137. crop in inv., no labor
	800	Lower Norfolk	Minutes, 1643–1646: 84. crop share
1646	2,220	Northampton	WD, 3: 145, 4: 52. crop in inv., no labor
	1,021	Northampton	WD, 3: 57. crop share
	950	Saint Mary's	*Archives of Md.,* I, 222. crop share
1647	1,500	York	DOW, 2: 372. crop in ground, no labor
	1,652	Lower Norfolk	Deeds, B: 55. inmate's crop, no labor
	600	Saint Mary's	*Archives of Md.,* IV, 501. servant's crop
1648	888	York	DOW, 2: 334. crop of best and seconds
	800	Northampton	WD, 3: 145. crop share
1649	2,000	All Va.	*Force Tracts,* II, 8. crop of one man
	1,500	All Va.	Bullock, *Virginia,* 63. one servant's crop
	850	Saint Mary's	*Archives of Md.,* IV, 466. crop share
	428	Lower Norfolk	Deeds, B: 114. inmate's crop, no labor
	1,429	Lower Norfolk	Deeds, B: 114. inmate's crop, no labor
1650	200	Northumberland	DO, 1650–1652: 28. inmate's crop
	1,011	Northumberland	DO, 1650–1652: 49. crop share
	987	Northampton	WD, 3: 26. crop share
	1,080	Northampton	WD, 3: 24. crop share
1651	500	Northumberland	Records, 1652–1658: 13. inmate's crop
	1,079	Northampton	WD, 4: 24. crop share
1652	1,440	Northumberland	Records, 1652–1658: 23. crop of landowner, no labor
1653	1,000	Northampton	WD, 5: 29–30. crop in inv., no labor
1654	1,900	Lancaster	Deeds, 2: 85. inmate's crop in inv.
	1,299	Lower Norfolk	Deeds, C: 93. crop in inv., no labor
1655	1,500	All Chesapeake	*Force Tracts,* 3, 13. man's crop commonly
	724	Northampton	WD, 5: 101. leaseholder's crop, no labor
	1,349	Lancaster	Deeds, 2: 28–30. crop share
	536	Lower Norfolk	Deeds, C: 113. inmate's share

Full citations for short titles and abbreviations in Appendixes 1, 2, and 3 appear at the end of Appendix 3.

YEAR	POUNDS	COUNTY	SOURCE, COMMENTS
	1,170	Lower Norfolk	Deeds, C: 153. crop in inv., no labor
	526	Lower Norfolk	Deeds, C: 154. inmate's crop
	1,600	Saint Mary's	*Archives of Md.,* XLI, 595. servant's crop share
1656	1,400	York	DOW, 3: 22. one man's crop, age 28
	826	York	DOW, 3. crop share
	770	Northumberland	Records, 1652–1658. inmate's crop
	1,390	Northumberland	Records, 1652–1658. crop in inv., no labor
	600	Northampton	WD, 5: 98. crop in inv., no labor
	1,373	Northampton	WD, 7–8: 32. crop share
	1,200–1,400	Saint Mary's	*Archives of Md.,* XLI, 595. servant's crop share
1658	1,100	York	DOW, 3: 44r. crop in inv., no labor
1659	1,093	York	DOW, 3: 109. crop share
	775	York	DOW, 3: 69r. share est. from inv.
	1,500	Saint Mary's	*Archives of Md.,* XLI, 595. share for seasoned hands
1660	1,213	York	DOW, 3: 109. crop share
	300	Northumberland	Records, 1658–1666: 46. crop share
1661	1,260–1,399	York	DOW, 3: 156. share est. from inv.
1662	1,574	York	DOW, 3: 156. share est. from inv.
	1,810	Saint Mary's	R. Cole Accts. crop share
1663	1,013	Northampton	WD, 9: 182. crop share
	1,037	Saint Mary's	*Archives of Md.,* LVII, 46. crop share
	1,291	Saint Mary's	R. Cole Accts. crop share
1664	906	York	DOW, 4: 265. crop share
	1,621	Saint Mary's	*Archives of Md.,* LVII, 49. crop share
	1,635	Saint Mary's	R. Cole Accts. crop share
1665	875	York	DOW, 4: 56. crop share
	1,190	York	DOW, 4: 6–8. share est. from inv.
	1,430	Northampton	Orders, 9: 1. crop share
	1,300	Northampton	Orders, 9: 24. crop in inv., no labor
	1,800	Charles	*Archives of Md.,* LIII, 617–618. crop share
	1,861	Md.	Test. Proc., 1E: 158. crop share
	1,241	Calvert	Test. Proc., 1F: 63. crop share
	1,627	Saint Mary's	R. Cole Accts. crop share
1666	860	Northampton	Orders, 9: 22. leaseholder's crop, no labor
	1,073	York	DOW, 4: 100. crop in inv., no labor
	2,520	Calvert	Test Proc., 1F: 75. crop in inv.
	1,186	Saint Mary's	R. Cole Accts. crop share
1667	720	York	DOW, 4: 136. female servant's crop share
	1,200	All Chesapeake	Wertenbaker, *Planters,* 64. size of common crop
	1,313	Saint Mary's	Test. Proc., 2: 395. crop in inv.
	1,272	Saint Mary's	R. Cole Accts. crop share
	1,760	Anne Arundel	Test. Proc., 2: 254. crop share
1668	1,394	Northampton	Orders, 9: 64. landowner's crop with 1 servant

YEAR	POUNDS	COUNTY	SOURCE, COMMENTS
	1,421	Lower Norfolk	Deeds, E: 69. single man's crop in inv., no labor
	1,142	Saint Mary's	Test. Proc., 3: 103. crop in inv.
	2,006	Saint Mary's	R. Cole Accts. crop share
1669	1,053	Northampton	Orders, 9: 64. crop share
	1,092	Northampton	Orders, 9: 61. servant's crop share
	1,258	Saint Mary's	R. Cole Accts. crop share
1670	1,052	Saint Mary's	R. Cole Accts. crop share
1671	620	York	DOW, 4: 318. crop in inv./tithables
	900	Talbot	Test. Proc., 5: 277. crop in inv.
	1,000	Anne Arundel	IA, 1: 77a. crop in inv.
	1,600	Saint Mary's	R. Cole Accts. crop share
1672	954 ½	York	DOW, 4: 14. crop share
	1,210	York	DOW, 4: 318, 377. crop in inv./tithables
	1,074	Lower Norfolk	Deeds, E: 124–125. crop in inv./tithables
	1,378	Saint Mary's	R. Cole Accts. crop share
1673	1,600	Dorchester	Test. Proc., 5: 471. crop in inv.
1674	3,000	Westmoreland	WD, 1665–1671: 187. crop share
	1,270	Calvert	Test. Proc., 6: 224. crop in inv.
	1,000	Anne Arundel	IA, 1: 247. crop in inv.
1675	2,600	York	DOW, 5: 136. crop in inv., no labor
	1,334	Lancaster	Wills, 5: 7–8. crop share est. from inv.
1676	374	Lancaster	Wills, 5: 19–22. crop/4 tithables in inv.
	2,627	Lancaster	Wills, 5: 30–31. crop share
	1,702	Northampton	Orders, 10. crop share
	700	York	DOW, 6: 34. slave's crop share
	2,359	Saint Mary's	IA, 4: 402. crop in inv.
	1,550	Saint Mary's	IA, 4: 207. crop in inv.
	769	Somerset	Deeds, L: 76. crop share
1677	840	York	DOW, 6: 34. slave's share of crop
	1,241	Lancaster	Wills, 5: 41. crop share est. from inv.
	1,750	Talbot	IA, 4: 630. crop in inv.
1678	1,264	York	DOW, 6: 390. crop share est. from inv.
	1,232	York	DOW, 6: 172. crop share est. from inv.
	1,362	York	DOW, 6: 391. crop share est. from inv.
	2,600	Charles	Chas. Co. Ct., I1: 40. crop share
1679	2,500–3,500	Cecil	Danckaerts, *Journal,* 133. crop of seasoned servant or slave
	250	Lower Norfolk	Deeds, 4: 54b. prob. crop in inv., no labor
	1,200	Lancaster	Wills, 5: 57–58. crop in inv., no labor
	654	Northampton	Orders, 11: 27. crop in inv., no labor
	2,881	York	DOW, 6: 341. crop in inv., no labor
1681	766	Lower Norfolk	Deeds, 4: 102b. prob. crop in inv., no labor
1682	2,000	Charles	Chas. Co. Ct., K1: 201. crop share

YEAR	POUNDS	COUNTY	SOURCE, COMMENTS
	1,812	Anne Arundel	IA, 8: 45, 325. servant's share of crop
1683	1,865	York	DOW, 6: 486. crop share est. from inv.
1684	2,000	Lower Norfolk	Deeds, 4: 183a. prob. crop in inv., no labor
1684–1696	1,585	6 Md. counties	Main, *Tobacco Colony,* 39. est. from invs.
1685	2,743	Anne Arundel	IA, 10: 51. servant's share of crop
	2,200	Anne Arundel	IA, 10: 51. single planter's crop, no labor
1686	300	Lower Norfolk	Deeds, 5, part 1: 2. prob. crop in inv., no labor
	1,888	York	DOW, 7: 159. crop in inv., no labor
	2,584	Cecil	IA, 9: 191. crop in inv., no labor
	2,240	Saint Mary's	IA, 10: 179. crop in inv., no labor
	1,238	Md.	IA, 9: 157. crop in inv., no labor
1687	2,500	Gloucester	[Durand], *Huguenot Exile,* 120–121. crop share
	1,238	Lower Norfolk	Deeds, 5, part 1: 2. prob. crop in inv., no labor
	714	Lower Norfolk	Deeds, 5, part 1: 42b. prob. crop in inv., no labor
	1,347	York	DOW, 7: 310. freedman's share of crop
	2,323	York	DOW, 7: 331. crop in inv., no labor
1688	1,959	York	DOW, 8: 176. crop in inv., no labor
1689	1,582	Lancaster	Wills, 5: 129. crop share
1692	817	Lower Norfolk	Deeds, 5, part 1: 169. crop in inv., no labor
1694	1,400	York	DOW, 9: 332. crop in inv., no labor
	1,093	York	DOW, 10: 22, 84. crop in inv., no labor
1695	917	York	DOW, 10: 244. crop share est. from inv.
1696	594	York	DOW, 10: 346. inmate's crop
1697	1,000	York	DOW, 10: 456. crop share est. from inv.
	2,000	All Va.	*Present State of Virginia,* 18. typical crop share
	1,000–1,500	All Chesapeake	*CSP,* XV, 591. typical crop share
1697–1704	1,522	6 Md. counties	Main, *Tobacco Colony,* 39. share est. from invs.
1698	1,938	Calvert	IA, 16: 30. crop in inv., no labor
1699	550	York	DOW, 10: 177. crop in inv., no labor
	2,233	Charles	Chas. Co. Ct., X1: 310. crop share per hand
To 1699	1,207	Middlesex	Rutman & Rutman, *Explicatus,* 16. share est. from invs.
1660–1699	1,914	Anne Arundel	Earle, *Evolution,* 27. share est. from invs.
1700–1709	969	Middlesex	Rutman & Rutman, *Explicatus,* 16. share est. from invs.
1700–1709	2,239	Anne Arundel	Earle, *Evolution,* 27. share est. from invs.
1701	1,676	York	DOW, 11: 455. crop in inv., no labor
	1,867	Lancaster	Wills, 8: 104–105. crop share est. from inv.
	800–1,000	Charles	Chas. Co. Ct., Y1: 339. poor man's "small" crop

YEAR	POUNDS	COUNTY	SOURCE, COMMENTS
	1,370	Charles	Chas. Co. Ct., Y1: 143. servant's share of crop
1701–1713	1,570	Talbot	Clemens, *Atlan. Econ.,* 173. share est. from invs.
1702	1,500–2,000	All Va.	Michel, *VMHB,* XXIV, 31. average worker's crop
	616	York	DOW, 11: 504. crop in inv./tithables in inv.
	1,600	Charles	Chas. Co. Ct., A2: 144. slave's share of crop
	2,464	Charles	Chas. Co. Ct., A2: 149. inmate's share of crop
1703	2,880	Talbot	R. Goldsborough Accts. share per hand
1704	1,387	Lancaster	Wills, 8: 130. crop share est. from inv.
	759	Talbot	R. Goldsborough Accts. est. share per hand
1705–1712	1,609	6 Md. counties	Main, *Tobacco Colony,* 39. share est. from invs.
1705	1,234	York	DOW, 12: 290. crop in inv., no labor
	1,333	All Va.	Beverley, *History,* 263. clergyman's crop per hand
1706	600	Charles	Chas. Co. Ct., B2: 465. slave's share of crop
	1,142	Charles	Chas. Co. Ct., B2: 493. slave's share of crop
	1,400	Charles	Chas. Co. Ct., B2: 530. share per hand
	1,522	Talbot	R. Goldsborough Accts. share per hand
1707	2,424	Charles	Pr. Geo. Accts., JB1: 21. share per hand
	2,366	Charles	Pr. Geo. Accts., JB1: 21. share per hand
	1,579	Talbot	R. Goldsborough Accts. share per hand
	661	Stafford	Loyd/Tayloe Accts. slave's share of crop
1708	1,568	Talbot	R. Goldsborough Accts. est. share per hand
	1,400	Charles	Chas. Co. Ct., B2: 530. share per hand
1709	709	York	Guardian Accts. share per hand
	1,390	Talbot	R. Goldsborough Accts. share per hand
1710	1,044	Talbot	R. Goldsborogh Accts. est. share per hand
	469	Stafford	Loyd/Tayloe Accts. share per hand
1710–1719	881	Middlesex	Rutman & Rutman, *Explicatus,* 16. share est. from invs.
1710–1719	1,688	Anne Arundel	Earle, *Evolution,* 27. share est. from invs.
1711	3,303	Talbot	R. Goldsborough Accts. slave's share of crop
	1,220	Talbot	R. Goldsborough Accts. est. share per hand
1712	1,680	King William	Byrd, *Diary,* 532. slave's share of crop
	737	York	DOW, 14: 232. share of crop est. from inv.
	902	York	DOW, 14: 214. crop in inv./tithables in inv.
	1,316	Pr. George's	Pr. Geo. Accts. JB1: 21. freedman's crop
	2,656	Talbot	R. Goldsborough Accts. share per hand
	1,221	Talbot	R. Goldsborough Accts. share per hand
1713	1,306	Talbot	R. Goldsborough Accts. share per hand
	2,144	Talbot	R. Goldsborough Accts. share per hand

YEAR	POUNDS	COUNTY	SOURCE, COMMENTS
1713–1719	1,515	6 Md. counties	Main, *Tobacco Colony,* 39. share est. from invs.
1714	492	Stafford	Loyd/Tayloe Accts. share per hand
	3,528	Charles	IA 36B: 25. crop share
	635	Talbot	R. Goldsborough Accts. est. share per hand
1715	400	Caroline/King & Queen	J. Baylor Accts. share per hand
	1,383	Talbot	R. Goldsborough Accts. share per hand
	2,615	Talbot	R. Goldsborough Accts. share per hand
	1,129	Stafford	Loyd/Tayloe Accts. share per hand
1716	1,936	Talbot	R. Goldsborough Accts. share per hand
	1,253	Talbot	R. Goldsborough Accts. share per hand
	1,008	Stafford	Loyd/Tayloe Accts. share per hand
	780	Anne Arundel	J. Carroll Accts. share per hand
1717	791	York	DOW, 15: 72. share of crop est. from inv.
	726–817	Stafford	Loyd/Tayloe Accts. share per hand
	1,018	Anne Arundel	J. Carroll Accts. share per hand
	1,392	Talbot	R. Goldsborough Accts. share per hand
	1,853	Talbot	R. Goldsborough Accts. share per hand
1718	470	York	DOW, 15: 553. share of crop est. from inv.
	1,062	York	DOW, 15: 263. crop in inv., no labor
	664	York	DOW, 15: 409–412. share of crop est. from inv.
	1,150	Caroline/King & Queen	J. Baylor Accts. prob. crop share
	1,444	Anne Arundel	J. Carroll Accts. prob. crop share
	1,226	Talbot	R. Goldsborough Accts. share per hand
	1,038	Talbot	R. Goldsborough Accts. share per hand
1719	715	York	DOW, 15: 497. crop share
	584	York	DOW, 15: 553. share of crop est. from inv.
	574	York	DOW, 15: 554. share of crop est. from inv.
	1,052	York	DOW, 15: 418. share of crop est. from inv.
	446	York	DOW, 15: 435. share of crop est. from inv.
	918	James City	DOW, 15: 497. slave's share of crop
	480	Caroline/King & Queen	J. Baylor Accts. share per hand
	1,617	Caroline/King & Queen	J. Baylor Accts. share per hand
	594	Caroline/King & Queen	J. Baylor Accts. share per hand
	683	Caroline/King & Queen	J. Baylor Accts. est. share per hand
	1,004	Stafford	Loyd/Tayloe Accts. share per hand
	2,101	Talbot	R. Goldsborough Accts. est. share per hand
	2,033	Talbot	R. Goldsborough Accts. share per hand
1720	751	York	DOW, 15: 651. share of crop est. from inv.
	1,140	York	DOW, 15: 680. crop in inv., no labor
	627	York	DOW, 15: 683. share of crop est. from inv.
	850	York	DOW, 15: 577. share of crop est. from inv.
	378	York	DOW, 15: 645. crop in inv., no labor

YEAR	POUNDS	COUNTY	SOURCE, COMMENTS
	916	Essex	P. Morgan, diss. share of crop est. from inv.
	1,235	Caroline/King & Queen	J. Baylor Accts. share per hand
	792	Caroline/King & Queen	J. Baylor Accts. share per hand
	465	Caroline/King & Queen	J. Baylor Accts. est. share per hand
	1,429	Stafford	Loyd/Tayloe Accts. share per hand
	905	Talbot	R. Goldsborough Accts. est. share per hand
	1,364	Talbot	R. Goldsborough Accts. share per hand
1720–1724	1,460	Talbot	Clemens, *Atlan. Econ.,* 173. share est. from invs.
1720–1724	1,280	Kent	Clemens, *Atlan. Econ.,* 173. share est. from invs.
1720–1729	988	Middlesex	Rutman & Rutman, *Explicatus,* 16. share est. from invs.
1720–1729	1,613	Anne Arundel	Earle, *Evolution,* 27. share est. from invs.
1721	743	York	DOW, 16: 64. share of crop est. from inv.
	1,113	Caroline/King & Queen	J. Baylor Accts. prob. share per hand
	987	Caroline/King & Queen	J. Baylor Accts. share per hand
	1,235	Caroline/King & Queen	J. Baylor Accts. share per hand
	627	Caroline/King & Queen	J. Baylor Accts. share per hand
	740	Stafford	Loyd/Tayloe Accts. share per hand
	390	Talbot	R. Goldsborough Accts. share per hand
	1,158	Talbot	R. Goldsborough Accts. est. share per hand
1722	340	Caroline/King & Queen	J. Baylor Accts. share per hand
	683	Stafford	Loyd/Tayloe Accts. share per hand
	768	Talbot	R. Goldsborough Accts. share per hand
	774	Talbot	R. Goldsborough Accts. share per hand
1723	627	York	DOW, 16: 245. share of crop est. from inv.
	520	Stafford	Loyd/Tayloe Accts. share per hand
	996	Talbot	R. Goldsborough Accts. share per hand
	1,095	Talbot	R. Goldsborough Accts. est. share per hand
	1,745	Queen Anne's	R. Tilghman Accts. share per hand
	1,140	Queen Anne's	R. Tilghman Accts. share per hand
1724	909	York	DOW, 16: 292. share of crop est. from inv.
	250	York	DOW, 16: 320. inmate's crop
	750–1,000	Lancaster	R. Carter Diary. at 6,000 plants per hand
	476	Queen Anne's	R. Tilghman Accts. share per hand
	1,520	Queen Anne's	R. Tilghman Accts. share per hand
1725	1,278	Talbot	R. Goldsborough Accts. share per hand
	2,766	Queen Anne's	R. Tilghman Accts. share per hand
	990	Queen Anne's	R. Tilghman Accts. share per hand
1726	1,386	York	DOW, 16: 417. crop in inv., no labor
	555 ½	Essex	E. Bagge Accts. share per hand

YEAR	POUNDS	COUNTY	SOURCE, COMMENTS
	1,256	Talbot	R. Goldsborough Accts. share per hand
	1,605	Talbot	R. Goldsborough Accts. share per hand
	2,088	Queen Anne's	R. Tilghman Accts. share per hand
	1,087	Queen Anne's	R. Tilghamn Accts. share per hand
1727	525	York	DOW, 16: 453. share of crop est. from inv.
	594	York	DOW, 16: 456. share of crop est. from inv.
	822	York	DOW, 16: 464. share of crop est. from inv.
	1,757	Talbot	R. Goldsborough Accts. share per hand
	870	Talbot	R. Goldsborough Accts. est. share per hand
	2,316	Queen Anne's	R. Tilghman Accts. share per hand
1728	314	Essex	E. Bagge Accts. share per hand
	1,315	Somerset	Inventories, 13: 342. one slave's crop
	1,868	Talbot	R. Goldsborough Accts. share per hand
	941	Talbot	R. Goldsborough Accts. share per hand
	1,448	Talbot	R. Goldsborough Accts. prob. share per hand
	1,981	Queen Anne's	R. Tilghman Accts. share per hand
1729	321	Essex	E. Bagge Accts. share per hand
	346	Essex	E. Bagge Accts. share per hand
	522	Essex	Essex Co. Loose Papers. share per hand
	576	Essex	Essex Co. Loose Papers. share per hand
	769	York	DOW, 16: 587. share of crop est. from inv.
	385	York	DOW, 16: 587. share of crop est. from inv.
	709	York	Guardian Accts. share per hand
	885	Talbot	R. Goldsborough Accts. share per hand
	636	Talbot	R. Goldsborough Accts. share per hand
	318	Talbot	R. Goldsborough Accts. prob. share per hand
	691	Talbot	R. Goldsborough Accts. free black's share of crop
	954	Queen Anne's	R. Tilghman Accts. share per hand
	908	Queen Anne's	R. Tilghman Accts. share per hand
1730	500	All Va.	R. Carter LB. est. share per hand
	538	Essex	E. Bagge Accts. share per hand, slave overseer
	1,101	Talbot	R. Goldsborough Accts. share per hand, overseer's = 1
	861	Talbot	R. Goldsborough Accts. est. share per hand
	1,426	Queen Anne's	R. Tilghman Accts. share per hand
	1,290	Queen Anne's	R. Tilghman Accts. share per hand
	1,034	Queen Anne's	R. Tilghman Accts. share per hand
1730–1739	602	Middlesex	Rutman & Rutman, *Explicatus,* 16. share est. from invs.
1730–1739	1,707	Anne Arundel	Earle, *Evolution,* 27. share est. from invs.
1731	308	Essex	E. Bagge Accts. share per hand
	341	Essex	E. Bagge Accts. share per hand
	1,094	Talbot	R. Goldsborough Accts. share per hand, overseer's = 1

YEAR	POUNDS	COUNTY	SOURCE, COMMENTS
	1,255	Queen Anne's	R. Tilghman Accts. share per hand
	1,323	Queen Anne's	R. Tilghman Accts. share per hand
	930	Queen Anne's	R. Tilghman Accts. share per hand
	1,132	Queen Anne's	J. Hollyday Accts. share per hand, overseer's = 1
	1,212	Queen Anne's	J. Hollyday Accts. share per hand, overseer's = 1
1732	514	Essex	E. Bagge Accts. share per hand
	548	Essex	E. Bagge Accts. share per hand
	811	York	DOW, 17: 312. crop in inv., no labor
	86	York	Guardian Accts. share per hand
	294	New Kent	Bassett FP. share per hand
	947	New Kent	Bassett FP. share per hand
	674	New Kent	Bassett FP. share per hand
	1,000	King & Queen	"Hugh Grove." share per hand
	1,263	Queen Anne's	R. Tilghman Accts. share per hand
	1,977	Queen Anne's	R. Tilghman Accts. share per hand
1733	380	New Kent	Bassett FP. share per hand
	520	New Kent	Bassett FP. share per hand
	660	New Kent	Bassett FP. share per hand
	733	York	Essex Will Book, 5: 312. share per hand
	623	Essex	E. Bagge Accts. share per hand
	2,083	Somerset	Inventories. share per hand
	1,347	Queen Anne's	R. Tilghman Accts. share per hand
	287	Queen Anne's	R. Tilghman Accts. share per hand
1733–1737	1,440	Talbot	Clemens, *Atlan. Econ.,* 173. est. share per hand
	1,100	Kent	Clemens, *Atlan. Econ.,* 173. est. share per hand
1734	571	New Kent	Bassett FP. share per hand
	1,219	New Kent	Bassett FP. est. share per hand
	439	Gloucester	R. Carter LB. est. share per hand
	750	Prince George's	B. Waring Accts. prob. share per hand
	1,538	Queen Anne's	R. Tilghman Accts. 1 or 1 ½ shares
	2,073	Queen Anne's	R. Tilghman Accts. 1 or 1 ½ shares
1735	365	New Kent	Bassett FP. share per hand
	1,221	New Kent	Bassett FP. est. share per hand
	272	York	Guardian Accts. net share after tax + overseer's share
	1,033	Essex	Essex Will Book, 5: 402. crop in inv., no labor
	292	King William	Jones FP. share per hand, overseer's = 1
	821	King William	Jones FP. share per hand, overseer's = 1
	900	King William	Jones FP. share per hand, overseer's = 1
	1,047	King William	Jones FP. share per hand, overseer's = 1
	864	King William	Jones FP. share per hand, overseer's = 1
	536	King William	Jones FP. share per hand, overseer's = 1
	442	Gloucester	Carter FP, *VMHB,* XXIII, 162. share per hand

YEAR	POUNDS	COUNTY	SOURCE, COMMENTS
	766	Queen Anne's	R. Tilghman Accts. prob. share per hand
	1,525	Queen Anne's	R. Tilghman Accts. 1 or 1 ½ shares
	1,231	Kent	J. Gresham Accts. share per hand, overseer's = 1
1736	606	New Kent	Bassett FP. share per hand
	952	New Kent	Bassett FP. share per hand
	373	New Kent	Bassett FP. share per hand
	359	York	Guardian Accts. net share after tax + overseer's share
	1,255	Queen Anne's	R. Tilghman Accts. share per hand
	1,527	Kent	J. Gresham Accts. share per hand, overseer's = 1
1737	684	New Kent	Bassett FP. share per hand
	760	New Kent	Bassett FP. share per hand
	328	New Kent	Bassett FP. share per hand
	585	York	Guardian Accts. share per hand
	800	Caroline/King & Queen	J. Baylor Accts. est. share per hand, overseer's = 1
	1,566	Queen Anne's	R. Tilghman Accts. 1 or 1 ½ shares
	2,089	Queen Anne's	R. Tilghman Accts. 1 or 1 ½ shares
1738	880	New Kent	Bassett FP. share per hand
	1,080	New Kent	Bassett FP. share per hand
	1,031	New Kent	Bassett FP. share per hand
	568	York	Guardian Accts. share per hand
	672	York	Guardian Accts. share per hand
	112	York	Guardian Accts. share per hand
	888–1,183	York/James City	C. Burwell Accts. est. share per hand
	506–648	York/James City	C. Burwell Accts. est. share per hand
	594–792	York/James City	C. Burwell Accts. est. share per hand
	705	Caroline/King & Queen	J. Baylor Accts. share per hand
	2,337	Queen Anne's	R. Tilghman Accts. 1 or 1 ½ shares
	1,377	Kent	J. Gresham Accts. share per hand, overseer's = 1
1739	830	New Kent	Bassett FP. share per hand
	906	New Kent	Bassett FP. share per hand
	848	New Kent	Bassett FP. share per hand
	796 ½	New Kent	Bassett FP. share per hand
	222	York	Guardian Accts. net share after tax + overseer's share
	329	York	Guardian Accts. share per hand
	345	York	Guardian Accts. share per hand
	564–752	York/James City	C. Burwell Accts. est. share per hand
	221–294	York/James City	C. Burwell Accts. est. share per hand
	671–894	York/James City	C. Burwell Accts. est. share per hand
	961	Caroline/King & Queen	J. Baylor Accts. share per hand
	587	Caroline/King & Queen	J. Baylor Accts. share per hand
	572	Caroline/King & Queen	J. Baylor Accts. share per hand
1740	788	New Kent	Bassett FP. share per hand
	664	New Kent	Bassett FP. share per hand

YEAR	POUNDS	COUNTY	SOURCE, COMMENTS
	704	New Kent	Bassett FP. share per hand
	285	York	Guardian Accts. share per hand
	921	York	Guardian Accts. share per hand
	991–1,320	York/James City	C. Burwell Accts. est. share per hand
	699–932	York/James City	C. Burwell Accts. est. share per hand
	863	Caroline/King & Queen	J. Baylor Accts. share per hand
	513	Caroline/King & Queen	J. Baylor Accts. share per hand
	1,098	Queen Anne's	R. Tilghman Accts. share per hand
	1,042	Queen Anne's	R. Tilghman Accts. share per hand
1740–1745	1,420	Talbot	Clemens, *Atlan. Econ.,* 173. share est. from invs.
	1,440	Kent	Clemens, *Atlan. Econ.,* 173. share est. from invs.
1740–1747	1,863	Anne Arundel	Earle, *Evolution,* 27. share est. from invs.
1740–1749	778.1	Middlesex	Rutman & Rutman, *Explicatus,* 16. share est. from invs.
1741	788	New Kent	Bassett FP. share per hand
	842	New Kent	Bassett FP. share per hand
	810	New Kent	Bassett FP. share per hand
	314	York	Guardian Accts. net share after tax + overseer's share
	435	York	Guardian Accts. share per hand
	342	York	Guardian Accts. share per hand
	449	York	J. Bray Accts. share per hand, overseer's = 1
	851–1,134	York/James City	C. Burwell Accts. est. share per hand
	632–842	York/James City	C. Burwell Accts. est. share per hand
	620	Caroline/King & Queen	J. Baylor Accts. share per hand
	794	Caroline/King & Queen	J. Baylor Accts. share per hand
	501	Caroline/King & Queen	J. Baylor Accts. share per hand
	1,652	Saint Mary's	Accounts, 18: 375. servant's share of crop
	1,229	Queen Anne's	R. Tilghman Accts. share per hand
1742	875	New Kent	Bassett FP. share per hand
	889	New Kent	Bassett FP. share per hand
	162	York	Guardian Accts. net after taxes + overseer's share
	376	York	Guardian Accts. share per hand
	631	York	Guardian Accts. share per hand
	1,444	York	J. Bray Accts. share per hand, overseer's = 1
	909–1,212	York/James City	C. Burwell Accts. est. share per hand
	594–792	York/James City	C. Burwell Accts. est. share per hand
	550	Caroline	Beverley Family Accts. share per hand, overseer's = 1
	868	Caroline	Beverley Family Accts. share per hand, overseer's = 1
	1,125	Caroline/King & Queen	J. Baylor Accts. share per hand
	835	Caroline/King & Queen	J. Baylor Accts. share per hand
	647	Westmoreland	W. Robinson Accts. share per hand, overseer = 1

YEAR	POUNDS	COUNTY	SOURCE, COMMENTS
	526	Queen Anne's	R. Tilghman Accts. share per hand
1743	867	New Kent	Bassett FP. share per hand
	782	New Kent	Bassett FP. share per hand
	596	New Kent	Bassett FP. share per hand
	956	York	J. Bray Accts. share per hand, overseer's = 1
	802	York	J. Bray Accts. est. share per hand
	689	York	J. Bray Accts. share per hand, overseer's = 1
	282 ½	York	Guardian Accts. net after taxes + overseer's share
	843–1,124	York/James City	C. Burwell Accts. est. share per hand
	573–764	York/James City	C. Burwell Accts. est. share per hand
	566–754	York/James City	C. Burwell Accts. est. share per hand
	1,250–1,666	York/James City	C. Burwell Accts. est. share per hand
	571	Caroline	Beverley Family Accts. share per hand
	785	Caroline	Beverley Family Accts. share per hand
	854	Caroline	Beverley Family Accts. est. share per hand
	545	Caroline/King & Queen	J. Baylor Accts. share per hand
	844	Caroline/King & Queen	J. Baylor Accts. share per hand
	394	Caroline/King & Queen	J. Baylor Accts. share per hand
	1,256	King & Queen	Berkeley FP. share per hand, overseer's = 1
	1,214	Queen Anne's	R. Tilghman Accts. share per hand
	1,092	Queen Anne's	R. Tilghman Accts. share per hand
1744	1,163	New Kent	Bassett FP. share per hand
	982	New Kent	Bassett FP. share per hand
	518	New Kent	Bassett FP. est. share per hand
	706	York	J. Bray Accts. est. share per hand
	894–1,192	York/James City	C. Burwell Accts. est. share per hand
	573–764	York/James City	C. Burwell Accts. est. share per hand
	584	Caroline	Dixon Accts. share per hand, overseer's = 1
	663	Caroline	Beverley Family Accts. share per hand
	593	Caroline	Beverley Family Accts. share per hand
	1,010	Caroline/King & Queen	J. Baylor Accts. share per hand
	705	Caroline/King & Queen	J. Baylor Accts. share per hand
	1,062	Middlesex	Berkeley FP. est. share per hand
	154	Northumberland	Jones FP. est. share per hand
	893	Westmoreland	W. Robinson Accts. est. share per hand, overseer's = 1
	1,010	Westmoreland	W. Robinson Accts. share per hand, overseer's = 1
	834	Westmoreland	W. Robinson Accts. share per hand, overseer's = 1
	584	Prince William	Dixon Accts. share per hand, overseer's = 1
	954	Queen Anne's	R. Tilghman Accts. share per hand
	647+	Queen Anne's	R. Tilghman Accts. share per hand, omits ground leaves
1745	1,149	New Kent	Bassett FP. share per hand
	1,425 ½	New Kent	Bassett FP. est. share per hand
	915	New Kent	Bassett FP. share per hand

YEAR	POUNDS	COUNTY	SOURCE, COMMENTS
	1,482	New Kent	Bassett FP. est. share per hand
	750	York	J. Bray Accts. est. share per hand
	989–1,319	York/James City	C. Burwell Accts. est. share per hand
	1,004	Gloucester	Berkeley FP. share per hand, overseer's = 1
	1,323	Middlesex	Berkeley FP. share per hand, overseer's = 1
	663	King & Queen	Berkeley FP. share per hand, overseer's = 1
	667	Caroline/King & Queen	J. Baylor Accts. share per hand
	838	Caroline/King & Queen	J. Baylor Accts. share per hand
	963	Caroline	Beverley Family Accts. share per hand
	374	Caroline	Beverley Family Accts. share per hand
	683	Caroline	Beverley Family Accts. share per hand
	461	Caroline	Beverley Family Accts. share per hand
	883	Caroline	Beverley Family Accts. share per hand
	1,100	Northumberland	Jones FP. est. share per hand
	700	Prince George's	Pr. Geo. Accts., D: 6. slave's share of crop
	800	Queen Anne's	R. Tilghman Accts. share per hand
1746	1,740	New Kent	Bassett FP. est. share per hand
	1,395	New Kent	Bassett FP. est. share per hand
	1,135	New Kent	Bassett FP. share per hand
	507	York	Guardian Accts. share per hand
	1,222	King William	Berkeley FP. est. share per hand
	1,131	Gloucester	Berkeley FP. share per hand, overseer's = 1
	1,380	Middlesex	Berkeley FP. share per hand, overseer's = 1
	281	King & Queen	Berkeley FP. share per hand, overseer's = 1
	763	Caroline/King & Queen	J. Baylor Accts. share per hand
	624	Caroline	Beverley Family Accts. share per hand
	883	Caroline	Beverley Family Accts. share per hand
	693	Northumberland	Jones FP. est. share per hand
	881	Stafford	J. Mercer Accts. share per hand
	900	Prince George's	Pr. Geo. Accts., D: 6. prob. share per hand
	800	Queen Anne's	R. Tilghman Accts. share per hand
	1,828	Queen Anne's	R. Tilghman Accts. share per hand
1747	1,236	New Kent	Bassett FP. est. share per hand
	744	New Kent	Bassett FP. share per hand
	652	New Kent	W. Massie Accts. share per hand
	868	New Kent	W. Massie Accts. share per hand, overseer's = 1
	526	New Kent	W. Massie Accts. share per hand
	1,016	New Kent	W. Massie Accts. share per hand, overseer's = 1
	741	York	Guardian Accts. share per hand
	583–777	York/James City	C. Burwell Accts. est. share per hand
	632–842	York/James City	C. Burwell Accts. est. share per hand
	660	King William	Berkeley FP. est. share per hand
	869	Gloucester	Berkeley FP. share per hand, overseer's = 1

YEAR	POUNDS	COUNTY	SOURCE, COMMENTS
	689	Middlesex	Berkeley FP. share per hand, overseer's = 1
	287	King & Queen	Berkeley FP. est. share per hand, overseer's = 1
	601	Caroline	Dixon Accts. share per hand
	817	Northumberland	Jones FP. est. share per hand
	714	Stafford	Bristow LB. est. share per hand
	931	Stafford	J. Mercer Accts. share per hand
	858	Prince William/Fairfax	J. Mercer Accts. share per hand
	900	Prince William	Dixon Accts. share per hand, overseer's = 1
	704+	Queen Anne's	R. Tilghman Accts. share per hand, omits ground leaves
	1,037	Queen Anne's	R. Tilghman Accts. share per hand
	1,121	Dorchester	Ennalls Accts. est. share per hand
1748	640	New Kent	Bassett FP. share per hand
	573	New Kent	Bassett FP. share per hand
	900	New Kent	Bassett FP. share per hand
	1,228	New Kent	W. Massie Accts. share per hand, overseer's = 1
	511	York	Guardian Accts. share per hand
	939–1,252	York/James City	C. Burwell Accts. est. share per hand
	1,580	King William	Berkeley FP. est. share per hand, overseer's = 1
	1,263	Gloucester	Berkeley FP. share per hand, overseer's = 1
	931	Middlesex	Berkeley FP. share per hand, overseer's = 1
	698	Caroline	Dixon Accts. share per hand
	589	Caroline	Beverley Family Accts. share per hand
	775	King & Queen	Berkeley FP. share per hand, overseer's = 1
	756	Stafford	J. Mercer Accts. share per hand
	754	Prince William/Fairfax	J. Mercer Accts. est. share per hand
	832	Prince William/Fairfax	J. Mercer Accts. est. share per hand
	1,099	Prince William	Dixon Accts. est. share per hand, overseer's = 1
	1,600	Queen Anne's	R. Tilghman Accts. est. share per hand
	853	Dorchester	Ennalls Accts. est. share per hand
1749	675	New Kent	Bassett FP. share per hand
	900	New Kent	Bassett FP. share per hand
	499	York	Guardian Accts. est. share per hand
	564–752	York/James City	C. Burwell Accts. est. share per hand
	980–1,306	York/James City	C. Burwell Accts. est. share per hand
	975	Gloucester	Berkeley FP. est. share per hand, overseer's = 1
	1,249	Middlesex	Berkeley FP. share per hand, overseer's = 1
	1,070	King & Queen	Berkeley FP. share per hand, overseer's = 1
	1,007	Stafford	J. Mercer Accts. share per hand
	1,050	King George/ Prince William	Dixon Accts. share per hand
	597	King George/ Prince William	Dixon Accts. share per hand
	1,030	Prince William/Fairfax	J. Mercer Accts. share per hand

YEAR	POUNDS	COUNTY	SOURCE, COMMENTS
	978	Prince William/Fairfax	J. Mercer Accts. share per hand
	1,372	Queen Anne's	R. Tilghman Accts. share per hand
1750–1759	913	Anne Arundel	Earle, *Evolution,* 27. est. share per hand
1750	850	Stafford	J. Mercer Accts. share per hand
	1,205	Anne Arundel	S. Galloway Accts. share per hand
	1,000	Baltimore	Jesuit Accts. est. share per hand, overseer's = 1
	1,470	Queen Anne's	R. Tilghman Accts. share per hand
1751	889	Baltimore	Jesuit Accts. est. share per hand, overseer's = 1
	975	Queen Anne's	R. Tilghman Accts. share per hand
1752	820	King William	Berkeley FP. est. share per hand, overseer's = 1
	962	Gloucester	Berkeley FP. est. share per hand, overseer's = 1
	1,073	Middlesex	Berkeley FP. share per hand, overseer's = 1
	667	Baltimore	Jesuit Accts. est. share per hand, overseer's = 1
1753	800	Gloucester/Lancaster	Bristow LB. expected share per hand
	620	Gloucester/Lancaster	Bristow LB. est. share per hand
	639	King George/ Prince William	Dixon Accts. share per hand
	511	Baltimore	Jesuit Accts. est. share per hand, overseer's = 1
	751	Talbot	R. Goldsborough Accts. share per hand, overseer's = 1
1754	678	York/James City	C. Burwell Accts. share per hand
	769	Gloucester	Berkeley FP. share per hand, overseer's = 1
	639	Caroline	Dixon Accts. share per hand
	222	Baltimore	Jesuit Accts. est. share per hand, overseer's = 1
	537	Queen Anne's	J. Hollyday Accts. share per hand
1755	862	York/James City	C. Burwell Accts. share per hand
	645	King William	Jones FP. share per hand, overseer's = 1
	964	Middlesex	Berkeley FP. share per hand
	645	Caroline	Jones FP. share per hand, overseer's = 1
	1,006	Northern Neck	Jones FP. share per hand, overseer's = 1
	573	Queen Anne's	J. Hollyday Accts. share per hand
1756	658	Middlesex	R. Corbin Papers. slave's share of crop
	541	Middlesex	R. Corbin Papers. slave's share of crop
	1,327	Middlesex	Berkeley FP. share per hand
	944	King William	Jones FP. share per hand, overseer's = 1
	444	King William/Caroline	Jones FP. share per hand, overseer's = 1
	725	King George/ Prince William	Dixon Accts. share per hand
	667	Baltimore	Jesuit Accts. est. share per hand, overseer's = 1
	537	Queen Anne's	J. Hollyday Accts. share per hand
	1,436	Queen Anne's	R. Lloyd Accts. share per hand, overseer's = 1
	656	Queen Anne's	R. Lloyd Accts. share per hand, overseer's = 1
	737	Queen Anne's	R. Lloyd Accts. share per hand, overseer's = 1

YEAR	POUNDS	COUNTY	SOURCE, COMMENTS
	1,280	Queen Anne's	R. Lloyd Accts. share per hand, overseer's = 1
	1,171	Queen Anne's	R. Lloyd Accts. share per hand, overseer's = 1
1757	528	Henrico	Byrd Memo. Book. net crop after overseer's share
	95	Nansemond	A. Meade Accts. share per hand, overseer's = 1
	136	Nansemond	A. Meade Accts. share per hand, overseer's = 1
	711	King William	Jones FP. share per hand, overseer's = 1
	243	King William/Caroline	Jones FP. share per hand, overseer's = 1
	1,003	King William/Caroline	Jones FP. share per hand, overseer's = 1
	596	King William/Caroline	Jones FP. share per hand, overseer's = 1
	782	Caroline	Jones FP. share per hand, overseer's = 1
	984	Caroline	Dixon Accts. share per hand
	288	King George/ Prince William	Dixon Accts. share per hand
	684	King George/ Prince William	Dixon Accts. share per hand
	778	Baltimore	Jesuit Accts. est. share per hand, overseer's = 1
	613	Queen Anne's	J. Hollyday Accts. share per hand
	895	Queen Anne's	R. Lloyd Accts. est. share per hand
	810	Queen Anne's	R. Lloyd Accts. share per hand
	906	Queen Anne's	R. Lloyd Accts. share per hand
	1,178	Queen Anne's	R. Lloyd Accts. share per hand
	966	Queen Anne's	R. Lloyd Accts. share per hand
1758	618	Henrico	Byrd Memo. Book. est. share per hand
	667	King William/Caroline	Jones FP. share per hand, overseer's = 1
	781	Caroline	Jones FP. share per hand, overseer's = 1
	288	Caroline	Dixon Accts. share per hand
	684	Caroline	Dixon Accts. share per hand
	115	Caroline	Dixon Accts. share per hand
	879	Prince George's	C. Hill Accts. share per hand
	889	Baltimore	Jesuit Accts. est. share per hand, overseer's = 1
	489	Queen Anne's	J. Hollyday Accts. share per hand
	943	Queen Anne's	R. Lloyd Accts. est. share per hand
	604	Queen Anne's	R. Lloyd Accts. share per hand
	1,662	Queen Anne's	R. Lloyd Accts. share per hand
	708	Queen Anne's	R. Lloyd Accts. share per hand
1759	806	New Kent	D. P. Custis Estate. share per hand, overseer's = 1
	846	New Kent	D. P. Custis Estate. share per hand, overseer's = 1
	673	King William	D. P. Custis Estate. share per hand, overseer's = 1
	1,464	King William	D. P. Custis Estate. share per hand, overseer's = 1
	1,098	King William	R. Corbin Papers. share per hand
	676	King William	Jones FP. share per hand, overseer's = 1
	1,112	Middlesex	R. Corbin Papers. share per hand
	1,341	Middlesex	R. Corbin Papers. share per hand

YEAR	POUNDS	COUNTY	SOURCE, COMMENTS
	602	Caroline	Jones FP. share per hand, overseer's = 1
	596	King George	Dixon Accts. est. share per hand, overseer's = 1
	827	King George	Dixon Accts. est. share per hand, overseer's = 1
	584	King George	Dixon Accts. est. share per hand, overseer's = 1
	385	King George	Dixon Accts. est. share per hand, overseer's = 1
	931	Saint Mary's	R. Boarman Accts. prob. share per hand, overseer's = 1
	814	Prince George's	C. Hill Accts. est. share per hand
	908	Anne Arundel	J. Hall Accts. share per hand, overseer's = 1
	959	Baltimore	Jesuit Accts. est. share per hand, overseer's = 1
	864	Queen Anne's	J. Hollyday Accts. share per hand
	1,109	Queen Anne's	R. Lloyd Accts. share per hand
	1,234	Queen Anne's	R. Lloyd Accts. share per hand
1760–1769	1,032	Anne Arundel	Earle, *Evolution,* 27. est. share per hand
1760–1769	880	Talbot	Clemens, *Atlan. Econ.,* 173. share per slave
1760–1769	530	Kent	Clemens, *Atlan. Econ.,* 173. share per slave
1760	436	Middlesex	R. Corbin Papers. share per hand, overseer's = 1
	489	Middlesex	R. Corbin Papers. share per hand
	486	Middlesex	R. Corbin Papers. share per hand
	605	Middlesex	R. Corbin Papers. share per hand
	484	Middlesex	R. Corbin Papers. share per hand
	704	Middlesex	Jones FP. est. share per hand, overseer's = 1
	838	King William	R. Corbin Papers. share per hand
	704	King William	Jones FP. share per hand, overseer's = 1
	729	Caroline	Jones FP. share per hand, overseer's = 1
	791	Caroline	R. Corbin Papers. share per hand
	774	Caroline	R. Corbin Papers. share per hand
	725	Caroline	Dixon Accts. share per hand
	546	King & Queen	R. Corbin Papers. share per hand
	561	King & Queen	Berkeley FP. share per hand, overseer's = 1
	1,256	Prince William	D. Payne Accts. prob. share per hand
	826	King George	Dixon Accts. prob. share per hand, overseer's = 1
	582	King George	Dixon Accts. prob. share per hand, overseer's = 1
	725	King George	Dixon Accts. share per hand, overseer's = 1
	862	King George	Dixon Accts. prob. share per hand, overseer's = 1
	582	King George	Dixon Accts. prob. share per hand, overseer's = 1
	444	King George	Dixon Accts. share per hand, overseer's = 1
	1,721	Saint Mary's	R. Boarman Accts. prob. share per hand, overseer's = 1
	1,381	Queen Anne's	R. Tilghman Accts. prob. share per hand
	774	Queen Anne's	J. Hollyday Accts. share per hand
	954	Queen Anne's	R. Lloyd Accts. est. share per hand

YEAR	POUNDS	COUNTY	SOURCE, COMMENTS
	791	Queen Anne's	R. Lloyd Accts. share per hand
	1,552	Queen Anne's	R. Lloyd Accts. share per hand
	832	Queen Anne's	R. Lloyd Accts. share per hand
	1,381	Queen Anne's	R. Lloyd Accts. share per hand
1761	679	King William	Jones FP. share per hand, overseer's = 1
	930	King William	Jones FP. est. share per hand, overseer's = 1
	824	King William	R. Corbin Papers. share per hand
	792	Middlesex	R. Corbin Papers. share per hand
	728	Middlesex	R. Corbin Papers. share per hand
	758	Middlesex	R. Corbin Papers. share per hand
	607	Middlesex	R. Corbin Papers. share per hand
	734	Middlesex	R. Corbin Papers. share per hand
	566	Middlesex	R. Corbin Papers. share per hand
	1,054	Caroline	Jones FP. share per hand, overseer's = 1
	927	Caroline	R. Corbin Papers. share per hand
	765	Caroline	R. Corbin Papers. share per hand
	756	King & Queen	R. Corbin Papers. share per hand
	538	King & Queen	R. Corbin Papers. share per hand
	671	King & Queen	Berkeley FP. share per hand, overseer's = 1
	638	King George	Dixon Accts. prob. share per hand, overseer's = 1
	991	King George	Dixon Accts. prob. share per hand, overseer's = 1
	831	King George	Dixon Accts. prob. share per hand, overseer's = 1
	674	Northumberland	Jones FP. est. share per hand, overseer's = 1
	626	Prince William	D. Payne Accts. prob. share per hand
	788	Prince George's	C. Hill Accts. share per hand
	699	Queen Anne's	J. Hollyday Accts. share per hand
	1,209	Queen Anne's	R. Lloyd Accts. est. share per hand
	516	Queen Anne's	R. Lloyd Accts. share per hand
	1,831	Queen Anne's	R. Lloyd Accts. share per hand
	1,000	Queen Anne's	R. Lloyd Accts. share per hand
1762	169	York	Guardian Accts. est. share per hand
	868	King William	Jones FP. share per hand, overseer's = 1
	841	King William	Jones FP. est. share per hand, overseer's = 1
	1,030	King William	Jones FP. est. share per hand, overseer's = 1
	805	Caroline	Jones FP. est. share per hand, overseer's = 1
	659	Prince William	D. Payne Accts. prob. share per hand
	768	King George	Dixon Accts. prob. share per hand, overseer's = 1
	485	King George	Dixon Accts. share per hand, overseer's = 1
	402	King George	Dixon Accts. share per hand, overseer's = 1
	403	King George	Dixon Accts. est. share per hand, overseer's = 1
	849	King George	Dixon Accts. est. share per hand, overseer's = 1
	903	King George	Dixon Accts. est. share per hand, overseer's = 1
	831	King George	Dixon Accts. est. share per hand, overseer's = 1

YEAR	POUNDS	COUNTY	SOURCE, COMMENTS
	599	Northumberland	Jones FP. est. share per hand, overseer's = 1
	551	Saint Mary's	N. L. Sewall Accts. share per hand, overseer's = 1
	466	Saint Mary's	R. Boarman Accts. share per hand
	872	Queen Anne's	J. Hollyday Accts. share per hand
	617	Queen Anne's	R. Lloyd Accts. est. share per hand
	457	Queen Anne's	R. Lloyd Accts. share per hand
	1,427	Queen Anne's	R. Lloyd Accts. share per hand
	672	Queen Anne's	R. Lloyd Accts. share per hand
1763	818	York	Custis FP. share per hand
	287	York	Guardian Accts. est. share per hand
	769	King William	Jones FP. share per hand, overseer's = 1
	1,138	Middlesex	Berkeley FP. prob. share per hand
	485	Caroline	Dixon Accts. share per hand
	453	Northumberland	Jones FP. est. share per hand, overseer's = 1
	769	Northumberland	Jones FP. est. share per hand, overseer's = 1
	892	King George	Fitzhugh Papers. share per hand, overseer's = 1
	872	King George	Dixon Accts. est. share per hand, overseer's = 1
	844	King George	Dixon Accts. share per hand, overseer's = 1
	902	King George	Dixon Accts. prob. share per hand, overseer's = 1
	822	King George	Dixon Accts. prob. share per hand, overseer's = 1
	420	Stafford	R. W. Carter Diary. est. share per hand
	466	Saint Mary's	R. Boarman Accts. share per hand
	600	Anne Arundel	J. Hall Accts. share per hand, overseer's = 1
	606	Queen Anne's	R. Lloyd Accts. share per hand

APPENDIX 2 *Corn Crop Shares per Laborer*

YEAR	BUSHELS	COUNTY	SOURCE, COMMENT
1640	2.00	Saint Mary's	*Archives of Md.,* IV, 92. mate's share of crop
	4.00	Kent	*Archives of Md.,* IV, 92. share per hand
1642	5.50	Saint Mary's	*Archives of Md.,* IV, 330. share per hand
1647	16.00	Saint Mary's	*Archives of Md.,* IV, 501. servant's share of crop
1649	5.50	Saint Mary's	*Archives of Md.,* IV, 466. share per hand
1650	5.50	Northampton	DW, 4: 26. share per hand
	8.00	Northampton	DW, 4: 222. share per hand
1655	4.50	Northampton	DW, 5: 117. share est. from inv.
1656	9.00	York	DOW, 3: 9. share est. from inv.
1657	7.00	Calvert	*Archives of Md.,* X, 478–479. share per hand
1659	11.80	York	DOW, 3: 69r. share est. from inv.
1661	7.90	York	DOW, 4: 191. share est. from inv.
1665	4.00	York	DOW, 4: 56. share per hand
	10.30	York	DOW, 3. share per hand
	9.00	Northampton	Orders, 9: 24. crop in inv., no labor
1667	9.00	Lower Norfolk	Deeds, E: 22. crop in inv., no labor
1669	2.50	Northampton	Orders, 9: 61. share per hand
1671	4.50	York	DOW, 4: 318. crop in inv./tithables
1674	18.00	Westmoreland	WD, 1665–1677: 187. share per hand
1679	7.00	Northampton	Orders, 11: 27. crop in inv., no labor
1682	5.00	Charles	Test. Proc., 12B: 92–95. servant's share of crop
1685	7.80	Charles	Chas. Co. Ct., M1: 48. share per hand
	5.70	Anne Arundel	IA, 10: 51. servant's share of crop
1686	5.00	Lower Norfolk	Deeds, 5: 1–2. crop in inv., no labor
1687	13.30	Gloucester	[Durand], *Huguenot Exile,* 120–121. share per hand
1696	5.60	York	DOW, 10: 346. inmate's share of crop
1699	4.50	Lancaster	Wills, 8: 89. share est. from inv.
1701	4.50	Charles	Chas. Co. Ct., Y1: 143. share per hand
1702	6–8	All Va.	Michel, *VMHB,* XXIV, 31. average laborer's crop
	9.00	York	DOW, 12: 59. crop in inv./tithables
	6.30	York	DOW, 12: 59. share est. from inv.
	10.70	Charles	Chas. Co. Ct., A2: 149. share per hand
1705	14.00	York	DOW, 12: 401. crop in inv., no labor
	10.00	York	DOW, 12: 440. share per hand
1706	9.50	Charles	Chas. Co. Ct., B2: 465. share per hand
	9.00	Charles	Chas. Co. Ct., B2: 493. share per hand
1707	4.00	Stafford	Loyd/Tayloe Accts. share per hand
1708	5.00	York	DOW, 13: 123. share est. from inv.
	4.00	Charles	Chas. Co. Ct., B2: 493. share per hand
1712	9.00	York	DOW, 14: 232. share est. from inv.

YEAR	BUSHELS	COUNTY	SOURCE, COMMENT
1714	6.00	Charles	IA, 36B: 25. share per hand
1716	4.30	Stafford	Loyd/Tayloe Accts. share per hand
1717	8.00	Anne Arundel	J. Carroll Accts. share per hand, slave overseer
1718	11.60	York	DOW, 15: 409. share est. from inv.
	2.30	York	DOW, 15: 553. share est. from inv.
	7.00	Caroline/King & Queen	J. Baylor Accts. share per hand
1719	2.30	York	DOW, 15: 435. share est. from inv.
	2.30	York	DOW, 15: 553. share est. from inv.
	4.90	York	DOW, 15: 554. share est. from inv.
	4.00	York	DOW, 15: 683. share est. from inv.
	12.00	Lower Norfolk	Orders, 1719, b51 c1. share est. from inv.
	11.00	Caroline/King & Queen	J. Baylor Accts. share per hand
	7.00	Caroline/King & Queen	Baylor FP. share per hand
1720–1729	9.00	Talbot	Clemens, *Atlan. Econ.*, 175. share per taxable male in invs.
1720–1729	4.00	Kent	Clemens, *Atlan. Econ.*, 175. share per taxable male in invs.
1720	8.50	York	DOW, 15: 645. share est. from inv.
	4.70	York	DOW, 15: 648. share est. from inv.
	9.00	York	DOW, 15: 680. crop in inv., no labor
	4.00	York	DOW, 15: 683. share est. from inv.
	11.50	Caroline/King & Queen	J. Baylor Accts. share per hand
1721	8.25	Anne Arundel	J. Carroll Accts. share per hand
	5.50	Anne Arundel	J. Carroll Accts. share per hand
	7.00	Talbot	R. Goldsborough Accts. share per hand
	9.00	Talbot	R. Goldsborough Accts. share per hand
1722	12.00	Queen Anne's	R. Tilghman Accts. share per hand
	17.69	Talbot	R. Goldsborough Accts. share per hand
1723	10.00	York	DOW, 16: 292. share est. from inv.
	12.00	Queen Anne's	R. Tilghman Accts. share per hand
	6.30	Talbot	R. Goldsborough Accts. share per hand
1724	6.00	York	DOW, 16: 320. inmate's crop
	12.00	Queen Anne's	R. Tilghman Accts. share per hand
1726	8.30	Essex	E. Bagge Accts. share per hand
	11.60	Talbot	R. Goldsborough Accts. share per hand
1727	13.25	York	DOW, 16: 453. share est. from inv.
	5.10	York	DOW, 16: 463. share est. from inv.
	15.30	Talbot	R. Goldsborough Accts. share per hand
1728	8.40	York	DOW 16: 570. share est. from inv.
	7.60	York	DOW 16: 509. share est. from inv.
	6.00	York	Guardian Accts. share per hand
	15.90	Essex	E. Bagge Accts. share per hand, overseer's = 1
	20.00	Talbot	R. Goldsborough Accts. share per hand

YEAR	BUSHELS	COUNTY	SOURCE, COMMENT
	12.20	Talbot	R. Goldsborough Accts. share per hand
1729	10.00	York	DOW, 16: 587. share est. from inv.
	9.60	York	DOW, 16: 570. share est from inv.
	7.70	York	Guardian Accts. share per hand
	11.60	Essex	E. Bagge Accts. share per hand
	8.80	Essex	Essex Co. Loose Papers. share per hand
	11.80	Essex	Essex Co. Loose Papers. share per hand
	13.00	Talbot	R. Goldsborough Accts. share per hand
	23.00	Talbot	R. Goldsborough Accts. share per hand
1730–1739	9.00	Talbot	Clemens, *Atlan. Econ.,* 175. share per taxable male
1730–1739	7.00	Kent	Clemens, *Atlan. Econ.,* 175. share per taxable male
1730	4.60	York	Guardian Accts. share per hand
	6.90	Essex	E. Bagge Accts. share per hand
	11.00	Saint Mary's	Accounts, 10: 320–328. share est. from inv.
	22.40	Talbot	R. Goldsborough Accts. share per hand
	30.00	Queen Anne's	R. Tilghman Accts. share per hand
1731	4.60	York	Guardian Accts. share per hand
	23.30	Talbot	R. Goldsborough Accts. share per hand
1732	5.70	York	Guardian Accts. share per hand
	8.20	Essex	E. Bagge Accts. share per hand
	8.33	Essex	E. Bagge Accts. share per hand
	7.00	New Kent	Bassett FP. share per hand
1733	9.00	New Kent	Bassett FP. share per hand
	8.50	Prince George's	B. Waring Accts. share est. from inv.
	8.00	Somerset	Inventories, 20: 175. inmate's crop
1734	6.00	New Kent	Bassett FP. share per hand
1735	6.67	York	Guardian Accts. prob. surplus per hand
	9.00	New Kent	Bassett FP. share per hand
1736	3.86	York	Guardian Accts. share per hand
	17.00	Anne Arundel	Earle, *Evolution,* 127. share per hand, overseer's = 1
1737	5.00	York	Guardian Accts. prob. surplus per hand
	11.00	New Kent	Bassett FP. share per hand
	10.00	Caroline/King & Queen	J. Baylor Accts. share per hand
	10.00	Caroline/King & Queen	J. Baylor Accts. share per hand
1738	3.75	York	Guardian Accts. prob. surplus per hand
	8.33	York	Guardian Accts. share per hand
	9.00–12.00	York/James City	C. Burwell Accts. est. share per hand
	15.00	New Kent	Bassett FP. share per hand
	20.00	New Kent	Bassett FP. share per hand
	12.00	Caroline/King & Queen	J. Baylor Accts. share per hand
	14.60	Kent	J. Gresham Accts. share per hand

YEAR	BUSHELS	COUNTY	SOURCE, COMMENT
1739	4.00	York	Guardian Accts. share per hand
	10.00–13.60	York/James City	C. Burwell Accts. est. share per hand
	10.00–13.30	York/James City	C. Burwell Accts. est. share per hand
	17.00	New Kent	Bassett FP. share per hand
	7.00	Caroline/King & Queen	J. Baylor Accts. share per hand
	15.00	Caroline/King & Queen	J. Baylor Accts. share per hand
1740–1749	18.00	Talbot	Clemens, *Atlan. Econ.,* 175. share per taxable male
1740–1749	8.00	Kent	Clemens, *Atlan. Econ.,* 175. share per taxable male
1740	3.25	York	Guardian Accts. prob. surplus per hand
	2.67	York	Guardian Accts. prob. surplus per hand
	11.50–15.60	York/James City	C. Burwell Accts. est. share per hand
	7.50–10.00	York/James City	C. Burwell Accts. est. share per hand
	8.40–11.20	York/James City	C. Burwell Accts. est. share per hand
	17.50	New Kent	Bassett FP. share per hand
	14.00	Caroline/King & Queen	J. Baylor Accts. share per hand
	6.70	Caroline/King & Queen	J. Baylor Accts. share per hand
1741	2.75	York	Guardian Accts. prob. surplus per hand
	4.33	York	Guardian Accts. prob. surplus per hand
	11.50–15.30	York/James City	C. Burwell Accts. est. share per hand
	7.50–10.30	York/James City	C. Burwell Accts. est. share per hand
	18.50	New Kent	Bassett FP. share per hand
	16.50	New Kent	Bassett FP. share per hand
	10.30	Caroline/King & Queen	J. Baylor Accts. share per hand
	6.70	Caroline/King & Queen	J. Baylor Accts. share per hand
	15.50	Caroline/King & Queen	J. Baylor Accts. share per hand
	10.00	King & Queen	Berkeley FP. share per hand
1742	3.13	York	Guardian Accts. prob. surplus per hand
	22.30	York	J. Bray Accts. share per hand, overseer's = 1
	5.50–7.00	York/James City	C. Burwell Accts. est. share per hand
	3.00–4.00	York/James City	C. Burwell Accts. est. share per hand
	18.00	New Kent	Bassett FP. share per hand
	17.50	New Kent	Bassett FP. share per hand
	10.50	Caroline/King & Queen	J. Baylor Accts. share per hand
	9.00	Caroline/King & Queen	J. Baylor Accts. share per hand
	14.00	King & Queen	Berkeley FP. share per hand, overseer's = 1
	9.50	Queen Anne's	R. Tilghman Accts. net share less consumption
1743	9.50–12.70	York/James City	C. Burwell Accts. est. share per hand
	8.25–11.00	York/James City	C. Burwell Accts. est. share per hand
	6.70–8.90	York/James City	C. Burwell Accts. est. share per hand
	8.00	York	J. Bray Accts. est. share per hand, overseer's = 1
	17.20	New Kent	Bassett FP. share or share and ½

YEAR	BUSHELS	COUNTY	SOURCE, COMMENT
	14.00	King & Queen	Berkeley FP. share per hand, overseer's = 1
	6.70	Caroline/King & Queen	J. Baylor Accts. share per hand
	23.00	Caroline	Beverley Family Accts. share per hand, overseer's = 1
	12.00	Caroline	Beverley Family Accts. share per hand, overseer's = 1
	12.00	Caroline	Beverley Family Accts. share per hand, overseer's = 1
	13.00	Queen Anne's	R. Tilghman Accts. net share less consumption
	15.00	Queen Anne's	R. Tilghman Accts. net share less consumption
1744	2.00–2.30	York/James City	C. Burwell Accts. est. share per hand
	6.90–9.20	York/James City	C. Burwell Accts. est. share per hand
	5.40	York	J. Bray Accts. share per hand, overseer's = 1
	14.10	New Kent	Bassett FP. share per hand
	16.30	Middlesex	Berkeley FP. est. share per hand, overseer's = 1
	12.00	Caroline	Beverley Family Accts. share per hand, overseer's = 1
	8.00	Caroline/King & Queen	J. Baylor Accts. share per hand
	8.80	King & Queen	Berkeley FP. share per hand, overseer's = 1
	17.00	Northumberland	Jones FP. share per hand
	7.50	Queen Anne's	R. Tilghman Accts. net share less consumption
	8.00	Queen Anne's	R. Tilghman Accts. net share less consumption
	8.00	Queen Anne's	R. Tilghman Accts. net share less consumption
1745	10.00–13.60	York/James City	C. Burwell Accts. est. share per hand
	11.10–14.80	York/James City	C. Burwell Accts. est. share per hand
	18.70	New Kent	Bassett FP. share or share and ½
	39.60	Gloucester	Berkeley FP. share per hand, overseer's = 1
	10.00	Caroline	Beverley Family Accts. share per hand, overseer's = 1
	20.50	Queen Anne's	R. Tilghman Accts. share per hand
	8.00	Queen Anne's	R. Tilghman Accts. net share less consumption
1746	1.25	York	Guardian Accts. prob. surplus per hand
	7.00	York	Guardian Accts. share per slave
	10.20–13.60	York/James City	C. Burwell Accts. est. share per hand
	8.35–11.10	York/James City	C. Burwell Accts. est. share per hand
	5.00–6.60	York/James City	C. Burwell Accts. est. share per hand
	15.31	New Kent	Bassett FP. share or share and ½
	24.00	New Kent	W. Massie Accts. share per hand, overseer's = 1
	5.50	Stafford	J. Mercer Accts. share per hand
	6.50	Queen Anne's	R. Tilghman Accts. net share less consumption
	34.50	Queen Anne's	R. Tilghman Accts. share per hand
1747	6.40–8.50	York/James City	C. Burwell Accts. est. share per hand
	7.00–9.30	York/James City	C. Burwell Accts. est. share per hand
	19.50	New Kent	Bassett FP. est. share per hand
	21.30	New Kent	W. Massie Accts. share per hand, overseer's = 1
	18.00	New Kent	W. Massie Accts. share per hand
	5.70	New Kent	W. Massie Accts. share per hand
	15.90	Middlesex	Berkeley FP. share per hand, overseer's = 1

YEAR	BUSHELS	COUNTY	SOURCE, COMMENT
	4.50	Stafford	J. Mercer Accts. share per hand
	7.50	Prince William/Fairfax	J. Mercer Accts. share per hand
	2.20	Queen Anne's	R. Tilghman Accts. net share less consumption
	18.50	Queen Anne's	R. Tilghman Accts. share per hand
1748	11.00–14.70	York/James City	C. Burwell Accts. est. share per hand
	25.50	New Kent	Bassett FP. est. share per hand
	20.50	New Kent	Bassett FP. share per hand
	14.00	New Kent	W. Massie Accts. share per hand, overseer's = 1
	13.90	Gloucester	Berkeley FP. share per hand, overseer's = 1
	2.00–3.00	Lancaster	Bristow LB. average share per hand
	7.50	Prince William/Farifax	J. Mercer Accts. share per hand
	10.00	Prince William/Farifax	J. Mercer Accts. share per hand
	32.50	Queen Anne's	R. Tilghman Accts. share per hand
	9.50	Queen Anne's	R. Tilghman Accts. net share less consumption
1749	2.57	York	Guardian Accts. prob. surplus per hand
	17.00–22.70	York/James City	C. Burwell Accts. est. share per hand
	23.25	New Kent	Bassett FP. est. share per hand
	11.50	New Kent	Bassett FP. share per hand
	28.00	Queen Anne's	R. Tilghman Accts. share per hand
1750	15.00	New Kent	Bassett FP. est. share per hand
	12.30	Anne Arundel	S. Galloway Accts. share per hand, overseer's = 1
	16.00	Baltimore	Jesuit Accts. est. share per hand, overseer's = 1
	3.00	Queen Anne's	R. Tilghman Accts. net share less consumption
1751	9.40	New Kent	Bassett FP. share per hand
	13.30	Baltimore	Jesuit Accts. est. share per hand, overseer's = 1
	27.50	Queen Anne's	R. Tilghman Accts. net share less consumption
1752	10.00	Norfolk area	Stewart LB. est. crop per hand
	16.20	Baltimore	Jesuit Accts. est. share per hand, overseer's = 1
	9.00	Queen Anne's	R. Tilghman Accts. net share less consumption
1753	14.70	Baltimore	Jesuit Accts. est. share per hand, overseer's = 1
	12.60	Talbot	R. Goldsborough Accts. share per hand, overseer's = 1
1754	14.50	York/James City	C. Burwell Accts. est. share per hand
	15.10	Baltimore	Jesuit Accts. est. share per hand, overseer's = 1
	18.00	Talbot	R. Goldsborough Accts. share per hand, overseer's = 1
1755	12.00	York/James City	C. Burwell Accts. est. share per hand
	15.40	Middlesex	Berkeley FP. share per hand, overseer's = 1
	13.70	Baltimore	Jesuit Accts. est. share per hand, overseer's = 1
	10.20	Talbot	R. Goldsborough Accts. share per hand, overseer's = 1
1756	15.00	King William	Jones FP. share per hand, overseer's = 1
	15.00	Middlesex	Berkeley FP. share per hand, overseer's = 1
	10.00	Caroline	Jones FP. share per hand, overseer's = 1
	13.10	Baltimore	Jesuit Accts. est. share per hand, overseer's = 1
	28.66	Queen Anne's	R. Lloyd Accts. share per hand

YEAR	BUSHELS	COUNTY	SOURCE, COMMENT
1757	15.40	Nansemond	A. Meade Accts. share per hand, overseer's = 1
	11.70	King William	Jones FP. share per hand, overseer's = 1
	18.00	Caroline	Jones FP. share per hand, overseer's = 1
	16.30	Caroline	Jones FP. share per hand, overseer's = 1
	14.90	Baltimore	Jesuit Accts. est. share per hand, overseer's = 1
1758	35.75	King William/Caroline	Jones FP. est. share per hand, overseer's = 1
	16.58	Caroline	Jones FP. est. share per hand, overseer's = 1
	18.00	Caroline	Jones FP. est. share per hand, overseer's = 1
	12.00	King & Queen	Berkeley FP. share per hand, overseer's = 1
	14.30	Baltimore	Jesuit Accts. est. share per hand, overseer's = 1
	20.50	Queen Anne's	R. Lloyd Accts. share per hand
	22.00	Queen Anne's	R. Tilghman Accts. prob. share per hand
1759	8.66	York	Guardian Accts. est. share per hand
	18.30	York	Guardian Accts. est. share per hand
	7.00	York	D. P. Custis Estate. share per hand, overseer's = 1
	6.40	York	D. P. Custis Estate. share per hand, overseer's = 1
	11.10	New Kent	D. P. Custis Estate. share per hand, overseer's = 1
	12.40	New Kent	D. P. Custis Estate. share per hand, overseer's = 1
	12.50	New Kent	D. P. Custis Estate. share per hand, overseer's = 1
	17.60	King William	D. P. Custis Estate. share per hand, overseer's = 1
	24.00	King William	Jones FP. est. share per hand, overseer's = 1
	18.10	King William	Jones FP. est. share per hand, overseer's = 1
	26.00	Caroline	Jones FP. est. share per hand, overseer's = 1
	14.00	Caroline	Jones FP. est. share per hand, overseer's = 1
	8.50	Anne Arundel	H. Hall Accts. share per hand, overseer's = 1
	15.80	Baltimore	Jesuit Accts. est. share per hand, overseer's = 1
	23.00	Queen Anne's	R. Lloyd Accts. share per hand
	17.00	Queen Anne's	R. Lloyd Accts. share per hand
	39.50	Queen Anne's	R. Lloyd Accts. share per hand
1760–1769	14.00	Talbot	Clemens, *Atlan. Econ.,* 175. share per taxable male
	9.00	Kent	Clemens, *Atlan. Econ.,* 175. share per taxable male
1760	18.30	York	Guardian Accts. est. share per hand
	33.30	King William	Jones FP. est. share per hand, overseer's = 1
	32.00	King William	Jones FP. share per hand, overseer's = 1
	4.70	King William	R. Corbin Papers. share per hand
	7.50	King William	R. Corbin Papers. share per hand
	12.63	Caroline	R. Corbin Papers. share per hand
	3.00	Caroline/King George	R. Corbin Papers. share per hand
	18.10	Caroline	Jones FP. est. share per hand, overseer's = 1
	17.30	King & Queen	R. Corbin Papers. share per hand
	4.30	Middlesex	R. Corbin Papers. share per hand
	10.60	Middlesex	R. Corbin Papers. share per hand

YEAR	BUSHELS	COUNTY	SOURCE, COMMENT
	9.40	Middlesex	R. Corbin Papers. share per hand
	6.10	Middlesex	R. Corbin Papers. share per hand
	12.20	Saint Mary's	R. Boarman Accts. share per hand, overseer's = 1
	19.20	Baltimore	Jesuit Accts. est. share per hand, overseer's = 1
	16.00	Queen Anne's	R. Tilghman Accts. net share less consumption
	19.00	Queen Anne's	R. Lloyd Accts. share per hand
	41.50	Queen Anne's	R. Lloyd Accts. share per hand
1761	14.50	York	Guardian Accts. est. share per hand
	18.30	York	Guardian Accts. est. share per hand
	9.80	King William	R. Corbin Papers. share per hand
	5.40	King William	R. Corbin Papers. share per hand
	16.30	Caroline	R. Corbin Papers. share per hand
	16.70	Caroline	R. Corbin Papers. share per hand
	24.70	Caroline	Jones FP. share per hand, overseer's = 1
	8.00	King & Queen	R. Corbin Papers. share per hand
	8.50	King & Queen	R. Corbin Papers. share per hand
	8.00	Middlesex	R. Corbin Papers. share per hand
	10.20	Middlesex	R. Corbin Papers. share per hand
	9.70	Middlesex	R. Corbin Papers. share per hand
	4.50	Middlesex	R. Corbin Papers share per hand
	10.50	Middlesex	R. Corbin Papers. share per hand
	9.10	Middlesex	R. Corbin Papers. share per hand
	25.40	Prince George's	B. Waring Accts. est. share per hand
	12.20	Baltimore	Jesuit Accts. est. share per hand, overseer's = 1
	21.80	Queen Anne's	R. Lloyd Accts. share per hand, overseer's = 1
1761 or 1762	10.30	Middlesex	Grymes Estate. share per hand
	9.20	Middlesex	Grymes Estate. share per hand
	10.40	Middlesex	Grymes Estate. share per hand
	11.00	Middlesex	Grymes Estate. share per hand
	6.20	Middlesex	Grymes Estate. share per hand
	5.60	Middlesex	Grymes Estate. share per hand
	12.60	Middlesex	Grymes Estate. share per hand
1762	8.80	York River Basin	Custis FP. share per hand
	25.50	York	Guardian Accts. est. share per hand
	32.00	King William	Jones FP. share per hand, overseer's = 1
	17.50	King William	Jones FP. share per hand, overseer's = 1
	32.30	Prince George's	B. Waring Accts. est. share per hand
	19.10	Baltimore	Jesuit Accts. share per hand, overseer's = 1
	31.00	Queen Anne's	R. Lloyd Accts. share per hand
	30.10	Queen Anne's	R. Lloyd Accts. share per hand
1762–1765	10.30	Middlesex	Grymes Estate. share per hand

YEAR	BUSHELS	COUNTY	SOURCE, COMMENT
	9.20	Middlesex	Grymes Estate. share per hand
	10.40	Middlesex	Grymes Estate. share per hand
	11.00	Middlesex	Grymes Estate. share per hand
	6.20	Middlesex	Grymes Estate. share per hand
	5.60	Middlesex	Grymes Estate. share per hand
1763	16.00	York	Guardian Accts. est. share per hand
	8.80	York	Custis FP. share per hand, overseer's = 1
	19.00	King William	Jones FP. share per hand, overseer's = 1
	15.30	Middlesex	Berkeley FP. est. share per hand, overseer's = 1
	15.10	Middlesex	Berkeley FP. est. share per hand, overseer's = 1
	14.30	Middlesex	Berkeley FP. share per hand, overseer's = 1
	15.00	King & Queen	Berkeley FP. share per hand
	9.00	King George	Fitzhugh FP. share per hand, overseer's = 1
	16.25	King George	Fitzhugh FP. share per hand, overseer's = 1
	21.30	Northumberland	Jones FP. est. share per hand, overseer's = 1
	17.80	Saint Mary's	N. L. Sewall Accts. share per hand, overseer's = 1
	29.90	Saint Mary's	N. L. Sewall Accts. share per hand, overseer's = 1
	30.90	Prince George's	B. Waring Accts. est. share per hand
	17.80	Baltimore	Jesuit Accts. est. share per hand, overseer's = 1

APPENDIX 3 *Wheat Crop Shares per Laborer*

YEAR	BUSHELS	COUNTY	SOURCE, COMMENT
1719	3.5	Essex	Essex Co. Loose Papers. share per hand
1730–1739	5	Talbot	Clemens, *Atlan. Econ.*, 175. output per taxable male
1730–1739	19	Kent	Clemens, *Atlan. Econ.*, 175. output per taxable male
1730	19.1	Talbot	R. Goldsborough Accts. share per hand, overseer's = 1
1736	2.29	York	Guardian Accts. share per hand
	3.4	Kent	J. Gresham Accts. share per hand
1737	6.8	Kent	J. Gresham Accts. share per hand
1738	13.8	York	Guardian Accts. share per hand
	3.57	York	Guardian Accts. share per hand
	13.3	Kent	J. Gresham Accts. share per hand
1740	2	York	Guardian Accts. net less consumption
	4	York	Guardian Accts. share per hand
1740–1749	4	Talbot	Clemens, *Atlan. Econ.*, 175. output per taxable male
	37	Kent	Clemens, *Atlan. Econ.*, 175. output per taxable male
1741	4.83	York	Guardian Accts. share per hand
1743	8	Caroline	Beverley Family Accts. share per hand, overseer's = 1
	5	Caroline	Beverley Family Accts. share per hand, overseer's = 1
	4	Queen Anne's	R. Tilghman Accts. net less consumption
1744	3.6	York	J. Bray Accts. share per hand, overseer's = 1
1746	1.25	York	Guardian Accts. share per hand
	11.5	Queen Anne's	R. Tilghman Accts. net less consumption
1747	1.11	York	Guardian Accts. share per hand
	3.5	Queen Anne's	R. Tilghman Accts. net less consumption
1748	11	Queen Anne's	R. Tilghman Accts. net less consumption
1749	5.57	York	Guardian Accts. est. share per hand
1750	4.8	Anne Arundel	S. Galloway Accts. share per hand, overseer's = 1
	5.6	Baltimore	Jesuit Accts. est. share per hand, overseer's = 1
1751	4.4	Baltimore	Jesuit Accts. est. share per hand, overseer's = 1
1752	4.4	Baltimore	Jesuit Accts. est. share per hand, overseer's = 1
1753	4.4	Baltimore	Jesuit Accts. est. share per hand, overseer's = 1
1754	6.2	Baltimore	Jesuit Accts. est. share per hand, overseer's = 1
1755	3.3	Baltimore	Jesuit Accts. est. share per hand, overseer's = 1
	14	Talbot	R. Goldsborough Accts. share per hand
1756	3.3	Baltimore	Jesuit Accts. est. share per hand, overseer's = 1
	76	Queen Anne's	R. Lloyd Accts. est. share per hand
	12.5	Queen Anne's	R. Lloyd Accts. share per hand
	49.5	Queen Anne's	R. Lloyd Accts. share per hand

YEAR	BUSHELS	COUNTY	SOURCE, COMMENT
	22	Queen Anne's	J. Hollyday Accts. share per hand
1757	4.8	Nansemond	A. Meade Accts. share per hand, overseer's = 1
	31	Queen Anne's	R. Lloyd Accts. share per hand
	20.3	Queen Anne's	J. Hollyday Accts. share per hand
	20	Queen Anne's	R. Lloyd Accts. share per hand
1758	6.1	King William	Jones FP. share per hand, overseer's = 1
	15	King William/Caroline	Jones FP. share per hand, overseer's = 1
	4	Caroline	Jones FP. share per hand, overseer's = 1
	20	Queen Anne's	R. Lloyd Accts. share per hand
	16.3	Queen Anne's	J. Hollyday Acct. share per hand
1759	9.4	New Kent	D. P. Custis Estate. share net of seed, overseer's = 1
	8.9	New Kent	D. P. Custis Estate. share net of seed, overseer's = 1
	9.1	King William	D. P. Custis Estate. share net of seed, overseer's = 1
	12	King William	Jones FP. share per hand, overseer's = 1
	11.2	Caroline	Jones FP. share per hand, overseer's = 1
	4	Saint Mary's	R. Boarman Accts. share per hand, overseer's = 1
	3.1	Baltimore	Jesuit Accts. est. share per hand, overseer's = 1
	54	Queen Anne's	R. Lloyd Accts. share per hand
	87	Queen Anne's	R. Lloyd Accts. share per hand
	59.3	Queen Anne's	J. Hollyday Accts. share per hand
1760–1769	24	Talbot	Clemens, *Atlan. Econ.,* 175. share per taxable male
1760–1769	54	Talbot	Clemens, *Atlan. Econ.,* 194. share per slave
1760–1769	89	Kent	Clemens, *Atlan. Econ.,* 175. share per taxable male
1760–1769	93	Kent	Clemens, *Atlan. Econ.,* 194. share per slave
1760	6	King William	Jones FP. share per hand, overseer's = 1
	13	King William	Jones FP. share per hand, overseer's = 1
	10	King William	Jones FP. share per hand, overseer's = 1
	8.6	Middlesex	R. Corbin Papers. share per slave
	5.15	Caroline	Jones FP. share per hand, overseer's = 1
	9	Caroline	Jones FP. share per hand, overseer's = 1
	7	Saint Mary's	R. Boarman Accts. share per hand, overseer's = 1
	37	Queen Anne's	R. Lloyd Accts. share per hand
	27.4	Queen Anne's	R. Lloyd Accts. est. share per hand
	14.7	Queen Anne's	J. Hollyday Accts. share per hand
1761	1.5	York	Guardian Accts. est. share per hand
	13	Middlesex	R. Corbin Papers. share per slave
	2.9	Middlesex	R. Corbin Papers. est. share per slave
	5.6	Middlesex	R Corbin Papers. est. share per slave
	7.3	Caroline	Jones FP. share per hand, overseer's = 1

YEAR	BUSHELS	COUNTY	SOURCE, COMMENT
	25	Queen Anne's	R. Lloyd Accts. share per hand
	40	Queen Anne's	J. Hollyday Accts. share per hand
1762	0.7	York	Guardian Accts. est. share per hand
	7	King William	Jones FP. est. share per hand, overseer's = 1
	13	King William	Jones FP. est. share per hand, overseer's = 1
	7	Northumberland	Jones FP. est. share per hand, overseer's = 1
	13	Saint Mary's	N. L. Sewall Accts. share per hand, overseer's = 1
	7.9	Prince George's	B. Waring Accts. est. share per hand
	43	Queen Anne's	R. Lloyd Accts. share per hand
	25.9	Queen Anne's	R. Lloyd Accts. share per hand
	30.1	Queen Anne's	R. Lloyd Accts. share per hand
	31.3	Queen Anne's	J. Hollyday Accts. share per hand
1763	9	Northumberland	Jones FP. est. share per hand, overseer's = 1
	12.5	Fairfax	G. Washington Papers. share per hand, overseer's = 1
	7.8	Fairfax	G. Washington Papers. share per hand, overseer's = 1
	17.3	Fairfax	G. Washington Papers. share per hand, overseer's = 1
	14.5	Saint Mary's	N. L. Sewall Accts. share per hand, overseer's = 1
	6.4	Prince George's	B. Waring Accts. est. share per hand
	22.2	Baltimore	Jesuit Accts. est. share per hand, overseer's = 1
	36	Queen Anne's	J. Hollyday Accts. est. share per hand

Sources

A. Meade Accts. Andrew Meade Account Book, 1758–1795, MS 3961, University of Virginia, Charlottesville, microfilm, M-62.2, John D. Rockefeller, Jr., Library, Colonial Williamsburg Foundation, Williamsburg, Va.

Accounts. Prerogative Court Accounts, 1718–1777, Maryland State Archives, Annapolis.

Ames, *County Court Records.* Susie M. Ames, *County Court Records of Accomack-Northampton, Virginia, 1640–1645* (Charlottesville, Va., 1973).

Archives of Maryland. William Hand Browne et al., eds., *Archives of Maryland,* 72 vols. (Baltimore, 1883–1972).

B. Waring Accts. Basil Waring Notebook, 1732–1743, Basil Waring Account Book, 1761–1765, Waring Collection, Maryland State Archives, Annapolis.

Bassett FP. William Bassett Account Book, 1728–1755, MS1 B2944 1a, and Elizabeth Bassett Accounts, 1744–1754, MS1 B2944 a8–16, Bassett Family Papers, Virginia Historical Society, Richmond.

Berkeley FP. Berkeley Family Papers, MS 38–113, microfilm, Records of Ante Bellum Southern Plantations, Series E, pt. 2, University of Virginia, Charlottesville.

Beverley, *History.* Robert Beverley, *The History and Present State of Virginia* (1705), ed. Louis B. Wright (Chapel Hill, N.C., 1947).

Beverley Family Accts. William Beverley Letterbook, 1737–1744/5, Account Book of Estate of Henry Beverley, 1743–1744, Rents, 1742, 1748, Accounts, 1743, Beverley Family Papers, 1654–1901, MS, New York Public Library, microfilm, M-1334, John D. Rockefeller, Jr., Library, Colonial Williamsburg Foundation, Williamsburg, Va.; Account Book, 1747 (Mss1B4678a 4782), Beverley Family Papers, Virginia Historical Society, Richmond.

Bristow LB. Robert Bristow Account and Letterbook, 1688–1750, MS, the Library of Virginia, Richmond, microfilm, M-1164, John D. Rockefeller, Jr., Library, Colonial Williamsburg Foundation, Williamsburg, Va.; Robert Bristow Letterbook, 1747–1770, PRO/A.O. 13/84, microfilm, M-494, John D. Rockefeller, Jr., Library, Colonial Williamsburg Foundation, Williamsburg, Va.

Bullock, *Virginia.* William Bullock, *Virginia Impartially Examined* . . . (London, 1649).

Byrd, *Diary.* Louis B. Wright and Marion Tinling, eds., *The Secret Diary of William Byrd of Westover, 1709–1712* (Richmond, Va., 1941).

Byrd Memo. Book. William Byrd III, Memorandum Book, 1757–1758, Library of Congress, Washington, D.C., microfilm, M-97, John D. Rockefeller, Jr., Library, Colonial Williamsburg Foundation, Williamsburg, Va.

C. Burwell Accts. Carter Burwell Ledger, 1738–1755, Burwell Family Papers, MS 64.4, John D. Rockefeller, Jr., Library, Colonial Williamsburg Foundation, Williamsburg, Va.

C. Hill Accts. Clement Hill Memorandum Book, 1756–1762, Clement Hill Papers, MS 446, Maryland Historical Society, Baltimore.

Calvert Papers. *The Calvert Papers,* 3 vols., Maryland Historical Society Fund Publication, nos. 28, 34, 35 (Baltimore, 1889).

Carter Diary. Robert Carter Diary, 1722–1727, Robert ("King") Carter Papers, MS 3807, University of Virginia, Charlottesville.

Carter FP, *VMHB.* Carter Family Papers, in "Letters and Other Papers, 1735–1829, from the Collections of the Virginia Historical Society," *Virginia Magazine of History and Biography,* XXIII (1915), 162–192.

Carter LB. Robert Carter Letterbook, 1728–1731, Special Collections, University of Virginia, Charlottesville.

Chas. Co. Ct. Charles County, Maryland, Court Records, I1, K1, M1, X1, Y1, A2, B2, Maryland State Archives, Annapolis.

Clarendon Manuscripts. Clarendon Manuscripts [Edward Hyde, first Earl of Clarendon], Bodleian Library, University of Oxford, cited in Edmund S. Morgan, *American Slavery, American Freedom: The Ordeal of Colonial Virginia* (New York, 1975), 142–143.

Clemens. *Atlan. Econ.* Paul G. E. Clemens, *The Atlantic Economy and Colonial Maryland's Eastern Shore: From Tobacco to Grain* (Ithaca, N.Y., 1980).

CSP. Great Britain, Public Record Office, *Calendar of State Papers,* Colonial Series, *America and West Indies,* ed. W. Noel Sainsbury et al. (London, 1864–).

Custis FP. Custis Papers, 1711–1792, Virginia Historical Society, Richmond, microfilm, M-58.1, M-1073.3, John D. Rockefeller, Jr., Library, Colonial Williamsburg Foundation, Williamsburg, Va.

D. P. Custis Estate. Donald Jackson, ed., *The Diaries of George Washington,* I (Charlottesville, Va., 1976); Joseph E. Fields, comp., *"Worthy Partner": The Papers of Martha Washington* (Westport, Conn., 1994); Custis Papers, 1711–1792, Virginia Historical Society, Richmond, microfilm, M-58.1, M-1073.3, John D. Rockefeller, Jr., Library, Colonial Williamsburg Foundation, Williamsburg, Va.

D. Payne Accts. Daniel Payne Account Book, Dumfries, Virginia, MS 635, Maryland Historical Society, Baltimore, microfilm, M-1580, John D. Rockefeller, Jr., Library, Colonial Williamsburg Foundation, Williamsburg, Va.

Danckaerts, *Journal.* Jasper Danckaerts, *Journal of Jasper Danckaerts, 1679–1680,* ed. Bartlett Burleigh James and J. Franklin Jameson, Original Narratives of Early American History (New York, 1913).

Deeds. Lancaster County, Virginia, Deeds, 2, 1654–1702, microfilm, the Library of Virginia, Richmond.

Deeds. Lower Norfolk County, Virginia, Deeds and Wills, B, 1646–1651, C, 1651–1656, E, 1666–1675, 4, 1675–1686, 5, 1686–1695, microfilm, the Library of Virginia, Richmond.

Deeds. Somerset County, Maryland, L, 1691–1697, Maryland State Archives, Annapolis.

Dixon Accts. Dixon Accounts. Edward Dixon Papers, 1743–1808, Library of Congress, Washington, D.C., microfilm, M-1310.1–8, John D. Rockefeller, Jr., Library, Colonial Williamsburg Foundation, Williamsburg, Va.

DO. Northumberland County, Virginia, Deeds and Orders, 1650–1652, microfilm, the Library of Virginia, Richmond.

DOW. York County, Virginia, Deeds, Orders, Wills, 2–16, 1645–1729, microfilm, the Library of Virginia, Richmond.

[Durand], *Huguenot Exile.* [Durand, of Dauphine], *A Huguenot Exile in Virginia; or, Voyages of a Frenchman Exiled for His Religion; with a Description of Virginia and Maryland, from the Hague Edition of 1687,* ed. Gilbert Chinard (New York, 1934).

E. Bagge Accts. Edmund Bagge Account Book, 1726–1733, John D. Rockefeller, Jr., Library, Colonial Williamsburg Foundation, Williamsburg, Va.

Earle, *Evolution.* Carville V. Earle, *The Evolution of a Tidewater Settlement System: All Hallow's Parish, Maryland, 1650–1783* (Chicago, 1975).

Ennalls Accts. Joseph Ennalls Account Book, 1729–1762, MS 937, Maryland Historical Society, Baltimore.

Essex Co. Loose Papers. Essex County, Virginia, Loose Papers, Essex County Court, Tappahannock, Va., box 107.

Essex Will Book. Essex County, Virginia, Will Book, 5, 1730–1735, microfilm, the Library of Virginia, Richmond.

Fitzhugh Papers. Henry Fitzhugh Ledger, 1747–1789, Henry Fitzhugh Papers, 1746–1786, Duke University Library, Durham, N.C., microfilm, M-1886.29, John D. Rockefeller, Jr., Library, Colonial Williamsburg Foundation, Williamsburg, Va.

Force Tracts. Peter Force, comp., *Tracts and Other Papers, Relating Principally to the Origin, Settlement, and Progress of the Colonies in North America, from the Discovery of the Country to the Year 1776,* 4 vols. (1836; rpt. Gloucester, Mass., 1963).

G. Washington Papers. George Washington Papers, 1741–1799, Library of Congress, Washington, D.C.

Grymes Estate. Phillip Grymes Estate Accounts, Harrison Family Papers, Virginia Historical Society, Richmond, microfilm, 1073.3, John D. Rockefeller, Jr., Library, Colonial Williamsburg Foundation, Williamsburg, Va.

Guardian Accts. York County, Virginia, Guardian Accounts, microfilm, the Library of Virginia, Richmond.

H. Hall Accts. Henry Hall Account Book, 1735–1777, William Henry Hall Papers, 1736–1862, MS 2267, Special Collections, Duke University Library, Durham, N.C.

"Hugh Grove." Gregory A. Stiverson and Patrick H. Butler III, eds., "Virginia in 1732: The Travel Journal of William Hugh Grove," *Virginia Magazine of History and Biography,* LXXXV (1977), 18–44.

IA. Prerogative Court Inventories and Accounts, 1674–1718, Maryland State Archives, Annapolis.

Inventories. Prerogative Court Inventories, 1718–1777, Maryland State Archives, Annapolis.

J. Baylor Accts. John Baylor Ledgers, 1719–1755, Baylor Family Papers, 1653–1915, MS 2257, University of Virginia, Charlottesville, microfilm, M-62.5, M-62.6, John D. Rockefeller, Jr., Library, Colonial Williamsburg Foundation, Williamsburg, Va.

J. Bray Accts. James Bray Ledger, 1736–1746, Burwell Family Papers, MS 64.4, John D. Rockefeller, Jr., Library, Colonial Williamsburg Foundation, Williamsburg, Va.

J. Carroll Accts. James Carroll Daybook, 1714–1721, Jesuit Provincial Archives, Special Collections, Georgetown University Library, Washington, D.C.

J. Gresham Accts. John Gresham, Sr., and John Gresham, Jr., Account Book, 1716–1750, SC 1999-1-1, Maryland State Archives, Annapolis.

J. Hall Accts. John Hall Account Book, 1765–1788, William Henry Hall Papers, 1736–1862, MS 2267, Special Collections, Duke University Library, Durham, N.C.

J. Hollyday Accts. James Hollyday Account Book, 1746–1784, MS 454.1, Maryland Historical Society, Baltimore.

J. Mercer Accts. John Mercer Ledgers, 1725–1732, [1741]–1750, Bucks County Historical Society, Doylestown, Pa., microfilm, M-164, John D. Rockefeller, Jr., Library, Colonial Williamsburg Foundation, Williamsburg, Va.

Jesuit Accts. Bernard Diderick, Baltimore Town Account Book, MS 1133, Maryland Historical Society, Baltimore.

Jones FP. Roger Jones Family Papers, 1649–1889, MS 18,063, Library of Congress, no. 263, Washington, D.C., microfilm, M-1397.1–15, John D. Rockefeller, Jr., Library, Colonial Williamsburg Foundation, Williamsburg, Va.

Loyd/Tayloe Accts. Stephen Lyde Account Book, 1708–1711, and Accounts and Letters of John Tayloe I and John Tayloe II, 1714–1778, Tayloe Family Papers, MSS1T2118, b1, Virginia Historical Society, Richmond.

Main, *Tobacco Colony.* Gloria L. Main, *Tobacco Colony: Life in Early Maryland, 1650–1720* (Princeton, N.J., 1982).

Michel, *VMHB,* XXIV. William J. Hinke, ed. and trans., "Report of the Journey of Francis Louis Michel, from Berne, Switzerland, to Virginia, October 2, 1701–December 1, 1702," *Virginia Magazine of History and Biography,* XXIV (1916), 1–43, 113–141, 275–303.

Minutes. Lower Norfolk County, Virginia, Minutes, 1637–1646, microfilm, the Library of Virginia, Richmond.

N. L. Sewall Accts. Nicholas Lewis Sewall Account Book B, 1760–1800, Special Collections, Georgetown University Library, Washington, D.C.

Orders. Lower Norfolk County, Orders, 1719, microfilm, the Library of Virginia, Richmond.

Orders. Northampton County, Virginia, Orders, 9, 1664–1674, 10, 1674–1679, 11, 1678–1683, microfilm, the Library of Virginia, Richmond.

P. Morgan, diss. Philip David Morgan, "The Development of Slave Culture in Eighteenth Century Plantation America (Ph.D. diss., University College, London, 1977).

Pr. Geo. Accts. Prince George's County, Maryland, Accounts, D, JB1, Maryland State Archives, Annapolis.

Present State of Virginia. Henry Hartwell, James Blair, and Edward Chilton, *The Present State of Virginia, and the College,* ed. Hunter Dickinson Farish (Williamsburg, Va., 1940).

R. Boarman Accts. Richard Boarman Account Book, 1755–1782, Thomas Papers, MSG176, Maryland State Archives, Annapolis.

R. Carter LB. Robert Carter Letterbook, 1728–1731, Robert ("King") Carter Papers, MS 3807, University of Virginia, Charlottesville, microfilm, M-113, John D. Rockefeller, Jr., Library, Colonial Williamsburg Foundation, Williamsburg, Va.

R. Cole Accts. Robert Cole Plantation Accounts, Testamentary Proceedings, 6: 118–147, Maryland State Archives, Annapolis.

R. Corbin Papers. Corbin Tobacco Book, 1746–1795, and Corbin Family Loose Papers, microfilm, M-55.1, M-55.2, John D. Rockefeller, Jr., Library, Colonial Williamsburg Foundation, Williamsburg, Va.; and Edmund Wilcox Ledger, 1760–1776 (vol. 3), Wilcox Farm Journal, 1761–1762 (vol. 4), and Wilcox Account Book, 1761–1764 (vol. 6), in the E. W. Hubard Papers, 1741–1907, University of North

Carolina, Chapel Hill, microfilm, M-24.1 and M-24.2, John D. Rockefeller, Jr., Library, Colonial Williamsburg Foundation, Williamsburg, Va.

R. Goldsborough Accts. Robert Gouldesborough Memorandum Book, 1696–1771, Robert Goldsborough Ledgers, 1687–1719, 1718–1742, 1751–1755, all privately owned, microfilm, Maryland State Archives, Annapolis.

R. Lloyd Accts. Robert Lloyd's Plantation Crop Book, Lloyd Papers, MS, 2001, microfilm, reel 6, Maryland Historical Society, Baltimore.

R. Tilghman Accts. Ledger A, 1718–1741, Ledger B, 1725–1765, Tilghman Family Papers, 1712–1843, Special Collections, Georgetown University Library, Washington, D.C.

R. W. Carter Diary. Robert Wormeley Carter Diary, 1764, William L. Clements Library, University of Michigan, transcript, John D. Rockefeller, Jr., Library, Colonial Williamsburg Foundation, Williamsburg, Va.

Records. Northumberland County, Virginia, Records, 1652–1658, 1658–1666, microfilm, the Library of Virginia, Richmond.

Relation of Maryland. *A Relation of Maryland* . . . (London, 1635), reprinted in Clayton Colman Hall, ed., *Narratives of Early Maryland, 1633–1684,* Original Narratives of Early American History (New York, 1910), 63–112.

Rutman and Rutman, *Explicatus.* Darrett B. Rutman and Anita H. Rutman, *A Place in Time: Explicatus* (New York, 1984).

S. Galloway Accts. Samuel Galloway Ledger, 1748–1758, Cheston-Galloway Papers, MS 1994, Maryland Historical Society, Baltimore; Samuel Galloway Ledgers, 1748–1751, 1753–1779, Galloway / Maxcy / Markoe Papers, I–VI, Library of Congress, Washington, D.C.

Stewart LB. Charles Stewart Letterbook, 1751–1753, Historical Society of Pennsylvania, Philadelphia, microfilm, M-32, John D. Rockefeller, Jr., Library, Colonial Williamsburg Foundation, Williamsburg, Va.

Test. Proc. Testamentary Proceedings, Maryland State Archives, Annapolis.

W. Massie Accts. William Massie Account Book, 1747–1748, MS 21, Virginia Historical Society, Richmond.

W. Robinson Accts. William Robinson Account Book, Bunker Hill, Westmoreland County, 1736–1823, Robert E. Lee Memorial Foundation, Stratford Hall, Va., microfilm, M-141, John D. Rockefeller, Jr., Library, Colonial Williamsburg Foundation, Williamsburg, Va.

WD. Northampton County, Virginia, Wills and Deeds, 3, 1645–1651, 4, 1651–1654, 5, 1654–1655, 7 and 8, 1655–1668, 9, 1657–1666, microfilm, the Library of Virginia, Richmond.

WD. Westmoreland County, Virginia, Wills and Deeds, 1665–1671, microfilm, the Library of Virginia, Richmond.

Wertenbaker, *Planters.* Thomas J. Wertenbaker, *The Planters of Colonial Virginia* (1922), rpt. in Wertenbaker, *The Shaping of Colonial Virginia* (New York, 1958).

Wills. Lancaster County, Virginia, Wills, 5, 1674–1689, 8, 1690–1709, microfilm, the Library of Virginia, Richmond.

INDEX